Physical Methods in Bioinorganic Chemistry

SPECTROSCOPY AND MAGNETISM

LIST OF CONTRIBUTORS

Numbers in parentheses indicate the pages on which the authors' contributions begin.

N. DENNIS CHASTEEN (187)
Department of Chemistry
University of New Hampshire
Durham, NH 03824

ROMAN S. CZERNUSZEWICZ (59)
Department of Chemistry
University of Houston
Houston, TX 77204

JEAN-JACQUES GIRERD (321)
Laboratoire de Chimie Inorganique
URA CNRS 420, Bat. 420
Université Paris-Sud
F-91405 Orsay
FRANCE

MICHAEL K. JOHNSON (233)
Department of Chemistry
University of Georgia
Athens, GA 30602

YVES JOURNAUX (321)
Laboratoire de Chimie Inorganique
URA CNRS 420, Bat. 420
Université Paris-Sud
F-91405 Orsay
FRANCE

DAVID R. McMILLIN (1)
Department of Chemistry
Purdue University
West Lafayette, IN 47907

LI-JUNE MING (375)
Department of Chemistry
University of South Florida
4202 Fowler Avenue
Tampa, FL 33620

ECKARD MÜNCK (287)
Department of Chemistry
Carnegie Mellon University
4400 Fifth Avenue
Pittsburgh, PA 15213

GRAHAM PALMER (121)
Department of Biochemistry and Cell Biology
Rice University
P. O. Box 1892
Houston, TX 77251-1892

LAWRENCE QUE, JR. (513)
Department of Chemistry
University of Minnesota
207 Pleasant St. SE
Minneapolis, MN 55455

JOANN SANDERS-LOEHR (505)
Department of Chemistry, Biochemistry and Molecular Biology
Oregon Graduate Institute of Science & Technology
Portland, OR 97291-1000

ROBERT A. SCOTT (465)
Department of Chemistry
University of Georgia
Athens, GA 30601

PENNY A. SNETSINGER (187)
Department of Chemistry
Sacred Heart University
Fairfield, CT 06432

THOMAS G. SPIRO (59)
Department of Chemistry
Princeton University
Princeton NJ 08544

Physical Methods in Bioinorganic Chemistry

SPECTROSCOPY AND MAGNETISM

EDITED BY

Lawrence Que, Jr.

UNIVERSITY OF MINNESOTA

University Science Books

Sausalito, California

University Science Books
55D Gate Five Road
Sausalito, CA 94965

Fax: (415) 332-5393
www.uscibooks.com

Production manager: *Susanna Tadlock*
Manuscript editor: *Jeannette Stiefel*
Designer: *Robert Ishi*
Compositor: *Asco Typesetters*
Printer & Binder: *Edwards Brothers*

This book is printed on acid-free paper.

Library of Congress Cataloging-in-Publication Data

Physical methods in bioinorganic chemistry : spectroscopy and magnetism / edited by Lawrence Que.

p. cm.

Includes bibliographical references and index.

ISBN 1-891389-02-5 (alk. paper)

1. Bioinorganic chemistry—Methodology. 2. Metalloproteins—Research—Methodology. 3. Spectroscopy. 4. Nuclear magnetic resonance spectroscopy. 5. Physical biochemistry—Methodology. I. Que, Lawrence, 1949–

QP531.P47 2000
572′.51′028—dc21 99-051651

Printed in the United States of America
10 9 8 7 6 5 4 3 2 1

Contents

Preface vii

1 Electronic Absorption Spectroscopy 1
David R. McMillin, Purdue University

2 Resonance Raman Spectroscopy 59
Thomas G. Spiro, Princeton University and
Roman S. Czernuszewicz, University of Houston

3 Electron Paramagnetic Resonance of Metalloproteins 121
Graham Palmer, Rice University

4 ESEEM and ENDOR Spectroscopy 187
N. Dennis Chasteen, University of New Hampshire
Peggy A. Snetsinger, Sacred Heart University

5 CD and MCD Spectroscopy 233
Michael K. Johnson, University of Georgia

6 Aspects of ^{57}Fe Mössbauer Spectroscopy 287
Eckard Münck, Carnegie Mellon University

7 Molecular Magnetism in Bioinorganic Chemistry 321
Jean-Jacques Girerd and Yves Journaux, Université Paris-Sud

8 Nuclear Magnetic Resonance of Paramagnetic Metal Centers in Proteins and Synthetic Complexes 375
Li-June Ming, University of South Florida

9 X-Ray Absorption Spectroscopy 465
Robert A. Scott, University of Georgia

10 Case Study: The Cu_A Site of Cytochrome *c* Oxidase 505
Joann Sanders-Loehr, Oregon Graduate Institute of Science & Technology
Case Study: Isopenicillin N Synthase 513
Lawrence Que, Jr., University of Minnesota

APPENDIX: Problems from Selected Chapters 529

Index 539

Preface

The development of bioinorganic chemistry as a discipline over the last forty years is due in no small part to the availability of a variety of physical methods with which to probe the nature of the metal ion environment in the biological system of interest. These tools have been quite useful in providing insight into the electronic structure of the metal ion and the ligands that surround it. By piecing together the various clues derived from the physical methods, bioinorganic chemists have been able to form a coherent picture of the metal binding site and to deduce the role of the metal ion in a number of biological processes.

This book is aimed at providing the novice in bioinorganic chemistry with a fundamental understanding of the physical methods employed. It is intended not only for beginning graduate students, but also as a source of information for scientists not previously trained in spectroscopy and magnetism who find their research projects evolving in a direction that requires such physical methods. Thus, many examples are provided to illustrate the types of physical data that can be obtained so that the reader can attempt to match the experimental data with data published in the book.

The first nine chapters cover individual methods that are useful in bioinorganic chemistry. Included are spectroscopic methods such as electronic absorption spectroscopy, resonance Raman spectroscopy, electron paramagnetic resonance and associated double resonance methods, circular dichroism and magnetic circular dichroism spectroscopy, Mössbauer spectroscopy, nuclear magnetic resonance spectroscopy of paramagnetic molecules, and x-ray absorption spectroscopy. There is also one chapter on magnetism. Excluded were discussions on nuclear magnetic resonance in general and x-ray crystallography, both of which are topics that are covered in entire books. Finally, the concluding chapter provides two case studies that illustrate how data from physical methods have been used to assemble a picture of a particular metal center in biology.

Lawrence Que, Jr.

Physical Methods in Bioinorganic Chemistry

SPECTROSCOPY AND MAGNETISM

1

Electronic Absorption Spectroscopy

DAVID R. McMILLIN

Purdue University
W. Lafayette, IN 47907

CONTENTS

I. Introduction

The method of absorption spectroscopy is one of the most convenient and useful tools for probing electronic structure. Molecules possess not one but several accessible electronic energy levels, and transitions between states can often be induced by the absorption of a photon. However, to appreciate the electronic absorption spectrum of a system, it is necessary to take account of important electron–electron interactions that are often neglected in elementary treatments of electronic structure. Consider, for example, the simple diatomic molecule O_2 (dioxygen), which is characterized by the familiar molecular orbital (MO) scheme in Figure 1. According to this scheme, the ground electronic configuration is

$$\sigma_{2s}{}^{2}\sigma_{2s}{}^{*2}\sigma_{2p}{}^{2}\pi_{2p}{}^{4}\pi_{2p}{}^{*2}$$

where, as usual, σ_{2s} denotes the bonding combination of the 2s atomic orbitals centered on each oxygen atom, and so on. This model adequately describes many important properties of the dioxygen molecule including the fact that the bond order is 2 and the fact that the molecule has two unpaired electrons. Figure 1 also correctly implies that $\pi_{2p} \rightarrow \pi_{2p}{}^{*}$ orbital excitation is possible; in fact, one such transition begins in the far ultraviolet (UV) at a wavelength of around 200 nm and maximizes at about 145 nm. The process involves the injection of a great deal of energy into the molecule, enough to cleave it into oxygen atoms, and atomic oxygen is the dominant form in the upper atmosphere where much of the UV radiation from the sun is absorbed. However, the electronic structure of dioxygen is much richer than this. The scheme in Figure 1 ignores important repulsive interactions that operate among the negatively charged electrons. As will become clearer below, there are actually three distinct states associated with the

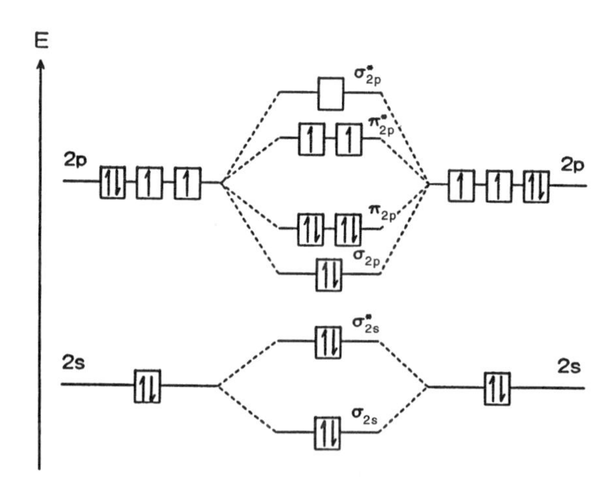

Figure 1
Molecular orbital scheme of O_2. The energy levels are derived from a one-electron model in which the electron–electron repulsive interactions are not explicitly considered. The antibonding orbitals are labeled with an asterisk, and the nuclei lie along the *z* axis.

ground electronic configuration. When the two electrons in the π_{2p}^* level have parallel spins, the system is in the lowest energy state, but there are two other states in which the electrons have paired spins. These states occur at higher energies because the electron–electron repulsions are greater for these arrangements. Although the transitions to each of these states are formally spin forbidden, they are observed in the atmospheric absorption spectrum, albeit with low intensities, because there are so many dioxygen molecules in the path. These excited states are not just curiosities; one of them is responsible for the luminescence and the important photochemistry that is associated with dioxygen in solution. A completely different picture emerges when the dioxygen molecule undergoes a change in oxidation state and binds to a metalloprotein (see below).

Absorption spectroscopy is a very versatile method in that measurements can be made using solid, liquid, or gaseous samples. The spectrum obtained is characteristic of a particular chromophore in a specific environment. Therefore, because of the possibility of solvolysis reactions, and so on, it is very important to establish exactly what species is (are) present for studies that are carried out in solution.

This chapter is organized such that the theoretical framework is sketched before the applications. The development assumes a working knowledge of group theory. It is not necessary to master the theory in its entirety before moving on to applications. Indeed, some sections, such as the one on vibronic coupling (Section VI), might be skipped on the first reading. For most students, the theoretical principles become clearer once they are applied. Finally, a few specialized topics are introduced at the end of the chapter.

II. Atomic Structure

According to quantum theory, the energy states of an atom can be derived from the Hamiltonian operator, which will be signified by the symbol $\mathscr{H}$. The lead terms in the Hamiltonian are those that account for the kinetic energy of the electrons and those that account for the Coulombic attraction between the nucleus and the individual electrons. Solving for the energy levels of even relatively simple systems can be extremely difficult; fortunately, explicit solutions are not really necessary for many purposes. A review of some of the formalism will suffice for the present purposes. In the formulation due to Schrödinger, the ith available energy, E_i, is an eigenvalue of the Hamiltonian operator, and the associated wave function ψ_i is an eigenfunction of $\mathscr{H}$. In an alternative formulation, introduced by Heisenberg, a set of basis functions is selected and used to construct a Hamiltonian matrix. When the proper eigenfunctions (also known as the stationary-state wave functions) are used, the diagonal terms:

$$\langle\psi_i|\mathscr{H}|\psi_i\rangle = E_i \tag{1}$$

correspond to the energies of the stationary states. Note that several steps are involved in evaluating this matrix element including (1) application of the Hamiltonian operator to the function ψ_i, (2) left multiplication by ψ_i (or the complex conjugate if the function involves a complex variable), and (3) integration over all relevant coordinates of the system.

When all wave functions and energies are known for a system, the interpretation of the electronic absorption spectrum should be a straightforward process. Usually, however, only qualitative or semiquantitative information is available about the energy states. A notable exception is the case of the hydrogen atom where the ψ_i correspond to the well-known atomic orbitals: 1s, 2s, 2p, and so on. Although detailed theoretical calculations are available for many other atoms by now, for general purposes it will be most helpful to review basic approximations that are applicable to almost any system. For an atom with N electrons, the total wave function is often approximated as a product of N one-electron orbitals that have the same orbital angular momentum values and labels (s, p, d, etc.) as the orbitals of the hydrogen atom. However, in the many electron atom, orbitals with the same principal quantum number n typically do not have the same energy. Within a given shell, the energies vary as $E(n\mathrm{d}) > E(n\mathrm{p}) > E(n\mathrm{s})$ in accordance with the ability to penetrate underlying shells. In line with the Pauli exclusion principle, the atom is assigned an electronic configuration by the Aufbau method (e.g., $1s^2 2s^2 2p^2$ for the ground state of the carbon atom).

III. Term Energies

A. Russell–Saunders Coupling

Any closed-shell configuration of a many electron atom generates a single energy level. However, when there is a partially filled shell, the electronic repulsion energy and magnetic interactions may differ depending on the way in which electrons are distributed within the shell. Except for heavy atom systems, which will not be considered here, the electron–electron repulsions are the dominant interactions. These repulsions give rise to a series of energy levels known as Russell–Saunders terms where the spacings between terms can be of the order of electron volts (eV). In the strictest sense, a term corresponds to an energy level, which may be degenerate, and a state corresponds to one particular wave function. In common practice, however, and in this chapter, the two words are sometimes used interchangeably. The number of Russell–Saunders terms is easily determined because each term has a well-defined net orbital angular momentum and a net spin angular momentum, characterized by the quantum numbers L and S, respectively. Note that in a multielectron atom the net, or total, orbital angular momentum and the net spin angular momentum of the whole set of electrons are the important quantities. The momenta associated with individual electrons are *not* conserved because the exchange of momentum is so facile between electrons.

As an example, consider a 2p shell containing two electrons. The two electrons can have their spins parallel or antiparallel, that is, the net spin can be $S = 1$ or $S = 0$. An $S = 1$ state is said to be a triplet state because it has threefold $(2S + 1)$ degeneracy. In this case, the degeneracy arises because the magnetic quantum number M_s can range through the values 1, 0, and −1; in general the M_s quantum number can be equal to $S, S-1, \ldots,$ or $-S$. Straightforward procedures can be used to determine the total orbital angular momentum values (L values) that occur with each S value.[1] Once the values of the angular momenta are known, each term can be designated by a letter that codes for the L quantum number and a superscript that reveals the multiplicity. For example, the terms associated with the $2p^2$ configuration are a ^3P, a ^1D and a ^1S, for which $L = 1, 2,$ and 0, respectively. See Appendix I for the derivation of these terms. These particular terms are associated with a doubly excited configuration of the helium atom, but the same terms arise for the ground electronic configuration of the carbon atom even though the configuration is $1s^22s^22p^2$. The reason is that the electrons in filled shells add no new degrees of freedom to the system.

From a semiclassical perspective, the energies of the terms differ because the average distance between any pair of electrons and the average distance between the nucleus and each given electron vary depending on the way in which the motion of the electrons is coupled, that is, on the composite orbital and spin angular momenta of the system. Due to magnetic interactions, that is, spin–orbit coupling, the ^3P ground term of the carbon atom splits further into three distinct sublevels, but this splitting is small for carbon and most of the lighter elements and will be ignored here.

B. Terms in the O_2 Molecule

Terms for molecules can be identified by focusing on the electrons in the unfilled shells just as in atoms. Consider the ground electronic configuration of the O_2 molecule that has two electrons in the partially occupied $\pi_{2p}{}^*$ shell (Fig. 1). There are actually six independent ways to assign two electrons to the shell. To see this, let x and y designate the π-antibonding MOs derived from the $2p_x$ and the $2p_y$ atomic orbitals, respectively. Then, one possible way to assign the electrons would be symbolized x^+x^-, where the superscripts indicate that one electron is assigned the $m_s = +\frac{1}{2}$ quantum number and the other the $m_s = -\frac{1}{2}$ quantum number as required by the Pauli principle when two electrons occupy the same spatial orbital. Note that x^-x^+ is an alternative representation for the same function because the two electrons are indistinguishable particles. In this notation, six independent ways of assigning the two electrons to the $\pi_{2p}{}^*$ shell are

$$x^+y^+, x^-y^-, x^+x^-, y^+y^-, x^+y^-, \quad \text{and} \quad x^-y^+$$

When electron–electron repulsions are properly taken into account, appropriate linear combinations of these functions can be used do describe stationary states (terms) of the O_2 molecule. (The reader will be able to derive the term symbols later after appropriate methodology is developed in the chapter; see Problem 8.) For now, note that there are functions with parallel spins, so there is a triplet ($S = 1$) term, which is the ground term. This term explains why O_2 is paramagnetic even though it has an even number of electrons. Two higher energy terms with energies in the visible and near-infrared (IR) are also spanned by the

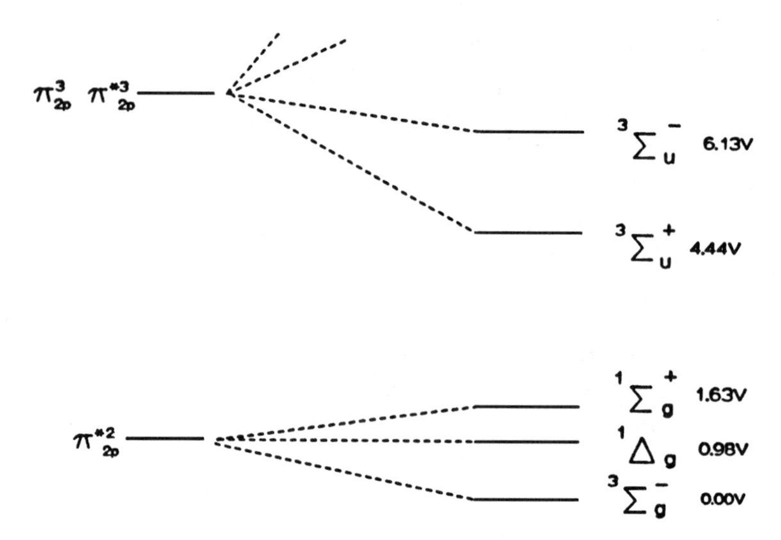

Figure 2
Terms associated with the two lowest lying electronic configurations of O_2. For linear molecules Greek letters are used to indicate the amount of orbital angular momentum about the internuclear axis. Relative energies are given in units of volts (V).

configuration as indicated in Figure 2. The physical basis for the energy differences between terms can be understood in terms of the relative disposition of these two electrons.[2] In the ground-state triplet, each electron is, on the average, distributed in an orbital that is rotated 90° from the orbital containing the other electron. This disposition is mandated by the Pauli principle because electrons with the same spin have to reside in different orbitals. The next lowest energy term is a singlet ($S = 0$), but it has a twofold orbital degeneracy and exhibits a net orbital angular momentum about the internuclear axis. In this state, all values of the azimuthal angle between the two electrons are equally probable. Finally, the third term is a nondegenerate singlet state in which the two electrons tend to be in the same π orbital. When the electrons are on the same atom, therefore, they must be in the same atomic orbital, and this is clearly a high-energy arrangement. There are several other open-shell systems in which important excited states have the same electronic configuration as the ground state. Examples are the methylene radical, cyclobutadiene, and octahedral complexes of Cr(III). In the case of Cr(III), the lowest energy doublet excited state generally stems from the ground-state configuration, and the output of the ruby laser is due to stimulated emission from one such doublet.

IV. Electric Dipole Transitions

Transitions between energy levels are usually induced with electromagnetic radiation. Energy transfer from the radiation field is possible whenever the atom or molecule can resonate with the perpendicularly oscillating electric or magnetic fields of a light wave. Consider the hydrogen atom. There is no net charge on the atom, but the charge distribution within the atom can exhibit an instantaneous electric dipole moment, magnetic dipole moment, or electric quadrupole moment. Coupling with the electric dipole moment is usually by far the most important effect in absorption spectroscopy. From the viewpoint of classical mechanics, the atom absorbs energy by acting like an antenna. Oscillating electric dipole character arises when the perturbing radiation field causes the atom to begin to take on the character of an excited-state wave function Φ_k of appropriate symmetry. (For more details, see the treatment of time-dependent superpositions in Appendix III.) The excited-state wave functions, which provide the oscillating dipole character, are those for which an integral of the type:

$$\langle \Phi_k | d | \Phi_0 \rangle \tag{2}$$

is nonzero, where d is one of the components of the dipole operator, that is, the functions x, y or z. In practice, Eq. (2) determines whether or not a particular transition occurs with any intensity. When the integral vanishes, the transition is forbidden to occur by the electric dipole mechanism.

Since the electric field of a light wave has no direct influence on the electron spin, the S quantum number is preserved during an electric dipole transition. Therefore, an important selection rule for an allowed transition is

$$\Delta S = 0$$

When $\Delta S \neq 0$ for any transition, it is said to be *spin forbidden*. As we will see in Section VIB, it is also possible for a transition to be orbitally forbidden.

V. Polyatomic Molecules

A. Born–Oppenheimer Approximation

Molecular systems are much more complicated that atoms because of the additional degrees of freedom connected with the nuclear framework. However, considerable simplification is achieved if the Born–Oppenheimer approximation is invoked. The basis of the approximation is that nuclei move much more slowly than electrons due to the difference in mass. Accordingly, the Schrödinger equation can be solved for the electrons under the assumption that the nuclei are stationary centers of positive charge. In principle, the electronic energy can then be calculated for all possible configurations of the nuclei that preserve the center of mass to obtain the potential energy function that constrains the nuclei. In turn, this potential can be used to solve for the vibrational wave functions associated with the electronic state. The ladder of vibrational energy levels so obtained is conveniently displayed within the potential energy surface (Fig. 3). The rotational and translational degrees of freedom can be treated, too, but will not be considered here.

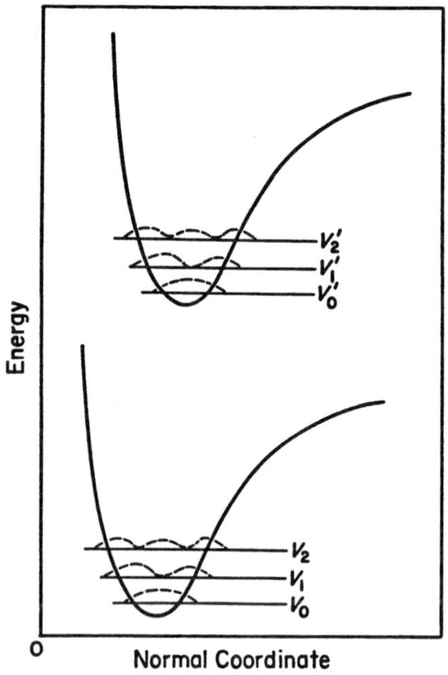

Figure 3
Potential energy surfaces for the ground state and the first excited state as a function of a specific nuclear displacement. Vibrational energy levels are presented as horizontal lines. The dashed lines represent the probability distributions defined by the vibrational wave functions.

B. The Role of Molecular Symmetry

The total orbital angular momentum can be used to classify the electronic states of an atom, but the orbital angular momentum is usually quenched in molecules so the symmetry properties of the wave function are used instead. Most electronic states exhibit a minimum energy for some particular nuclear arrangement called the equilibrium geometry, which can be assigned a point group. Since the molecular Hamiltonian must be invariant with respect to any operation that results in an indistinguishable arrangement of particles, the electronic wave functions can be required to be eigenfunctions of the molecular symmetry operators, and therefore can be classified by symmetry type. The symmetry operators form a mathematical group, and the symmetry properties of the wave functions are conveniently derived by the application of group representation theory.

C. Orbital Symmetry

Once the point group is known, the first step in the symmetry analysis is to choose the basis functions. Because approximate MOs are often constructed from linear combinations of atomic orbitals, the valence orbitals of the constituent atoms generally serve as a convenient basis set. As an elementary example, consider the $2p\pi$ orbitals of four N atoms arranged at the corners of a square. It can be shown that the four atomic orbitals span three irreducible representations of the D_{4h} point group:

$$\Gamma = a_{2u} + b_{2u} + e_g$$

If the basis set is expanded to include the $2p\pi$ orbitals of all non-metal atoms in the metalloporphyrin framework in Figure 4, 18 irreducible representations in all

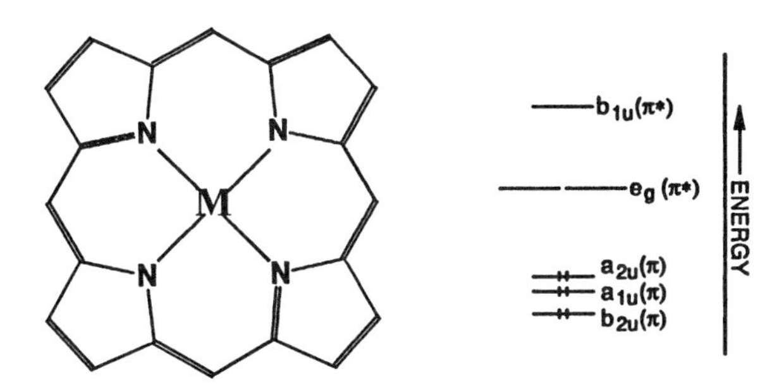

Figure 4
A metalloporphyrin with hydrogen atoms omitted. Some of the π energy levels are depicted in the scheme on the right.

are subtended

$$\Gamma = 4a_{2u} + 3b_{2u} + 2a_{1u} + 3b_{1u} + 6e_g$$

Each representation corresponds to an energy level within the π system of the metalloporphyrin. One-half of these representations are π-bonding orbitals, and the other one-half are π^* (or π-antibonding) orbitals. In the ground state, the π orbitals are filled, and the π^* orbitals are all empty. The low-lying π–π^* electronic transitions of the metalloporphyrin can be assigned once the relative energies of the orbitals are established. Many years ago, Longuet–Higgins and co-workers employed Hückel theory to predict that in the ground state the highest occupied molecular orbitals (the two HOMOs) would be close-lying a_{1u} and a_{2u} orbitals and that the lowest unoccupied molecular orbital (LUMO) would be a pair of $e_g{}^*$ orbitals.[3] Some of the relevant orbitals are shown schematically in Figure 4. More sophisticated calculations and experimental results have confirmed this scheme.[4]

D. Electronic States and Transition Energies

To begin the analysis of the electronic absorption spectrum of a molecule such as a metalloporphyrin, the terms associated with the various electronic configurations must be identified. In general, the term symmetry, that is, the symmetry of the wave function that encompasses all of the electrons of the system, is derived from the direct product of the symmetry species of the populated one-electron orbitals. For the system in Figure 4, the relevant direct product is

$$b_{2u} \times b_{2u} \times a_{1u} \times a_{1u} \times a_{2u} \times a_{2u} = a_{1g}$$

With this or any other closed-shell configuration, the term symmetry is necessarily totally symmetric. Since $S = 0$, the ground term is labeled as ${}^1A_{1g}$, where a capital letter is used to designate the state symmetry, and the superscript is used to indicate the spin multiplicity.

The excited-state terms are derived in the same way. Note that when the direct products are taken, the occupied orbitals of any filled shells can be ignored in this exercise since they always span the totally symmetric representation. The symmetry of the excited state associated with the $a_{2u} \rightarrow e_g$ transition of the metalloporphyrin is, therefore,

$$a_{2u} \times e_g = e_u$$

Since the spins can be either parallel or antiparallel, two terms are spanned, namely, ^1E_u and ^3E_u. Similarly, the $a_{1u} \rightarrow e_g$ transition gives rise to another pair of ^1E_u and ^3E_u terms. Within the spirit of Hückel theory, the total energy of any given electronic configuration can be calculated by summing the energies of all of the electrons. The theory predicts that two nearly degenerate, spin-allowed transitions occur, and both turn out to be dipole allowed (see below). However, the spectrum presented at the bottom of Figure 5 reveals that the two lowest energy band systems are separated by several thousand wavenumbers and that the higher energy transition has a much greater intensity. The transition to the higher energy ^1E_u state occurs at about 420 nm and is known as the Soret band or sometimes as the B band. Excitation to the other ^1E_u state is realized as a pair of bands centered around 550 nm. Two bands, known as Q_o and Q_v or as α and β, occur because vibrational motion is simultaneously excited as discussed below. However, the reason for the rather large energy gap between the two transitions will be discussed first.

E. Configuration Interaction

The main problem with the Hückel theory is that it essentially ignores the electron–electron interactions and assumes that each electron moves completely independently of the others. In this view, the ^1E_u excited states of the metalloporphyrin can unambiguously be assigned specific electronic configurations with the hole in either the a_{2u} or the a_{1u} level. However, the electron–electron interactions of the full Hamiltonian operator must be taken into account. If the effects are not too great, the interactions can be viewed as perturbations on the original system in which case they have the effect of mixing wave functions that have the same symmetry. When the two ^1E_u states of the metalloporphyrin mix together, the system is said to undergo a configuration interaction, and the resulting states can no longer be ascribed to specific electronic configurations. In the state diagram in Figure 5, the dashed lines correspond to the energies the two ^1E_u states would have in the absence of the electron–electron interactions. The heavy lines are indicative of the relative energies when configuration interaction is accounted for.

The physics behind the mixing is not always very transparent, but in this case some insight can be gained from a semiclassical picture of the system. The fundamental problem is that the zero-order states, represented by the two configurations $a_{1u}{}^2a_{2u}{}^1e_g{}^1$ and $a_{1u}{}^1a_{2u}{}^2e_g{}^1$, are not stationary states of the isolated system. To see this, suppose the system could be prepared as the pure $a_{1u}{}^2a_{2u}{}^1e_g{}^1$ configuration. This state would not persist because the system can evolve into a

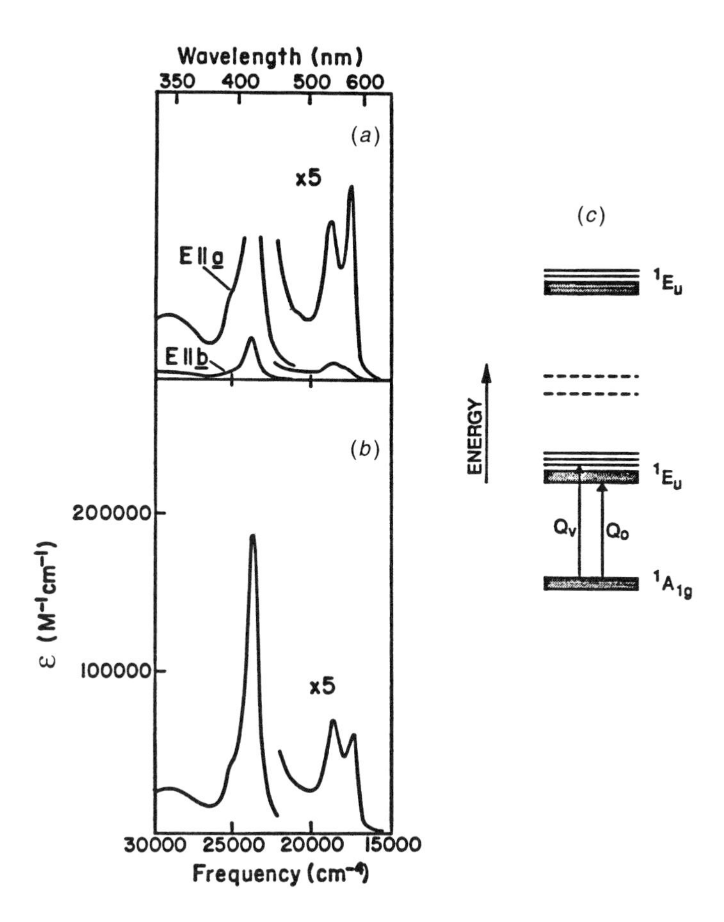

Figure 5
Absorption spectra of the carbonyl adduct of sperm whale myoglobin. (*a*): Single-crystal absorption spectra obtained with the electric vector of the incident plane-polarized light aligned parallel to the *a* and *b* axes. The *b* axis is nearly perpendicular to the plane of the porphyrin. (*b*): solution spectrum. (*c*): The thick lines represent the energies of the observed electronic states. The spectral data are from Makinen and Chung.[5]

state with a lower average energy by forming a superposition with the $a_{1u}{}^1a_{2u}{}^2e_g{}^1$ configuration. In the superposition state, an electron jump from the a_{2u} orbital to an e_g orbital occurs in synchrony with an electron jump from the e_g level down to the a_{1u} level. The reason that the system achieves a lower energy is that the electrons are better able to avoid each other by synchronously alternating between orbitals. The coupling is effective so long as the gain in stabilization due to the decrease in the electron–electron repulsion energy outweighs the cost of moving the hole from the a_{2u} orbital to the a_{1u} orbital. The other stationary state represents a higher energy admixture of the same two electronic configurations. This phenomenon is discussed in more detail in Appendix III.

VI. Vibronic Coupling

The fine structure, which appears in an absorption band due to concurrent excitation of vibrational motion, can be explained in terms of wave functions that depend on the vibrational coordinates as well as the electronic coordinates. A diatomic molecule with a potential energy surface like that in Figure 3 has only one vibrational degree of freedom, but a nonlinear molecule containing N atoms has $3N - 6$ degrees of freedom associated with the nuclear framework. In the simplest approximation, a wave function that includes the vibrational and electronic degrees of freedom is the product function:

$$|\psi_i> |\chi_{v1}> |\chi_{v2}> \cdots \tag{3}$$

where ψ_i is the electronic wave function of the ith electronic state, χ_{v1} is the vibrational function for normal coordinate ξ_1 with quantum number $v1$, and so on. Purely electronic transitions are possible, but there may also be a vibronic transition in which there is simultaneous vibrational and electronic excitation. Examination of Figure 5 reveals that there is vibronic structure associated with both ^1E_u transitions of the metalloporphyrin. Since electronic excitation typically occurs in 10^{-15} s or less and a typical vibration requires about 10^{-14} s, the nuclei do not have time to move during the excitation. This notion is the basis of the Franck–Condon principle, which states that only "vertical" transitions occur. To do some justice to the treatment of vibronic transitions, orbitally allowed and orbitally forbidden transitions will be considered separately.

A. Electronically Allowed Transitions

In view of the form of Eq. (3), the matrix element of the electronic dipole moment operator that connects a vibrationally excited level of the ith electronic state to the ground vibrational level of the ground electronic state takes the form:

$$\langle\psi_i|d|\psi_0\rangle\langle\chi'_{v1}|\chi_0\rangle\langle\chi'_{v2}|\chi_0\rangle\cdots \tag{4}$$

where the product of vibrational overlap integrals extends over all $3N - 6$ degrees of vibrational freedom. The primes are introduced to designate the vibrational wave functions of the excited state since it typically has a different potential energy function than the ground state. If Eq. (4) is nonzero when at least one of the vk values is not equal to zero, simultaneous electronic and vibrational excitations are allowed.

Since the transition is assumed to be electronically allowed, the lead integral involving the dipole moment operator is nonzero, and the focus is on the vibrational overlap integrals. Group theory reveals that an overlap integral will be nonzero when the direct product:

$$\Gamma(\chi'_{vk}) \times \Gamma(\chi_0)$$

contains the totally symmetric representation where $\Gamma(\chi'_{vk})$ and $\Gamma(\chi_0)$ denote the representations spanned by the excited- and ground-state vibrational wave functions, respectively. Since the ground vibrational function χ_0 is always totally symmetric, the transition to the vkth vibrational state is only allowed if it is also totally symmetric. Even-order harmonics of asymmetric vibrations can also be excited if the harmonic is totally symmetric. Note that even-order harmonics of other fundamentals can be excited so long as they represent totally symmetric vibrations. A similar treatment can give the selection rules for a less common type of transition known as a hot band. Hot bands originate from molecules that are already vibrationally excited when the transition from the ground electronic state occurs.

In the absence of hot bands, the lowest energy transition can be labeled as the 0–0 transition, which corresponds to the transition from the ground vibrational state of the ground electronic state to the ground vibrational level of the electronic excited state. If one or more harmonics of a mode are excited, they are said to form a progression with the 0–0 transition as the origin. The length of a progression varies from case to case as the individual transition intensities depend on the vibrational overlap integrals, or Franck–Condon factors, indicated in Eq. (4). As an illustration of a vibronic transition, consider the B band in Figure 5. Some 1300 cm^{-1} above the main transition, there is a shoulder that has been assigned as a vibronic transition. Since the associated electronic transition is strongly allowed, a totally symmetric mode must be involved, and this frequency is reasonable for a stretching vibration of the porphyrin framework. The fact that the 0–1 transition is observed shows that the ^1E_u excited state does not present the same potential energy surface for vibrations as the $^1A_{1g}$ ground state. If it did, χ'_1 would be orthogonal to χ_0 since the vibrational wave functions of any particular potential well form an orthonormal set. One difference in the surfaces is likely to be in the average lengths of the carbon–carbon and carbon–nitrogen bonds of the ring since a π^* orbital is populated in the ^1E_u state. In terms of Figure 3, the excited-state minimum should be horizontally displaced relative to the minimum of the ground state. Matching the experimentally observed vibronic envelope with a calculated spectrum based on a model potential is one way of obtaining information about the structure of excited states.

B. Orbitally Forbidden Transitions

In conjunction with Eq. (2), group representation theory reveals that a transition from the ground-state ψ_0 to an electronic excited-state ψ_1 is orbitally forbidden if the direct product:

$$\Gamma(\psi_1) \times \Gamma(d) \times \Gamma(\psi_0) \tag{5}$$

does not contain the totally symmetric representation. However, the transition is vibronically allowed if the direct product spans one of the normal coordinates, say ξ_k, that describes a fundamental vibration of the molecule. The reason is that the symmetry constraint is removed during the vibration, and the ever-present zero-point motions of the nuclei generally suffice to insure a finite absorption intensity. Of course, even larger nuclear displacements occur when vibrationally

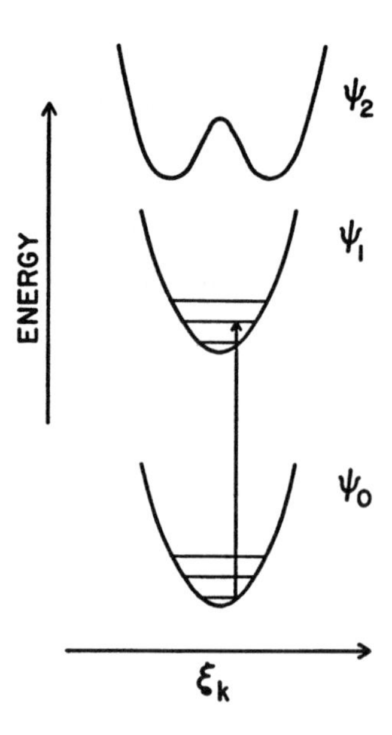

Figure 6
Potential energy surfaces for the ground and two excited states of a molecule. The normal coordinate ξ_k is assumed to be a nontotally symmetric vibration.

excited levels of the ground electronic state are populated; therefore, the intensities of vibronic transitions are characteristically temperature dependent. As a rule of thumb, the intensity of a vibronically allowed transition is only about 0.1% that of an orbitally allowed transition. Next, the essential ideas are presented, but a more detailed treatment of vibronic interactions will be given in Appendix II.

Consider a molecule that contains three well-separated electronic states: ψ_0 and two excited states ψ_1 and ψ_2. Furthermore, let the transition from the ground state to ψ_1 be electronically forbidden but the transition to ψ_2 be electronically allowed. Figure 6 contains representative potential energy surfaces where the nuclear displacement ξ_k is assumed to be along a nontotally symmetric coordinate. Note the contrast with Figure 3. Since ξ_k is a nontotally symmetric coordinate, the excited-state energy surface has to have a minimum or a local maximum at the position that corresponds to the equilibrium geometry of the ground state.[6] The transition from the ground-state ψ_0 to the first excited-state ψ_1 is vibronically allowed if movement along ξ_k induces a mixing between the wave functions ψ_1 and ψ_2. As shown in Appendix II, this is possible if the direct product:

$$\Gamma(\psi_1) \times \Gamma(\psi_2) \tag{6}$$

forms a basis for the same irreducible representation as the normal coordinate ξ_k. Because of the way in which the absorption intensity of the vibronic transition arises, the transition is said to become allowed by "borrowing" or "stealing" intensity from the $\psi_0 \rightarrow \psi_2$ transition. The vibronically allowed transition involves the excitation of one quantum of ξ_k, which is referred to as the promoting mode. Excitation of various totally symmetric vibrations, so-called acceptor modes, may occur as well. However, in the spectrum they converge on a false origin displaced from the 0–0 transition by the energy of the ξ_k fundamental.

As an illustration, consider the Q_v band in the spectrum of the metalloporphyrin discussed above (Fig. 5). Even though the parent electronic transition is not symmetry forbidden, Gouterman[4] showed that the Q_v band steals intensity from the B band via a vibronic coupling mechanism. In accord with Eq. (6), the normal modes that can theoretically mix the two excited states are determined by the direct product:

$$e_u \times e_u = a_{1g} + a_{2g} + b_{1g} + b_{2g}$$

The stretching modes of the ring framework of the porphyrin ligand include several examples of each of the four types of vibrations, and the separation of about 1200 cm^{-1} between the Q_v and Q_0 bands is consistent with this type of vibration.

As implied above, there are two criteria that can be used to determine if an electronically forbidden transition is vibronically allowed. If either the direct product in Eq. (5) or (6) forms the basis for the irreducible representation of some normal coordinate ξ_k, which describes one of the fundamental vibrations of the system, the transition $\psi_0 \rightarrow \psi_1$ is vibronically allowed. As stated, the criterion based on Eq. (5) is the stronger one because the direct product of $\Gamma(d)$ with $\Gamma(\psi_0)$ spans the symmetries of all states to which symmetry-allowed transitions occur. Of course, the irreducible representation for any allowed transition can be used in place of $\Gamma(\psi_2)$ in Eq. (6).

VII. Polarization of Absorption

The electric dipole moment that is responsible for the absorption event may develop along a particular molecular axis or it may be confined within a plane of the molecule. Determining the polarization generally involves working with oriented single crystals, although sometimes special effects can be exploited. For example, DNA molecules can be oriented in fluid solution by imposition of an electric field. Group theory can be used to explain the polarizations.

A. Pure Electronic Transitions

A pure electronic transition $\psi_0 \rightarrow \psi_i$ is allowed by the electric dipole mechanism if the direct product:

$$\Gamma(\psi_i) \times \Gamma(d) \times \Gamma(\psi_0) \tag{7}$$

contains the totally symmetric representation, or, equivalently, if the direct product:

$$\Gamma(\psi_i) \times \Gamma(\psi_0) \tag{8}$$

contains $\Gamma(d)$, where $\Gamma(d)$ is an irreducible representation spanned by one or more of the Cartesian axes. This symmetry argument is valid because the components of the electric dipole operator have the same symmetry, that is, they span the same irreducible representation(s), as the Cartesian axes x, y, and z.

To continue with the example of the metalloporphyrin, consider the π–π^* transitions. For this and all other planar aromatic molecules, the allowed π–π^* transitions are polarized within the molecular plane. If the the xy plane is taken to be the molecular plane, no π–π^* transition can be z polarized because all one-electron integrals of the form

$$\langle \psi_i | z | \psi_j \rangle \tag{9}$$

vanish, where ψ_i and ψ_j are one-electron MOs of the π system. The reason is that the π orbitals and the function z are each antisymmetric with respect to reflection through the molecular plane. Consequently, the integrand in Eq. (9) is also an antisymmetric function, and the integral over all space must vanish. It is easy to show that the **B** and the **Q** bands of the metalloporphyrin are allowed. Since the ground state has A_{1g} symmetry, Eq. (8) reveals that a transition to an E_u state occurs with dipole operators of E_u symmetry. This representation is spanned by the x and y axes. The strong degree of polarization that occurs is evident in the data presented in Figure 5(a).

B. Vibronically Allowed Transitions

For vibronically allowed transitions, the symmetry of the promoting mode ξ_k helps determine the polarization. Consider the transition from the lowest vibrational level of the ground electronic state ψ_0 to the first excited vibrational level of ψ_i. The arrow in Figure 6 indicates an example of this type of transition. It follows from the previous discussion that the transition is vibronically allowed for excitation by an electric field oscillating along axis d if Eq. (5) forms a basis for the representation spanned by normal mode ξ_k.

VIII. Absorption Intensity

Experiment reveals that the fraction of the incident light that is transmitted by an absorbing solution decreases exponentially with an increase in the concentration of the absorber or with an increase in the pathlength. This relationship is embodied in Beer's law:

$$A = -\log_{10}(I/I_0) = \varepsilon c l \tag{10}$$

where A is the absorbance, I is the intensity of the transmitted light, I_0 is the incident light intensity, ε is the molar extinction coefficient at the wavelength of investigation, c is the molar concentration of the sample, and l is the pathlength in centimeters. As defined, the absorbance is an additive function in that if two species absorb at the wavelength of interest:

$$A_{\mathrm{T}} = \varepsilon_1 c_1 l + \varepsilon_2 c_2 l \tag{11}$$

where A_T denotes the total absorbance. When there are two absorbing species, the fraction of light absorbed by species 1 is given as

$$\frac{\varepsilon_1 c_1}{\varepsilon_1 c_1 + \varepsilon_2 c_2}(1 - 10^{-A_T})$$

While the molar absorptivity is useful in quantifying the absorption intensity at a single wavelength, the overall absorption intensity is best described in terms of a dimensionless quantity f, which is termed the oscillator strength.[7] Because the intensity is distributed over a band with a finite width, f is calculated by means of the equation:

$$f = 4.32 \times 10^{-9} \int \varepsilon(\bar{\nu})\, d\bar{\nu} \tag{12}$$

where the bar indicates a wavenumber scale. If the absorption band can be approximated as a Gaussian distribution with half-width $\Delta\bar{\nu}_{1/2}$:

$$f \simeq 4.61 \times 10^{-9} \varepsilon_{max} \Delta\bar{\nu}_{1/2} \tag{13}$$

Within the spirit of this approximation, representative values of ε_{max} for different types of transitions are quoted in Table I.

Table I
Electronic Transitions and Band Intensities

Type of Transition	Range of ε
Orbitally and spin forbidden	$\varepsilon < 1$
Orbitally forbidden	$1 \leq \varepsilon \leq 10^2$
Spin forbidden	$\varepsilon \leq 10^2$ (depends on atomic number)
Allowed	$10^2 \leq \varepsilon \leq 10^6$

IX. Metal Complexes

An MO of a transition metal complex can usually be assigned predominantly metal or ligand character according to the participation of the constituent atomic orbitals. A qualitative MO scheme, appropriate for an octahedral complex, is presented in Figure 7, where the dashed lines indicate participation in the resulting MO. The *d* orbitals mainly participate in two of the MOs. One of these, the t_{2g} level, is formally nonbonding when strictly σ bonding ligands are present. If, on the other hand, the ligands are capable of π bonding, the t_{2g} level can have metal–ligand bonding or antibonding character, depending on the relative energies

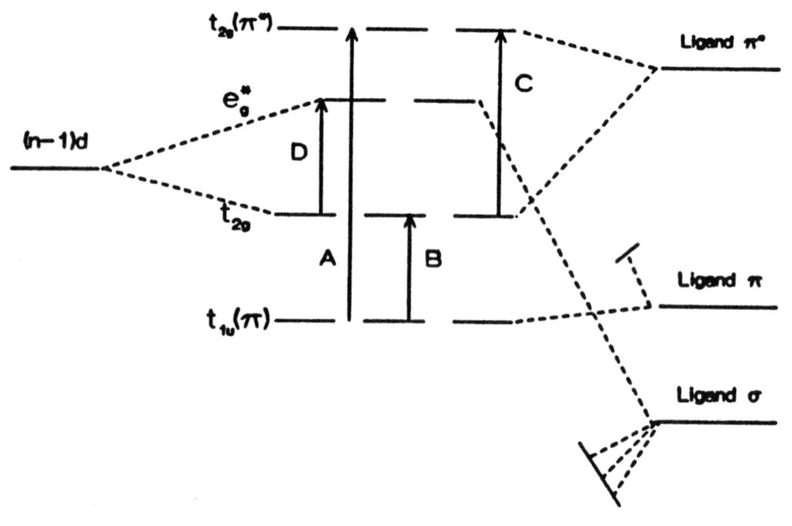

Figure 7
Selected transitions within an octahedral metal complex. Only a fraction of the total number of MOs are illustrated. For example, in the case of a cyano complex there are $6 \times 2 = 12\pi$ orbitals associated with the ligands, but only the t_{1u} molecular orbitals are included. Transitions are as follows: A is intraligand, B is ligand–metal CT, C is metal–ligand CT, and D is ligand field or d–d.

of the ligand π and π^* orbitals. The other MO with predominantly d orbital character, the $e_g{}^*$ level, is σ antibonding between the metal and the ligands.

Four types of transitions, labeled A–D in Figure 7, are apparent. Transition A represents an intraligand transition, analogous to the Soret band of the metalloporphyrin discussed above. Because of the makeup of the participant orbitals, transition B is called a ligand–metal charge transfer, or LMCT, transition, and transition C is an example of a metal–ligand charge transfer, or MLCT, transition. The charge-transfer transitions are specific for the complex and do not appear in the spectrum of either the free ligand or the free metal ion. Finally, transition D is formally regarded as a metal-centered or d–d transition and is Laporte forbidden when the complex has a center of symmetry. The Laporte selection rules are based on the symmetry with respect to the inversion center. A function can be gerade (symmetric) or ungerade (asymmetric) with respect to this operation. Since the dipole moment function is ungerade, only gerade–ungerade or ungerade–gerade transitions are symmetry allowed. Even in lower symmetries lacking an inversion center, transitions that are essentially d–d in nature tend to be weak. In contrast, intraligand transitions can be highly allowed; for example, Soret bands are very intense, and the molar extinction coefficient at the absorption maximum typically exceeds $10^5\ M^{-1}\,\text{cm}^{-1}$. Finally, electronically allowed charge-transfer transitions tend to have intermediate intensities. Because the intraligand transitions of porphyrins have been discussed in detail above, the development that follows will focus on charge-transfer and d–d transitions.

X. Ligand–Metal Charge Transfer

The electronic spectra of complexes involving high-valent metal ions and electron-rich ligands commonly reveal LMCT transitions. Classic examples of LMCT transitions arise in d^0 complexes. A partial MO diagram for a tetrahedral oxo anion is presented in Figure 8, which depicts only eight of the atomic orbitals of the ligand oxygens and the five d orbitals of the metal. The metal d orbitals are primarily distributed in two MOs in a tetrahedral complex, specifically the $t_2^*(M)$ and the $e^*(M)$ levels. In contrast with Figure 7, however, the doubly degenerate pair of levels falls at lower energy due to the difference in the type of orbital overlap that occurs with a tetrahedral ligand field. The lowest energy orbital excitation involves a transition from the HOMO, which is the ligand-localized, strictly nonbonding $t_1(L)$ level, to the LUMO, which is the $e^*(M)$ level. The term symmetries are determined by the direct product of e with t_1, which spans $t_1 + t_2$. Since either singlet or triplet multiplicity is possible, the terms associated with this excited-state configuration are

$$^1T_1,\ ^3T_1,\ ^1T_2, \quad \text{and} \quad ^3T_2$$

Since the ground electronic configuration involves only filled shells, the ground state is a $^1A_1$ term. Hence, the transitions to the $^1T_1$ and $^1T_2$ terms are spin allowed, but only the $^1T_2$ transition is orbitally allowed because the dipole moment operators form a basis for the t_2 representation of the tetrahedral point group. The next higher energy allowed transition can be associated with either $t_1(L) \rightarrow t_2^*(M)$ or $t_2(L) \rightarrow e^*(M)$ excitation since both excited configurations give rise to a $^1T_2$ term.

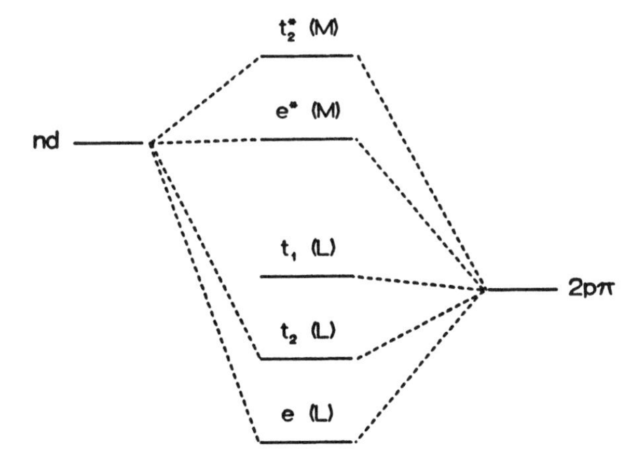

Figure 8
Partial MO scheme for a tetrahedral oxoanion. (M) denotes a MO with predominantly metal d character, and (L) denotes a MO composed of predominantly oxygen-centered atomic orbitals. The oxygen atom orbitals are the $2p_x$ and $2p_y$ orbitals, where the local axis system at each ligand is chosen such that the z axis points toward the metal.

A. Permanganate

As a specific example, consider the permanganate ion, which exhibits the absorption spectrum depicted in Figure 9. The first allowed transition is centered at about 530 nm and has been assigned to the $^1T_2$ term associated with the $t_1(L) \rightarrow e^*(M)$ excitation.[8,9] The vibronic structure that is resolved is attributed to the totally symmetric Mn–O stretching vibration within the excited state. The low-energy shoulder at about 640 nm is probably associated with the orbitally forbidden $^1T_1$ term, which derives from the same excited configuration. To higher energy there is a strong absorbance maximizing in the near-UV at about 310 nm, and this probably represents two overlapping transitions involving $t_1(L) \rightarrow t_2^*(M)$ and $t_2(L) \rightarrow e^*(M)$ excitations. If the influence of the electron–electron repulsions is ignored, the energies of the transitions originating in $t_1(L)$ imply that the energy difference between the $t_2^*(M)$ and $e^*(M)$ orbitals is approximately 14,000 cm^{-1}. In the case of the chromate ion, which involves a less oxidizing metal center, analogous transitions occur but at higher energies. Thus, a solution containing the permanganate ion is purple, but solutions of the chromate ion are yellow because the long-wavelength transition maximizes at 350 nm and merely tails into the visible. The next higher energy transition of the chromate ion is observed just below 260 nm. In crude terms, the reason for the shift in energy is the fact that Mn(VII) has a greater electron affinity than Cr(VI).

As an aside, it is worth noting that the designation "charge transfer" may be more meaningful in regard to the change in configuration rather than the net distribution of the electron density. Recent theoretical calculations confirm that the dominant configuration involved in the lowest energy $^1T_2$ excited state is consistent with $t_1(L) \rightarrow e^*(M)$ excitation; however, they suggest that there is little or no *net* transfer of electron density.[10,11] Although the calculations show that electron density is introduced into an orbital with metal character and removed

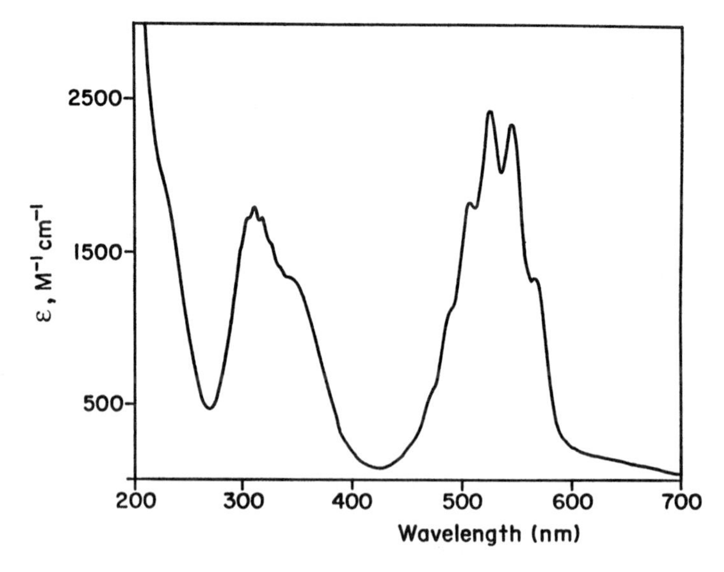

Figure 9
Visible and near-UV absorption spectrum of MnO_4^- in aqueous solution.

from a ligand-centered orbital, there is concomitant orbital relaxation that influences the composition of all other occupied orbitals. The net result of the orbital relaxation is a compensating shift of electron density from the metal back to the ligands.

B. Other Examples of LMCT Transitions

The data from the oxo anion systems reveal that the energy of a LMCT transition shifts when there is a change of the central metal. A similar effect is observed in the absorption spectra of a series of metal-substituted derivatives of azurin, a bacterial protein that normally contains a copper ion. The X-ray structure shows that in the native protein, copper forms relatively short bonds to a thiolate sulfur and two imidazole nitrogens from protein side chains. There is a rather long bond to a methionine sulfur as well, and the coordination geometry can be roughly described as pseudotetrahedral as in structure **1**. In solution, the copper-containing

(His)N — Cu — N (His), S(Cys); (Met)S

1

protein exhibits a deep blue color due to a strong absorption band that maximizes at 626 nm. As can be seen in Figure 10, when Cu(II) is replaced by Ni(II), the absorption maximum shifts to 440 nm. And when Co(II) is bound instead, the

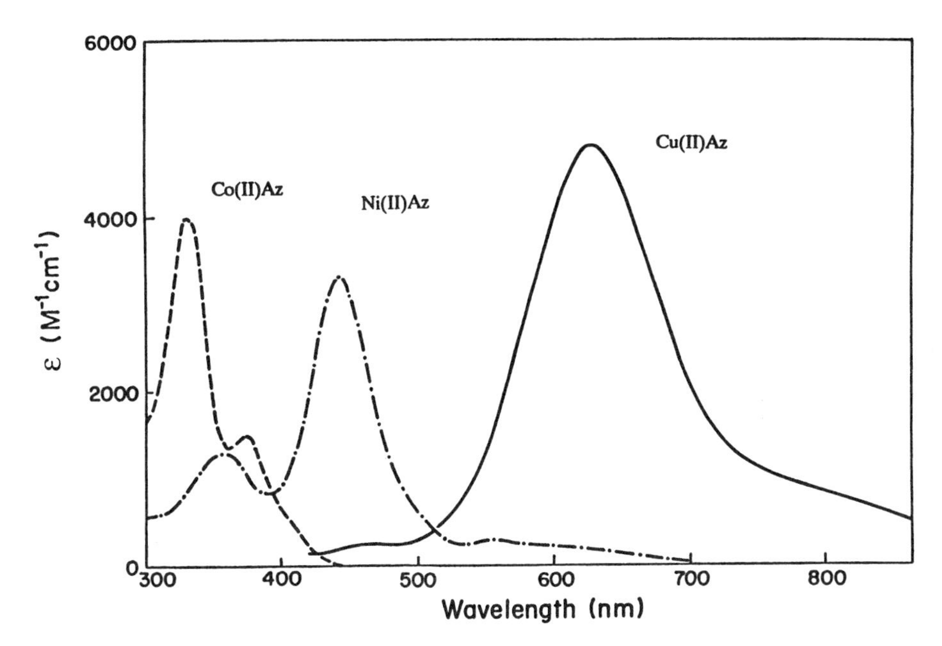

Figure 10
Visible and near-UV absorption spectra of azurin derivatives in aqueous solution. [The data are taken with permission from Tennent and McMillin.[12]] [In this figure – – – is Co(II) azurin; – · – is Ni(II) azurin, and —— is Cu(II) azurin].

maximum is at 330 nm. These results are consistent with a LMCT assignment because of the increase in electron affinity of the metal ion with a shift to the right within the periodic table.[12]

An empirical correlation proposed by Jorgensen[13] can be used to treat these observations in a more quantitative fashion. In this scheme, for a given coordination geometry, each metal ion is assigned an optical electronegativity χ_M that reflects its ability to accept the electron in a charge-transfer transition. Similarly, each ligand is assigned an optical electronegativity χ_L in accordance with its ability to release the electron. The transition energy can then be calculated as in Eq. (14):

$$\bar{\nu} = 30\,(\chi_L - \chi_M) \tag{14}$$

where the energy is in units of kilokaisers (1 kK = 1000 cm^{-1}). Optical electronegativity parameters are assigned empirically by correlating data from a variety of systems, and a selected sample of values is compiled in Table II. Note that the χ_M value is tabulated for a specific coordination geometry and that the χ_L value differs depending on whether the excitation originates in an orbital with σ or π symmetry with respect to the metal–ligand bond.

Equation (14) is valid as written for a d^0 system, but certain corrections must be applied for other ions. One of the effects is illustrated in Figure 11 where, for

Table II
Optical Electronegativities

Metal Ion	O_h	T_d	Ligand	π	σ^a
V (IV)	2.6		F^-	3.9	(4.4)
V (III)	1.9		Cl^-	3.0	3.4
Cr (III)	1.8–1.9		Br^-	2.8	3.3
Fe (III)	2.4–2.5		I^-	2.5	3.0
Fe (III)	2.1[b]		H_2O	3.5	
Os (III)	1.95[b]		NH_3		3.3
Fe (II)		1.8	CN^-	2.8	
Co (III)	2.3[b]		N_3^-	2.8	
Co (II)		1.8–1.9	RS^-	2.5	
Ni (II)		2.0–2.1			
Cu (II)		2.3–2.4			
Zn (II)		1.2			

[a] The values in parentheses are estimated.
[b] Low-spin complexes.

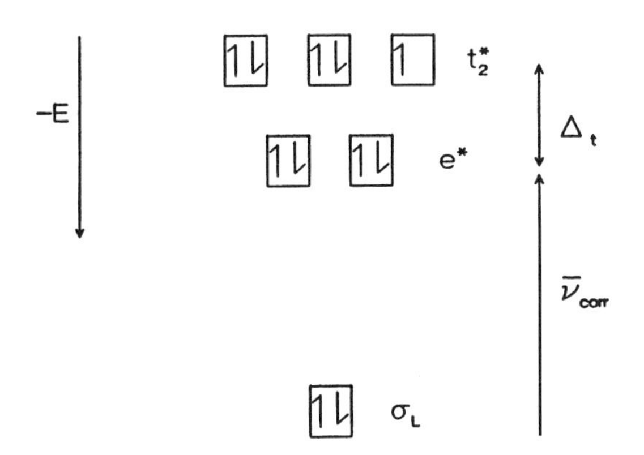

Figure 11
Definition of $\bar{\nu}_{corr}$. If the transition terminates in the $t_2{}^*$ orbital of a tetrahedral complex, Δ_t must be subtracted from the transition energy.

convenience, the d orbitals of a Cu(II) center are split as in strict tetrahedral symmetry. Jorgensen originally assigned the reference level to be $e^*(M)$ for a tetrahedral complex, but the transition terminates in $t_2{}^*(M)$ for a d^9 system; hence, the observed transition energy must be corrected for the d level splitting:

$$\bar{\nu}_{corr} = \bar{\nu}_{obs} + \Delta_t$$

When the metal center contains more than one unpaired electron, Jorgensen found that it is necessary to correct for the change in the spin pairing energy (SPE) as well. For an ion with n electrons in the d shell and a total spin S, the SPE is given as

$$\mathrm{SPE} = 7B\left[\frac{3}{4}n\left(1 - \frac{n-1}{9}\right) - S(S+1)\right]$$

where B is an experimentally determined parameter that is a measure of the electron–electron interactions. This parameter is discussed in more detail below in the section on ligand field transitions.

Consider, for example, a tetrahedral Ni(II) complex with four electrons in the t_2 shell. The most stable arrangement is the one with two unpaired electrons. However, one of the latter is paired with the incoming electron as a result of charge transfer, and the increase in the local electron–electron repulsion interaction is reflected in the transition energy. An even larger correction is required for Co(II), which has three unpaired d electrons in the ground state. The data relevant to the azurin derivatives are compiled in Table III, and the results are consistent with the intense absorption being assigned to a S(cysteine)$\rightarrow M^{II}$ absorption. Equation (14) is useful, but for most transition metal systems quantitative applications depend on the availability of reliable estimates of Δ_t and B, which come from the analysis of the d–d transitions (see below).

Another example of a biological system with important LMCT transitions is the oxygenated form of hemocyanin. Hemocyanins are the oxygen transport proteins in arthropoda and mollusca. The dioxygen-binding site within the pro-

Table III
Energy Quantities ($\times 10^{-3}$ cm^{-1}) for LMCT Transitions in Azurins

Metal Derivative	$\bar{\nu}_{obs}$	SPE	Δ_t	$\bar{\nu}_{corr}$
Cu^{II}azurin	16.0	0.0	−8.2	7.8
Ni^{II}azurin	22.7	3.3	−5.9	13.5
Co^{II}azurin	32.3	6.8	−4.9	20.6

tein involves two Cu(I) ions bound to the protein by imidazole groups from the side chains of various histidine residues. When dioxygen binds, an intense transition appears in the UV region at about 345 nm ($\varepsilon \approx 20{,}000\ M^{-1}\,cm^{-1}$) and a less intense transition appears in the visible region at about 575 nm ($\varepsilon \approx 1000\ M^{-1}\,cm^{-1}$). Dioxygen binding involves formal oxidation of the copper centers with formation of a peroxide moiety that binds to each copper in an edge-on fashion. Although at each copper center there are three histidine ligands, one forms a relatively weak bond. Hence, Solomon et al.[14] used the idealized D_{2h} structure in Figure 12 to assign the transition of the oxygenated protein. Figure 12 also contains a partial MO scheme derived from the filled π^* orbitals of the bridging peroxide and the half-filled d_{xy} orbital of each copper center. The out-of-

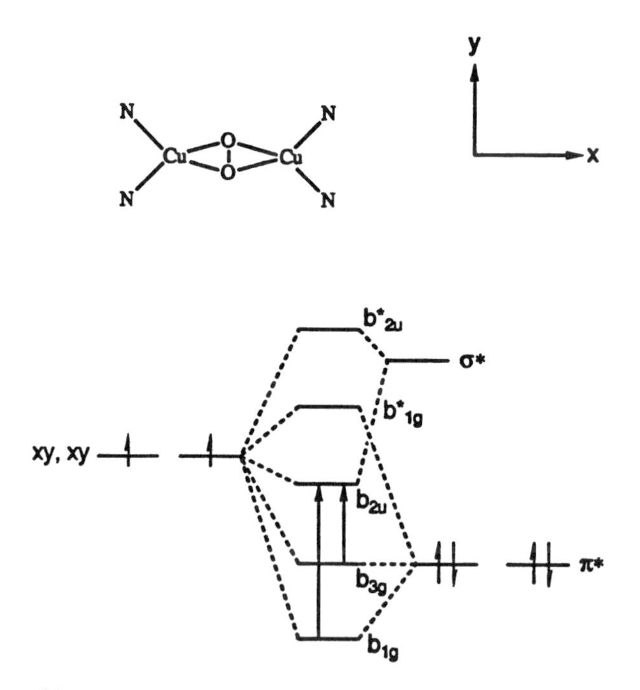

Figure 12
Schematic MO diagram for relevant orbitals of oxygenated hemocyanin. The z axis is orthogonal to the plane of the Cu_2O_2 unit.

plane π^* orbital of the peroxide moiety forms a basis for the b_{3g} representation and has no net overlap with the d_{xy} orbitals of the metal centers. In contrast, what was the other π^* orbital of the ligand engages in sigma bond formation with the metal centers and, hence, moves to lower energy. At first glance, one anticipates a $b_{1g}{}^2b_{3g}{}^2b_{2u}{}^2$ configuration for the ground state, consistent with the fact that the oxygenated protein is diamagnetic. However, this configuration does not provide a good description of the $^1A_{1g}$ ground state because the two electrons in the HOMO would spend too much time together in the same d_{xy} orbital.[15] In reality, the motion of the two electrons is correlated, and when one electron is at the copper center on the left, the other electron tends to be on the right. To account for the correlation, a proper description of the ground state includes an admixture of the $b_{1g}{}^2b_{3g}{}^2b_{1g}{}^{*2}$ configuration, and low-lying CT excited states involve mixed configurations, too. The upshot is that there are two symmetry-allowed peroxide-to-copper CT transitions, which terminate in the $^1B_{3u}$ and $^1B_{1u}$ excited states derived from $b_{1g} \rightarrow b_{2u}$ and $b_{3g} \rightarrow b_{2u}$ excitations, respectively. On energy grounds, the transition to the $^1B_{3u}$ state can be assigned to the band at 345 nm.

XI. Metal–Ligand Charge Transfer

A. Cu(I) Diimines

The opposite, or inverted, type of CT often occurs in complexes involving low-valent, oxidizable metal ions and π acid ligands. As an example, consider the pseudotetrahedral $Cu(GLI)_2{}^+$ complex, where GLI is N,N'-di-*tert*-butyl glyoxaldiimine, also known as N,N'-di(*tert*-butyl)-ethylenediimine, which is shown in structure **2**, where *t*-Bu denotes *tert*-butyl. The absorption spectrum, presented in

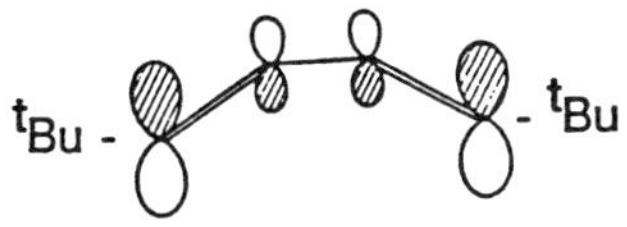

^{t}Bu - … - ^{t}Bu

2

Figure 13, consists of one major band in the visible with a maximum at about 19,000 cm^{-1} along with a shoulder and another weaker transition toward higher energies.[16] The spectral analysis begins with the one-electron energy level scheme, a part of which is presented in Figure 14. In D_{2d} symmetry, the d orbitals that would span the t_2 representation in tetrahedral symmetry are split into two levels, b_2 and e. Back-bonding is present because the filled $e(xz, yz)$ level of the metal ion interacts with the $e(\psi)$ orbitals of the ligands. The basis set for $e(\psi)$ consists of the LUMO from each of the GLI ligands. The LUMO of the GLI ligand is a π^* orbital, which is sometimes called the ψ orbital because it is antisymmetric with respect to the C_2 axis that passes through the ligand. It is depicted schematically in the structure **2** drawn for the ligand.

According to Figure 14, the lowest energy MLCT transitions are expected to be associated with the $b_2 \rightarrow e^*$ and $e \rightarrow e^*$ excitations of the complex. The for-

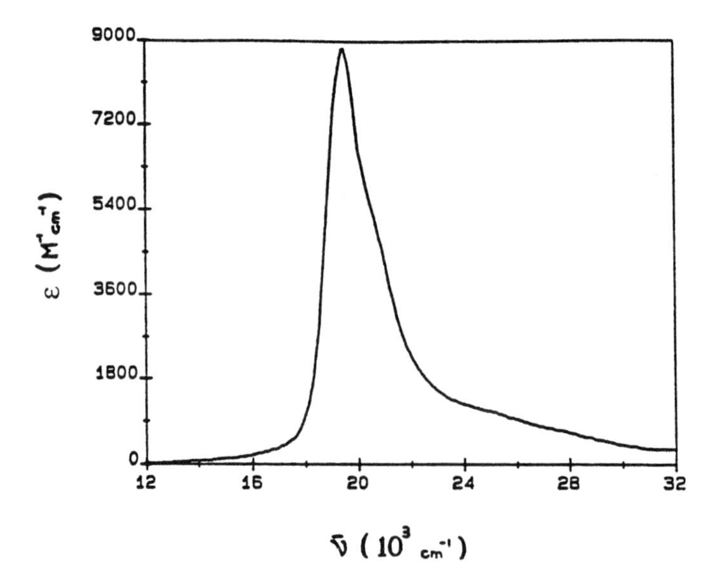

Figure 13
Absorption spectrum of $[Cu(GLI)_2]^+$ in ethanol at room temperature. [The spectrum has been adapted from one drawn in Daul et al.[16]]

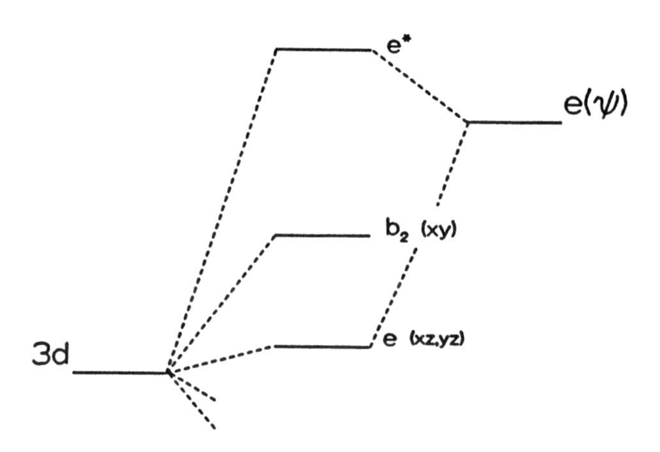

Figure 14
Partial MO scheme for $[Cu(GLI)_2]^+$. The LUMOs of the ligands form a basis for the e representation of the D_{2d} point group, and they are capable of π back-bonding with the d_{xz} and d_{yz} orbitals of the metal. The placement of the energy levels is schematic.

mer excitation gives rise to ^1E and ^3E terms, and the transition to the ^1E term is allowed with xy polarization since the ground state is a $^1A_1$. The singlet states associated with $e \rightarrow e^*$ excitation are

$$^1A_1, ^1A_2, ^1B_1, \quad \text{and} \quad ^1B_2$$

and the transition to $^1B_2$ is z polarized. The band intensities give a clue to one of the assignments. Mulliken used perturbation theory to show that CT bands have relatively high intensities when the acceptor orbital(s) mixes (mix) with the donor orbital(s).[17] When this is the case, what Mulliken called the CT term makes an important contribution to the transition dipole moment. For $Cu(GLI)_2^+$, the CT term contributes to the transition moment for the $e \rightarrow e^*$ transition because $e(\psi)$ mixes with $e(xz, yz)$, but it does not appear in the transition moment for the $b_2 \rightarrow e^*$ excitation. The reason is the b_2 orbital is orthogonal to the $e(\psi)$ set. Note that the CT term characteristically generates a dipole moment along the axis that connects the metal and the ligand centers.[17,18] In this case, it is the z-axis. Therefore, on intensity grounds the band at 19,000 cm^{-1} can be assigned to the $^1B_2$ term. In view of the nodal pattern within the orbital, injecting an electron into the ψ orbital would be expected to alter the equilibrium internuclear distances among the atoms of the ligand. Indeed, the shoulder that is displaced about 1500 cm^{-1} to the high-energy side of the MLCT maximum can be assigned as a vibronic transition associated with the totally symmetric C=N stretching motion within the excited state.

Similar MLCT transitions are observed in Cu(I) complexes of 2,2′-bipyridines (bpy), 1,10-phenanthrolines (phen), and other unsaturated π acid ligands. Since these transitions tend to fall in the visible and below the onset of intraligand absorption, the ligands can be used for the colorimetric determination of copper. In contrast, the Ag(I) analogues are colorless compounds because the Ag(II) oxidation state is much less accessible.

B. MLCT Transitions in Metal Carbonyls

Carbon monoxide is a classic π acceptor ligand, hence MLCT transitions are expected to occur in the spectra of metal carbonyls as well. The Group 6 (VIB) carbonyls serve as a useful case study. A partial MO scheme, modeled after the calculations of Beach and Gray,[19] is presented in Figure 15. In the case of an

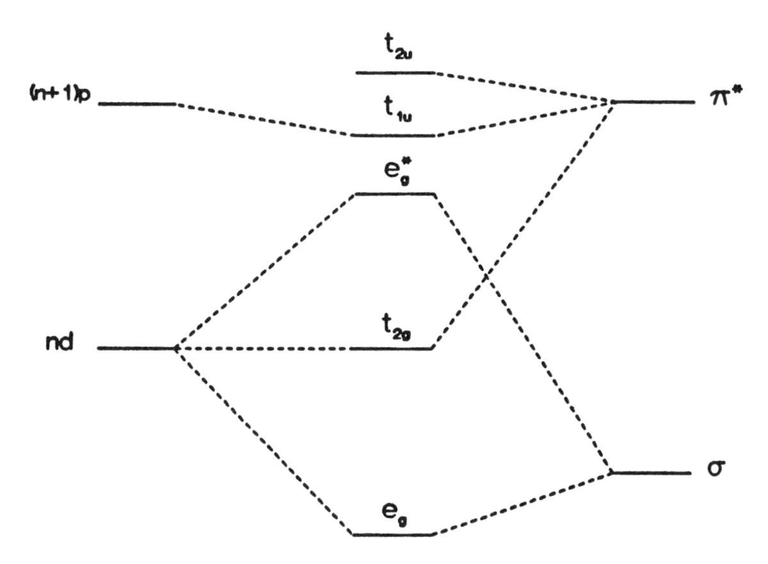

Figure 15
Partial MO scheme for a $M(CO)_6$ system. Only a few of the levels involving the π^* orbitals of the CO ligands are depicted.

octahedral d^6 system, the t_{2g} level and all lower energy levels are filled, hence the ground state is classified as $^1A_{1g}$. In a CT excited state an electron resides in a MO derived from the π^* acceptor orbitals of the six CO ligands, which span the representations:

$$t_{1g} + t_{2g} + t_{1u} + t_{2u}$$

Since the dipole operators transform as t_{1u}, the only allowed transitions are those that terminate in a $^1T_{1u}$ term. Consideration of the possible direct products shows that $t_{2g} \rightarrow t_{1u}$ or $t_{2g} \rightarrow t_{2u}$ excitation gives rise to T_{1u} terms with singlet and triplet multiplicities. Accordingly, two relatively low-lying MLCT states are predicted to be allowed, and the spectrum of $Mo(CO)_6$ in Figure 16 reveals a moderately intense transition in the UV at about 290 nm and another at about 225 nm, which have been assigned to MLCT states. Because the $t_{2g}(d)$ orbitals of the metal mix with the $t_{2g}(\pi^*)$ orbitals of the ligands, contributions from the CT term are possible. (The shoulders on either side of the lower energy CT transition in Fig. 16 have been assigned to formally forbidden d–d states, which are discussed in more detail below.)

The t_{1u} level, which is mainly derived from the π^* orbitals of the ligands, mixes with the $(n+1)$p orbitals of the metal and is calculated to occur at lower energy than the ligand-centered t_{2u} level. Indeed, Gray and Beach[19] assigned the higher energy of the two MLCT transitions as the one that terminates in the t_{2u} level. It is interesting to note that the relative intensities of the two transitions are reminiscent of the porphyrin system with its intense Soret band and much weaker Q_0 band. Configuration interaction may be an important effect in the excited state manifold of the metal carbonyls as well.

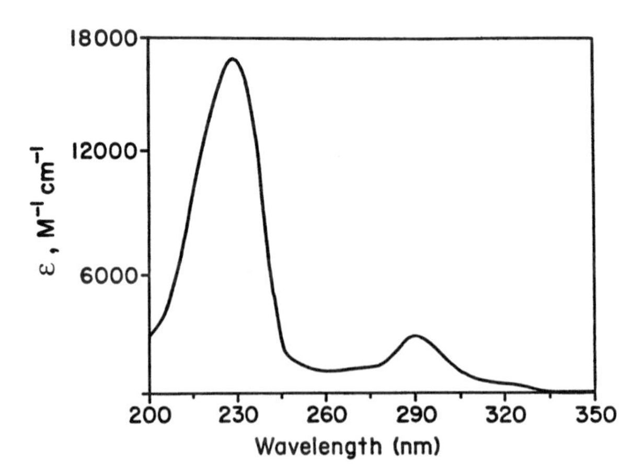

Figure 16
Electronic spectrum of $Mo(CO)_6$ in the vapor phase. The shoulders at around 270 and 320 nm are believed to be d–d transitions. [This spectrum has been constructed from data presented in Gray and Beach.[19] Reprinted with permission from Gray, H. B.; Beach, N. A. *J. Am. Chem. Soc.* **1963**, *85*, 2922–2927. Copyright © 1963 American Chemical Society.]

XII. Ligand Field Transitions

A. A T_d Example

This section begins with a treatment of the metal-centered transitions of a tetrahedral complex. These transitions are often called ligand field transitions or d–d transitions because the MOs involved are largely based at the metal. As indicated above, these orbitals are often metal–ligand antibonding orbitals; however, so long as attention is confined to this group of orbitals there is no confusion if the symmetry labels do not carry the asterisks used previously. The relative energies of the "d orbitals" in a tetrahedral ligand field can be rationalized by recognizing that the four ligands occupy alternate corners of a cube centered about the metal ion. The convention is that the *x, y,* and *z* axes pass through the faces of the cube. Consequently, the lobes of the three t_2 orbitals are directed at the edges of the cube, while the lobes of the two e orbitals are directed at the faces. With a tetrahedral distribution of ligand orbitals better overlap occurs with the t_2 set, and these orbitals are destabilized to a greater degree than the e set (Fig. 17). If Co(II) is the central ion, there are seven d electrons to be distributed among the d orbitals, and it is clear that the lowest energy configuration is $e^4t_2{}^3$. There are several terms associated with the ground configuration that have yet to be determined, but the e^4 part of the configuration can be ignored during the exercise since it is a closed-shell configuration. Therefore, the problem reduces to determining the terms associated with the $t_2{}^3$ configuration. At the outset, it is clear that, with allowance for the m_s quantum number, there are $(6 \times 5 \times 4)/(3 \times 2 \times 1) = 20$

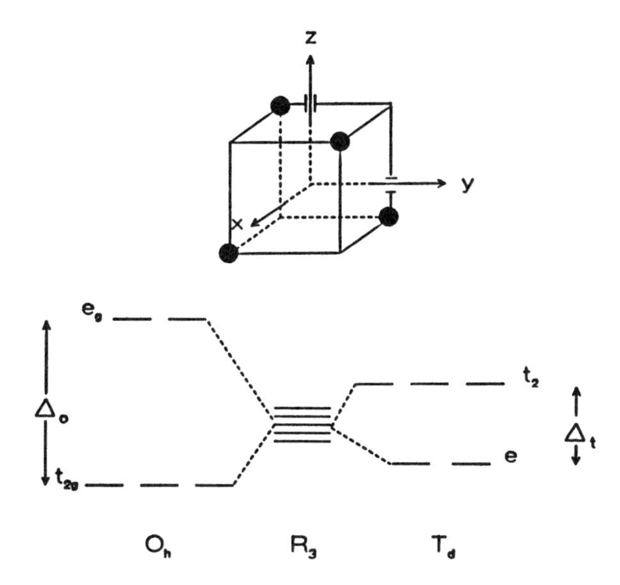

Figure 17
Splitting diagrams for tetrahedral and octahedral ligand fields. In the case of an octahedral complex, the ligands would lie along the Cartesian axes in the faces of the cube.

possible wave functions associated with the t_2^3 configuration. To see this, note that there are six possible orbital assignments for the first electron because there are three spatial orbitals and two possible spin orientations within each. That leaves five choices for the second electron and four for the third. Multiplication gives the total number of possibilities once we divide by 3! to exclude those assignments that are simply permutations of electrons within a given set of three orbitals. Elucidation of the terms is not quite as straightforward a process as it has been in the examples above. Thus, the triple direct product of t_2 with itself yields

$$a_1 + a_2 + 2e + 3t_1 + 4t_2 \tag{15}$$

Now some of these symmetry species have to be spurious since spanning Eq. (15) requires 27 functions, and as shown above, there are only 20 independent wave functions associated with the $e^4t_2^3$ configuration. The theorem stated in Section XII.B provides a way to identify the terms.

B. Terms for the $\gamma_1{}^n\gamma_2{}^m$ Configuration

The theorem can be stated as follows: Suppose a system presents two shells with symmetries γ_1 and γ_2 that contain n and m electrons, respectively. Let ${}^{2S_1+1}\Gamma_1$ be one of the terms associated with the $\gamma_1{}^n$ configuration, and let ${}^{2S_2+1}\Gamma_2$ be a term associated with the $\gamma_2{}^m$ configuration. Then the direct product $\Gamma_1 \times \Gamma_2$ spans the symmetry species of corresponding terms associated with the combined $\gamma_1{}^n\gamma_2{}^m$ configuration, and each symmetry type occurs with all possible values of the total spin S which are

$$S_1 + S_2, S_1 + S_2 - 1, \ldots, |S_1 - S_2| \tag{16}$$

Surprising as it may seem, by judicious use of symmetry it is possible to apply the same theorem to determine the terms of the $\gamma_1{}^n$ and $\gamma_2{}^m$ configurations.

To illustrate the use of the theorem, return to the t_2^3 configuration. Violations of the Pauli principle could easily be avoided if each of the three d orbitals had a unique symmetry designation. Since this is true in a lower symmetry point group where orbital degeneracy is removed, it is convenient to ignore some of the symmetry and to work in a subgroup of the actual point group for bookkeeping purposes. Therefore, the objective is to find a subgroup of T_d such that d_{xz}, d_{yz}, and d_{xy} are each the basis of a separate one-dimensional representation. A solution lies in Table IV, which illustrates how irreducible representations correlate with one another in a series of related point groups. Inspection of Table IV shows that in the C_{2v} subgroup t_2 orbitals transform as $a_1 + b_1 + b_2$. In C_{2v} symmetry, the simplest configurations are the six wherein two electrons are paired up in one of the three orbitals. They generate six doublets:

$$2\,{}^2A_1 + 2\,{}^2B_1 + 2\,{}^2B_2 \tag{17}$$

Table IV
Correlations between Selected Point Groups

R_3	O_h	T_d	D_{4h} [a]	C_{4v} [a]	C_{2v}
s	a_{1g}	a_1	a_{1g}	a_1	a_1
p [b]	t_{1g}	t_1	$a_{2g} + e_g$	$a_2 + e$	$a_2 + b_1 + b_2$
	e_g	e	$a_{1g} + b_{1g}$	$a_1 + b_1$	$a_1 + a_2$
d	t_{2g}	t_2	$b_{2g} + e_g$	$b_2 + e$	$a_1 + b_1 + b_2$
	a_{2g}	a_2	b_{1g}	b_1	a_2
f [b]	t_{1g}	t_1	$a_{2g} + e_g$	$a_2 + e$	$a_2 + b_1 + b_2$
	t_{2g}	t_2	$b_{2g} + e_g$	$b_2 + e$	$a_1 + b_1 + b_2$
	a_{1g}	a_1	a_{1g}	a_1	a_1
	e_g	e	$a_{1g} + b_{1g}$	$a_1 + b_1$	$a_1 + a_2$
g	t_{1g}	t_1	$a_{2g} + e_g$	$a_2 + e$	$a_2 + b_1 + b_2$
	t_{2g}	t_2	$b_{2g} + e_g$	$b_2 + e$	$a_1 + b_1 + b_2$
	e_g	e	$a_{1g} + b_{1g}$	$a_1 + b_1$	$a_1 + a_2$
h [b]	$(2)t_{1g}$	$(2)t_1$	$(2)a_{2g} + (2)e_g$	$(2)a_2 + (2)e$	$(2)a_2 + (2)b_1 + (2)b_2$
	t_{2g}	t_2	$b_{2g} + e_g$	$b_2 + e$	$a_1 + b_1 + b_2$
	a_{1g}	a_1	a_{1g}	a_1	a_1
	a_{2g}	a_2	b_{1g}	b_1	a_2
i	e_g	e	$a_{1g} + b_{1g}$	$a_1 + b_1$	$a_1 + a_2$
	t_{1g}	t_1	$a_{2g} + e_g$	$a_2 + e$	$a_2 + b_1 + b_2$
	$(2)t_{2g}$	t_2	$b_{2g} + e_g$	$b_2 + e$	$a_1 + b_1 + b_2$

[a] The D_{4h} and C_{4v} point groups are not subgroups of T_d. All three point groups have subgroups in common, however.

[b] The gerade representations are appropriate for terms derived from d-electron configurations. The p, f, and h orbitals have, however, ungerade symmetry.

For example, the $b_1{}^2a_1{}^1$ and the $b_2{}^2a_1{}^1$ configurations each generate a $^2A_1$ term. The other possible configuration is $a_1{}^1b_1{}^1b_2{}^1$. The $a_1{}^1b_1{}^1$ component gives rise to $^1B_1 + ^3B_1$, while the $b_2{}^1$ configuration gives rise to a $^2B_2$ term.

Since the direct product of B_1 with B_2 is A_2, according to Eq. (16) the resultant terms associated with the $a_1{}^1b_1{}^1b_2{}^1$ configuration are $2^2A_2 + {}^4A_2$. Therefore, in C_{2v} symmetry the terms spanned by the three electrons are

$$2^2A_1 + 2^2A_2 + 2^2B_1 + 2^2B_2 + {}^4A_2 \tag{18}$$

As expected, the set of irreducible representations in Eq. (18) accounts for exactly 20 wave functions. The task that remains is to use Table IV once more to relate these representations to the ones appropriate for the real symmetry, which is T_d. Since there is only one quartet term, it is convenient to begin with the $^4A_2$ term. The only one-dimensional representation in Eq. (15) that correlates with a_2 in C_{2v} symmetry is the species a_2; therefore, one of the terms associated with the $t_2{}^3$ configuration in T_d symmetry is a $^4A_2$. In other words, the a_2 representation in

Eq. (15) does occur, and it occurs as a quartet. Other obligatory correlations are also evident. Thus, Table IV reveals that the $2^2B_1 + 2^2B_2$ terms in Eq. (18) have to be associated with t_1 and/or t_2 species from Eq. (15). Now there cannot be 2^2T_2 terms because there would be two A_2 terms left unaccounted for in Eq. (18). There cannot be 2^2T_1 terms either because that would mean that a $^2A_1$ term in Eq. (18) could not be correlated with Eq. (15). Hence, there must be a $^2T_1$ and a $^2T_2$ term associated with the tetrahedral ion. The two terms from Eq. (18) that remain to be correlated are $^2A_1 + {}^2A_2$. Inspection shows that they must correlate with a ^2E term. Therefore, in T_d symmetry the terms associated with the $t_2{}^3$ configuration are

$$^4A_2 + {}^2E + {}^2T_1 + {}^2T_2 \tag{19}$$

The unused symmetry species from Eq. (15) are irrelevant. The following scheme summarizes the correlation between the terms of the two point groups:

C_{2v}	T_d
$^4A_2$	$^4A_2$
$^2A_1 + {}^2A_2$	^2E
$^2A_1 + {}^2B_1 + {}^2B_2$	$^2T_2$
$^2A_2 + {}^2B_1 + {}^2B_2$	$^2T_1$

Terms associated with other excited-state configurations of the tetrahedral ion are presented in Table V.

Table V
Terms of a d^7 Ion in T_d Symmetry

Configuration	Terms
$e^1t_2{}^6$	^2E
$e^2t_2{}^5$	$^4T_1, 2^2T_1, 2^2T_2$
$e^3t_2{}^4$	$^4T_1, {}^4T_2, 2^2T_1, 2^2T_2, 2^2E, {}^2A_1, {}^2A_2$
$e^4t_2{}^3$	$^4A_2, {}^2T_1, {}^2T_2, {}^2E$

From Hund's rules, the ground term in Eq. (19) must be the $^4A_2$. Accordingly, the only spin-allowed transitions are those that terminate in quartet states derived from the higher lying configurations. A crude diagram, portraying the quartet energies, is presented on the right-hand side of Figure 18. A sketch of the vis–near-IR absorption spectrum of $[CoCl_4]^{2-}$ in solution is given in Figure 19. Two transitions are plainly resolved in the solution spectrum, and a third has been observed in the IR region from solid samples. The bands at about 6000 and

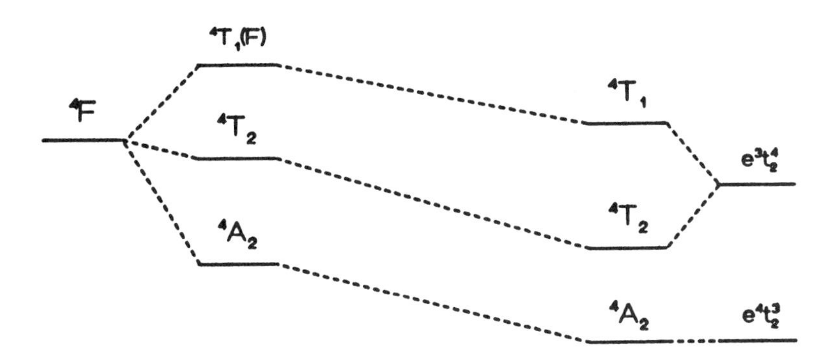

Figure 18
A correlation diagram relating the weak- and strong-field limits for a d^7 ion in a tetrahedral ligand field. The ligand field strength increases from left to right in the diagram.

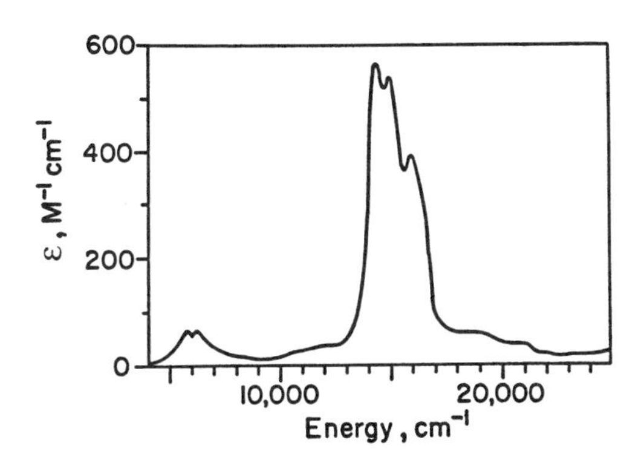

Figure 19
Absorption spectrum of $CoCl_4^{2-}$ in 10 *M* HCl.

about 15,000 cm^{-1} have been assigned as transitions to the two $^4T_1$ terms, while the transition in the IR has been assigned to the $^4T_2$ term. In the absense of quantitative information about the relative energies of the terms, however, it is difficult to justify these assignments. The assignments will become more obvious shortly when the analysis is couched in terms of a different set of zero-order states.

XIII. The Weak Field Versus the Strong Field Limit

The spectral assignments for $[CoCl_4]^{2-}$ are more transparent when the so-called weak field limit is adopted. Before this method can be used, however, it is necessary to understand the term structure of the free Co^{2+} ion.

A. Free Ion Terms

The term levels of the various d^n configurations found for transition metal ions are presented in Table VI. The reader can develop this table with the method

Table VI
Terms Associated with the d^n Configurations

Configuration	Terms
d^1, d^9	^2D
d^2, d^8	$^3F, {}^3P, {}^1G, {}^1D, {}^1S$
d^3, d^7	$^4F, {}^4P, {}^2H, {}^2G, {}^2F, (2)^2D, {}^2P$
d^4, d^6	$^5D, {}^3H, {}^3G, (2)^3F, {}^3D, (2)^3P, {}^1I, (2)^1G, {}^1F, (2)^1D, (2)^1S$
d^5	$^6S, {}^4G, {}^4F, {}^4D, {}^4P, {}^2I, {}^2H, (2)^2G, (2)^2F, (3)^2D, {}^2P, {}^2S$

described in Appendix I. The relative energies of the terms have been determined experimentally and can be expressed analytically as functions of the Racah parameters *A*, *B* and *C* (all positive quantities), which are determined by integrals involving the electron–electron repulsion operator e^2/r_{ij}.[20] For example, the terms associated with the d^7 configuration have the energies:

$$^4F = 3A - 15B$$

$$^4P = 3A$$

$$^2H = {}^2P = 3A - 6B + 3C$$

$$^2G = 3A - 11B + 3C$$

$$^2F = 3A + 9B + 3C$$

$$^2D = 3A + 5B + 5C \pm (193B^2 + 8BC + 4C^2)^{1/2}$$

For our purposes, however, the analytical forms are not needed, because the parameters will be evaluated empirically.

Note that the d^n and d^{10-n} configurations always possess the same terms. This can be understood by appealing to what is known as the hole formalism. The idea is that the same degrees of freedom are available to n electrons occupying the d subshell as are available to the n vacancies or "holes" associated with the d^{10-n} configuration. In general, gaps between terms spanned by the same d^n configuration are independent of the parameter A. Moreover, for any d^n configuration that has two terms with the maximum multiplicity (e.g., ^4F and ^4P in the case of the d^7 ion), the energy difference between the two does not depend on the value of C. The energy difference between the other terms does, however, depend on C, and both the B and the C parameters vary from ion to ion.

B. The Weak Field Limit

In the weak field limit, the ligand field is viewed as a perturbation on the free ion term structure. The energy expressions listed above reveal that the ^4F term is the ground term of the Co^{2+} ion, as predicted by Hund's rules. The other quartet term (^4P) occurs at higher energy. The separation is $15B$, and B has been experimentally determined to be 970 cm^{-1} for this ion. When the ligand field due to the four chloride ligands is introduced as a perturbation on the system, the point symmetry is lowered from spherical (R_3) to tetrahedral (T_d), and the orbital degeneracy is reduced. As can be deduced from Table IV, the F term splits into three new terms, $^4T_1$, $^4T_2$, and $^4A_2$. The ^4P term does not split but becomes another $^4T_1$ term, often denoted $^4T_1(P)$. (Note that the spin is unaffected because the ligand field operator does not couple to the internal degrees of freedom of the electron.)

The energy differences among the $^4T_1(F)$, $^4T_2$, and $^4A_2$ terms depend on the strength of the ligand field. A semiquantitative way of modeling the system is to assume that the ligand field interaction is basically an electrostatic perturbation, that is, a crystal field effect. In this case, a center of gravity is preserved in Figure 17 when the splitting occurs, and the e orbitals are stabilized by $\frac{3}{5}\Delta_t$ while the t_2 orbitals are destabilized by $\frac{2}{5}\Delta_t$. The dependence on the ligand field strength can then be represented with the diagram in Figure 18. The extreme left side of the diagram represents the weak field limit in which the effect of the ligand field is almost nil as, for example, when the ligands are far removed from the metal center. As the ligands move closer to the metal center, the strength of the perturbation increases, and the terms derived from ^4F steadily diverge from each other. These terms ultimately correlate with the quartet terms in Table V. If only terms from the free ion states are considered, there are no other terms with the same multiplicity and the same symmetry as the $^4A_2$ and the $^4T_2$ terms. Each, therefore, experiences only a first-order perturbation and shifts linearly with the field strength. Moreover, the slope is directly proportional to the ligand field splitting energy Δ_t in Figure 17. In fact, since in the strong field limit the $^4A_2$ and the $^4T_2$ terms correlate with the $e^4t_2{}^3$ and the $e^3t_2{}^4$ configurations, the ligand field energies are $-\frac{6}{5}\Delta_t$ and $-\frac{1}{5}\Delta_t$, respectively. In contrast, the $^4T_1(F)$ and the $^4T_1(P)$ terms have the same symmetry and multiplicity; hence, they interact, or mix, with each other under the influence of the ligand field perturbation. However, in the limit of a very weak ligand field, the interaction with the ^4P term is negligible, and the terms that split out of the free ion ^4F term retain a center-of-gravity relationship with each other. Accordingly, on the far left side of Figure 18, the $^4T_1(F)$ term is destabilized by $+\frac{3}{5}\Delta_t$ relative to the unperturbed ^4F term. With the aid of Figure 18 the electronic spectrum of $[CoCl_4]^{2-}$ is readily assigned. The low-energy, spin-allowed transition, which is observed in the IR, is assigned as the transition to the $^4T_2$ term; its energy serves as a direct measure of Δ_t. The transition at 6000 cm^{-1} is assigned to the $^4T_1(F)$ term, and the transition at 15,000 cm^{-1} is assigned to the $^4T_1(P)$.

If V_{el} denotes the Hamiltonian operator for the interelectron repulsions among the d electrons and V_{lf} denotes the ligand field operator that splits the d orbitals, the two limits in Figure 18 are described by the following inequalities:

$$\text{Weak field} \qquad V_{el} > V_{lf} \tag{20}$$

$$\text{Strong field} \qquad V_{lf} > V_{el} \tag{21}$$

The two effects act somewhat in opposition to each other. On the one hand, the electron–electron repulsions favor spreading the electrons around as much as possible within the d shell. On the other hand, the effect of the ligand field is to destabilize electron density in selected d orbitals, namely, those with lobes extending toward the ligands. Another way of saying the same thing is that in the weak field limit the term energies are closely related to those of the d^7 ion while in the strong field limit, the term energies are better described in the context of the various $e^n t_2^{7-n}$ configurations.

C. Extension to Other Ions

Even though a d^3 ion has the same term structure as a d^7 ion, Figure 18 cannot be used to describe the energy levels of a d^3 ion in a tetrahedral field, but, surprising as it may seem, the diagram is qualitatively correct for a d^3 ion in an octahedral ligand field. If a tetrahedral ligand field is applied to the d^3 ion, the ^4F terms split into $^4T_1$, $^4T_2$ and $^4A_2$ terms as before; however, the energy contribution from the ligand field has the opposite sign. Thus, the $^4T_1(F)$ term becomes the ground state because the most stable distribution of seven holes in the field of the ligands is the least stable distribution of seven electrons and vice versa. In octahedral symmetry, however, there is a compensating change in sign because the d-level splitting is inverted, and the triply degenerate set of d_{xz}, d_{yz}, and d_{xy} orbitals falls below the doubly degenerate set formed by the $d_{x^2-y^2}$ and d_{z^2} orbitals (Fig. 17). The only change required in Figure 18 for use with a d^3 ion in an octahedral field is that a subscript g must be added to the labels for the irreducible representations to designate the inversion symmetry of the d orbitals.

Figure 20 contains the same type of correlation diagram appropriate for a d^6 ion in an octahedral ligand field. The most striking aspect of the diagram is the predicted change in the ground state at higher ligand field strengths. In a weak field, the ground state is a quintet with four unpaired electrons, but in a strong field there are no unpaired electrons. A detailed analysis reveals that the switch to the low-spin form occurs when the ligand field splitting Δ_0 exceeds the energy

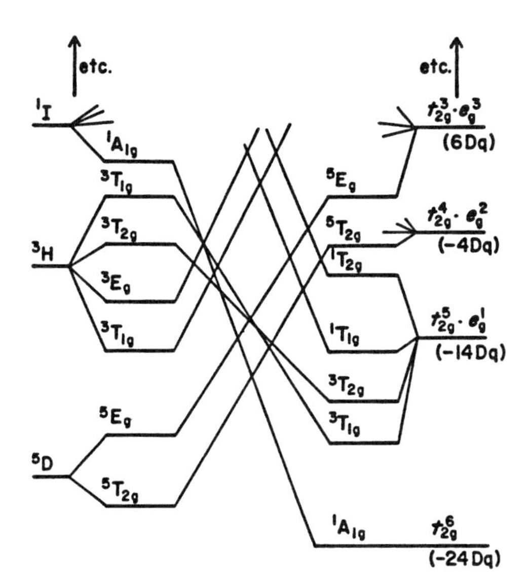

Figure 20
Schematic correlation diagram for a d^6 ion in an octahedral ligand field. The diagram is also relevant for a d^4 ion in a tetrahedral field except that the g and the u subscripts must be dropped from the symmetry labels.

required to pair two electrons in the same d orbital. Of course, the magnitude of the ligand field splitting Δ_O varies with the nature of the ligand. For example, the splitting is comparatively small for halide ligands, so $[CoF_6]^{3-}$ is high spin and has four unpaired electrons. On the other hand, the ligand field splitting is much larger for $[Co(NH_3)_6]^{3+}$, and the complex is diamagnetic. In keeping with Figure 20, two spin-allowed d–d transitions are observed in the spectrum of the yellow-orange hexaammine complex. The $^1A_{1g} \rightarrow {}^1T_{1g}$ transition is assigned to a band that maximizes at 472 nm, and the $^1A_{1g} \rightarrow {}^1T_{2g}$ transition is assigned to a band that maximizes at 338 nm. The ligand field splitting energy is even greater for the cyano complex, $[Co(CN)_6]^{3-}$, which is colorless because the lowest energy, spin-allowed d–d band maximizes in the near UV at 313 nm.

XIV. Tanabe–Sugano Diagrams

Correlation diagrams like those presented in Figures 18 and 20 are useful in a qualitative sense, but more detailed energy level schemes, known as Tanabe–Sugano diagrams, are available. In the latter, the theoretical energies, including interelectronic repulsions, of virtually all important terms are plotted as a function of the strength of the ligand field. Certain conventions are followed in order to allow for a two-dimensional presentation. One is that the ligand field strength is expressed as 10 Dq rather than Δ_O. Another convention is that the transition energy E and the ligand field splitting parameter are plotted in units of the interelectronic repulsion parameter B. Since the energies also depend on the inter-electronic repulsion parameter C, the plot is drawn for a stated value of the C/B ratio. In truth, this ratio varies from ion to ion, but it is usually around 4. Finally, the Tanabe–Sugano diagram only describes the energy differences between terms, and the ground state is therefore always assigned an energy of zero.

A. Application to the d^3 Configuration

A Tanabe–Sugano diagram for the d^3 configuration is presented in Figure 21. The diagram can be used for any d^3 ion in octahedral symmetry or for any d^7 ion in tetrahedral symmetry. However, the results hinge on the particular choice of the C/B ratio which, of course, varies from configuration to configuration. The diagram shows that the d^3 ion is expected to show three spin-allowed transitions from the $^4A_{2g}$ ground state. The use of the diagram can be illustrated with spectral data for the $[Cr(H_2O)_6]^{3+}$ ion, which shows weak absorptions

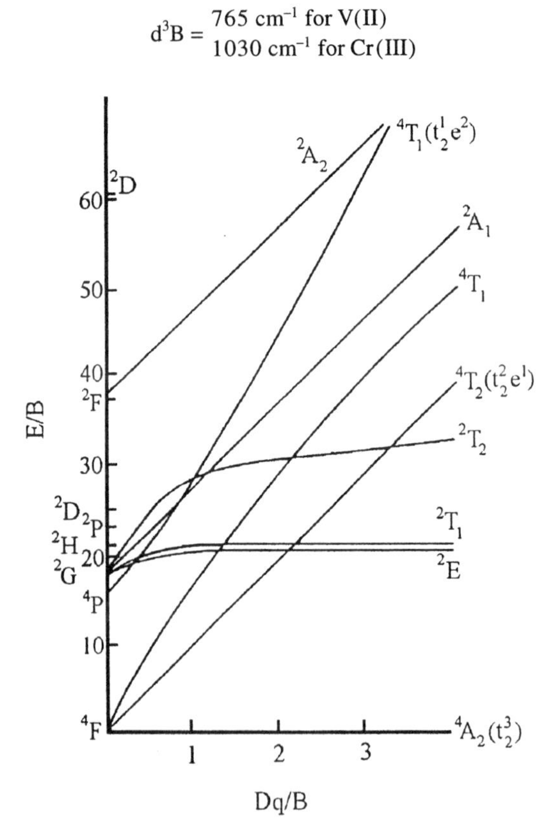

Figure 21
Tanabe–Sugano diagram for a d^3 ion in an octahedral ligand field or a d^7 ion in a tetrahedral ligand field. A C/B ratio of 4.5 is assumed. The superscript g should be added to the symmetry labels for octahedral ions. [Reprinted with permission from Tanabe, Y.; Sugano, S. *J. Phys. Soc. Jpn.* **1954**, *9*, Fig. 5, p. 769 and Fig. 8, p. 770.]

($\varepsilon \approx 15\ M^{-1}\,cm^{-1}$) at about 17,400 and at 24,500 cm^{-1} in solution. The lowest energy transition should occur to the $^4T_{2g}$ state, and since $B = 1030\ cm^{-1}$ for the gaseous Cr^{3+} ion, the E/B ratio of the term can be estimated to be 16.9. From the graph of the $^4T_{2g}$ term energy in Figure 21, a value of $E/B = 16.9$ on the ordinate corresponds with a Dq/B value on the abscissa of about 1.75. Hence, 10 Dq can be estimated to be $17.5 \times 1030\ cm^{-1}$, or about 18,000 cm^{-1}. At the same Dq/B value, the E/B value for the next highest spin-allowed transition to the $^4T_{1g}$ level is estimated to be about 25. The second spin-allowed transition is therefore expected at about 25,800 cm^{-1} in fair agreement with the experimental spectrum.

It is also possible to proceed by fitting the ratio of the transition energies in which case B can be treated as a parameter that is determined empirically. The ratio of the experimental transition energies is 1.41 for the $[Cr(H_2O)_6]^{3+}$ ion, and it can be shown by trial and error (use a ruler!) that the displacements of the $^4T_{1g}$ and $^4T_{2g}$ terms from the ground state satisfy this ratio when $Dq/B \approx 2.41$ and the E/B ratios are 33.6 and 23.7. The E/B ratios for the $^4T_{1g}$ and $^4T_{2g}$ transitions predict B values of 729 and 734 cm^{-1}, respectively. If B is assumed to be 730 cm^{-1}, the selected Dq/B ratio predicts 10 $Dq \approx 17{,}600\ cm^{-1}$. This analysis suggests that the interelectronic repulsion parameter B is significantly reduced in the complex in comparison with the free ion. One important difference is that in the complex the electrons are not found in pure d orbitals, since there is some degree of covalent bonding with the ligands. This has been called the nephelauxetic effect (see below).

B. Application to the d^6 Configuration

Figure 22 depicts the Tanabe–Sugano diagram for a d^6 ion. It is important to realize that the discontinuities in the curves are purely artifacts of the method of presentation. As demonstrated in Figure 20, the $^1A_{1g}$ term becomes the ground state at some value of Dq/B, and at this point the $^1A_{1g}$ term is assigned an energy of 0 for all higher values of Dq/B. Consequently, there is a sudden change in the slopes of the curves representing all higher energy terms. High-spin complexes are described by the left-hand side of the Tanabe–Sugano diagram, and low-spin complexes are described by the right-hand side. Inspection reveals there should be one spin-allowed d–d transition for a high-spin d^6 system and two low-lying spin-allowed d–d transitions for low-spin complexes.

The use of the diagram can be illustrated by the analysis of the spectrum of the low-spin complex $[Rh(NH_3)_6]^{3+}$ (Fig. 23). The spin-allowed d–d bands appear at 32,800 and 39,200 cm^{-1}, so the ratio of the energies is 1.20. Although a large extrapolation of the curves is required, one can show that the energies of the $^1T_{2g}$ and $^1T_{1g}$ terms are predicted to be in a ratio of 1.20 at a Dq/B value of about 7.5. From the corresponding E/B values, B can be estimated to be 460 cm^{-1}. Since $Dq/B = 7.5$, it follows that 10 $Dq \approx 34{,}200$. According to the Tanabe–Sugano diagram, there should be a transition to a $^3T_{1g}$ term at lower energies; however, the transition is spin forbidden and difficult to locate in the absorption spectrum. The phosphorescence signal of the complex probably originates from this term.

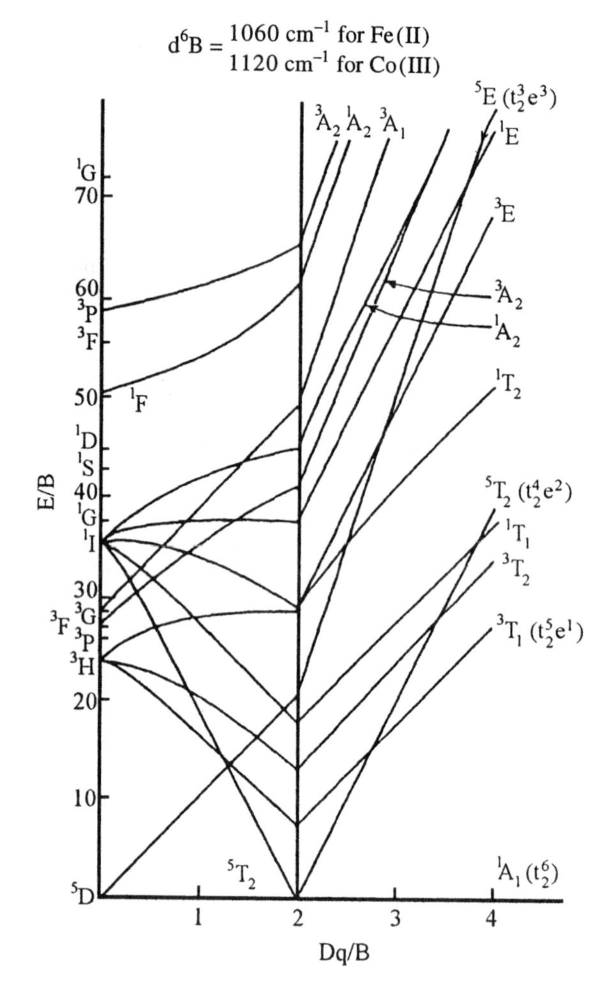

Figure 22
Tanabe–Sugano diagram for a d^6 ion in an octahedral ligand field. A C/B ratio of 4.8 has been assumed. The subscript g should be added to the symmetry labels for octahedral ions. [Reprinted with permission from Tanabe, Y.; Sugano, S. *J. Phys. Soc. Jpn.* **1954**, *9*, Fig. 5, p. 769 and Fig. 8, p. 770.]

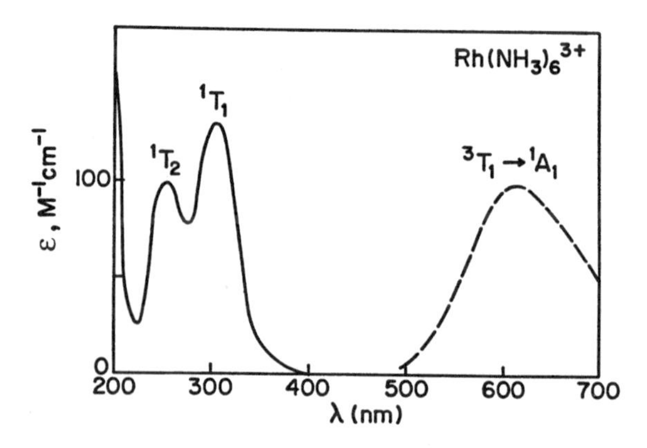

Figure 23
Absorption spectrum of $[Rh(NH_3)_6]^{3+}$. The low-temperature phosphorescence spectrum, which maximizes at around 610 nm, is shown as the dashed line. [The data are reprinted with permission from Ford.[21]]

XV. Overview of Spectral Parameters

Data from a series of d^3 systems are found in Table VII. The entries show that the magnitude of 10 Dq depends on a number of factors. For example, all else the same, 10 Dq increases with the oxidation state of the metal; hence, the splitting is larger for a M^{4+} ion than a M^{3+} ion. In addition, for a given oxidation state and a given set of ligands, 10 Dq is larger for a second-row transition ion than a first-row ion, and it is larger still for a third-row ion. Although the spin-allowed bands of these d^3 ions move to higher energy in the heavy metal complexes, the doublet states occur at lower energies because the electron–electron repulsion integrals are reduced in metal ions having more diffuse d orbitals.

Table VII also illustrates how 10 Dq varies with the ligand. In general, the same trend is observed independent of the d-electron configuration; therefore, ligands can be ordered according to the ability to induce an increase in the value of 10 Dq:

$$Cl^- < F^- < OH^- < H_2O < NH_3 < en < CN^- < CO \tag{22}$$

which is known as the spectrochemical series and where en-ethylenediamine. Finally, the spectroscopically determined B value is invariably smaller than the free ion value, and ligands can also be ordered according to the ability to decrease B:

$$F^- < H_2O < NH_3 < en < CN^- \sim Cl^- < Br^- \tag{23}$$

which is called the nephelauxetic series, where the word nephelauxetic derives from the Greek for "cloud expanding". The original idea was that the reduction

Table VII
Spectral Data and Ligand Field Parameters for Some d^3 Complexes[22]

Complex	$^4T_{1g}(P)$	$^4T_{1g}(F)$	$^4T_{2g} \approx 10\ Dq$	$^2T_{2g}, {}^2E_g$	B
$[V(H_2O)_6]^{2+}$	27,900	18,500	12,350	13,100	690
$[CrCl_6]^{3-}$			13,800		510
$[CrF_6]^{3-}$	34,400	22,700	14,900	15,700	820
$[Cr(H_2O)_6]^{3+}$	37,800	24,600	17,400	15,000	725
$[Cr(NH_3)_6]^{3+}$		28,500	21,550	15,300	650
$[Cr(CN)_6]^{3-}$		32,600	26,700		
$[MoCl_6]^{3-}$		24,000	19,200	9,650	440
$[Mo(H_2O)_6]^{3+}$		31,750	26,110		505
$[MnF_6]^{2-}$		28,090	21,800	16,020	600
$[ReF_6]^{2-}$		37,500?	32,800	10,000?	500

in B could be attributed to covalency and expansion of the d orbitals. Although this interpretation is intuitively appealing, the results of ab initio calculations are not very easily reduced to such a simple physical picture.[23]

XVI. Bandshapes and Miscellaneous Effects

A. The Effect of Reduced Symmetry

Figures 18 and 20 show how a reduction symmetry increases the number of observable, spin-allowed transitions. A further reduction in symmetry compounds the effect. In Figure 24, the low-energy absorption of $[Co(CN)_6]^{3-}$ is compared with that of $[Co(CN)_5(OH_2)]^{2-}$. What was the $^1T_{1g}$ band of the hexacyanocobaltate(III) complex suffers several changes. Specifically, the band splits into two components, the bulk of the absorption shifts to lower energy, and there is an increase in net absorption intensity. All three effects can be rationalized in terms of the reduction in symmetry, which will be approximated as C_{4v} in the mixed-ligand complex.

One way to analyze the absorption spectrum of the $[Co(CN)_5(OH_2)]^{2-}$ ion is to construct a ligand field splitting diagram, like the one depicted in Figure 25, and to establish all possible terms. However, it is much more convenient to take advantage of what is known about the energy level scheme of the octahedral $[Co(CN)_6]^{3-}$ system and to recognize that it must be closely related to that of the $[Co(CN)_5(OH_2)]^{2-}$ ion. The correlation diagram provided in Figure 26 has been constructed with the help of Table IV. This diagram reveals that the $^1T_{1g}$ transition of the O_h complex splits into two transitions in C_{4v} symmetry. The net increase in absorption intensity can be explained by the fact that the transition to

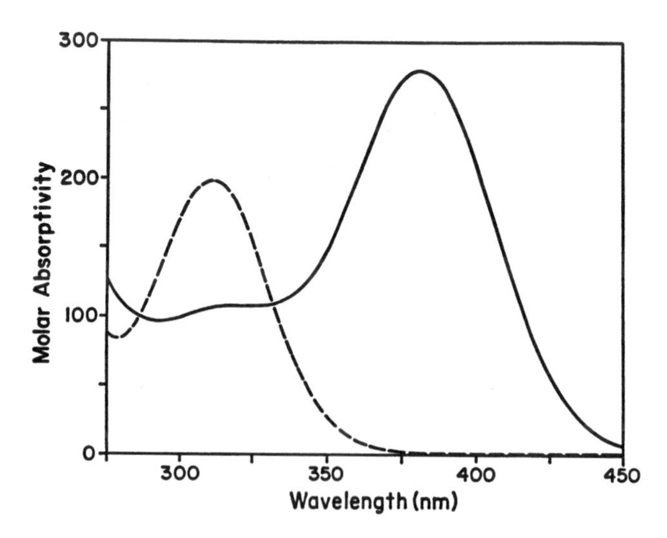

Figure 24
Near-UV absorption spectra of $[Co(CN)_6]^{3-}$ (----) and $[Co(CN)_5(OH_2)]^{2-}$ (——) in aqueous solution. [Reprinted with permission from Stacy et al. *J. Photochem. Photobiol. A.* **1989**, *47*, p. 87, Fig. 3.]

Figure 25
Ligand field splitting diagram in C_{4v} symmetry with strong field cyanide ligands in the equatorial plane.

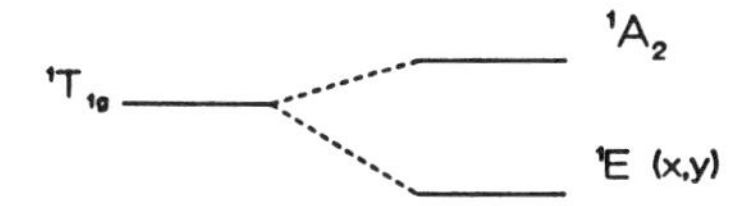

Figure 26
Effect of a C_{4v} perturbation on two terms of a d^6 complex originally in an octahedral ligand field.

the ^1E term is formally allowed with *xy* polarization in C_{4v} symmetry. On this basis, the more intense of the two components in the spectrum of the mixed-ligand complex can be assigned to the ^1E state. The transition to the $^1A_2$ term remains forbidden and can be assigned to the weaker component apparent at higher energy in Figure 24. Even though the transition to the ^1E term is symmetry allowed, it is still not very intense because the transition retains for the most part d–d character. While there is the potential for 4p orbitals to mix with 3d orbitals in the absence of the center of symmetry, the magnitude of the intensity enhancement suggests that the mixing is not very extensive. Finally, the overall shift of absorption intensity to lower energy is sensible because the replacement of a cyanide ligand by a water ligand reduces the average strength of the ligand field.

B. The Jahn–Teller Effect

Low-symmetry effects can also be seen in the spectra of certain homoleptic complexes, such as $[Cu(H_2O)_6]^{2+}$, which are Jahn–Teller active. (A homoleptic complex is one in which there is only one kind of ligand present in the coordina-

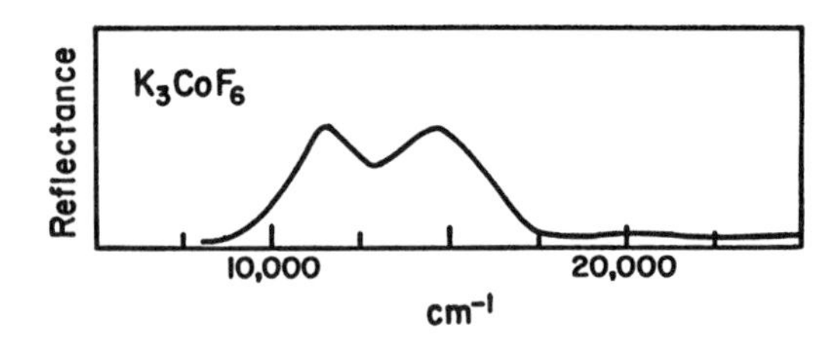

Figure 27
Room temperature reflectance spectrum of $K_3CoF_{6(s)}$ in the visible region.

tion sphere.) The Jahn–Teller effect is a vibronic interaction that destabilizes symmetrical forms of nonlinear molecules whenever an electronic state arises that is orbitally degenerate. In such cases, the system achieves a lower energy by distorting to a lower symmetry structure in which the orbital degeneracy is removed. In the case of $[Cu(OH_2)_6]^{2+}$, which would have a ^2E_g ground state in octahedral symmetry, the complex exists as a tetragonally distorted form with two long Cu–O bonds and four short Cu–O bonds. Therefore, a ligand field splitting diagram like the one in Figure 25 is appropriate. Instead of a single $^2E_g \rightarrow {}^2T_{2g}$ transition, three distinct, but overlapping transitions appear in the spectrum of the $[Cu(H_2O)_6]^{2+}$ ion.

The $^5T_{2g}$ ground state of the high-spin complex $[CoF_6]^{3-}$ is also Jahn–Teller active, but the distortion is believed to be small because the orbital degeneracy is associated with the t_{2g} subshell rather than the strongly antibonding e_g subshell. Nevertheless, contrary to the prediction of the Tanabe–Sugano diagram, the absorption spectrum of $[CoF_6]^{3-}$ contains a pair of transitions split by about 2000 cm^{-1} (Fig. 27). This is attributed to a Jahn–Teller splitting of the ^2E_g excited state in which the orbital degeneracy is, once again, associated with the e_g subshell. Although the Jahn–Teller effect can have important consequences in select systems, the unfortunate fact is that the magnitude of the splitting is difficult to calculate from theory and is often too small to be resolved. Note that no splitting is apparent in the $^1T_{2g}$ transition of the low-spin $[Co(CN)_6]^{3-}$ complex (Fig. 24) even though the excited state is orbitally degenerate and is formally associated with a configuration having an odd electron in the e_g subshell.

C. Factors Influencing Absorption Intensities

Ligand field transitions often obtain intensity by a vibronic mixing mechanism. When this is the case the transition intensity is expected to be temperature dependent due to thermal population of low-frequency vibrational modes of the ground state. Figure 28 shows the temperature dependence of two ligand field transitions of the $CuCl_4^{2-}$ complex in a crystal where the site symmetry is approximately D_{4h}. Note that the absorption intensity decreases as the temperature is lowered. It never vanishes because zero-point vibrational motion is enough to guarantee some vibronic intensity.

In appropriate symmetries, a ligand field, or "d–d", transition can also have intrinsic electric dipole strength if there is d–p and/or d–f character involved in

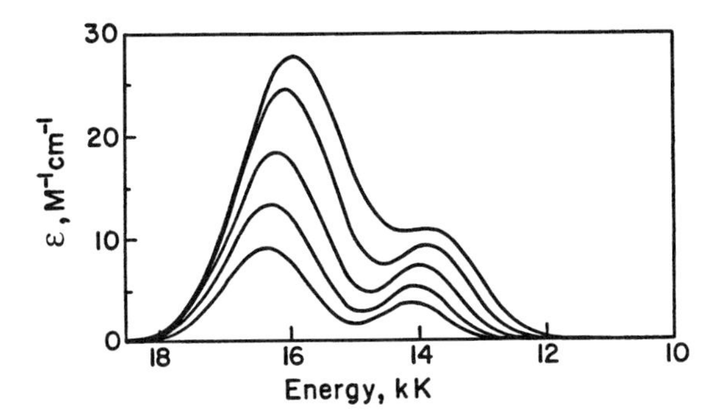

Figure 28
Single-crystal absorption spectrum of the planar $[CuCl_4]^{2-}$ ion as a function of temperature. The crystal was oriented such that the electric vector of the incident light was perpendicular to the plane of the ion. In order of increasing intensity, the spectra were measured at 10, 50, 90, 140, and 180 K. [This figure is based on data from McDonald.[24] Reprinted with permission from McDonald, R. G.; Hitchman, M. A. *Inorg. Chem.* **1986**, *25*, 3273–3281. Copyright © 1986 American Chemical Society.]

the transition.[25] This is possible, for example, in mixed-ligand complexes that lack a center of symmetry. The intensity generated in this way can vary with the coordination number. Consider, for example, high-spin Co(II) systems, which are of interest because Co(II) is often used as a spectroscopic probe of metal-binding sites in proteins, especially Zn(II) proteins. High-spin Co(II) systems generally exhibit a ligand field transition in the visible region somewhere around 500 or 600 nm; however, the intensity varies greatly depending on the coordination number. Data from a few representative systems are compiled in Table VIII. Although it would be risky to predict the coordination number from this type of data alone, a molar absorptivity of 400 M^{-1} cm^{-1} or greater is probably a good indication of a pseudotetrahedral coordination geometry. Likewise, if the molar absorptivity is less than about 50 M^{-1} cm^{-1}, chances are the site is six coordinate.

Formally, spin-forbidden transitions can also appear in the absorption spectrum. This occurs when the spin–orbit interaction mixes states with different spin

Table VIII
Absorptivity Data from Cobalt(II) Substituted Proteins

Chromophore[a]	λ_{max} (nm)	ε(M^{-1} cm^{-1})	Coordination Number	Reference
Co^{II}ADH	650	650	4	26
Adduct with 1,10-phen	618	300	5	
Co^{II}CPD	575	150	5	27
Adduct with pyrophosphate	580	<50	6	
Co^{II}ovotransferrin	550	20	6	28

[a] Abbreviations: ADH = alcohol dehydrogenase and CPD = carboxypeptidase.

quantum numbers and is common in heavy metal systems where the spin–orbit coupling constants are large. The energy difference between the states being mixed is also important. Some of the weak absorption that is observed near the $^4T_1(P)$ transition of $[CoCl_4]^{2-}$ in Figure 19 may be due to spin-forbidden transitions to doublet states that borrow intensity from the $^4A_2 \rightarrow {}^4T_1(P)$ transition.

D. Bandwidths

Ligand field transitions tend to be rather broad and unstructured for several reasons. For example, transitions to orbitally degenerate states can be broadened because of an unresolved Jahn–Teller splitting or unresolved spin–orbit structure. Very often there is unresolved vibronic structure due to simultaneous excitation of multiple low-frequency vibrations. Consider, for example, an octahedral Cr(III) complex. Population of the $^4T_{2g}$ state requires promotion of an electron from the nonbonding t_{2g} subshell to the σ-antibonding e_g subshell. Consequently, the metal–ligand bond lengths are altered in the excited state, and vertical transitions to a host of vibrationally excited levels of the $^4T_{2g}$ state are possible. What is typically observed in such a case is the envelope of a vibronic progression (Fig. 29). The problem is made more difficult by the fact that several different vibrations may be capable of acting as promoting modes in which case there may be several different overlapping progressions, each with a distinct origin. On top of this, there may be progressions built on vibrationally excited levels of the ground state. Sometimes, reducing the temperature alleviates problems and improves resolution.

In select systems, relatively narrow transitions are observed when the complications associated with the vibronic states vanish because the potential energy surfaces of the ground and excited states parallel each other. This is nearly true for the ^2E_g state of an octahedral Cr(III) complex because formation of the ^2E_g state

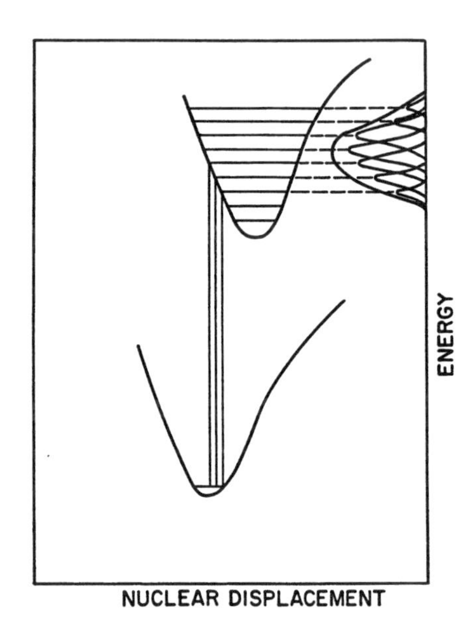

Figure 29
Franck–Condon transitions from the ground to an excited state with a different equilibrium structure. The absorption envelope is indicated on the upper right.

involves only a spin flip of an electron in the t_{2g} subshell. Since this level is essentially nonbonding in most cases, there is very little change in the metal–ligand bond order in the excited state. Thus, the ligand field exerts practically the same effect on the ^2E_g term and the $^4A_{2g}$ ground state, and the lines representing these states are virtually parallel in the Tanabe–Sugano diagram. Transitions to the doublet states are therefore relatively narrow and well resolved in these systems. See, for example, the spectrum of $[Cr(CN)_6]^{3-}$ in Problem 10 at the end of the chapter.

XVII. Metal–Metal Bonded Systems

A. Orbital Scheme and Transitions

Complexes that contain metal–metal bonds can be highly colored and exhibit interesting electronic spectra. For example, $Mn_2(CO)_{10}$ contains a metal–metal single bond and is bright yellow while the mononuclear carbonyls of the neighboring metals Mo and Ru are colorless compounds. Metal–metal multiple bonds are also known, particularly among heavy metals, the classic example being the $[Re_2Cl_8]^{2-}$ system, which is depicted in Figure 30. Despite the stereoelectronic constraints posed by the eight chloride ligands, this dirhenate complex exhibits an eclipsed geometry. The metal–metal bonds inherent in the D_{4h} structure are described in the MO diagram in Figure 30. There is a σ bond involving the two d_{z^2} orbitals, two π bonds involving the d_{xz} and d_{yz} orbitals, and a δ bond involving the d_{xy} orbitals of each metal atom. (The $d_{x^2-y^2}$ orbitals are involved in bonds to

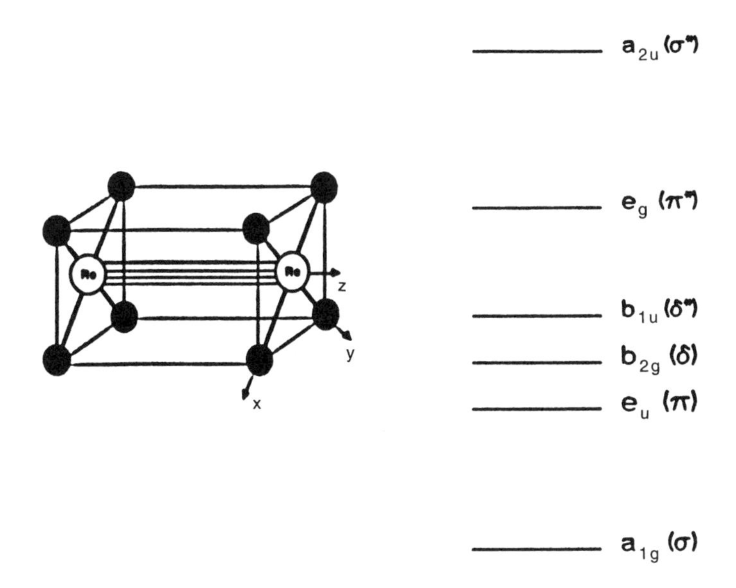

Figure 30
Structure and MO scheme of the D_{4h} ion $[Re_2Cl_8]^{2-}$. The z axis is along the Re–Re axis, and the x and y axes lie in the vertical symmetry planes containing chlorine atoms.

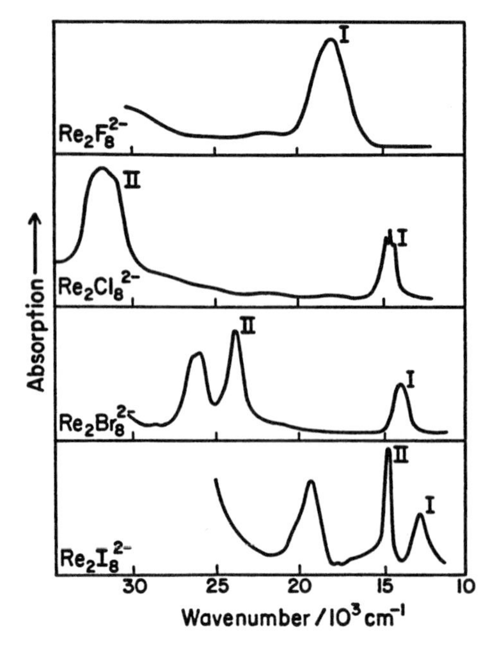

Figure 31
Electronic spectra of $[Re_2X_8]^{2-}$ systems at 14 K. [This drawing was adapted from Clark.[29] Reproduced by permission from "Synthesis, Structure, and Spectroscopy of Metal–Metal Dimers, Linear Chains, and Dimer Chains", by Clark, R. J. H., Royal Society of Chemistry, Cambridge, 1990.]

the chloride ligands and have been omitted from the diagram.) The δ bond enforces the eclipsed conformation. Spectra for a series of octahalodirhenate(III) complexes are presented in Figure 31. Two, apparently analogous bands, labeled I and II, have been identified for almost every member of the series. Band I, reasonably intense and always the lowest energy transition, is predicted to be the δ–δ^* transition on the basis of the MO scheme. Consistent with this assignment, the transition is polarized along the metal–metal bond axis and, at low temperatures, often exhibits vibronic structure that can be assigned to metal–metal stretching in the excited state. Even though the δ–δ^* transition is a type of d–d transition, it is dipole allowed because the center of symmetry lies between the metal atoms. Thus, there are ungerade combinations of d orbitals. The basis of the absorption intensity will be discussed in Section XVIIB.

In contrast to band I, band II shifts dramatically with a change in halide, and it has been assigned as a LMCT transition. This band appears to be perpendicularly polarized and therefore is assigned to a ^1E_u state. In line with theoretical estimates of orbital energies,[29] it has been assigned to $e_g(X\pi) \rightarrow b_{1u}(\delta^*)$ excitation where $e_g(X\pi)$ denotes a level involving MOs, which are mainly comprised of combinations of the halide pπ orbitals that overlap with the d_{xy} orbitals of the metal atoms. The third intense band resolved in the spectra of the bromide and iodide derivatives may represent a π–π^* transition within the metal–metal bond system. Assignments of other weaker bands in the spectrum of $[Re_2Cl_8]^{2-}$ have also been proposed but will not be discussed here.

B. The $\delta \rightarrow \delta^*$ Transition

Since the d_{xy} orbitals of the rhenium atoms lie in parallel planes, there is limited orbital overlap between them and, consequently, a relatively small energy difference between the δ and the δ^* orbitals. The interesting questions are why then does the $\delta \rightarrow \delta^*$ transition show up in the visible region and why is the transition as intense as it is. Gray and co-workers[30] pointed out that an increase in interelectronic repulsion within the excited state contributes to the transition energy. In the ground state, the metal–metal bond is covalent, and in valence bond terms the δ bond can be described in terms of a covalent structure that might be written:

$$Re^{III}\text{–}Re^{III} \tag{24}$$

However, the ground-state singlet and excited state triplet wave functions exhaust the group of covalent wave functions that can be derived from a pair of atomic orbitals,[31] and the lowest energy singlet excited state must be described as a resonance hybrid of two mixed-valence structures:

$$Re^{IV}Re^{II} \leftrightarrow Re^{II}Re^{IV} \tag{25}$$

In other words, in the excited state the two electrons formally associated with the δ and δ^* orbitals are constrained to spend most of their time paired in one or the other of the d_{xy} orbitals, and the system is destabilized accordingly. A clue to the absorption intensity comes from the fact that the molar extinction coefficient of the $\delta \rightarrow \delta^*$ transition increases from chloride to bromide to iodide. This is also the order of decreasing energy of the halide–metal CT excited states. Thus, Gray and co-workers[30] also proposed that the oscillator strength of band I is "borrowed" from a higher lying LMCT state. The idea is that the $\delta \rightarrow \delta^*$ excited state is mixed with another ${}^1B_{1u}$ excited state that stems from a LMCT excited-state configuration. This mixing is allowed because the excited states have the same spin multiplicity and the same symmetry. With the heavier halogens, the energy gap between the two excited states should be smaller, and the interaction should be more effective. Covalent mixing of ligand orbitals with the metal d orbitals also enhances the oscillator strength by increasing the dipole length of the transition.

APPENDIX I

Russell–Saunders Coupling

What follows is an outline of the derivation of the Russell–Saunders terms for the $2p^2$ configuration. The m_l and m_s values for each electron serve as labels for the relevant functions (microstates). In the notation used here, the m_s quantum number appears as a superscript. Thus, (1^+1^-) designates a microstate with $m_l = 1$ and $m_s = +\frac{1}{2}$ for electron 1 and $m_l = 1$ and $m_s = -\frac{1}{2}$ for electron 2. The first step is to write down the independent microstates and to organize them according to the total quantum numbers:

$$M_L = m_l(1) + m_l(2)$$

$$M_S = m_s(1) + m_s(2)$$

For the $2p^2$ configuration, the highest possible M_L value is 2. The highest possible M_S value is 1, when both electons have $m_s = +\frac{1}{2}$. In Table IX, the microstate with the maximum M_L has $M_S = 0$. This microstate must be part of a ^1D term since the maximum M_L and M_S values for every term are L and S, respectively. Because there are $2L + 1$ possible values of M_L, the ^{1}D term is fivefold degenerate. To account for these microstates, cross out or remove one microstate from each row of the table in the $M_S = 0$ column. Since this is a purely bookkeeping exercise, delete any $M_S = 0$ microstate when there is more than one with the same M_L value.

Of the remaining microstates, the maximum M_S value is 1 when M_L assumes the maximal value $M_L = 1$. It follows that this microstate is part of a ^{3}P term that is ninefold degenerate. In other words, there are nine possible combinations of M_L and M_S. Remove nine microstates from the appropriate positions in Table IX, and there is one microstate left in the $M_L = 0$, $M_S = 0$ position. It necessarily represents a ^{1}S term. According to Hund's rules, the ground term is the ^{3}P.

Table IX
Microstates for the $2p^2$ Configuration

		M_S		
		+1	0	−1
	2		(1^+1^-)	
M_L	1	(1^+0^+)	$(1^+0^-)(1^-0^+)$	(1^-0^-)
	0	$(1^+ - 1^+)$	$(1^+ - 1^-)(1^- - 1^+)(0^+0^-)$	$(1^- - 1^-)$
	−1	$(0^+ - 1^+)$	$(0^+ - 1^-)(0^- - 1^+)$	$(0^- - 1^-)$
	−2		$(-1^+ - 1^-)$	

APPENDIX II

More About Vibronic Coupling

Recall the system in Figure 6 in which the transition from the ground state ψ_0 to the first excited state ψ_1 is electronically forbidden, whereas the transition from ψ_0 to ψ_2 is allowed. The following development shows how the forbidden transition can borrow intensity from the allowed transition by vibronic mixing. The discussion ends with an illustration that should make the process seem more intuitive.

The symmetry restriction in question is based on the properties of the wave functions ψ_0 and ψ_1 at the equilibrium geometry $\xi_k{}^0$. However, the symmetry changes when the system is displaced along ξ_k, where ξ_k is assumed to be a non-totally symmetric vibrational coordinate. For small displacements along ξ_k, the new wave function can be described in terms of an expansion of the original functions via perturbation theory. Since the $\psi_0 \rightarrow \psi_2$ transition is allowed, a sufficient condition for a nonzero electric dipole transition moment is for ψ_1 to evolve into a new function $\psi_1{}^{(1)}$, which is of the form:

$$\psi_1{}^{(1)} = \psi_1 + \alpha\psi_2 + \cdots \qquad \text{(A.II.1)}$$

where α is a mixing coefficient. The part of the Hamiltonian operator that describes the vibronic perturbation can be approximated by the first term in a Taylor expansion along the coordinate ξ_k:

$$\mathscr{H}^{(1)} \approx V_0'\xi_k \qquad \text{(A.II.2)}$$

where V_0' is the partial derivative (with respect to ξ_k) of the potential energy term in the Hamiltonian that relates the electrons to the nuclear framework. Since the expansion is centered about the equilibrium geometry, $\mathscr{H}^{(1)}$ must form a basis for the totally symmetric representation of the original point group. Consequently, V_0', the part of the operator that depends on the electronic coordinates, must be a basis for the same symmetry species as ξ_k. The mixing coefficient α is defined by the integral:

$$\alpha = \langle \psi_2 | V_0' | \psi_1 \rangle \qquad \text{(A.II.3)}$$

and is nonzero only if the direct product:

$$\Gamma(\psi_1) \times \Gamma(\psi_2)$$

forms a basis for the same irreducible representation as the normal coordinate ξ_k. Ultimately, the magnitude of the transition moment depends on α, on the dipole strength of the transition from ψ_0 to ψ_2, and on integrals of the form:

$$\langle \chi_{vk}' | \xi_k | \chi_0 \rangle \qquad \text{(A.II.4)}$$

where χ'_{vk} is a vibrational wave function associated with the normal coordinate ξ_k. For a standard harmonic oscillator the ground state vibrational wave function is totally symmetric, and the $vk = 1$ state transforms like ξ_k. Therefore, Eq. (A.II.4) reveals that the transition is allowed when it occurs with the concomitant excitation of one quantum of ξ_k.

To solidify the ideas, consider the $PtCl_4{}^{2-}$ ion. Although the metal–ligand bonds are often placed along the x and y axes in D_{4h} molecules, for the present purposes it is convenient to rotate the molecule by 45° such that the lobes of the d_{xy} orbital point toward the ligands. With this convention the d_{xy} orbital is the LUMO of the d^8 system instead of the $d_{x^2-y^2}$ orbital. (This alignment makes the impending correlation with the D_{2d} point group a more straightforward task.) The square planar complex $[PtCl_4]^{2-}$ is a closed-shell system and the ground state is $^1A_{1g}$ in D_{4h} symmetry. Excitation of an electron from d_{z^2}, which has a_{1g} symmetry, to d_{xy}, which has b_{2g} symmetry, generates a $^1B_{2g}$ state, hence the transition is Laporte forbidden because of the gerade–gerade character. However, displacement along an ungerade normal coordinate destroys the center of symmetry and makes it possible for the transition to be observed. In particular, it can be shown that the $a_{1g} \rightarrow b_{2g}$ transition is vibronically allowed with z polarization if it occurs in conjunction with a b_{1u} vibration. One reason is that in D_{2d} symmetry the d_{xy} orbital and the p_z orbital can mix together because each forms a basis for the b_2 representation. In fact, the mixing is a natural consequence of the out-of-plane distortion because the lobes of the d_{xy}/p_z hybrid orbital are directed more toward the ligands in the D_{2d} structure (Fig. 32). To the extent that this mixing occurs, the transition from the d_{z^2} orbital becomes allowed because of its $d_{z^2} \rightarrow p_z$ character.

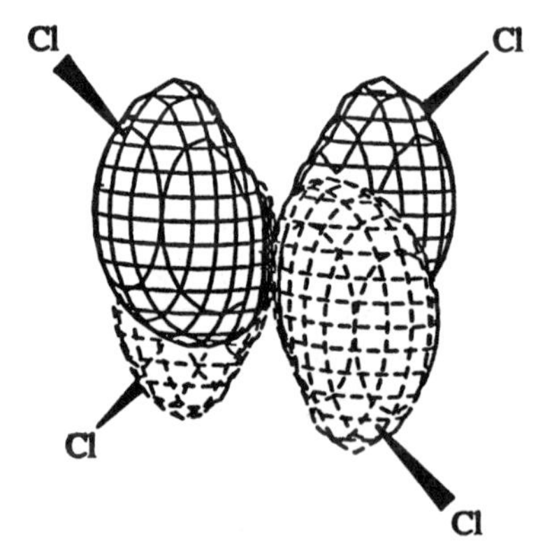

Figure 32
Schematic view of a hybrid orbital of $[PtCl_4]^{2-}$. The orbital contains 60% d_{xy} character and 40% p_z character, and is appropriate for D_{2d} symmetry. The x, y, and z axes bisect the Cl–Pt–Cl angles. Positive and negative amplitudes are indicated by solid and dashed lines, respectively.

APPENDIX III

Superpositions and Time-Dependent Phenomena

An earlier section on electric dipole transitions stated that in terms of classical mechanics an atom has to develop an oscillating electric dipole moment in order to absorb energy by the electric dipole mechanism. This section describes the oscillations in more detail, and it illustrates why a 1s → 2p transition is allowed for an atom while a 1s → 2s transition is not. Finally, the concepts are used to provide more insight into the configuration interaction that has been invoked to explain the relative energies of the two lowest energy singlet excited states in metalloporphyrins.

A. Electric Dipole Transitions

The discussion begins with a description of the time dependence of the wave function. In the Bohr model of the hydrogen atom the electron travels in a fixed orbit, and its position in the atom varies with time. According to quantum theory, however, the electron has a fixed spatial distribution relative to the nucleus and a fixed energy so long as the atom is in a stationary state. The time dependence shows up in the phase of the wave function. Thus, if the complication of electron spin is ignored, the total electronic wave function for the jth state of the hydrogen atom can be written:

$$\Phi_j(\tau)\exp(-2\pi\mathbf{i}E_jt/h) \tag{A.III.1}$$

where τ denotes the spatial coordinates of the electron, t denotes the time, Φ_j is an atomic orbital, E_j is the energy of the state, $\mathbf{i}$ is the square root of -1, and h denotes Planck's constant. Note that the time-dependent phase factor has no effect on the charge density since the density depends on the square of the amplitude. If Eq. (A.III.1) looks unfamiliar, the reason is that the phase factor is often ignored because, for the isolated atom, the function Φ_j contains all the relevant information. However, time-dependent forces come into play in the presence of an oscillating electric field, and the time-dependent part of the wave function becomes extremely important.

The electric field associated with light energy perturbs the atom. As in the case of the vibronic interaction described above, the perturbation causes the wave function to evolve into a superposition, or admixture, of the ground and excited states:

$$\psi(\tau,t) = \Sigma a_k(t)\Phi_k(\tau)\exp(-2\pi\mathbf{i}E_kt/h) \tag{A.III.2}$$

where the mixing coefficients, the $a_k(t)$ values, are also time-dependent quantities.

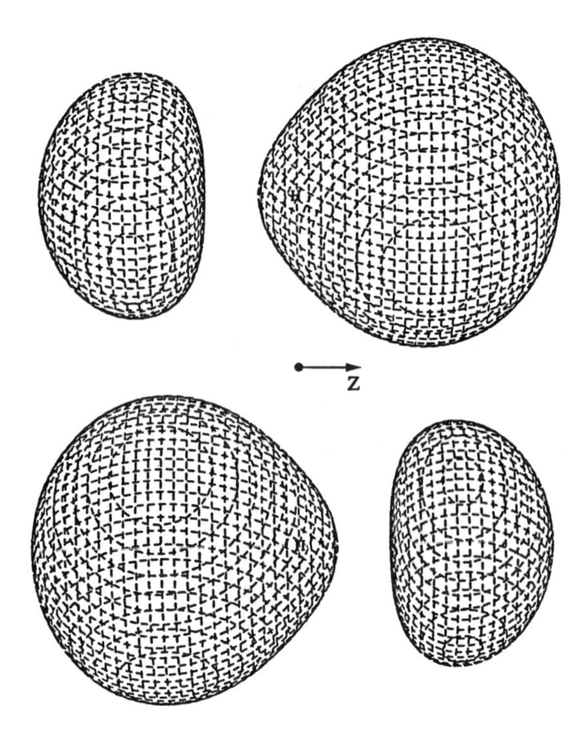

Figure 33
Electron density contours at two instants in time for a superposition of the $1s$ and $2p_z$ orbitals of the hydrogen atom. The times have been chosen to illustrate the maximum degree of charge displacement in the positive and negative z directions. In each plot, the nucleus is located at $z = 0$, and the contour level is 0.01 $e\text{A}^{-3}$. The coefficients for the $1s$ and $2p_z$ orbitals in the linear combination are are 0.5 and 0.866, respectively.

The charge distribution associated with Eq. (A.III.2) attempts to oscillate in step with the perturbing electric field. When the energy difference between the ground state and one of the excited states involved matches the photon energy, energy transfer from the radiation field to the atom is possible and Eq. (A.III.2) evolves into the excited-state wave function.

Suppose the radiation field is tuned to the right frequency to induce a transition from the ground state to the state Φ_k. Then, excitation to Φ_k can occur by the electric dipole mechanism if superposition with Φ_0 induces oscillating electric dipole character into the system. As discussed in the references, superposition of Φ_0 and Φ_k necessarily induces oscillatory character into the charge density $\psi^*\psi$ with a frequency given by $(E_k - E_0)/h$.[32,33] The only question is does the superposition provide for oscillating dipole character. This depends on the nature of the interference that occurs when the wave functions overlap each other. For two different points in time Figure 33 depicts the shape of the charge cloud when the $2p_z$ orbital is superposed with the $1s$ orbital. When the $2p_z$ orbital overlaps the 1s orbital, there is constructive interference on one side of the nucleus and destructive interference on the other. Because the interference pattern is modulated by a cosine function,[31] the electron density sways from side to side about the nucleus. Consequently, this transition is allowed when there is an electric field oscillating along the z axis with an appropriate frequency. In contrast, the overlap between the 2s orbital and the 1s orbital produces an interference that results in a spherically symmetric displacement of the electron density. Here there is no charge polarization, and the transition is forbidden to occur by the electric dipole mechanism.

B. Application to a Type of Configuration Interaction

The text states that the zero-order states associated with the two excited configurations $a_{1u}{}^2a_{2u}{}^1e_g{}^1$ and $a_{1u}{}^2a_{2u}{}^1e_g{}^1$ do not represent stationary states of the metalloporphyrin. To see why, note that if each transition is orbitally allowed, there is a skewing of the charge distribution during the transition due to interference between the initial and final orbitals. Now when two such transitions are coupled properly, the buildup in the electron density, that is, the constructive interference that occurs during one of the transitions, falls where the deficit in electon density is created during the other, and vice versa. Thus, in this semiclassical view the system spontaneously achieves a lower energy by means of a dynamic correlation between the motions of the electrons. Note that this type of coupling is possible between any two transitions that are dipole allowed to states of the same symmetry, but the mechanism loses its effectiveness when the two transitions have greatly different frequencies and do not synchronize well. From the point of view of the dipole moments the two transition dipoles align favorably so as to cancel each other in the combination state described above. This combination is consistent with a lower energy state. When the cancellation is very effective, however, there is almost no net dipole moment associated with the transition to the combination state. Thus, the $\mathbf{Q}_0$ band in metalloporphyrins is a relatively weak transition. On the other hand, there is a higher energy combination state in which the dipole moments are in phase and parallel. The transition to this state carries the vast majority of the dipole strength and is strongly allowed.

The dipole–dipole interaction explains why the two lowest energy singlet states of the metalloporphyrin have quite different transition energies and transition intensities. In contrast, calculations show that the energy gap between the corresponding ^3E_u states is comparatively small.[4] The reason is that the same kind of correlation is not possible in the triplet manifold because the analogous electron jumps would each require a concomitant spin flip, and in the absence of a strong spin-orbit interaction this is not very probable. In other words, the dipole–dipole coupling is rendered ineffective because the individual transitions are spin forbidden.

References

1. Hyde, K. E. "Methods for Obtaining Russell–Saunders Term Symbols from Electronic Configurations" *J. Chem. Educ.* **1975**, *52*, 87–89.
2. Salem, L. *Electrons in Chemical Reactions*; Wiley: New York, 1982; pp 67–72.
3. Longuet–Higgins, H. C.; Rector, C. W.; Platt, J. R. "Molecular Orbital Calculations on Porphine and Tetahydroporphine" *J. Chem. Phys.* **1950**, *18*, 1174–1181.
4. Gouterman, M. "Optical Spectra and Electronic Structure of Porphyrins and Related Rings" In *The Porphyrins*; Dolphin, D., Ed.; Academic: New York, 1978; Vol. III, Part A; pp 1–165.
5. Makinen, M. W.; Chung, A. K. "Structural and Analytical Aspects of the Electronic Spectra of Hemeproteins" In *Physical Bioinorganic Chemistry Series. Iron Porphyrins*, Part I; Lever, A. B. P.; Gray, H. B., Eds.; Addison-Wesley: Reading, MA 1983; pp 141–234.
6. Riley, M. J.; Hitchman, M. A. "Temperature Dependence of the Electronic Spectrum of the Planar $CuCl_4^{2-}$ Ion: Role of the Ground- and Excited-State Potential Surfaces" *Inorg. Chem.* **1987**, *26*, 3205–3215.
7. Lever, A. B. P. *Inorganic Electronic Spectroscopy*; Elsevier: New York, 1984; 2nd ed.; pp 161–163.
8. Wrobleski, J. T.; Long, G. J. "The Electronic Spectra of the Tetrahedral $[MnO_4]^{m-}$ Chromophore" *J. Chem. Educ.* **1977**, *54*, 75–79.
9. Ballhausen, C. J.; Gray, H. B. "Electronic Structures of Metal Complexes" In *Coordination Chemistry*; Martell, A. E., Ed.; Van Nostrand Reinhold: New York, 1971; pp 3–83.
10. Johansen, H. "Analysis of a Configuration Interaction Calculation for a Charge Transfer Transition in MnO_4^-" *Mol. Phys.* **1983**, *49*, 1209–1216.
11. Miller, R. M.; Tinti, D. S.; Case, D. A. "Comparison of the Electronic Structures of Chromate, Halochromates, and Chromyl Halides by the Xα Method" *Inorg. Chem.* **1989**, *28*, 2738–2743.
12. McMillin, D. R.; Tennent, D. L. "Physical Studies of Azurin and Some Metal Replaced Derivatives" In *ESR and NMR of Paramagnetic Species in Biological and Related Systems*; Bertini, I., Drago, R. S., Eds.; Reidel: Dordrecht, Holland, **1979**; pp 369–379.
13. Jorgensen, C. K. "Electron Transfer Spectra" *Prog. Inorg. Chem.* **1970**, *12*, 101–158.
14. Solomon, E. I.; Baldwin, M. J.; Lowery, M. D. "Electronic Structures of Active Sites in Copper Proteins: Contributions to Reactivity" *Chem. Rev.* **1992**, *92*, 521–542.
15. Hay, P. J.; Thibeault, J. C.; Hoffmann, R. "Orbital Interactions in Metal Dimer Complexes" *J. Am. Chem. Soc.* **1975**, *97*, 4884–4899.
16. Daul, C.; Schläpfer, C. W.; Goursot, A.; Pénigault, E.; Weber, J. "The Electronic Structure of $Cu(ethanediimine)_2^+$" *Chem. Phys. Lett.* **1981**, *78*, 304–310.
17. Mulliken, R. S. "Molecular Compounds and their Spectra II" *J. Am. Chem. Soc.* **1952**, *74*, 811–824.
18. Phifer, C. C.; McMillin, D. R. "The Basis of Aryl Substituent Effects on Charge-Transfer Absorption Intensities" *Inorg. Chem.* **1986**, *25*, 1329–1333.
19. Gray, H. B.; Beach, N. A. "The Electronic Structures of Octahedral Metal Complexes. I. Metal Hexacarbonyls and Hexacyanides" *J. Am. Chem. Soc.* **1963**, *85*, 2922–2927.
20. Griffith, J. S. *The Theory of Transition-Metal Ions*; Cambridge University Press: London, 1961; pp 83–96.
21. Ford, P. C. "The Photosubstitution Reactions of Rhodium(III) Ammine Complexes" *J. Chem. Educ.* **1983**, *60*, 829–833.
22. Jørgensen, C. J. "Spectroscopy of Transition-Group Complexes" *Adv. Chem. Phys.* **1963**, *5*, 33–146.
23. Gerloch, M.; Slade, R. C. *Ligand-Field Parameters*; Cambridge University Press: London, 1973; pp 197–228.

24. McDonald, R. G.; Hitchman, M. A. "Electronic "*d–d*" spectra of the Planar $CuCl_4^{2-}$ Ions in Bis(methadonium) Tetrachlorocuprate(II) and Bis(creatinium) Tetrachlorocuprate(II): Analysis of the Temperature Dependence and Vibrational Fine Structure" *Inorg. Chem.* **1986**, *25*, 3273–3281.
25. Fenton, N. D.; Gerloch, M. "Intensity Distributions within the "*d–d*" spectra of Tetrakis(diphenylmethylarsine oxide)(nitrato)cobalt(II) and Tetrakis(diphenylmethylarsine oxide)(nitrato)nickel(II) Nitrates" *Inorg. Chem.* **1989**, *28*, 2975–2983.
26. Sartorius, C.; Dunn, M. F.; Zeppezauer, M. "The Binding of 1,10-phenanthroline to Specifically Active-site Cobalt(II)-substituted Horse-liver Alcohol Dehydrogenase" *Eur. J. Biochem.* **1988**, *177*, 493–499.
27. Bertini, I.; Donaire, A; Messori, L.; Moratal, J. "Interaction of Phosphate and Pyrophosphate with Cobalt(II) Carboxypeptidase" *Inorg. Chem.* **1990**, *29*, 202–205.
28. Bertini, I.; Luchinat, C.; Messori, L.; Monnanni, R.; Scozzafava, A. "The Metal-binding Properties of Ovotransferrin" *J. Biol. Chem.* **1986**, *261*, 1139–1146.
29. Clark, R. J. H. "Synthesis, Structure, and Spectroscopy of Metal–Metal Dimers, Linear Chains, and Dimer Chains" *Chem. Soc. Rev.* **1990**, *19*, 107–131.
30. Hopkins, M. D.; Gray, H. B.; Miskowski, V. M. "$\delta \rightarrow \delta^*$ Revisited: What the Energies and Intensities Mean" *Polyhedron* **1987**, *6*, 705–714.
31. Salem, L.; *Electrons in Chemical Reactions*; Wiley: New York, **1982**; pp 60–63.
32. McMillin, D. R. "Fluctuating Electric Dipoles and the Absorption of Light" *J. Chem. Educ.* **1978**, *55*, 7–11.
33. Henderson, G. "How a Photon is Created or Absorbed" *J. Chem. Educ.* **1978**, *56*, 631–635.

General References

BASIC TREATMENTS

Harris, D. C.; Bertolucci, M. D. *Symmetry and Spectroscopy*, Dover: New York, 1978. This is a good introductory text. Chapter 4 deals with molecular orbital theory, and Chapter 5 discusses electronic spectroscopy. The book is especially good in its treatment of diatomic molecules.

Pavia, D. L.; Lampman, G. M.; Kriz, G. S. *Introduction to Spectroscopy*, 2nd ed.; Harcourt Brace: New York 1996. Chapter 6 is an introductory discussion of the electronic spectroscopy of organic chromophores.

DISCUSSIONS OF ORGANIC CHROMOPHORES

Jaffè, H. H.; Orchin, M. *Theory and Applications of Ultraviolet Spectroscopy*, Wiley: New York, 1962. A classic, readable overview of electronic spectroscopy with an emphasis on organic chromophores.

Suzuki, H. *Electronic Absorption Spectra and Geometry of Organic Molecules*, Academic: New York, 1967. Rigorous exposition.

DISCUSSIONS OF INORGANIC COMPLEXES

Shriver, D. F.; Atkins, P.; Langford, C. H. *Inorganic Chemistry*, 2nd ed.; W. H. Freeman: New York, 1994. Chapter 14 provides a thumbnail sketch of the electronic spectra of metal complexes.

Figgis, B. N. "Ligand Field Theory" In *Comprehensive Coordination Chemistry*, Wilkinson, G.; Gillard; R. D.; McCleverty, J. A., Eds.), Vol. 1, Chapter 6. Pergamon: Oxford, 1987. A concise treatment of the spectra of metal complexes. This chapter also deals with magnetic properties.

Lever, A. B. P. *Inorganic Electronic Spectroscopy*, 2nd ed.; Elsevier: New York, 1984. Thorough and detailed; an excellent reference. It also includes a good discussion of many metalloprotein systems.

Ballhausen, C. J. *Introduction to Ligand Field Theory*. McGraw-Hill, New York, 1962. A rigorous treatment of ligand field theory.

Jorgensen, C. K. *Modern Aspects of Ligand Field Theory*, North-Holland: Amsterdam, The Netherlands, 1971. An overview written by a long-time practitioner.

MOLECULAR SYMMETRY AND GROUP THEORY

Cotton, F. A. *Chemical Applications of Group Theory*, 3rd ed; Wiley: New York, 1990. An introduction to group theory; Chapter 9 deals with ligand field theory.

Douglas, B. E.; Hollingsworth, C. A. *Symmetry in Bonding and Spectra. An Introduction* Academic: New York, 1985. Chapter 9 provides a treatment of the electronic spectroscopy of metal complexes with a strong emphasis on symmetry considerations.

2

Resonance Raman Spectroscopy

THOMAS G. SPIRO

Department of Chemistry
Princeton University
Princeton, New Jersey 08544

ROMAN S. CZERNUSZEWICZ

Department of Chemistry
University of Houston
Houston, Texas 77204

CONTENTS

I. Introduction

Resonance Raman (RR) spectroscopy is capable of providing a useful structure probe of metal complexes and of metal centers in biological systems. When the Raman excitation laser is tuned to the wavelength of an electric-dipole allowed electronic transition, certain Raman bands, those associated with vibrational modes that mimic the excited-state distortion, are greatly enhanced. Since vibrational frequencies are sensitive to changes in molecular bonding and conformation, the positions of the enhanced Raman bands can be used to monitor the chromophoric structure. Metal centers are frequently involved in allowed electronic transitions, due to ligand–metal charge-transfer (LMCT) transitions, or due to the influence of the metal on π–π^* transitions of a ligand (e.g., porphyrin), and consequently they give wide scope to the application of resonance Raman spectroscopy. (The d–d electronic transitions that often give simple coordination compounds their characteristic colors are electric-dipole forbidden, and are ineffective in resonance enhancement.)

In this chapter, we review the principles of vibrational spectroscopy, emphasizing aspects of molecular structure that determine vibrational frequencies, especially those associated with metal–ligand bonds. We then discuss the nature of the resonance Raman effect, in order to clarify the kind of information that can be obtained, and the rules for optimal resonance enhancement. These principles are illustrated with specific applications to bioinorganic systems from recent research. For more thorough exposure to Raman spectroscopy, the reader is referred to the books in Ref. 1. Biological applications of Raman spectroscopy have been reviewed.[2]

II. Infrared and Raman Spectroscopy

A. Physical Principles

Molecular vibrational frequencies lie in the infrared (IR) region of the electromagnetic spectrum. Transitions to vibrationally excited states can therefore be accessed by direct absorption of light in the IR region (IR spectroscopy). Alternatively, the same transition can be accessed via an inelastic scattering process, in which the requisite energy is transferred from a photon of higher energy, the scattered photon emerging with correspondingly lowered energy ("Stokes" radiation). This Raman process is shown diagrammatically in relation to IR absorption in Figure 1. A Raman spectrum contains a series of scattering peaks, whose frequencies are shifted from that of the incident photons (or of the elastic scattering Rayleigh peak) by the characteristic vibrational frequencies. If the incident frequency is offset to zero, then Raman peaks occur at the same frequencies as peaks in the IR spectrum, as illustrated in Figure 2 for the molecule methyldithioacetate. ["Anti-Stokes" Raman peaks also appear, on the high-energy side of the Rayleigh peak, due to energy transfer from vibrationally excited molecules. Since the population of such molecules depends on the Boltzmann factor

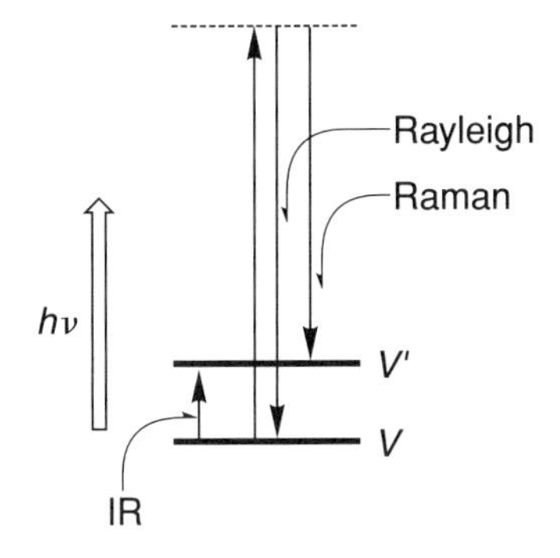

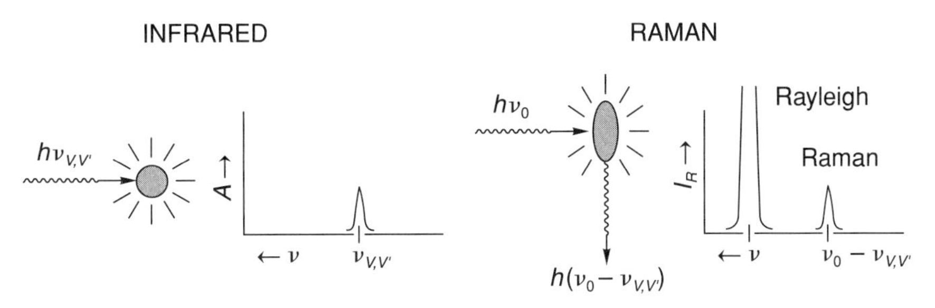

Figure 1
Infrared and Raman processes: Absorption of an IR photon raises the molecule from the ground (v) to an excited (v') vibrational level, and produces a peak at $\nu_{v,v'}$ in the IR spectrum. Scattering of visible photons produces an intense Rayleigh peak, and weaker Raman peaks, displaced from the Rayleigh peak by $\nu_{v,v'}$. These arise from energy transfer from the incident photon to the molecule, which is raised to an excited vibrational level.

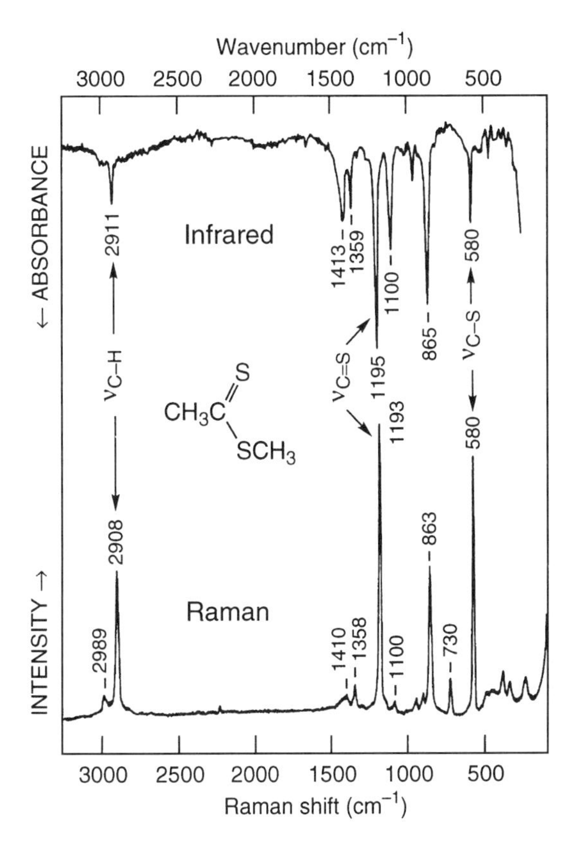

Figure 2
Comparison of IR and Raman spectra of methyldithioacetate. The Raman shifts are the same as the IR frequencies, although the relative intensities of the bands differ. Bands that mainly involve the stretching of C–H, C=S, and C–S bonds are labeled. [Adapted from Carey,[1a] p. 20.]

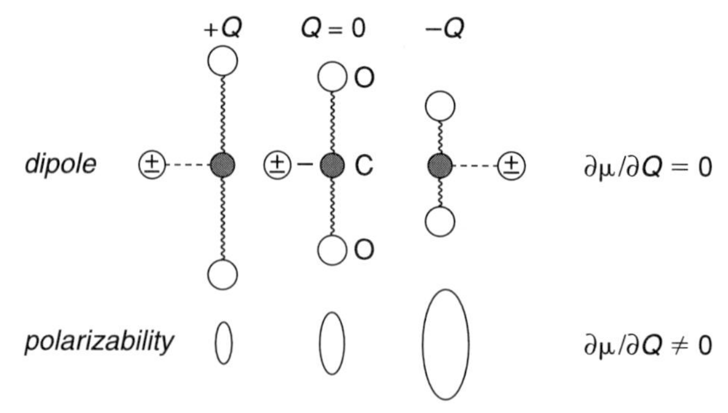

(*a*) ν_1, symmetric stretching mode; IR - inactive, Raman - active

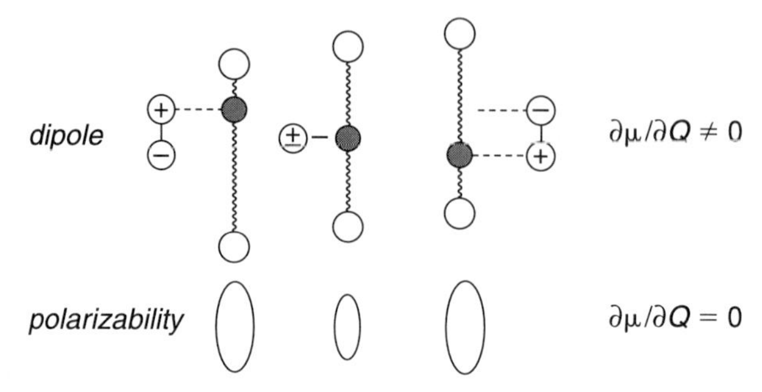

(*b*) ν_3, antisymmetric stretching mode; IR - inactive, Raman - inactive

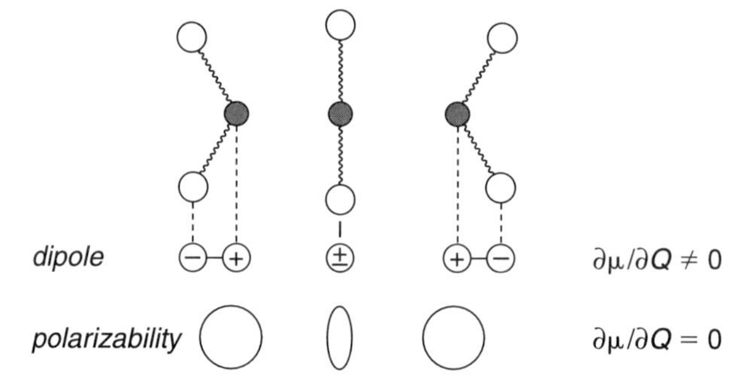

(*c*) ν_2, bending mode (degenerate); IR - inactive, Raman - inactive

Figure 3
Changes in the electric dipole moment (μ) and molecular polarizability (α) produced by three fundamental vibrations of the centrosymmetric molecule carbon dioxide. The center column shows the equilibrium position of the molecule ($Q = O$) while to right and left are the extremes of each vibrations ($\pm Q$) (the amplitudes are greatly exaggerated for clarity). In order to be IR active, that is, in order to absorb IR radiation, the vibration must cause a change in the molecular dipole moment μ. In order to be Raman (R) active, that is, in order to inelastically scatter (usually) ultraviolet–visible (UV–vis) photons, it is necessary that the vibration causes a change in the induced dipole moment P or molecular polarizability α ($P = \alpha E$). In the equilibrium position ($Q = O$), $\mu = 0$ for CO_2 (required by the symmetry of the molecule), whereas $P \neq 0$ (the molecule suffers some distortion in the applied field E, the positively charged nuclei being attracted toward the negative pole of the field, the electrons to the positive pole). (*a*) The symmetrical, in-phase stretching vibration ν_1 does not change the symmetry in the distorted molecules [both C–O bonds stretch ($+Q$) or contract ($-Q$) simultaneously], and

($kT/hc = 209$ cm^{-1} at $T = 300$ K), the intensities of the anti-Stokes peaks fall off rapidly with increasing vibrational frequency.]

Although the frequencies of Raman and IR peaks are the same, their relative intensities can differ greatly, because of the different physical mechanisms involved. Absorption of IR photons requires a change in the dipole moment, μ, of the molecule associated with the vibration under investigation. The IR intensity is given by the square of the dipole moment derivative with respect to the vibrational coordinate, Q.

$$I_{\mathrm{IR}} \propto (\partial\mu/\partial Q)^2 \tag{1}$$

Raman scattering, on the other hand, depends on the fluctuating dipole moment, P, induced by the incident photon. The moment P is the product of the photon electric field, E, and the change in the polarizability of the molecule, α'. The Raman intensity is given by

$$I_{\mathrm{R}} \propto P^2 = (\alpha' E)^2$$

where

$$\alpha' = (\partial\alpha/\partial Q)_0 Q \tag{2}$$

As an example, Figure 3 shows the nature of the dipole and polarizability changes for the three fundamental vibrations of carbon dioxide, ν_1 (symmetric stretch), ν_2 (bending, degenerate), and ν_3 (antisymmetric stretch). A vibration producing a large dipole change might not produce a large polarizability change, and vice versa. Indeed, for centrosymmetric molecules the two processes are mutually exclusive, since a change in the dipole moment requires a loss of the symmetry center, while a change in the polarizability requires its preservation (see ν_3 and ν_2 vs. ν_1 in Fig. 3). Moreover, symmetric bond stretching vibrations, which produce large polarizability changes, usually dominate the Raman spectra, whereas antisymmetric stretching and deformation modes, involving large dipole

consequently $\partial\mu/\partial Q = 0$ for this vibration. The ν_1 is thus IR *inactive* (the E field of the IR photon cannot pull the two negatively charged O atoms in opposite directions as required during this vibration). In contrast, since the CO_2 molecule changes size during the ν_1 motion, there is a corresponding fluctuation in the size of the ellipsoids; the $\partial\alpha/\partial Q \neq 0$ and the motion is thus Raman *active*. (*b*) and (*c*) Both the antisymmetrical, out-of-phase stretching vibration ν_3 (one C–O bond stretches while the other contracts) and the bending vibration ν_2 change the symmetry in the distorted molecules (by destroying the center of inversion), and cause the dipole moment to oscillate along (ν_3) or across (ν_2) the molecular axis. The $\partial\mu/\partial Q \neq 0$ for ν_3 and ν_2, and both are thus IR *active* (the E field of the IR photon pulls the positively charged C atom in one direction and the negatively charged O atoms in the opposite direction along or across the molecular axis as required to excite ν_3 and ν_2). In contrast, although the ν_3 and ν_2 motions also change the polarizability ellipsoides, the respective change is exactly the same for both positive and negative Q. Consequently, $\partial\alpha/\partial Q = 0$ for ν_3 and ν_2, and both vibrations are Raman *inactive*.

changes, are strong in IR spectra. For an associated liquid like H_2O, the many vibrations associated with the H-bond network produce stronger IR than Raman signals; consequently, H_2O interferes much less with Raman than IR spectra, giving Raman spectroscopy an advantage for aqueous solution studies.

B. Polarization

The dipole moment is a vector quantity, and the absorption of IR light depends on the relative orientation of the molecule and the photon electric vector, as illustrated in Figure 4. If all the molecules are aligned, as in a crystal or a stretched film, and the photon vector points along a molecular axis, (e.g., z), then absorption occurs for those vibrations that displace the dipole along z. Vibrations that are purely x or y polarized would be absent. Thus extra information about the

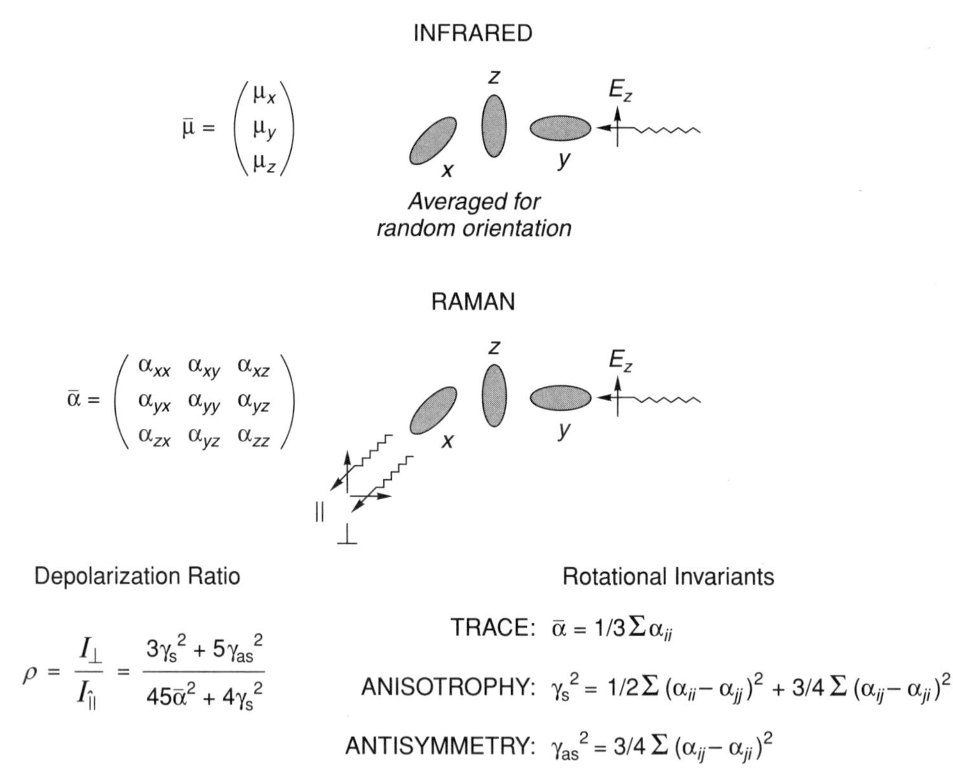

Figure 4

Polarization properties of IR absorption and Raman scattering. The dipole moment associated with the vibration is a vector quantity, and polarization information is lost if the molecules are randomly oriented. The polarizability change, however, is a tensor quantity and even for randomly oriented molecules, symmetry information can be obtained by analyzing the scattered light into components parallel and perpendicular to the incident light vector. This information is contained in the rotational invariants of the tensor, which determine the depolarization ratio.

character of the vibration is available for oriented samples. If the sample is unordered, all polarization information is lost.

This is not the case for Raman scattering, however, since the incident and scattered photons both involve a polarization, and their mutual orientation is important for the scattering intensity, even if the sample is unordered. Because of the two polarization directions, the polarizability is a tensor quantity with nine elements, α_{xx}, α_{xy}, α_{xz}, ..., and so on. If the molecules in the sample were all aligned in a given direction, then the scattering intensity would be determined by α_{zz} if the incident photon vector were aligned with the molecular z axis and the scattered light was analyzed along z. Only those vibrations having nonzero α_{zz} would be represented in this spectrum. On the other hand, if the scattered light were analyzed along x, then vibrations with nonzero α_{xz} would be represented. This situation is illustrated graphically in Figure 4 for the usual scattering geometry in which scattered light is collected at 90° to the incident beam.

When the sample is unoriented, the direction of the incident polarization is unrelated to the molecular axes, but the scattered light can be analyzed parallel ($\parallel$) or perpendicular ($\perp$) to the incident light. The ratio of the two intensities ($I_{\perp}/I_{\parallel}$) is called the depolarization ratio ρ, and it can be expressed in terms of the rotational invariants of the scattering tensor, as defined in Figure 4. Under most circumstances, the scattering tensor is symmetric (i.e., $\alpha_{ij} = \alpha_{ji}$, and consequently $\gamma_{as} = 0$). In that case, the depolarization ratio must be $\frac{3}{4}$ (when $\bar{\alpha} = 0$) or less ($\bar{\alpha} \neq 0$). Only for vibrations that preserve the symmetry of the molecule (totally symmetric vibrations) are diagonal polarizability tensor elements nonzero ($\bar{\alpha} \neq 0$); these modes have $\rho < \frac{3}{4}$, and are said to be polarized. Nontotally symmetric vibrations ($\bar{\alpha} = 0$) give rise to depolarized bands ($\rho < \frac{3}{4}$) ordinarily. In special cases involving the resonance Raman effect, however, the scattering tensors are not symmetric, $\alpha_{ij} \neq \alpha_{ji}$, in which case $\gamma_{as} \neq 0$, and $\rho > \frac{3}{4}$, a situation described as anomalous polarization. If the symmetry is high enough, the scattering tensor may be antisymmetric, and $\alpha_{ij} = -\alpha_{ji}$, in which case $\gamma_{s} = 0$ and $\rho = \infty$ (the band intensity in parallel scattering is zero), a situation termed inverse polarization.

C. Experimental Aspects

Historically, Raman spectroscopy provided our first catalog of molecular vibrational frequencies in the years following the discovery of the Raman effect by Raman and Krishnan (1928).[3] Commercial IR spectrophotometers became available in the 1940s. However, they quickly eclipsed Raman spectrometers as a method for obtaining vibrational spectra, since Raman scattering is weak. The total intensity in the Raman peaks is generally 1 millionfold smaller than the intensity of the Rayleigh peak. In prelaser days, heroic measures were required to obtain a Raman spectrum. Typically the sample, held in a long tube, was illuminated coaxially by filtered monochromatic radiation from gas discharge lamps, and the scattered photons were collected through an optical flat at the end of the tube. A substantial volume ($\sim$5 mL) of a concentrated, optically transparent sample, was required.

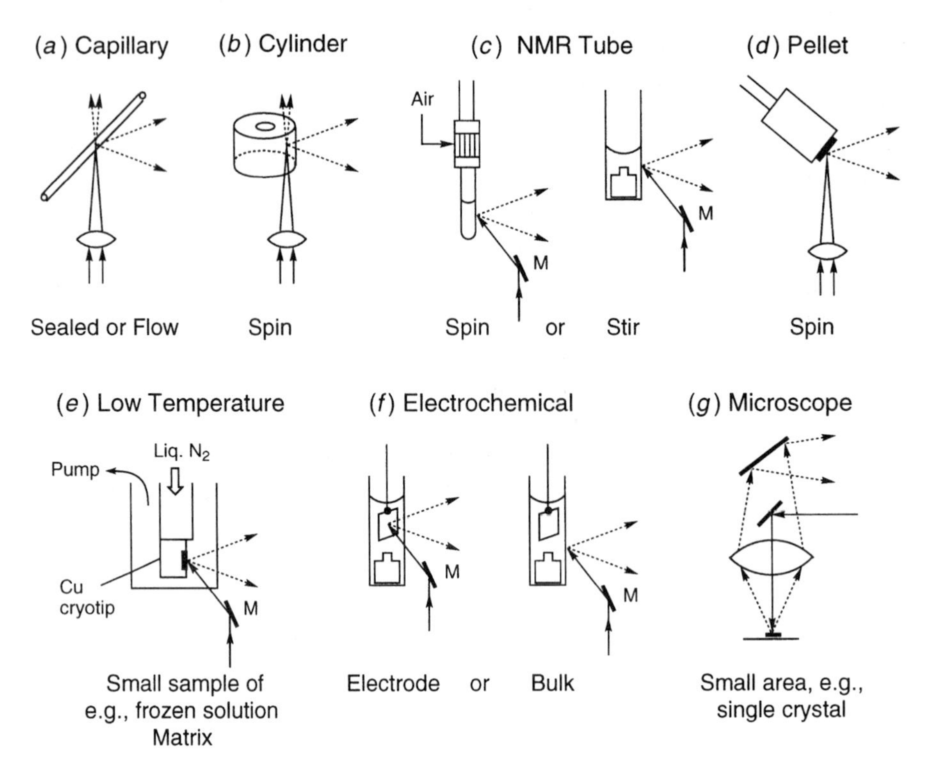

Figure 5

A variety of sampling devices are compatible with the flexible, directional properties of the laser beam. The simplest is a transverse capillary (*a*), in which the sample can be sealed (milliliter amounts can be examined if the sample is not damaged by the laser beam) or flowed, in order to avoid thermal or photo damage, or to examine a kinetic product of a flow reactor. Sample damage can also be avoided via a spinning cylinder [90° geometry (*b*)] or a NMR tube in front scattering, either spun or stirred (*c*). A solid sample can be examined in a pellet via front scattering, and spun to avoid damage by the incident laser beam (*d*). A cryotip (*e*) is especially useful for examining scarce materials. A drop of frozen solution (e.g., of a protein) can provide a Raman spectrum, whose quality is generally improved by the low temperature, the cryotip also serving to inhibit thermal damage. Electrochemical cells (*f*) are readily adapted to the Raman experiment with the laser scattered either from the solution phase or directly from the electrode surface. Finally, microscope optics can be used to examine very small areas of a sample, including, for example, protein single crystals (*g*).

The advent of the laser revolutionized the practice of Raman spectroscopy. Because it can be focused to a spot, the laser reduced the effective scattering volume from milliliters to microliters, and eliminated restrictions on the sampling geometry. Nowadays, spectra can be obtained in the conventional 90° geometry, or approximately equal to 135° front- or 180° backscattering geometries to accommodate special sample requirements. Figure 5 illustrates some of the sampling arrangements that have been found to be useful. The simplest sample holder is a capillary tube held transverse to the laser beam, from which scattered light can be collected at 90°. The sample can be stationary, in which case only milliliters of the material are required, or it can be flowed through the capillary tube in a recirculating arrangement with a reservoir. Motion of the sample through the laser beam may be needed in order to eliminate the buildup of decomposition products. This aspect is particularly important in resonance Raman studies when the excitation wavelength is deliberately tuned to an absorption band, so that photo and thermal effects are to be expected. The flowing arrangement is also useful for kinetic Raman studies, in which two solutions are mixed upstream of the laser, and spectra are obtained of chemical intermediates formed within the transit time from the point of mixing to the laser focus.

Alternatively, the sample can be spun in the laser beam so that the radiation is spread over the sample volume; this arrangement is convenient for small volumes and/or for air-sensitive samples requiring a confined space. The spinning cylinder, typically used for transparent liquids, allows for 90° scattering while front scattering from a spinning NMR tube provides a particularly convenient experimental arrangement when the sample is optically dense. In most applications, the angle of incidence is about 135°, as shown in Figure 5, although 180° backscattering may also be obtained if desired. A disadvantage of the spinning NMR tube is that vertical mixing of the solution is ineffective and the irradiated volume tends to be confined to a circle around the tube. This problem can be obviated by inserting a stirrer in the tube.

Solid samples are readily examined with a laser excitation source. A spinning pellet is a particularly useful arrangement. The directional properties of the laser make it easy to record spectra of frozen solution samples with a liquid nitrogen cooled finger (77 K) [or a cryotip attached to a liquid He refrigerator (10 K)]. This method is particularly valuable for examining scarce protein solutions, since only a drop is required. In addition, the absence of a broad Raman band (300–550 cm^{-1}) arising from the glass sample container makes this technique particularly suitable for the study of low-frequency vibrational modes. Heat removal by the cold tip minimizes laser-induced damage, and moreover, the spectra are generally of higher quality because of the sharpening of the Raman bands at low temperature.[4a,b] Electrochemical cells are likewise readily adapted to laser excitation for Raman spectroscopy of electrogenerated products, either in the bulk solution or directly at the electrode surface.[4c,d] Finally, the laser can be directed through a microscope and the backscattered light collected by the microscope objective in order to excite Raman spectra of microscopic samples in a Raman microprobe arrangement. Thus the Raman experiment can be adapted to a wide range of experimental conditions.

Figure 6 is a block diagram of a Raman spectrometer. The scattered photons are collected by a lens or mirror system and focused on the entrance slit of a spectrometer, consisting of a monochromator and detector system. The spectrum may be scanned, by rotating the monochromator gratings and monitoring the dispersed light with an exit slit and a photomultiplier tube, which is generally equipped with photon counting electronics for maximum sensitivity at the low light levels encountered in Raman spectroscopy. Alternatively, the monochromator can be operated as a spectrograph by eliminating the exit slit and capturing the dispersed light with a multichannel detector.[4e] In this way an entire

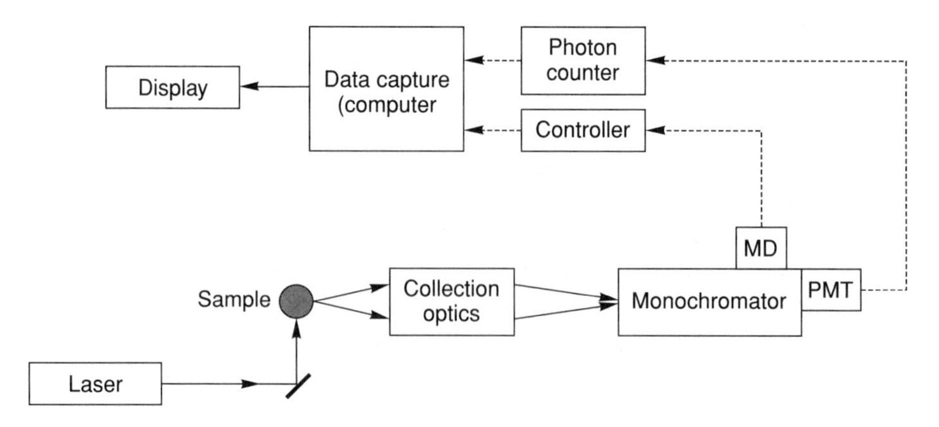

Figure 6
Diagram of the elements of a Raman spectrometer. A laser beam is focused onto the sample, from which scattered photons are collected and focused onto the entrance slit of a monochromator. The monochromator may be operated in scanning mode, the analyzed light at the exit slit being monitored by a photomultiplier (PMT), usually with photon-counting electronics, or it may be operated as a spectrograph, the disbursed spectrum being recorded with a multichannel detector (MD) (usually an intensified diode array, or a charge-coupled device).

section of the spectrum is recorded simultaneously, the coverage depending on the resolution of the spectrometer and the width of the detector target. Because of the large number of resolution elements (typically 500–1000) multichannel detection can produce a spectrum in much shorter times than scanning. This characteristic is particularly important in time-resolved experiments, when transient species are being sought. Also, if the sample is undergoing a chemical alteration during spectral acquisition, better results are obtained with multichannel detection since the scanned spectrum will be distorted by the time dependent change. There are, however, disadvantages with multichannel detection. Only coarse changes in resolution are possible, via grating replacements, whereas with a scanning instrument one can examine a portion of the spectrum at whatever resolution is required to bring out selected features of interest. Stray light is a more difficult problem with multichannel detectors, especially in the low-frequency region of the spectrum, since the exit slit of the spectrometer is eliminated. For this reason, triple monochromators are often used in multichannel applications, but their low throughput removes some of the speed advantage of multichannel detection. Thus, there are trade-offs between scanning and multichannel instruments, which vary from one application to another.

The response of the system depends critically on the wavelength and other operating characteristics of the laser. The most common and reliable Raman sources are continuous wave (CW) gas lasers. Argon and krypton ion lasers can operate at a series of wavelengths in the visible and near-UV region, as shown diagrammatically in Figure 7. In addition, the helium–neon and helium–cadmium lasers have useful wavelengths, albeit at lower power levels. The argon and krypton lasers can also be used to pump CW dye lasers, for continuous wavelength tuning.

In recent years, pulsed lasers have been introduced into Raman spectroscopy. The very high photon flux in the laser pulses produces efficient dye laser pumping and also the possibility of extensive frequency shifting through doubling and mixing crystals, and Raman shifting in media with high cross section for the stimulated Raman effect. A particularly useful device is the H_2 Raman shift cell, a pipe filled with high-pressure hydrogen gas and fitted with appropriate windows. When a laser pulse is focused into this cell, a series of new laser lines appear, with frequencies that are the input laser frequency plus and minus multiple quanta of the H_2 stretching frequency, 4115 cm^{-1}. This device has been used to generate deep UV Raman excitation lines by H_2–Raman shifting of the harmonic output of the pulsed Nd:YAG laser.[5] Recently, frequency doubling of pulsed dye lasers, pumped by Nd:YAG or excimer lasers, has come to the fore as a means of generating UV Raman excitation.[6] An approach using a modelocked Nd:YAG laser to create a quasi-CW UV source has also been introduced,[7] while the Coherent Laser Group (division of Coherent, Inc.) developed an intracavity frequency-doubled CW Ar^+ laser with five excitation wavelentghts in the UV region (229–257 nm) of moderate output powers (30–400 mW). The latter laser permitted a major improvement in spectral signal to noise (S/N) and, for the first time, allowed UV-RR studies to be performed on solid-state samples.[8] Ultraviolet Raman spectroscopy is gaining importance in biological applications since it can provide selective information about the environment of aromatic residues in proteins and nucleic acids.[9] The wide range of laser frequencies available with pulsed lasers is illustrated in Figure 7. Pulsed lasers are also useful in time-resolved Raman studies, a rapidly emerging technique for the study of chemical dynamics.

Also in recent years, the dramatic upsurge in application of near-IR Raman spectroscopy has been stimulated, chiefly thanks to the development of commercially available Fourier transform (FT) Raman spectrometers in conjunction with a CW Nd:YAG 1064-nm excitation.[10] Fourier transform Raman spectroscopy offers the multiplex advantage of the Michelson interferometer to increase the spectral S/N that is limited by IR detector background noise, and has the advantage of excitation in a spectral region that is frequently free of fluorescence interferences. Several promising FT Raman studies on selected proteins,[11a] photobiological systems,[11b] and photoactive proteins[11c] have appeared in the literature, and a general application to biological materials has been discussed.[11d] An important disadvantage of FT Raman spectroscopy at the moment is the lack of variable excitation required for RR studies. The recent advent of a tunable Ti–saphire dye laser in the near-IR region (~700–1100 nm) may allow one to circumvent this limitation, however.

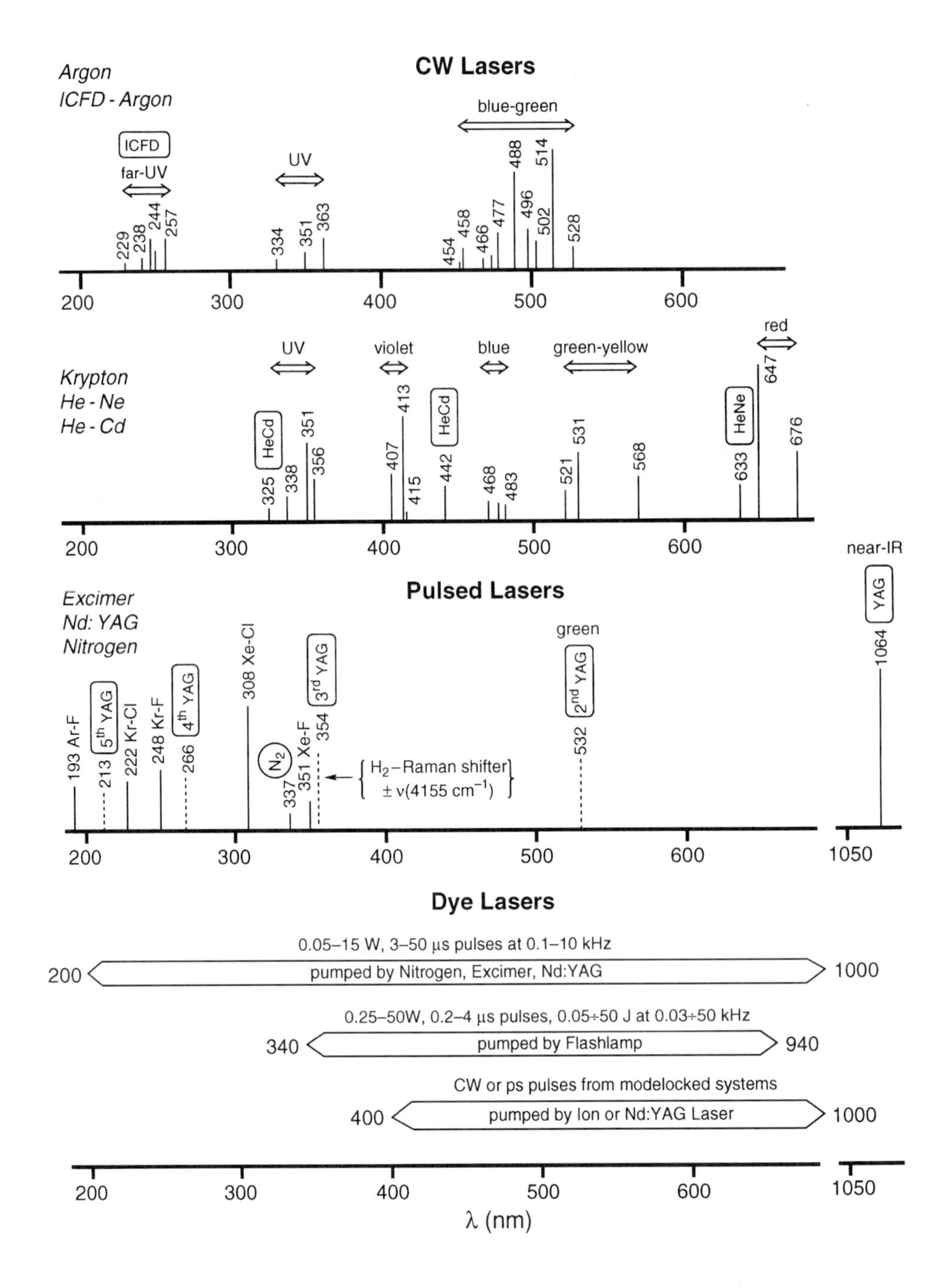

Figure 7
Diagram of laser sources suitable for Raman spectroscopy. The best quality spectra are provided by CW lasers, Ar^+, Kr^+, He-Ne, and He-Cd, operating at fixed frequencies (the length of the lines indicate the relative output for a given laser) throughout the visible and near-UV region. An intracavity frequency-doubled (ICFD) Ar^+ laser has recently been developed with five useful CW excitation wavelengths in the far-UV region (257, 248, 244, 238, and 228.9 nm). The high-powered Ar^+ and Kr^+ lasers can also be used to pump dye lasers that are tunable throughout the visible region. A wider tuning range is afforded by pulsed lasers, N_2, excimer, or Nd:YAG. The N_2 and excimer lasers are not used directly, because of large line

III. Molecular Vibrations

In this section, we outline some fundamental considerations about molecular vibrations that are important in understanding vibrational spectra and applying them to structural problems. The main points emerge from the vibrational characteristics of diatoms and triatoms, which can be considered as building blocks for the vibrations of more complex molecules.

A. Diatoms

To a first approximation, the frequency of the stretching vibration of a diatomic molecule, X–Y, is described by Hooke's law

$$\nu(\mathrm{Hz}) = \frac{1}{2\pi}\left[F\left(\frac{1}{m_X}+\frac{1}{m_Y}\right)\right]^{1/2} \quad \text{or} \quad \tilde{\nu}(\mathrm{cm})^{-1} = \frac{\nu}{c} = 1303.1[F(\mu_X+\mu_Y)]^{1/2} \tag{3}$$

in which μ_X and μ_Y are reciprocals of the masses of the atoms, and F is the spring constant for the bond between them. Vibrational spectroscopists use the second form of the equation, with frequency expressed as wavenumbers (cm^{-1}), the reciprocal of the wavelength; the factor 1303.1 gives ν in reciprocal centimeters, when F is expressed in millidynes per angstrom (mdyn $Å^{-1}$). The stiffer the spring and the lighter the atoms, the higher the bond stretching frequency. Because the reciprocals of the masses are added to produce the reduced mass of the diatom, the lighter atom dominates the frequency. Table I shows that the vibrational frequencies of the hydrogen halides are in direct proportion to the square of the force constant, since the reduced mass is essentially unity for all of them. Between HF and HI the frequency falls by about $\sqrt{3}$, corresponding to the approximate threefold decrease in the force constant. In contrast, the halogens, which also show a factor of approximately three variations in force constant, have a much wider frequency range, reflecting the much greater reduced mass of I_2 than F_2.

Hooke's law describes a harmonic oscillator, whose potential is parabolic in the internuclear displacement, as illustrated in Figure 8 (dotted curve) for the molecule HBr.[12] Quantum mechanics tells us that only certain energy levels on

widths, but are used to pump dye lasers in the visible region, which can be doubled and/or mixed with nonlinear generating crystals to produce useful power deep into the UV region. The 1064 nm Nd:YAG output can yield usable power at the second, third, fourth, and fifth harmonics, generated in crystals; these outputs have sufficient power to pump dye lasers or to generate stimulated Raman scattering in H_2–Raman shift cells, producing a cone of lines, separated by the 4155 cm^{-1} H_2 vibrational frequency, built on the incident wavelength. The YAG laser is also available in a CW version; this version can be mode locked to produce a picosecond pulse train, which can be frequency doubled or tripled and used to pump picosecond dye lasers synchronously.

Table I
Mass and Force Constant Dependence of Diatomic Frequencies

Molecule	ν (cm^{-1})[a]	F (mdyn $Å^{-1}$)[b]
F_2	892	4.68
Cl_2	546	3.11
Br_2	319	2.39
I_2	215	1.73
HF	4138	9.68
HCl	2991	5.16
HBr	2650	4.12
HI	2309	3.14

[a] Frequencies from Nakamoto.[31]
[b] The F values were calculated from Eq. (3).

the parabola correspond to stationary states, and the spacing between these levels is the energy corresponding to the classical Hooke's law of frequency. The transition from the $v = 0$ to the $v' = 1$ vibrational level is the fundamental transition, while the transitions to higher levels are the harmonics, or overtones.

The harmonic oscillator potential is obviously inadequate to represent a real molecule, since it requires infinite energy for infinite internuclear separation. A more realistic description is given by the Morse potential (Fig. 8, solid curve), which incorporates a dissociation limit, the energy at which the atoms fly apart. The Morse potential force constant (curvature at the bottom of the potential well) is given by

$$F = \left(\frac{d^2V}{dQ^2}\right)_{Q=0} = 2D_e\beta^2 \tag{4}$$

where $Q = r - r_e$ is the vibrational coordinate, D_e is the dissociation energy, and β is the spreading parameter of the curve. This relationship provides a theoretical basis for our intuitive feeling that the force constant should be directly related to the bond energy. Figure 9 shows that experimental force constants for a wide range of diatoms do vary directly with the bond enthalpy.[12] There are, however, significant deviations from the best straight line passing through the intercept, which presumably reflect variations in the Morse potential β. There are a variety of empirical relationships among force constant, bond energy, and bond distance, which have recently been critically reviewed by Burgi and Dunitz.[13] The most useful is Badger's rule,[14a] or its more current variants,[14b,c] which relates the force constant to bond distance with empirical parameters that depend on the row of

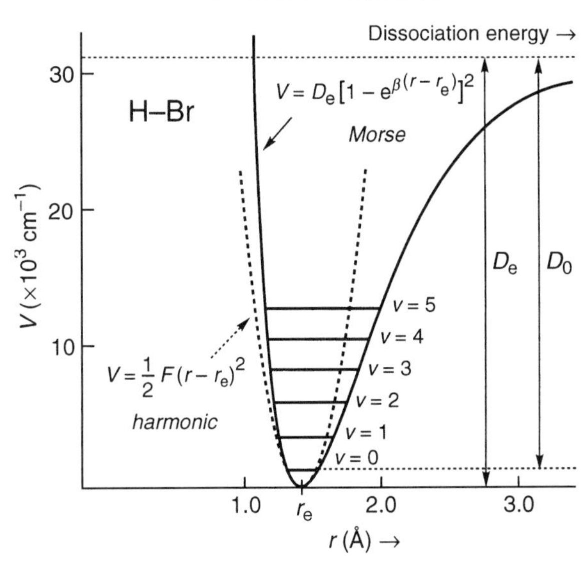

Figure 8
The potential curve for a diatomic oscillator, HBr,[12] is closely approximated by the Morse curve. As the vibrational quantum number increases, the level spacings decrease, and collapse at the dissociation energy. Near the bottom of the well, the potential is approximated by that of a harmonic oscillator, obeying Hooke's law. The force constant, F, is the curvature at the potential minimum.

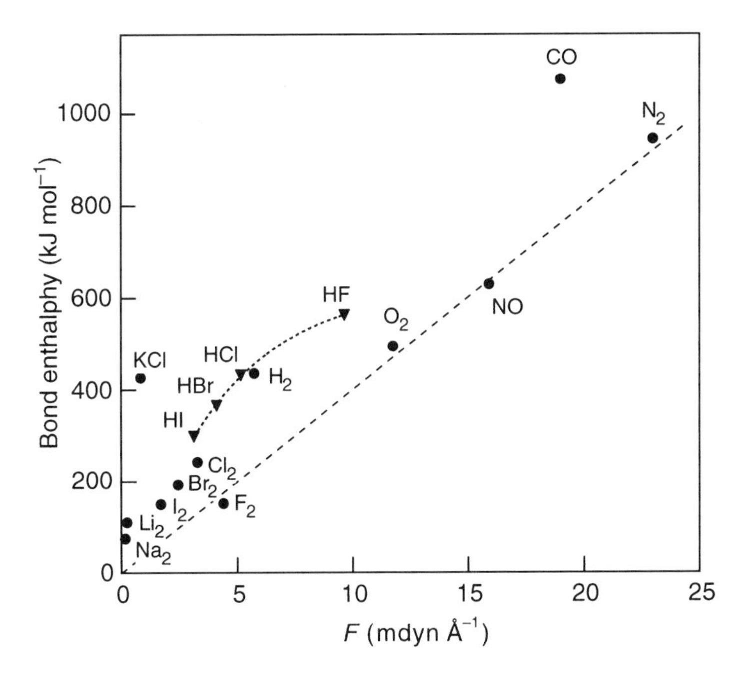

Figure 9
Plot of bond enthalpy versus force constant (corrected for anharmonicity), for several diatomic molecules.[12] Most points fall on a straight line passing through the intercept, consistent with the curvature of the potential well at its minimum being proportional to the well depth (constant β in the Morse equation), but there are significant deviations, reflecting the breakdown of this approximation.

the periodic table containing the connected atoms. These relations can be used to obtain starting approximations of force constants from structural data, or to constrain the ratios of force constants of bonds connecting similar atoms, in vibrational analyses of complex molecules.

B. Triatoms

We next consider the stretching vibrations of a linear triatom, Y–X–Y. There are two such vibrations, one for each of the bonds, and it is clear from the diagram in Figure 10 that the two bond stretching motions interact strongly. When the two

Coupled Bond Stretches in Triatomic Molecules

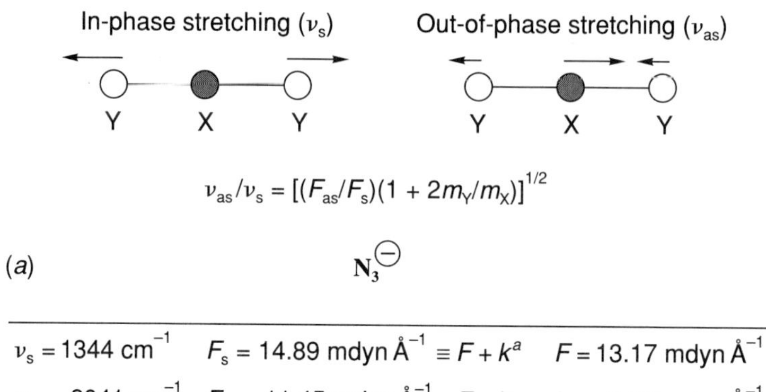

$\nu_s = 1344\ \text{cm}^{-1}$	$F_s = 14.89$ mdyn Å^{-1} $\equiv F + k^a$	$F = 13.17$ mdyn Å^{-1}
$\nu_{as} = 2041\ \text{cm}^{-1}$	$F_{as} = 11.45$ mdyn Å^{-1} $\equiv F - k$	$k = 1.72$ mdyn Å^{-1}
$\nu_{as}/\nu_s = 1.52$	(vs. $\sqrt{3}$ if $k = 0$)	

[a] *F* and *k* are principal and stretch–stretch interaction force constants, respectively. *k* = + means stretching one bond contacts the other:

$N{=}N{=}N^{\ominus} \longrightarrow N{\cdots\cdots}N{\equiv}N^{\ominus}$

(*b*)

O=U^{2+}=O $\quad \nu_{as} = 931\ \text{cm}^{-1}$ $\quad \nu_s = 856\ \text{cm}^{-1}$

$\nu_{as}/\nu_s \rightarrow$ *for light* Y

porphyrin Fe–O–Fe $\quad \nu_{as} = 850\ \text{cm}^{-1}$ $\quad \nu_s = 360\ \text{cm}^{-1}$

$\nu_{as}/\nu_s \rightarrow$ large *for heavy* Y

Figure 10
In- and out-of-phase stretching modes of a linear triatomic molecule, showing how the stretching motions oppose or reinforce one another, leading to a higher frequency for the out-of-phase mode due to the kinematic coupling. (*a*) If all three masses and the two force constants, are the same the ratio of the frequencies should be $\sqrt{3}$. The ratio is less than this for the azide ion[17] because the electronic structure lowers the energy of the out-of-phase motion, giving a positive stretch–stretch interaction force constant. (*b*) If the mass of the central atom is much greater than that of the outer atoms, for example, UO_2^{2+},[18] the two frequencies approach one another (mass uncoupling), while in the reverse situation, for example, $(Fe^{III}$ porphyrin$)_2O$,[19] the ratio becomes very large.

bonds are stretched in-phase, the displacements are in opposition and motion of the central X atom is canceled. The frequency for this motion is given by

$$\nu_s = 1303.1\{F_s[\mu_Y + \mu_X(1 + \cos\alpha)]\}^{1/2} \tag{5}$$

where the subscript "s" stands for symmetric (the motion retains the molecular symmetry), $\mu_Y = 1/m_Y$ and $\mu_X = 1/m_X$, and α is the Y–X–Y angle. Since $\alpha = 180°$ for a linear triatom, Eq. (5) reduces to

$$\nu_s = 1303.1(F_s\mu_Y)^{1/2} \tag{6}$$

and the frequency for the symmetric stretching vibration does not depend on the mass of the central atom; the reduced mass for this motion is just the reciprocal of the mass of Y. When the two bonds stretch out of phase (one stretches while the other contracts), the displacements reinforce one another and the X atom moves twice as far as it did in the X–Y diatom (assuming the same force constant). The frequency for this motion is given by

$$\nu_{as} = 1303.1\{F_s[\mu_Y + \mu_X(1 - \cos\alpha)]\}^{1/2} \tag{7}$$

or, for $\alpha = 180°$, simply by

$$\nu_{as} = 1303.1[F_{as}(\mu_Y + 2\mu_X)]^{1/2} \tag{8}$$

(The subscript "as" stands for antisymmetric; the motion destroys the symmetry of the molecule.) The X atom contribution to the reduced mass is doubled. If X and Y are the same atoms, and if the force constant is the same for the in- and out-of-phase stretches (i.e., $F_s = F_{as}$), their frequencies are predicted to be in the ratio $1 : \sqrt{3}$, while the X–Y diatomic frequency for the same force constant would scale as $\sqrt{2}$. As an example, the C=C stretching frequency for ethylene[15] is 1623 cm^{-1} while the in-phase and out-of-phase C=C frequencies of allene[16] are 1071 and 1980 cm^{-1}. These values are not too far from the expected $1 : \sqrt{2} : \sqrt{3}$ intervals; closer agreement cannot be expected because the motions of the H atoms also contribute somewhat to these frequencies.

The frequencies of the azide ion, N_3^-,[17] can be treated exactly in our simple scheme (Fig. 10). The in-phase and out-of-phase frequency ratio, 1.52, is significantly less than $\sqrt{3}$, showing that the force constant does not have the same value for the two motions. This inequality is due to an electronic interaction between the two bond displacements, which introduces a cross term in the potential energy second derivative, called an interaction force constant, k. This term adds to the bond stretching constant (the principal valence force constant) for the in-phase motion ν_s but subtracts from it for the out-of-phase motion ν_{as}. The positive interaction constant for N_3^-, can be understood on the basis of its electronic structure. Since there are two double bonds in line, stretching one of them makes the other one harder to stretch but easier to contract, because π electron density can be transferred from the first to the second bond. Thus the frequency is raised for the in-phase but lowered for the out-of-phase motion. Interaction force constants are generally smaller than 10% of the valence force constant, but can be somewhat larger for interacting π bonds.

The ratio of the atomic masses also affects the in-phase and out-of-phase frequency separation. The frequencies approach one another when the outer atoms are much lighter than the central atom, which then makes little contribution to either motion. Conversely, very large separations are seen when the outer atoms are much heavier than the central atoms. These cases are illustrated in Figure 10 with UO_2^{2+}[18] and the μ-oxo dimer of Fe^{III} porphyrin.[19,20]

The in-phase and out-of-phase motions interact less for unsymmetrical linear triatoms if the force constants of the two bonds or the masses of the two outer atoms are very different. This is illustrated in Figure 11 for a series of X–C≡N molecules.[21] The natural frequency for the C≡N triple bond is about 2100 cm^{-1}.

Energy Factoring of Bond Stretches in X–C≡N Molecules

Frequencies (cm^{-1}) and Force Constants (mdyn $Å^{-1}$)

	ν_{as}	ν_s	F_{as}	F_s
^1HCN	3312	2089	5.8	17.9
^2HCN	2629	1906	5.8	17.9
^3HCN	2460	1724	5.8	17.9
FCN	2290	1077	9.2	16.3
ClCN	2201	729	5.2	16.7
BrCN	2187	580	4.2	16.9
ICN	2158	470	3.0	16.7

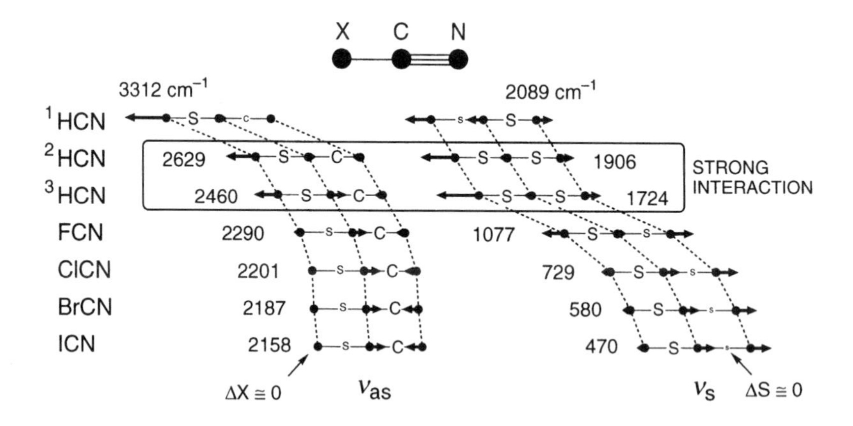

Figure 11
Uncoupling due to energy separation: For X–C≡N molecules,[21] the two stretching modes can be classified as in-phase (lower) and out-of-phase (upper) (S = bond stretching, C = bond contraction). But the interaction is strong only when the frequencies are not too far apart (^2HCN and ^3HCN). For ^1HCN, the upper frequency, 3312 cm^{-1}, involves stretching of the H–C bond, exclusively, while the lower frequency, 2089 cm^{-1}, is mostly C≡N stretching. Coversely, for the halogen cyanides, the upper frequency ($\sim$2200 cm^{-1}) is mostly C≡N stretching, while the lower frequency is mostly X–C stretching.

One of the two stretching frequencies is close to this value for HCN (lower frequency) as well as the halogen cyanides (upper frequency). Calculation of the atomic motions (Fig. 11) shows that in these cases C≡N stretching is a good approximate description of the motion, whether it is in phase or out of phase, because the remaining bond is displaced very little. Only when the two frequencies approach one another, in ^{2}HCN and ^{3}HCN, do the two bond displacements interact strongly, producing shifts in the natural frequencies and thoroughly mixed motions. As the frequencies move apart, the interaction of the displacements decreases, an effect known as energy factoring. Approximate energy factoring is widely encountered in complex molecules and is of great assistance in permitting qualitative band assignments.

Uncoupling of adjacent bond stretches is also encountered if the triatom (or triatomic fragment of a more complex molecule) is allowed to bend, as illustrated in Figure 12. The mutual opposition or reinforcement of the two bond displacements is now modified by the cosine of the bond angle. Thus coupling is maximal when the bond angle is 180° but vanishes when it is 90°. The equations given in Figure 12 predict equal frequencies at 90° for the in-phase and out-of-phase

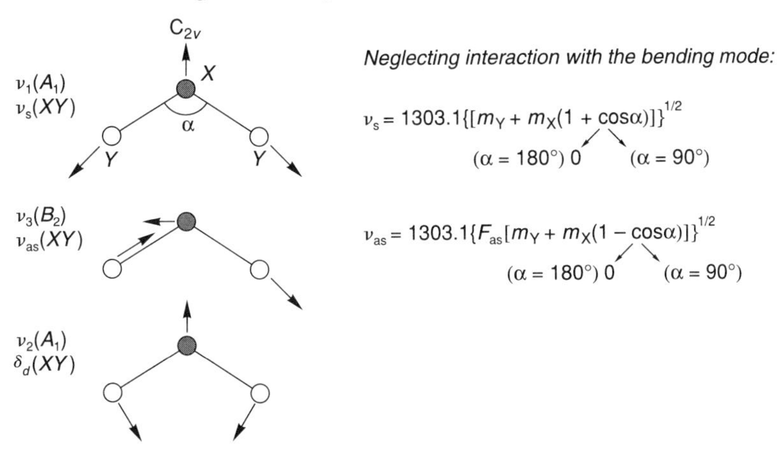

Bond Angles (°) and Vibrational Frequencies (cm^{-1})

	α	ν_s	ν_{as}	δ_d
F_2O	102°	925	821	461
Cl_2O	–111°	631	671	296

Figure 12
Modes of a bent triatomic molecule. Interaction between the symmetric and antisymmetric stretches decreases to zero as the bond angle a decreases to 90°. The symmetric stretch interacts with the bend, but this is a small effect since the bending frequency is typically about one-half of the stretching frequency. For angles near 90°, however, this effect is sufficient to raise ν_s above ν_{as} (F_2O).[22]

stretches if the force constants are the same (negligible interaction constant), and the same frequency as for an X–Y molecule with the same force constant. However, the equations do not account for mixing of the in-phase stretch with the angle bending motion of the molecule. These motions each contribute to the two symmetric modes of the molecule, but not to the one antisymmetric mode (the out-of-phase stretch) since modes of different symmetry cannot mix. (This is why the bending mode could be left out of the discussion of the bond stretches of a linear triatomic molecule; mixing is precluded because angle bending destroys the symmetry axis of the molecule, while the bond stretches preserve it.) The effect of mixing angle bending into the in-phase stretching mode of a bent triatom is not large, because the bending frequency is typically only about one-half that of the stretching frequency (energy factoring). The table in Figure 12 shows that the effect is sufficient to push the in-phase above the out-of-phase mode for bond angles near 90°, as in the case of F_2O ($\alpha = 102°$).[22] For Cl_2O, the angle is slightly larger ($\alpha = 111°$) and the in-phase frequency is below the out-of-phase frequency.[23]

C. Metal–Ligand Vibrations

These issues of vibrational coupling are important in deciphering metal–ligand modes in the Raman and IR spectra of inorganic and bioinorganic complexes. Because the masses are relatively large and the bonds relatively weak, metal–ligand stretching vibrations are found in the low-frequency region of the spectra, usually below 500 cm^{-1} unless the ligands are involved in multiple bonding with the metal (e.g., M=O bonds). Figure 13 lists metal–ligand frequencies for the

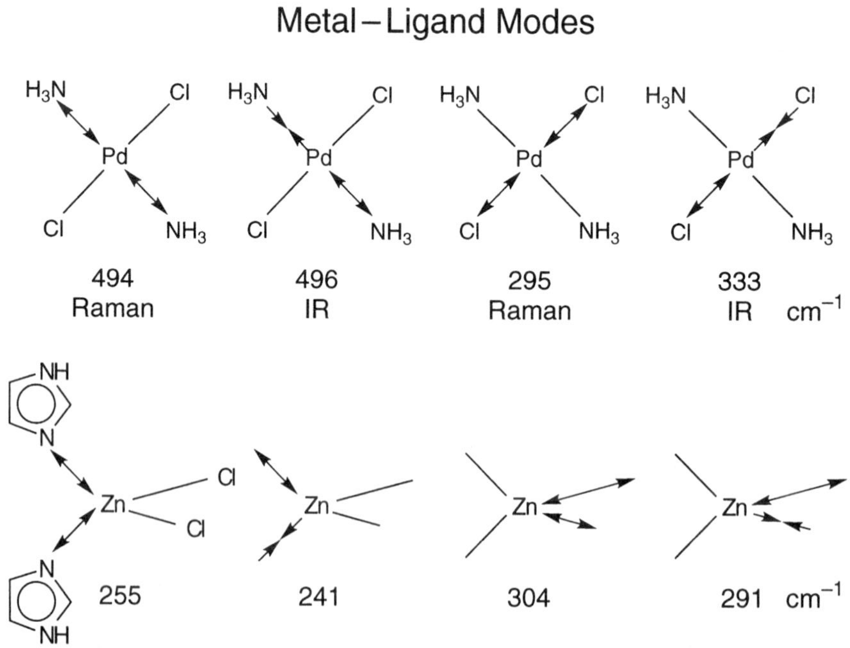

Metal – Ligand Modes

Figure 13
Representations of metal–ligand stretching modes in the complexes *trans*-$Pd(NH_3)_2Cl_2$ [24] and $Zn(ImH)_2Cl_2$,[25] showing differential effects of ligand masses and bond angles.

complexes *trans*-$Pd(NH_3)_2Cl_2$[24] and $Zn(ImH)_2Cl_2$[25] (ImH = imidazole) and illustrates the vibrations to which they are assigned. Because the bonds to Pd are stronger for NH_3 than for Cl^-, and the ligand mass is smaller, the Pd–NH_3 frequencies (~ 500 cm^{-1}) are much higher than the Pd–Cl frequencies (~ 300 cm^{-1}). The complex is tetragonal and can be thought of as being made up of two orthogonal PdX_2 vibrating units, each with an in-phase and out-of-phase stretch; the H atoms of the NH_3 ligands can be ignored, because their vibrations are all at much higher frequencies (the H atoms being so light). For the Pd–Cl bonds, the in-phase stretch is lower than the out-of-phase stretch, as expected. The former is found in the Raman spectrum and the latter in the IR spectrum, since the former preserves the symmetry center while the latter destroys it. For the Pd–NH_3 bonds, however, the two stretches are at essentially the same frequency, because the Pd atom is so much heavier than the NH_3 ligands. The apparent violation of the mutual exclusion rule (IR-active modes do not appear in the Raman spectrum of centrosymmetric molecules, and vice versa) is due to accidental degeneracy; the IR and Raman bands arise from different modes that happen to have the same frequency.

The situation is significantly different for the zinc complex. Even though NH_3 and ImH are both nitrogenous ligands, and have similar bonding properties, ImH has a much larger effective mass because the atoms of the rigid ring all vibrate in phase, relative to the metal atom. (This does not happen in the case of a trialkyl amine, for example, whose skeleton is flexible; its M–L frequencies are not far from those of NH_3). The ImH effective mass is close to the sum of the atomic masses (68 amu), and depresses the M–L frequencies below those of the Cl^- ligands. Because the bond angles are close to tetrahedral in the Zn complex, interaction between in-phase and out-of-phase bond stretches is small, and the in-phase stretches are at slightly higher frequencies because of interactions with bond bending. On the other hand, since the angle between the Zn–ImH and Zn–Cl bonds is not 90°, and the frequencies are comparable, their stretches can interact somewhat; the modes are not as pure as the representations in Figure 13 suggest. Finally, since the symmetry is so low, all four modes are active in both the Raman and IR spectra.

D. Band Assignment

How does one assign the bands in a Raman or IR spectrum to particular modes? This question is the central problem of a vibrational analysis, and experience and judgment are frequently required for complex molecules. Empirical correlations leading from simple to more complicated molecules have shown that certain groupings of atoms give rise to vibrational frequencies in fairly narrow ranges. The best guess from these correlations can be tested by a calculation of the frequencies using the techniques of normal coordinate analysis, as discussed in Section 3.E. The most reliable tool for assignment is isotopic substitution, however. When one isotope is replaced by another, a frequency shift is produced for any vibration in which the atom in question moves significantly. Even in quite complicated spectra a particular feature of interest can frequently be detected by appropriate isotope substitution. A case in point is the Fe_2O symmetric stretch of

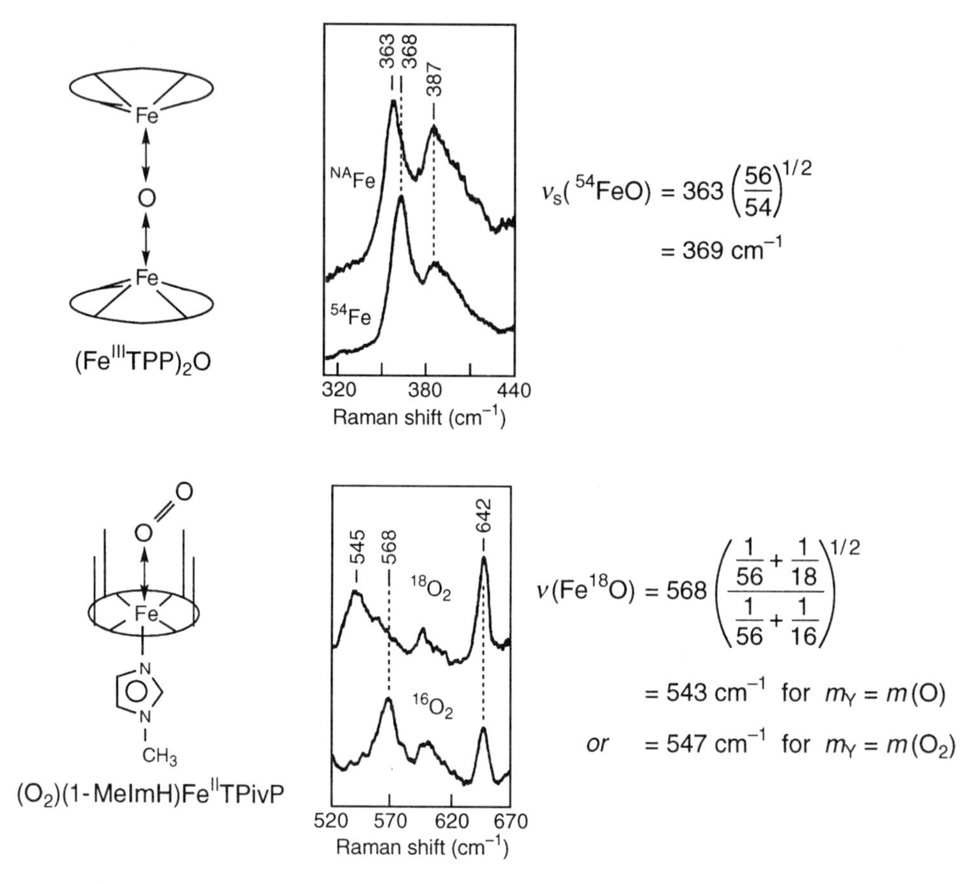

Figure 14
Examples of isotope substitution to identify metal–ligand modes in complex spectra: The Fe–O–Fe symmetric stretching mode in $(Fe^{III}TPP)_2O$ (TPP = tetraphenylporphyrin), via its ^{54}Fe shift,[19] and the Fe–O_2 stretch in the "picket fence" porphyrin hemoglobin model, via its ^{18}O shift.[26] The remaining bands in the spectra are due to the prophyrins. Calculation of the isotope shifts using simple triatom and diatom relations are shown.

the μ-oxo porphyrin dimer mentioned above. Figure 14 shows how this vibration was identified[19] with a Raman band at 363 cm^{-1} via its 5 cm^{-1} shift to higher frequency when the molecule was prepared with the ^{54}Fe isotope. The isotope shift clearly distinguishes this mode from the nearby porphyrin mode at 387 cm^{-1}. A simple triatom calculation (see above) predicts a 6-cm^{-1} shift on the basis of the 54 : 56 Fe mass ratio. The accuracy of this calculation implies that the Fe_2O motion does not couple significantly to the porphyrin motions; the lack of coupling is because the Fe–O and Fe–N(porphyrin) bonds are nearly orthogonal. The ^{18}O substitution would have been useless in this experiment because the central atom does not move in the symmetric stretch of a linear symmetric triatom. A large ^{18}O shift is expected for the antisymmetric stretch, however, and has been used to identify this mode at 885 cm^{-1} ($\Delta \nu = 30$ cm^{-1}) in the IR spectrum.[20]

Another example in Figure 14 is the identification by $^{18}O_2$ substitution of the Fe–O_2 stretch, at 568 cm^{-1}, in the O_2 adduct of Fe "picket fence" porphyrin with a 1-methylimidazole axial ligand,[26] a model for oxyhemoglobin. The shift is calculated quite accurately just by using Eq. (3) for a diatomic molecule, even though the O–Fe–N bond forms a linear system. The lack of significant coupling with the Fe–N(1-MeImH) stretch is attributable to energy factoring, associated with the low frequency (~250 cm^{-1}) expected for the latter stretch because of the large effective mass of the ligand. There is an ambiguity as to whether the mass of one or two O atoms should be used as the ligand mass in this simple calculation, but the predicted difference between the two assumptions is only 4 cm^{-1} (Fig. 14).

E. Normal Mode Analysis

A molecule containing N atoms has $3N$ degrees of freedom. Six of these (five for linear molecules) are the translations and rotations of the molecule as a whole, leaving $3N - 6$ ($3N - 5$) internal degrees of freedom that comprise the molecular vibrations. These can be calculated by solving Newton's equations of motion, provided that the geometry of the molecule, the masses of the atoms, and the force constants (collectively called the force field) are all known. The result of this calculation is a set of $3N - 6$ eigenvalues, which are the vibrational frequencies, and the corresponding eigenvectors, which describe the motion. A proper solution leads to periodic motions, in which the atoms move along recurring paths. These are called the normal coordinates, Q, which describe the normal modes. For example, the arrows in Figure 12 represent the eigenvectors for the three normal modes of a typical bent triatom. Each normal mode involves motion on a separate parabolic potential (to the extent that the harmonic approximation is valid, which it is near the bottom of the wells), like the diatomic potential in Figure 8, but with the normal coordinate as the abscissa. For methods of carrying out the calculations, the reader is referred to Herzberg,[27] Wilson et al.,[28] Schachtschneider,[29] Gans,[30] and Nakamoto.[31]

The main problem in normal coordinate analysis is that the force field is not known in advance. Moreover, it cannot be determined unambiguously from the vibrational spectra because, except for diatoms, the number of force constants always exceeds the number of modes. Even for a bent Y–X–Y triatom there are three modes but four valence force constants (two principal constants, stretch and bend, and two interaction constants, stretch–stretch and stretch–bend), and a fifth constant (called a Urey–Bradley constant) should be added to take into account the nonbonded repulsion between the terminal atoms. Thus the problem is overdetermined. Rapid progress is being made on ab initio electronic structure calculations of the force field,[32] even for moderate to complex molecules, such as porphyrins.[32c] The usual empirical approach is to transfer force constants from small molecules having bonding features of the larger molecule under study, and then to adjust their values incrementally in order to minimize the discrepancy between calculated and experimental frequencies. While this procedure cannot lead to a unique solution, it is possible to constrain the force constants to physically reasonable values, on the basis of the extensive body of empirical analyses, and the increasing number of ad initio studies. The purpose of the exercise is to provide a reliable guide to the assignment of the vibrational spectra. In this situation it is important to have isotopic data, since the calculation of isotope shifts is a stringent test of the adequacy of the force field, as well as the assignments. An eigenvector is accurately calculated if the isotope shifts of the atoms participating in the motion are reproduced, whether or not the force field is unique.

As an example of a vibrational analysis, we cite a study of the nickel(II)-dithiooxalato (dto) complex, $K_2Ni(dto)_2$,[33] whose IR and Raman spectra are compared in Figure 15. The high symmetry of this molecule (D_{2h}) can be used to

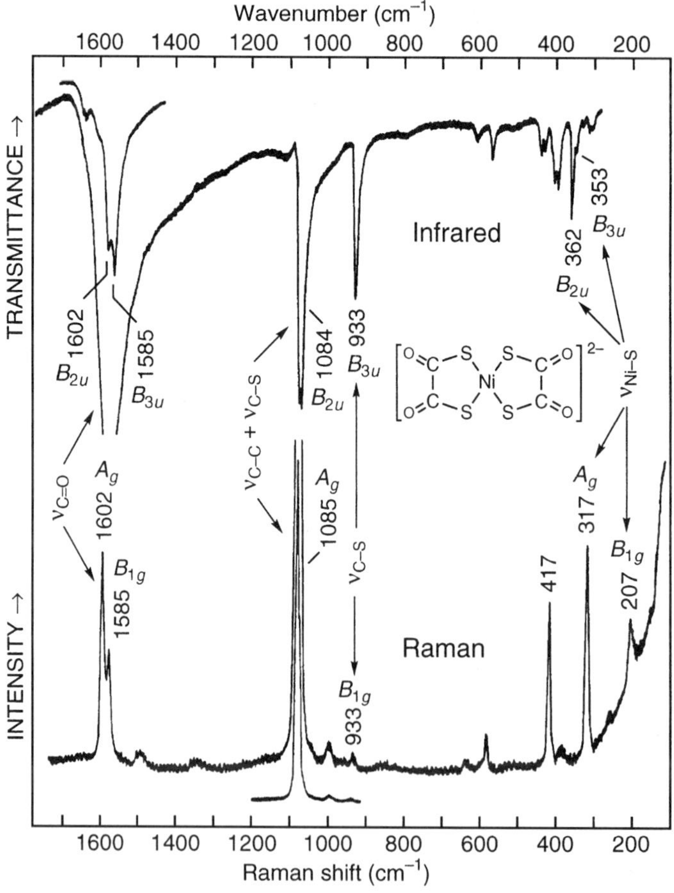

Figure 15
Infrared and Raman spectra for $K_2Ni(dto)_2$ (dto = dithiooxalato anion).[33] Band designations represent the major bond contributions to the mode as determined from a normal mode analysis. Because of the high symmetry, D_{2h}, the Ni–S stretches fall into different representations. The antisymmetric stretches, in which the symmetry center is destroyed, are IR active, and, because the Ni atom participates in the motion, are at higher frequencies than the symmetric modes, which are Raman active. In the high frequency region, one finds modes localized on the chelate rings. The IR and Raman components show essentially the same frequencies (accidental degeneracy) because the motions are localized on the individual rings, and interaction between the rings is small.

advantage. The in-plane modes fall into two different representations for the Raman (A_{1g} and B_{1g}) as well as the IR (B_{2u} and B_{3u}) spectra. By symmetry, the out-of-plane modes cannot mix with the in-plane modes; moreover, they are expected to be weak (small dipole and polarizability changes) and to lie at quite low frequencies, since the motions involve bond deformations and the atoms are relatively heavy. The out-of-plane modes are therefore conveniently ignored. Assignment of the spectral bands to the expected in-plane modes was based on an accurate calculation of the frequencies using physically reasonable force constants.[33] The compound was prepared with ^{62}Ni and ^{58}Ni, and the isotope shifts of the IR bands (the Raman bands do not shift since the Ni is on the symmetry center) were also calculated accurately. Two of these, 362 and 353 cm^{-1}, showed large shifts, 7 and 5 cm^{-1}, identifying them with Ni–S stretching motions primarily. Their Raman counterparts were identified with the 317- and 207-cm^{-1} bands by the calculation. The IR modes lie above the Raman modes because the Ni atom participates in the IR-active modes (and the ligand–metal mass disparity is much less than in, e.g., the Pd–NH_3 modes discussed above.) The reason for the large difference in the Raman frequencies is differential mixing with deformation motions of the chelate rings.[33]

While the IR and Raman Ni–S stretching bands occur at distinctly different frequencies, this is not true for the bands above 900 cm^{-1}, all of which show accidental degeneracies. The motions in this region are localized on the chelate rings. The IR- and Raman-active modes differ only in the phasing of the local motions on the two different rings. Consequently, their frequencies differ only to the extent that there is electronic communication between the rings, across the Ni atom; this is evidently very small.

IV. Resonance Enhancement

The intensity of Raman scattering depends strongly on the excitation wavelength. When the laser is tuned to the vicinity of an allowed electronic transition (resonance), the scattering cross section increases markedly for some of the Raman bands. This happens because the polarizability, and its dependence on the molecular motions, is enhanced via the electronic transition. Not all Raman bands intensify, however, but only those associated with normal coordinates that carry the molecule into its excited state geometry. For example, resonance with a π–π^* transition enhances stretching modes of the π bonds, which alter their length in the excited state, while charge-transfer (CT) transitions of metal complexes generally enhance metal–ligand stretching modes, and sometimes internal modes of the ligands as well.

The laser wavelength is an important consideration in Raman applications because resonance enhancement greatly increases both the sensitivity and the selectivity of the technique. Nonresonance Raman scattering is weak enough that sample concentrations in the molar range are often required to obtain good quality spectra, but resonance enhancement can lower the required concentration to the millimolar (mM) to micromolar (μM) ranges. Selectivity is an equally important feature since vibrational spectroscopy frequently suffers from crowding and overlap of bands when complex molecules or mixtures are investigated. Because only Raman modes associated with the excited-state distortion are enhanced, the RR spectrum is simplified and one can be sure that the enhanced bands are associated with the chromophore; interferences from other parts of the molecule or other constituents of the mixture are eliminated. Of course, this selectivity means a loss of information if the nonchromophoric parts of the sample are of interest. Wide tunability of the laser source is desirable so that different chromophores in the sample (or different excited states of the same chromophore, which may enhance different Raman bands) can be accessed. While most RR studies have relied on CW lasers[34] and have therefore been limited to the visible and near-UV region, deep UV capabilities have recently become available[5,6,35] via pulsed lasers (see Fig. 7). There are many more UV than visible chromophores, including all the aromatic residues of proteins and nucleic acids.

Another advantage of wide laser tunability is the ability to minimize undesirable side effects of laser irradiation, namely, photodegradation and fluorescence. It is often possible to reduce photodegradation by tuning the laser to the long-wavelength region of a molecule's absorption spectrum, while interference from fluorescence can be reduced by tuning to the blue side of the fluorescence envelope. The RR scattering and fluorescence are intimately connected since in both phenomena the emitted photon results from excitation in an absorption band. Their relationship is shown diagrammatically in Figure 16. The two Morse curves represent the potential energy of the molecule in its ground and excited state, along a particular vibrational coordinate. Excitation into the absorption band (upward arrow) is followed by emission processes that are illustrated in the schematic spectrum: the Rayleigh peak, a Raman peak, and a broad fluorescence envelope. This broad envelope, properly called relaxed fluorescence, follows rapid vibrational relaxation of the photoexcited molecule to the lowest level of the excited state, from which emission occurs to many levels in the ground state. If vibrational relaxation is not rapid, a situation that can be encountered in small molecules with widely spaced vibrational levels, then it is possible to produce unrelaxed or "resonance" fluorescence, at the same position as the Raman peak, provided the laser is in exact resonance with one of the excited state vibrational levels. Time resolution of the emission can distinguish the two processes,[36] since resonance fluorescence decays with the lifetime of the excited state while RR scattering is instantaneous, coinciding with the excitation pulse. Of course, if the laser is tuned away from the resonant level (outside its bandwidth) the resonance fluorescence disappears, while the RR peak is still observed, at the same frequency shift. In complex molecules, vibrational relaxation is generally rapid and resonance fluorescence is rarely seen. Indeed, it is common for vibrational relaxation to continue from the bottom of the excited-state potential to high vibrational levels of the ground state and then to the ground vibrational level (wavy arrows in the diagram), thereby quenching the fluorescence (radiationless relaxation). Relatively few molecules fluoresce strongly, and this is what makes RR spectroscopy possible at all, since RR scattering, even though enhanced, is intrinsically much weaker than fluorescence.

High backgrounds due to fluorescence are a common frustration for Raman spectroscopists. As mentioned above, the interference can sometimes be reduced

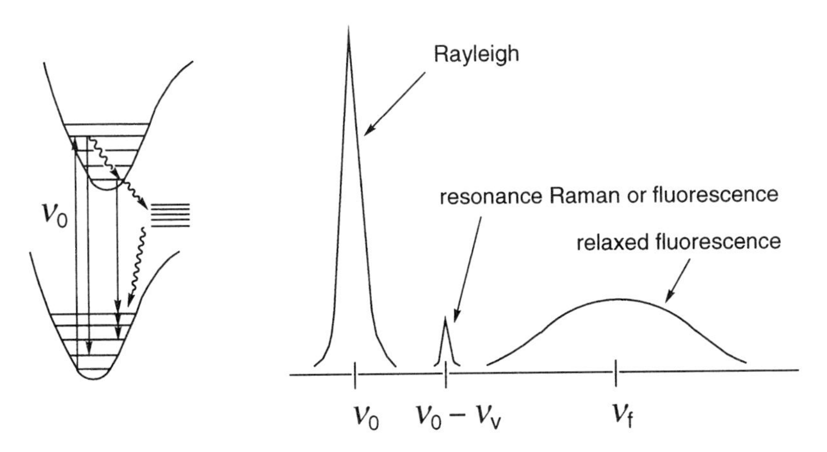

Figure 16
Schematic representation of processes leading to relaxed fluorescence and to Rayleigh and resonance Raman (or fluorescence) re-emission. Excitation with light that is resonant with the excited state (upward arrow) can be followed by emission that is relaxed or unrelaxed. The former produces a broad envelope of emission representing transitions from the bottom of the excited-state potential to the various levels of the ground state; the intensity of the envelope diminishes to the extent that radiationless relaxation occurs to high-lying vibrational levels of the ground state (wavy arrows). The distinction between resonance Raman and fluorescence (unrelaxed) is that the latter involves actual population of the excited state, the emission decaying with the excited-state lifetime, whereas the RR process is simultaneous with the excitation pulse. Off exact resonance with the upper state level, only RR scattering is seen. The parameters ν_0, ν_v, and ν_f = frequencies of the excitation radiation, vibrational transitions, and fluorescence radiation, respectively.

by increasing the laser wavelength, and external heavy-atom quenching agents can sometimes be added to the sample (a KI salt is often quite effective in this regard). Coherent Raman techniques can be applied to fluorescent samples[37] because the directional character of the signal lends itself to spatial filtering from the isotropic fluorescence; these techniques are technically difficult to implement, however. Often the most effective remedy is to purify the sample, since in many cases the interference is not from the intrinsic fluorescence of the molecule under study but from strongly fluorescing impurities.

A. Enhancement Mechanisms

With the use of second-order perturbation theory, the polarizability can be expressed in terms of a sum of contributions from the wave functions of all the excited states. The result is the Kramers–Heisenberg dispersion equation[38]

$$\alpha_{\rho\sigma} = \frac{1}{\hbar}\sum_{\mathrm{e}} \frac{\langle \mathrm{f}|\mu_\rho|\mathrm{e}\rangle\langle \mathrm{e}|\mu_\sigma|\mathrm{g}\rangle}{\nu_{\mathrm{eg}} - \nu_0 + \mathrm{i}\Gamma_{\mathrm{e}}} + \frac{\langle \mathrm{f}|\mu_\sigma|\mathrm{e}\rangle\langle \mathrm{e}|\mu_\rho|\mathrm{g}\rangle}{\nu_{\mathrm{ef}} + \nu_0 + \mathrm{i}\Gamma_{\mathrm{e}}} \tag{9}$$

Here the subscripts ρ and σ are Cartesian directions of the dipole moment operator μ and the polarizability tensor element α, while g, e, and f are wave functions of the ground, excited and final states. The denominators contain the frequencies of the g $\rightarrow$ e and e $\rightarrow$ f transitions, ν_{eg} and ν_{ef}, the laser frequency, ν_0, and a damping constant Γ_e, corresponding to the half-bandwidth of the excited state. Resonance enhancement arises from the first term, since its denominator decreases rapidly as ν_0 approaches ν_{eg}.

To consider the implications for RR scattering, we apply the Born–Oppenheimer approximation of separability of electronic and vibrational wave functions, since the nuclei move much more slowly than the electrons:

$$\alpha = \frac{1}{\hbar} M_e{}^2 \sum_v \frac{\langle j|v\rangle\langle v|i\rangle}{\nu_{vi} - \nu_0 + i\Gamma_v} \tag{10}$$

where M_e is the pure electronic transition moment

$$M_e = \langle e_0|\mu|g_0\rangle \tag{11}$$

evaluated at the ground-state equilibrium geometry (symbolized by the "0" subscripts). In Eq. (10), the polarization directions, ρ and σ, are dropped for simplicity, and we assume the laser to be close to resonance with a given excited state, e, so that only a single term in the sum over excited states in Eq. (9) needs to be considered, and the nonresonant component can be dropped. The resonant term is now a sum over the vibrational levels (v) of the excited state, and the denominator takes into account that the laser can be tuned to resonance with each individual level v, with a frequency from the ground state ν_{vi} and a damping constant Γ_v.

The integrals $\langle v|i\rangle$ and $\langle j|v\rangle$ are overlap integrals between the vibrational wave functions of the initial, i, and final, j, vibrational levels of the ground state with the intermediate vibrational level, v, of the excited state. They are called Franck–Condon factors. For fundamental Raman transitions $j = i + 1$, and the wave functions differ by one node. Consequently, either $\langle v|i\rangle$ or $\langle j|v\rangle$ must be zero unless there is a shift in the excited-state potential along the vibrational coordinate; then both overlaps can be nonzero, as illustrated in Figure 17. This requirement means that only totally symmetric modes can have RR intensity, according to Eq. (10), since vibrations that fail to preserve the symmetry of the molecule must have symmetric displacements with respect to the equilibrium geometry, and therefore cannot undergo a shift in the excited state. Another peculiarity of Eq. (10), is that the Raman intensity is predicted to vanish off resonance. When $\nu_0 \ll \nu_{vi}$ the denominator becomes independent of v and can be factored out. Then the sum over the numerators can be shown to reduce to $\langle j|i\rangle$. But this overlap is zero, unless $i = j$, implying that only Rayleigh scattering survives off resonance.

These limitations of Eq. (10) can be remedied by recognizing that the electronic transition moment is itself a function of the vibrational coordinate:

$$M_e = M_e{}^0 + M_e{}' \qquad \text{where} \qquad M_e{}' = (\partial M_e/\partial Q)\, Q \tag{12}$$

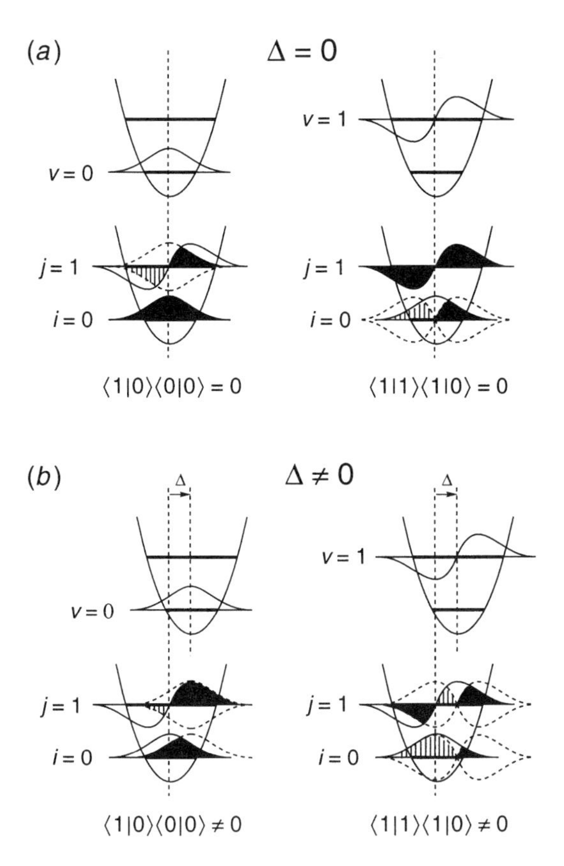

Figure 17
Representation of Franck–Condon overlap products (opposite signs indicated by different shading patterns) involved in the A term RR scattering of a vibrational fundamental, from the ground vibrational level ($i = 0, j = 1$) via the first two vibrational levels of the excited state ($v = 0, 1$). If there is no shift of the excited-state potential (*a*) then $\langle 0|0\rangle$ and $\langle 1|1\rangle$ are unity, because of exact overlaps, but $\langle 1|0\rangle = 0$ because the $v = 0$ and 1 wave functions are orthogonal and positive and negative overlaps cancel. If there is a shift (*b*) this cancellation is no longer exact and $\langle 1|0\rangle \neq 0$, so that the product is nonzero.

As the molecule departs from the ground-state equilibrium geometry, via the coordinate Q, the transition moment is altered. When Eq. (12) is inserted in Eq. (9), the polarizability becomes the sum of three terms:

$$\alpha = \mathrm{A} + \mathrm{B} + \mathrm{C} \tag{13}$$

$$\mathrm{A} = (M_e{}^0)^2 \sum_v F_v/D_v \quad \text{where} \quad F_v = \langle j|v\rangle\langle v|i\rangle \quad \text{and}$$

$$D_v = (\Delta\nu_v - \Gamma_v)/\hbar \tag{13a}$$

$$\mathrm{B} = M_e{}^0 M_e{}' \sum_v F_v{}'/D_v \quad \text{where} \quad F_v{}' = \langle j|Q|v\rangle\langle v|i\rangle + \langle j|v\rangle\langle v|Qi\rangle \tag{13b}$$

$$\mathrm{C} = (M_e{}')^2 \sum_v F_v{}''/D_v \quad \text{where} \quad F_v{}'' = \langle j|Q|v\rangle\langle v|Q|i\rangle \tag{13c}$$

The A term is the same as the original Eq. (10), and it is the leading term for fully allowed transitions. These have large M_e^0, by definition, and M_e' is relatively small in comparison. Thus totally symmetric modes do dominate RR spectra, when in resonance with strongly allowed transitions. When the transition is weakly allowed, however, M_e' can be of the same magnitude as M_e^0, or even exceed it. The value of M_e' depends on the degree to which the excited-state gains absorption strength from other excited states by mixing with them via the mode Q (called vibronic mixing). In the Herzberg–Teller formalism,[38]

$$M_e' = M_s^0 \frac{\langle s|\partial M_e/\partial Q|e\rangle}{\nu_s - \nu_e} \tag{14}$$

where s is another excited state and ν_s is its frequency above the ground state. The stronger the transition to the state s, and the closer it is in energy to e, the larger is M_e'. Thus B term scattering becomes especially important in resonance with a weak electronic transition that is mixed vibronically with a nearby strong one.

The B term denominator is the same as for the A term, but its numerator contains the Q-dependent integrals $\langle v|Q|i\rangle$ and $\langle j|Q|v\rangle$. These integrals connect ground- and excited-state vibrational levels that differ by one quantum. When they are multiplied by Franck–Condon factors, $\langle j|v\rangle$ or $\langle v|i\rangle$, having the same quantum numbers in the ground and excited states (e.g., $\langle 1|1\rangle$ and $\langle 0|0\rangle$), the numerator does not vanish even if there is no excited-state shift of the potential. Consequently, nontotally symmetric modes can be, and usually are enhanced. The magnitude of the B term enhancement depends on the effectiveness of the vibration in mixing the excited states. Moreover, under nonresonance conditions, when the denominator can be factored out, the sum over the numerator reduces to $\langle j|Q|i\rangle$, which is nonzero for $j = i + 1$, that is, fundamental modes. Thus off-resonance Raman scattering is preserved by the B term, although under these conditions any specific resonance effects are lost in the sum over all excited states, Eq. (13b).

B. Illustration: Metalloporphyrins

The interplay between A and B term scattering is nicely illustrated by the RR spectra of metalloporphyrins.[39] For these highly symmetric (effectively D_{4h}) aromatic chromophores, the first two excited states are well described by Gouterman's four-orbital model,[40] illustrated in Figure 18, along with the absorption spectrum of nickel octaethylporphyrin (NiOEP). The two LUMOs are a degenerate pair (e_g^*), while the two HOMOs, a_{2u} and a_{1u}, have nearly the same energy. Consequently, the excitations $e_g^* \leftarrow a_{2u}$ and $e_g^* \leftarrow a_{1u}$, which have the same symmetry ($a_{1u} \times e_g = a_{2u} \times e_g = E_u$), interact strongly. The transition dipoles add up for the higher energy transition but nearly cancel for the lower energy transition (cancellation is exact if a_{1u} and a_{2u} have the same orbital energy). The result is a very strong band at about 392 nm, called the B or Soret band, and a much weaker band at 552 nm called the Q_0 or α band. The latter steals back approximately 10% of the intensity of the former via vibronic mixing, producing a side band at 516 nm, called Q_1 or β, which is the envelope of the vibronic inten-

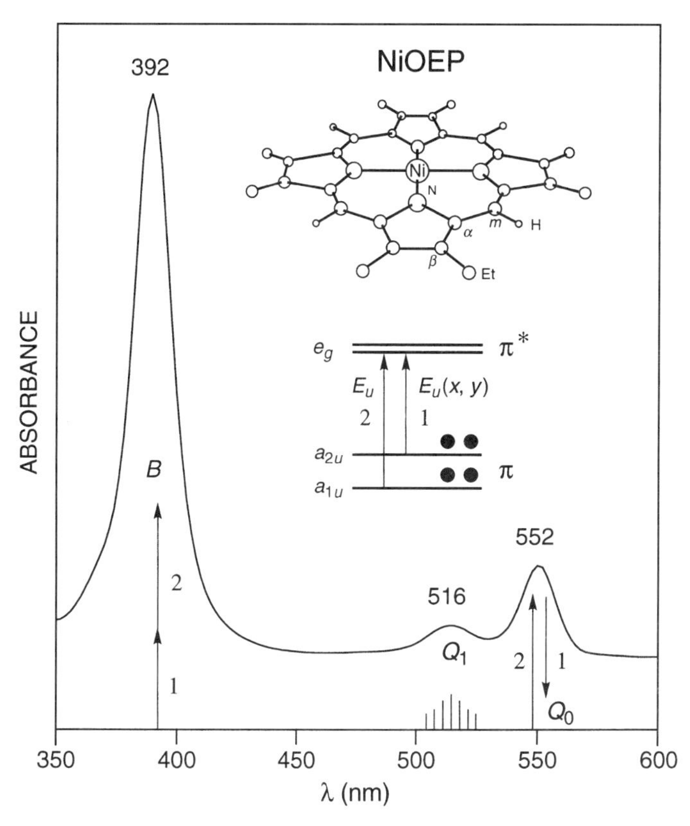

Figure 18
Absorption spectrum of NiOEP (OEP = octaethylporphyrin) and its interpretation via Gouterman's four-orbital model.[40] The $e_g \leftarrow a_{2u}$, a_{1u} orbital excitations (labeled 1 and 2), being of the same symmetry (E_u) and nearly degenerate, interact strongly, the transition dipoles adding up for the intense B transition at 392 nm, and nearly cancelling for the weaker Q_0 transition at 552 nm. About 10% of the B band intensity is borrowed by the Q transition producing the Q_1 vibronic side band 516 nm, above Q_0 by $\approx 1300\ cm^{-1}$, the average frequency of the vibronically effective modes.

sity, the most effective mixing vibrations having an average frequency of about $1300\ cm^{-1}$ (the Q_0–Q_1 separation).

The symmetry of the mixing vibrations is given by the direct product of the excited-state electronic symmetries, $E_u \times E_u = A_{1g} + A_{2g} + B_{1g} + B_{2g}$. However, the A_{1g} modes are ineffective in mixing the transition moments, because of the high molecular symmetry. Consequently, RR spectra with excitation in the Q bands are dominated by nontotally symmetric modes. The B_{1g} and B_{2g} modes give depolarized bands, while the A_{2g} modes give anomalously polarized bands.[39] The A_{2g} modes have rotational symmetry, for example, they involve alternating stretching and contraction of equivalent bonds around the ring, and therefore they can rotate the transition moments, giving an antisymmetric tensor $\alpha_{xy} = -\alpha_{yx}$. Meanwhile, RR spectra excited near the B band show much higher enhancements, via the A term, and are dominated by polarized bands from totally

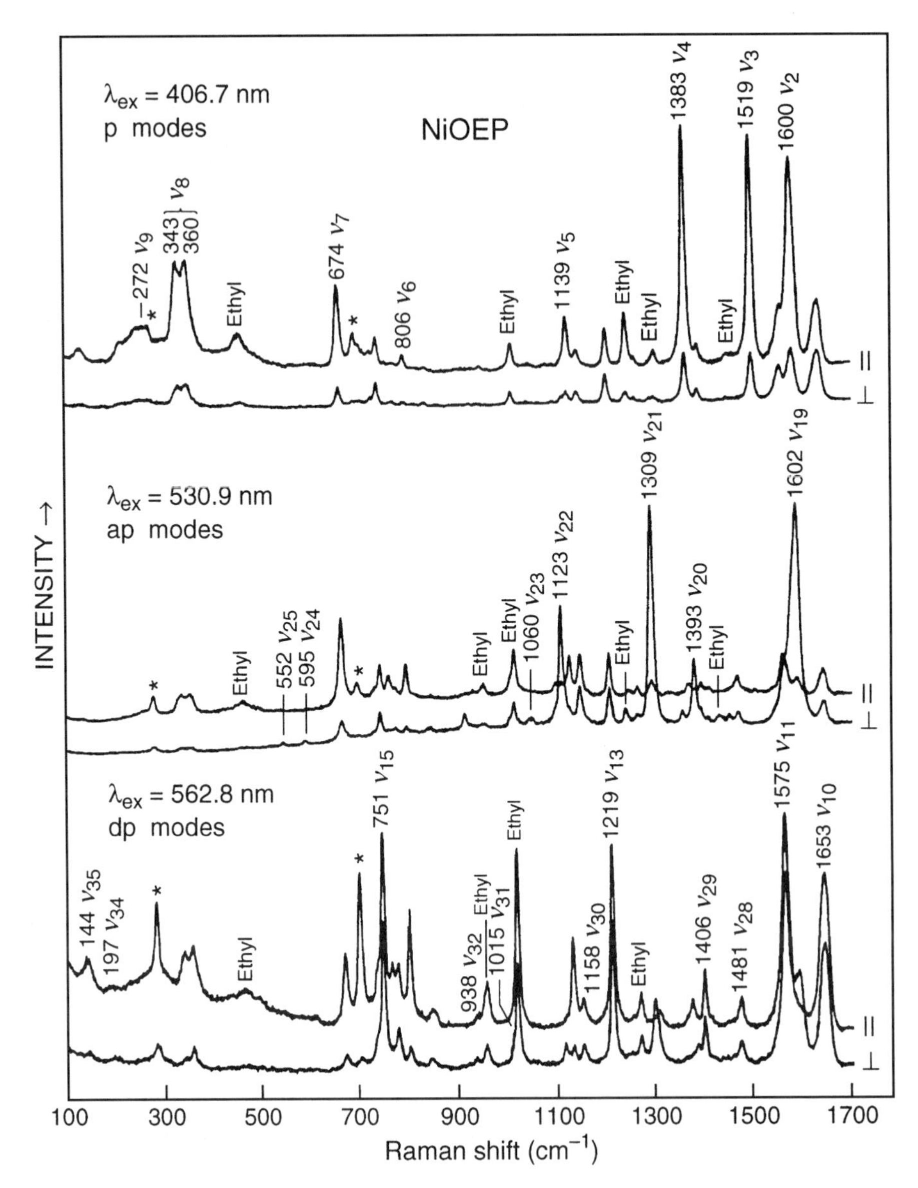

Figure 19
The RR spectra in parallel (||) and perpendicular (⊥) scattering for NiOEP,[41] showing the selective enhancement of different modes with different excitation wavelengths. Resonance with the B absorption band (406.7 nm) enhances mainly polarized bands, arising from totally symmetric modes, A_{1g}, although appreciable enhancement is also seen for Jahn–Teller active B_{1g} modes (see ν_{10} and ν_{11}), which are depolarized. Resonance in the Q band region enhances depolarized (B_{1g}, B_{2g}) and anomalously polarized (A_{2g}) modes that are effective in vibronic mixing of the Q and B electronic transitions. Due to interference effects the A_{2g} modes are brought out most strongly via excitation between Q_0 and Q_1 (530.9 nm), while the B_{1g} and B_{2g} modes are brought out more strongly with excitation outside the Q bands (568.2 nm). The mode labels ν_i refer to porphyrin skeletal modes;[39] modes assignable to the ethyl substituents are also seen, as indicated.

symmetric vibrations (A_{1g}). Figure 19, showing RR spectra in resonance with the B and Q bands of NiOEP,[41] illustrates the point that quite different vibrational modes are enhanced in the two regimes. This is very helpful in structural applications because variable wavelength excitation can be used to enhance selected bands, and to disentangle them from nearby bands of different symmetry. Also, excitation at different wavelengths in the Q band region produces selectivity because of interference effects (see below). The two terms in F_v' of the B term [Eq. (13b)] add for B_{1g} and B_{2g} modes, but subtract for A_{2g} modes, because of the tensor symmetries. Consequently, excitation between the Q_0 and Q_1 transition, where the energy denominators for the Q_0 and Q_1 resonances are opposite in sign, selects for A_{2g} modes, while excitation on either side selects for B_{1g} and B_{2g} modes.

The B-band excited spectra also show appreciable intensity for certain B_{1g} modes (see, e.g., the bands marked ν_{10} and ν_{11} in the top spectrum of Fig. 19). Their enhancement is attributed to a dynamical Jahn–Teller effect in the degenerate excited state (E_u).[42] The two components of this state can be mixed by B_{1g} or B_{2g} vibrations; this is a limiting case of vibronic mixing, in which the states being mixed are actually degenerate. (The above equations, which are derived from perturbation theory, no longer apply however, and a quantitative treatment requires direct diagonalization of the molecular Hamiltonian.)

C. C Term Scattering: Overtones

The numerator of the C term [Eq. (13c)] contains products of two Q-dependent integrals. Since each of these connects vibrational levels differing by one quantum, the final level, j, must differ from i by two quanta, as diagrammatically shown in Figure 20. Consequently, only overtones are enhanced by the C term.

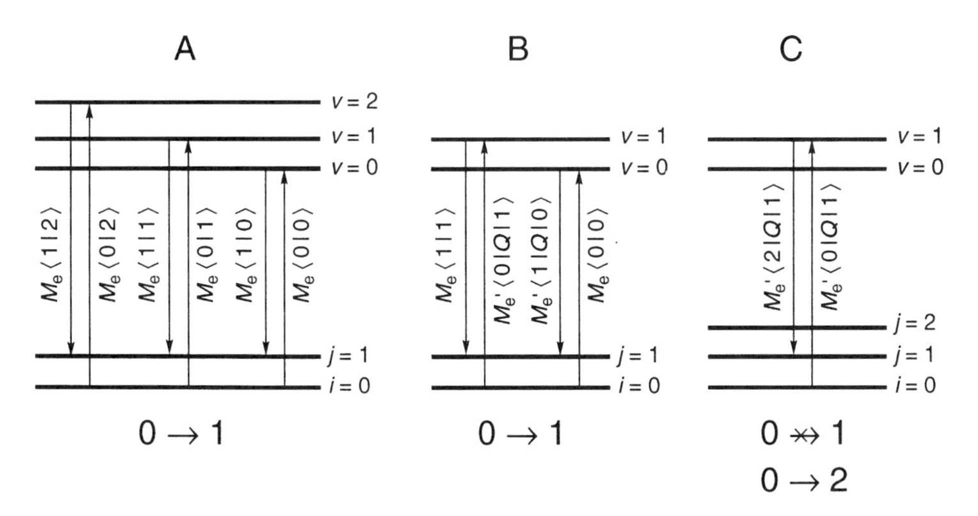

Figure 20

Diagrams showing the Raman processes responsible for A, B, and C term resonance scattering for fundamentals (A and B) and overtones (C). The applicable transition moment and overlap integral are indicated for each of the contributing transitions for $v = 0, 1$, and 2 intermediate levels of the resonant excited state.

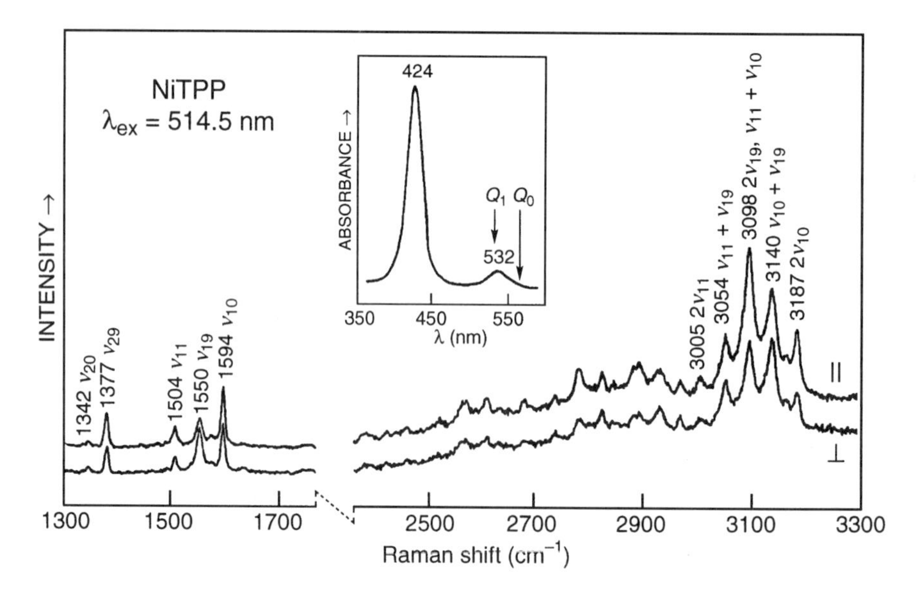

Figure 21
The RR spectrum of NiTPP with 514.5-nm (Q_1) excitation, showing strong enhancement of overtone and combination modes involving ν_{19}.[43] The Q_0 band is very weak in this case, giving large overtone/fundamental ratios via the larger C than B term scattering (see text).

Since the electronic factor is $(M_e')^2$, the C term can outweigh the B term when $M_e' > M_e^0$. An example is shown in Figure 21, where overtone and combination bands involving the mode ν_{19} in the Q-band spectrum of NiTPP are seen to be substantially stronger than the fundamentals.[43] The Q_0 absorption band is very weak in this case, indicating near degeneracy of the a_{1u} and a_{2u} orbitals, while the Q_1 band retains about 10% of the intensity of the B band. Consequently, $M_e^0 < (M_e^1)$ and B < C. In the case of the benzene molecule,[35a] the first two electronic transitions, B_{2u} and B_{1u}, are forbidden by symmetry, but the latter is strongly mixed vibronically with the nearby allowed transition, E_{1u}. The overtones of the mixing modes are observed in RR spectra excited in resonance with B_{1u}, via the C term, but the fundamentals are missing since M_e^0, and therefore the B term, is zero by symmetry.[35a]

Overtones can also be enhanced by the A term, via the Franck–Condon factors, when in resonance with allowed transitions. Indeed, it is possible to enhance overtones, which are themselves symmetric, of nontotally symmetric vibrations, provided there is a change in the force constant of the mode in the excited state. This produces a change in the curvature of the potential well, and leads to non-zero Franck–Condon products for the overtone transition. Thus overtones can sometimes be stronger than fundamentals, even in resonance with allowed transitions; such instances, for example, have recently been reported for the phenyl ring of phenylalanine,[44] and for the μ-oxo-bridged dinuclear iron(III) center in the hemerythrin model complex $[Fe_2O(O_2CCH_3)_2(HBpz_3)_2]$ [$HBpz_3^-$ = hydrotris(1-pyrazolyl)borate anion].[45]

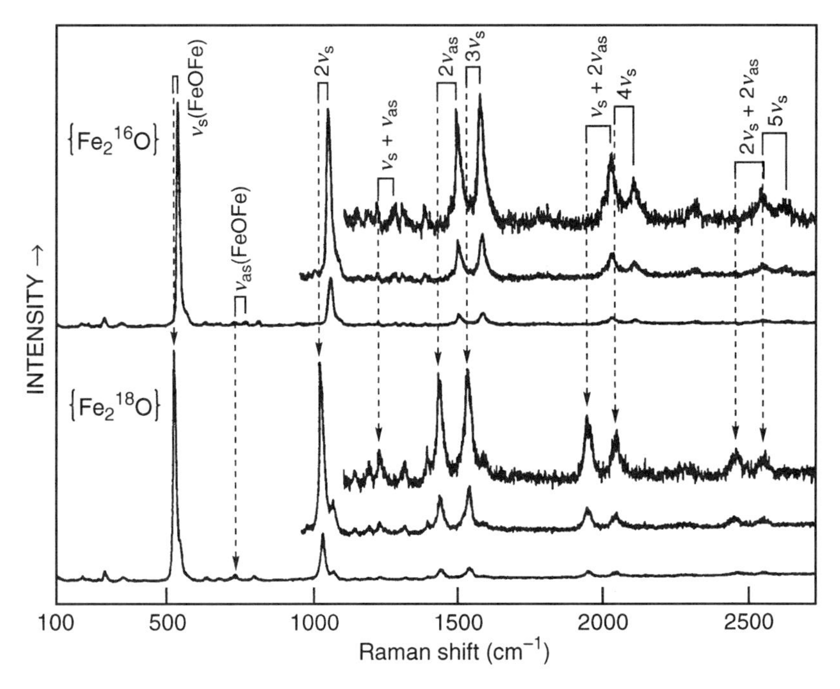

Figure 22
Low-temperature (77 K) Fe–O–Fe overtone region RR spectra of crystalline $[Fe_2O(O_2CCH_3)_2(HBpz_3)_2]$ and its μ-^{18}O analogue obtained with 406.7-nm excitation (150 mW) and 7-cm^{-1} slit width.[45]

Figure 22 shows a crystalline RR spectrum for the latter compound in the 100–2700-cm^{-1} region obtained with 406.7-nm excitation, near resonance with the O → Fe CT transition.[45] Besides a striking series of overtones of the Fe–O–Fe symmetric stretch, ν_s (530 cm^{-1}), out to $5\nu_s$, Figure 22 shows that the first overtone, $2\nu_{as}$ (1504 cm^{-1}), of the corresponding antisymmetric Fe–O–Fe stretch, ν_{as} (754 cm^{-1}), is substantially more intense than the fundamental. In this case, the fundamental is of B_2 symmetry (under the effective C_{2v} point group of the bent Fe–O–Fe bridge) and requires a vibronic mechanism (generally much weaker than a Franck–Condon A-term mechanism) for activation, whereas its overtone is of $B_2 \times B_2 = A_1$ symmetry and is Franck–Condon allowed. Not only does $2\nu_{as}$ show appreciable intensity ($\simeq$ to $3\nu_s$) but it forms a subsidiary progression with ν_s, out to $2\nu_s + 2\nu_{as}$. The combination of ν_s with the ν_{as} fundamental is also detectable. All of these features can readily be sorted out by their ^{18}O shifts, as shown in Table II.

The observation of a number of overtones of the ν_s(FeOFe) in the RR spectra of $[Fe_2O(O_2CCH_3)_2(HBpz_3)_2]$ makes it possible to determine the harmonic wavenumber, ω_s, and anharmonicity constant, X_{11}. In general, the observed wavenumber $\nu(n_1)$, of any overtone of the n_1 fundamental, is given by the

Table II
Observed Overtone and Combination Wavenumbers in the RR Spectrum of $[Fe_2O(O_2CCH_3)_2(HBpz_3)_2]$ (406.7-nm Excitation).[a]

Band[b]	^{16}O (cm^{-1})	^{18}O (cm^{-1})	Band[b]	^{16}O (cm^{-1})	^{18}O (cm^{-1})
		$n\nu_s$(FeOFe) Progression			
ν_s	530	513	$4\nu_s$	2105	2036
$2\nu_s$	1058	1024	$5\nu_s$	2627	2539
$3\nu_s$	1583	1532			
		$n\nu_s$(FeOFe) + $2\nu_{as}$(FeOFe) Progression[c]			
$\nu_s + 2\nu_{as}$	2028	1941	$2\nu_s + 2\nu_{as}$	2550	2444
		Other Overtone and Combination Modes			
$\nu_s - \delta$(FeOFe)[d]	428	410	$\nu_s + \nu_{as}$	1284	1230
$\nu_s + \delta$(FeOFe)	633	620[e]	$2\nu_{as}$	1504	1432
$\nu_s + \nu(FeN_{trans})$[f]	803	784			

[a] See Czernuszewicz et al.[45]
[b] Symmetric and antisymmetric stretching modes = ν_s and ν_{as}, respecitively; δ = bending mode.
[c] ν_{as}(FeOFe) = 754 cm^{-1} (719 cm^{-1} in ^{18}O).
[d] Difference combination mode with δ(FeOFe) = 104 cm^{-1} (103 cm^{-1} in ^{18}O).[45]
[e] Overlapped with ligand mode.
[f] $\nu(FeN_{trans})$ = 275 cm^{-1} (272 cm^{-1} in ^{18}O); stretching mode of the Fe–N(pyrazole) bond trans to the Fe–O bridge bond.

expression[27]

$$\nu(n_1) = G(n_1) - G(0) = n_1\omega_1 + X_{11}(n_1{}^2 + n_1 d_1) + \cdots \quad (15)$$

where $G(n_1)$ is the term value of the n_1th vibrational level, and n_1 and d_1 are the vibrational quantum number and degeneracy, respectively, of this fundamental. With the simplification that $d_1 = 1$ (since ν_1 is totally symmetric and consequently nondegenerate), a plot of $\nu(n_1)/n_1$ versus n_1 will then give a straight line with a slope equal to X_{11} and an intercept equal to $\omega_1 + X_{11}$. Figure 23 shows that the $n\nu_s$ progression is well behaved, a plot of frequncy against vibrational quantum number being accurately linear for both ^{16}O and ^{18}O. These plots[46] provide harmonic frequencies, $\omega_s = 532.5 \pm 0.6$ cm^{-1} for ^{16}O and 515.5 ± 0.7 cm^{-1} for ^{18}O, and an anharmonicity constant, $X_{11} = -1.25 \pm 0.05$ cm^{-1}, which is the same, within experimental error, for ^{16}O and ^{18}O.[45]

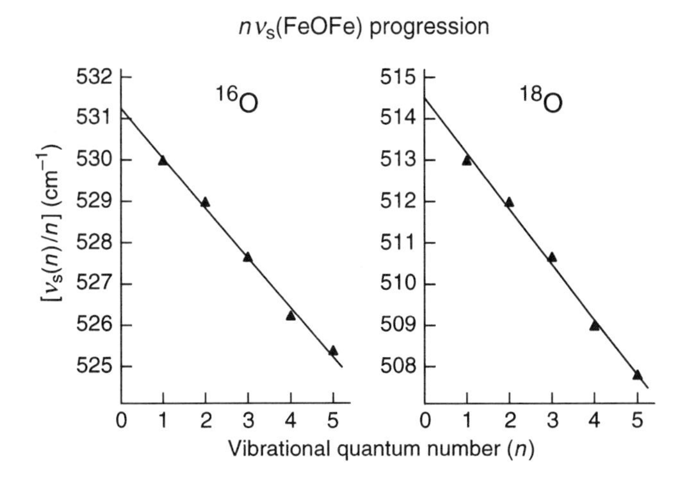

Figure 23
Plots of $\nu(n)/n$ versus n for $n\nu_s$(FeOFe) progressions for $[Fe_2O(O_2CCH_3)_2(HBpz_3)_2]$ and its μ-^{18}O analogue.[45]

D. Excitation Profiles, Multimode Effects, Time-Dependent, and Transform Theories

A plot of RR intensity against the excitation wavelength is called an excitation profile (EP). Such plots track the absorption spectrum in a general way, but differ from it in detail, frequently showing more structure. There are three basic reasons for such differences: (1) The absorption dipole change depends on the vibrational overlap integrals for the transition from the ground to the excited state, but the Raman polarizability change depends on the product of overlap integrals for the up and down transitions (Fig. 20); the latter can have a distinctly different wavelength dependence. (2) The absorption oscillator strength depends on the sum of the squares of the overlap integrals for different modes, whereas the Raman intensity is proportional to the square of the polarizability change, so that the sum over the terms in Eq. (13) is formed before squaring. Consequently, the scattering contributions from the individual vibrational levels of the excited state can either add or subtract (the denominators change signs as the laser frequency is tuned through the local resonance, and the numerators can be positive or negative); contributions from nearby excited states can also interfere negatively as well as positively. (3) The EP emphasizes electronic factors that are particularly sensitive to the mode under consideration, while the absorption shows no such selectivity. Consequently, the EPs can provide information about the excited state that may be hidden in the absorption spectrum.

A case in point is the EP of the 351-cm^{-1} RR band of the enzyme cytochrome P_{450},[47] shown in Figure 24, which was assigned to the Fe–S stretching mode of the bond from the heme Fe atom to the cysteine axial ligand on the basis of its isotope shift in enzyme prepared from microorganisms grown on ^{34}S.[48] The

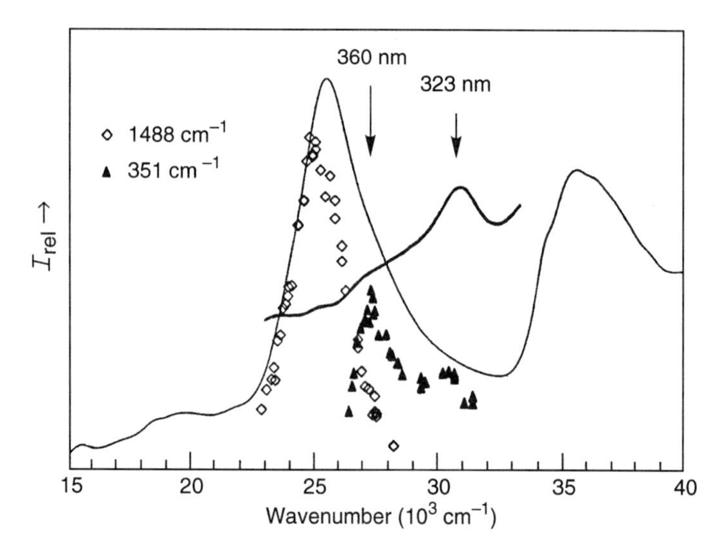

Figure 24
Cytochrome P_{450} EPs for the 1488-cm^{-1} porphyrin mode (◇) and the 351-cm^{-1} Fe–S stretching mode (△). The former maximizes in the B absorption band of the heme (thin solid line) but the latter shows shorter wavelength maxima, which are attributed to S → Fe charge-transfer transitions, for which evidence can be seen in the z-polarized single-crystal absorption spectrum (thick solid line). [From Champion.[47]]

EP shows maxima that are at distinctly higher energy than the dominant absorption band in the region, the B band of the heme group, while the EP of a porphyrin mode at 1488 cm^{-1} is seen to maximize in the B band. The resonances of the Fe–S mode are therefore attributed to other electronic transitions, involving S → Fe charge transfer, which are buried in the absorption spectrum in the high-energy tail of the B band. As shown in Figure 24, these transitions can, however, be seen in the single-crystal absorption spectrum when the electric vector is aligned perpendicular to the plane of the porphyrin so that the in-plane π–π^* transitions are suppressed.

The quantitative treatment of RR intensities and EPs, as well as of the absorption spectrum, is of considerable current interest, because it can provide interesting information about the resonant excited states. If the system is well behaved, it is possible to calculate EPs from the absorption spectrum, and vice versa, with only the coupling strengths of the individual normal modes as adjustable parameters, using transform theory.[49] This methodology has been applied extensively to heme proteins by Champion.[50]

Another approach is to model both the EPs and the absorption spectrum using the sum-over-states formalism introduced above. The equations as written are inadequate for this purpose, however, since they tacitly assume that only a single normal mode contributes to the intensity of a given RR band. In reality, the vibrational wave functions are multidimensional, and each one contains all the normal modes in some quantum state. Only one normal mode at a time is excited in the vibrational fundamentals, but the numerators of the terms in Eq. (13) should all be multiplied by Franck–Condon products $\Pi\langle j|v\rangle\langle v|i\rangle$, one for each normal mode that is not excited. These products are close to unity only if the excited state is not significantly displaced along the mode in question. Otherwise they depart from unity and become significant determinants of the RR intensity and the EP. Multimode effects considerably complicate the sum-over-states calculation,[51] although computer programs are available that can handle a substantial number of modes.

Another alternative for the investigation of excited-state properties from RR intensities is the time-dependent theory developed by Holler and co-workers.[52] In this approach, the sum-over-states expressions are transformed from a frequency to a time basis, and the resulting equations can be interpreted as reflecting the evolution of the overlap of a wave packet moving along the excited-state potential surface with the initial and final vibrational wave functions of the ground state. This approach has the conceptual advantage of providing a dynamical description of the excited state, and also has computational advantages for systems with many active modes, or when there are large distortions in the excited state. A full description of applications to RR intensities and EPs has been given by Myers and Mathies.[51]

V. Application: Iron–Sulfur Proteins

We now discuss applications of RR spectroscopy to a particular area of bioinorganic chemistry, namely, iron–sulfur proteins. This subject has been reviewed, as have RR studies of several other bioinorganic systems including proteins containing heme, chlorophyll, dinuclear Cu and Fe centers, "blue" Cu sites, and tyrosinate-Fe sites.[53] The Fe–S proteins provide a set of structural variations on a common theme that offer particularly instructive illustrations of the principles developed in the preceding sections. All of them have Fe coordinated by sulfur atoms of cysteine side chains (in a few cases one or more cysteine ligands are replaced by non-S-containing ligands), and most of them have sulfide ions (labile sulfur) bridging the Fe atoms in clusters containing two, three, four, and possibly more Fe atoms. Figure 25 gives absorption spectra in oxidized (solid lines) and one-electron reduced (dashed lines) forms for proteins representing the classical structural types, containing 1-Fe, [2Fe–2S], and [4Fe–4S] clusters. The visible absorption bands are associated with CT transitions from sulfur (both bridging and terminal) to Fe(III). For the oxidized proteins, Fe(III) is the oxidation level for the one Fe in rubredoxin (Rd), both Fe atoms in the [2Fe–2S] proteins, two of the four Fe atoms in bacterial ferredoxin (Fd), and three of the four Fe atoms in high-potential iron protein (HiPIP). Oxidized Fd and reduced HiPIP are at the same average oxidation level, from which Fd can be reduced by one electron, and from which HiPIP can be oxidized by one electron (at a much higher potential). The loss in visible band absorption intensity is consistent with there being fewer Fe(III) atoms in the reduced forms: zero for Rd, one for the [2Fe–2S] proteins and for bacterial Fd, and two for HiPIP.

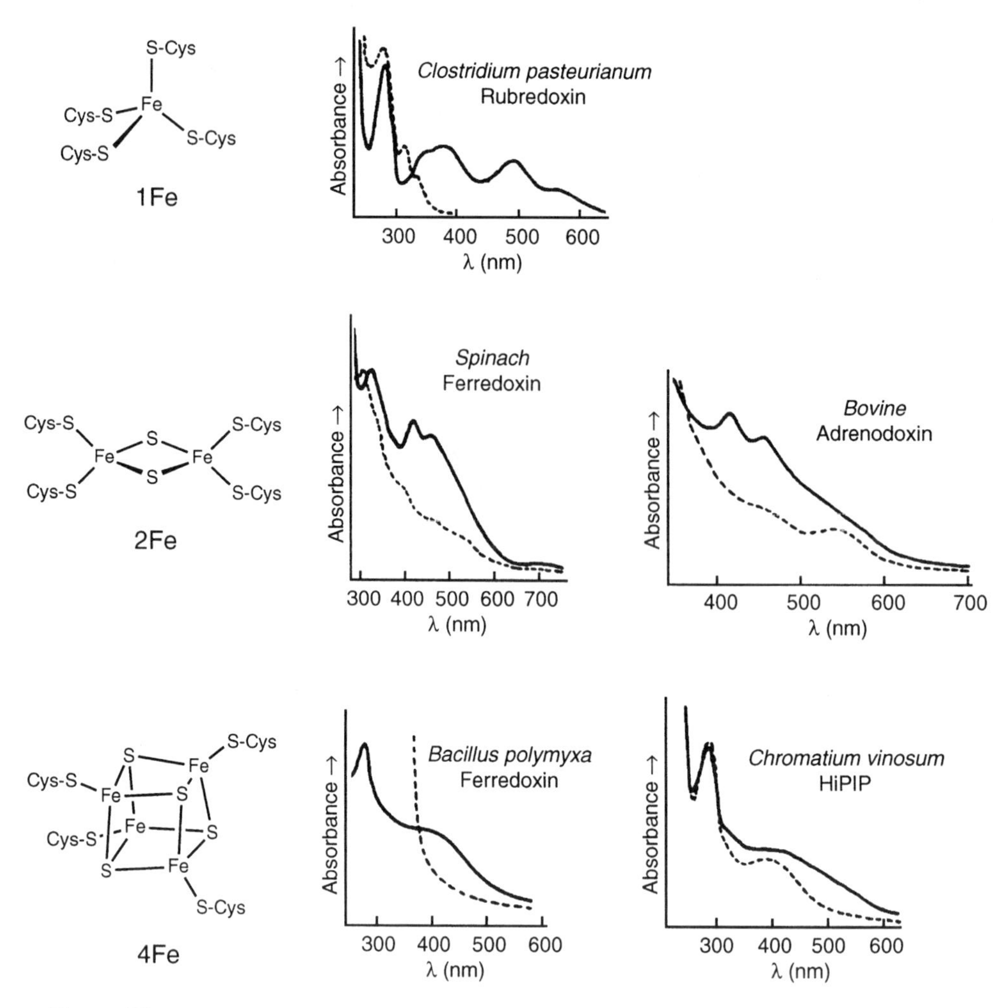

Figure 25
Optical absorption spectra of iron–sulfur proteins (oxidized —— and reduced ----) examplifying the three classic structural types shown on the left-hand side: 1-Fe (rubredoxin),[54] [2Fe–2S] (plant ferredoxin[55] and adrenodoxin[56]), and [4Fe–4S] (bacterial ferredoxin[57] and high-potential iron protein[58]). The visible absorption bands are attributed to S → Fe^{III} charge-transfer transitions.

A. Rubredoxin

Rubredoxin was one of the very first biological molecules to which RR spectroscopy was applied. Figure 26 shows the spectra reported in 1971 by Long et al.[59] with 488.0-nm excitation. Four bands were identified in the metal–ligand stretching and bending region, just the number of normal modes expected for a tetrahedral complex.[31] The eigenvectors and symmetry labels of these modes[60] are also shown in Figure 26. Only one of these is totally symmetric (A_1). It is the

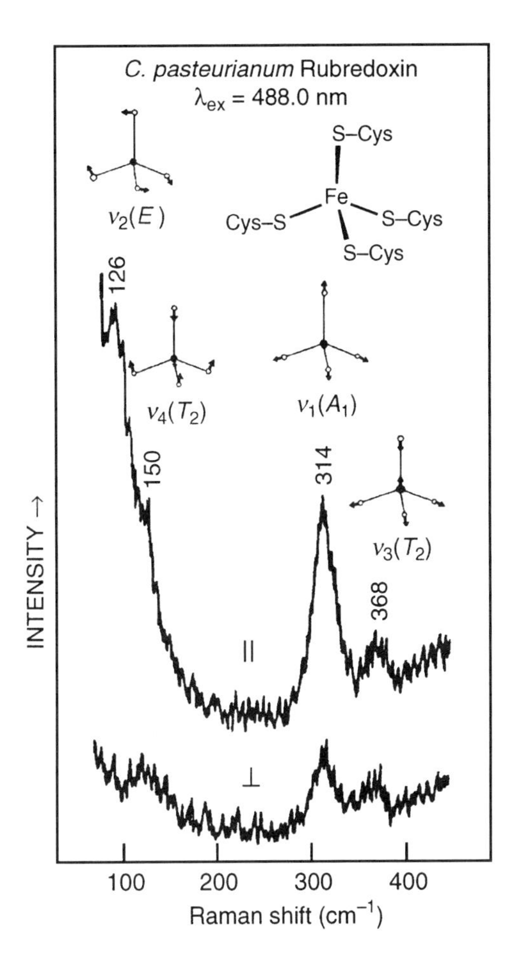

Figure 26
The RR spectra of *C. pasteurianum* rubredoxin obtained in 1971 by Long et al.[59] using 488.0-nm excitation. The suggested assignments to modes of the FeS_4 tetrahedron are indicated by the labels and the eigenvectors of the modes.

breathing mode, ν_1, involving the in-phase stretching of all four Fe–S bonds. This mode was assigned to the strong, polarized band at 314 cm^{-1}. The other three stretches (there must be four altogether, one for each bond) comprise the triply degenerate mode $\nu_3(T_2)$; this was assigned to the weak, depolarized 368-cm^{-1} band. The six angle bending coordinates make up a triply degenerate mode, $\nu_4(T_2)$, and a doubly degenerate mode, $\nu_2(E)$ (for a total of five combinations. The sixth mode has zero frequency since it is impossible to expand all of the angles at the same time; this is called a redundancy. There are a total of nine vibrational degrees of freedom, $\Gamma_{vib} = A_1 + E + 2T_2$, as required for a five-atom molecule). The ν_2 and ν_4 modes were assigned to the shoulders at 125 and 150 cm^{-1} on the rising background of the Rayleigh wing. These four bands are close in frequency to the modes of the isoelectronic complex $FeCl_4^- = 385(\nu_3)$,

330(ν_1), 133(ν_4), and 108 cm^{-1} (ν_2).[61] Thus the analysis of the protein RR spectrum was satisfyingly straightforward.

Subsequent work has shown the situation to be more complex, however.[62] Figure 27 shows an interesting variation with laser wavelength of the ν_2 and ν_4 bending modes. The EPs, constructed by ratioing the band intensities of the frozen protein sample to the intensity of the nearby ice band at 227 cm^{-1}, showed ν_2 to be strongly resonant with the strong absorption band at 495 nm, while ν_4 is weakly resonant with the 565-nm shoulder. These results were interpreted on the basis of a previous electronic assignment by Eaton and Lovenberg,[63] also shown in Figure 27. Two CT transitions are expected in the visible region, from the sulfur π to the Fe dπ and dσ orbitals, but these are split by a lowering of the effective symmetry of the chromophore, from T_d to at least D_{2d}, which can be envisioned as a compression (or elongation) of the tetrahedron along the S_4 symmetry axis. The 495-nm band and its 565-nm shoulder were assigned, respectively, to the ^6E and $^6B_2$ components of the Fe(d$_\pi$) $\leftarrow$ S(π) transition. These components can be coupled vibronically via a mode of $B_2 \times E = E$ symmetry. The T_2 vibrational modes also have E and B_2 components in D_{2d} symmetry. Therefore the E component of the T_2 mode can couple the weak 565-nm transition with the strong nearby 500-nm transition, thereby accounting for the ν_4 enhancement via the B-term scattering mechanism. On the other hand, the ν_2 enhancement in resonance with the strong 495-nm transition can be attributed to Franck–Condon scattering, since the E (T_d) mode has A_1 and B_1 components in D_{2d} symmetry, and the A_1 component is subject to A-term enhancement. (We note that the second components of each mode, B_2 of ν_4 and B_1 of ν_2, are not expected to be observed because they lack effective enhancement mechanisms.)

Further complexity was discovered when the Fe–S stretching region was reexamined at higher resolution in frozen solution,[62] as shown in Figure 28. Instead of one broad band above ν_1, three were now seen, at 376, 363, and 348 cm^{-1}, and, in addition, a 324-cm^{-1} shoulder became apparent on the ν_1 band. To investigate the origins of these bands, a difference RR spectrum was obtained using protein that was reconstituted with ^{54}Fe, in order to obtain isotope shifts. For a tetrahedral FeS_4 complex the $^{56/54}$Fe shift is predicted by the normal mode frequency relationships to be zero for ν_1, because the Fe atom does not move in the breathing mode, and 2.7 cm^{-1} for ν_3. The experimental shifts were 2.5, 1.4, and 1.1 cm^{-1} for the 376-, 363-, and 348-cm^{-1} bands, while no isotope shift was detected for the 314-cm^{-1} band (as expected for ν_1), or its 324-cm^{-1} shoulder. This experiment indicated that the ^{54}Fe sensitive bands could be identified with the three expected components of ν_3, but the 30-cm^{-1} splitting was larger than expected from a symmetry-lowering distortion, and the depressed isotope shifts for the 363- and 348-cm^{-1} components indicated that they were coupled kinematically to other modes, which were omitted from the tetrahedral model. The likeliest candidate for a coupling vibration was considered to be the SCC bend of the coordinated cysteine ligands, since this motion and Fe–S stretching have similar natural frequencies and should be linked kinematically (closing the SCC angle helps stretch the Fe–S bond and vice versa). The SCC bending might account for the 324-cm^{-1} shoulder.

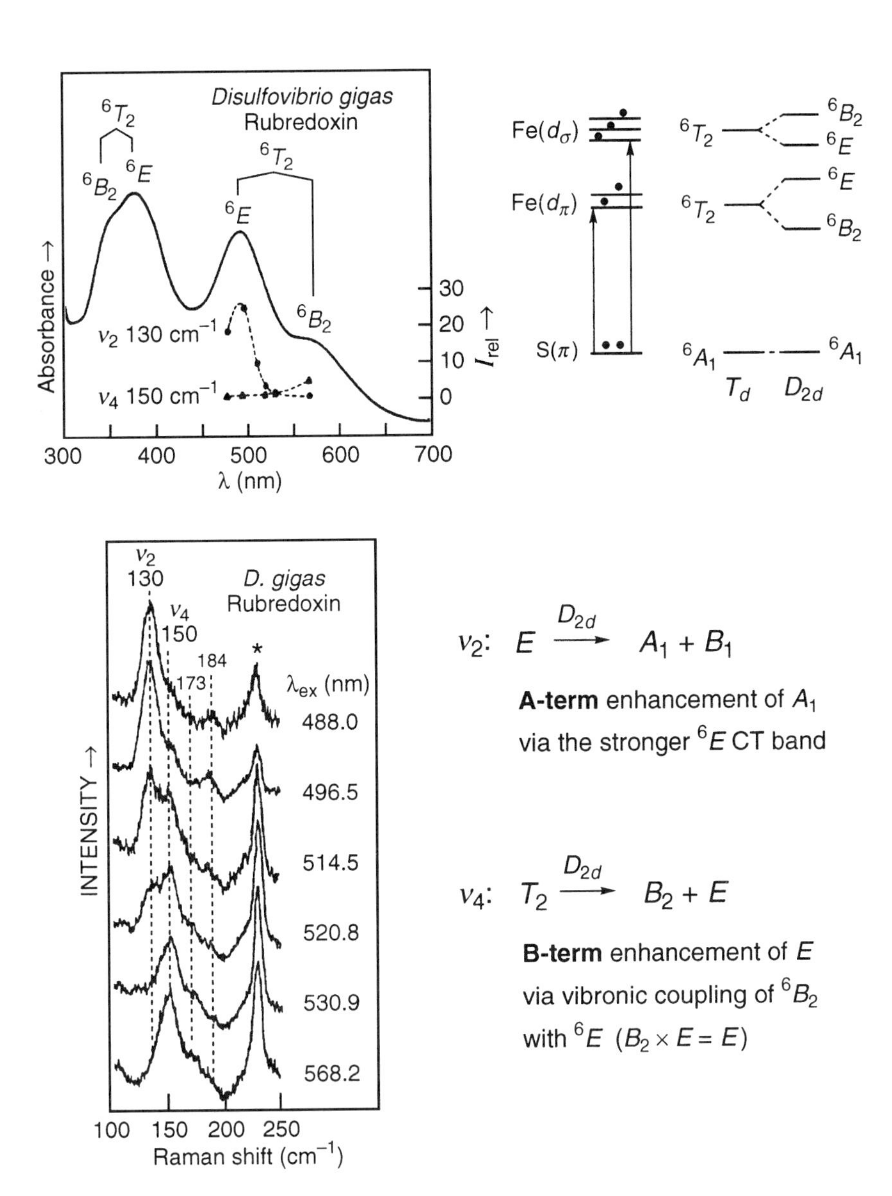

Figure 27
Variable wavelength RR spectra of a frozen *D. gigas* rubredoxin solution in the bending mode region.[62] The intensities of ν_2 and ν_4 were determined by ratioing them with the 227-cm^{-1} ice band (marked with an asterisk) intensity. The resulting excitation profiles are compared with the absorption spectrum, showing that ν_2 is strongly resonant with the 495-nm absorption band, while ν_4 is weakly resonant with the 565-nm shoulder. These transitions are assigned,[63] respectively, to the ^{6}E and $^6B_2$ components of the S(π) $\rightarrow$ Fe(dπ) charge-transfer transition, split by symmetry lowering from T_d to D_{2d}. The former electronic component provides Franck–Condon enhancement of the A_1 vibrational component of the E bending mode (ν_2) while the latter provides B term enhancement to the E component of the T_2 mode (ν_4), via vibronic coupling with the nearby ^{6}E electronic transition.

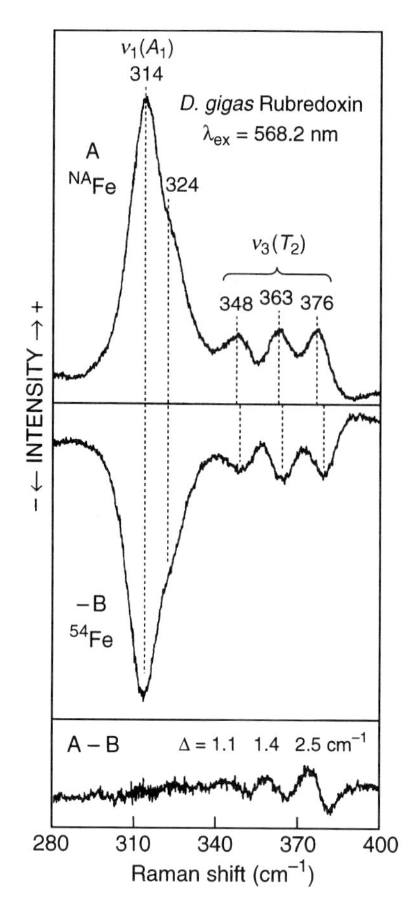

Figure 28
The 568.2-nm excited RR spectra in the Fe–S stretching region for natural abundance rubredoxin (A) and of the protein reconstituted with ^{54}Fe (B).[62] The frozen solution spectra were obtained with a low-temperature difference Raman cell,[63] permitting accurate subtraction to determine the isotope shifts from the amplitudes of the difference signals (A – B).[65] On this basis the three bands at 348, 363, and 376 cm^{-1} were assigned to components of the tetrahedral $\nu_3(T_2)$ stretching vibration, although isotope shifts for the first two are lower than expected (2.5 cm^{-1}). These lowerings and the frequency spread were attributed to vibrational coupling, probably with SCC bending motions.

This coupling proposal was also advanced to explain an otherwise puzzling difference between the RR spectra of Rd and an analogue complex $[Fe(S_2\text{-}o\text{-}xyl)_2]^-$ (S_2-*o*-xyl = *o*-xylene-α, α′-dithiolate).[66] As seen in Figure 29, the ν_1 band is found to be 16 cm^{-1} lower in the analogue than in the protein. This shift is large, 5% of the frequency, and therefore 10% of the force constant, which varies as the square of the frequency, [see Eq. (3)] if the shift were due to a force constant change. Yet, the Fe–S distances are indistinguishable between the analogue and the protein, as judged by X-ray crystallography[68] and by extended X-ray absorption fine structure (EXAFS)[69] measurements. The main structural

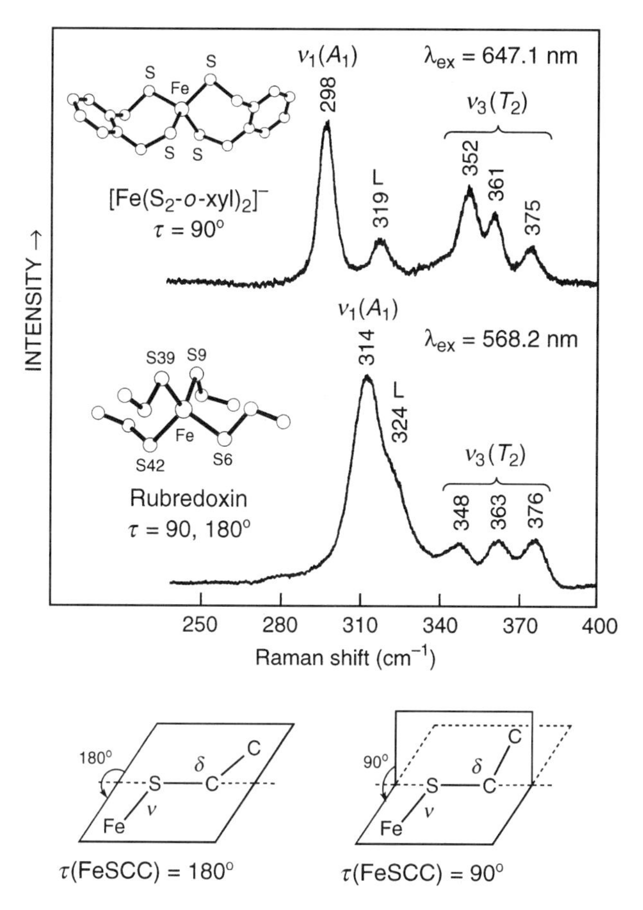

Figure 29
The 568.2 (protein)- and 647.1 (analogue)-nm excited RR spectra in the Fe–S stretching region and structural drawings for rubredoxin and for the xylenedithiolate analogue complex.[62,67] Similar frequencies are seen for the ν_3 components, but the ν_1 breathing mode is substantially lower for the analogue. This difference may possibly be associated with SCC bending interactions that are expected to differ between protein and analogue due to the different distribution of τ(FeSCC) dihedral angles. These are 90° for the analogue but two of them are 180° for the protein. Coupling between FeS stretching and SCC bending is maximal for $\tau = 180°$, when the coordinates are in line, and minimal at $\tau = 90°$. The band marked L (ligand) may be due to SCC bending.[62]

difference between the two species lies in the τ(FeSCC) dihedral angles. In the analogue complex, the constraint of the chelate ring forces this angle to be 90°,[68b] whereas in the protein two of the angles are found to be 90° but the other two are close to 180°.[68a] This difference is important because the proposed kinematic coupling of Fe–S stretching with SCC bending is a strong function of τ. As illustrated in Figure 29, coupling is maximal for $\tau = 180°$, where the coordinates are in line, but minimal for 90°, where they are orthogonal. It was therefore proposed that the coupling is restricted to the protein, where it serves to shift ν_1 up relative to its unperturbed frequency in the analogue.[62]

B. [2Fe–2S] Proteins

A rather complete analysis of the vibrational modes of [2Fe–2S] clusters has been carried out[70] with the aid of the analogue complexes $[Fe_2S_2(SMe)_4]^{2-}$, $[Fe_2S_2(SEt)_4]^{2-}$, and $[Fe_2S_2(S_2\text{-}o\text{-xyl})]^{2-}$. Figure 30 shows RR and IR spectra of these complexes, while Figure 31 displays eigenvectors and frequencies calculated from a normal mode analysis. This calculation was carried out for a $[Fe_2S_2(SEt)_4]^{2-}$ model using the structural parameters[71] of $[Fe_2S_2(S_2\text{-}o\text{-xyl})_2]^{2-}$, since extensive isotopic data was available for the latter, including ^{54}Fe and ^{34}S in both bridging and terminal positions, and also ^{2}H in the methylene position. Having the IR spectra was essential for the analysis since the mutual exclusion rule applies to these centrosymmetric molecules. The point group is D_{2h} for the central $Fe_2S_2{}^{b}S_4{}^{t}$ cluster (S^b, S^t = bridging and terminal sulfur atoms), and C_{2h} if the carbon framework is included. Symmetry labels appropriate to both point groups are indicated in Figure 31. The modes selected for display are those that have mainly Fe–S stretching character, as well as one of the SCC bending modes

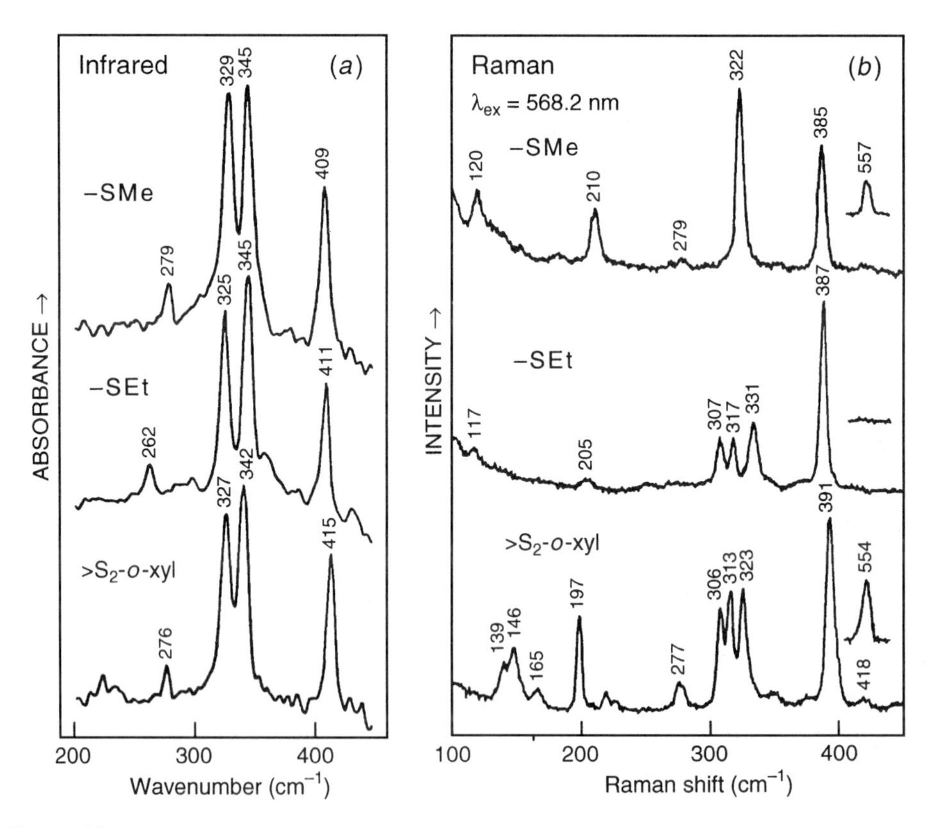

Figure 30

The IR (*a*) and RR (*b*) spectra of the methylthiolate (top), ethylthiolate (middle), and xylenedithiolate (bottom) [2Fe–2S] analogue complexes.[70]

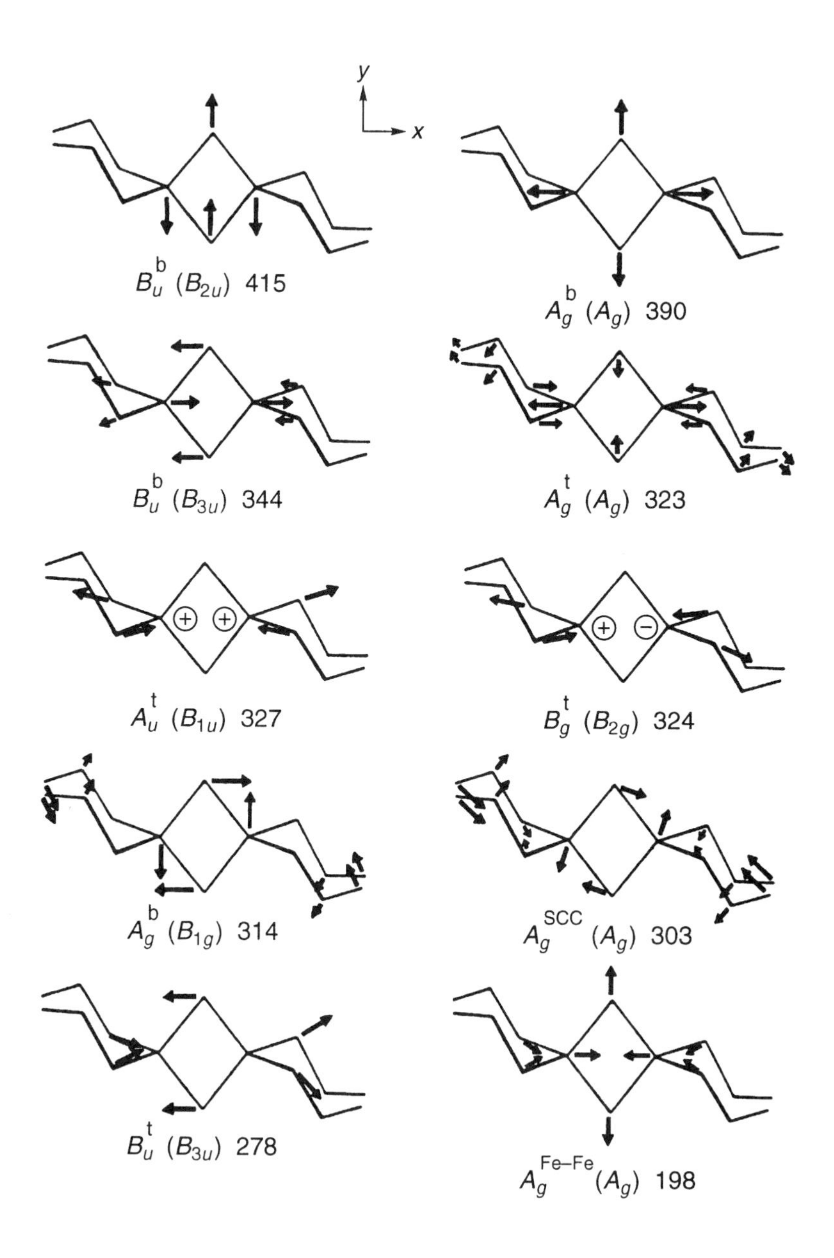

Figure 31
Eigenvectors for selected normal modes of $[Fe_2S_2(SEt)_4]^{2-}$, calculated using the geometry and the isotope shifts of the xylenedithiolate analogue.[70] The eight modes with mainly Fe–S stretching character are identified according to the dominant terminal (t) or bridging (b) contribution (although some have major contributions from both). Calculated frequencies and mode symmetries in the C_{2h} point group are given as well as the parent symmetry (parentheses) in the D_{2h} symmetry of the $Fe_2S_2{}^bS_4{}^t$ core. Also shown is the symmetric ring deformation, which modulates the Fe–Fe interaction, as well as the mainly SCC bending mode (303 cm^{-1}) that interacts strongly with the nearby $A_g{}^b$ Fe–S stretching mode (314 cm^{-1}) sharing its intensity and isotope shift.

and the "Fe–Fe stretching" mode. This last is a symmetric deformation of the Fe_2S_2 ring, whose frequency carries no implication about Fe–Fe bonding per se, since the force constant can be made up from angle bending as well as Fe–Fe stretching contributions. The complexes are all diamagnetic, despite the high-spin Fe(III) character of the individual atoms,[72] but the required antiferromagnetism can be mediated by the bridging S atoms (superexchange mechanism). Direct overlap of Fe orbitals may play some role at the observed separation of 2.75 Å, however, and the substantial RR enhancement seen for a mode near 200-cm^{-1} identifiable (via its ^{54}Fe shift) with "Fe–Fe stretching" implies that modulation of this interaction is significant in the resonant electronic transition.

The four Fe–S^b and the four Fe–S^t stretching coordinates each contribute to two IR-active (symmetry label u) and two Raman-active (g) modes. The eight mainly Fe–S eigenvectors in Figure 31 are labeled with b or t superscripts to indicate mainly bridging or terminal character, although in some cases there are substantial contributions from both coordinates, as indicated by mixed-isotope patterns.[59] The four u modes are clearly observed as four well-resolved IR bands in the 250–440-cm^{-1} region for all three complexes (Fig. 30). The methanethiolate anologue shows two strong bands in the 568.2-nm excited RR spectrum, at 385 and 322 cm^{-1}, and a weaker one at 314 cm^{-1} (not shown), which is found only with longer wavelength excitation (647.1 nm). The former two were assigned to the two breathing modes (A_g) of the $Fe_2S_2{}^bS_4{}^t$ cluster (D_{2h}), while the 314-cm^{-1} band was assigned to the B_{1g} bridging mode,[69] which requires vibronic activation. For the ethanemono- and xylenedithiolate analogues, however, this band is seen quite strongly with 568.2-nm excitation. This activation was attributed to the B_{1g} mode becoming A_g in the C_{2h} point group. The actual symmetry in the methyl analogue is likewise no higher than C_{2h}, but the involvement of SCC bending seems to be required for effective symmetry lowering. The final Raman mode, B_{2g}, does not become totally symmetric in C_{2h} symmetry, and remains weak for all three complexes. It was observed at 331 cm^{-1} in the xylenedithiolate analogue when the stronger A_g band at 323 cm^{-1} was shifted out of the way by deuterating the methylene groups, as shown in Figure 32.

The deuteration experiment was carried out in order to investigate the origin of still another band, at 306 cm^{-1}, seen only in the ethanemono- and xylenedithiolate analogues, and to test the hypothesis that this band is due to intensity borrowing by a SCC bending mode from the nearby $A_g{}^b$ band at 315 cm^{-1}. Consistent with this hypothesis deuteration, which was expected to lower the SCC bending frequency, collapsed the 306/313 cm^{-1} doublet into a single band at 312 cm^{-1}, with the intensity of the two bands combined. Moreover, the 312-cm^{-1} band was found, (using a doubly labeled compound) to have a ^{34}S isotope shift close to the total shift of the 313- and 306-cm^{-1} bands in the natural abundance material. These observations established that the 306-cm^{-1} band does indeed represent a SCC bending mode, whose coordinate mixes with the $A_g{}^b$ mode due to their near coincidence. The calculated eigenvectors of the two modes (Fig. 31) are quite similar.

This coupling served to allow the SCC bending force constant to be determined quite accurately, making it possible to carry out model calculations to test

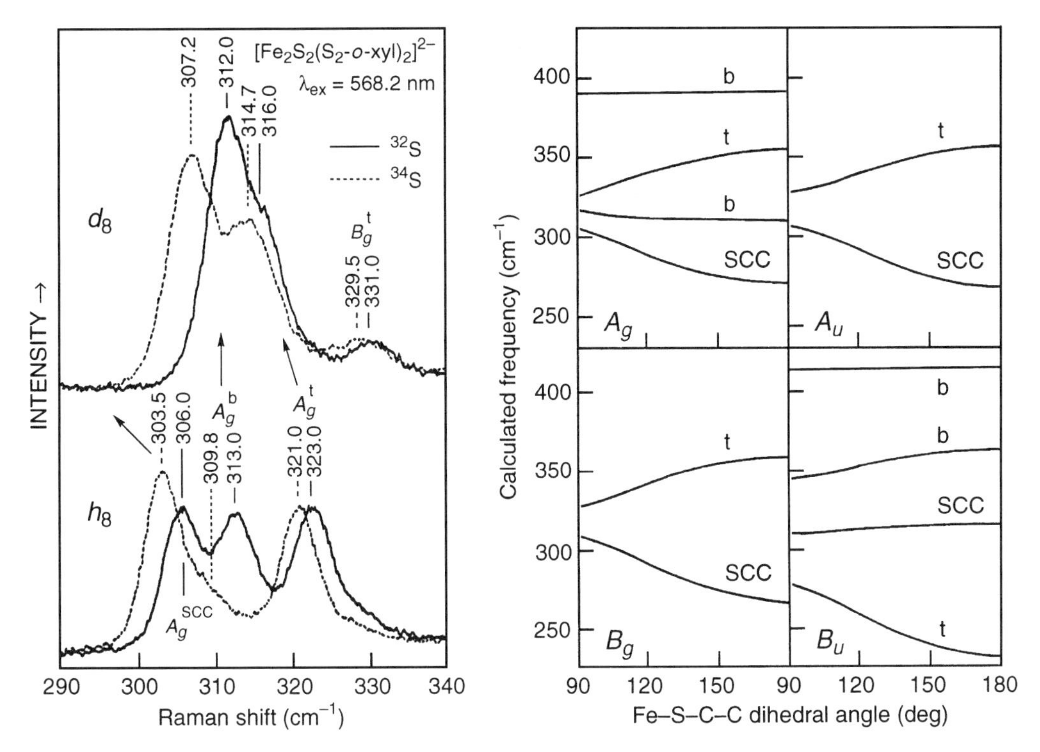

Figure 32
A detailed view of the 300-cm^{-1} region of the $[Fe_2S_2(S_2\text{-}o\text{-xyl})_2]^{2-}$ RR spectrum showing the effects of methylene deuteration and, for both ^1H and ^2H species, the effect of ^{34}S substitution in the bridging positions.[70] The mode assignments are also shown. Deuteration collapses the 306- and 313-cm^{-1} bands and increases the ^{34}S shift of the resulting 312-cm^{-1} band to nearly the total of the 306- and 313-cm^{-1} ^{34}S shifts in the ^1H species. This effect was interpreted as reflecting loss of the interaction between $A_g{}^b$ and $A_g{}^{SSC}$ as the latter is moved to lower frequency by deuteration of the methylene group. Also shown are graphs of the predicted pattern of frequency changes for the various FeS stretching modes when the FeSCC dihedral angle is increased from 90 to 180° in a concerted manner for all four ligands, maintaining the C_{2h} symmetry.

the effects of varying τ(FeSCC), and therefore the kinematic coupling with the Fe–S^t stretches.[70] The results are shown in Figure 32. As expected, coupling increased as τ increased from 90° toward 180° (in concert for the four ligands, in order to preserve the C_{2h} symmetry). The result was to force upward the Fe–S^t modes that are higher in frequency than SCC bending, but to force downward the one lower lying FeS^t mode, $B_u{}^b$.

These coupling effects suggested an interpretation of the RR spectra of the [2Fe–2S] proteins plant Fd and adrenodoxin (Ado),[73] which display distinct differences (Fig. 33), despite their chromophores having nominally the same structure. (An X-ray crystal structure is available for Fd[73] but not for Ado, although the chemical and spectroscopic properties of the latter all point to the same

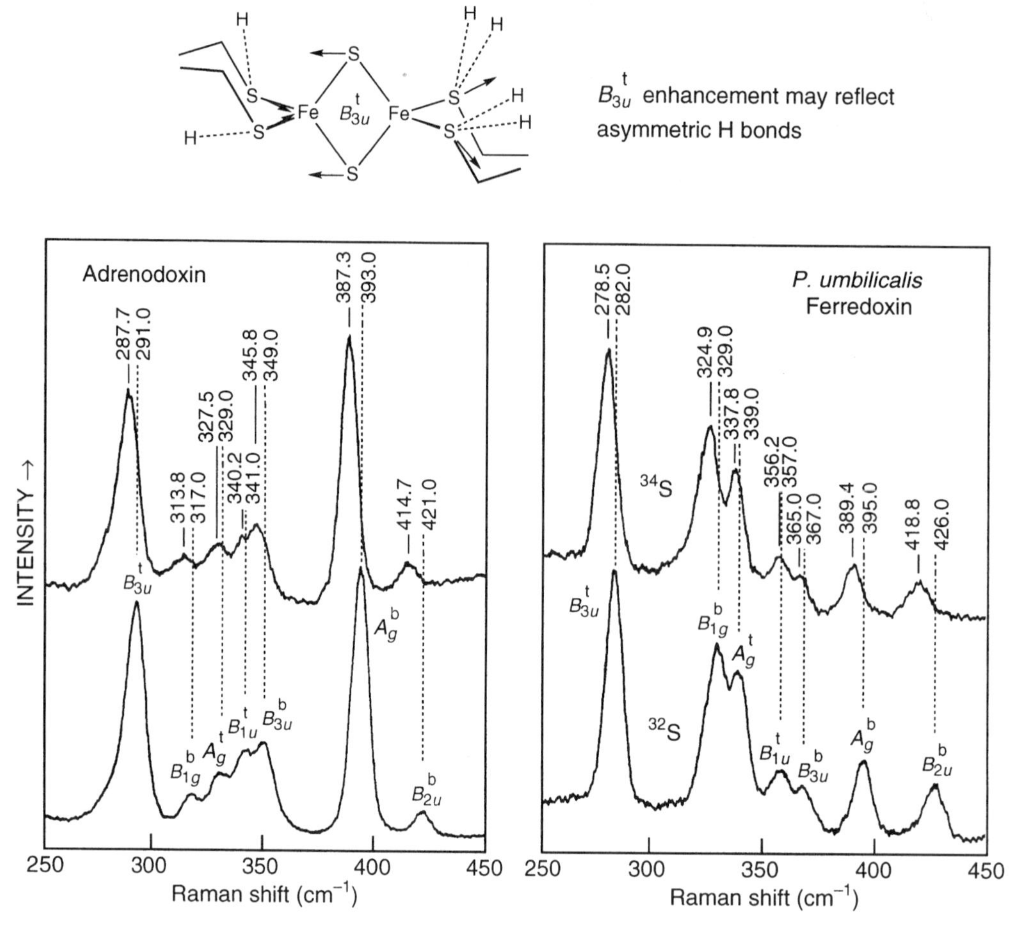

Figure 33
The 406.7-nm excited RR spectra of plant ferredoxin and adrenodoxin[73a] showing isotope shifts on reconstitution with $^{34}S^{2-}$. The mode assignments were made by comparison with the analogue complex vibrational spectra.[62] The frequency differences between proteins were suggested to arise in part from different FeSCC dihedral angles. The remarkable intensification of the approximate 280-cm^{-1} mode assigned to $B_{3u}{}^{t}$ was attributed to asymmetry in the H-bond pattern that destroys the phase cancellation of the FeS_4 breathing motions; the diagram shows the H-bond pattern indicated by the crystal structure of *Spirulina platensis* ferredoxin.[74]

$Fe_2S_2{}^{b}S_4{}^{t}$ center.) The bands were assigned via their $^{34}S^{b}$ shifts, in protein reconstituted with $^{34}S^{2-}$, by comparison with the modes of the analogue complexes. Although IR data are not available for the protein chromophores, all the u modes were nevertheless located (Fig. 33), indicating loss of the electronic symmetry center by asymmetric influences of the protein. Remarkably, the $B_{3u}{}^{t}$ mode gives rise to one of the strongest bands in the spectra. The eigenvector of this mode (Fig. 31) shows it to consist of an out-of-phase breathing motion of the two linked FeS_4 tetrahedra. Since FeS_4 breathing produces the strongest band in the RR spectra of 1-Fe complexes, it is not surprising that the $B_{3u}{}^{t}$ mode gains substantial RR intensity once the phase cancellation is overcome by environmental asymmetry. This asymmetry was suggested[73a] to stem from the pattern of H bonds to the coordinated cysteine S atoms, since the crystal structure of plant Fd indicates that there are two such bonds on one side of the cluster but four on the other (Fig. 33).[74]

In other respects, the spectra of the two proteins differ significantly, both in relative intensities and band frequencies. Most frequencies are higher for Fd than Ado, substantially so in the 310–360-cm^{-1} region, but the $B_{3u}{}^{t}$ frequency is 10 cm^{-1} lower. This pattern is exactly that expected if coupling with SCC bending is increased via increases in τ(FeSCC) (Fig. 29). Calculations with the C_{2h} model gave a good account of the observed frequencies with $\tau = 90°$ for Ado but 105° for Fd, and a slightly higher Fe–S^b force constant for the latter.

C. [4Fe–4S] Proteins

Figure 34 compares RR spectra in the Fe–S stretching region for the proteins HiPIP (from *C. vinosum*) and bacterial Fd (from *C. pasteurianum*), both at the same oxidation level (reduced for the former and oxidized for the latter), and also for the synthetic analogue complex $[Fe_4S_4(SCH_2Ph)_4]^{2-}$,[75] in frozen solution and in crystals with tetraethylammonium (TEA) counterions. In these crystals, the Fe_4S_4 cube is known to undergo a compression along the S_4 axis (Fig. 34) leaving a D_{2d} structure with four short and eight long bonds.[76] The frozen solution spectrum is noticeably simpler than that of the crystals. It can be analyzed in terms of the expected modes of a tetrahedral molecule; as indicated in Figure 34, the four Fe–S^t stretches contribute to two tetrahedral modes, A_1 and T_2, while the 12 Fe–S^b stretches contribute to five modes: $A_1 + E + T_1 + 2T_2$. The indicated assignments were made on the basis of ^{34}S (both t and b) and ^{54}Fe shifts, and a normal coordinate calculation. The additional complexity of the crystal spectrum is attributable to mode splitting in the lower symmetry, as indicated in Figure 34. The perturbed frequencies and isotope shifts were quantitatively reproduced with a D_{2d} calculation, using Fe–S force constants scaled to the different bond distances via Badger's rule.[14a,b] Thus the spectral pattern is quite well understood.

It can be seen from Figure 34 that the RR spectrum of *C. vinosum* HiPIP resembles that of the analogue in frozen solution while that of *C. pasteurianum* Fd resembles the spectrum of the TEA crystals. It is clear that the *C. vinosum* HiPIP spectrum has a more tetrahedral-like pattern, while that of Fd is D_{2d}-like, indicating a greater degree of cluster distortion in Fd than HiPIP. The $Fe_4S_4(Cys)_4$ clusters in HiPIPs can adopt less symmetrical structures, however, as revealed by RR studies on proteins from different organisms (*Rhodopila globiformis, Rhodocyclus tenuis*, and *Ectothiorhodospira halophila*).[77] These subtle differences are not at all apparent in the crystal structures,[78] all of which have been analyzed in terms of a low-symmetry distribution of bond lengths. Protein X-ray diffraction measurements cannot reliably discriminate bond distance differences on the order of 0.1 Å or less, however, while such differences have marked effects on the vibrational spectrum. Thus the RR spectra are probably more reliable indicators of chromophore symmetry than the X-ray diffraction measurements.

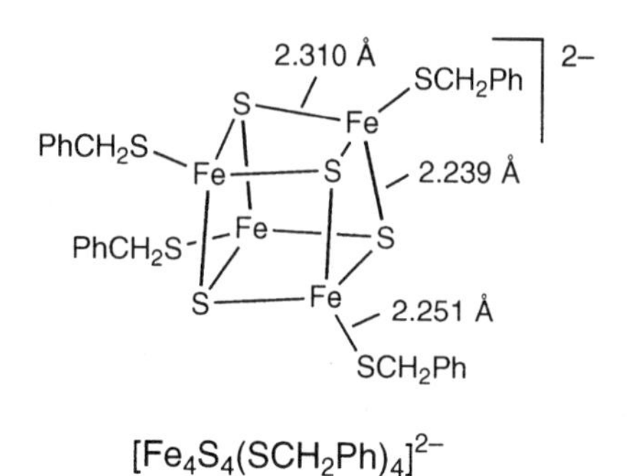

Bond	T_d	D_{2d}	cm^{-1}
4×Fe–S^t	A_1	A_1	~390
	T_2	$B_2 + E$	~360
12×Fe–S^b	T_2	$B_2 + E$	~385
	A_1	A_1	~335
	E	$A_1 + B_1$	~280
	T_1	$A_2 + E$	~275
	T_2	$B_2 + E$	~250

- $\boldsymbol{T_d \rightarrow D_{2d}}$ distortion seen in T_2 mode splittings and $E(A_1)$ mode activation
- Fd_{ox} shows greater distortion than $HiPIP_{red}$

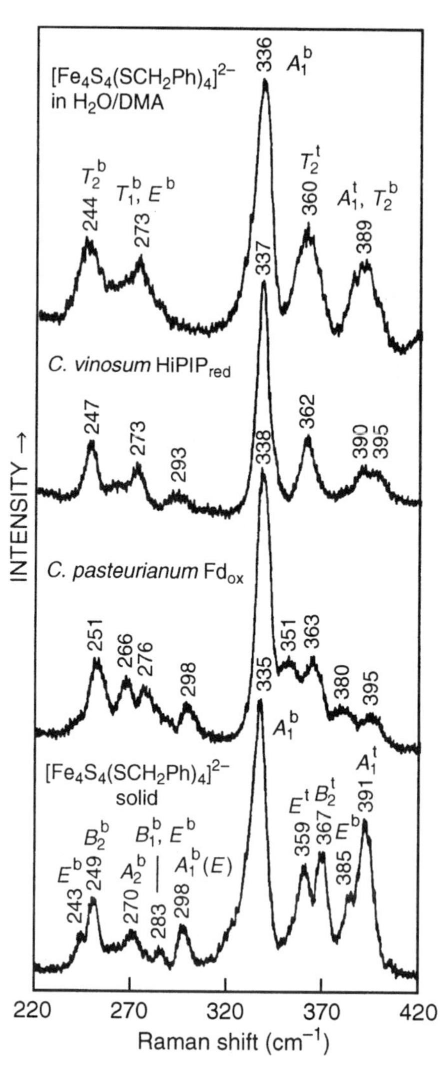

Figure 34

The 457.9 (proteins)- and 514.5 (analogue)-nm excited RR spectra of [4Fe–4S] proteins and their benzyl thiolate analogue.[75] The $[Fe_4S_4(SCH_2Ph)_4]^{2-}$ spectra are assignable on the basis of a tetrahedral structure in frozen solution, but the known D_{2d} distortion in tetraethylammonium crystals. Assignments are indicated in the figure, while the table gives the expected $T_d \rightarrow D_{2d}$ correlations. The RR spectrum of oxidized bacterial ferredoxin is very similar to the analogue crystals, indicating a similar D_{2d} distortion, while the spectrum of reduced HiPIP (same oxidation level) is similar to the analogue spectrum in solution, indicating a structure that is more tetrahedral. (*N,N*-Dimethylactamide is DMA.)

D. [3Fe–4S] Proteins

A number of proteins have been shown to contain [3Fe–4S] clusters. For one of these, Fd I from Azotobacter *vinelandii*, an early X-ray crystal structure[79a] was reported to establish a nearly planar Fe_3S_3 hexagon. This structure was demonstrated, however, to be incompatible with the RR spectrum of *A. vinelandii* Fd I. or of proteins with other [3Fe–4S] centers[67] (Fig. 35). The critical evidence is the finding of cluster modes, identified via their $^{34}S^b$ shifts in reconstituted protein, at

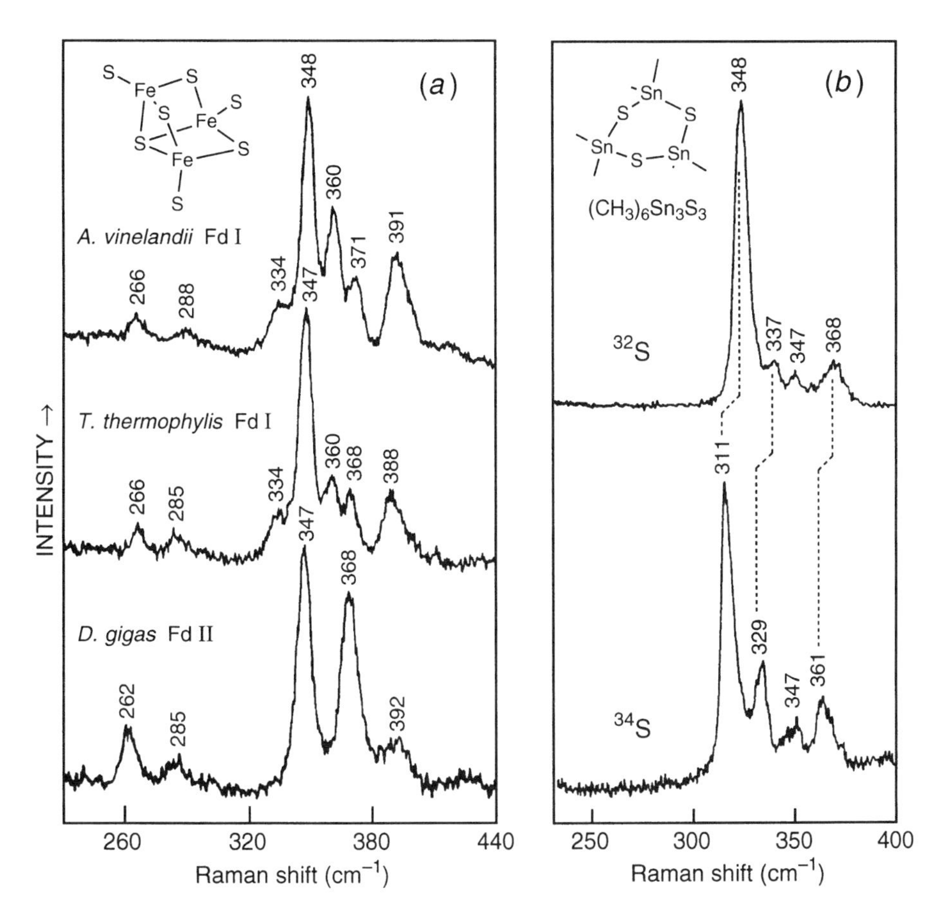

Figure 35
(*a*) The 488.0-nm excited RR spectra for the [3Fe–4S] centers of the indicated proteins.[67] Two of the three proteins *A. vinelandii* Fd I and *Thermus thermophylis* Fd, have [4Fe–4S] as well as [3Fe–4S] centers, but their RR spectra can be suppressed by appropriate choice of excitation wavelength; only the strongest band of the [4Fe–4S] centers is seen as a weak shoulder at 334 cm^{-1}. (*b*). Raman spectra of $(CH_3)_6Sn_3S_3$[67a] showing the expected ^{34}S shifts of the Sn–S stretching modes. The hexagonal ring structure of $(CH_3)_6Sn_3S_3$[80] can be ruled out for the [3Fe–4S] centers because antisymmetric cluster modes (265, 285 cm^{-1}, identified by their $^{34}S^b$ shifts)[67a] are found below the breathing mode (347 cm^{-1}) indicating acute Fe–S–Fe angles, consistent with the cube-derived Fe_3S_4 structure (*a*) proposed by Beinert et al.,[81] and since confirmed by X-ray crystallography.[65]

frequencies substantially below the breathing mode, assigned to the strongest cluster band, at 347 cm^{-1}. Nontotally symmetric modes of a Fe_3S_3 hexagon must lie above the breathing mode, because the Fe–S–Fe angles ($\sim 120°$) are substantially greater than 90°. The situation is the same as for a triatom (see Section III.B) with a large bond angle; the antisymmetric stretch must lie above the symmetric stretch because of the reinforcement of the displacements. In the context of trinuclear clusters, this point was illustrated[66a] with the Raman spectrum of the molecule $(CH_3)_6Sn_3S_3$, which has a Sn_3S_3 hexagon.[79] As shown in Figure 35, the antisymmetric ring modes (^{34}S sensitive) of this molecule do lie above the intense breathing mode.

The RR spectra of the [3Fe–4S] proteins were shown by model calculations[66a] to be consistent with the proposal of Beinert et al.[81] of a Fe_3S_4 structure, based on a Fe_4S_4 cube with one corner removed. (This structure has Fe–S–Fe angles $<90°$.) This structure is in fact correct for the *A. vinelandii* Fd I [3Fe–4S] center, the original X-ray analysis having been shown to be in error.[79b–d] X-ray molecular parameters of the Fe_3S_4 cube-derived structure were also determined for a [3Fe–4S] cluster in *D. gigas* Fd II.[79a]

References

1. (a) Carey, P. R. *Biochemical Applications of Raman and Resonance Raman Spectroscopies;* Academic: New York, 1982. (b) Tu, A. T. *Raman Spectroscopy in Biology;* Wiley: New York, 1982. (c) Parker, F. S. *Application of Infrared, Raman and Resonance Raman Spectroscopy in Biochemistry;* Plenum: New York, 1983. (d) Moore, C. B., Ed., *Chemical and Biochemical Applications of Lasers;* Academic: New York, Vol. 1 to present. (e) Szymanski, H. A., Ed., *Raman Spectroscopy: Theory and Practice;* Plenum: New York, 1967 and 1970; Vols. 1 and 2. (f) Koningstein, J. A. *Introduction to the Theory of the Raman Effect;* Reidel, Dordrecht: The Netherlands 1973, (g) Long, D. A. *Raman Spectroscopy;* McGraw-Hill: New York, 1977. (h) Strommen, P. P.; Nakamoto, K. *Laboratory Raman Spectroscopy;* Wiley: New York 1984. (i) Clark, R. J. H.; Hester, R. E., Eds. *Advances in Infrared and Raman Spectroscopy;* Vols. 1–12; and *Advances in Spectroscopy;* Wiley: Vol. 13 to present; Chichester, UK, 1970 to present. (j) Gardiner; D. J.; Graves, P. R., Eds., *Practical Raman Spectroscopy;* Springer-Verlag: Heidelberg, 1989.
2. Spiro, T. G., Ed., *Biological Applications of Raman Spectroscopy;* Wiley-Interscience: New York, 1988; Vols. 1–3.
3. (a) Raman, C. V.; Krishnan, K. S. "A New Type of Secondary Radiation" *Nature* (*London*) **1928**, *121*, 501. (b) Raman, C. V. Krishnan, K. S. "A Change of Wave-Length in Light Scattering" *Nature* (*London*) **1928**, *121*, 619.
4. (a) Czernuszewicz, R. S.; Johnson, M. K. "A Simple Low-Temperature Cryostat for Resonance Raman Studies of Frozen Protein Solutions" *Appl. Spectrosc.* **1983**, *37*, 297. (b) Czernuszewicz, R. S. "Closed-Cycle Refrigerator Solution and Rotating Solid Sample Cells for Anaerobic Resonance Raman Spectroscopy" *Appl. Spectrosc.* **1986**, *40*, 571. (c) Hildebrandt, P.; Macor, K. A.; Czernuszewicz, R. S. "A Novel Cylindrical Rotating Electrode for Anaerobic Surface Enhanced Raman Spectroscopy" *J. Raman Spectrosc.* **1988**, *19*, 65. (d) Czernuszewicz, R. S., Macor, K. A. "Low-Temperature Bulk Electrolysis Cell for In Situ Resonance Raman Spectroelectrochemistry: Observation of the Fe(IV)=O Stretching Frequency for Electrogenerated Ferryl Tetramesitylporphyrin" *J. Raman Spectrosc.* **1988**, *19*, 553. (e) McCreery, R. L. "Instrumentation for Dispersive Raman Spectroscopy," in *Modern Techniques in Raman Spectroscopy*, Laserna, J. J., Ed., Wiley: New York, 1996, pp 41–72.
5. Fodor, S. P. A.; Rava, R. P.; Copeland, R. A.; Spiro, T. G. "H_2 Raman-Shifted YAG Laser Ultraviolet Raman Spectrometer Operating at Wavelengths Down to 184 nm" *J. Raman Spectrosc.* **1986**, *17*, 471.
6. (a) Asher, S. A.; Johnson, C. R.; Murtaugh, J. "Development of a New UV Resonance Raman Spectrometer for the 217–400-nm Spectral Region" *Rev. Sci. Instrum.* **1983**, *54*, 1657. (b) Jones, C. M.; DeVito, M. L.; Harmon, R. A.; Asher, S. A. "High-Repetition-Rate Excimer-Based UV Laser Excitation Source Avoids Saturation in Resonance Raman Measurements of Tyrosinate and Pyrene" *Appl. Spectrosc.* **1987**, *41*, 1268.
7. (a) Williams, K. P. J.; Klenerman, D. J. "UV Resonance Raman Spectroscopy Using a Quasi-Continuous-Wave Laser Developed as an Industrial Analytical Technique" *J. Raman Spectrosc.* **1992**, *23*, 191. (b) Gustafson, T. L. "The Generation of Quasi-Continuous, Tunable 200 nm Excitation Using the MHz Amplified Synchronously Pumped Dye Laser. Application to UV Resonance Raman Spectroscopy" *Opt. Commun.* **1988**, *67*, 53.
8. Asher, S. A.; Bormett, R. W.; Chen, X. G.; Lemmon, D. H.; Cho, N.; Peterson, P.; Arrigoni, M.; Spinelli, L.; Cannon, J. "UV Resonance Raman Spectroscopy Using a New cw Laser Source: Convenience and Experimental Simplicity" *Appl. Spectrosc.* **1993**, *47*, 628.
9. Asher, S. A. "UV Resonance Raman Spectroscopy for Analytical, Physical, and Biological Chemistry" *Anal. Chem.* **1993**, *65*, 59A and 201A and references cited therein.
10. (a) Chase, B. "Fourier Transform Raman Spectroscopy" *Anal. Chem.* **1987**, *59*, 881A. (b) Cutler, D. J. "Fourier Transform Raman Instrumentation" *Spectrochim. Acta* **1990**, *46A*, 131. (c) Schrader, B.; Hoffman, A.; Simon, A.; Podschadiowski, R.; Fisher, M. "NIR–

FT–Raman Spectroscopy, State of the Art" *J. Mol. Struct.* **1990**, *217*, 207. (d) Chase, D. B.; Rabolt, J. F., Eds., "Fourier-Transform Raman Spectroscopy: From Concept to Experiment", Academic Press: San Diego, 1994.

11. (a) Barnett, S. M.; Dicaire, F.; Ismail, A. A. "Application of FT-Raman Spectroscopy to the Study of Selected Organometallic Complexes and Proteins" *Can. J. Chem.* **1990**, *68*, 1196. (b) Sawatzki, J.; Fischer, R.; Scheer, H.; Siebert, F. "Fourier Transform Raman Spectroscopy Applied to Photobiological Systems" *Proc. Natl. Acad. Sci. USA* **1990**, *87*, 5903. (c) Johnson, C. K.; Rubinovitz, R. "Fourier Transform Raman Spectroscopy of Photoactive Proteins with Near-Infrared Excitation" *Appl. Spectrosc.* **1990**, *44*, 1103. (d) Levin, I. W.; Lewis, E. N. "Fourier Transform Raman Spectroscopy of Biological Materials" *Anal. Chem.* **1990**, *62*, 1101A.
12. Harris, D. C.; Bertolucci, M. D. *Symmetry and Spectroscopy;* Oxford University Press: New York, 1978, Chapter 3.
13. H.-B. Burgi, H.-B.; Dunitz, J. D. "Fractional Bonds: Relations Among Their Lengths, Strengths, and Stretching Force Constants" *J. Am. Chem. Soc.* **1987**, *109*, 2924.
14. (a) Badger, R. M. "The Relation Between the Internuclear Distance and Force Constants of Molecules and Its Application to Polyatomic Molecules" *J. Chem. Phys.* **1935**, *3*, 710. (b) Herschbach, D. R.; Laurie, V. W. "Anharmonic Potential Constants and Their Dependence upon Bond Length" *J. Chem. Phys.* **1961**, *35*, 458. (c) Conradson, S. P.; Sattelberger, A. P.; Woodruff, W. M. "X-ray Absorption Study of Octafluorodichromate(III): EXAFS Structures and Resonance Raman Spectroscopy or Octahalodichromates" *J. Am. Chem. Soc.* **1988**, *110*, 1309.
15. Dollish, F. R.; Fateley, W. G.; Bentley, F. F. *Characteristic Raman Frequencies of Organic Compounds;* Wiley-Interscience: New York, 1974; p 60.
16. Dollish, F. R.; Fateley, W. G.; Bentley, F. F. *Characteristic Raman Frequencies of Organic Compounds;* Wiley-Interscience: New York, 1974; p 143.
17. Gray, P.; Waddington, T. C. "Fundamental Vibration Frequencies and Force Constants in the Azide Ion" *Trans. Faraday Soc.* **1956**, *53*, 901.
18. Jones, L. H. "Infrared Spectra and Structure of the Crystalline Sodium Acetate Complexes of U(VI), Np(VI), Pu(VI) and Am(VI). A Comparison of Metal–Oxygen Bond Distance and Bond Force Constant in this Series" *J. Chem. Phys.* **1955**, *23*, 2105.
19. Burke, J. M.: Kincaid, J. R.; Spiro, T. G. "Resonance Raman Spectra and Vibrational Modes of Iron(III) Tetraphenylporphine μ-Oxo-Dimer. Evidence for Phenyl Interaction and Lack of Dimer Splitting" *J. Am. Chem. Soc.* **1978**, *100*, 6077.
20. Fleischer, E.; Srivastava, T. S. "The Structure and Properties of μ-Oxo-bis (tetraphenylporphine iron(III)) *J. Am. Chem. Soc.* **1969**, *91*, 2403.
21. Colthup, N. B.; Daley, L. H.; Wiberley, S. E. *Introduction to Infrared and Raman Spectroscopy;* Academic: New York, 1990.
22. Ogden, J. S.; Tumer, J. J. "Photolytic Fluorine Reactions, $F_2 + N_20$ at 4K" *J. Chem. Soc. A* **1967**, 1483.
23. Rochkind, M. M.; Pimentel, G. C. "Infrared Spectrum and Vibrational Assignment for Chlorine Monooxide, Cl_2O" *J. Chem. Phys.* **1965**, *42*, 1361.
24. Perry, C. H.; Athans, D. P.; Young, E. F.; During, J. R.; Mitchell, B. R. "Far Infrared Spectra of Palladium Compounds—III. Tetrahalo, Tetraammine and Dihalodiammine Complexes of Palladium" *Spectrochim. Acta* **1967**, *23A*, 1137.
25. (a) Cornilsen, B.; Nakamoto, K. "Metal Isotope Effect on Metal–Ligand Vibrations-XII. Imidazole Complexes with Co(II), Ni(II), Cu(II), and Zn(II)" *J. Inorg. Nucl. Chem.* **1974**, *36*, 2467. (b) Hodgson, J. B.; Percy, G. C.; Thornton, D. A. "The Infrared Spectra of Imidazole Complexes of First Transition Series Metal(II) Nitrates and Perchlorates" *J. Mol. Struct.* **1980**, *66*, 81.
26. Burke, J. M.; Kincaid, J. R.; Peters, S.; Gagne, R. R.; Collman, J. P.; Spiro, T. G. "Structure-Sensitive Resonance Raman Bands of Tetraphenyl and 'Picket Fence' Porphyrin–Iron Complexes, Including an Oxyhemoglobin Analogue" *J. Am. Chem. Soc.* **1978**, *100*, 6083.
27. Herzberg, G. *Molecular Spectra and Molecular Structure.* Vol. II: *Infrared and Raman Spectra of Polyatomic Molecules;* Van Nostrand: Princeton, NJ, 1945.

28. Wilson, E. B.; Decius, J. C.; Gross, P. C. *Molecular Vibrations;* McGraw-Hill: New York, 1955.
29. Schachtschneider, J. H.; Shell Development Co., Technical Report No. 51–65, and 231–264, 1962.
30. Gans, P. *Vibrating Molecules;* Chapman & Hall: London, 1971.
31. Nakamoto, K. *Infrared and Raman Spectra of Inorganic and Coordination Compounds;* Wiley-Interscience: New York, 1997; Part A, Chapter 1.
32. (a) Fogarasi, G.; Pulay, P "*Ab initio* Vibrational Force Fields", *Annu. Rev. Phys. Chem.* **1984**, *35*, 191. (b) Hess, B. A., Jr.; Schaad, L. J.; Carsky, P.; Zahradnik, R. "An *ab initio* Calculations of Vibrational Spectra and Their Use in the Identification of Unusual Molecules" *Chem. Rev.* **1986**, *86*, 709. (c) Kozlowski, P. M.; Zgierski, M. Z.; Pulay, P. "An Accurate In-Plane Force Field for Porphine: A Scaled Quantum Mechanical Study," *Chem. Phys. Lett.* **1995**, *247*, 379.
33. Czernuszewicz, R. S.; Nakamoto, K.; Strommen, D. P. "Resonance Raman Spectra and Vibrational Assignments of 'Red' and 'Black' Forms of Bis(dithiooxalato)nickel(II)" *J. Am. Chem. Soc.* **1982**, *104*, 1515.
34. Czernuszewicz, R. S. "Resonance Raman Spectroscopy of Metalloproteins Using CW Laser Excitation" In *Methods in Molecular Biology;* Jones, C.; Mulloy, B.; Thomas, A. H., Eds.; Humana Press: Totawa, 1993; Vol. 17, Chapter 15, p 345.
35. (a) Ziegler, L. D.; Hudson, B. "Resonance Raman Scattering of Benzene and Benzene-d_6 with 212.8 nm Excitation" *J. Chem. Phys.* **1986**, *74*, 982. (b) Hudson, B.; Mayne, L "Ultraviolet Resonance Raman Spectroscopy of Biopolymers" *Methods Enzymol.* **1986**, *130*, 331. (c) Johnson, C. R.; Ludwig, M.; O'Donnell, S.; Asher, S. A. "UV Resonance Raman Spectroscopy of the Aromatic Amino Acids and Myoglobin" *J. Am. Chem. Soc.* **1984**, *106*, 5008. (d) Asher, S. A.; Ludwig, M.; Johnson, C. R. "UV Resonance Raman Excitation Profiles of the Aromatic Amino Acids" *J. Am. Chem. Soc.* **1986**, *108*, 3186. (e) Rava, R. P.; Spiro, T. G. "Selective Enhancement of Tyrosine and Tryptophan Resonance Raman Spectra via Ultraviolet Laser Excitation" *J. Am. Chem. Soc.* **1984**, *106*, 4062. (f) Rava, R. P.; Spiro, T. G. "Resonance Enhancement in the Ultraviolet Raman Spectm of Aromatic Amino Acids" *J. Phys. Chem.* **1985**, *89*, 1856.
36. Rousseau, D. L.; Williams, D. F. "Resonance Raman Scattering of Light from a Diatomic Molecule" *J. Chem. Phys.* **1976**, *64*, 3519.
37. (a) Dutta, P. K.; Nestor, J. R.; Spiro, T. G. "Resonance Coherent Anti-Stokes Raman Scattering Spectra of Fluorescent Biological Chromophores: Vibrational Evidence for Hydrogen Bonding of Flavin to Glucose Oxidase and for Rapid Solvent Exchange" *Proc. Natl. Acad. Sci. USA* **1077**, *74*, 4146. (b) Dutta, P. K.; Dallinger, R.; Spiro, T. G. "Resonance CARS (Coherent Anti-Stokes Raman Scattering) Line Shapes via Franck–Condon Scattering: Cytochrome *c* and *β*-Carotene" *J. Chem. Phys.* **1980**, *73*, 3580. (c) Morris, M. D.; Wallan, D. J.; Ritz, G. P.; haushalter, J. P. "AC-Coupled Inverse Raman Spectrometry" *Anal. Chem.* **1978**, *50*, 1796. (d) Clark, R. J. H.; Hester, R. E., Eds.; *Advances in Spectroscopy, Vol. 15: Advances in Non-Linear Spectroscopy;* Wiley: Chichester, 1988.
38. Albrecht, A. C. "On the Theory of Raman Intensity" *J. Chem. Phys.* **1961**, *34*, 1476.
39. Spiro, T. G.; Li, Y.-Y. "Resonance Raman Spectroscopy of Metalloporphyrins" In *Biological Applications of Raman Spectroscopy;* Spiro, T. G. Ed.; Wiley-Interscience: New York, 1988, Vol. 3, Chapter 1, pp 1–38.
40. Gouterman, M. "Optical Spectra and Electronic Structure of Porphyrin and Related Ring" In *The Porphyrins;* Dolphin, D. Ed.; Academic: New York, 1979, Vol. 3, Part A, pp 1–156.
41. Li, X.-Y.; Czernuszewicz, R. S.; Kincaid, J. R.; Stein, P.; Spiro, T. G. "Consistent Porphyrin Force Field. 2. Nickel Octaethylporphyrin Skeletal and Substituent Mode Assignments from ^{15}N, Meso-d_4, and Methylene-d_{16} Raman and Infrared Isotope Shifts" *J. Phys. Chem.* **1990**, *94*, 47.
42. Cheung, L. D.; Yu, N.-T.; Felton, R. H. "Resonance Raman Spectra and Excitation Profiles of Soret-Excited Matalloporphyrins" *Chem. Phys. Lett.* **1978**, *55*, 527.

43. Li, X.-Y.; Czernuszewicz, R. S.; Kincaid, J. R.; Su, Y. O.; Spiro, T. G. "Consistent Porphyrin Force Field. 1. Normal-Mode Analysis for Nickel Porphine and Nickel Tetraphenylporphine from Resonance Raman and Infrared Spectra and Isotope Shifts" *J. Phys. Chem.* **1990**, *94*, 31.
44. Fodor, S. P. A.; Copeland, R. A.; Grygon, C. A.; Spiro, T. G. "Deep-Ultraviolet Raman Excitation Profiles and Vibronic Scattering Mechanism of Phenylalanine, Tyrosine and Tryptophan", *J. Am. Chem. Soc.* **1989**, *111*, 5509.
45. Czernuszewicz, R. S.; Sheats, J. E.; Spiro, T. G. "Resonance Raman Spectra and Excitation Profile for $[Fe_2O(O_2CCH_3)_2(HBpz_3)_2]$, a Hemerythrin Analogue" *Inorg. Chem.* **1987**, *26*, 2063.
46. Clark, R. J. H.; Stewart, B. "The Resonance Raman Effect. Review of the Theory and of Applications in Inorganic Chemistry" *Struct. Bonding (Berlin)* **1979**, *36*, 1.
47. Champion, P. M. "Elementary Electronic Excitations and the Mechanism of Cytochrome $P450_{cam}$" *J. Am. Chem. Soc.* **1989**, *111*, 3433.
48. Champion, P. M.; Stallard, B. R.; Wagner, G. C.; Gunsalus, I. C. "Resonance Raman Detection of an Fe–S Bond in Cytochrome $P450_{cam}$" *J. Am. Chem. Soc.* **1982**, *104*, 5469.
49. (a) Tonks, D. L.; Page, J. B. "First-Order Resonance Raman Profile Lineshapes from Optical Absorption Lineshapes—a Consistency Test of Standard Theoretical Assumptions" *Chem. Phys. Lett.* **1979**, *66*, 449. (b) Page, J. B.; Tonks, D. L. "On the Separation of Resonance Raman Scattering into Orders in the Time Correlator Theory" *J. Chem. Phys.* **1981**, *75*, 5694.
50. Champion, P. M. "Cytochrome P450 and the Transform Analysis of the Heme Protein Raman Spectra" In *Biological Applications of Raman Spectroscopy;* Spiro, T. G. Ed.; Wiley-Interscience: New York, 1986, Vol. 3, Chapter 6, pp 249–292.
51. Meyers, A. B.; Mathies, R. A. "Resonance Raman Intensities: A Probe of Excited-State Structure and Dynamics" In *Biological Applications of Raman Spectroscopy;* Spiro, T. G. Ed.; Wiley-Interscience: New York, 1988, Vol. 2, Chapter 1, pp. 1–58.
52. (a) Lee, S.-Y.; Heller, E. J. "Time-Dependent Theory of Raman Scattering" *J. Chem. Phys.* **1979**, *71*, 4777. (b) Heller, E. J.; Sundberg, R. L.; Tannor, D. "Simple Aspects of Raman Scattering" *J. Phys. Chem.* **1982**, *86*, 1822. (c) Tannor, D. J.; Heller, E. J. "Polyatomic Raman Scattering for General Harmonic Potentials" *J. Chem. Phys.* **1982**, *77*, 202.
53. Spiro, T. G., Ed., *Biological Applications of Raman Spectroscopy;* Wiley-Interscience: New York, 1988, Vol. 3.
54. Sobel, B. E.; Lovenberg, W. "Characteristics of *Clostridium pasteurianum* Ferredoxin in Oxidation–Reduction Reactions" *Biochemistry* **1966**, *5*, 6.
55. Palmer, G.; Brintzinger, H.; Estabrook, R. W. "Spectroscopic Studies on Spinach Ferredoxin and Adrenodoxin" *Biochemistry* **1967**, *6*, 1658.
56. Estabrook, R. W.; Suzuki, K.; Mason, J. I.; Baron, J.; Taylor, W. E.; Simpson, E. R.; Purvis, J.; McCarthy, J. "Adrenodoxin: An Iron-Sulfur Protein of Adrenal Cortex Mitochondria" In *Iron–Sulfur Proteins;* Lovenberg, W., Ed., Academic: New York, 1973; Vol. 1, Chapter 8, p 199.
57. Shethna, Y. I.; Stombaugh, N. A.; Burris, R. H. "Ferredoxin from *Bacillus polymyxa*" *Biochem. Biophys. Res. Commun.* **1971**, *42*, 1108.
58. K. Dus, K.; DeKlerk, H.; Stetten, K.; Bartsch, R. G. "Chemical Characterization of High Potential Iron Proteins from *Chromatium* and *Rhodopseudomonas gelatinasa*" *Biochim. Biophys. Acta* **1967**, *40*, 291.
59. Long, T. V.; Loehr, T. M.; Alkins, J. R.; Lovenberg, W. "Determination of Iron Coordination in Nonheme Iron Using Laser-Raman Spectroscopy. II. *Clostridium pasterianum* Rubredoxin in Aqueous Solution" *J. Am. Chem. Soc.* **1971**, *93*, 1809.
60. Nakamoto, K. *Infrared and Raman Spectra of Inorganic and Coordination Compounds;* Wiley-Interscience: New York, 1986.
61. Woodward, L. A.; Taylor, M. J. "Raman Effect and Solvent Extraction. Part II. Spectra of Tetrachloroiodate and Tetrachloroferrate Ions" *J. Chem. Soc.* **1960**, 4473.

62. Czernuszewicz, R. S.; J. LeGall, J.; Moura, I; Spiro, T. G. "Resonance Raman Spectra of Rubredoxin: New Assigments and Vibrational Coupling Mechanism from Iron-54/Iron-56 Isotope Shifts and Variable-Wavelength Excitation" *Inorg. Chem.* **1986**, *25*, 696.
63. Eaton, W. A.; Lovenberg, W. "The Iron-Sulfur Complex in Rubredoxin" In *Iron Sulfur Proteins;* W. Lovenberg, Ed.; Academic Press: New York, 1973; Vol. 12, Chapter 3.
64. Eng, J. F.; Czernuszewicz, R. S.; Spiro, T. G. "Raman Difference Spectroscopy via Backscattering from a Spinning Tube and from a Low-Temperature Tuning Fork" *J. Raman Spectrosc.* **1985**, *16*, 432.
65. (a) Laane, J.; Kiefer, W. "Determination of Frequency Shifts by Raman Difference Spectroscopy" *J. Chem. Phys.* **1980**, *72*, 5305. (b) Rousseau, D. L. "Raman Difference Spectroscopy as a Probe of Biological Molecules" *J. Raman Spectrosc.* **1981**, *10*, 94.
66. Yachandra, V. K.; Hare, J.; Moura, I.; Spiro, T. G. "Resonance Raman Spectra of Rubredoxin, Desulforedoxin, and the Synthetic Analogue $[Fe(S_2\text{-}o\text{-xyl})_2]^-$: Conformational Effects" *J. Am. Chem. Soc.* **1983**, *105*, 6455.
67. (a) Johnson, M. K.; Czernuszewicz, R. S.; Spiro, T. G.; Fee, J. A.; Sweeney, W. V. "Resonance Raman Spectroscopic Evidence for a Common [3Fe–4S] Structure among Proteins Containing Three-Iron Clusters" *J. Am. Chem. Soc.* **1983**, *105*, 6671. (b) Johnson, M. K.; Czeruszewicz, R. S.; Spiro, T. G.; Ramsay, R. R.; Singer, T. P. "Resonance Raman Studies of Beef Heart Aconitase and a Bacterial Hydrogenase" *J. Biol. Chem.* **1983**, *258*, 12771.
68. (a) Watenpaugh, K. D.; Sieker, L. C.; Jensen, L. H. "The Structure of Rubredoxin at 1.2 Å Resolution" *J. Mol. Biol.* **1979**, *131*, 509. (b) Lane, R. W.; Ibers, J. A.; Frankel, R. B.; Holm, R. H.; Papaefthymiou, G. C. "Synthetic Analogues of the Active Sites of Iron–Sulfur Proteins. 14. Synthesis, Properties and Structures of Bis(*o*-xylyl-α,α'-dithiolato)-ferrate(II,III) Anions, Analogues of Oxidized and Reduced Rubredoxin Sites" *J. Am. Chem. Soc.* **1978**, *99*, 84.
69. Shulman, R. G.; Eisenberger, P. I.; Teo, B. K.; Kincaid, B. M.; and Brown, G. S. "Fluorescence X-ray Absorption Studies of Rubredoxin and its Metal Compounds" *J. Mol. Biol.* **1978**, *124*, 305.
70. Han, S.; Czernuszewicz, R. S.; Spiro, T. G. "Vibrational Spectra and Normal Mode Analysis for [2Fe–2S] Protein Analogues Using ^{34}S, ^{54}Fe and ^{2}H Substitution: Coupling of Fe–S Stretching and S–C–C Bending Modes" *J. Am. Chem. Soc.* **1989**, *111*, 3496.
71. Mayerle, J. J.; Denmark, S. E.; DePamphilis, B. V.; Ibers, J. A.; Holm, R. H. "Synthetic Analogues of the Active Sites of Iron–Sulfur Proteins. XI. Synthesis and Properties of Complexes Containing the Fe_2S_2 Core and the Structures of Bis[*o*-xylyl-α,α'-dithiolato-μ-sulfido-ferrate(III)] and Bis[*p*-tolylthiolato-μ-sulfido-ferrate(III)] Dianions" *J. Am. Chem. Soc.* **1975**, *97*, 1032.
72. Noodleman, L.; Case, D. A.; Aizman, A. "Broken Symmetry Analysis of Spin Coupling in Iron–Sulfur Clusters" *J. Am. Chem. Soc.* **1988**, *110*, 1001 and references cited therein.
73. (a) Han, S.; Czernuszewicz, R.; T. Kimura, T.; Adams, M. W. W.; and Spiro, T. G. "Fe_2S_2 Protein Resonance Raman Spectra Revisited: Structural Variations Among Adrenodoxin, Ferredoxin, and Red Paramagnetic Protein" *J. Am. Chem. Soc.* **1989**, *111*, 3505. (b) Fu, W.; Drozdzewski, P. M.; Davies, M. D.; Sligar, S. G.; Johnson, M. K. "Resonance Raman and Magnetic Circular Dichroism Studies of Reduced 2Fe–2S Proteins" *J. Biol. Chem.* **1992**, *267*, 15502.
74. Tsukihara, T.; Fukuyama, T.; Nakamura, M.; Katsube, Y.; N. Tanaka, N.; Kakudo, M.; Wada, K.; Matsubara, H. "X-ray Analysis of a [2Fe–2S] Ferredoxin from *Spirulina platensis*. Main Chain Fold and Location of Side Chains at 2.5 Å Resolution" *J. Biochem. (Tokyo)* **1981**, *90*, 1763.
75. Czernuszewicz, R. S.; Macor, K. A.; Johnson, M. K.; A. Gewirth, A.; Spiro, T. G. "Vibrational Mode Structure and Symmetry in Proteins and Analogs Containing Fe_4S_4 Clusters: Resonance Raman Evidence for Different Degree of Distortion in HiPIP and Ferredoxin" *J. Am. Chem. Soc.* **1987**, *109*, 7178.
76. Averill, B. A.; T. Herskovitz, T.; Holm, T. H.; Ibers, J. A. "Synthetic Analogs of the

Active Sites of Iron–Sulfur Proteins. II. Synthesis and Structure of the Tetra(mercapto-μ3-sulfidoiron) Clusters, $[Fe_4S_4(SR)_4]^{2-}$ *J. Am. Chem. Soc.* **1973**, *95*, 3523.

77. Backes, G.; Yoshiki, M.; T. Loehr, T. M.; Meyer, T. E.; M. A. Cusanovich, M. A.; Sweeney, W. V.; Adman, E. T.; Sanders-Loehr, J. "The Environment of Fe_4S_4 Clusters in Ferredoxins and High-Potential Iron Proteins. New Information from X-ray Crystallography and Resonance Raman Spectroscopy" *J. Am. Chem. Soc.* **1991**, *113*, 2055.
78. (a) Adman, E. T.; Sieker, L. C.; and Jensen, L. H. "The Structure of a Bacterial Ferredoxin. Structure of *Peptococcus aerogenes* Ferredoxin" *J. Biol. Chem.* **1976**, *248*, 3987; *251*, 3801. (b) Carter, C. W., Jr.; Kraut, J.; Freer, S. T.; Xuong, N. H.; Alden, R. A.; and Bartsch, R. G. "Two-Angstrem Crystal Structure of Oxidized *Chromatium* High Potential Iron Protein" *J. Biol. Chem.* **1974**, *249*, 4212. (c) Carter, C. W., Jr.; Kraut, J.; Freer, S. T.; Alden, R. A. "Comparison of Oxidation-Reduction site Geometries in Oxidized and Reduced *Chromatium* High Potential Iron Protein and Oxidized *Peptococcus aerogenes* Ferredoxin" *J. Biol. Chem.* **1974**, *249*, 6339. (d) Freer, S. T.; Alden, R. A.; Carter, C. W., Jr.; Kraut, J. "Crystallographic Structure Refinement of *Chromatium* High Potential Iron Protein at 2 Å Resolution" *J. Biol. Chem.* **1975**, *250*, 46. (e) Fukuyama, K.; Nagahara, Y.; Tsukihama, T.; Katsube, Y. "Tertiary Structure of *Bacillus thermoprotedyticus* [4Fe–4S] Ferredoxin. Evolutionary Implications for Bacterial Ferredoxins" *J. Mol. Biol.* **1988**, *199*, 183. (f) Fukuyama, K.; Matsubara, H.; Tsukihama, T.; Katsube, Y. "Structure of [4Fe–4S] Ferredoxin from *Bacillus thermoprotedyticus* Refined at 2.3 Å Resolution—Structural Comparison of Bacterial Ferredoxins" *J. Mol. Biol.* **1989**, *210*, 383.
79. (a) Ghosh, D.; O'Donnell, S.; Furey, W., Jr.; Robbins, A.; Stout, C. D. "Iron–Sulfur Clusters and Protein Structure of *Azotobacter* Ferredoxin at 2.0 Å Resolution" *J. Mol. Biol.* **1982**, *158*, 73. (b) Stout, G. H.; Turley, S.; Sieker, L. C.; Jensen, L. H. "Structure of Ferredoxin I from *Azotobacter vinelandii*" *Proc. Natl. Acad. Sci. USA* **1988**, *85*, 1020. (c) Stout, C. D. "7-Iron Ferredoxin Revisited" *J. Biol. Chem.* **1988**, *263*, 9256. (d) Stout, C. D. "Refinement of the 7-Fe Ferredoxin from *Azotobacter Vinelandii* at 1.9 Å Resolution" *J. Mol. Biol.* **1989**, *205*, 545. (e) Kissinger, C. R.; Adman, E. T.; Sieker, L. C.; Jensen, L. H. "Structure of the 3Fe–4S Cluster in *Desulfovibrio gigas* Ferredoxin II" *J. Am. Chem. Soc.* **1088**, *110*, 8721.
80. Minzebach, P.; Blackman, P. "The Crystal Structure of Hexamethyltristannathiane" *J. Organomet. Chem.* **1975**, *91*, 291.
81. Beinert, H.; Emptage, M. H.; Dryer, Y.-L.; Scott, R. A.; Hahn, J. E.; Hodgson, K. O.; Thompson, A. J. "Iron–Sulfur Stoichiochemistry and Structure of Iron–Sulfur Clusters in Three-Iron Proteins: Evidence for [3Fe–4S] Clusters" *Proc. Natl. Acad. of Sci. USA* **1983**, *80*, 393.

General References

Ferraro, J. R.; Nakamoto, K. "Introductory Raman Spectroscopy" Academic: New York, 1994. A well-balanced Raman text on introductry level that explains basic theory, instrumentation and experimental techniques (including special techniques), and a wide variety of applications in structural chemistry, biochemistry, biology and medicine, solid-state chemistry and industry. Will serve well as a quide for beginners.

Harris, D. C., Bertolucci, M. D. "Symmetry and Spectroscopy: An Introduction to Vibrational and Electronic Spectroscopy", Oxford University Press: New York, 1979. Easy-to-follow, instructive textbook on vibrational and electronic spectroscopy with numerous helpful figures, line drawings, and exercises illustrating important concepts. Excellent choice for anyone teaching or studying molecular spectroscopy at junior to beginning graduate level. Also available in paperback from Dover, New York.

Nakamoto, K. "Infrared and Raman Spectra of Inorganic and Coordination Compounds", Part A and B, 5th ed., Wiley-Interscience: New York, 1997. Unsurpassed resource textbook available for researchers and graduate students in the field of vibrational spectroscopy, inorganic chemistry, organometallic chemistry, and bioinorganic chemistry—now fully revised and updated.

Long, D. A. "Raman Spectroscopy", McGraw-Hill: New York, 1977. Clearest, unified and fully illustrated treatment of the basic theory and physical principles of Raman, resonance Raman and nonlinear Raman scattering.

Decius, J. C.; Cross, P. C. Wilson, E. B. Jr., "Molecular Vibrations: The Theory of Infrared and Raman Vibrational Spectra", McGraw-Hill: New York, 1955. Classic graduate text; contains rigorous yet easily followed exposition of the mathematics involved in detailed vibrational analyses of polyatomic molecules. Also available in paperback from Dover, New York.

Spiro, T. G. Ed., "Biological Applications of Raman Spectroscopy", Wiley-Interscience: New York, 1988 Vols. 1–3. Through summaries of representative areas by authorities in biological Raman applications, this set of three volumes provides in-depth coverage of the theoretical and experimental aspects of modern bio-Raman spectroscopy.

3

Electron Paramagnetic Resonance of Metalloproteins

GRAHAM PALMER*

Department of Biochemistry and Cell Biology
Rice University
6100 Main
Houston, TX 77005-1892

CONTENTS

* With a contribution by Michael P. Hendrich, Department of Chemistry, Carnegie Mellon University, Pittsburgh, PA.

I. Introduction

Electron paramagnetic resonance spectroscopy (EPR)[a] is a technique that probes certain properties and the environment of a paramagnetic center by characterizing the interaction of that center with an applied magnetic field. In principle, the method is applicable to any species containing one-or-more unpaired electrons. In chemistry and biochemistry, this includes inorganic and organic free radicals, triplet states (e.g., dioxygen, O_2), and systems that contain transition metal ions. Although the spectroscopy of radical and triplet species are of considerable importance in the chemical and biological sciences, the focus of this chapter is directed at molecules containing metal ions; Carrington[1] has written a review of the EPR of aromatic radicals, inorganic radicals are reviewed in the book by Atkins and Symons,[2] radicals in enzyme systems by Edmondson[3], and paramagnetic species in photosynthesis by Miller and Brudwig.[4]

Table I summarizes those metals that are of particular interest to bioinorganic chemists and rcvcals some of their relevant properties. The purpose of this chapter is to provide the reader with the skills necessary to conduct a basic interpretation of the features present in an EPR spectrum, and from that information draw useful conclusions about the environment and electronic geometry of the metal system under study, whether it be incorporated into a protein, nucleic acid, or some small molecule. Because the material draws heavily on information from the physical sciences certain background concepts may be unfamiliar to the reader. To avoid interrupting the text, some of these background concepts have been reviewed in the appendices to this chapter.

There are three basic equations. These are

$$\mu = -g\beta\mathbf{S} \tag{1}$$

$$E = -\mu \cdot \mathbf{B} \tag{2a}$$

$$E = h\nu \tag{3}$$

The first of these equations relates μ, the magnitude of the magnetic moment of the electron[b] to the electron spin via the quantities g and β. The quantity β is the Bohr magneton. It is defined as

$$\beta = \frac{e}{2m}\frac{h}{2\pi} \tag{4}$$

The first ratio in this definition arises from a derivation of the magnetic moment expected from a *classical* particle of atomic charge e and mass m circulating

[a] Also known as electron spin resonance (ESR).

[b] A moving charged particle generates a magnetic field. In EPR, the motion of interest is the "spin" (S) of the electron and, as will become apparent later, the circulation of the electron about its nucleus (the orbital motion). This magnetic field is associated with a magnetic moment. The symbol for the magnetic moment (μ) is often written with a subscript (such as μ_e) to avoid confusion with the magnetic moment arising from the nucleus. In this chapter, the confusion will be minimized by always writing μ_n for the nuclear magnetic moment.

Table I
EPR Properties of Metal Ions Found in Biochemical and Related Bioinorganic Systems

Metal Ion	Electronic Configuration	Spin State	Example	*g*-Values[a]	Observation Temperature (K)
V(IV)	$3d^1$	$\frac{1}{2}$	Vanadyl-substituted proteins	$\sim 2(O_h)$	
Mn(II)	$3d^5$	$\frac{5}{2}$	Concanavalin A	2–6	<300
Fe(III) (heme)	$3d^5$	$\frac{1}{2}$	Cytochrome *c*	3.8–0.25	<100
Fe(III) (heme)	$3d^5$	$\frac{5}{2}$	Cytochrome $P450_{cam}$ + camphor	8–1.8	<100
Fe(III) (non-heme)	$3d^5$	$\frac{5}{2}$	Lipoxygenase	10–0.25	<100
Fe(III), Fe(II) (pair)	$3d^5$, $3d^6$	$\frac{1}{2}$	Spinach ferredoxin purple acid phosphatase	1.25–2.1	<50
3Fe(III) (triad)	$(3d^5)_3$	$\frac{1}{2}$	Aconitase	2–2.1	<50
3Fe(III), Fe(II) (tetrad)	$(3d^5)_3 3d^6$	$\frac{1}{2}$	HiPIP[b]	2–2.2	<50
Fe(III), 3Fe(II) (tetrad)	$3d^5(3d^6)_3$	$\frac{1}{2}$	Bacterial ferredoxin	1.7–2.1	<50
Fe(II), Fe(II) (pair)	$3d^6$, $3d^6$	2, 2 weakly coupled	Methane mono-oxygenase	16	<50
Co(II)	$3d^7$	$\frac{1}{2}$	Cobalamin (B_{12r})	2.0–2.3	<120
Co(II)	$3d^7$	$\frac{3}{2}$	Cobalt-substituted p-roteins	1.8–6	<40
Ni(III)	$3d^7$	$\frac{1}{2}$	Hydrogenase	2–2.2	<50
Cu(II)	$3d^9$	$\frac{1}{2}$	Plastocyanin	2–2.4	<300
Mo(V)	$4d^1$	$\frac{1}{2}$	Xanthine oxidase	1.95–2.0	<300

[a] The ranges shown encompass both *g* anisotropy in a given system and the variations among systems.
[b] High-potential non protein = HiPIP.

in a Bohr orbit at the velocity of light. The second ratio, Planck's constant h divided by 2π, is the unit of quantum mechanical angular momentum and is the only correction needed for an orbiting, charged, *quantum mechanical* particle. The parameter β has the magnitude 9.285×10^{-21} erg G^{-1} in cgs units and 9.285×10^{-24} C T^{-1} in MKS units ($1\ T = 10^4$ G).

The second quantity, g, is called the g factor or spectroscopic splitting factor. Subsequent to the classical development of the relationship between an electron's magnetic moment and its orbital motion about the nucleus, it was discovered that the electron also had an intrinsic magnetic moment; this is usually visualized as arising from the electron spinning about its own axis. The spin magnetism was found to be twice as intense as that anticipated from the initial theory and this fact was accommodated by introducing g into the expression for μ. For our purposes, g can be taken as having the value of 2.0; a more precise value for the free electron is 2.0023. As we shall see, electrons in metal systems typically exhibit g-values quite different from 2.0.

The last quantity in Eq. (1) is **S**, the symbol for the total spin associated with the electron (i.e. spin quantum number); the spin angular momentum is $\mathbf{S}\,(h/2\pi)$. In elementary discussions **S** is usually visualized as a vector; in more sophisticated treatments **S** is taken as a quantum mechanical operator. The same is true for **I**, the nuclear spin. Note that both operators and vectors are written with a bold face typeface.

The expression for μ contains the minus sign; β is inherently positive (because e is positive) and the minus sign is needed to accommodate the negative charge of the electron.[c]

Equation (2a) describes the change in energy experienced by an electron exposed to an external magnetic field, **B**.[d] Because **B** has both magnitude (its intensity) and direction it is a vector quantity. The use of the dot product implies that the magnetic moment is also a vector quantity, which follows from its direct proportionality to **S**. Equation (2a) is the classical expression for the change in energy. To proceed to the quantum mechanical equivalent we replace μ by its equivalent operator and E by $\mathscr{H}$, the Hamiltonian. Equation (2) is simplified as

$$\mathscr{H} = g\beta\mathbf{S}\cdot\mathbf{B} \tag{2b}$$

$$E = g\beta m_S B \tag{2c}$$

In Eq. (2c), the dot product has been explicitly replaced by m_S, the projection of **S** onto **B**,[e] multiplied by the magnitude of **B**.

For the $S=\frac{1}{2}$ electron m_S is allowed to have two values, namely, $\pm\frac{1}{2}$ (corresponding to normal and inverted cones, see Kevan and Bowman[6]). Thus

$$E_{1/2} = \tfrac{1}{2}g\beta B$$

$$E_{-1/2} = -\tfrac{1}{2}g\beta B$$

and

$$\Delta E = E_{1/2} - E_{-1/2} = g\beta B \tag{5}$$

---

[c] There is an alternative way of writing Eq. (1), namely, $\mu = \gamma hS/2\pi$ By comparison with Eq. (1), we see that γ must have the definition $-g_e/2mc$. The parameter γ is called the magnetogyric ratio; it is a signed quantity (being positive for the proton). This representation for the magnetic moment is usually used when one is dealing with the precession picture for magnetic resonance.[5]

[d] The literature uses the symbols H or B to represent magnetic field; B is used in this book.

[e] It is common practice to represent m_s as the height of the cone described by the precession of **S** about **B**.

Consequently, upon exposure to a magnetic field, B_L,[f] the electron can be either stabilized ($m_S = -\frac{1}{2}$) or destabilized ($m_S = +\frac{1}{2}$), and the magnitude of this effect varies linearly with the intensity of B_L; the interaction of these magnetic moments with B_L is called the *Zeeman effect* or *Zeeman interaction*. For magnetic fields usually available in the laboratory (1–10 kG, 0.1–1.0 T) the change in energy, ΔE, is 5.94×10^{-17} erg molecule^{-1} ($\sim$1 cal mol^{-1})[g]. Hence, at room temperature, where thermal energy, $k_B T = (1.381 \times 10^{-16}\ \text{erg deg}^{-1})(300\ {}^\circ\text{K}) = 4.14 \times 10^{-14}$ erg, the ratio of the populations of the spin-up to spin-down orientations is obtained from the Boltzmann formula:

$$\frac{N^+}{N^-} = \exp(-\Delta E/k_B T) = \exp\left(-\frac{1}{700}\right) \approx 1 - \frac{1}{700} \approx 0.999 \tag{6}$$

that is, there is about a 0.1% net excess of spins in the more stable, spin-down, orientation.

This net difference in population is a necessary requirement for EPR absorption, which requires that the number of transitions from the more-stable to the less-stable states (by flipping the spin) exceed those in the opposite direction. The condition for "flipping the spin" is governed by Eq. (3), namely, we must provide a quantum of energy, $h\nu$, equal to the separation in energy between the $\pm\frac{1}{2}$ states as revealed by the formulas:

$$h\nu = \Delta E = g\beta B_0 = g\beta B_R \tag{7}$$

The condition defined by Eq. (7) is called the *resonance condition*. The parameter B_R or alternatively B_0, specifies the intensity of the magnetic field, which satisfies Eq. (7) and is required for resonance.

Clearly, ΔE, the separation in energy between the two spin states is variable and depends on the intensity of B_L; this is illustrated in Figure 1 (*a*). It follows that two alternative instrumental configurations are conceivable. In the first configuration, B_L is set to a fixed value (e.g., 300 mT) and the energy source is varied over the appropriate frequency range. For $g = 2$, this might mean varying the frequency from 9 to 10 GHz[h] in search of the resonant frequency. In the second configuration, the frequency is set to a fixed value (e.g., 9.2 GHz) and the intensity of the magnetic field varied over the relevant range. For $g = 2$, this might mean varying B_L from 300 to 350 mT. The former configuration is to be preferred because the abscissa of the spectrum is frequency, and hence linearly proportional to energy. Unfortunately, technical considerations make this preferred configuration very difficult to implement, especially when large variations in g are

[f] Throughout this chapter, we will be referring to a variety of magnetic fields. The most obvious is provided by the electromagnet that is an integral component of an EPR spectrometer. This magnetic field will be alluded to using the notation $\mathbf{B_L}$ (the laboratory field). When no subscript is employed the origin of the magnetic field is not important to the discussion.

[g] More precisely, at 3300 G, the magnetic field corresponding to $g = 2$ in most X-band EPR spectrometers, the energy difference between the spin-up and spin-down states is 0.87 cal mol^{-1}.

[h] Pronounced gigahertz: The prefix giga means 10^9.

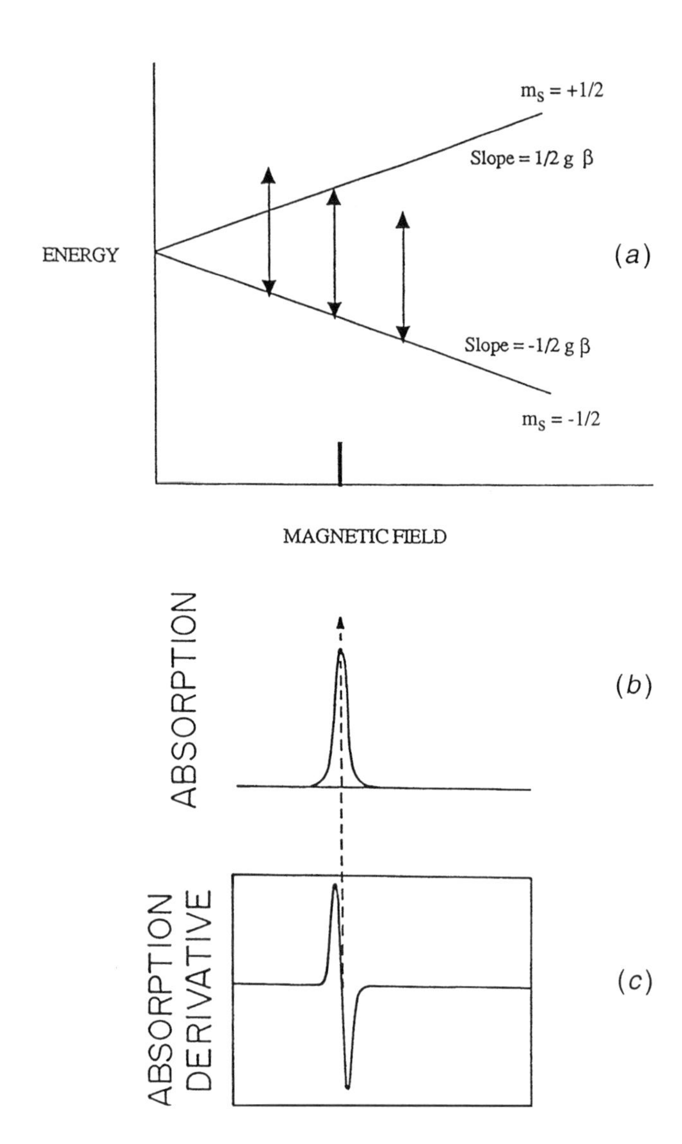

Figure 1
Simple energy level diagram with absorption and first-derivative modes of presentation. (*a*) This graph shows the dependence on magnetic field of the energies of the $m_s = \pm\frac{1}{2}$ states of an electron. The vertical double-headed arrows signify the energy associated with the microwave quantum and imply that there is a unique value of the magnetic field at which absorption of the microwave energy can occur. This absorption results in the "spike" located on the abscissa, which because of line-broadening phenomena results in a typical absorption curve as shown in (*b*). Because of the details of the electronic circuitry, this absorption curve is routinely displayed as its first derivative (*c*).

to be studied. Consequently, the standard instrumentation utilizes a fixed frequency and a variable magnetic field; the abscissa is then proportional to reciprocal energy [see Eq. (8)]. This instrument is called a continuous wave (CW) spectrometer.

The frequencies that are needed are in the microwave region of the spectrum, commonly 2–36 GHz, and are provided by stable oscillators such as klystrons or Gunn diodes. Most spectrometers operate at 9 GHz (called X band), a frequency that provides a compromise between sensitivity and ease of sample handling.

Measurements are sometimes made at other frequencies. Because EPR spectral line widths are often field dependent, spectra taken at lower frequencies (e.g., 3 GHz) can enhance the resolution of otherwise unresolved hyperfine structure. Higher frequency data (e.g., 35 GHz) are of value in increasing the resolution associated with g anisotropy and for distinguishing spectral features due to g anisotropy from those due to hyperfine and other interactions; the former is field dependent while the latter are not. These aspects of EPR spectra will be discussed in detail in subsequent sections of this chapter.

In addition to CW techniques, a variety of pulsed EPR techniques are available; a recent compendium of these methods can be found in the volume edited by Kevan and Bowman.[6] Electron–nuclear double resonance (ENDOR) and electron spin echo envelope modulation (ESEEM) are covered in Chapter 4 of this book by Chasteen and Snetsinger.

When $h\nu = g\beta B$ the energy present in the microwave quantum matches the energy separation between the two spin orientations and absorption of the microwave energy occurs as required by the *selection rule*, which states that during an EPR transition ΔM_S can only change by ± 1 unit. The resultant EPR spectrum might therefore have the conventional absorption lineshape depicted in Figure 1 (*b*). In practice the details of the electronic circuitry used to acquire the EPR signal leads to a spectrum that is the first derivative of the conventional curve, as shown in Figure 1 (*c*). The peak of the absorption curve, and equally the zero crossing of the first derivative, is found at the magnetic field for maximum absorption, B_R.

From Eq. (7), $B_R = h\nu/g\beta$. More usefully

$$g = h\nu/\beta B_R \tag{8}$$

Thus, from a knowledge of the operating frequency of one's spectrometer and the intensity of the magnetic field at which maximum EPR absorption occurs, it is trivial to calculate g.[i] When necessary, g can be calculated with very high accuracy, the usual limitation being the intrinsic width of the EPR line, and hence the difficulty in locating the relevant spectral feature. The microwave frequency is readily measured with an appropriate counter while a measurement of the magnetic field is more tedious but not normally a limitation.

As the temperature is lowered, the argument to the Boltzmann expression [Eq. (6)] increases and so does the difference in populations. With more spins in

[i] A convenient simplification is that $g = 714.48[(\upsilon(\mathrm{GHz}))/(B(G))]$.

the stable orientation the probability of absorbing microwaves is increased. Thus, all other things being equal, the sensitivity of an EPR measurement increases with decreasing temperature. As long as the argument to the exponent of Eq. (6) is sufficiently small that the exponential term can be adequately approximated as $(1 - x)$, the intensity of the EPR signal is linearly dependent on $1/T$. This linear dependence is called the Curie law. This dependence and the additional circumstance that the line widths of metal EPR spectra frequently increase rapidly with increasing temperatures are the reasons that metal EPR spectra are most commonly recorded at low temperatures, 4–77 K. Suitable temperatures for various metal centers are noted in Table I.

An important attribute of EPR is that it is relatively straightforward to determine the number of paramagnetic centers present in a sample. One simply compares the EPR intensity of the sample with that of a suitable standard. The procedure involves the double integration of the EPR spectrum of both standard and unknown, correction of each double integral for certain instrumental parameters (only when they are different in the two measurements) and use of the relationship

$$\text{Concn of spins in unknown} = \text{concn of spins in standard} \times \frac{\text{CDBLI(unknown)}}{\text{CDBLI(standard)}}$$

where CDBLI stands for corrected double integral. A detailed account of the procedure is given by Palmer.[7] The method is very straightforward for $S = \frac{1}{2}$ systems and is particularly true in the frozen state when possible differences in the propensity of the solvent to scavenge the microwaves is minimized. With care, accuracies of about 5% can be obtained, especially if the EPR parameters of the standard are not too different from those of the unknown. Otherwise it is necessary to make an additional correction to the CDBLI, which takes into account that the transition probabilities depend on the g values and their anisotropies.[8] When $S > \frac{1}{2}$, then the method is much more difficult because (a) the spectra tend to be highly anisotropic and (b) the observed EPR is exhibited by only one of the several spin states, which are present, as in Section II.D. Then, even with careful integrations, control of temperature, and matching of the properties of unknown and standard accuracies better than 50%, requires the attention of an experienced spectroscopist and data that is favorable. Even so, such measurements are worthwhile because they will show whether the paramagnet being studied is a trivial or substantial component of the system being studied. The latter tends to be more interesting than the former.

II. The *g*-Factor is Anisotropic

Because the measurable experimental quantity appears to be the simple number, 2.0, at first sight it might appear that EPR is a very limited technique. If this were true, then this chapter would not have been written! In fact, we will see that there are several important structural interactions that dramatically affect the appear-

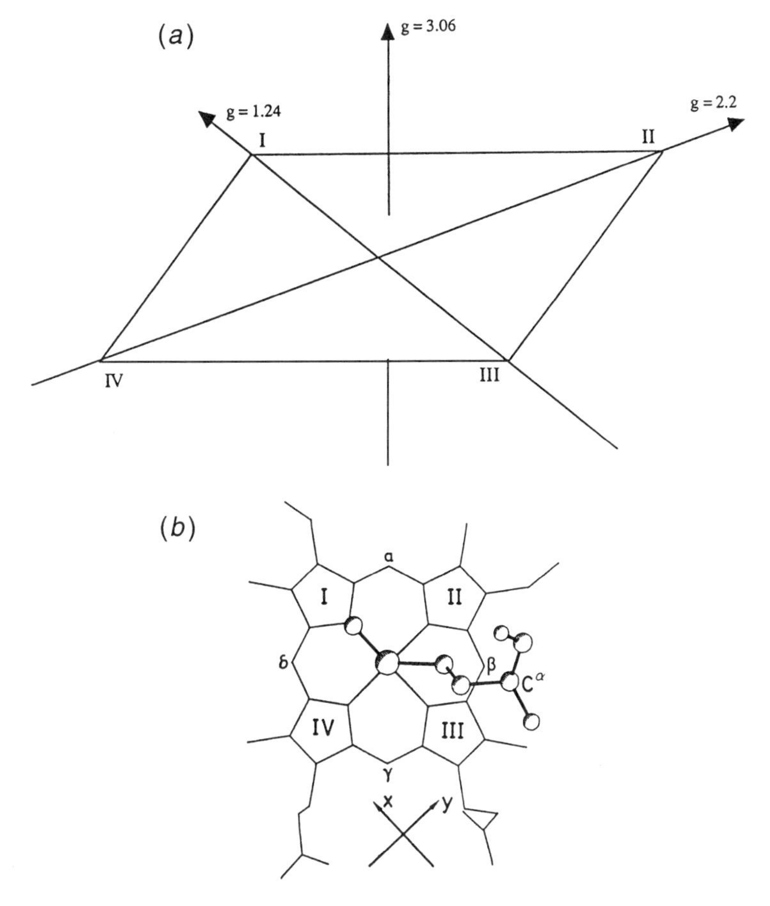

Figure 2
The principal g values of cytochrome c. Panel (b) shows the heme prosthetic group of cytochrome c together with the numbering for each of the four pyrrole groups (pyrroles I and II bear the two thioether functions); the orientation of the axial methionine is also shown. Panel (a) indicates the g values observed when EPR measurements are made with the magnetic field aligned along each of the three principal structural axes.

ance of the EPR spectrum. The first of these, the spin–orbit interaction, is introduced with a simple experiment.

Ferricytochrome c is a low-spin hemoprotein and thus contains a single unpaired electron (present in the $3d_{yz}$ orbital of the heme iron). This protein can be obtained in crystalline form and these crystals are large enough so that they can be studied by EPR.

When a single crystal of cytochrome c is mounted between the poles of the magnet of an EPR spectrometer and this crystal is oriented so that the heme normal is parallel to the magnetic field (Fig. 2) then, for a spectrometer frequency of 9.2 GHz, the value of the field required for maximum EPR absorption is found at 214.8 mT. By using Eq. (8), the g value is calculated to be 3.06. If this crystal is now rotated so that B_L is parallel to the line connecting pyrroles I and III, the EPR measurement reveals that maximum absorption occurs at 530 mT; the g value is 1.24. Finally, when the crystal is oriented so that B_L is parallel to the line connecting pyrroles II and IV, the EPR measurement shows that maximum absorption is at 298.8 mT; the g value is 2.2. Clearly, the g value for cytochrome c is orientation dependent. The implication, therefore, is that the magnetic moment associated with the single unpaired electron of the heme iron is not a simple number; it exhibits the property of being *anisotropic*.

When an anisotropic body (the paramagnet) is subjected to a stimulus (B_L) it may take as many as nine numbers to fully characterize the response (R), the deviation of g from 2. These numbers, R_{ij}, form a 3×3 square array called a tensor and reflect the fact that the stimulus and the response are both vector quantities. For example, R_{xy} and R_{xz} represent the y and z components of the response due to the x component of the stimulus, and so on; in an arbitrary Cartesian coordinate system there are nine such components. For systems of interest to us, these tensors are usually symmetric about the diagonal; then it is possible to find a coordinate system such that only the diagonal components of the tensor, R_{xx}, R_{yy}, and R_{zz}, are nonzero. This coordinate system is called a principal axis system and these diagonal values are called the principal values of the response. The principal axis system is related to the original arbitrary system by rotations in Cartesian space.

Frequently, the principal axes are the same as obvious structural axes of the paramagnet; for example, for cytochrome c, the principal axes of the g tensor are (almost) the same as those axes used in the single-crystal experiment just described.

The question now arises: If a single crystal of cytochrome c exhibits an orientation dependent EPR spectrum, what will be the spectrum of a solution of cytochrome c, in which all molecular orientations are present[j]? The answer is relatively straightforward and can be conveniently explained by outlining the steps in a computer program that would calculate the EPR spectrum of such a solution.

A molecule of cytochrome c oriented at some particular direction with respect to B_L will exhibit a g value, $g_{\theta\phi}$, given by the equation

$$g\theta_\phi = ({g^2}_{xx} \cos^2\phi \sin^2\theta + {g^2}_{yy} \sin^2\phi \sin^2\theta + {g^2}_{zz} \cos^2\theta)^{1/2} \tag{9}$$

This equation says that the g value observed in some particular direction (specified by the angles θ and ϕ) is the weighted mean of the three principal values, g_x, g_y, and g_z. The weighting factors are the projections of the direction of the magnetic field onto the principal axes of the g tensor[k] and are given in terms of θ and

[j] A typical EPR sample has a volume of 0.3 mL and a concentration of about 0.1 mM. This sample contains 10^{16} molecules.

[k] A useful aid here is to consider gB as representing the internal magnetic field experienced by the paramagnet: $B_{\text{int}} = ({g_{xx}}^2 B^2 + {g_{yy}}^2 B^2 + {g_{zz}}^2 B^2)^{1/2}$. When B_L is directed along a principal axis, for example, x, $B_{\text{int}} (= g_{xx} B_L)$ is collinear with B_L but possibly of changed magnitude. When B_L is not directed along a principal axis then B_{int} may be changed both in direction and magnitude relative to B_L.

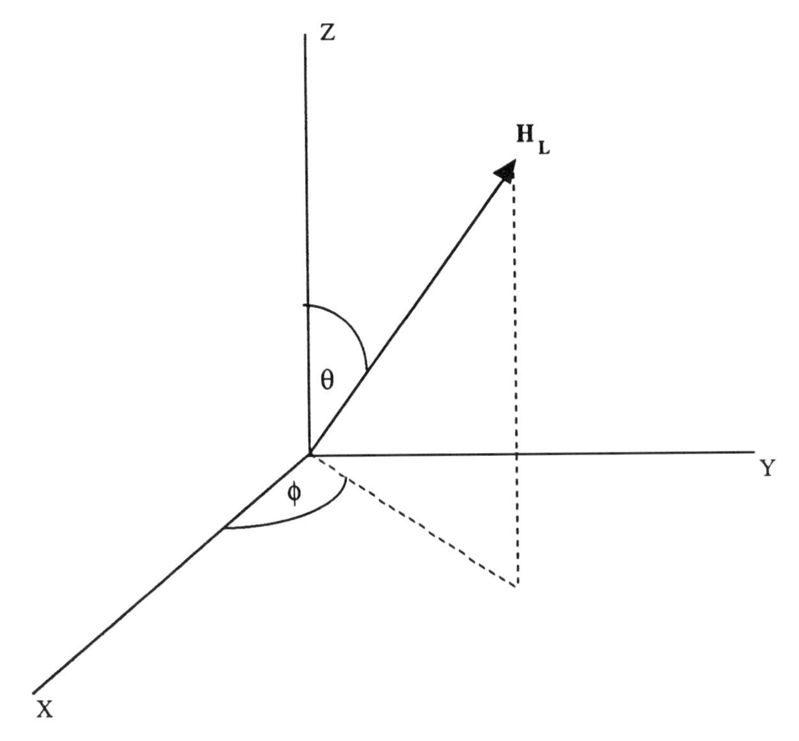

Figure 3
The definition of the axial and polar angles conventionally employed in describing the angular dependence of magnetic resonance parameters.

ϕ, the polar and azimuthal angles between the principal axis of the g-tensor and the direction of B_L (Fig. 3). The fraction of molecules at this orientation is proportional to sin θ.

By using Eq. (9), it is a relatively straightforward matter to calculate the EPR spectrum. The calculation consists of two loops:

1. An outer loop that increments θ from 0° to 90° in steps of $\delta\theta$.[l]
2. An inner loop that increments ϕ from 0° to 90° in steps of $\delta\phi$.
3. Each time through the inner-loop Eq. (9) is evaluated to obtain $g_{\theta\phi}$ and the corresponding magnetic field for resonance, B_R, is calculated via Eq. (7) using the appropriate value for the microwave frequency.
4. An absorption curve (in first derivative form) of suitable width is now centered at B_R with an amplitude proportional to $\sin\theta\, d\theta\, d\phi$.[m] This latter quantity is proportional to the number of molecules subtended by the solid angle specified by the current values of θ, ϕ, $d\theta$, and $d\phi$.
5. Each time through the inner loop the calculated absorption curves are accumulated.

[l] For certain low-symmetry geometries, the integration must be done over 180°.
[m] The lineshape is usually either Gaussian or Lorentzian in form, depending on the nature of the problem. In those cases, where unresolved structure is anticipated, a Gaussian formula is used.

6. When all passes through the two loops are complete the accumulated absorption curve yields the desired EPR spectrum.[n]

The EPR spectra calculated according to this recipe exhibit four limiting cases:

a. when $g_x = g_y = g_z$. Then the magnetic moment is independent of orientation; it is said to be *isotropic* and a single symmetric EPR absorption is obtained. The paramagnet can be represented by a sphere [Fig. 4(*a*)].
b. when $g_x = g_y < g_z$. The paramagnet can be represented by a football (prolate ellipsoid) and exhibits a minor feature at low field (from g_z) and

[n] An additional correction due to variations in transition probability with orientation has been left out of the above description; its inclusion requires a formula similar to Eq. (9).[8]

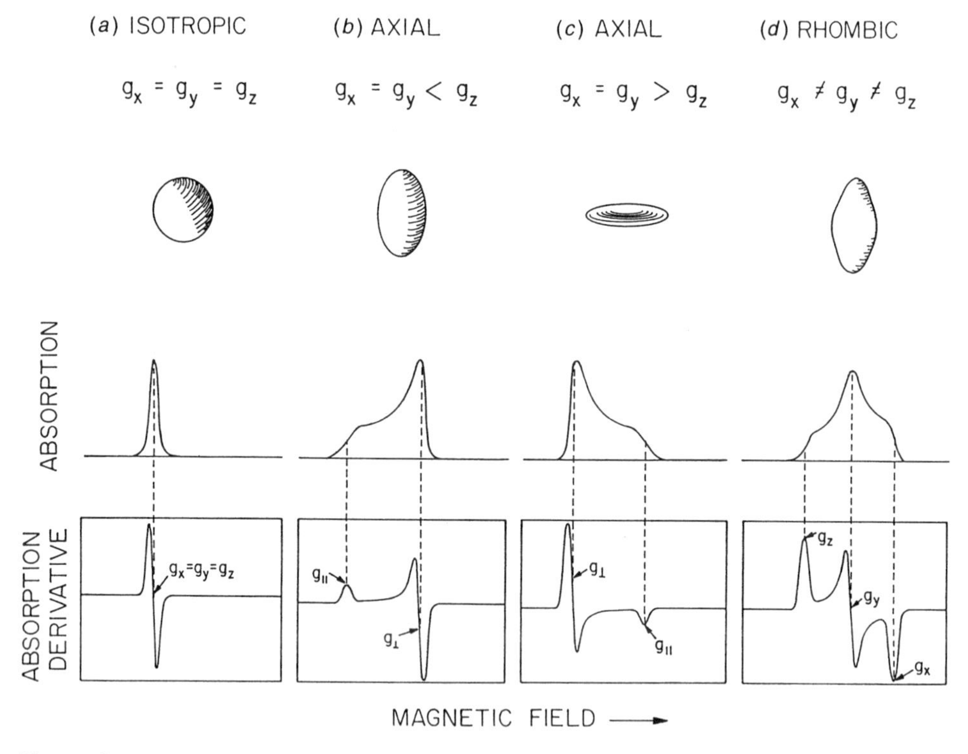

Figure 4
The basic spectral envelopes to be found with $S = \frac{1}{2}$ paramagnets. Columns (*a–d*) name each spectral type and immediately below show the associated pattern of *g* anisotropy, which is presented both numerically and graphically as solid bodies. Note that axial symmetry can be either prolate (an American football) or oblate (a discus). The rhombic body (*d*) was harder to draw but can be thought of as a partially inflated football, especially if you subject it to an uniaxial compression at its equator. Beneath each solid is the idealized lineshape, first in absorption presentation and subsequently in derivative form. The broken vertical lines serve to correlate the features shown in the absorption and derivative presentations.

a major feature at high field (from g_x and g_y) [Fig. 4(*b*)]. The spectrum is said to be *axial*. In axial spectra the two common g values are often referred to as $g_\perp$ (pronounced g perpendicular) while the unique g value is called $g_\parallel$ (pronounced g parallel).[o]

c. when $g_x = g_y > g_z$. The paramagnet can be represented by a discus (oblate ellipsoid) and exhibits a minor feature at high field (from g_z) and a major feature at low field (from g_x and g_y) [Fig. 4(*c*)]. This spectrum is also *axial*.

d. when $g_x \neq g_y \neq g_z$. Now the spectrum shows three major features reflecting the unique values of the three components of the g tensor [Fig. 4(*d*)]. The spectrum is said to be *rhombic* and the magnetic moment has the shape of a partially inflated football.

While the shapes of the isotropic and two axial spectral types are somewhat intuitive that of the rhombic spectrum is not. In particular, the characteristic peaking of the intensity at g_y might be a surprise for g_y is characteristic of those molecules oriented along y and such molecules should be no more probable than x-oriented or z-oriented molecules. However, the value of g_y is exhibited both by y-oriented molecules and by all those other molecules that happen to have values for θ and ϕ that will yield g_y when inserted into Eq. (9).[p] Finally, note that for molecules tumbling rapidly [more rapidly than $(g_z - g_x)2\pi\beta B/h$)] one observes an average g value $= (g_x + g_y + g_z)/3$. Metal centers in proteins are unable to tumble this rapidly.

It is often the case that the observed spectrum has contributions from several species. Then it is important to be able to identify the features in the spectrum associated with each of these species. One systematic way is as follows:

1. Does the data show a derivative lineshape? If not then only axial species are present and one looks for the peaks and troughs associated with the oblate and prolate axial forms.
2. If a derivative feature is present then one looks for pairs of peaks and troughs contributed by rhombic species. If these are absent then only isotropic species are present.

If the species that are present have the same general envelope, then additional data is usually needed; for example, samples in which the relative amounts of the

---

[o] The notation for g values is a little unsatisfactory because there is an implicit assumption that one knows the relationship between the x, y, and z components of the g tensor and the x, y, and z directions of the molecular axes of the paramagnet. For that reason, the notations g_1, g_2, g_3, and g_{min}, g_{mid}, and g_{max} are sometimes used. The latter seems the most satisfactory.

[p] Imagine standing at the north pole ($+z$) of an ellipsoid whose principal axes are proportional to g_x, g_y, and g_z. If you walk toward the equator along the arc defined by the yz plane the "radius" of the ellipsoid changes continuously from g_z to g_y. However, when you walk from the north pole to the equator along the arc defined by the xz plane then the radius changes continuously from g_z to g_x and somewhere on the journey you will encounter a spot where the "radius" is proportional to g_y. In fact, you can walk on any arc from the north pole to the equator and you will always encounter a location where the "radius" is proportional to g_y. These "g_y locations" comprise two ribbons that are orthogonal to one another and that both pass through $\pm y$.

various forms are different. Furthermore, one needs to examine the data very carefully to be sure that weak and/or broad features are not missed. This is particularly important when there are features at high field where the effects of the most common line-broadening mechanism (g strain) is most apparent.

A. What Is the Physical Origin of g-values Different from 2.0?

Deviations of g from 2.0 can be considered under five categories:

1. $S = \frac{1}{2}$, $\Delta g < 0.4$
2. $S = \frac{1}{2}$, $\Delta g > 0.4$
3. $S = \frac{1}{2}$, coupled spin systems
4. $S > \frac{1}{2}$, half-integer spin systems
5. $S > \frac{1}{2}$, integer-spin systems

where $\Delta g = |g - g_e|$. The distinction between the first two classes is rather artificial and reflects the way in which one thinks about the system; the fundamental interaction is the same. This interaction is the spin–orbit interaction, a phenomenon whereby the relative motion of the electron and its charged nucleus results in the electron being exposed to a local magnetic field arising from the nuclear charge (see Appendix I). The spin–orbit interaction is written as $\lambda\mathbf{L} \cdot \mathbf{S}$, where $\lambda = \zeta/2S$ (see Appendix I); it is a correction to the energy of the electron. By analogy to Eq. (2a), λL can be interpreted as the intensity of this local magnetic field. Thus, speaking anthropomorphically, one might say that the electron has a strong incentive to acquire orbital motion so that it experiences the magnetic field of nuclear origin, and hence acquires an additional lowering of its energy. The spin–orbit coupling constant, ζ, has the dimensions of reciprocal centimeters; its value depends on Z, the atomic number of the nucleus and is about 150 cm^{-1} for Ti^{3+} increasing monotonically to a value of 830 cm^{-1} for copper ($\lambda = 150$ to -830 cm^{-1} reflecting the sense of the magnetic field that is produced, see below).

B. Category I: $S = \frac{1}{2}$, Small Deviations of g from 2 [e.g., Ni(III), Ni(I), Cu(II), and Mo(V)]

The basic phenomenon can be illustrated by the example of a transition metal ion containing a single unpaired electron. This example is provided by molybdenum(V) and the splitting of the d orbitals of this metal ion in two typical geometries are shown in Figure 5. In a compressed octahedron, the single electron is located in d_{xy}. This orbital is related to the other four d orbitals by rotations about one or other of the Cartesian axes. These rotational relationships are summarized in Table II where the individual entries represent the consequences of rotating one of the d orbitals listed in the left-hand column about an axis identified in the top row. For example, rotation of d_{xy} about x yields d_{xz}, rotation of d_{xy} about y yields d_{yz} while rotation of d_{xy} about z yields $d_{x^2-y^2}$ (the significance of the associated small numeric factors will be clarified shortly). Thus the unpaired electron in d_{xy} can exploit the spin–orbit coupling in all three directions for rotation about all three axes converts d_{xy} into one of the other d orbitals.

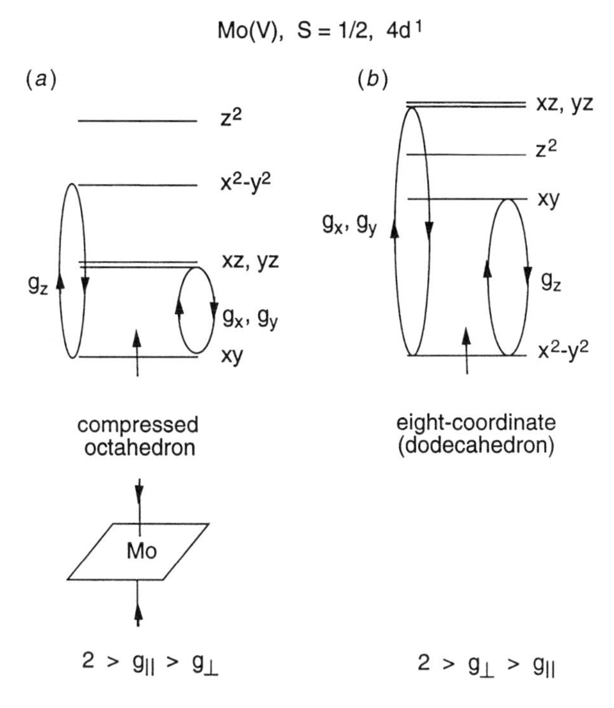

Figure 5
The relative order of the energy levels of the five 3d orbitals for two common geometries of molybdenum(V) ($4d^1$). (*a*): A compressed octahedron; (*b*): eight-coordinate molybdenum. The electron circulations associated with the orbital contributions to the three g values are depicted by the loops. The contribution of a loop to the orbital magnetism is inversely related to its amplitude but the identity of the participating orbitals is also important (Table II).

Table II
Rotational Relationships of the Real *d*-Orbitals

	For Rotations about		
	X	*Y*	*Z*
Initial d Orbital		Final d Orbital	
$d_{x^2-y^2}$	$-id_{yz}$	$-id_{xz}$	$2id_{xy}$
d_{z^2}	$-i\sqrt{3}d_{yz}$[a]	$i\sqrt{3}d_{xz}$[a]	0[b]
d_{xy}	id_{xz}	$-id_{yz}$	$-2id_{x^2-y^2}$
d_{xz}	$-id_{xy}$	$id_{x^2-y^2} - i\sqrt{3}d_{z^2}$	id_{yz}
d_{yz}	$id_{x^2-y^2} + i\sqrt{3}d_{z^2}$	id_{xy}	$-id_{xz}$

[a] Equate d_{z^2} with $d_{x^2-z^2} + d_{y^2-z^2}$.
[b] No other orbital reached.

Two additional facts are now needed: (1) there is a barrier to rotation that is provided by the energy separation between d_{xy} and the orbital into which it is being rotated. In our example, this separation is provided by $\Delta_x (= \Delta_y)$ for rotations about x and y, and by Δ_z for rotations about z. (2) If the rotation moves the electron into a vacant orbital, as is the case here, then the sense of the current resulting from the circulation leads to a magnetic field that opposes the applied field. Consequently, a larger laboratory field is required to meet the Zeeman condition and the calculated g value will be smaller than that of the free electron via Eq. (8) (i.e., <2.0). Conversely, if the electron can be rotated into an occupied orbital then the magnetic field produced by the circulation aids the applied field. Now a smaller laboratory field is required to meet the Zeeman condition and the calculated g value will be greater than that of the free electron (i.e., >2.0)

With these concepts together with Table II and the anticipated ordering of d orbitals for the given metal ion, it is possible to predict the qualitative form of the EPR spectrum for any $S = \frac{1}{2}$ transition metal ion. Thus $3d^9$ in an elongated octahedron (Fig. 6) yields an epr envelope similar to that of Fig. 4b with all g-values > 2; $3d^7$ in an compressed octahedron also has all g-values > 2 but the epr envelope will depend upon the energy gap between d_{xy} and $d_{xz,yz}$. However $3d^9$ in an compressed octahedron and $3d^7$ in an elongated octahedron yield Fig. 4c with $g_z = 2.00$ and the remaining 2 g-values > 2.

A more quantitative approach requires using a result from quantum mechanical perturbation theory, namely,

$$g_i = g_e\left(1 + \frac{n_i \lambda}{\Delta_i}\right) \qquad i = x,\, y, \text{ and } z \tag{10}$$

where $n_i = \langle \text{matrix element} \rangle^2$, and $\langle \text{matrix element} \rangle$ is the relevant entry in Table II, g_e is the g value of the free electron (2.0023), and Δ_i is the relevant energy separation.

For example, for Cu(II) (elongated d^9, see Fig. 6) the unpaired electron is present in $d_{x^2-y^2}$. A rotation about z generates d_{xy} with a matrix element $= 2i$. So $n = -4$ and

$$g_z = g_e - \frac{8\lambda}{\Delta_z} \tag{10a}$$

Likewise

$$g_x = g_e - \frac{2\lambda}{\Delta_x} \tag{10b}$$

$$g_y = g_e - \frac{2\lambda}{\Delta_y} \tag{10c}$$

As $\lambda = -830\ \text{cm}^{-1}$ all three g values will be greater than 2. The values for Δ_x, Δ_y, and Δ_z can be obtained from the optical spectrum of the compound for they represent the $d \rightarrow d$ optical transitions. For typical Cu(II) complexes, they are found at about 600 nm (16,000 cm^{-1}). Insertion of these numbers into Eq. (10a–c) yields $g_z = 2.4$, $g_y = g_x = 2.1$. It is usually found that the g values calculated

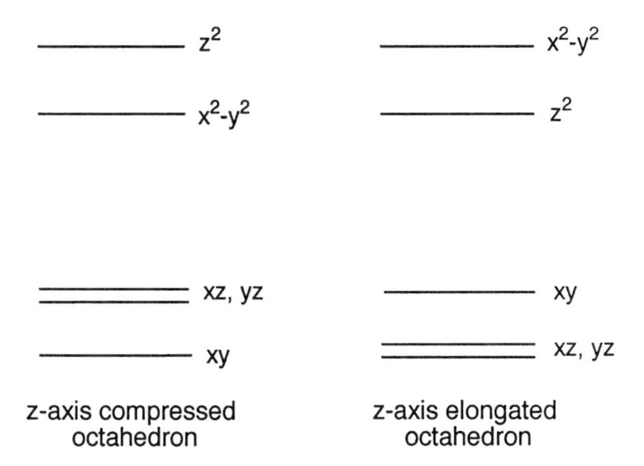

Figure 6
The relative order of the energy levels of the five 3d orbitals for two common distorted octahedral geometries of Ni(I), Ni(III), and Cu(II).

from the optical transitions are larger than those observed experimentally. This discrepancy is usually interpreted as evidence for covalent bonding, which results in the unpaired electron spending a fraction of its time on one-or-more ligand atoms with the consequence that the effective spin–orbit coupling constant is smaller than that for the free metal ion. Reductions in λ ranging from 20–50% are common.

More formal treatments of this approach can be found in Chapters 9 and 10 of Carrington and McLachlan[9] and Section 11-6 of Wertz and Bolton[10] (who write n/Δ as Λ).

C. Category II: $S = \frac{1}{2}$, Large Deviations of g from 2 [e.g., Low-Spin Fe(III)]

Use of the above method for low-spin d^5 systems such as the heme containing cytochromes would suggest that Δg_z = large and positive, Δg_y = small and positive, and Δg_x = very small and negative. Indeed some systems do appear to fit this pattern. However, there are a number of low-spin systems that clearly do not conform to this prediction and exhibit, for example, Δg_y = small and negative and Δg_x = large and negative; potassium ferricyanide and met-myoglobin cyanide are examples. Clearly, a more general treatment is needed.

Inspection of Eq. (10) immediately reveals that the perturbation approach must fail if $\Delta = 0$. In fact, the method is only valid when the second term in Eq. (10) is small and any corrections to the shape of the parent orbital containing the unpaired electron is equally small. When Δ is small relative to λ the orbitals that are related by rotation lie energetically close to one another and the above requirement of the perturbation treatment is not met. In the more general approach, the "true" orbital is described explicitly as a (linear) combination of the parent d orbitals with the extent of contribution of the participating orbitals being established by the relative magnitudes of λ and Δ.

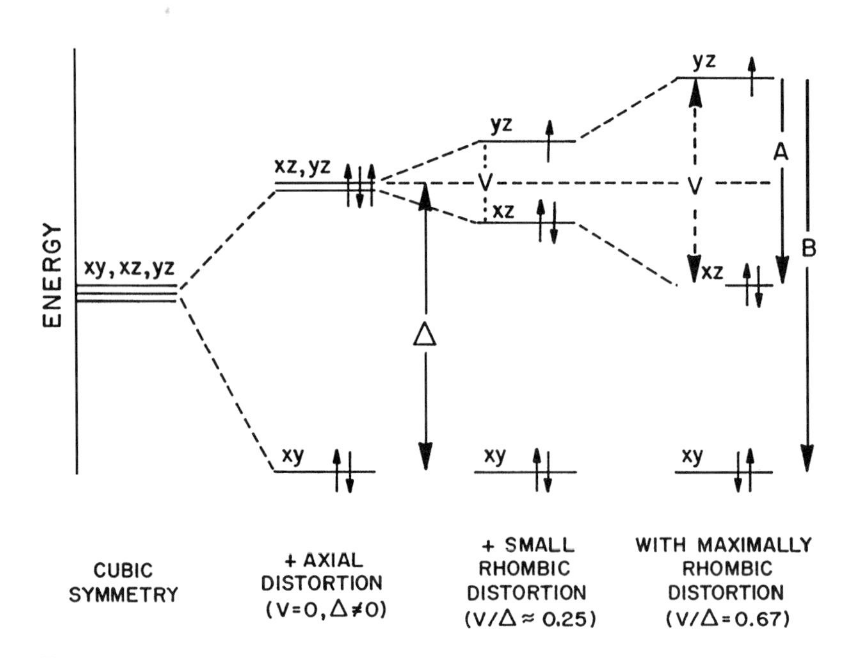

Figure 7
The relative order of the energy levels for three of the five 3d orbitals of low-spin Fe(III) ($S = \frac{1}{2}$); the remaining two orbitals are normally taken to be at a sufficiently high energy as to be unimportant. In cubic symmetry (left), all three orbitals are equi-energetic. An axial compression destabilizes d_{xz} and d_{yz} equally relative to d_{xy}. This destabilization is quantified by the parameter Δ, which is specified in units of the spin–orbit coupling constant (λ). Inequivalence in the xy plane leads to a raising of d_{yz} relative to d_{xz} quantified by the parameter V. The direction of this rhombic distortion is labeled so as to yield this arrangement. At a sufficiently large value of V the ratio $V/\Delta = \frac{2}{3}$. In a proper coordinate system, this is called a maximally rhombic system for with larger values of V one is free to reassign the labeling of the axes so that this pattern of behavior is maintained. The absolute energies of d_{xz} and d_{xy} relative to d_{yz} are specified by A and B.

This situation is found with the low-spin iron(III) hemes that have the formal orbital arrangement shown in Figure 7, with a single unpaired electron present in d_{yz}. Note that the contribution to $d_{x^2-y^2}$ and d_{z^2} are omitted because it is assumed that spin–orbit effects within the d_{yz}, d_{xz}, and d_{xy} trio will totally dominate the phenomenon. From Table II, we see these orbitals can be rotated into each other. The spin–orbital combinations that are mixed to one another by the spin–orbit interaction ($\mathbf{L} \cdot \mathbf{S}$) are $d_{yz}{}^{+}$, $d_{xz}{}^{+}$, and $d_{xy}{}^{-}$ while $d_{yz}{}^{-}$ is mixed with $d_{xz}{}^{-}$ and $d_{xy}{}^{+}$ (see Weissbluth,[11] Table 5.13); the superscripts ($\pm$) denote the spin-up or spin-down electron. Instead of the electron being in d_{yz} we represent the spin-up and spin-down orientations of the electron as $|+\rangle$ and $|-\rangle$ and then the paramagnetic orbitals have the form

$$|+\rangle = a \cdot d_{yz}{}^{+} - ib \cdot d_{xz}{}^{+} - c \cdot d_{xy}{}^{-} \tag{11a}$$

$$|-\rangle = -a \cdot d_{yz}{}^{-} - ib \cdot d_{xz}{}^{-} - c \cdot d_{xy}{}^{+} \tag{11b}$$

where a, b and c are real constants and $a^2 + b^2 + c^2 = 1$.

The way in which the coefficients shown in Eqs. (11) are chosen is not arbitrary. Some latitude is allowed in selecting the coefficients in Eq. (11a) and their choice is governed primarily by the convenience of the solutions that ultimately result. Once those of Eq. (11a) are selected those for Eq. (11b) are chosen so that the spin–orbit coupling energies of $|+\rangle$ and $|-\rangle$ are the same; specific formulas exist to convert the coefficients in Eq. (11a) to those in Eq. (11b).[11]

The particular form of coefficients shown in Eqs. (11) are those introduced by Taylor[12] for they led to particularly convenient expressions for g_x, g_y, and g_z in terms of a, b, and c, and, conversely, expressions for a, b, and c in terms of g_x, g_y, and g_z (Appendix II).

Now the values of a, b and c depend on both the value of λ and the *relative* energies of the three d orbitals (A and B Fig. 7). Knowing the composition of $|+\rangle$ (i.e., the values of a, b, and c, which can be obtained from the three g values) it is possible to calculate the values of A and B.

There is an alternative way of specifying the energy parameters. These alternative parameters are Δ [the difference in energy of d_{xy} and the mean of the energies of d_{xz} and d_{yz}; $\Delta = (B - A/2)\lambda$] and V (the difference in energy between d_{xz} and d_{yz}; $V = A\lambda$). These latter quantities are called the axial and rhombic components of the ligand field.[q]

The ability to characterize the strength and geometry of the ligand field from the measured g values has proven to be a powerful tool for analyzing the ligation of low-spin hemes by their respective proteins. This technique, which was introduced by Blumberg and Peisach in 1971,[13] became particularly useful with the introduction of the Taylor relationships just reviewed (see Appendix II). The procedure is straightforward. One records the EPR spectrum and determines the g values. These values are inserted into Eq. (A.II.8) and (A.II.9) and the values of V/λ and Δ/λ calculated. These calculated values are then inserted into a graph of V/Δ versus Δ/λ (see Fig. 8). In this graph the abscissa is a measure of the strength of the axial ligand field while the ordinate gauges the deviation from axial symmetry: the rhombicity. Points that fall on the same vertical line represent species that experience the same axial field but suffer varying amounts of rhombic distortion; points that fall on the same horizontal line represent species that experience varying amounts of axial field but deviate from axiality to the same degree.

Blumberg and Peisach[13] carried out this manipulation on a large number of low-spin heme centers and noted that the data clustered into five domains, designated C, H, B, O, and P. The first four of these domains experience comparable amounts of rhombic distortion but increasing axial ligand field; they are assigned to heme centers that all have histidine as the fifth (proximal) iron ligand but differ

[q] As is often the case the definition of the coordinate system to be used in this system is somewhat arbitrary. However, many people feel comfortable with the so-called "proper" coordinate system in which the major distortion from cubic symmetry is assigned to the axial direction and labeled z, while any subsequent inequivalence in the xy plane is attributed to the rhombic component. If the rhombic component is increased to the point that its magnitude exceeds the axial component, then the axes are relabeled using the same recipe. In this proper coordinate system, the maximum value of $V/\Delta = \frac{2}{3}$ (see Fig. 7).

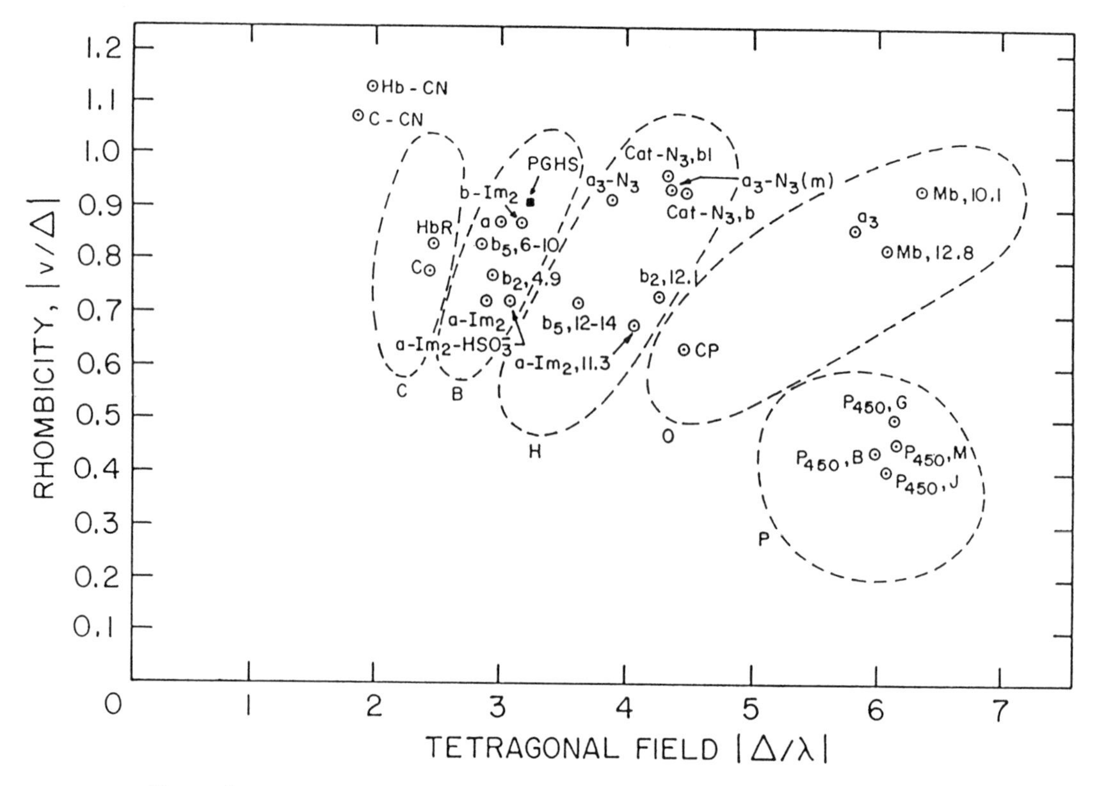

Figure 8
The correlation of the axial and rhombic ligand field parameters for a large number of low-spin heme centers, both protein and non-protein. The abscissa measures the strength of the axial field in units of the spin–orbit coupling constant. The ordinate measures the ratio of the rhombic to axial components. Both axes are dimensionless. Note that there are values on the ordiante larger than 0.67; consequently a proper coordinate system was not used in developing this graph (see text and Appendix II).

in that the sixth (distal) ligand is methionine (C), neutral histidine (H), anionic histidine (B), and oxide (O) respectively. The fifth family (P) is clearly different and represents the P_{450} heme enzymes; these have cysteine and water as the ligands to the heme iron.

This procedure has been widely used in defining the axial coordination of newly discovered low-spin heme centers. There are, however, three caveats in its use. First, the particular form of the truth diagram shown in Figure 8 uses an unconventional assignment of g values, namely, $g_x = g_{\mathrm{mid}}$, $g_y = -g_{\mathrm{max}}$, and $g_z = g_{\mathrm{min}}$ (the basis for these choices is described in Palmer).[14] Second, a consistency check should be applied to the g values used. This check is

$$g_x^2 + g_y^2 + g_z^2 - g_x g_z + g_y g_z - g_x g_y + 4(g_x - g_y - g_z) = 0 \tag{12}$$

which is true provided that the normalization condition ($a^2 + b^2 + c^2 = 1$) holds.

The inability to satisfy this consistency check[r] implies either that one or more of the measured values is incorrect or that the normalization condition is not obeyed (implying that covalency effects or higher excited states are important). When g_x is broad, and its precise location consequently uncertain, the former explanation is the more likely. Alternatively one can assume that the normalization condition is true and use Eq. (12) to determine the value of the third g value when only two of them can be measured. This proves to be quite useful in practice.

Finally, it appears that the Blumberg correlations only work well when V/Δ is quite large; ambiguities creep in when this ratio is small. These ambiguities are particularly serious when d_{yz} and d_{xz} have the same energy (they are degenerate, $V = 0$ and $a^2 = b^2$), for the two orbitals must then make equal contributions to $|\pm\rangle$. Under these conditions, g_z is determined only by the value of a, which in turn is controlled by the value of Δ. As Δ approaches infinity $a^2 \rightarrow 0.5$, $c^2 \rightarrow 0.0$, $g_z \rightarrow 4.0$ [Eq. (A.II.10)] and g_x and g_y both approach zero. When Δ is less than infinite the contribution of d_{xy} to $|\pm\rangle$ is not zero, the value of a^2 is necessarily smaller than 0.5 and $g_z = 12a^2 - 2.0$ while g_x and g_y become finite [Eq. (A.II.11)].

Such behavior is found in a number of cytochrome species and these are often referred to as the highly anisotropic low-spin (HALS) species, though as the origin of the phenomenon is a *raising* of the symmetry ($V = 0$) a more appropriate ascription would be the HALS *spectra*, or, more usefully, the axially symmetric low-spin hemes. Particularly important examples of this behavior are to be found in the b and $c_1(= f)$ cytochromes present in the electron-transfer complexes of mitochondria and of photosynthesis (both bacterial and plant).[15] With these systems values of g_z approaching 3.8 have been observed while g_x and g_y are often so small, and their EPR absorption sufficiently broad, that reliable estimates of their values do not exist.

The important structural question raised by this phenomenon is the nature of the ligation that is responsible for these HALS species. The most likely explanation is to be found in the relative orientation of the two ligands that occupy the two axial coordination sites in these cytochromes. In normal cases ($g_z < 3.1$), for which substantial X-ray data exists, it is found that the planes of the two axial ligands are nearly parallel to one another when viewed down the z axis. However, their projection onto the heme plane can be parallel either to the axis connecting opposed pyrrole rings or opposed methine bridges; thus the projection on the heme ring appears not to be a significant factor when the two ligands are coplanar.

Coplanar ligands clearly impose an asymmetry on what is otherwise a fourfold symmetric porphyrin ring, and it is this asymmetry that leads to a differentiation in the energies of d_{yz} and d_{xz}. However, when the planes of the two axial

---

[r] An alternative representation of this consistency check is to use equations a, b, and c of Appendix II to see if the normalization condition holds. For g_{max} values greater than 3, a quick though not totally reliable check is that $g_x^2 + g_y^2 + g_z^2 = 16.0$.

[s] The planes of the two ligands are readily apparent when they are both histidines, as in the b cytochromes. In the c cytochromes, the relevant planes are that of the histidine present in the proximal location and the bonding lone pair on the methionine sulfur in the distal coordination site. This lone pair is perpendicular to the plane defined by the methyl-thioether sulfur-methylene components of the methionine sidechain; hence, this latter plane and that of the histidine will be coplanar.

ligands are perpendicular[s] there is no differentiation between these two orbitals and the required axial behavior is obtained. In model systems studied thus far, there is no example of a HALS system in which the planes of the ligands lie over the pyrrole nitrogens; in all cases they lie over the methine bridges. This orientation is presumably a result of steric hindrance that cannot be relieved by heme ruffling when the ligands are perpendicular, but can be relieved when they are parallel.

One consequence of HALS behavior is that the lineshape of the low-field g_z envelope might be markedly asymmetric. Because the value of g_z, which is obtained in the axial limit has the maximum value possible for the particular value of Δ then any deviations from axiality, regardless of the direction (x vs. y), will lead to a smaller value for g_z. Now a major contribution to the EPR line width of many metalloproteins is "g strain", a variation in g reflecting variations in the ligand field, which result from a distribution of microconformations [let Δ of Eq. (10) have a distribution rather than a fixed value]. Normally, such g-strain leads to a Gaussian distribution of intensity symmetrically disposed about the "true" g value; hence, the EPR lineshape at g_z in non-HALS systems is quite reasonably reproduced with a Gaussian formula. But at the axial limit it is not possible to have this distribution lead to a value for g_z, which is larger than the maximum value and thus Gaussian behavior is only found on the low side of the g value (i.e., the high-field side); the low-field side of the EPR signal rises much more rapidly than can be represented by a Gaussian formula and, in extreme cases, this increase in EPR intensity is almost vertical and referred to as the "ramp". There are two consequences of this phenomenon.[16] First, it is not possible to simulate the g_z region of the spectrum using normal lineshape techniques; early attempts to do so found it necessary to invoke multiple species to reproduce the spectral shape and hence led, erroneously, to the proposal of multiple species of these cytochromes. Second, the presence of the ramp allows one to obtain an independent estimate of Δ; for example, for bis-histidine hemes it appears that Δ is approximately 3.2λ.

The important biological question, which has yet to be answered, is whether-or-not such a high-energy conformation of the ligands serves a biological role. It may be relevant to this question that many of the HALS proteins[15] are part of the energy conserving electron transport systems found in membranes. In these proteins, the hemes are believed to be coordinated to amino acid residues present in approximately parallel, trans-membrane helices, and one might speculate that it is a consequence of this arrangement that such a "strained" geometry is produced.

Finally, one should note that alternative methods exist to identify the axial ligand to the heme; of these near-infrared (IR) magnetic circular dichroism (**MCD**) is particularly good for identifying methionine. An application of the method, its utility, and its limitations can be found in recent articles.[17,18]

D. Category III: $S = \frac{1}{2}$, Coupled Spin Systems

Another category of behavior is that found in systems in which two or more paramagnetic centers are present. When the paramagnetic centers are sufficiently

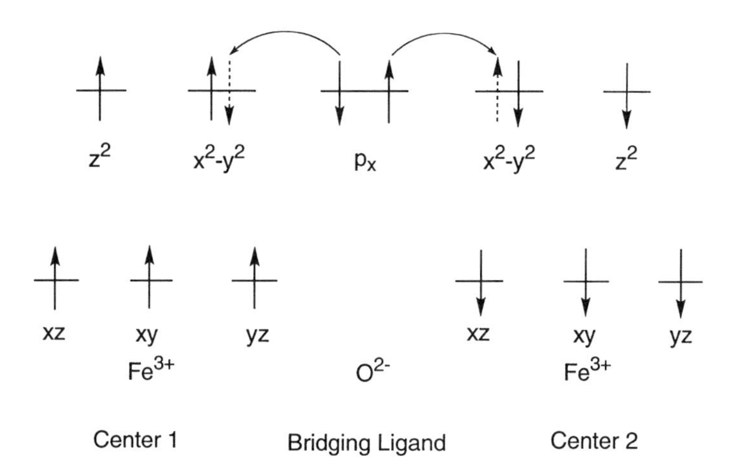

Figure 9
A schematic representation of the mechanism of antiferromagnetic exchange between two high-spin iron(III) ions mediated by a pair of electrons from a bridging ligand atom.

far apart so that the dominant interaction is through space, one class of behavior is obtained and is described at the end of Section III.E. Here we focus on the case where the metal centers are sufficiently close that there is the possibility of bonding, either real or incipient between them. Of particular interest is the case where there are two metal centers together with one-or-more bridging ligands. This situation is found in iron–sulfur proteins, and nonheme diiron proteins such as hemerythrin, purple acid phosphatase (uteroferrin), ribonucleotide reductase, and methane monooxygenase.

The scenario can be understood by reference to a system composed of two high-spin iron centers connected by the oxygen atom of a bridging ligand. For convenience, we arrange these three atoms on a line that we will call the x axis (Fig. 9).

We begin by letting both iron atoms be high-spin iron(III); thus in each case the five 3d orbitals contain a single electron. The bridging oxygen atom contains two electrons in a single orbital, p_x, for example. The bond between the left-hand iron center and the ligand is formed by delocalizing one of the oxygen p_x electrons into $d_{x^2-y^2}$ of the metal. From the Pauli principle, this partially transferred electron is required to "point down", as shown. Consequently, the fraction of the electron remaining on the ligand also points down and its spin-paired partner must "point up". But this second electron is partially transferred onto the right-hand iron center and, again from the Pauli principle, this forces the five 3d electrons of the second center to point down. We thus have the circumstance that the five 3d electrons on each of the two iron atoms have their spin oriented in opposite directions; the spin of the two-iron system is thus zero and we have a *molecular diamagnet*. This phenomenon is called antiferromagnetic exchange. An event that causes the spins to flip on one metal center will simultaneously flip the spins on the second iron atom, via this antiferromagnetic coupling of the two metal centers.[t]

From the point of view of EPR, the more interesting case is when one of the two iron atoms (e.g., the left-hand iron) is Fe(II), and hence has six 3d electrons. Now this center has four unpaired electrons and can "cancel" the paramagnetism of four of the five electrons on the second, Fe(III), center. But this still leaves a single electron on the Fe(III) uncompensated. This "single" electron can now be visualized by EPR. We label the spins on the Fe(III) and Fe(II) as S_1 and S_2 and the net spin resulting from the antiferromagnetic interaction as S. When $S_1 = S_2 = \frac{5}{2}$, then $S = 0$; when $S_1 = \frac{5}{2}$ and $S_2 = \frac{4}{2}$, then $S = \frac{1}{2}$.

The spin, S, clearly has contributions from S_1 and S_2 and these contributions determine the magnetic behavior of the net spin. As it is S that aligns along any applied magnetic field then so does S_1; consequently, S_2 must align itself against the magnetic field. The relative contributions of S_1 and S_2 to S are established by constructing a triangle whose sides are proportional to S_1, S_2, and S and then calculating the vector projections of S_1 and S_2 onto S. As described in Appendix IV, this leads to an expression for the g values:

$$g_i = \frac{7}{3}\, g_{1i} - \frac{4}{3} g_{2i} \qquad i = x, y, \text{ and } z \tag{13}$$

That is, the g value observed in a particular direction (i) is given by the weighted average of the g values of S_1 and S_2, in that direction with the weighting factors being $\frac{7}{3}$ and $-\frac{4}{3}$ (the minus sign reflects the fact that S_2 points in the opposite direction to S). An expression similar to Eq. (17) also exists for hyperfine interactions (Appendix IV).

For mononuclear high-spin Fe(III), the intrinsic g factor is close to 2.0 and is isotropic; however, mononuclear high-spin Fe(II) does have some orbital magnetism and its g values are anisotropic, falling in the range 2–2.4. Thus g values in the range of 1.6–2.0 are expected for the spin-coupled system and this range of values is observed in practice.

Examples of such Fe(II)Fe(III) sites in proteins are shown in Figure 10.[19,20] Interestingly, the 2 Fe ferredoxins exhibit one g value (g_z) greater than 2.0. Such behavior is not predicted by the above theory and is attributed to substantial delocalization of the iron electrons onto the sulfur ligands, either terminal or bridging, that are present in this class of proteins. Because of its high atomic number, sulfur has a large spin–orbit coupling constant and that fraction of the paramagnetism present on sulfur experiences an orbital contribution that opposes

---

[t] Although the process as described requires the participation of a bridging ligand, there is no fundamental requirement that such a bridge be present. It is sufficient that (a) the participating metal orbitals have the same symmetry so that they can overlap with each other and (b) that they are sufficiently close that the incipient bond responds to the Pauli principle. At the same time, the Pauli principle specifies that those orbitals that are orthogonal, for example $d_{x^2-y^2}$ on center 1 and d_{yz} on center II, will have parallel spins and is called a *ferromagnetic* interaction. These latter interactions are generally much weaker than an antiferromagnetic interaction and will be obscured when antiferromagnetism is present.

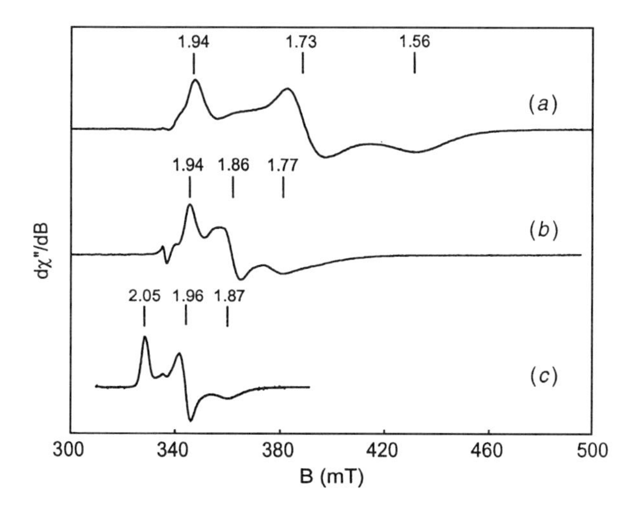

Figure 10
The EPR spectra of proteins containing mixed-valence diiron(II,III) sites: (*a*) purple acid phosphatase, (*b*) hydroxylase component of methane monooxygenase, and (*c*) reductase component of methane monooxygenase. The exchange coupling between the Fe(II) and Fe(III) atoms of the mixed-valence state of the proteins gives a system spin $S = \frac{1}{2}$ state low in energy. The sample temperatures were approximately 4 K and the microwaves were 0.02 mW at 9.43 GHz. [Adapted from Debrunner et al.[18] (spectrum *a*) and Fox et al.[19] (spectra *b* and *c*) with permission from the publishers.]

that due to the iron(II) center; this shifts the g values closer to 2.0 and actually leads to one g value being larger than 2.0.

Complications to this picture arise when one considers that each metal ion will have a zero-field splitting. These complications are discussed by Guigliarelli et al.[21]

E. Category IV: $S > \frac{1}{2}$, Half-Integer Spin Systems

When there is more than one unpaired electron in the paramagnetic center a new phenomenon is encountered. This phenomenon is called zero-field splitting (zfs). It is the separation in energy of the various m_S states in the absence of an applied magnetic field and is a result of interelectronic interactions and ligand fields of low symmetry. The Hamiltonian for zfs is indicated below:

$$H_{\mathrm{zfs}} = D[\mathbf{S}_z^2 - \tfrac{1}{3}\mathbf{S}^2 + E/D(\mathbf{S}_x^2 - \mathbf{S}_y^2)] \tag{14}$$

where D is the axial zfs parameter and E/D indicates the degree of rhombic distortion in the electronic environment.

The zfs is quantified as a correction to the energies of the individual spin states arising from spin–orbit coupling." (see Fig. 7 of Palmer).[14] A purely axial distortion ($E/D = 0$) leads to a differential stability of the six spin configurations, which is proportional to m_S^2 in units of D, the axial zfs parameter. The relative energies of the various levels is given by

$$\text{Energy}(m_S) = D\{m_S^2 - \tfrac{1}{3}[S(S+1)]\} \tag{15}$$

For $S = \frac{5}{2}$, Eq. (15) yields values of $-\frac{8}{3}$, $-\frac{2}{3}$, and $\frac{10}{3}$ D for the $m_S = \pm\frac{1}{2}$, $\pm\frac{3}{2}$, and $\pm\frac{5}{2}$ states, respectively. When D is positive the $\pm\frac{1}{2}$ levels are the lowest in energy and vice-versa. The D values for $S = \frac{5}{2}$ ions range from 0.1 cm^{-1} [for Mn(II)] to 10 cm^{-1} [for Fe(III) porphyrins]. The parameter D is a directed quantity in that it characterizes the magnitude and direction of the axial distortion; in high-spin hemes this direction is normal to the heme plane.

Figure 11 illustrates three cases of interest in an $S = \frac{5}{2}$ system. When $D = 0$ [Fig. 11(*a*)] there is no zfs and the six spin configurations diverge linearly with magnetic field with a slope that is directly proportional to m_S. The separation between any adjacent pair of levels is always the same and thus all levels come into resonance at the same value for B_L. A single EPR line is expected. With high-spin Fe(III) and Mn(II) both of which have a half-filled shell of electrons there is no orbital angular momentum and the g value for this line equals 2.0. This has never been observed in any biological system nor, to my knowledge, has it been observed with paramagnetic molecules in the frozen state.

When $0 < D < h\nu$, the three pairs of levels are split in zero field, but the separation between these pairs is smaller than the microwave quantum [Fig. 11(*b*)]. Application of B_L again produces a linear change in the energy of each level that is proportional to m_S. But as the levels are already partially split in zero field certain pairs of levels come into resonance before others. Explicitly, for the transition between m_S and $m_S - 1$ the resonant field (B_R) is related to the resonant field in the absence of zfs (B_0) by the magnitude of D and the angle (θ) between the direction of the applied magnetic field and the zfs axis,[22] that is,

$$B_R = B_0 - (m_S - \tfrac{1}{2})[D(3\cos^2\theta - 1)] \tag{16}$$

The pattern of splittings shown in Figure 11 describe the behavior when the magnetic field is aligned along the axis of the distortion and the resulting EPR

---

[a] The means whereby an electrical field affects the energy of a magnetic state is not necessarily obvious and arises from Pauli's principle that electrons of the same spin are less likely to be found in the same space than are electrons of opposite spin (hence electrons in the same orbital are always paired). This principle affects the motions of the individual electrons such that, on average, spin configurations in which the electrons have the smallest number of parallel spins (e.g., $S = \frac{5}{2}$, $m_s = \pm\frac{1}{2}$; see Fig. 9) are more compact than spin configurations in which the number of parallel spins is larger (e.g., $S = \frac{5}{2}$, $m_s = \pm\frac{5}{2}$). In the more compact configuration, the paramagnetic electrons will suffer a larger interelectron electrostatic repulsion. Thus these spin-opposed configurations have more to gain by an expansion of their environment (such as would occur by removing one of the axial ligands) and more to lose by an axial compression. The former removes negative charge associated with the liganding atom and thus allows the paramagnetic electrons to expand along the axial direction; the latter produces the inverse affect. For a purely axial distortion, this leads to a differential stability of the six spin configurations that is proportional to m_{s^2}.

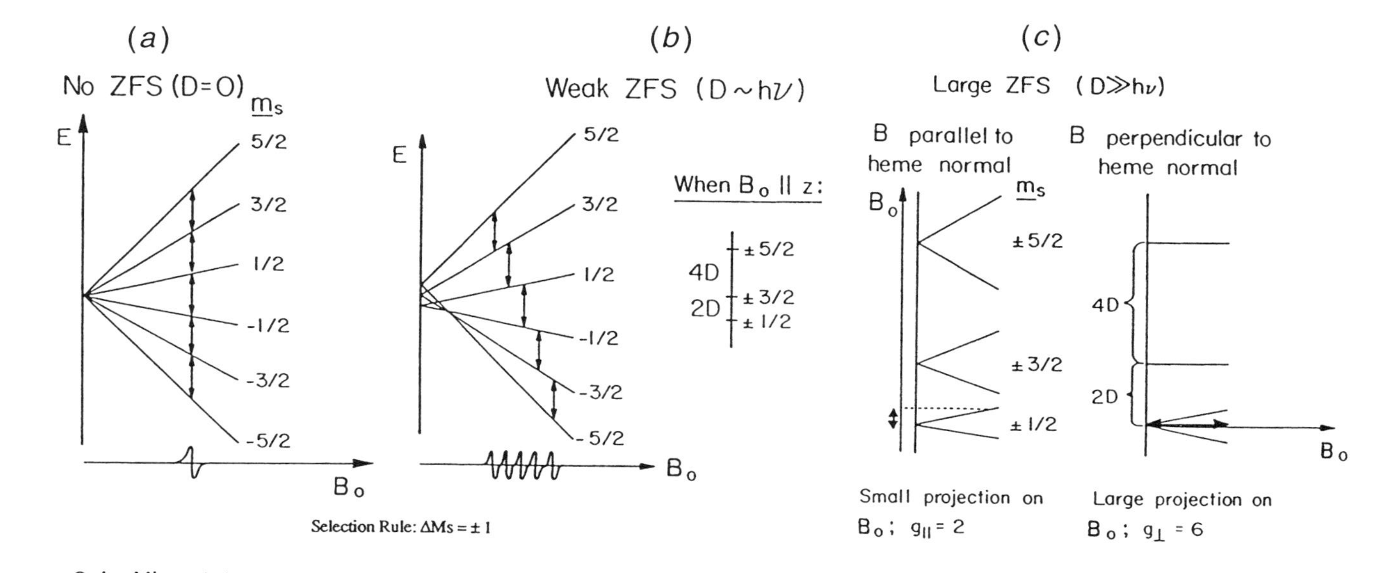

Figure 11
The behavior of paramagnets with $S = 5/2$. (*a*) This panel illustrates the case when the paramagnet is in high symmetry; each m_s level converges to a common origin at zero field and each level comes into resonance at the same value of the applied field. A single EPR line is observed. In panel (*b*) the paramagnet experiences a small asymmetry in its environment which separates the m_s levels according to their absolute value. The levels now converge to different origins. When the zero-field separation is small compared to the Zeeman interaction (which is about 0.3 cm^{-1} at X band) the levels can still come into resonance with available values of the laboratory field and five separate resonances are now observed (the zero-field splitting brings some levels into resonance sooner than they otherwise might while other levels come into resonance later than they might). Only the transition from the $\pm\frac{1}{2}$ levels is unaffected. Panel (*c*) shows the behavior at very large values of the zero-field splitting where the separation between the $m_s \pm\frac{3}{2}$ and $\pm\frac{5}{2}$ is already sufficiently large that there is no hope of seeing transitions between these levels. Only transitions between the $\pm\frac{1}{2}$ states can be observed and these transitions are very sensitive to the orientation of the paramagnet with respect to the applied field (panel c and text). The bottom of the figure provides a simple picture of the interpretation of the various m_s states; similar pattern exists for $S = -\frac{5}{2}$. Be aware that these states are not necessarily unique. For example, the $m_s = \frac{5}{2}$ state is unique but the $m_s = \frac{3}{2}$ configuration can be achieved in five different ways by simply moving the "down" spin.

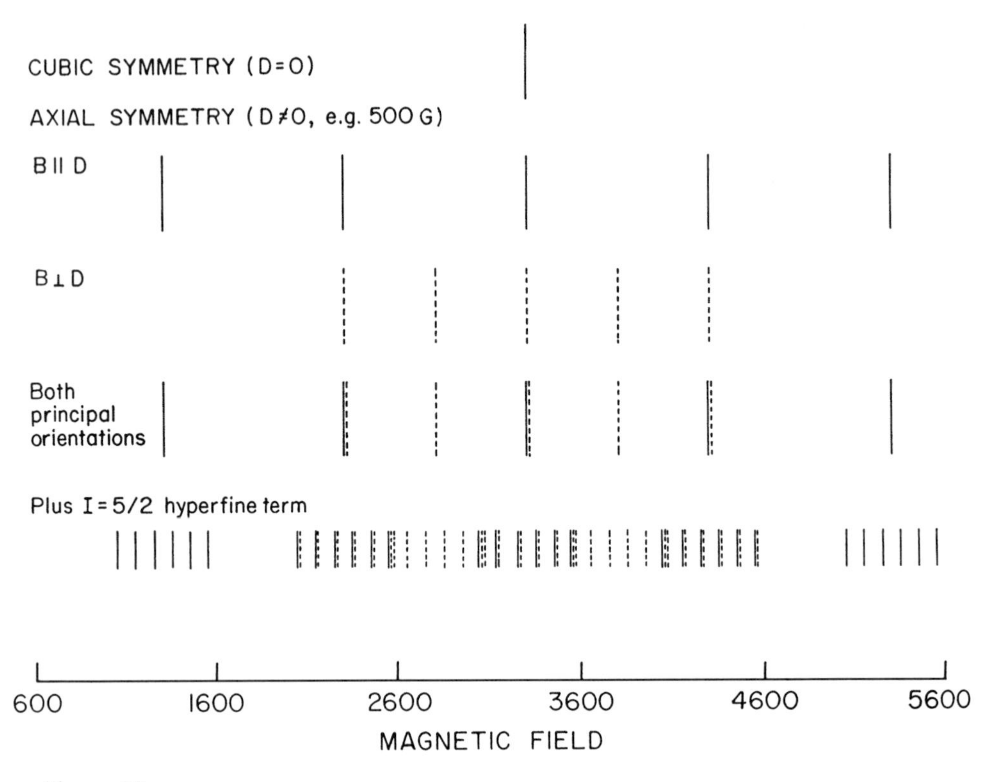

Figure 12
The complexity in the EPR of Mn(II) ($S = \frac{5}{2}$). For a single crystal, the small zfs converts the parent single resonance obtained in high symmetry to a quintet with a separation that depends on whether the D axis of the paramagnet is aligned parallel or perpendicular to the applied field (lines two and three). The summation of these two orientations is given in the fourth line. Each of these transitions is now split into six by virtue of the interaction of the electrons and the nuclear moment of the manganous ion (see later in chapter). When combined with a finite spectral envelope and including all other "off-axis" orientations as would be found in a frozen solution, this results in an absorption that is broad and difficult to resolve.

spectrum observed at this orientation consists of five EPR lines with adjacent lines being separated by $2D$. Perpendicular to this orientation ($\theta = 90°$) the splittings are one-half as large while at intermediate angles the magnitude of the splittings vary from 0 ($\theta = 54°$) to D. Consequently, the typical EPR spectrum is broad and unresolved. This problem is compounded in Mn(II) (Fig. 12; $D = -50$ mT) because of the nuclear moment of Mn ($I = \frac{5}{2}$, see below), which splits each EPR line into six satellites. For this case of small D, the resolution of the spectrum can be improved by recording data at higher magnetic fields for the magnitudes of D relative to B_L decreases and a condition closer to Figure 11(a) is approached; such measurements require a higher frequency spectrometer and thus experiments at 35 GHz are particularly desirable for Mn(II).[22]

In the third case $D \gg h\nu$. Now the individual pairs of m_S levels are sufficiently separated that transitions between the pairs of levels cannot occur with normal laboratory fields for the usual selection rule ($\Delta m_S = \pm 1$). This is the circumstance in many high-spin iron(III) complexes, particularly for hemes where the square planar array of four porphyrinic nitrogens gives rise to D values in the range of 5–10 cm^{-1}; thus only transitions within the $m_S = \pm\frac{1}{2}$ levels can be observed [Fig. 11(c)]. These transitions correspond to $g_{\parallel}$ of 2.0 and $g_{\perp}$ of 6.0 ($2S + 1$ with $S = \frac{5}{2}$).

For positive D, the $m_S = \pm\frac{3}{2}$, $\pm\frac{5}{2}$ are excited states. The $g_{\parallel}$ values for these states are 6 and 10, respectively. However, in the perpendicular orientation the levels are not affected by the magnetic field and have $g_{\perp} = 0$. In a conventional EPR spectrometer B_1, the magnetic field associated with the microwave quantum that is responsible for inducing the EPR transition, is oriented perpendicular to B_L and it is the g values in the plane containing B_1 that determines the transition probability. Consequently, for these two excited states the transition probability in the parallel orientation is zero. In the perpendicular orientation, the transition probability is not zero but, as $g_{\perp} = 0$, the Zeeman interaction is zero and the levels do not split in the magnetic field.[v] Thus these excited states do not exhibit EPR signals under any condition.

The next complication is the introduction of a rhombic distortion to the ligand field, making all three axes distinct from each other. In this circumstance, a second parameter, E, is used to characterize the additional asymmetry; it is analogous to V in the low-spin iron(III) ($S = \frac{1}{2}$) case described earlier. The nonzero value for E activates the $\mathbf{S}_x$ and $\mathbf{S}_y$ operators in the zfs Hamiltonian [Eq. (14)] and allows for the mixing of the various m_s states. The zero-field states can thus no longer be described with simple m_S labels. As was noted for V/Δ the ratio E/D has a limited range, which in this case equals $0 - \frac{1}{3}$. When E/D is small (< 0.1), small deviations from the axial EPR spectrum just described are observed; this shows up as a splitting of $g_{\perp}$ about its axial value of 6 and a shift of $g_{\parallel}$ to values smaller than 2. When $g_x - g_y < 4$ the following relationships are approximately correct (Slappendel et al.,[23]):

$$\begin{aligned} g_x &= 6 + 24\,E/D - 79(E^2/D^2) \\ g_y &= 6 - 24\,E/D - 79(E^2/D^2) \\ g_z &= 2 - 74(E^2/D^2) \\ g_x - g_y &= 48(E/D) \end{aligned} \tag{17}$$

Figure 13 illustrates graphically how the g values of the three doublets of the $S = \frac{5}{2}$ system change as a function of E/D. It should be noted that as E/D grows in value, the extent of mixing between the m_s states increases and it is now possible to observe EPR signals arising from the previously EPR silent doublets.

[v] The resonance field is infinite at which point the condition $D \gg h\nu$ no longer applies.

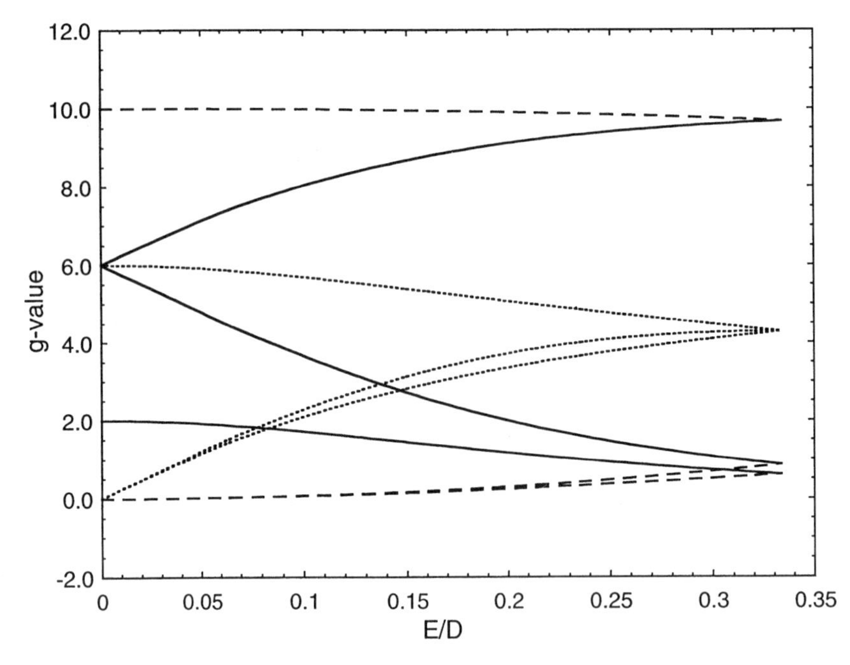

Figure 13
The g values for a $S = \frac{5}{2}$ paramagnet as a function of E/D. The solid lines represent the x, y, z components of the $m_s = \pm\frac{1}{2}$ states, the dotted lines represent the $M_s = \pm\frac{3}{2}$ states and the dashed lines the $M_s = \pm\frac{5}{2}$ states. Note that these spin-state labels are only valid when $E = 0$ but that the states become admixed as E/D increases.

The EPR spectra of several high-spin iron(III) systems are shown in Figure 14. Metmyoglobin fluoride exhibits the typical axial heme spectrum [Fig. 14(a)] with ground doublet g values at 5.8, 5.8, and 2 ($E/D = 0$). The excited-state doublets do not give rise to any signals. Examples of heme systems with increased rhombicity ($E/D > 0$) include horseradish peroxidase [$E/D = 0.02$, Fig. 14(b)] and camphor-bound cytochrome P_{450} [$E/D = 0.08$, Fig. 14(c)].[24] The increased rhombicity observed for cytochrome P_{450} is presumably due to the substitution of the axial imidazole with a thiolate, a notion corroborated by suitable model complexes. In addition, the cytochrome P_{450}–camphor complex exhibits low-spin Fe(III) signals at $g = 2.41$, 2.23, and 1.79.

Figure 14(d) shows the EPR spectrum of protocatechuate 3,4-dioxygenase with g values at 9.6 and 4.3, corresponding to the maximum E/D value of 0.33.[25] The absorption-like signal at $g = 9.6$ arises from a ground doublet with companion signals expected at $g = 0.6$ and 0.9, but the latter signals are seldom observed due to the high fields required. The $g = 4.3$ signal derives from the middle doublet and is independent of orientation, that is, $g_x = g_y = g_z = 4.3$ (Fig. 13). The isotropic nature of this signal allows it to be readily observed, since all the signal intensity is confined to a very small field range, in contrast to the signals from the ground doublet, which span a range of 1 tesla. Thus the $g = 4.3$ signal in

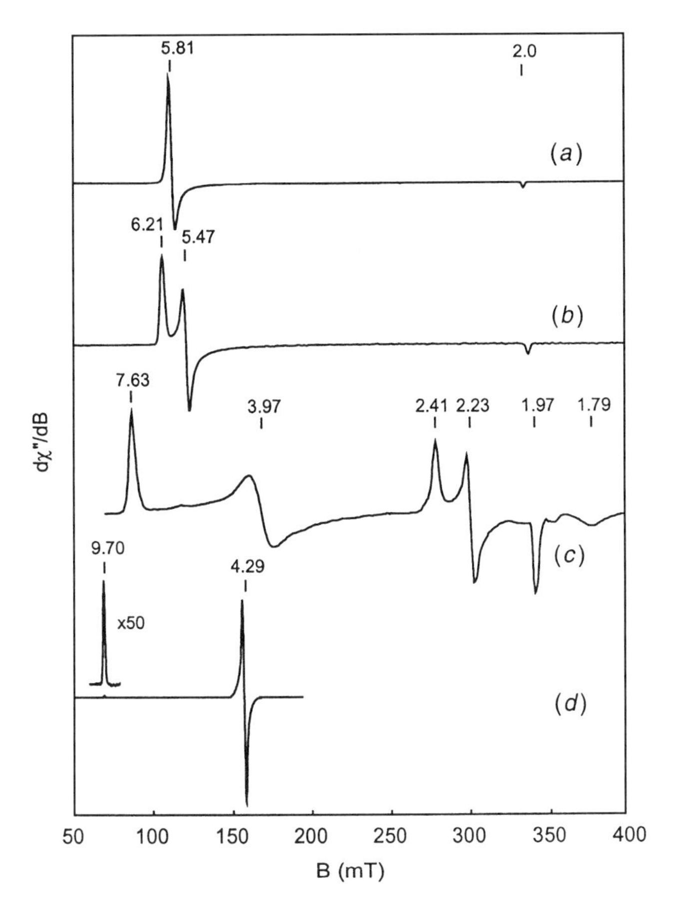

Figure 14
The EPR spectra of various iron proteins in their high-spin iron(III) state: (*a*) aquometmyoglobin, (*b*) horseradish peroxidase, (*c*) cytochrome P_{450}–camphor complex, (*d*) protocatechuate 3,4-dioxygenase. The sample temperatures were all less than 10 K and the microwave powers were less than 1 mW at 9.44 GHz. [Adapted from Tsai et al.[22] (spectrum *c*) and Whittaker et al.[23] (spectrum *d*) with permission from the publishers.]

Fig. 14(*c*) appears about three orders of magnitude taller than the $g = 9.7$ signal. The $g = 4.3$ resonance is infamous in biological samples, because its isotropic nature makes it readily observed even when only minute quantities of iron(III) contamination is present. Other proteins that also exhibit an intrinsic $g = 4.3$ signal include rubredoxin and transferrin.

The reader may wonder how, in a complicated spectrum, the g values can be assigned to appropriate doublets and the E/D value determined. The choice is limited by the shapes of the various signals; a suitable set of signals should have an absorption-like component, a derivative-like component, and an emission-like component (see Fig. 4). The temperature dependence of the signals can also distinguish signals that arise from the ground doublet and those that arise from

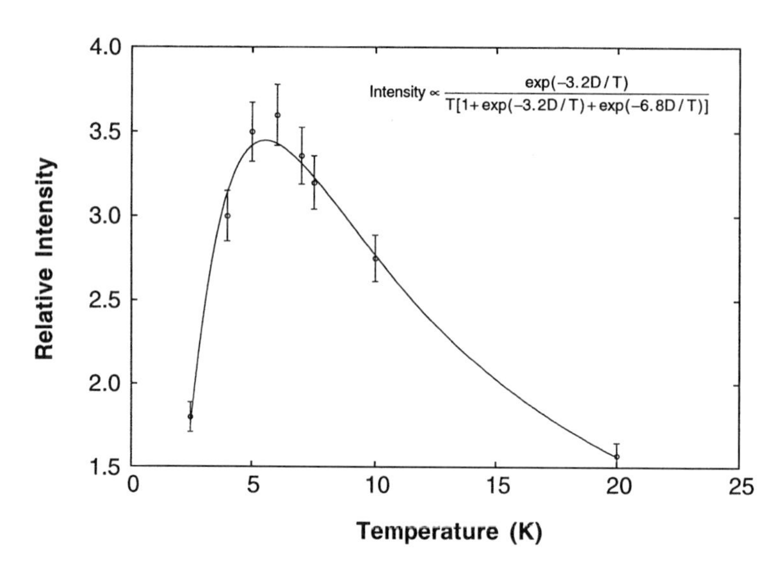

Figure 15
Temperature dependence of the intensity of EPR signal at $g = 4.3$ in protocatechuate 3,4-dioxygenase. Theoretical line calculated using the parameters $E/D = 0.29$, $D = 1.7\ \text{cm}^{-1}$. [Data points are obtained from Blumberg.[24]]

excited states. Indeed such measurements allow D to be measured. In the case of protocatechuate 3,4-dioxygenase, the $g = 9.6$ signal from the lowest Kramer doublet increases in intensity as the temperature is lowered, while the $g = 4.3$ signal achieves a maximum intensity at 6 K and then diminishes (Fig. 15). The temperature dependence of the $g = 4.3$ signal is governed by two factors, the magnitude of D, which determines to what extent the middle doublet is populated at a particular temperature $[\exp(-3.2\ D/kT)]$ and the population difference within the doublet $(1/T)$. Hence the observed intensity maximum in the temperature dependence plot (Fig. 15).[26]

Another commonly encountered spin state is the $S = \frac{3}{2}$ system. Figure 16 shows the variation of g values as a function of E/D for such a system. Figure 17 displays bioinorganic examples of such spectra, including (*a*) the NO complex of isopenicillin N synthase $(E/D = 0.01)$,[27] (b) the FeMo cluster in nitrogenase $(E/D = 0.03)$,[28] and (c) an antiferromagnetically coupled $Fe^{III}Ni^{II}$ complex $(E/D = 0.3)$.[29]

F. Category V: $S > \frac{1}{2}$, Integer Spin Systems

(*Contributed by* Michael P. Hendrich)

Section II.E focused on centers containing an odd number of unpaired electrons or Kramers centers; this section discusses those with an even number of unpaired electrons or non-Kramers centers. Zero-field splitting for Kramers centers results in a separation of m_s levels with degenerate $\pm m_s$ levels. These pairs of states, $\pm m_s$, only differ by the sense of rotation of the electron, either clockwise or counterclockwise, and can only be split by an external magnetic field. This is

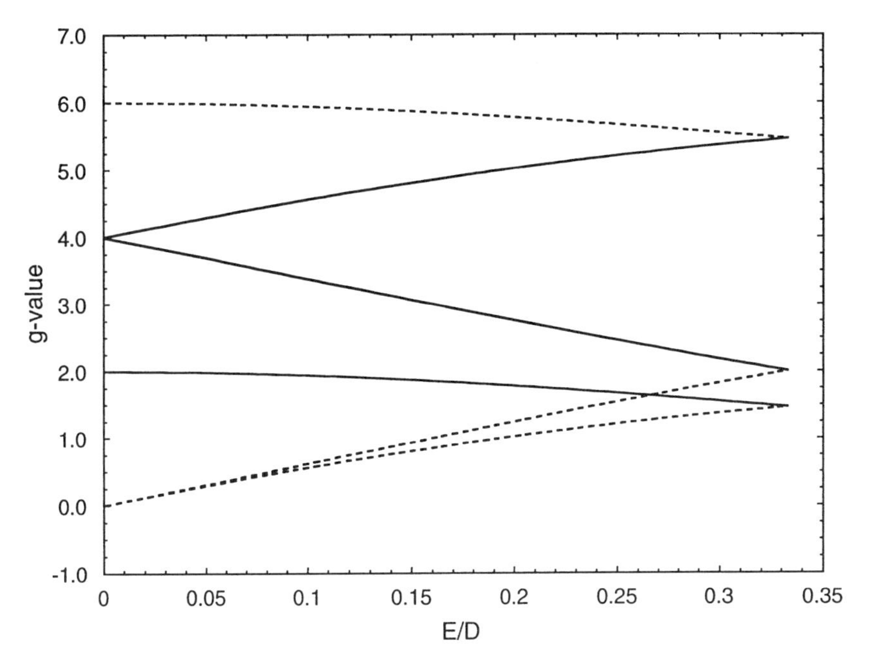

Figure 16
The g values and energy levels for a $S = \frac{3}{2}$ paramagnet as a function of E/D.

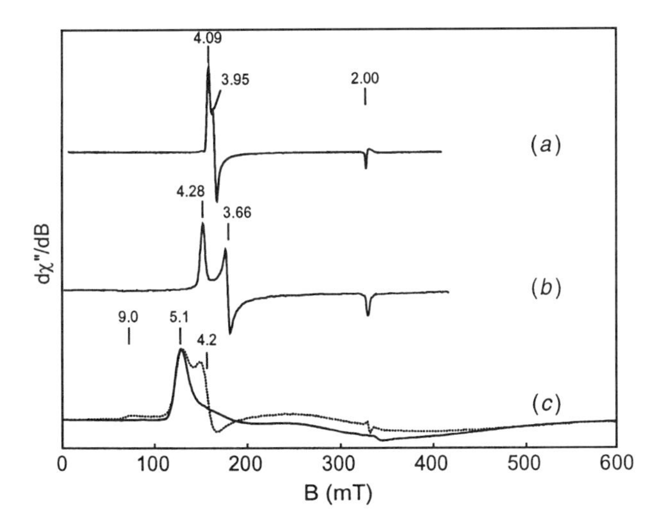

Figure 17
X-band (9.23 GHZ) EPR spectra of metal centers and clusters having a system spin $S = \frac{3}{2}$ low in energy. (a) Isopenicillin N synthase, $E/D = 0.015$; (b) the FeMo cluster of nitrogenase, $E/D = 0.03$; and (c) $Fe^{III}Ni^{II}BPMP(OPr)_2$ in CH_3CN, $E/D = 0.32$, $T = 3$ K (solid line), $T = 44$ K (dashed line). The dashed line in (c) shows the growth of an EPR Signal ($g = 9, 4.3$) from a low-lying excited $S = \frac{5}{2}$ states of the same FeNi cluster. [Adapted from Chen et al.[25] (spectrum a), Rawlings et al.[26] (spectrum b), and Holman et al.[28] (spectrum c) with permission from the publishers.]

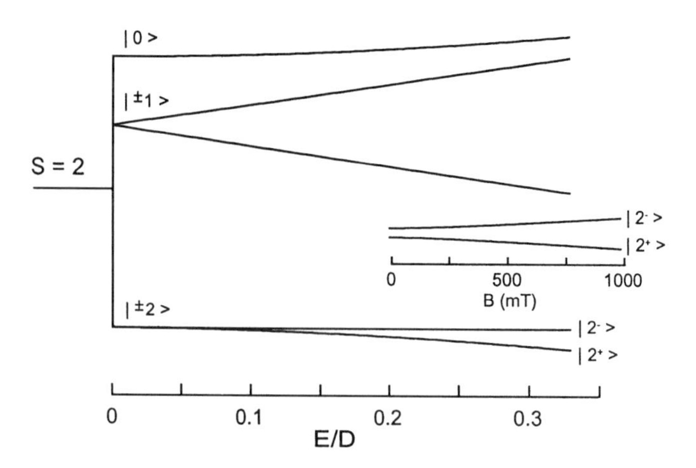

Figure 18
Energy levels as a function of E/D of an $S = 2$ system for $B = 0$ and $D < 0$ according to Eq. (14). The inset shows the effect of a magnetic field the states $|2\,\pm\rangle$ for $E/D = 0.2$.

formally known as Kramers theorem and so centers that obey it are known as Kramers centers.

For non-Kramers or integer spin systems, the m_s levels are also split by zfs according to Eq. (14). However, since they do not obey Kramers theorem, the zfs can separate all the m_s levels, including the $\pm m_s$ pairs. Figure 18 illustrates the zfs for an $S = 2$ system. Note that the $\pm m_s$ pairs are degenerate when $E/D = 0$. In this case, no signal is observed when D is greater than $h\nu$. Transitions within ± 1 and ± 2 m_s states are forbidden by the EPR selection rule ($\Delta m_S = \pm 1$) while $m_S = 0$ is nonmagnetic. Only when the intensity of B_L is large enough to reduce the energy gap to within hν between the $m_S = 0$ level and one of the $m_S = \pm 1$ levels (or between one of the $m_S = \pm 2$ levels and one of the $m_S = \pm 1$ levels) will EPR be possible. Frequently, this requires a value for B_L, which is beyond the range of the instrument and thus, in a practical sense, the material is not EPR active.

However, the degeneracy of the $m_s = \pm n$ levels is lifted as E/D deviates from 0, even in the absence of a magnetic field (Fig. 18). In this case, the $\pm m_s$ levels are split in energy by Δ, which causes dramatic changes in the EPR spectra relative to those of Category IV (Section II.E). Figure 19 shows several representative examples. Integer spin EPR signals ahave been observed for the P clusters of nitrogenase [Fig. 19(*a*)],[30] beef heart cytochrome *c* oxidase [Fig. 19(*b*)],[31] and the reduced hydroxylase component of methane monooxygenase [Fig. 19(*c*)].[32] Note that these signals are typically found at low field and that their signal intensities increase significantly when the EPR spectrometer configuration is changed from the conventional mode (microwave magnetic field B_1 oscillating perpendicular to B_L) to a parallel mode (B_1 oscillating parallel to B_L). Often the integer spin EPR signal is the only convenient spectroscopic probe for a sample in this particular spin state.

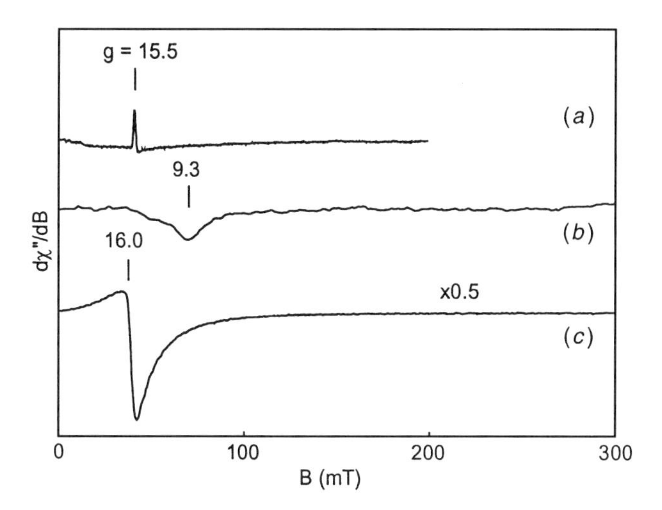

Figure 19
The X-band (9.1 GHz) EPR spectra of integer spin systems (*a*) P clusters of nitrogenase, (*b*) beef heart cytochrome *c* oxidase, and (*c*) the reduced hydroxylase component of methane monooxygenase. The spectra were recorded for B_1 parallel to B and plotted for equal sample concentrations (0.2 mM) and instrumental parameters ($T = 4$ K). [Adapted from Surerus et al.[30] (spectrum *a*), Hendrich and Debrunner[29] (spectrum *b*), and Hendrich et al.[38] (spectrum *c*) with permission from the publishers.]

How do these signals come about? We shall consider an $S = 2$ system and draw on the discussion in Section 3.14 of Abragam and Bleaney, 1970.[32a] Figure 18 shows the zero-field energies from Eq. (14) as a function of E/D for $D < 0$. The states $|m_s{}^{\pm}\rangle$ are defined as linear combinations of $m_S = \pm 1$ or of ± 2.

$$\begin{aligned} |m_{s^+}\rangle &= \cos\alpha| + m_S\rangle + \sin\alpha| - m_S\rangle \\ |m_{s^-}\rangle &= \sin\alpha| + m_S\rangle - \cos\alpha| - m_S\rangle \end{aligned} \tag{18}$$

where α is a mixing coefficient whose value is determined by the magnitudes of D and E relative to $g\beta B$. The difference in energy between $|m_{s^+}\rangle$ and $|m_{s^-}\rangle$ is

$$\Delta E = [(g'\beta B\cos\theta)^2 + \Delta^2]^{1/2} \tag{19}$$

where θ is the angle between B and the axis of D.

$$\text{For } m_S = \pm 1 \qquad g' = 2g_Z; \quad \Delta = 6E$$

$$\text{For } m_S = \pm 2 \qquad g' = 4g_Z; \quad \Delta = 3E^2/D$$

When the first term in the energy expression is much larger than the second, α approaches zero and $|m_{s^+}\rangle$ and $|m_{s^-}\rangle$ simplify to ± 1 or ± 2 as appropriate; when the first term is much smaller than the second α approaches 45° and $|m_{s^+}\rangle$ and

$|m_{s^-}\rangle$ become

$$|m_{s^+}\rangle = 2^{1/2}(|+m_S\rangle + |-m_S\rangle)$$
$$|m_{s^-}\rangle = 2^{1/2}(|+m_S\rangle - |-m_S\rangle) \tag{20}$$

These $|m_{s^+}\rangle$ and $|m_{s^-}\rangle$ states cannot be interconverted in a conventional EPR spectrometer with B_1 perpendicular to B_L for the standard selection rule of $\Delta m_S = \pm 1$ cannot be met. However, when the EPR instrument is configured so that B_1 is parallel to B_L, the selection rule changes to $\Delta m_S = 0$. Then transitions between states $|m_{s\pm}\rangle$ become possible. Such transitions are often referred to as $\Delta m_S = 2$ ($m_S = \pm 1$) or 4 ($m_S = \pm 2$); this is a misnomer, since these are not multiple quantum transitions. For states $|m_{s\pm}\rangle$, m_s is not a good quantum number and so any selection rule which uses m_s may be misleading. A better statement of the selection rule is $\Delta m_S = 0$, since the microwave interaction with the spin system derives from the S_z component in parallel mode (note that $\langle m_{s^-}|S_z|m_{s^+}\rangle = 1$, while $\langle m_{s^-}|S_x|m_{s^+}\rangle = \langle m_{s^-}|S_y|m_{s^+}\rangle = 0$).

The probability of these $\Delta m = 0$ transitions is proportional to $\Delta^2/[(g'\beta B\cos\theta)^2 + \Delta^2]$; consequently, one condition for the observation of these transitions is that E be nonzero. The transition requires a quantum $h\nu = \Delta E = \{(g_{\text{eff}}\beta B\cos\theta)^2 + \Delta\}^{1/2}$ and thus resonance always occurs at applied fields lower than expected from g_{eff} by an amount specified by Δ. For $m_S = \pm 2$, $g_{\text{eff}} = 8$ and thus signals are to be expected at applied fields less than 80 mT when measurements are made at 9 GHz. This magnetic field leads to observed "g values", which are unexpectedly large. In fact, as Δ becomes equal to or larger than $h\nu$, the resonances can be shifted to $B_L = 0$ and out of range of the spectrometer.

A good example of these effects is found in the spectra of reduced ferredoxin II ($S = 2$) of *Desulfovibrio gigas*. Figure 20 shows two EPR spectra of reduced ferredoxin II recorded with modes B_1 parallel and perpendicular to B_L, respectively.[33] Several features distinguish these signals as arising from an integer spin system. First, signals are observed in both modes. This immediately indicates that the signals derive from an integer spin center. Signals are present in both modes, because the random molecular orientations of solution samples allow for nonzero projections of B_1 along S_z for both modes. Half-integer spin signals, like the small impurity resonance at $g = 2$ in Figure 20(*c*), which is absent in Figure 20(*b*), are easily distinguished from integer spin signals. Second, integer spin signals have nonzero intensity at very low fields as is evident from the first integral, Figure 20(*a*), which shows absorption at $B_L = 0$. For most proteins, the value of Δ has a considerable spread. Thus, the fraction of molecules with $\Delta = h\nu$, in accordance with Eq. (19), resonate at $B_L = 0$. Third, owing to the distribution in Δ values, the spectra are broad and lack identifiable features that correlate directly with, molecular properties. Note that the minima of the spectra in Figure 20 do not occur at the same field, showing that one cannot obtain the g_{eff} value by simply marking the minima of the curves.

As a result of these complications, we must utilize computer simulations to obtain zero-field parameters and g_{eff} values. The dashed lines in Figure 20 are the result of simulations using $S = 2$ and Eq. (14), with $D = -2.5$ cm^{-1}, $E/D = 0.23$,

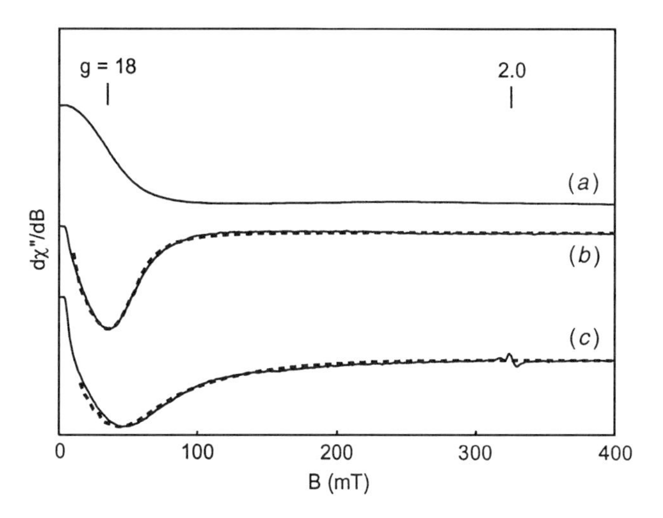

Figure 20
The X-band EPR spectra of the $S = 2$ state of the Fe_3S_4 cluster of reduced *D. gigas* ferredoxin II recorded at 4 K with B_1 parallel (b) or perpendicular (*c*) to B. The dashed curves are theoretical spectra computed for $D = -2.5\ cm^{-1}$ and $E/D = 0.23$. The absorption spectra (a) is a digital integration of spectrum *B* demonstrating presence of signal intensity at zero field; a common feature of integer-spin centers but not observed for half-integer-spin centers. [Adapted from Münck et al.[31] with permission from the publishers]

and $g_{eff} = 8$. These simulations determine the spread in the values of Δ, and, most importantly, the simulations are required to determine the spin concentration of the sample. These topics are discussed in detail in Hendrich and Debrunner[31] Hendrich et al.,[34] and Münck et al.[33]

III. The Origins of Structure in the EPR Spectrum

The discussion to this point has dealt with those phenomena that affect g values, and hence the general shape of the spectrum. We now turn to those processes that impose additional structure on the EPR envelope. These can be divided into two classes. The more common are the consequences of the paramagnetic electron being associated with a nucleus that possesses a nuclear spin; these are the nuclear hyperfine and superhyperfine interactions. Less common are the magnetic effects due to the paramagnetic electron being situated close to a second paramagnetic center such as is found when two or more iron–sulfur clusters are present in the same protein.

A. Hyperfine Interactions

Hyperfine interactions arise when the paramagnetic electron of interest finds itself within the sphere of influence of a nucleus that possesses a nuclear spin and consequently a nuclear magnetic moment. The interactions are divided into two classes. When the relevant nucleus is part of the parent atom of the paramagnet, that is, the electron interacts with its own nucleus, the phenomenon is simply called the hyperfine interaction. When the relevant nucleus belongs to a different part of the molecule, most commonly the ligand to the metal, then the interaction is often called the superhyperfine interaction.

Table III summarizes some of the properties of those nuclei that are of interest. Notice that some of the nuclear isotopes indicated are not naturally abundant to any significant extent but they are of considerable value in the laboratory.

Since the nuclei listed in Table III possess both a nuclear spin ($\mathbf{I}$) and charge they also possess a nuclear moment

$$\mu_n = g_n \beta_n \mathbf{I} \tag{21}$$

where g_n and β_n are the nuclear g factors and the nuclear magneton respectively.[w] In the presence of an applied magnetic field, a nucleus can adopt $2\mathbf{I} + 1$ orientations, as specified by the allowed values of $m_\mathbf{I}$. Thus the proton ($\mathbf{I} = \frac{1}{2}$) has two orientations corresponding to $m_\mathbf{I} = \pm\frac{1}{2}$. As the nuclear magneton is very much smaller than the electron magneton the relative energies of these nuclear orientations is also much smaller with the consequence that, in magnetic fields typically used in an EPR spectrometer, the excess number of nuclei in the stable ($m_\mathbf{I} = +\frac{1}{2}$) orientation[x] is about $1:10^6$; for our purposes the populations of the two nuclear spin states are identical.

Because of its magnetic moment the nucleus is the source of a magnetic field and this nuclear hyperfine field (B_{HF}) can combine with the applied, laboratory field (B_{L}) to provide the magnetic field required to satisfy the resonance condition. This can be illustrated by reference to the simplest of all possible cases, a system with $S = \frac{1}{2}$ and $I = \frac{1}{2}$, namely, the hydrogen atom.

The basis requirement for EPR is that

$$h\nu = g\beta B_{\mathrm{R}} \tag{22}$$

where B_{R} specifies the resonance field, the magnetic field *at* the electron needed to satisfy Eq. (22). In the present situation, B_{R} has two possible contributions

$$B_{\mathrm{R}} = B_{\mathrm{L}} + B_{\mathrm{HF}} \tag{23}$$

The value of B_{HF} is dictated by the nucleus under consideration; for the hydrogen atom it is 50 $m_I\, mT$.

[w] Unlike the electron for which the intrinsic g factor is always 2.0, the nuclear g factor is characteristic of the nucleus; it can be either positive or negative. For example, the proton has a nuclear g value of approximately 6 while for ^{17}O it is approximately -1. The nuclear magneton is 5.05×10^{-24} erg^{-1} G^{-1} approximately 2000 × smaller than the electron magneton, which is a consequence of the difference in the masses of the proton and the electron.

[x] Note the change of sign, a consequence of Eqs. (2) and (18).

Table III
Nuclear Spins for Commonly Studied Nuclei

Nucleus	Spin	% Natural Abundance
^{1}H	$\frac{1}{2}$	ca. −100
^{2}D	1	0.01
^{13}C	$\frac{1}{2}$	1
^{14}N	1	ca. 100
^{15}N	$\frac{1}{2}$	0.4
^{17}O	$\frac{5}{2}$	0.04
^{19}F	$\frac{1}{2}$	100
^{33}S	$\frac{3}{2}$	0.7
^{31}P	$\frac{1}{2}$	100
^{50}V	$\frac{7}{2}$	ca. 100
^{55}Mn	$\frac{5}{2}$	100
^{57}Fe	$\frac{1}{2}$	2
^{59}Co	$\frac{7}{2}$	100
$^{63,65}Cu$	$\frac{3}{2}$	100 (combined)
^{61}Ni	$\frac{3}{2}$	1
^{77}Se	$\frac{1}{2}$	7.5
$^{95,97}Mo$	$\frac{5}{2}$	25 (combined)

Assume that the magnetic field needed for resonance in the absence of any nucleus (i.e., $B_{\mathrm{HF}} = 0$) is 330 mT ($g = 2$, $\nu = 9.2$ GHz). In our sample, essentially one-half of the hydrogen atoms have $m_I = \frac{1}{2}$ and for these atoms $B_{\mathrm{HF}} = +25$ mT. Then to achieve a value of 330 mT for B_{R} we need only provide a value of 305 mT with B_{L}. For the remaining 50% of the hydrogen atoms $m_I = -\frac{1}{2}$ and B_{HF} is consequently −25 mT. To establish a value of 3300 for B_{R} we must now provide a value of 355 mT with B_{L}. Thus the EPR spectrum consists of two lines of equal intensity split symmetrically about the free electron position of 330 mT. One line appears at 305 mT on the spectral abscissa and is due to those atoms for which $m_I = +\frac{1}{2}$; the second line appears at 355 mT and arises from those atoms for which $m_I = -\frac{1}{2}$. The distance between these two lines, 50 mT, is called the hyperfine splitting and is given the symbol a, the hyperfine splitting constant.

The magnetic field is not a rigorous unit to quantify this phenomenon. The fundamental effect of the nuclear magnetic field is to modify the energy of the electron and thus it is appropriate to express the hyperfine splitting in energy units. We know that the resonance expression [Eq. (7)] provides a relation between magnetic field and energy (frequency) and by analogy we write

$$hA = g\beta a \tag{24}$$

where A is the hyperfine coupling constant in hertz; for $g = 2$, $A = 2.8\ a$ (MHz). A common alternative is to further transform the units to reciprocal centimeters using the relation that 30 GHz = 1 cm^{-1}. Thus (at $g = 2$) a hyperfine splitting of 10 mT can be expressed equally as an hyperfine coupling constant of 280 MHz or 0.0093 cm^{-1}.

The hyperfine interaction of the hydrogen atom can also be visualized as shown in Figure 21 and described in its legend. The basis for this figure is to be found in the equation for the energy in the presence of the hyperfine interaction:

$$\begin{aligned} \mathcal{H} &= -\mu \cdot \mathbf{B_R} \\ E &= -\mu_z B_\mathrm{R} \end{aligned} \tag{25}$$

provided z is parallel to the direction of B_R. With B_HF very much smaller than B_L this expression can be simplified by equating B_HF with am_I

$$\begin{aligned} E &= -\mu_z(B_\mathrm{L} + am_I) \\ &= g\beta m_s B_\mathrm{L} + g\beta a m_s m_I \end{aligned}$$

As $g\beta a = hA$

$$E = g\beta m_s B_\mathrm{L} + A(h) m_s m_I$$

Note that h is enclosed in parentheses because it is often omitted, being accommodated in the units of the spin momenta. The resultant energy levels are shown in Table IV.

For each particular value of electron spin, the original energy level is split by $\pm A/4$ (Fig. 21). As the experiment consists of irradiating the sample with frequencies appropriate to the separation in the energy levels of the electron the allowed transitions involve flipping the electron spin without any change in the orientation of the associated nuclear spin, thus the selection rule for allowed

Table IV
Hyperfine Energies for the Hydrogen Atom

	$m_s + \frac{1}{2}$	$m_s = -\frac{1}{2}$
$m_I = +\frac{1}{2}$	$\frac{1}{2}g\beta B_\mathrm{L} + A/4$	$-\frac{1}{2}g\beta B_\mathrm{L} - A/4$
$m_I = -\frac{1}{2}$	$\frac{1}{2}g\beta B_\mathrm{L} - A/4$	$-\frac{1}{2}g\beta B_\mathrm{L} + A/4$

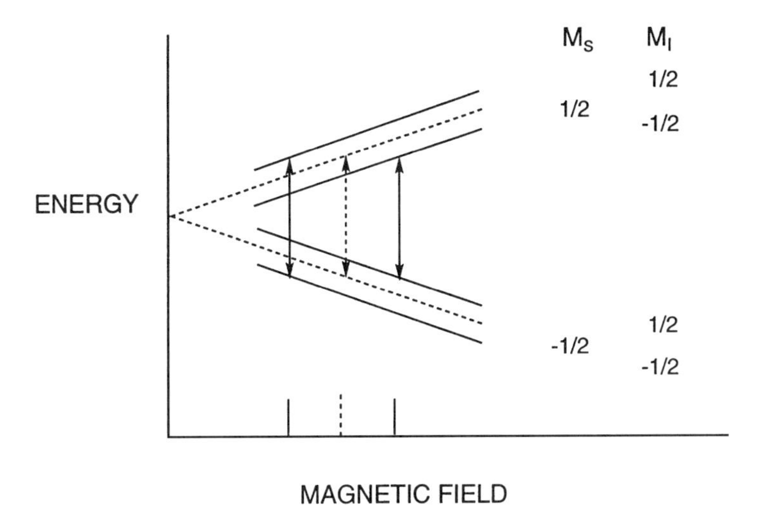

Figure 21
The behavior of the energy levels of the $M_s = \pm\frac{1}{2}$ states of a $S = \frac{1}{2}$ paramagnet upon interaction with a $I = \frac{1}{2}$ nuclear moment. The broken lines show the levels before interaction and the solid lines the levels after the interactions. The vertical arrows and the bars on the abscissa represent the location of the transitions in the absence and the presence of the hyperfine interaction.

transitions is $\Delta m_s = \pm 1$, $\Delta m_I = 0$. These transitions are represented by the double-headed arrows in Figure 21 and show that transitions that originate from $m_S = -\frac{1}{2}$, $m_I = +\frac{1}{2}$ are to lower magnetic fields by an amount equal to $A/2$ while transitions originating from $m_S = -\frac{1}{2}$, $m_I = -\frac{1}{2}$ are to higher applied fields by the same amount. The difference in magnetic field, which separate the two transitions, is consequently specified by A.

When the electron is only exposed to a single nuclear moment the number of hyperfine lines is $2I + 1$ and each of these hyperfine lines is of equal intensity; this is always the case for an electron interacting with its own nucleus.

B. Anisotropies in the Hyperfine Interaction

The hyperfine interaction has two general origins. The first is the so-called "contact" interaction. It is the isotropic component of the hyperfine interaction and arises because the unpaired electron has a finite probability of being in the same space as the nucleus. In transition metals, the electron is localized in a d orbital that has a node (zero crossing) at the nucleus and thus, at first sight, should not occupy any nuclear space. Various mechanisms are invoked to rationalize this apparent discrepancy; the easiest to visualize is to take the node as a point but the nuclear volume as finite so that there is net overlap between the electron and the nucleus (see Wertz and Bolton[9] for a detailed discussion).

The value of the isotropic contact term, A_{iso}, is given by $-\kappa P$ with the dimensionless κ measuring the extent to which the electron "occupies" the nucleus while P converts κ into the relevant energy units; $P = g_e g_n \beta_\varepsilon \beta_n \langle r^{-3} \rangle$, the term in pointed brackets quantifying how fast the d orbital decays as one moves away from the nucleus. The parameter P is the quantum mechanical equivalent to $\mu_e \mu_n / r^3$; it varies with the nucleus (via $\langle r^{-3} \rangle$).

The hyperfine interaction can also be anisotropic. This anisotropy arises because the electron occupies a d orbital of finite size, which is distributed in a limited region of space and can thus undergo a "through-space" dipolar interaction with the nuclear magnetic moment.

For convenience, this dipolar interaction can be broken down into two parts. The first part is between the spin component of the electron's magnetic moment and the nuclear moment and can be determined from the shape of the orbital in which the electron is confined and the expression that describes the spatial dependence of the dipole–dipole interaction (see Appendix V)

$$V_{12} = \frac{\mu_e \mu_n}{r^3}(1 - 3\cos^2\theta) \tag{26}$$

Equation (26) is for point dipoles. Now while it may be acceptable to treat the nucleus as a point in space it is certainly not a valid assumption for the electron that is distributed throughout its parent d orbital. Consequently, the interaction between the electron and nucleus must be averaged over the electron's spatial distribution. (An example of such a calculation can be found in Appendix III). For a single unpaired electron confined to a d orbital, the x, y, and z components of the interaction energy are listed in Table V. In this table, all entries are in units of P/7. From this table, it is clear that (a) the dipolar contribution is anisotropic (the entries depend on the direction); and (b) that it averages to zero (the sum across a line is zero). Thus these dipolar terms should only be observed under conditions where the sample is rigid (frozen liquids). In solution, they disappear because the rapid tumbling motion of the paramagnetic averages the directional variation unless this motion of the molecule is restricted, as in usually the case with proteins.

For traditional square planar Cu(II), the unpaired electron is located in $d_{x^2-y^2}$. Using Table V we can now deduce that $A_x = (\frac{2}{7} - \kappa)P$, $A_y = (\frac{2}{7} - \kappa)P$, and $A_z = (-\frac{4}{7} - \kappa)P$. Thus A_z is expected to be large (and negative) while A_x and A_y will be much smaller, and might even be zero (in the special circumstance that $\kappa = \frac{2}{7}$).

The second component of the dipolar interaction is due to the orbital contribution to the magnetic moment of the electron; it can be estimated using the deviations of the g factor from 2 for, as described earlier, these deviations reflect the orbital magnetism (at least for $S = \frac{1}{2}$). The full expressions for Cu(II) including this orbital term are[35]

$$\begin{aligned} A_x = A_y &= [\tfrac{2}{7} - \kappa + \tfrac{11}{14}(g_{x,y} - g_e)]P \\ A_z &= [-\tfrac{4}{7} - \kappa + (g_z - g_e) + \tfrac{3}{7}(g_{x,y} - g_e)]P \end{aligned} \tag{27}$$

Table V
Anisotropic Hyperfine Contributions Due to the Dipolar Interaction of a Nucleus with a Single *d* Electron[a]

	x	y	z
z^2	−2	−2	4
$x^2 - y^2$	2	2	−4
yz	−4	2	2
xz	2	−4	2
xy	2	2	−4

[a] All entries are in units of P/7 (see text).

For Cu(II) $P \approx 0.035\ \text{cm}^{-1}$, κ lies between 0.23 and 0.32 (depending on covalency), $g_x \approx g_y = 2.1$ and g_z is about 2.3; $g_e = 2.0023$. Thus $A_z = 0.02\ \text{cm}^{-1}$ while A_x and $A_y \approx 0.001\ \text{cm}^{-1}$ ($\sim$20 and 1 mT, respectively).

C. The Super-Hyperfine Interaction

The extension of these principles to the case where the magnetic nucleus is a satellite rather than the parent nucleus of the paramagnet is very straightforward. This effect is particularly striking in organic free radicals when the unpaired electron is often found in an orbital delocalized over a large aromatic framework. and thus the electron might be under the influence of a large number of nuclei. This topic is discussed in detail in Chapter 4 of Wertz and Bolton.[9] The effect can also be observed in metal complexes in cases where the metal ligands have a nuclear moment. For an electron exposed to n equivalent nuclei,[y] the number of lines is equal to $2nI + 1$ but the intensity of the individual lines varies. For $I = \frac{1}{2}$ nuclei, the intensity distribution can be predicted by the straightforward application of Pascal's triangle; for example, three equivalent $I = \frac{1}{2}$ nuclei yields a four line pattern with relative intensities 1 : 3 : 3 : 1. In general, for n equivalent nuclei of nuclear spin I in intensity distribution is given by the coefficients of the expansion of the formula:

$$\prod_{1}^{n} \sum_{i=0}^{2I} x^i \tag{28}$$

[y] In the present context, equivalent nuclei have the same values for both I and A. Chemically identical nuclei with different values for A are inequivalent for they will absorb energy at different applied fields.

For example, for three $\mathbf{I} = \frac{1}{2}$ nuclei this formula yields $(1 + x)^3 = 1 + 3x + 3x^2 + x^3 = 1 : 3 : 3 : 1$.

The only real complications arise when (a) $n > 1$, and (b) there may be several classes of nuclei. For example, in the nitrosyl complexes of iron(II) heme proteins there are two inequivalent axial nitrogen ligands. The first nitrogen couples strongly to the unpaired electron and yields a widely split triplet with each component of equal intensity and separated by 2.1 mT. The coupling to the second nitrogen is weaker and splits each component of the original triplet into a second triplet, though in this case the splitting is one-third the size, about 0.7 mT. Barring accidental overlap the result is a nine-line spectrum in which each component is of relative intensity one. (The procedure for constructing these splitting patterns is precisely the same as that used in analyzing the splitting patterns due to spin–spin interactions seen in the NMR spectra of AX systems.)

Even though superhyperfine interactions can also be anisotropic, by mechanisms similar to those described for hyperfine splittings, it is important to appreciate that superhyperfine splitting patterns from an isotropic interaction might only be observed on one of the three g features of the EPR spectrum. For example, in the heme–nitrosyl system just described the unpaired electron has z symmetry, being delocalized from the NO ligand into d_{z^2} of the iron atom. Thus only nitrogen atoms present on the z axis contribute in a significant way to the splitting pattern; the pyrrole nitrogens, which lie in the xy plane, interact only weakly with the electron. Likewise for axially elongated Cu(II) the electron is confined to $d_{x^2-y^2}$ and only the nuclei of ligands present in the xy plane can have a strong interaction with the electron. Even though ligand nuclei may be present on the z axis their splitting is usually too small to be resolved in CW spectra.

The question naturally arises "How does one tell if an observed splitting arises from a hyperfine or superhyperfine interaction?" The answer depends on the spectrum being considered. If the interaction is with a single nucleus, so that all hyperfine components are of equal intensity, then a clear potential for ambiguity exists and one must draw upon other information. Fortunately, Nature has conspired to be of major help for the most abundant ligand nucleus is nitrogen, typically originating in the imidazole sidechain of a histidine residue. For nitrogen, $I = 1$; there is no commonly available paramagnetic metal for which $I = 1$ (see Table III). Thus this potential ambiguity is not as serious as might be anticipated. When there is more than one nucleus the number of lines might be the same as those to be expected from a single metal nucleus. For example, with three $I = \frac{1}{2}$ nuclei we expect four lines, the same number as those anticipated from the Cu(II) nucleus; however, the hyperfine lines expected from the Cu(II) nucleus will have the relative intensity $1:1:1:1$ while those from the three $I = \frac{1}{2}$ nuclei will be in the ratio $1:3:3:1$

Most ambiguities are found in the situation where there are two (or more) inequivalent ligand nuclei and one wishes to establish which ligand is responsible for the major splitting and which is for the minor splitting. Resolution of this issue requires that either or both of the ligands in question be replaced with their homolog is which the liganding atom has been substituted with an appropriate isotope (see Table III). For nitrogen, this means replacing ^{14}N ($I = 1$) with

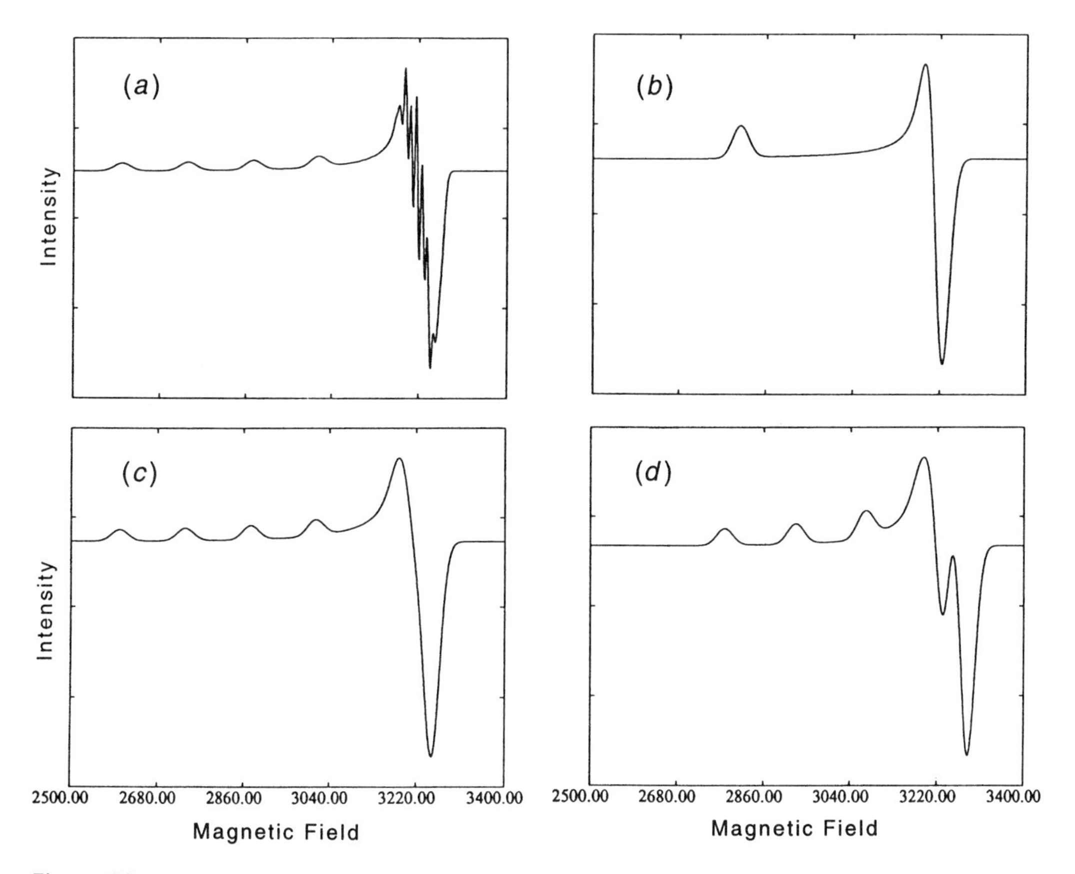

Figure 22
An illustration of the mental process involved in the analysis of an EPR spectrum (see the text for a description). The spectra were calculated using the following parameters; (*b*): $g_x = 2.05$, $g_y = 2.05$, $g_z = 2.35$; (*c*): a hyperfine coupling to a $I = \frac{3}{2}$ nucleus was added with $A_x = 0.001$, $A_y = 0.001$, and $A_z = 0.015\ \text{cm}^{-1}$. For (*a*) an additional coupling to two equivalent nitrogens was added with $A_x = 0.001$, $A_y = 0.001$, and $A_z = 0\ \text{cm}^{-1}$. For all three cases, the line widths were $W_x = 0.5$, $W_y = 0.5$, and $W_z = 2.0$ mT. The microwave frequency was 9.232 GHz. For (*d*), the parameters were $g_x = g_y = 2.06$, $g_z = 2.285$; $A_x = A_y = 0.001$, $A_z = 0.02\ \text{cm}^{-1}$; $W_x = W_y = W_z = 2.0$ mT.

^{15}N ($I = \frac{1}{2}$). For example, for heme nitrosyl compounds the use of ^{15}NO in place of ^{14}NO converts the triplet of splitting 2.1 mT into a doublet of splitting $1.4 \times 21 = 30$ mT.[z] The smaller splitting of 0.7 mT can then be attributed to the nitrogen nucleus of the proximal histidine.

D. An Example of Spectral Analysis

The information so far acquired can now be synthesized in the analysis of the EPR spectrum of a typical copper compound, shown in Figure 22(*a*).

---

[z] The factor of 1.4 reflects the fact that the ratio $\mu/\mathbf{I}$ for ^{15}N is 1.4 times that of ^{14}N. This ratio is reflected in the hyperfine coupling constants.

One's first task is to try and establish the nature of the g anisotropy that is present in this spectrum. This task is complicated by the hyperfine and superhyperfine structure that is present but, even so, one should be able to recognize that the overall spectral envelope is axial with $g_\perp$ smaller than $g_\parallel$, as illustrated in Figure 22(b). The next step is to recognize that the peak at $g_\parallel$ has been split into four lines of equal intensity (these are symmetrically disposed about the original peak) implying the presence of a nuclear spin of $\frac{3}{2}$ that immediately suggests Cu(II). This then leads to the spectrum shown in Figure 22(c). Finally, one hopefully realizes that the difference between Fig. 22(a and b) is the presence of the five-line splitting pattern present on $g_\perp$. As there is no nuclear spin of 2 that is present in high natural abundance, this five-line pattern must arise from the coupling to two-or-more nuclei. Two equivalent nitrogens with $I = 1$ will lead to a five-line pattern and this would undoubtedly be one's first prediction. These should be in the ratio 1:2:3:2:1. This ratio of intensities is not immediately obvious from the spectrum shown. Because the splitting falls on the steep portion of the EPR envelope it is often necessary to subtract out the background EPR to reveal the correct ratio of intensities. Actually, a better approach would be to use a computer program to calculate spectra for specific guesses of the nuclear spin and coupling constants; in this instance, the spectral parameters given in the legend to Figure 22 were used to produce these spectra. With this data one would undoubtedly conclude that this EPR spectrum was due to Cu(II) in a square planar complex with two equivalent nitrogen ligands present in the equatorial coordination sites.

Figure 22(d) shows an interesting phenomenon that occurs with certain values of $g_\parallel$. Then for some orientations of the copper with respect to B_L, the copper hyperfine line that arises from the $m_I = -\frac{3}{2}$ nuclear level is actually located to slightly higher field than is $g_\perp$ with the result that there is an apparent rhombic splitting of $g_\perp$, as shown in the figure. If one were not aware of this phenomenon one might well conclude that this is a copper complex with a marked rhombic distortion. The lesson here is that copper samples that yield apparently rhombic spectra should be reexamined at higher microwave frequencies (e.g., 35 GHz). At the higher frequency, the separation in millitesla between $g_\perp$ and $g_\parallel$ increases by about a factor of 4; as the magnitude of the hyperfine splittings do not change with magnetic field the $m_I = \frac{3}{2}$ line will now be found to much lower field in the higher frequency spectrometer and a spectral envelope resembling Figure 22(c) will once more be obtained. The feature found at high field in Figure 22(d) is usually referred to as the "overshoot" line. More examples of copper EPR spectra can be found in the books edited by Lontie.[36]

E. Long-Range Electron–Electron Interactions

We saw earlier that two paramagnets in close proximity can have a profound effect on the EPR spectrum. In fact, when two paramagnets are separated by distances that are sufficiently large that the possibility of direct or ligand mediated bonding seems remote we are still faced with the possibility of a through-space dipolar interaction. This interaction is described formally by precisely the same equation used to describe the dipolar part of the hyperfine interaction [Eq. (22)] with the

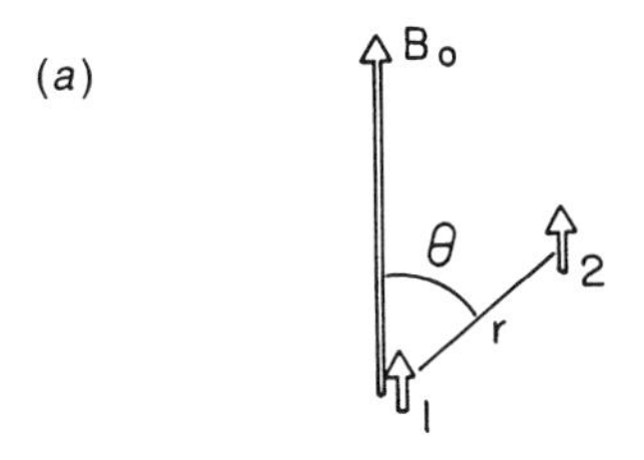

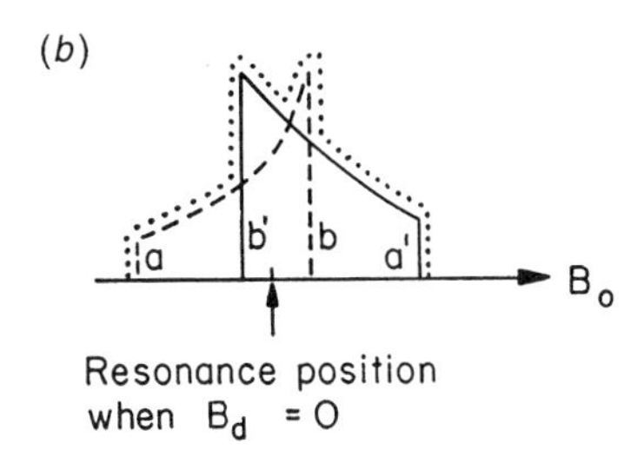

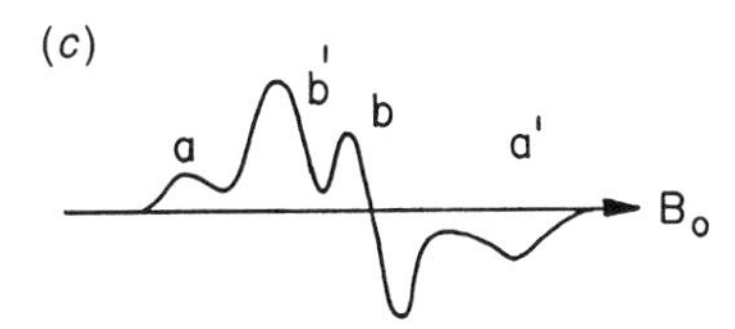

Figure 23
The EPR spectrum from two isotropic $S = \frac{1}{2}$ centers that interact through-space via a dipole–dipole interaction. The top of the figure shows one possible orientation for the two centers and the definition of the angle θ used in Eq. (29). See text for additional explanation.

simple change that μ_1 and μ_2 [Eq. (A.III.1)] are now the magnetic moments of the two paramagnets that are involved.

The phenomenon is best described in terms of the simplest possible model, namely, two $S = \frac{1}{2}$ magnetic moments with an isotropic g factor = 2.0 arranged as shown in Figure 23. Then the dipolar magnetic field (B_D, mT) produced by the spin-up No. 2 at spin No. 1 is

$$B_D(\mathrm{mT}) = \frac{\mu_e}{r^3}(1 - 3\cos^2\theta) \approx \frac{1{,}400}{r^3}(1 - 3\cos^2\theta) \tag{29}$$

(Figure 23 and the following text refer to spin Nos. 1 and 2. The reader should understand this to mean the magnetic moments associated with these spins.) When r, the distance between the two moments, is infinitely large the dipolar field is zero and EPR is obtained when $B_R = B_L$; say at 330 mT. To appreciate the consequences of finite r we let the two paramagnets be separated by a distance of

10 Å, whereupon r^3 is 10^3 Å^3. For certain molecules, θ is zero (spin No. 2 lying above or below spin No. 1). The parameter $B_D = -2.8$ mT; thus B_L must be 332.8 mT to establish resonance and these molecules will exhibit EPR at the location marked a′. The intensity will be relatively small because the polar volumes contain only a small number of molecules. When $\theta = 90°$ spin No. 2 lies on the equator with respect to spin No. 1. Now $B_D = +1.4$ mT and for these molecules B_L need only be 328.6 mT to meet the resonance condition. A relatively large number of molecules are found at the equator, and hence the intensity at location b′ will be more intense than at a′. Molecules oriented so that θ lies between 0° and 90° (as in a frozen solution) will exhibit intermediate values for B_D and these will require intermediate values of B_L. The total EPR contribution from the 50% of molecules in which spin No. 2 has $m_s = +\frac{1}{2}$ is given by the solid line in Figure 23(*b*). In the remaining molecules spin No. 2 has $m_s = -\frac{1}{2}$ and for these molecules the sign of B_D is reversed. This leads to the dashed line in Figure 23(*b*). The dotted line in the same figure is the combination of the solid and dashed lines and yields the EPR contribution due to spin 1. But in this example spin Nos. 1 and 2 are identical and the effects of spin No. 1 upon spin No. 2 are exactly the same as those of spin No. 2 on spin No. 1. By repeating the above arguments, we find that the dotted line also represents the EPR envelope due to spin No. 2 and the total EPR is twice the amplitude of the dotted line. The derivative of this dotted line is shown in Figure 23(*c*) and represents the EPR spectrum expected for this example. Note that the distance from *a* to *a*′ is measured readily; it will be 5.6 mT and represents $2B_D$ when $\theta = 0°$. Insertion of these values into Eq. (24) allows r to be quickly determined.

There are three important complications to the situation just described. The first exists when the g factor is anisotropic. Then the magnitude of the interaction in a particular direction will be modulated by the magnitude of the g factor of each spin in that direction. The second occurs when the paramagnets are not identical, most simply because their g tensors are not collinear. These two complications are readily handled by computer programs that can calculate the lineshape expected for any arbitrary mutual orientation of the g tensors. The final complication is best illustrated by an example.

Many bacterial ferredoxins contain two four–iron iron–sulfur clusters, which are located about 12 Å apart (center-to-center); when fully reduced both clusters are $S = \frac{1}{2}$ paramagnets. If the degree of reduction of the sample is adjusted so that, on average, only one cluster per protein is reduced, then the observed EPR spectrum has a simple rhombic shape at both 9 GHz [Fig. 24(*a*)][37] and 35 GHz (*b*). However, the EPR spectrum of fully reduced ferredoxin recorded at 9 GHz exhibits a most complicated lineshape [Fig. 24(*a*)], which clearly contains more features than is expected from one of the four basic EPR profiles (Fig. 4). Furthermore, it is not obvious how adding together two of the basic profiles would lead to the observed spectrum and so one might anticipate that in the fully reduced protein there is a dipolar interaction between the two clusters. However, when the spectrum of the same sample is recorded at 35 GHz the spectral envelope changes in a dramatic way [Fig. 24(*b*)].

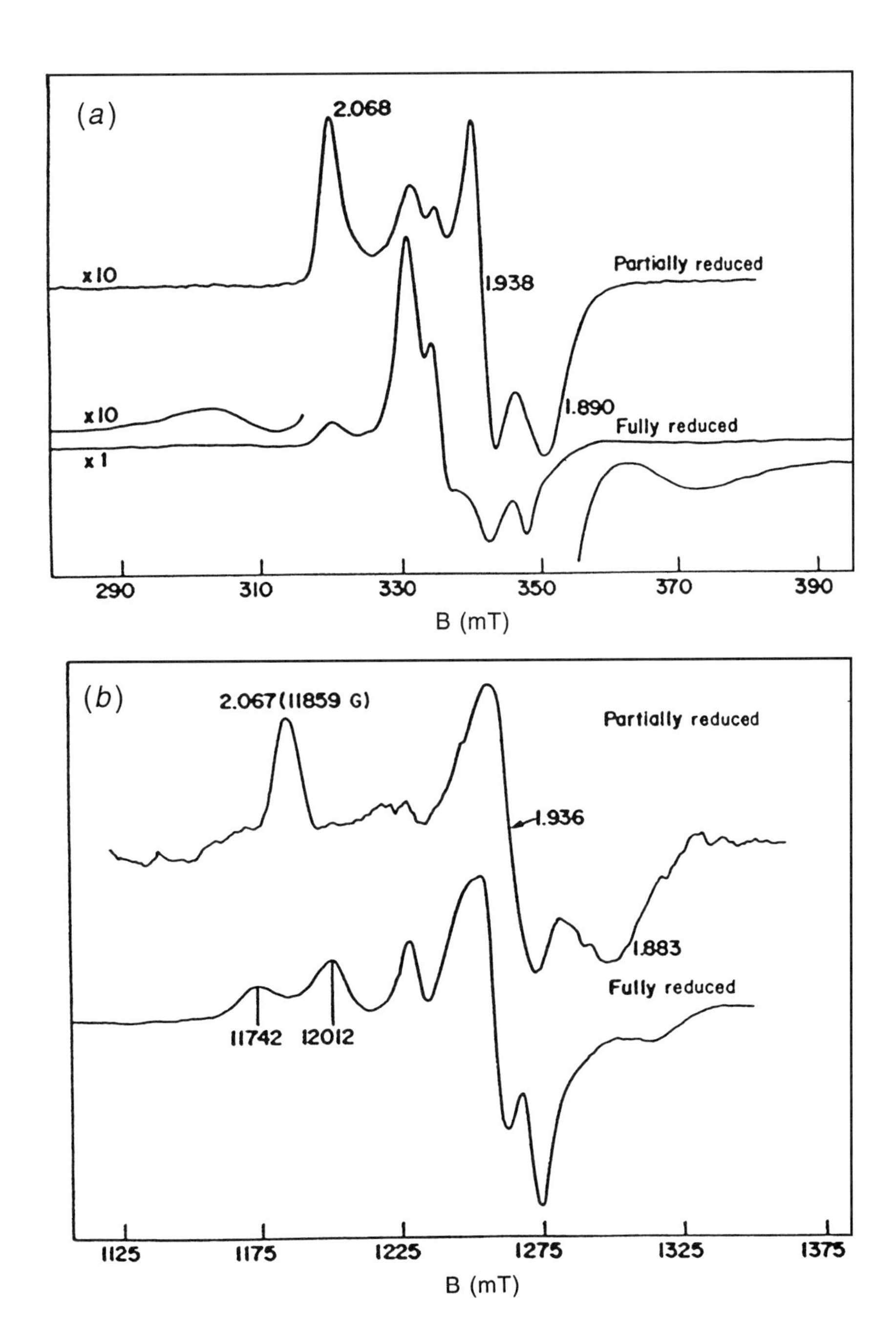

Figure 24
The EPR spectra of the ferredoxin from *Micrococcus lactilyticus* at 9 GHz (*a*) and 35 GHz (*b*). In each panel, the spectrum is shown of a sample in which both iron–sulfur clusters are reduced (fully reduced) and a sample in which only one of the two clusters is reduced (partially reduced). Note that the spectra of the partially reduced sample is essentially the same at the two frequencies while the spectra of the fully reduced sample are strikingly different. [Adapted from Mathews et al.[37] with permission from the publishers]

Now the dipolar interaction (like the hyperfine interaction described earlier) are small effects relative to the Zeeman interaction and should not be field dependent, that is, the spectra at 9 and 35 GHz should be qualitatively similar. (Indeed, were the g tensor isotropic they would look identical). The demonstration that the spectra are dramatically sensitive to spectrometer frequency implies that some other interaction is present.

This interaction has an analogue in NMR, called the AB case. In NMR, this occurs when a molecule has two nuclei with slightly different chemical shifts and there is a spin–spin coupling (parameterized by the quantity J) between the nuclei that is of a magnitude comparable to this difference. Cast into EPR language the phenomenon requires that the two paramagnets exhibit a difference in g factor (Δg) along a particular direction and that there be spin–spin coupling in the same direction that is comparable to Δg. When these conditions hold, the EPR spectrum at a particular orientation will consist of four lines the *locations and intensities* of which depend on the *relative* magnitudes of the Zeeman interaction (via Δg) and J at this orientation; this four-line pattern has to be summed over all orientations. Even if J is isotropic the variation in Δg with orientation will lead to a net EPR spectrum that is extremely complex. Furthermore, as the contribution of Δg is manifested through its interaction with B_L, spectra recorded at different frequencies will exhibit a field dependence with the spectra at higher fields becoming increasingly simplified as the Zeeman term increasingly dominates J. In NMR language, the AB spectrum is converted to an AX spectrum. However, even at an applied field sufficiently large to effectively suppress the contribution from J the complications due to the through-space dipolar interactions will still be present and the spectra "simplify" to one resembling that described in Section III.E. (For further discussion of this topic, see the article by Sands[38] and the book by Bencini and Gatteschi.[39])

IV. Coda

The material in this chapter has dealt primarily with the energetic interactions of unpaired electrons and their environments and how a study of these interactions can be of value in characterizing these microenvironments. The reader should be aware, however, that a number of important aspects of electron paramagnetic resonance have been omitted. The topics that have been omitted are those that were believed to be of little utility in furthering the objectives of this chapter. The reader should, however, not assume that these topics should be ignored. Indeed in other areas of contemporary research it is the topics that have been omitted that are of paramount importance. This finding is particularly true if one is interested in the effects of motion on EPR spectra, as would be studied in both conventional experiments with spin labels and with saturation transfer EPR. An up-to-date treatise on this aspect of EPR is provided in the book edited by Berliner and Reuben.[40] The reader should also be aware of the developments in pulsed EPR techniques; these are covered in Chapter 4 of this volume by Chasteen and Snetsinger.

ACKNOWLEDGMENTS

I wish to thank G. R. Eaton, S. S. Eaton, M. P. Hendrich, E. Münck, and L. Que for their input on this chapter. I would also like to acknowledge my longtime collaboration with Richard H. Sands who did his best to educate me in the physics of epr. The preparation of this chapter was supported by grants from the National Institutes of Health (GM 21337) and the Robert A Welch Foundation (C636).

APPENDIX I

Spin–Orbit Coupling

If an electron, of mass m, rotates clockwise about its nucleus with a velocity $\mathbf{v_e}$, the electron perceives the nucleus rotating clockwise about it at a velocity $-\mathbf{v_n}$ (from reversing the sign of $\mathbf{r}$). The magnetic field ($\mathbf{B}$) at the electron from the nucleus is then given by Jackson.[41]

$$\mathbf{B} = \frac{-1}{c^2}\mathbf{E} \times \mathbf{v_n} = \frac{1}{c^2}\mathbf{E} \times \mathbf{v_e}$$

As the force, $\mathbf{F}$, on the electron $= -e\mathbf{E}$ and as $\mathbf{F} = \frac{-Ze^2}{r^2} \cdot \frac{\mathbf{r}}{r}$ where $\mathbf{r}/r$ is a *unit* vector in the direction of $\mathbf{F}$ (i.e., along a radius) then

$$\mathbf{E} = \frac{Ze\mathbf{r}}{r^3}$$

So

$$\mathbf{B} = \frac{Ze}{c^2 r^2}\mathbf{r} \times \mathbf{v_n}$$

$\mathbf{L}$, the orbital angular momentum (OAM) of the electron $= \mathbf{r} \times m\mathbf{v_e}$, $\mathbf{L}/m = \mathbf{r} \times \mathbf{v_e}$ and

$$\mathbf{B} = \frac{Ze}{mc^2 r^3} \cdot \mathbf{L}$$

The potential energy, E, of the electron in this field $= -\mu \cdot \mathbf{B}$ and $\mu = -g\beta\mathbf{S}$

$$E = \frac{-Ze^2 gh}{2m^2c^2r^3\, 2\pi}\mathbf{L} \cdot \mathbf{S}$$

$$= \frac{-Ze^2}{2m^2c^2r^3} \cdot \mathbf{L} \cdot \mathbf{S}$$

where the definition of β has been used explicitly, g is taken as 2 and $\mathbf{S}$ represents the spin angular momentum (p. 122).

Clearly, the quantity $(-Ze^2/2m^2c^2r^3)\mathbf{L}$ is equivalent to a magnetic field; this ratio is our first approximation to a quantity called the *spin–orbit coupling constant* (ζ); it is a positive quantity. The approximation can be "improved" by

1. Adding a factor or two in the denominator; this arises when the relative motions of the electron and nucleus are correctly treated (the electron is in a circular orbit, which means that it is always accelerating).
2. Replacing the classical orbit of the electron. In practice, one must acknowledge that the electron is in a diffuse orbit with a complicated radial dependence. This is accommodated by replacing $1/r^3$ by $\langle 1/r^3 \rangle$, its average (or expectation) value, which is obtained by integrating the radial part of the wave function from zero to infinity.
3. Treating the nuclear charge correctly. For an electron in an $n = 2$, $\ell = 1$ orbit of hydrogen (the first orbit with any OAM) the "perceived" magnetic field is about 1 T. The value is about 400 T for iron and 800 T for copper; it increases rapidly with nuclear charge. Why does it increase with nuclear charge, that is, why do the core electrons not shield the valence electron from the nuclear charge? Well they do, in part. But all orbitals orginate at the nucleus and that part of, for example, a 3d orbital that is close to the nucleus is much less shielded than is the part that lies "outside" the core. Thus the full expression for the spin–orbit coupling constant takes care of all the above factors and becomes[42]

$$\zeta = \frac{\alpha^2 \mathrm{R_H} Z^4}{n^3 \ell(\ell + \frac{1}{2})(\ell + 1)}$$

α is the "fine structure constant" $e^2/(4\pi\varepsilon_0 \hbar c) = \frac{1}{137}$, and $\mathrm{R_H}$, the Rydberg constant[aa] for hydrogen $= m\mathrm{e}^4/(8\varepsilon_0{}^2 h^3 c) = 10^5\ \mathrm{cm}^{-1}$.

One usually sees the spin–orbit coupling represented by the symbol $\lambda (= \pm\zeta/2S)$, where the plus sign is used for shells that are less than one-half full ($\mathrm{d}^1 - \mathrm{d}^4$) and the minus sign is used for shells that are greater than one-half full ($\mathrm{d}^6 - \mathrm{d}^9$). The practical consequence of using λ is that one can take the theoretical expressions for Ti(III), for example, (having a single d electron) and use it for Cu(II), which contains nine d electrons, one short of a full shell, and which is thus equated to a single positively charged "electron hole".

In summary, because the energy of an electron is lowered when it is exposed to a magnetic field this stabilization serves to "motivate" the electron to acquire an orbital circulation with the attendant development of the spin–orbit magnetic field.[bb]

---

[a] The Rydberg constant is a parameter that defines the energy of an atomic orbital with a particular value for n as $E_n = -hcR_0/n^2$.

[b] If the above treatment is done in the MKS system the very first equation contains $1/c^2$ but the Bohr magneton lacks the c in the denominator, so that the final result is the same.

APPENDIX II

EPR of Low-Spin Hemes

By applying the Zeeman operator $[\beta H(\mathbf{L}+2S)]$ to the modified orbitals shown in Eqs. (11) Taylor[12] deduced

$$g_z = 2[(a+b)^2 - c^2] \tag{A.II.1}$$

$$g_y = z[(a+c)^2 - b^2] \tag{A.II.2}$$

$$g_x = 2[a^2 - (b+c)^2] \tag{A.II.3}$$

or equivalently

$$a = \frac{g_z + g_y}{D} \qquad b = \frac{g_z - g_x}{D} \qquad c = \frac{g_y - g_x}{D} \tag{A.II.4}$$

with $D = [8(g_z + g_y - g_x)]^{1/2}$.

The squared terms in Eqs. (A.II.1–A.II.3) reflect the orbitals that are related by rotation, and hence contribute to the orbital angular momentum about the pertinent axis. By assuming no contribution from covalency or participation of excited states, then $a^2 + b^2 + c^2 = 1$ and

$$g_x^2 + g_y + g_z^2 + g_y g_z - g_x g_z - g_x g_y + 4(g_x - g_y - g_z) = 0 \tag{A.II.5}$$

which provides a check for the internal consistency of measured g values.

The ligand field parameters A, B, C, V, and Δ can be expressed in terms of either a, b, and c or the g values:

$$\begin{aligned} A &= E(yz) - E(xz) \\ &= \frac{a+c}{2b} - \frac{b+c}{2a} = \frac{-g_x}{g_z + g_y} + \frac{g_y}{g_z - g_x} \end{aligned} \tag{A.II.6}$$

$$\begin{aligned} B &= E(yz) - E(xy) \\ &= \frac{a+b}{2c} - \frac{b+c}{2a} = \frac{g_x}{g_z + g_y} + \frac{g_z}{g_z - g_x} \end{aligned} \tag{A.II.7}$$

$$\frac{\Delta}{\lambda} = A \tag{A.II.8}$$

$$\frac{\Delta}{\lambda} = B - \frac{A}{2} = \frac{g_x}{2(g_z + g_y)} + \frac{g_z}{g_y - g_x} - \frac{g_y}{2(g_z - g_x)} \tag{A.II.9}$$

Note: When using Eqs. (A.II.8) and (A.II.9) to make entries on Blumberg–Peisach graphs, g_x, g_y, and g_z should be assigned as $-g_{\text{mid}}$, g_{max}, and $-g_{\text{min}}$, respectively, that is, an improper coordinate system was selected for the original analysis. For a proper coordinate system,[13] g_x, g_z, and g_x correspond to g_{min}, g_{mid}, and $-g_{\text{max}}$ (g_{min} and g_{max} are numerically the smallest and largest g values).

From Eqs. (A.II.1–A.II.3), the following special cases can be deduced:

1. When the unpaired electron is confined exclusively to d_{yz}, then $a = 1$, $b = c = 0$ and $g_x = g_y = g_z = 2$.
2. When the unpaired electron is shared equally by d_{yz}, d_{xz}, and d_{xy}, then $a = b = c = (0.33)^{1/2}$ and $g_z = g_y = g_z = 2$.
3. In the limit of axial symmetry then the contributions of d_{zy} and d_{xz} are equal. Then for finite values of $\Delta(a^2 < 0.5)$:

$$g_z = 12a^2 - 2 \tag{A.II.10}$$

$$g_x = g_y = 1 - 2a^2 - (4a^2 - 8a^4)^{1/2} = (4a^2 - 8a^4)^{1/2} \tag{A.II.11}$$

For an axial ligand field of sufficient strength (Δ approaching infinity) a^2 approaches 0.5, g_z approaches 4 and $g_x = g_y$ approach 0.

APPENDIX III

The Dipolar Contribution

To evaluate anisotropic contribution to the hyperfine interaction, one needs to solve integrals of the form

$$\int_{\psi^*} \mathscr{H}_d \psi \, d\tau \quad (d\tau \text{ is a volume element})$$

For real orbitals $\psi^* = \psi$, while $\mathscr{H}_d$, the recipe for the dipolar interaction (see Appendix V), can be written as

$$g_e g_n \beta_e \beta_n \left(\frac{3i^2 - r^2}{r^5} \right) \qquad i = x, y, \text{ or } z$$

The real orbitals are most commonly found represented in spherical coordinates and are the product of three terms A, B, C, where

A is a number $[(5/14\pi)^{1/2}$ for d orbitals]
B contains the r dependence and need not be explicitly evaluated.
C is a simple trigonometric term that contains the angular dependence of the orbital. For example, for d_{z^2} it is $3\cos^2\theta - 1$ and, for $d_{x^2-y^2}$ it is $\sin\theta\cos^2\phi$

Thus $d_{z^2} = (5/14\pi)^{1/2} f(r)\ (3\cos^2\theta - 1)$.

To calculate the z component of the hyperfine interaction for d_{z^2}, we evaluate

$$A_z = g_e g_n \beta_e \beta_n \int d_{z^2} \left(\frac{3z^2 - r^2}{r^5} \right) d_{z^2} d\tau$$

With $z = r\cos\theta$ in our coordinate system and $d\tau = r^2 \sin\theta \, d\theta \, d\phi$ we get by simple substitution and extraction of constants

$$\langle A_z \rangle = g_e g_n \beta_e \beta_n (5/14\pi) \int_0^\infty \int_0^\pi \int_0^{2\pi} [f^2(r)(3\cos^2\theta - 1)^2] \times \frac{(3r^2\cos^2\theta - r^2)}{r^5} r^2 \sin\theta \, d\theta \, d\Phi \, dr$$

$$= R \int_0^\pi \int_0^{2\pi} (3\cos^2\theta - 1)^2 (3\cos^2\theta - 1) \, dr \sin\theta \, d\theta \, d\Phi$$

The integral over $R = \int_0^\infty f(r)^2 (1/r^3) r^2 dr$ is not evaluated but parameterized

as $\langle 1/r^3 \rangle$. The integral over $\phi = \int_0^{2\pi} d\phi = 2\pi$ and the integral over θ is
The integral over θ, $I\theta$, can be written as

$$I\theta = \int_0^{\pi} (3\cos^2\theta - 1)^3 \sin\theta \, d\theta$$

(let $t = \cos\theta$; $dt = -\sin\theta \, d\theta$. When $\theta = \pi$, $t = -1$, when $\theta = 0$, $t = 1$) Therefore,

$$= \int_{-1}^{1} (3t^2 - 1)^3 \, dt$$

$$= \int_{-1}^{1} (27t^6 - 27t^4 + 9t^2 - 1) \, dt$$

$$= \Big|_{-1}^{1} \left[\frac{27}{7} t^7 - \frac{27}{5} t^5 + 3t^3 - t \right]$$

$$= \left\{ \frac{27}{7} - \frac{27}{5} + 3 - 1 \right\} - \left\{ -\frac{27}{7} + \frac{27}{5} - 3 + 1 \right\}$$

$$= \frac{32}{35}$$

So the grand result is

$$\langle A_z \rangle = g_e g_n \beta_e \beta_n \langle 1/r^3 \rangle (5/16\pi)(2\pi)(\tfrac{32}{35}) = \tfrac{4}{7} \; P$$

where P is shorthand for $g_e g_n \beta_e \beta_n \langle 1/r^3 \rangle$

If we repeat this whole operation solving for $A_x(i = x = r\sin\theta\cos\phi)$ or $A_y(i = y = r\sin\theta\sin\phi)$ we find $A_{x,y} = -\frac{2}{7} \; P$.

APPENDIX IV

g and A Values for Coupled $\frac{5}{2}$ and $\frac{4}{2}$ Centers

The individual spins with $S_1 = \frac{5}{2}$ and $S_2 = \frac{4}{2}$ precess about a common axis defined by $S(= \frac{1}{2})$, with a frequency $J = 30{,}000$ MHz cm^{-1} (Fig. 25). For $J = 200$ cm^{-1}, this precession frequency $= 6 \times 10^{12}$ Hz. Formally, the resultant spin $\mathbf{S} = \mathbf{S}_1 + \mathbf{S}_2$ while the associated magnetic moment $\mu = \mu_1 + \mu_2$

Because of their precession, the observable part of $\mathbf{S}_1$ (and $\mathbf{S}_2$) is the time independent part of $\mathbf{S}_1$, which is its projection of $\mathbf{S}$. This projection has a magnitude $\mathbf{S}_1 \cdot \frac{\mathbf{S}}{S}$, where $\frac{\mathbf{S}}{S} =$ unit vector in direction of S, and direction $\frac{\mathbf{S}}{S}$. The result $= \frac{(\mathbf{S}_1 \cdot \mathbf{S})\mathbf{S}}{S^2}$. In an applied field S, precesses about $\mathbf{B}_0$ at the Larmor frequency ω_L ((about 10 GHz, $\ll J$); thus $\mathbf{S}_1$ and $\mathbf{S}_2$ precess rapidly about $\mathbf{S}$, which is precessing much more slowly about $\mathbf{B}_0$.

Now

$$E = -\mu \cdot \mathbf{B}_0 = g\beta \mathbf{S} \cdot \mathbf{B}_0$$

$$\mathbf{S} = \mathbf{S}_1 + \mathbf{S}_2 = \frac{(\mathbf{S}_1 \cdot \mathbf{S})\mathbf{S}}{S^2} + \frac{(\mathbf{S}_2 \cdot \mathbf{S})\mathbf{S}}{S^2}$$

$$\text{or } \mu = -g\beta\mathbf{S} = -g_1\beta\frac{(\mathbf{S}_1 \cdot \mathbf{S})\mathbf{S}}{S^2} - g_2\beta\frac{(\mathbf{S}_2 \cdot \mathbf{S})\mathbf{S}}{S^2}$$

$$g = g_1\frac{(\mathbf{S}_1 \cdot \mathbf{S})}{S^2} + g_2\frac{(\mathbf{S}_2 \cdot \mathbf{S})}{S^2}$$

$$\text{As } \mathbf{S} = \mathbf{S}_1 + \mathbf{S}_2, (S)^2 = (\mathbf{S}_1 + \mathbf{S}_2)^2 = {S_1}^2 + {S_2}^2 + 2\mathbf{S}_1 \cdot \mathbf{S}_2$$

Therefore,

$$\mathbf{S}_1 \cdot \mathbf{S}_2 = \frac{S^2 - {S_1}^2 - {S_2}^2}{2} \quad \text{and as } \mathbf{S}_1 \cdot \mathbf{S} = {\mathbf{S}_1}^2 + \mathbf{S}_1 \cdot \mathbf{S}_2$$

$$\mathbf{S}_1 \cdot \mathbf{S} = {S_1}^2 + \frac{S^2 - {S_1}^2 - {S_2}^2}{2} = \frac{S^2 + {S_1}^2 - {S_2}^2}{2}$$

Thus

$$g_1\frac{\mathbf{S}_1 \cdot \mathbf{S}}{S^2} = g_1\frac{S^2 + {S_1}^2 - {S_2}^2}{2S^2}$$

$$= \frac{\frac{3}{4} + \frac{35}{4} - \frac{24}{4}}{\frac{3}{2}}$$

$$= \frac{7}{3}\, g_1$$

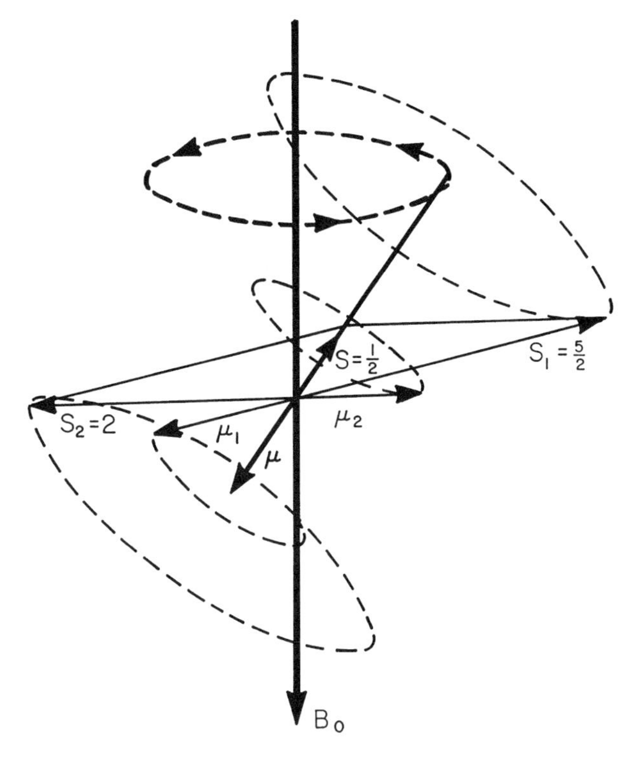

Figure 25
The coupling of two spins, $S_1 = \frac{5}{2}$ and $S_2 = \frac{4}{2}$ to yield a resultant $S = \frac{1}{2}$. Note that S_1 and S_2 precess around the axis of S, which in turn precesses about the axis of **B**. Also note that the magnetic moments are antiparallel to the spins; as it is μ that aligns along **B**, S points "up" while **B** points "down".

Similarly, for g_2 we get (by interchanging the subscripts 1 and 2)

$$g_2 \frac{\mathbf{S}_2 \cdot \mathbf{S}}{S^2} = g_2 \frac{S^2 + S_2{}^2 - S_1{}^2}{2S^2} = -\tfrac{4}{3} g_2$$

Thus

$$g = \tfrac{7}{3} g_1 - \tfrac{4}{3} g_2$$

The hyperfine interaction of the ion in the absence of coupling is $A_1 \mathbf{I}_1 \cdot \mathbf{S}_1$. In the coupled system, $\mathbf{S}_1$ is not a constant (only S is) and now the hyperfine interaction is $a\ \mathbf{I}_1 \cdot \mathbf{S}$ where a is a function of A_1 and S_1. We have seen that the time independent part of $\mathbf{S}_1 = [(\mathbf{S}_1 \cdot \mathbf{S})/S^2]\mathbf{S}$. So the free ion term $A_1 \mathbf{I}_1 \cdot \mathbf{S}_1$ becomes the coupled term $A[(\mathbf{S}_1 \cdot \mathbf{S})/S^2]\mathbf{I}_1 \cdot \mathbf{S}$. As $[(S_1 \cdot \mathbf{S}/S^2] = \frac{7}{3}$ for $\mathbf{S}_1$, $A_1\mathbf{I}_1$. Then $\mathbf{S}_1$ becomes $\frac{7}{3}$, $A_1 \mathbf{I}_1 \cdot \mathbf{S}$ or $a_1 = \frac{7}{3} A_1$. Likewise $a_2 = -\frac{4}{3} A_2$.

APPENDIX V

The Dipole–Dipole Interaction

For two point dipoles (electric or magnetic) oriented arbitrarily in space the expression for the energy of interaction (V_{12}) is

$$V_{12} = \frac{\mu_1 \cdot \mu_2}{r^3} - \frac{3(\mu_1 \cdot \mathbf{r})(\mu_2 \cdot \mathbf{r})}{r^5} \tag{A.V.1}$$

where μ_1 and μ_2 are the two dipoles (e.g., electron and nucleus, respectively), **r** is the vector between these two midpoints. The first dot product reflect the extent to which the two dipoles are colinear; the second and third dot products measure the extent to which μ_1 and μ_2 are parallel to the line connecting their midpoints.

In a strong magnetic field, when both magnetic moments are aligned along the field, and hence parallel, expression A.V.1 simplifies further to

$$V_{12} = \frac{\mu_1 \mu_2}{r^3}(1 - 3\cos^2\theta) \tag{A.V.2}$$

where θ is the angle subtended by the magnetic field and **r**. Equation (A.V.2) is axially symmetric and consequently has two limiting values. When $\theta = 0$ (μ_1 and μ_2 lie on the z axis):

$$V_{12} = \frac{-2\mu_1 \mu_2}{r^3} \tag{A.V.3}$$

When $\theta = 90$ (μ_1 and μ_2 lie in the xy plane):

$$V_{12} = \frac{\mu_1 \mu_2}{r^3} \tag{A.V.4}$$

When $\cos^2\theta = \frac{1}{3}$ ($\theta = 54°$), then $V_{12} = 0$; thus 54° is called the magic angle because the interaction between the two dipoles disappears.

As V_{12} is an energy, it is equivalent to the term $\mu \cdot \mathbf{B}$, which implies that

$$\frac{\mu_2(1 - 3\cos^2\theta)}{r^3}$$

is equivalent to a magnetic field. This field is

$$V_{12} = \frac{\frac{3}{2} g\beta^2}{r^3}(1 - 3\cos^2\theta)$$

$$= \frac{1{,}400}{r^3}(1 - 3\cos^2\theta)$$

The units are millitesla (mT). The three-halves arises from evaluating $gS^2 = gS(S+1)$. (When S_1 is not identical with S_2, the full expressions must be employed [in this context identical means identical g values, not identical chemical species)].

References

1. Carrington, A., "Electron-Spin Resonance Spectra of Aromatic Radicals and Radical Ions" *Quart. Rev.* **1963**, *17*, 67–89.
2. Atkins, P. W.; Symons, M. R. C., Eds., *The Structure of Inorganic Radicals*; Elsevier: Amsterdam, The Netherlands, 1967.
3. Edmondson, D. E. "ESR of Fine Radicals in Enzymatic Systems" In *Biological Magnetic Resonance*; Berliner, L. J.; Reuben, J. Eds.; Plenum: New York, 1978, Vol. 1.
4. Miller, A.-F.; Brudvig, G. W. "A Guide to Electron Paramagnetic Resonance Spectroscopy of Photosystem II Membranes" *Biochim. Biophys. Acta* **1991**, *1026*, 1–18.
5. Pake, G. E.; Estle, T. "*Paramagnetic Resonance*"; 2nd ed.; Benjamin: Reading, MA, 1973.
6. Kevan, L.; Bowman, M. K., Ed.; *Modern Pulsed and Continuous Wave Electron Spin Resonance*; Wiley: New York, 1990.
7. Palmer, G., "Electron Paramagnetic Resonance" *Methods Enzymol.* **1967**, *10*, 595–609.
8. Aasa, R.; Vanngard, T., "EPR Signal Intensities and Powder Shapes: A Reexamination" *J. Mag. Reson.* **1975**, *17*, 308–315.
9. Carrington, A.; McLachlan, A. D., "*Introduction to Magnetic Resonance*; Harper & Row: New York, 1967.
10. Wertz, J. E.; Bolton, J. R., "*Electron Spin Resonance: Elementary Theory and Practical Applications*; Chapman & Hall: New York, 1986.
11. *Hemoglobin: Cooperativity and Electronic Properties*; Weissbluth, M., Ed.; Springer-Verlag: Heidelberg, 1974. Structure and Bonding Vol. 15.
12. Taylor, C. P. S. "The EPR of Low-Spin Heme Complexes. Relation of the t_{2g} Model to the Directional Properties of the g-tensor, and a New Method for Calculating the Ligand Field Parameters" *Biochim. Biophys. Acta* **1977**, *491*, 137–149.
13. Blumberg, W. E.; Peisach, J. "Low-Spin Compounds of Heme Proteins" *Adv. Chem. Ser.* **1971**, *100*, 271–291.
14. Palmer, G. "Electron Paramagnetic Resonance of Hemoproteins" In *The Porphyrins*; Dolphin, D. Ed.; Academic: New York, 1979; Vol. 4; pp 313–354.
15. Palmer, G. "The Electron Paramagnetic Resonance of Metalloproteins" *Biochem. Soc. Trans.* **1985**, *13*, 548–559.
16. Salerno, J. "Electron Paramagnetic Resonance Lineshapes of Biological Molecules: Some Effects of Distributed Parameters in the Applications of Electron Spin/Paramagnetic Resonance in Biochemistry" *Biochem. Soc. Trans.* **1985**, *13*, 611–614.
17. Simpkin, D.; Palmer, G.; Devlin, F. J.; McKenna, M. C.; Jensen, G. M.; Stephens, P., "The Axial Ligands of Heme in Cytochromes: A Near-Infra red Magnetic Circular Dichroism Study of Yeast Cytochromes *c*, c_1 and *b* and Spinach Cytochrome *f*" *Biochemistry* **1989**, *28*, 8033–8039.
18. Cheesman, M. R.; Greenwood, C.; Thomson, A. J. "Magnetic Circular Dichroism of Hemoproteins" *Adv. Inorg. Chem.* **1991**, *36*, 201–255.
19. Debrunner, P. G.; Hendrich, M. P.; DeJersey, J.; Keough, D. T.; Sage, J. T.; Zerner, B., "Mössbauer and EPR Study of the Binuclear Iron Centre in Purple Acid Phosphatase" *Biochim. Biophys. Acta* **1983**, *745*, 103–106.
20. Fox, B. G.; Hendrich, M. P.; Surerus, K. K.; Andersson, K. K.; Froland, W. A.; Lipscomb, J. D.; Münck, E. "Mössbauer, EPR, and ENDOR Studies of the Hydroxylase and Reductase Components of Methane Monooxygenase from *Methylosinus trichosporium* OB3b" *J. Am. Chem. Soc.* **1993**, *115*, 3688–3701.
21. Guigliarelli, B.; Bertrand, P.; Gayda, J.-P. "Contributions of the Fine Structure Terms to the g-values of the Biological Fe(III)–Fe(II) Clusters" *J. Chem. Phys.* **1986**, *85*, 1689–1692.
22. Reed, G. H.; Ray, W. J., Jr. "Electron Paramagnetic Resonance of Manganese(II) Coordination in the Phosphoglucomutase System" *Biochemistry* **1971**, *10*, 3910–3917.
23. Slappendel, S., Veldink, G. A., Vliegenthart, J. F. G., Aasa, R and Malmstrom, B. EPR Spectroscopy of Soybean Lipoxygenase-1. Description and Quantification of the High-Spin Fe(III) Signals. *Biochim. Biophys Acta* **1981**, *667*, 77–86.

24. Tsai, R.; Yu, C. A.; Gunsalus, I. C.; Peisach, J.; Blumberg, W.; Orme-Johnson, W. H.; Beinert, H. "Spin-State Changes in Cytochrome P-450cam on Binding of Specific Substrates" *Proc. Natl. Acad. Sci. USA* **1970**, *66*, 1157–1163.
25. Whittaker, J. W.; Lipscomb, J. D.; Kent, T. A.; Münck, E. "*Brevibacterium fuscum* Protocatechuate 3,4-Dioxygenase. Purification, Crystallization, and Characterization" *J. Biol. Chem.* **1984**, *259*, 4466–4475.
26. Blumberg, W. E.; Peisach, J. "The Measurement of Zero Field Splitting and the Determination of Ligand Composition in Mononuclear Nonheme Iron Proteins" *Ann. N. Y. Acad. Sci.* **1973**, *222*, 539–560.
27. Chen, V. J.; Orville, A. M.; Harpel, M. R.; Frolik, C. A.; Surerus, K. K.; Münck, E.; Lipscomb. J. D. "Spectroscopic Studies of Isopenicillin N. Synthase" *J. Biol. Chem.* **1989**, *264*, 21677–21681.
28. Rawlings, J.; Shah, V. K.; Chisnell, J. R.; Brill, W. J.; Zimmermann, R.; Münck, E.; Orme-Johnson, W. H. "Novel Metal Cluster in the Iron-Molybdenum Cofactor of Nitrogenase" *J. Biol. Chem.* **1978**, *253*, 1001–1004.
29. Holman, T. R.; Juarez-Garcia, C.; Hendrich, M. P.; Que, L., Jr.; Münck, E. "Models for Iron-Oxo Proteins. Mössbauer and EPR Study of an Antiferromagnetically Coupled $Fe^{III}Ni^{II}$ Complex" *J. Am. Chem. Soc.* **1990**, *112*, 7611–7618.
30. Surerus, K. K.; Hendrich, M. P.; Christie, P. D.; Rottgardt, D.; Orme-Johnson, W. H.; Münck, E. "Mössbauer and Integer-Spin EPR of the Oxidized P-Clusters of Nitrogenase: P^{OX} is a Non-Kramers System with a Nearly Degenerate Ground Doublet" *J. Am. Chem. Soc.* **1992**, *114*, 8579–8590.
31. Hendrich, M. P.; Debrunner, P. "Integer-Spin Electron Paramagnetic Resonance of Iron Proteins" *Biophys. J.* **1989**, *56*, 489–506.
32. Hendrich, M. P.; Münck, E.; Fox, B. G.; Lipscomb, J. D. "Integer-Spin EPR Studies of the Fully Reduced Methane Monooxygenase Hydroxylase Component" *J. Am. Chem. Soc.* **1990**, *112*, 5861–5865.
32a. Abragam, A.; Bleaney, B., *Electron Paramagnetic Resonance of Transition Metal Ions*; Oxford University Press: Oxford, 1970, Section 3.14.
33. Münck, E.; Surerus, K. K.; Hendrich, M. P. "Combining Mössbauer Spectroscopy with Integer Spin EPR" *Methods Enzymol.* **1993**, *227*, 463–479.
34. Hendrich, M. P.; Pearce, L. L.: Que, L., Jr.; Chasteen, N. D.; Day, E. P. "Multifield Saturation and Multifrequency EPR Measurements of Deoxyhemerythrin Azide: A Unified Picture" *J. Am. Chem. Soc.* **1991**, *113*, 3039–3044.
35. Penfield, K. W.; Gewirth, A. A.; Solomon, E. I. "Bonding of the Blue Copper in Plastocyanin" *J. Am. Chem. Soc.* **1985**, *107*, 4519–4529.
36. Lontie, R., Ed.; *Copper Proteins and Copper Enzymes*; CRC Press: Boca Raton, FL, 1984; Vols. 1–3.
37. Mathews, R.; Charlton, S.; Sands, R. H.; Palmer, G. "On the Nature of the Spin-Coupling Between the Iron Sulfur Clusters in Eight-Iron Ferredoxins" *J. Biol. Chem.* **1974**, *249*, 4326–4328.
38. Sands, R. H. "ENDOR and ELDOR on Iron Sulfur Proteins" In *Multiple Electron Resonance Spectroscopy*; Dorio, M. M. Freed, J. H., Ed.; Plenum: New York, 1979; pp 331–374.
39. Bencini, A.; Gatteschi, D., "*EPR of Exchange Coupled Systems*; Springer-Verlag: Heidelberg, 1990.
40. Berliner, L. J.; Reuben, J., Eds.; *Biological Magnetic Resonance*; Plenum: New York, 1989; Vol. 8.
41. Jackson, J. D. "*Classical Electrodynamics*", 2nd ed.; Wiley: New York, 1975.
42. Atkins, P. W. "*Molecular Quantum Mechanics*; University Press: Oxford, 1983.

General References

The following references represent texts for EPR at different levels of sophistication. It is regrettable that many of them are out of print but should still be found in your university library. There are three books that have recently been published.

Symons, M. R. C. *Chemical and Biological Aspects of Electron Spin Resonance*; Wiley: New York **1978**. An introductory book that contains a lot of physical insight. A good place to begin especially for those with minimal physical chemistry.

Bersohn, M.; Baird, J. C. *An Introduction to Electron Paramagnetic Resonance*; Benjamin, New York, **1966**. Very similar to the book by Symons but covers a little more ground.

Atkins, P. W.; Symons, M. R. C. *The Structure of Inorganic Radicals*; Elsevier: Amsterdam, **1967**. Intuitive discussion; contains valuable data on small molecules of biological interest (e.g., superoxide)

One could then progress to the following, each of which is good though incomplete. My preference is for Wertz and Bolton.

Wertz, J. E.; Bolton, J. R. *Electron Spin Resonance: Elementary Theory and Practical Applications*; Chapman & Hall: New York, **1986**. It is also available in paperback. A second edition of this book is available: Weil, J. E.; Bolton, J. R.; Wertz, J. E. *Electron Paramagnetic Resonance: Elementary Theory and Practical Applications*; Wiley: New York, **1994**.

Carrington, A.; McLachlan, A. D. *Magnetic Resonance*; Harper & Row: New York **1967**. Treats EPR and NMR in a relatively general manner that emphasizes the similarities. The treatment of the AB NMR system is particularly useful. Also available in paperback.

The following are somewhat more advanced.

Atherton, N. M. *Electron Spin Resonance*; Halsted Press: London, **1973**. Complements (Wertz and Bolton) very nicely.

Pake, G. E.; Estle, T. *Paramagnetic Resonance* 2nd ed.; Benjamin: Reading, MA. **1973**. Written for physicists but the best description of the precession model of magnetic resonance, which we have not covered in this chapter.

Abragam, A.; Bleaney, B. *Electron Paramagnetic Resonance of Transition Metal Ions*; Oxford University Press: Oxford, **1970**. A masterful book that is the bible for the field. Assumes some knowledge of quantum mechanics but the early parts can be read by anyone with some background in physical chemistry

Pilbrow, J. *Transition Ion Electron Paramagnetic Resonance*; Oxford University Press: Oxford, **1990**. It provides a good introduction to the recent literature but is very expensive.

Bencini, A.; Gatteschi, D. *EPR of Exchange Coupled Systems*; Springer-Verlag, Heidelberg, **1990**. A comprehensive though rather advanced treatment that deal with interactions between paramagnetic centers such as are found in the iron–sulfur clusters.

Kevan, L.; Bowman, M. K. Eds., *Modern Pulsed and Continuous Wave Electron Spin Resonance*; Wiley: New York, **1990**. Covers recent developments in modern EPR techniques.

Reviews of the applications of EPR in biochemistry are

"The Applications of Electron Spin/Paramagnetic Resonance in Biochemistry." *Biochem. Soc. Trans.* **1985**, *13*, 541–641.

Hoff, A. J., Ed. *Advanced EPR: Applications in Biology and Biochemistry*; Elsevier: Amsterdam, **1989**.

Sigel, H., Ed., "Endor, EPR and Electron Spin Echo for Probing Coordination Spheres" In *Metal Ions in Biological Systems* Vol. 22, Marcel Dekker: NY, **1987**.

Specific topics are also covered in *Biological Magnetic Resonance*, a series edited by L. Berliner and J. Reuben and published by Plenum Press.

Edmonson, D. "Free Radicals in Enzymatic Systems" Vol. **1**.
Warden, J. T. "Paramagnetic Intermediates in Photosynthetic Systems" Vol. **1**.
Boas, J. F.; Pillbrow, J. R. "ESR of Copper in Biological Systems" Vol. **1**.
Bray, R. C. "EPR of Molybdenum Containing Enzymes" Vol. **2**.
Smith, T. D. "ESR of Iron Proteins" Vol. 2.
Chasteen, N. D. "Vandyl(IV) EPR Spin Probes: Inorganic and Biochemical Aspects" Vol. **2**.
Chien, J. C. W.; Dickinson, L. C. "EPR Crystallography of Metalloproteins and Spin-Labelled Enzymes" Vol. **3**.
Reed, G. H.; Markham, G. D. "EPR of Mn(II) Complexes with Enzymes and Other Proteins" Vol. **7**.

The following practical articles will all be found in *Methods in Enzymology* published by Academic Press.

Palmer, G. "Electron Paramagnetic Resonance" **1967**, *10*, 595–609.
Hyde, J. H. "Saturation Transfer Spectroscopy" **1978**, *49*, 480–511.
Fee, J. A. "Transition Metal Electron Paramagnetic Resonance Related to Proteins" **1978**, *49*, 512–528.
Beinert, H.; Orme-Johnson, W. H.; Palmer, G. "Special Techniques for the Preparation of Samples for Low-Temperature EPR Spectroscopy" **1978**, *54*, 111–132.
Beinert, H. "EPR Spectroscopy of the Components of the Mitochondrial Electron Transfer Chain" **1978**, *54*, 133–150.
Blumberg, W. E. "The Study of Hemoglobin by Electron Paramagnetic Resonance Spectroscopy" **1981**, *76*, 312.
Pilbrow, J. R.; Hanson, G. R. "Electron Paramagnetic Resonance" **1993**, *227*, 330.
Cammack, R.; Cooper, C. E. "Electron Paramagnetic Resonance Spectroscopy of Iron Complexes and Iron-Containing Proteins" **1993**, *227*, 353.
Paulsen, K. E.; Stankovich, M. T.; Orville, A. M. "Electron Paramagnetic Resonance Spectroelectro-chemical Titration" **1993**, *227*, 396.
Brudvig, G. W. "Electron Paramagnetic Resonance Spectroscopy" **1995**, *246*, 536.

4

ESEEM and ENDOR Spectroscopy

N. DENNIS CHASTEEN

Department of Chemistry
University of New Hampshire
Durham, NH 03824

PENNY A. SNETSINGER

Department of Chemistry
Sacred Heart University
Fairfield, CT 06432

I. Introduction

In this chapter, we discuss two related resonance techniques that derive from the basic EPR experiment, namely, electron spin echo envelope modulation (ESEEM) and electron–nuclear double resonance (ENDOR). These techniques have gained widespread use in chemistry and biochemistry. Hundreds of papers are published annually that make use of ESEEM and ENDOR to gain structural information about paramagnetic centers. In this chapter, we discuss the basic principles of ESEEM and ENDOR, emphasizing the information that can be readily obtained from an analysis of the spectrum. The reader should consult the references to the literature and several in-depth reviews given at the end of the chapter for further details and more rigorous treatments of these subjects.

Electron–nuclear hyperfine interactions (Chapter 3), which are too small to be resolved within the natural width of the EPR line, are generally amenable to study by either ESEEM or ENDOR. In transition metal complexes and metalloproteins, magnetic nuclei such as ^{1}H, ^{2}H, ^{13}C, ^{14}N, ^{15}N, ^{17}O, ^{31}P, and ^{33}S, in the vicinity of the paramagnetic metal ion can be detected by these techniques. Not only do ENDOR and ESEEM enable one to identify the presence of a particular ligand nucleus in a metal complex but in favorable circumstances metal–ligand nuclei distances and angles can be obtained as well.

ESEEM and ENDOR are complementary techniques. Broadly speaking, ESEEM is used to study "more distant" weakly coupled nuclei, having electron–nuclear coupling constants A less than about 10 MHz, depending on the nucleus. In contrast, ENDOR is used for investigations of more strongly coupled "closer" nuclei having A values typically in the range 2–40 MHz. Both techniques can be used to obtain quadrupole coupling constants of nuclei with spins $I \geq 1$, for example, ^{2}H ($I = 1$), ^{14}N ($I = 1$), and ^{17}O ($I = \frac{5}{2}$). In principle, the depth of the modulation in ESEEM can be used to determine the number of equivalent nuclei of a particular type coupling with the electron spin, although in practice this is often difficult (more later). In contrast, ENDOR spectroscopy cannot be used to count the number of equivalent nuclei. Before we discuss further the principles of ESEEM and ENDOR spectroscopy, let us consider in some detail the spin energy level diagram and transitions that are possible for two commonly encountered electron spin–nuclear spin systems, namely, the $S = \frac{1}{2}$, $I = \frac{1}{2}$ and $S = \frac{1}{2}$, $I = 1$ coupled spin systems. An understanding of the energy levels and transitions for these simple systems will form the basis for interpreting the ESEEM and ENDOR spectra that follow. The student is encouraged to work the problems in the Appendix, which serve to illustrate the use of the various equations that appear throughout.

II. The Energy Level Diagrams

A. The $S = \frac{1}{2}$, $I = \frac{1}{2}$ Case

Consider a simple $S = \frac{1}{2}$, $I = \frac{1}{2}$ spin system, that is, an unpaired electron magnetically coupled to an $I = \frac{1}{2}$ nucleus such as a proton. The energies (E) for this system are calculated from the solution to the Schrödinger equation from quantum mechanics,

$$\mathscr{H}\Psi = E\Psi \qquad (1)$$

Here Ψ is the *total* spin wave function for the system and consists of electron and nuclear spin functions. The spin Hamiltonian operator $\mathscr{H}$ given by

$$\mathscr{H} = \beta\hat{\mathbf{S}} \cdot \mathbf{g} \cdot \mathbf{B} - g_n\beta_n\hat{\mathbf{I}} \cdot \mathbf{B} + h\hat{\mathbf{S}} \cdot \mathbf{A} \cdot \hat{\mathbf{I}} \tag{2a}$$

or in expanded form by

$$\mathscr{H} = \beta(g_{xx}B_x\hat{\mathrm{S}}_x + g_{yy}B_y\hat{\mathrm{S}}_y + g_{zz}B_z\hat{\mathrm{S}}_z) - g_n\beta_n(B_x\hat{\mathrm{I}}_x + B_y\hat{\mathrm{I}}_y + B_z\hat{\mathrm{I}}_z)$$
$$+ hA_{xx}\hat{\mathrm{S}}_x\hat{\mathrm{I}}_x + hA_{yy}\hat{\mathrm{S}}_y\hat{\mathrm{I}}_y + hA_{zz}\hat{\mathrm{S}}_z\hat{\mathrm{I}}_z \tag{2b}$$

Symbols with circumflexes are spin operators. The terms $\beta\hat{\mathbf{S}} \cdot \mathbf{g} \cdot \mathbf{B}$ and $g_n\beta_n\hat{\mathbf{I}} \cdot \mathbf{B}$ in Eq. (2a) are the operators for calculating the field dependence of the energy levels for EPR and NMR spectroscopy, respectively. They represent the interaction of the electron spin magnetic moment $\boldsymbol{\mu} = -g\beta\mathbf{S}$ with the applied magnetic field $\mathbf{B}$ (the electron Zeeman interaction) and the interaction of the nuclear spin magnetic moment $\boldsymbol{\mu}_N = g_n\beta_n\mathbf{I}$ with the magnetic field (the nuclear Zeeman interaction), respectively. g and g_n are the electron and nuclear g factors, respectively. $\hat{\mathbf{S}} = \hat{\mathrm{S}}_x\mathbf{i} + \hat{S}_y\mathbf{j} + \hat{S}_z\mathbf{k}$ and $\hat{\mathbf{I}} = \hat{I}_x\mathbf{i} + \hat{\mathrm{I}}_y\mathbf{j} + \hat{\mathrm{I}}_z\mathbf{k}$ are the electron and nuclear spin operators and $\mathbf{B} = B_x\mathbf{i} + B_y\mathbf{j} + B_z\mathbf{k}$, where $\mathbf{i}$, $\mathbf{j}$, and $\mathbf{k}$ are the unit vectors along the x, y, and z axes of the coordinate system for the coincident $\mathbf{g}$ and nuclear hyperfine $\mathbf{A}$ matrices. Here, β $(= 9.27408 \times 10^{-24}\ \mathrm{J\,T^{-1}})$ and β_n $(= 5.05082 \times 10^{-27}\ \mathrm{J\,T^{-1}})$ are the Bohr and nuclear magneton constants, respectively, and h $(= 6.62618 \times 10^{-34}\ \mathrm{J\,s})$ is Planck's constant. The parameter A is the nuclear hyperfine coupling constant. For a free electron, the electron g factor is 2.00232 but in molecules it differs from this value due to spin–orbit coupling (see Chapter 3). The nuclear g_n factor is characteristic of the nucleus and determines its NMR frequency (Table I).

The last term $h\hat{\mathbf{S}} \cdot \mathbf{A} \cdot \hat{\mathbf{I}}$ in Eq. (2a) is the electron–nuclear hyperfine interaction and represents the coupling between the electron and nuclear magnetic moments. In general, the magnitudes of g and A depend on the direction of the magnetic field $\mathbf{B}$ in the molecular coordinate system and are said to be anisotropic (see Chapter 3). They are written as 3×3 matrices $\mathbf{g}$ and $\mathbf{A}$ in Eq. (2a).

To solve the Schrödinger equation for this $S = \frac{1}{2}$, $I = \frac{1}{2}$ spin system, one typically writes the Hamiltonian matrix using basis functions $\{\phi\}$, which are simple products of electron and nuclear spin functions, that is, $\phi = |m_s\rangle|m_I\rangle = |m_sm_I\rangle$, where $m_s = \pm\frac{1}{2}$ and $m_I = \pm\frac{1}{2}$. The energies to first order are given by the expression[a]

$$E(m_sm_I) = g\beta\mathrm{B}m_s - g_n\beta_n\mathrm{B}m_I + hAm_sm_I \tag{3}$$

For a system of axial symmetry, the angular dependence in g and A in Eq. (3) can

[a]The student should consult Chapter 2 of Carrington, A.; McLachlan, A. D. *Introduction to Magnetic Resonance*, Chapman & Hall: New York, 1967, for details of this type of calculation.

be calculated from the expressions

$$g(\theta) = ({g_\perp}^2 \sin^2\theta + {g_\parallel}^2 \cos^2\theta)^{1/2} \tag{4}$$

$$A(\theta) = ({g_\perp}^2 {A_\perp}^2 \sin^2\theta + {g_\parallel}^2 {A_\parallel}^2 \cos^2\theta)^{1/2}/g(\theta) \tag{5}$$

where θ is the angle between the direction of the applied field **B** and the sym-

Table I
Properties of Selected Magnetic Nuclei[a]

Nucleus	% Natural Abundance	I	g_n	NMR Frequency ν_n at $B_0 = 3.5$ KG (MHz)	Quadrupole Moment Q ($\lvert e \rvert \times 10^{-24}$ cm^2)
^{1}H	99.985	$\frac{1}{2}$	5.586	14.90	
^{2}H	0.0148	1	0.8574	2.288	0.002875
^{13}C	1.11	$\frac{1}{2}$	1.405	3.748	
^{14}N	99.63	1	0.4037	1.077	0.0193
^{15}N	0.366	$\frac{1}{2}$	−0.5664	1.511	
^{17}O	0.038	$\frac{5}{2}$	−0.7575	2.021	−0.026
^{19}F	100	$\frac{1}{2}$	5.258	14.03	
^{23}Na	100	$\frac{3}{2}$	1.4784	3.9442	0.108
^{27}Al	100	$\frac{5}{2}$	1.457	3.886	0.150
^{29}Si	4.67	$\frac{1}{2}$	−1.111	2.963	
^{31}P	100	$\frac{1}{2}$	2.263	6.038	
^{33}S	0.75	$\frac{3}{2}$	0.4291	1.145	−0.064
^{35}Cl	75.77	$\frac{3}{2}$	0.5479	1.462	−0.08249
^{37}Cl	24.23	$\frac{3}{2}$	0.4561	1.217	−0.06493
^{51}V	99.75	$\frac{7}{2}$	1.468	3.917	−0.0515
^{57}Fe	2.15	$\frac{1}{2}$	0.1806	0.4818	
^{63}Cu	69.2	$\frac{3}{2}$	1.484	3.959	−0.222
^{65}Cu	30.8	$\frac{3}{2}$	1.588	4.237	−0.195
^{95}Mo	15.9	$\frac{5}{2}$	−0.3656	0.9754	−0.019
^{97}Mo	9.6	$\frac{5}{2}$	−0.3734	0.9962	0.2
^{133}Cs	100	$\frac{7}{2}$	0.7378	1.9685	−0.003

[a] Data from the "Bruker Almanac", Bruker Instruments, 1990, p. 95.

metry axis and $g_{\parallel}$ $(= g_{zz})$, and $g_{\perp}$ $(= g_{xx} = g_{yy})$, and $A_{\parallel}$ $(= A_{zz})$, and $A_{\perp}$ $(= A_{xx} = A_{yy})$ are the principal values of the diagonalized **g** and **A** matrices. Here we assume that both the **g** and **A** matrices are diagonal in the same axis system. However, the principal axes of the hyperfine couplings of ligand nuclei are often noncoincident with those of the **g** matrix. The expression for the angular dependence in g for a rhombic system is given by Eq. (9) of Chapter 3.

Figure 1(a) shows the energy level diagram calculated from Eq. (3) and illustrates the splitting of the spin states by the successively smaller interactions for a magnetic field of approximately 330 mT (3300 G). Typically, the energies

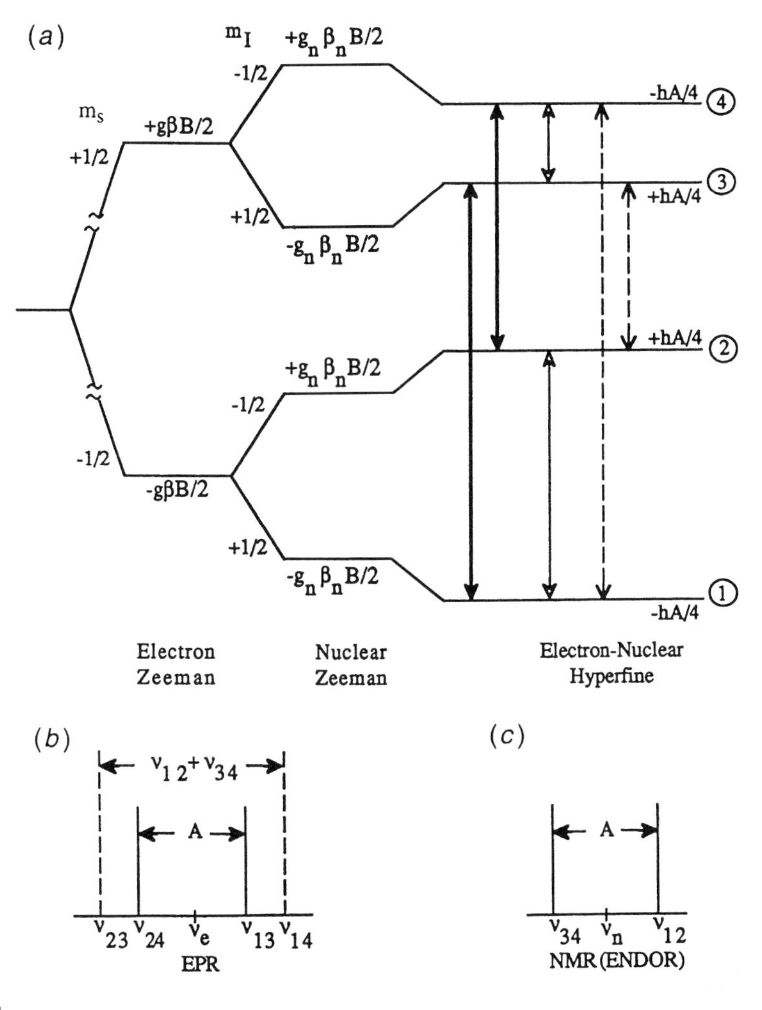

Figure 1
(a) Energy level diagram for an $S = \frac{1}{2}$, $I = \frac{1}{2}$ spin system described by the Hamiltonian given in Eq. (2) and calculated to first order. The g value and hyperfine coupling are assumed to be isotropic. The bold lines show the allowed EPR transitions, the solid lines the NMR (ENDOR, ESEEM) transitions, and the dashed lines the EPR "semiforbidden" transitions. (b) The stick EPR spectrum. (c) The stick ENDOR (NMR) spectrum. [Adapted from Snetsinger.[1]]

(in frequency units) follow the order: electron Zeeman ($\sim$10 GHz) > nuclear Zeeman (1–15 MHz) $\approx$ electron–nuclear hyperfine (0–10 MHz). As we shall see, the relative magnitude of the nuclear Zeeman ($g_n\beta_n \mathrm{B} m_I$) and nuclear hyperfine ($hAm_s m_I$) energy terms in Eq. (3) has a pronounced effect on ESEEM and ENDOR spectra.

Three types of transitions are indicated in Figure 1(*a*). First are the usual EPR transitions occurring at frequencies ν_{13} and ν_{24} and indicated by heavy solid lines between levels 1 and 3 and between levels 2 and 4. These transitions obey the usual EPR selection rule $\Delta m_s = \pm 1$, $\Delta m_I = 0$ where only electron spin flips but no nuclear spin flips are allowed. Second are the "semiforbidden" EPR transitions indicated by dashed lines in Figure 1(*a*) and occurring at frequencies ν_{23} and ν_{14}. These frequencies correspond to $\Delta m_s = \pm 1$, $\Delta m_I = \pm 1$ transitions where both the electron and nuclear spins flip simultaneously. To the first approximation, such transitions are forbidden; however, off-diagonal elements in the Hamiltonian matrix cause mixing of the states and lend some intensity to these transitions, hence the term "semiforbidden".[b] As we shall see, these transitions are key to the observation of ESEEM. The third transitions, indicated by thin solid lines, are NMR transitions ($\Delta m_s = 0$, $\Delta m_I = \pm 1$) and occur at the frequencies ν_{12} and ν_{34}. These frequencies are also known as the ENDOR or ESEEM frequencies. The NMR transition frequency ν_{12} corresponding to the lower β ($m_s = -\frac{1}{2}$) spin state of the electron is frequently denoted as ν_b. The frequency ν_{34} is denoted as ν_a, the NMR frequency for the upper α ($m_s = +\frac{1}{2}$) spin state of the electron. From the first-order Eq. (3), the ENDOR (or ESEEM) frequencies can be calculated

$$\nu_\mathrm{a} = \nu_{34} = \nu_\mathrm{n} - A/2 \tag{6a}$$

$$\nu_\mathrm{b} = \nu_{12} = \nu_\mathrm{n} + A/2 \tag{6b}$$

or more generally as

$$\nu_\mathrm{ENDOR} = \nu_\mathrm{ESEEM} = \nu_\mathrm{n} \pm A_i/2 \qquad i = x, y, \text{ or } z \tag{7}$$

where ν_n $(= g_\mathrm{n}\beta_\mathrm{n} B/h)$ is the Larmor or nuclear Zeeman (NMR) frequency at an applied magnetic field of strength B. The angular dependence of the anisotropic coupling constant A is given by Eq. (5) for an axial system.

The stick spectrum for a hypothetical frequency swept EPR spectrometer at fixed field B is shown in Figure 1(*b*). Both the allowed ($\Delta m_s = \pm 1$, $\Delta m_I = 0$) EPR transitions at frequencies ν_{13} and ν_{24} (solid lines) and the semiforbidden ($\Delta m_s = \pm 1$, $\Delta m_I = \pm 1$) transitions at frequencies ν_{14} and ν_{23} (dashed lines) are shown. The splitting of the allowed EPR lines by the coupling constant A is indicated. Here we assume that this splitting is much less than the natural width of the EPR line and would therefore be unresolved in the EPR spectrum. As will

[b]Normally forbidden $\Delta m_s = \pm 1$, $\Delta m_I = \pm 1$ transitions become allowed due to the mixing of states by the $A_{xx}\hat{S}_x\hat{I}_x + A_{yy}\hat{S}_y\hat{I}_y$ operator terms in Eq. (2b) and the normally forbidden $\Delta m_I = \pm 2$ transitions become allowed due to the mixing of states by the $\hat{I}_x^{\,2}$ and $\hat{I}_y^{\,2}$ operator terms in Eq. (9).

be seen, ENDOR and ESEEM are used to obtain this hyperfine coupling. Figure 1(*c*) shows the NMR (ENDOR) spectrum from which the coupling is measured directly. The number of ENDOR lines for a nucleus of spin I is given by the $4I$ rule ($2I$ NMR lines from each of the two m_s manifold of states). For $I = \frac{1}{2}$, two lines are predicted. The number of lines in the EPR spectrum is given by the $2I + 1$ rule which in this case also predicts two lines [Fig. 1(*b*)].

B. The $S = \frac{1}{2}$, $I = 1$ Case

The above discussion was limited to the simple case of $S = \frac{1}{2}$, $I = \frac{1}{2}$ such as ^{15}N, ^{1}H, ^{31}P, and ^{13}C. A more complete Hamiltonian is required to describe an $I \geq 1$ nucleus because of its quadrupole moment. Examples are ^{2}H and ^{14}N, both having $I = 1$. An additional operator term $\mathscr{H}_Q$ must be added to the Hamiltonian [Eq. (2a)] for a rhombic system. The operator $\mathscr{H}_Q$ takes on the form

$$\mathscr{H}_Q = h\hat{\mathbf{I}} \cdot \mathbf{P} \cdot \hat{\mathbf{I}} \tag{8}$$

Here $\mathbf{P}$ is the 3×3 quadrupole tensor represented by a matrix; its trace is zero, that is, $P_x + P_y + P_z = 0$. The z-component P_z is given by $P_z = e^2qQ/2hI(2I - 1)$, where e is the electronic charge and eq is the electric field gradient across the nucleus along the z axis ($eq = V_{zz} = d^2V/dz^2$). The quantity V is the electrostatic potential at the quadrupolar nucleus; this potential is generated by the surrounding charges of the electrons and nuclei in the molecule. Here Q is the nuclear quadrupole moment, which is a property characteristic of $I \geq 1$ nuclei (Table I). The quadrupole interaction is zero when there is cubic symmetry (tetrahedral or octahedral) about the nucleus since there is no electric field gradient in that case. The quantity e^2qQ is called the quadrupole coupling constant, not to be confused with the quadrupole coupling parameter Q', which is variously defined in the literature as $e^2qQ/2hI(2I - 1)$, $3e^2qQ/4hI(2I - 1)$, or $3e^2qQ/8hI(2I - 1)$. The parameter Q' is measured directly from the spectrum from which e^2qQ can be calculated [$Q' = P_z$ when Q' is defined as $e^2qQ/2hI(2I - 1)$].[c]

Since $\mathbf{P}$ is a traceless tensor, the quadrupolar Hamiltonian $\mathscr{H}_Q$ is often written in one of the following forms

$$\mathscr{H}_Q = [e^2Qq/4I(2I - 1)] \cdot [3\hat{\mathbf{I}}_{Z^2} - I(I + 1) + \eta(\hat{\mathbf{I}}_x{}^2 - \hat{\mathbf{I}}_y{}^2)] \tag{9a}$$

or

$$\mathscr{H}_Q = h\mathbf{P}_z/2[3\hat{\mathbf{I}}_{Z^2} - I(I + 1) + \eta(\hat{\mathbf{I}}_x{}^2 - \hat{\mathbf{I}}_y{}^2)] \tag{9b}$$

which are alternate ways of expressing Eq. (8). Here $\hat{\mathbf{I}}_x$ and $\hat{\mathbf{I}}_y$ are the operators for the x and y components of the nuclear spin angular momentum. The asymmetry parameter η is given by $0 \leq \eta = (P_x - P_y)/P_z \leq 1$. A value of $\eta = 0$ corresponds to complete axial symmetry and $\eta = 1$ to pure rhombic symmetry. The

[c] It is unfortunate that there is no common nomenclature for expressing the quadrupole interaction. One variously encounters usage of P, K, Q', and e^2qQ in the literature.

parameter η is a measure of the asymmetry of the electric field gradient within the xy plane. For a system of axial symmetry with $\eta = 0$, it follows that $P_x = P_y = -P_z/2$ since $\mathbf{P}$ is traceless.

The energy level diagram to first order for an $I = 1$ system with axial symmetry is shown in Figure 2(a) with the field along the symmetry z axis. In constructing Figure 2(a), we assume that the electron–nuclear hyperfine energy $hAm_s m_I$ with coupling constant A exceeds both the nuclear Zeeman energy $g_n\beta_n Bm_I$ and the quadrupole energy $hP_z[3m_I{}^2 - I(I+1)]/2$, for example, ^{14}N in a field of 330 mT (3300 G). The allowed EPR (heavy solid lines) and NMR (thin solid lines) transitions and the semiforbidden (dashed lines) transitions are indicated. The EPR spectrum [Fig. 2(b)] consists of three lines as expected for the coupling of the unpaired electron with an $I = 1$ nucleus ($2I + 1$ rule). The NMR (ENDOR) spectrum, however, reflects the quadrupole interaction and consists of four $\Delta m_I = \pm 1$ lines ($4I$ rule), two NMR transitions from each of the two m_S manifolds of states. The frequencies ν_{12}, ν_{23}, ν_{45}, and ν_{56} of these lines are given by the first-order equations

$$\nu_{12} = A_z/2 + \nu_n - 3P_z/2 \tag{10a}$$

$$\nu_{23} = A_z/2 + \nu_n + 3P_z/2 \tag{10b}$$

$$\nu_{45} = A_z/2 - \nu_n - 3P_z/2 \tag{10c}$$

$$\nu_{56} = A_z/2 - \nu_n + 3P_z/2 \tag{10d}$$

or more generally as

$$\nu_{\text{ENDOR}} = A_i/2 \pm \nu_n \pm 3P_i/2 \qquad i = x, y, \text{or } z. \tag{11}$$

The separation between lines ν_{12} and ν_{23} or ν_{45} and ν_{56} equals $3P_z$. The four ENDOR lines predicted by Eq. (10) are often not resolved. When P_z is less than the ENDOR line width, typically about 0.1–0.5 MHz for frozen solutions of transition metal ions, only two lines centered at $A_z/2$ and separated by $2\nu_n$ are observed.

Figure 3 shows the energy level diagram for a rhombic quadrupole interaction ($\eta \neq 0$) with both the nuclear hyperfine coupling constant a and g factor assumed to be isotropic and the magnetic field taken to be along the z axis. The energies shown in Figure 3 are obtained from solving the 3×3 secular determinant in the nuclear basis functions $|m_I\rangle$ ($m_I = 0, \pm 1$) for each of the $m_S = \pm \frac{1}{2}$ states. The parameter K is defined as $e^2qQ/4h$ and $\nu_n = g_n\beta_n B/h$. There are a total of six allowed and "semiforbidden" transitions, the frequencies of which are given by the equations:

$$\nu_{56} = 3K + \sqrt{(a/2 - \nu_n)^2 + \eta^2 K^2} \tag{12a}$$

$$\nu_{45} = -3K + \sqrt{(a/2 - \nu_a)^2 + \eta^2 K^2} \tag{12b}$$

$$\nu_{46} = 2\sqrt{(a/2 - \nu_n)^2 + \eta^2 K^2} \tag{12c}$$

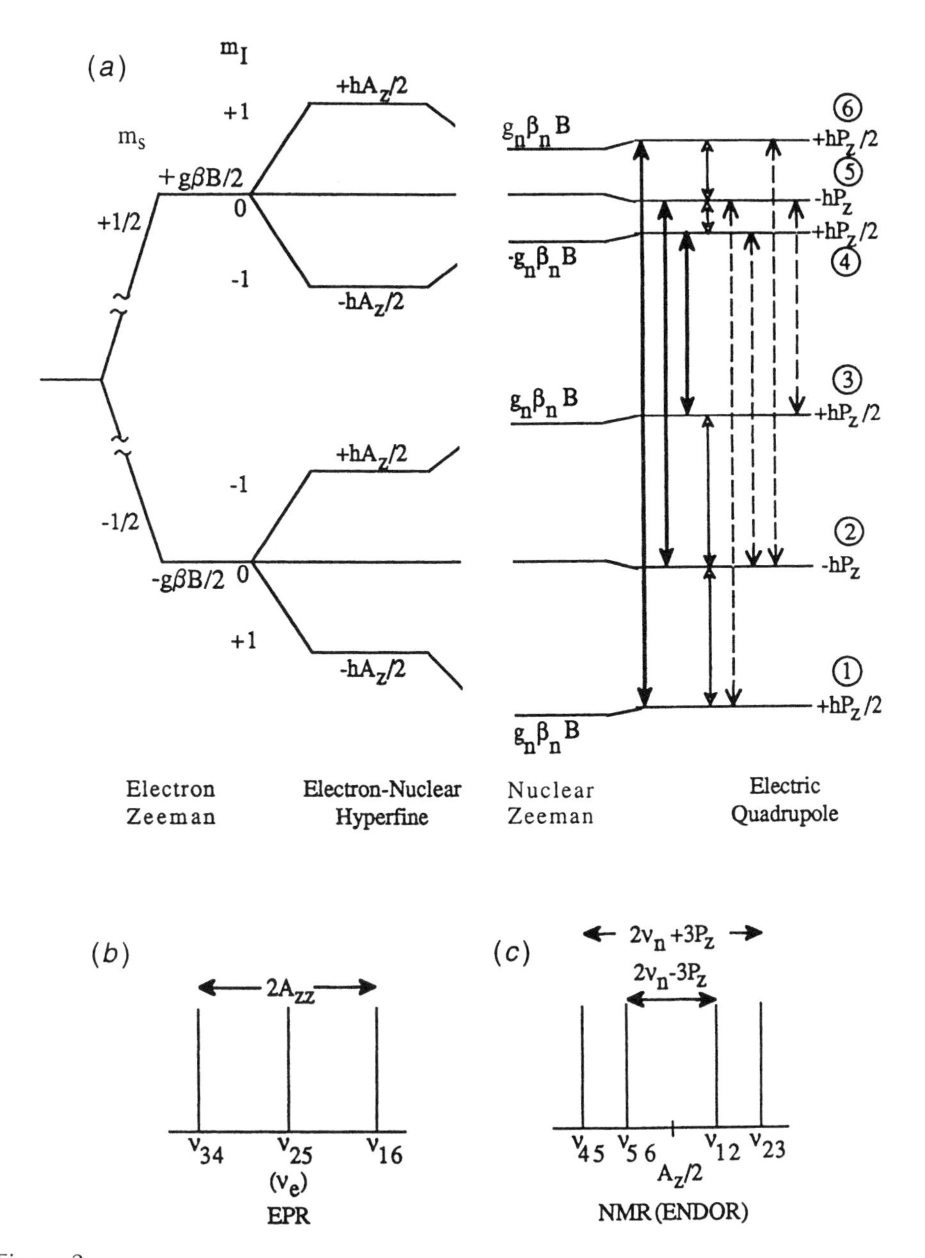

Figure 2
(a) The energy level diagram for an $S = \frac{1}{2}$, $I = 1$ spin system with an axial quadrupole coupling tensor and the magnetic field along the z axis. The g value and hyperfine coupling are assumed to be isotropic. The bold lines show the allowed EPR transitions, the solid lines the NMR (ENDOR, ESEEM) transitions, and the dashed lines the EPR "semiforbidden" transitions. (b) The stick EPR spectrum. (c) The stick ENDOR (NMR) spectrum. [Adapted from Snetsinger.[1]]

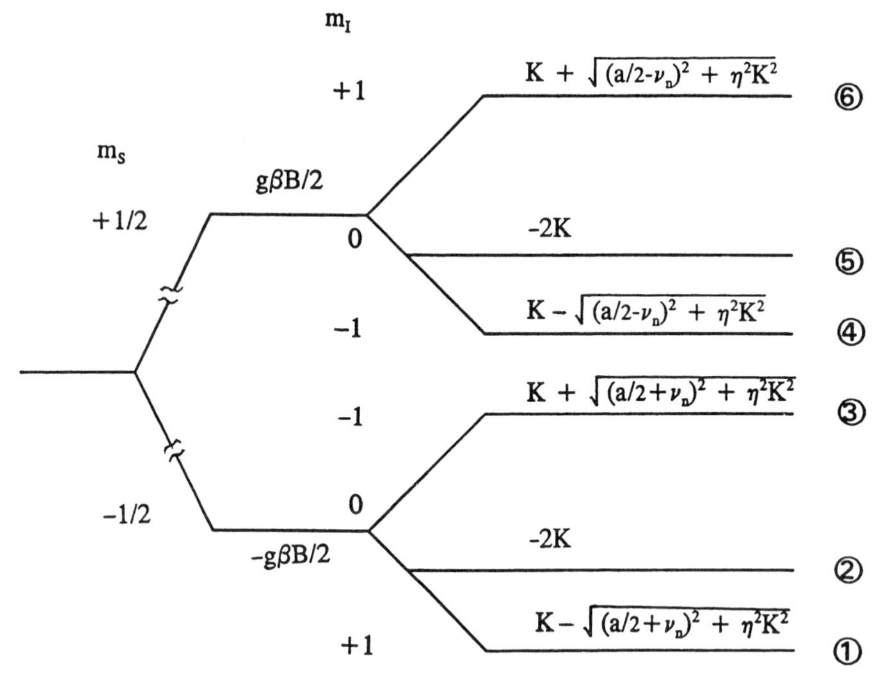

Figure 3
Energy level diagram for a $S = \frac{1}{2}$, $I = 1$ spin system with a rhombic quadrupole coupling tensor ($\eta \neq 0$), an isotropic hyperfine coupling constant a and isotropic g factor, and the magnetic field along the z axis. Here $K = P_z/2 = e^2qQ/4$. The energies are obtained from solving the nuclear secular determinants for the $m_S = +\frac{1}{2}$ and $-\frac{1}{2}$ electron spin states.

for the upper α ($m_s = +\frac{1}{2}$) electron spin manifold and

$$\nu_{23} = 3K + \sqrt{(a/2 + \nu_n)^2 + \eta^2 K^2} \tag{13a}$$

$$\nu_{12} = -3K + \sqrt{(a/2 + \nu_n)^2 + \eta^2 K^2} \tag{13b}$$

$$\nu_{13} = 2\sqrt{(a/2 + \nu_n)^2 + \eta^2 K^2} \tag{13c}$$

for the lower β ($m_s = -\frac{1}{2}$) electron spin manifold. The frequencies ν_{13} and ν_{46} correspond to double quantum $\Delta m_I = \pm 2$ semiforbidden transitions (Fig. 3). Whether or not all six of these transitions will be observed depends on the degree of mixing of the various spin states. The extent of mixing is determined by the magnitudes of the nuclear hyperfine and quadrupole terms in the Hamiltonian.

III. Electron Spin Echo Envelope Modulation Spectroscopy

ESEEM spectroscopy is a pulsed EPR technique that has much in common with pulsed Fourier transform NMR (FT–NMR) spectroscopy. The basic components of the spectrometer are a microwave bridge with a microwave pulse transmitter and receiver, a resonator for the sample, and an electromagnet. Microwave pulses of short duration (10–100 ns) are generated by fast Positive Internal Negative (PIN) diode modulation of the continuous microwave output from a microwave oscillator. The low power pulses are then amplified to around 1 kW. The pulse sequence is controlled by a pulse programmer. Signal averaging is done as in FT–NMR. Typically, 100–1000 transients (repetitions of the pulse sequence) at repetition rates in the range 100 Hz–2 KHz are required with transition metal samples at concentrations nominally 1–10 m*M* at liquid helium temperatures. The repetition rate has to be sufficiently slow for the electron spin magnetization to recover following each pulse. An interval of about $3T_{1e}$ between pulses is usually employed, where T_{1e} is the electron spin–lattice relaxation time.

A. The Electron Spin Echo

The Hahn spin echo 90°-τ-180° pulse sequence is used in the most basic ESEEM experiment (Fig. 4); we will therefore consider it in some detail. The basic Hahn spin echo experiment is also employed in NMR to measure the transverse or spin–spin relaxation time T_{2n} of the nucleus. The bulk magnetization $\mathbf{M}_0$ of the sample is the vector sum of the individual electron spin magnetic moments, which are at thermal equilibrium in an applied magnetic field of strength $\mathbf{B}_0$ [Fig. 4(*b*)]. In the rotating-frame coordinate system $(x'y'z)$,[d] the static magnetic field is along the z axis, and the x' and y' axes precess at the microwave frequency ν_0 of the EPR transition. The laboratory-frame coordinate axes (xyz) are fixed. The application of a microwave pulse of amplitude B_1 along $-x'$ for a duration t_p causes the magnetization vector $\mathbf{M}_0$ to precess about $\mathbf{B}_1$ and to tip through an angle θ given by Eq. (14).

$$\theta = \gamma_e B_1 t_p \tag{14}$$

Here γ_e ($= 1.7608 \times 10^{11}$ rad s^{-1} T^{-1}) is the magnetogyric ratio of the electron spin. A 90° pulse corresponds to a tip angle $\theta = \pi/2$ in Eq. (14). It is evident from Eq. (14) that the tip angle can be varied by either changing the strength of the field B_1 or the duration of the pulse t_p. Since it is easier instrumentally to control t_p, t_p is the parameter usually varied. For a 90° pulse, t_p is typically of the order of 9–18 ns for spectrometers producing pulses with B_1 values of 5–10 G (0.5–1 mT). In contrast, much longer t_p values in the range 5–40 μs are employed to obtain a 90° tip angle in FT–NMR. The difference of three orders of magnitude in t_p for NMR compared to EPR is due to the difference between the

[d] A description of the rotating frame can be found in most NMR books, for example, Abraham, R. J. and Loftus, P. *Proton and Carbon-13 NMR Spectroscopy*, Heyden: London, 1978; p 82.

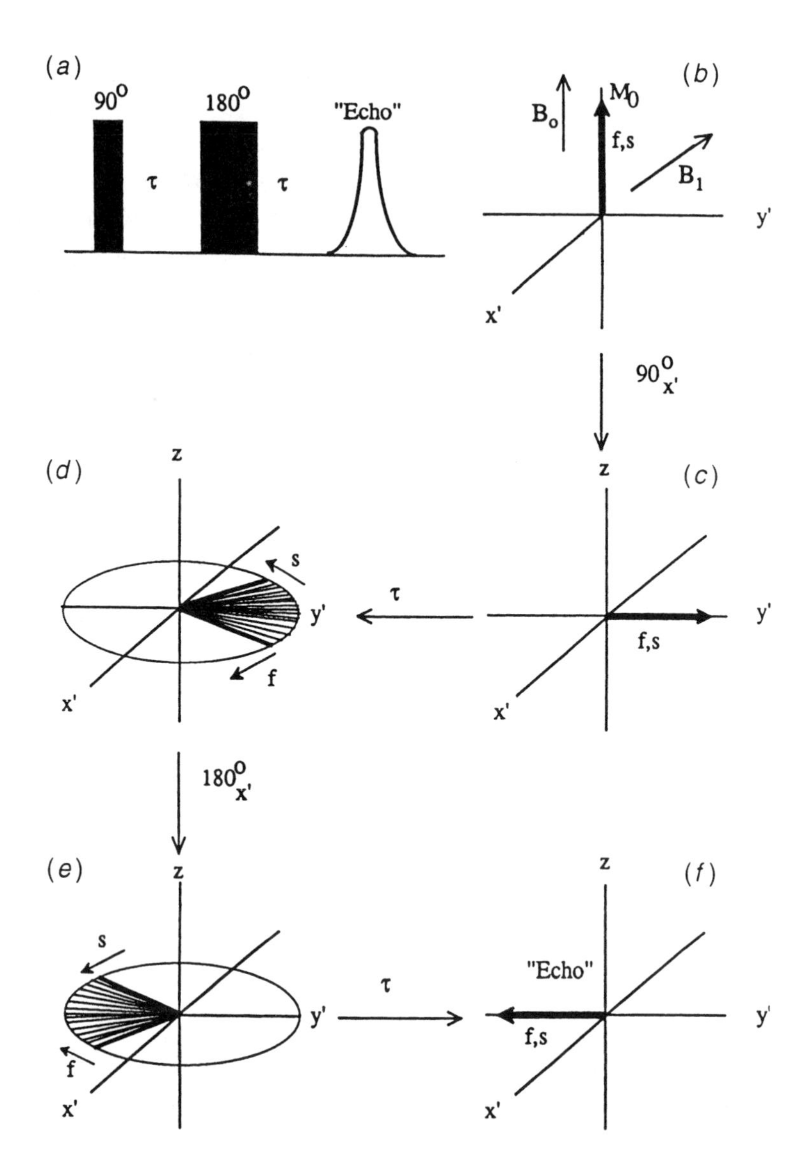

Figure 4
Vector diagram for the formation of a spin echo by the Hahn (90°-τ-180°) pulse sequence. (*a*) The pulse sequence. The echo is not to scale and is several orders of magnitude smaller than the heights of the pulses. (*b*) At $\tau = 0$, the sample magnetization $\mathbf{M}_0$ is aligned with the magnetic field $\mathbf{B}_0$ along the $+z$ axis. (*c*) The magnetization immediately after the 90° pulse is aligned along the y' axis. (*d*) The individual spin packets spread out after a time τ. (*e*) The magnetization after the 180° pulse. (*f*) After time τ, the magnetization vectors have refocused along the y' axis to form the echo. [Reproduced from Snetsinger.[1]]

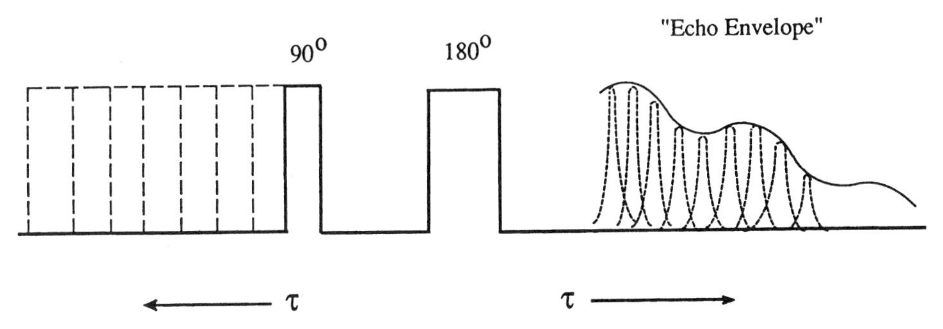

Figure 5
The modulated echo envelope as a result of the two-pulse spin echo experiment as the time between the first and second pulse, τ, is varied. In the absence of nuclear modulation, the echo would simply decay exponentially. Data collection commences following the 180° pulse after waiting for the instrument deadtime to elapse. The echo amplitude is not to scale and is several orders of magnitude smaller than the pulses. [Reproduced from Snetsinger.[1]]

magnetogyric ratios of the proton and the electron. Pulses in EPR and NMR spectroscopy are about equally strong, that is, B_1 (EPR) $\approx B_1$ (NMR).

Under the action of a 90° pulse, the magnetization vector $\mathbf{M}_0$ becomes directed along the y' axis as shown in Fig. 4(c). Since the EPR line is broadened by various magnetic interactions, the individual spin packets[e] within the bulk magnetization vector $\mathbf{M}_0$ precess at slightly different Larmor frequencies relative to the frequency ν_0 of the rotating-frame coordinate axes. This effect causes a "fanning out" of the magnetization after a time τ as shown in Fig. 4(d), with some spin packets moving slower (s) and others moving faster (f) relative to the rotating axis system. The resulting free-induction decay (FID) signal is the time variation in the magnetization detected along the y' axis.

Application of a second pulse of 180° after a time τ rotates the f and s magnetization components through 180° about the x' axis [Fig. 4(e)]. The "fan" of magnetization, although still in the $x'y'$ plane, closes because the fast spins catch up with the slow spins; an electron spin echo signal is observed at a time τ after the 180° pulse [Fig. 4(a and f)].

In the ESEEM experiment, the 90°-τ-180° series of pulses is repeated with different times τ between the first and second pulse, allowing the magnetization vectors to fan out for different periods of time. The echo is detected and recorded for each time τ before stepping to a new τ value. Figure 5 shows schematically the amplitude of the resultant echo envelope from an ESE experiment as τ is varied. An echo signal is observed that decays approximately exponentially with a characteristic decay time T_m called the "phase memory time". Superimposed on the decay is the nuclear modulation. The relaxation time T_m is a measure of the persistence of the precessional frequencies in the $x'y'$ plane of the individual

[e] A spin packet is a homogeneously broadened component of an overall inhomogeneous line and consists of those spins in the sample, which are precessing with the same Larmor frequency when the pulse is applied.

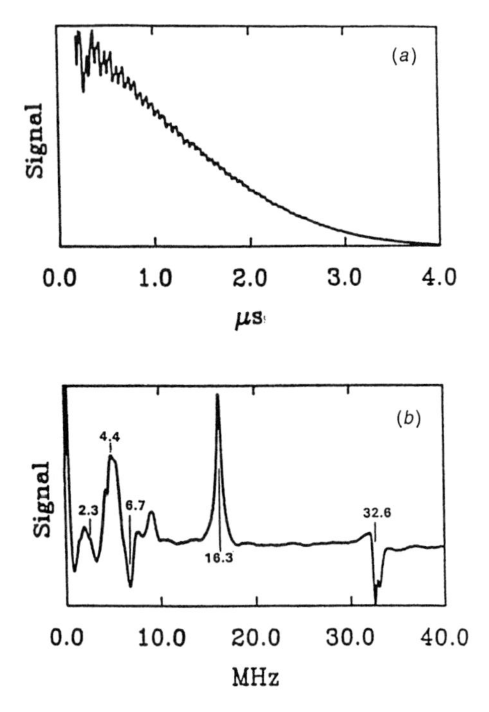

Figure 6
(*a*) Two-pulse ESEEM pattern at 4.2 K and (*b*) corresponding frequency spectrum of a VO^{2+}–apoferritin complex. The lack of data in the beginning of the ESEEM pattern is due to the deadtime of the instrument. [Reproduced with permission from Gerfen, G. J.; Hanna, P. M.; Chasteen, N. D.; Singel, D. J. *J. Am. Chem. Soc.* **1991**, *113*, 9513–9519. Copyright © 1991 American Chemical Society.]

electron spin packets of the sample. When loss of in-plane phase coherence is due solely to exchange of energy between electron spins, T_m is equivalent to the transverse or spin–spin relaxation time T_{2e} of the electron spin. Because phase coherence is also affected by nuclear spin flips in the vicinity of the unpaired electron, T_m is normally not identical to T_{2e}. For paramagnetic transition metal ions at liquid helium temperatures, about 4 K, ESEEM T_m values are typically around 1–10 μs.

In pulsed EPR, a modulation with well-defined periodicity is very often observed superimposed on the decaying echo envelope (Fig. 5). In the absence of modulation, the echo envelope would simply tail off exponentially or approximately so. The decay envelope modulation has its origin in the coupling of the resonant electron spin to nearby nuclear spins.

Figure 6 shows an experimental electron spin echo envelope for a VO^{2+} complex of the iron storage protein apoferritin and the corresponding frequency spectrum obtained from a cosine Fourier transformation of the modulation pattern.[2] As in NMR, the FT of the echo envelope converts time domain data to a frequency spectrum from which nuclear frequencies can more readily be obtained. The peaks below 10 MHz in the frequency spectrum are from ^{14}N ($I = 1$) and the higher frequency peaks from ^{1}H, as will be discussed later.

B. Requirements to Observe Nuclear Modulation

Once an echo is obtained, as determined by the electron relaxation properties of the sample, there are several conditions that must be met in order to observe modulation from a nucleus coupled to the electron spin.

1. Both Allowed and "Semiforbidden" Transitions Must Occur During the Microwave Pulse

This condition is very important. The transition moments for the semiforbidden transitions shown in Figs. 1 and 2 must be nonzero.

For the $S = \frac{1}{2}$, $I = \frac{1}{2}$ case, the mixing of states differing by $\Delta m_I = \pm 1$ is rigorously most effective when the magnitude of the nuclear Zeeman and hyperfine energies, that is, the second and third terms in the Hamiltonian given by Eq. (2a), are "matched". This situation gives good quantum mechanical mixing of the spin states and optimum modulation depths. For the $S = \frac{1}{2}$, $I = 1$ case, a rhombic nuclear quadrupole interaction gives rise to off diagonal matrix elements from the term $\eta(\hat{\mathbf{I}}_x{}^2 - \hat{\mathbf{I}}_y{}^2)$ in the Hamiltonian [Eq. 9(a)]. These elements enable mixing of states that differ by $|\Delta m_I| = 2$, which leads to semiforbidden "double quantum" ($\Delta m_I = \pm 2$) transitions, that is, transitions between levels 1 and 3 and between 4 and 6 in Figure 3.

Note that it is the differences between the EPR frequencies of the allowed and semiforbidden EPR transitions that correspond to the ENDOR frequencies (Figs. 1 and 2). In the ESEEM experiment, the simultaneous excitation of both allowed and semiforbidden transitions causes a beating of the EPR transition frequencies against one another, resulting in a modulation of the echo envelope with a periodicity of the ENDOR frequencies.

2. The Microwave Pulse Must be Sufficiently Strong to Excite all of the Transitions Between the Lower and Upper Electron Spin State Manifolds (Figs. 1 and 2)

The range of frequencies $\Delta\nu$ contained in a microwave pulse of strength B_1 is given by the approximate expression

$$\Delta\nu \approx \gamma_e B_1 / 2\pi \qquad (15)$$

Equation (15) predicts that for a microwave pulse of strength $B_1 = 10$ G all superhyperfine lines buried in the EPR line width within a frequency range $\Delta\nu$ of about 28 MHz will be excited. Thus, with a B_1 of about 10 G, a microwave field typically used in ESEEM measurements, the excitation of frequencies in the range of about $\Delta\nu \approx 30$ MHz ($\sim$1 G EPR line width) can be achieved. This places an approximate upper limit on the range of nuclear frequencies that can be probed by ESEEM with current microwave pulse technology.

3. The Electron Spin Phase Memory Time T_m of the Electron Spin Must be Sufficiently Long Relative to the Modulation Period of the Nucleus

In other words, an adequate number of nuclear modulation cycles must be observed before the echo signal decays away. For example, a 5 MHz NMR line has a modulation periodicity of 0.2 μs and would complete five cycles during an electron spin phase memory time of 1 μs. However, a nuclear modulation frequency less than 1 MHz does not complete even one cycle during the 1-μs observation time and, therefore, cannot be measured. Transition metal samples are usually run as magnetically dilute frozen solutions in order to minimize electron spin–spin interactions and, thereby, lengthen T_m. Measurements are usually made at liquid helium temperatures (2–10 K) to slow electron spin and nuclear spin relaxation rates, as done, for example, with VO^{2+}, Cu^{2+}, and high-spin Fe^{3+}. Some systems have been studied at 77 K such as the nitrosyl (NO) complexes of Fe(II) and Co(II) tetraphenyl porphyrins.[3]

4. The Instrument Deadtime Must be Short Relative to the Time Over Which the Nuclear Modulations Persist

The "deadtime" is the time required for the microwave energy remaining in the resonator after the pulse to dissipate before the receiver can be turned on and the electron spin magnetization in the $x'y'$ plane measured. Nuclear modulations that decay away within the instrument deadtime cannot be measured. For example, NMR lines having line widths greater than $\Delta\nu_{1/2} \sim 2$ MHz (width at half-height of the absorption curve) will be difficult to detect since the amplitudes of their nuclear modulations will die away with a lifetime $T_{2n} \approx 1/\pi\Delta\nu_{1/2} = 0.16$ μs, which is around the deadtime of most instruments. Therefore, broad NMR lines having $\Delta\nu_{1/2} \geq 1$ MHz cannot be measured by present ESEEM instrumentation.

The quality factor Q (= energy stored/energy dissipated per cycle) of the microwave resonator used for ESE measurements must be low in order to minimize the deadtime in the spin echo experiment. Typically ESEEM resonators have Q values around 200–400 and produce instrument deadtimes around 0.1–0.2 μs. Most microwave cavities for continuous wave EPR (CW EPR) have Q values an order of magnitude higher. Of course, deadtimes always result in some lost data at the beginning of the echo envelope. Efforts are often made to fill in the early part of the echo decay by various reconstruction techniques that attempt to extrapolate the measured data backward to zero time.[4] Such procedures minimize the impact of the lost data on the frequency spectrum obtained from Fourier transformation of the time domain data.

It should be evident from the foregoing that the failure to see couplings from a particular nucleus in the ESEEM experiment does not necessarily mean that the nucleus is not present. A similar limitation applies to ENDOR. In both forms of spectroscopy, a negative result must be interpreted with caution. In the case of ESEEM, the lack of modulation from a nucleus present means that its coupling constant must fall outside the range needed for stimulating both allowed and semiforbidden transitions, or that the modulation decays within the deadtime of the instrument.

C. The Two-Pulse Sequence

There are two commonly employed pulse sequences in ESEEM spectroscopy: the two-pulse 90°-τ-180° sequence already considered above (i.e., the Hahn spin echo) and the three pulse or stimulated echo sequence to be discussed in Section III.D. The quantum mechanical model for echo amplitude modulation was developed by Mims[5] using density matrix theory. Only the results are given here.

The time dependence of the echo amplitude can be written as a product of an exponential decay term V_{decay} and a nuclear modulation term V_{mod}.

$$V_{\mathrm{echo}} = V_{\mathrm{decay}}(\tau) \cdot V_{\mathrm{mod}}(\tau) \tag{16a}$$

where

$$V_{\mathrm{decay}}(\tau) = V_0 \exp(-\tau/T_{\mathrm{m}}) \tag{16b}$$

The echo amplitude V_{decay} in Eq. (16b) has a lifetime equal to the phase memory time T_{m} of the electron spin. The parameter V_0 is the amplitude at $\tau = 0$. Here exponential decay is assumed as is often the case experimentally. For the simplest case of an electron and a proton ($S = \frac{1}{2}$, $I = \frac{1}{2}$) in an axially symmetric system, the nuclear modulation V_{mod} is given by Eq. (17a). Mims[5a] gives the expressions for the $S = \frac{1}{2}$, $I = 1$ case.

$$V_{\mathrm{mod}}(2\tau, I = \tfrac{1}{2}) = 1 - \tfrac{1}{4}k[2 - 2\cos(\omega_{\mathrm{a}}\tau) - 2\cos(\omega_{\mathrm{b}}\tau) + \cos(\omega_{\mathrm{a}} - \omega_{\mathrm{b}})\tau + \cos(\omega_{\mathrm{a}} + \omega_{\mathrm{b}})\tau] \tag{17a}$$

where

$$k = (\omega_{\mathrm{n}} B/\omega_{\mathrm{a}}\omega_{\mathrm{b}})^2 \tag{17b}$$

$$\omega_{\mathrm{a}} = [(A/2 + \omega_{\mathrm{n}})^2 + (B/2)^2]^{1/2} \tag{17c}$$

$$\omega_{\mathrm{b}} = [(A/2 - \omega_{\mathrm{n}})^2 + (B/2)^2]^{1/2} \tag{17d}$$

$$A = F(3\cos^2\theta - 1) + 2\pi a \tag{17e}$$

$$B = 3F\cos\theta\sin\theta \tag{17f}$$

$$F = g_{\mathrm{e}} g_{\mathrm{n}} \beta_{\mathrm{e}} \beta_{\mathrm{n}}/r^3\hbar \tag{17g}$$

As is the customary practice, the quantities on the left-hand side of Eqs. (17a–g) are expressed in radians per second. The parameter ω_{n} is the free precessional (Larmor) frequency of the nucleus in radians per second ($\omega_{\mathrm{n}} = 2\pi\nu_{\mathrm{n}}$) at the applied magnetic field of the experiment, θ is the angle between the magnetic field and the electron–nuclear axis, and r is the effective electron–nuclear distance, which is usually assumed to be the same as the metal–ligand nucleus distance.

Planck's constant, $\hbar$ is $h/2\pi$. Equation (17g) is appropriate only when cgs units are used, namely, $\beta_e = 9.27402 \times 10^{-21}$ erg G^{-1} and $\beta_n = 5.05079 \times 10^{-24}$ erg G^{-1} and r is in centimeters. Equation (17h) is the correct expression for F in SI units

$$F(Hz) = \mu_0 g_e g_n \beta_e \beta_n / 4\pi r^3 h \tag{17h}$$

Here the additional factor $\mu_0 = 4\pi \times 10^{-7}$ $T^2 J^{-1} m^3$ is the permeability in a vacuum.

The distance r is employed as a parameter in simulating the ESEEM pattern. The frequencies ω_a $(= 2\pi\nu_a = 2\pi\nu_{34})$ and ω_b $(= 2\pi\nu_b = 2\pi\nu_{12})$ are the nuclear transition frequencies (the ENDOR frequencies) of the upper and lower electron α and β spin manifolds of Figure 1. The quantity a is the isotropic hyperfine coupling constant in hertz. The parameter k determines the depth of the modulation and is proportional to B^2 as well as various frequency factors. The parameter B is a dipolar term of the anisotropic electron–nuclear hyperfine coupling that varies as $1/r^3$ and has an angular dependence $\cos\Theta \sin\Theta$ [Eq. (17f and g)].

It is evident from Eq. (17a) that no modulation is possible when $B = 0$, since $k = 0$. For $I = \frac{1}{2}$ systems, observation of ESEEM is limited to solids where $F \neq 0$. Since the depth of modulation (k term) depends on r^{-6}, it falls off rather quickly with distance. As a practical matter, ligand $(I = \frac{1}{2})$ nuclei farther than about 4–5 Å from the metal cannot easily be observed by ESEEM.

Equation (17a) assumes that the electron and nuclear spins are quantized along the direction of the applied magnetic field and that the electron g factor is isotropic. The anisotropic part of the hyperfine coupling is assumed to be point-dipole in origin, which gives rise to the terms in F $(3\cos^2\theta - 1)$ and F $(3\cos\theta \sin\theta)$. Therefore, Eq. (17a) is expected to be valid for only weakly coupled protons at some effective distance (>2.5 Å) from the metal where the through-space magnetic point-dipole approximation is expected to hold reasonably well. Note that the $\cos\theta \sin\theta$ term is zero when $\theta = 0$ or 90° and, therefore, reduced modulation depth is expected when the field is approximately parallel or perpendicular to the principal axis of the nuclear hyperfine interaction, that is, the metal nucleus–ligand nucleus axis.

Another important result from Eq. (17a) is that the modulation pattern from the two-pulse sequence also contains, in addition to the ENDOR frequencies ω_a and ω_b, the sum and difference of these frequencies, $\omega_a + \omega_b$ and $\omega_a - \omega_b$, albeit at a twofold lower amplitude. Therefore, the spectrum can be complicated when several different nuclei are present, each contributing overlapping combination bands $(\omega_a \pm \omega_b)$ and fundamental frequencies (ω_a and ω_b) to the frequency spectrum.

In Figure 6(*b*), the counterphase peak at 6.7 MHz is a sum combination of fundamental peaks at 2.3 and 4.4 MHz. The fundamental frequency of matrix protons distant from the metal is observed at 16.3 MHz (i.e., the Larmor frequency ν_n). The first overtone (second harmonic) is observed at twice the frequency of the fundamental, 32.6 MHz, with a counterphase, intensity as predicted by Eq. (17a). (A third harmonic would be observed at three times the fundamental frequency.) In this case, the pairs of lines from proton dipolar couplings are not resolved within the line width of the fundamental peak at 16.3 MHz. The small peak to the high-frequency side of the 32.6-MHz overtone peak in Figure 6(*b*) arises from more strongly dipolar coupled protons in the first-coordination sphere.[2] Because of a combination of factors, proton couplings, large or small, are difficult to observe at the fundamental frequency in ESEEM. Therefore, ENDOR is usually better suited for their measurement.

The depth of modulation increases with the number of equivalent nuclei contributing to the modulation. In practice, it is difficult to determine accurately the number of equivalent nonhydrogen nuclei contributing to the modulation since both dipolar [i.e., Eqs. (17)] and pseudodipolar terms from electron spin density on the p and d orbitals of the observed nucleus contribute to the modulation.[6] Determination of the pseudodipolar terms requires a detailed knowledge of the electron spin density distribution onto the ligands, information that is usually not available. Nevertheless, simulations in conjunction with the use of model compounds has met with some success.[7] For example, it is easier to distinguish between one versus two nitrogen ligands being present than between two and three and so on.

When several inequivalent nuclei contribute to the modulation, the observed modulation pattern is simply a *product* of the modulations of the *n* individual nuclei.

$$V_{\text{mod}} = V_{\text{mod1}} \cdot V_{\text{mod2}} \cdot V_{\text{mod3}} \cdots V_{\text{mod}n} \tag{18}$$

Equation (18) can be used to good advantage in studies of isotope or ligand exchange in metal complexes in proteins. As an example, let us consider the ESEEM study of the Cu^{2+}–ovotransferrin–oxalate complex where the goal was to establish whether the oxalate is directly coordinated to the copper.[8] Figure 7(*a*) shows the ESEEM patterns of the complex prepared with ^{13}C-oxalate and with ^{12}C-oxalate, ^{13}C being a magnetic nucleus ($I = \frac{1}{2}$), whereas, ^{12}C is nonmagnetic.

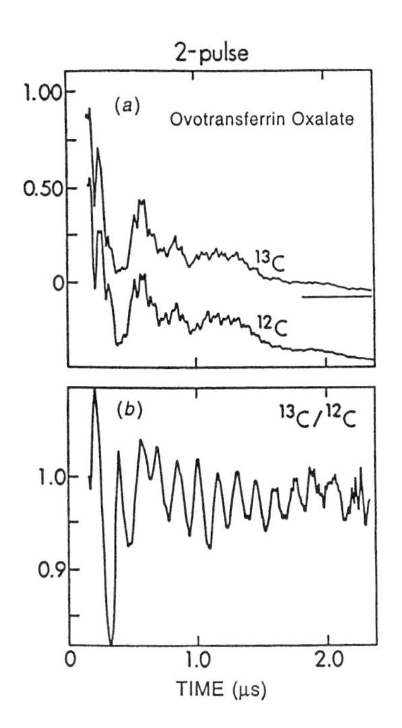

Figure 7
(*a*) ESEEM patterns for Cu^{2+}–ovotransferrin with [^{13}C] and [^{12}C] oxalate. (*b*) Ratio of the patterns in (*a*). The ^{13}C modulations are evident. [Adapted from Mims and Peisach[6] with permission.]

The ESEEM patterns obtained with the two isotopes are virtually identical. However, when the $^{13}C/^{12}C$ ratio of the two envelopes is taken, the modulation pattern for ^{13}C becomes evident as shown in Figure (7*b*). The strong modulations from protons and ^{14}N nuclei, which dominate the ESEEM patterns in Figure 7(*a*), effectively cancel since they appear as common factors in Eq. (18). The observation of ^{13}C ESEEM means that the oxalate is bound near the Cu^{2+} and almost certainly coordinated to it. Using a similar procedure, division of ESEEM patterns obtained with H_2O and D_2O solvents can be used to establish the presence of exchangeable protons, usually from water, in the first coordination sphere of metalloprotein complexes.[9]

D. The Three Pulse Sequence

The so-called stimulated echo or three-pulse, 90°-τ-90°-T-90°, sequence is illustrated in Figure 8. In this sequence, the time τ between the first and second pulses is held *constant* while the time T between the second and third pulses is *varied.* For the $S = \frac{1}{2}$, $I = \frac{1}{2}$ case, the stimulated echo modulation can be written as[5]

$$V_{\text{mod}}(2\pi + T, I = \tfrac{1}{2}) = 1 - k\{\sin^2(\omega_a\tau/2)\sin^2(\omega_b(\tau + T)/2) + \sin^2(\omega_b\tau/2)\sin^2[\omega_a(\tau + T)/2]\} \tag{19}$$

When $T = 0$, Eq. (19), after trigonometric substitution, reduces to Eq. (17a) for the two-pulse envelope modulation. In contrast to the two-pulse sequence, Eq.

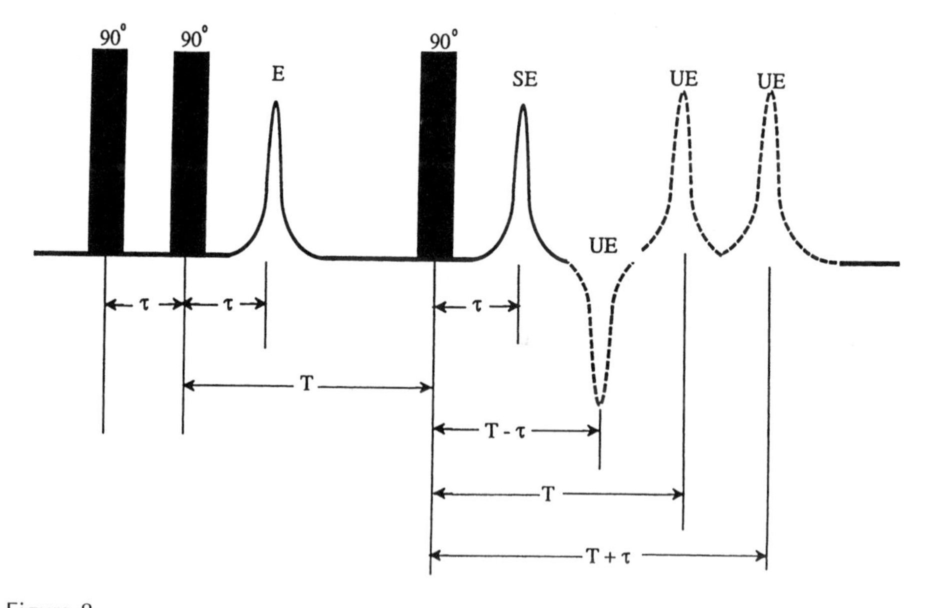

Figure 8
Three-pulse experiment showing all two-pulse echos: E (echo), SE (stimulated echo), and UE (unwanted echo). The amplitudes of the echos are not to scale. They are several orders of magnitude smaller than the pulse heights. Data collection commences following the third pulse. [Reproduced from Snetsinger.[1]]

(19) contains only the ENDOR frequencies ω_a and ω_b, not sums and differences. However, as illustrated in Figure 8 any combination of two pulses gives an echo. Thus in addition to the stimulated echo SE at time τ after the third pulse, unwanted echos UEs are also created. These unwanted echos can be suppressed by a procedure known as "phase cycling" between successive pulses.

It is apparent from Eq. (19), that if τ is set equal to a whole multiple of the number of the cycles of ω_a (period $= 2\pi/\omega_a$), then the $\sin^2(\omega_a\tau/2)$ term equals zero and we obtain

$$V_{\text{mod}}(2\tau + T, I = \tfrac{1}{2}) = 1 - k[\sin^2(\omega_b\tau/2)\sin^2[\omega_a(\tau + T)/2] \tag{20}$$

Since T is the experimental time variable in the $\sin^2[\omega_a(\tau + T)/2]$ term of Eq. (20), only the ω_a modulation will appear in the envelope but no ω_b modulation, the $\sin^2(\omega_b\tau/2)$ term being a constant. Although Eq. (20) was developed for a particularly simple case and is rarely strictly true, this "tau suppression" effect of the ω_b modulations is usually observed to some degree. The effect can be experimentally useful in eliminating modulations due to nuclei such as protons that are numerous and may mask modulations due to other nuclei.[5c] The tau suppression can, in some cases, actually complicate the frequency spectrum since it may cause one broad line to appear as two distinct lines if a frequency in the middle of the line is suppressed. This situation can be easily recognized by making measurements at more than one τ value.

The principal advantage of the three-pulse experiment is that the decay of the echo is no longer governed by the phase memory time T_m ($\approx T_{2e}$) as in the two-pulse sequence but rather by a time closely related to the spin–lattice or longitudinal relaxation time, T_{1e}, of the electron spin. Note that following the first two 90° pulses, the magnetization $\mathbf{M}_0$ is effectively along the negative z axis when on resonance. The third 90° pulse after the time T is a "sampling" pulse that measures the magnetization along -z at a time $\tau + T$. The z component of magnetization has a decay time T_{1e}. Since in general $T_{1e} \geq T_{2e}$, the envelope decays more slowly in the three-pulse experiment than in a two-pulse experiment. Therefore spectra have better resolution overall, which is especially important in the low-frequency regime. The three-pulse experiment enables observation of low-frequency nuclei that have long modulation periods and are thus not observable in the two-pulse experiment.

A three-pulse ESEEM and corresponding FT are illustrated in Figure 9. Note that longer time axis in Figure 9 (10 μs) for the three-pulse compared to that in Figure 6 (5 μs) for the two-pulse experiment. Weakly coupled nuclei most often studied by the three-pulse experiment include ^{14}N, ^{13}C, ^{2}H, and ^{27}Al. The presence of three low-frequency ^{14}N lines below 1.6 MHz in the frequency spectrum of Figure 9 makes it advantageous that the ESEEM measurement be made with the three-pulse sequence.

A disadvantage of the three-pulse sequence is that the time τ between the first and second pulses must be added to the "dead time" of the instrument and, therefore, rapidly decaying modulations may be missed. The two-pulse sequence is used to look for such modulations.

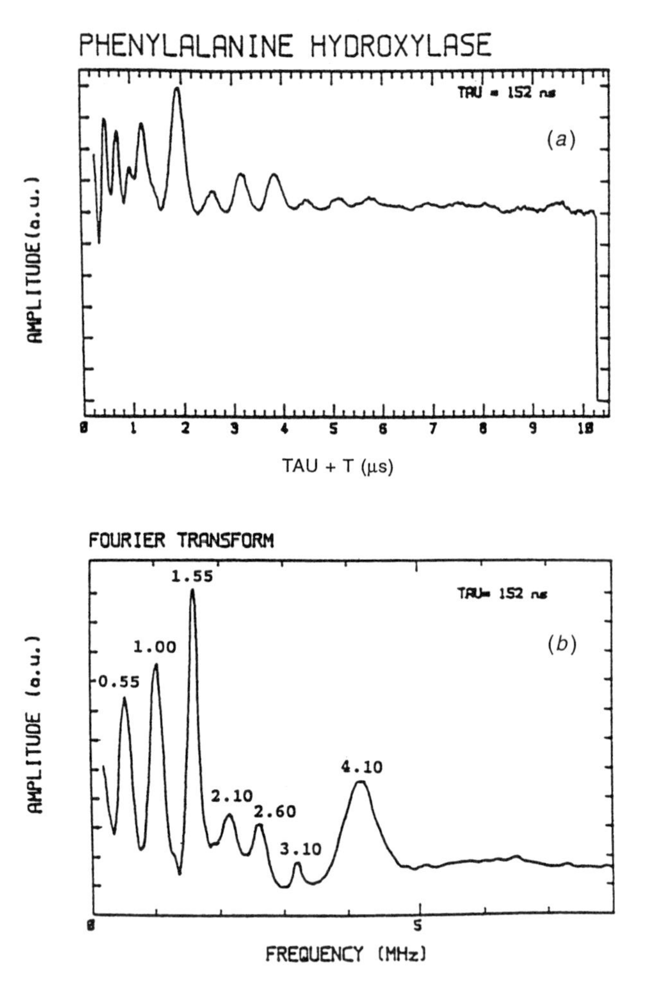

Figure 9
Three-pulse ESEEM (*a*) and FT (*b*) of phenylalanine hydroxylase at 4.2 K. The decay factor V_{decay} [Eq. (16)] has been factored out of the echo envelope. [Reproduced with permission from McCracken, J.; Pember, S.; Bernkovic S. J.; Villafraneo, J. J.; Miller, R. J.; Peesach, J. *J. Am. Chem. Soc.* **1988**, *110*, 1069–1074. Copyright © 1988 American Chemical Society.]

E. Examples of ESEEM Spectral Assignments

1. An $S = \frac{1}{2}$, $I = 1$ Example

The couplings from ^{14}N nuclei are commonly seen with metalloproteins due to the coordination of nitrogen containing ligands such as the imidazole group of histidine. It is therefore instructive to consider ESEEM from the ^{14}N nucleus in some detail.

The applicable energy level diagram for a rhombic $S = \frac{1}{2}$, $I = 1$ system with an isotropic hyperfine coupling constant a and a rhombic quadrupole interaction ($\eta \neq 0$) is given in Figure 3 and the frequencies of the transitions by Eqs. (12)

and (13). A special case is obtained when the nuclear Zeeman frequency ν_n ($= g_n\beta_n B/h$) is equal to one-half the hyperfine coupling constant $a/2$ in Eq. (12) for the upper electron spin manifold. The nuclear transitions within the upper manifold then become[f]

$$\nu_{56} = \nu_+ = K(3+\eta) \tag{21a}$$

$$\nu_{45} = \nu_- = K(3-\eta) \tag{21b}$$

$$\nu_{46} = \nu_0 = 2K\eta \tag{21c}$$

These frequencies are equivalent to the zero-field resonances observed in nuclear quadrupole resonance spectroscopy (NQR).[g] They are characterized by the fact that the sum of the two lower frequencies ν_0 and ν_- equal the third ν_+. It is desirable in ESEEM spectroscopy to carry out measurements at a microwave frequency (i.e., field strength) where the nuclear Zeeman and the hyperfine coupling terms in Eq. (12) cancel. Not only is the spectral analysis simplified, and the quadrupole parameters directly obtained from the frequency spectrum, but the amplitude of the modulation effect is maximized. The quadrupole lines tend to be sharp for frozen solutions of randomly oriented molecules because the quadrupole interaction is independent of the orientation of the magnetic field.

Figure 9 illustrates the three-pulse ESEEM pattern (*a*) and frequency spectrum (*b*) of the Cu^{2+} containing enzyme phenylalanine hydroxylase.[7] We now proceed to assign all of the peaks in the frequency spectrum. Three sharp peaks occur at 0.55, 1.00, and 1.55 MHz and are assigned to the pure quadrupole peaks, ν_0, ν_-, and ν_+ [Eqs. (21)], respectively, of the uncoordinated remote nitrogen of the imidazole ring of histidine. Note that the third peak at 1.55 MHz is the sum of the first two peaks as predicted by Eqs. (21) (i.e., $\nu_+ = \nu_- + \nu_0$).

The directly coordinated nitrogen of Cu^{2+} complexes cannot be observed by ESEEM at X-band microwave frequency because the ^{14}N hyperfine coupling constant of this nitrogen is too large relative to the Zeeman frequency, resulting in too shallow modulation. The directly coordinated nitrogen of the imidazole ring in Cu^{2+} protein complexes is usually the N3 nitrogen; the quadrupole parameters K and η measured by ESEEM are those of the remote N1 nitrogen.

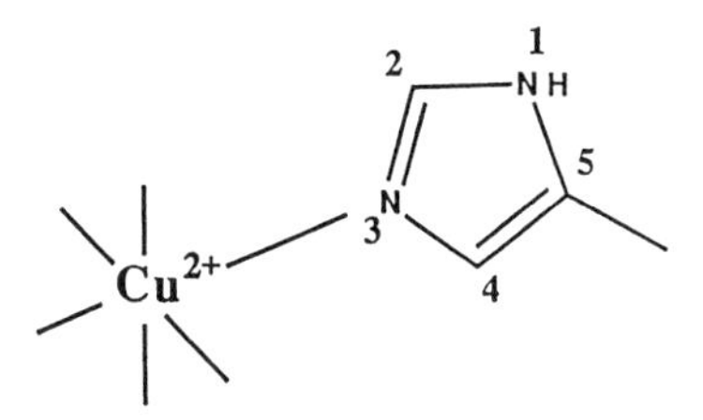

[f] The exact cancellation situation causes the sign of Eq. (21b) to be opposite that in Eq. (12b) due to the inversion of the energy levels 4 and 5 in Figure 11.

[g] The intensities of transitions corresponding to the NQR frequencies in the upper electron spin manifold are most favored when the ratio ν_{eff}/K is less than unity where $\nu_{eff} = g_n\beta_n B/h - A/2$, that is, the quadrupole term is the largest nuclear term in the Hamiltonian [See Eaton et al.[13]].

The parameter K can be calculated by combining Eqs. (21a) and (21c), $K = (\nu_+ - \nu_0/2)/3 = (1.55 - 0.55/2)/3 = 0.425$ MHz. The quadrupole coupling constant is thus $e^2qQ = 4K = 1.7$ MHz, a value typical of histidine nitrogen. The asymmetry parameter η can be calculated from Eq. (21c), $\eta = \nu_0/2K = 0.55/2(0.425) = 0.65$.[h] Thus the quadrupole tensor is approximately midway between axial ($\eta = 0$) and pure rhombic ($\eta = 1$). Since we are at a magnetic field strength B close to exact cancellation, namely, $a_N \approx 2\nu_n$ ($\nu_n = g_n\beta_n B/h$), it follows that the nitrogen coupling constant is approximately $a_N \approx 2(0.95) = 1.9$ MHz where 0.95 is the ^{14}N Zeeman frequency at the 3100-G field of the experiment. The uncertainty in the isotropic coupling constant a_N obtained by this procedure is around $\pm K$, or ± 0.4 MHz. To a good approximation, the hyperfine coupling constant can be assumed to be isotropic since the anisotropic or dipolar part is small for distant ^{14}N nuclei. In this case, the modulation depth derives from the quadrupolar term in the Hamiltonian in conjunction with the isotropic hyperfine interaction.[10]

In addition to the three pure quadrupole frequencies 0.55, 1.00, and 1.55 MHz in Figure 9, lines are observed at 2.1 ($= 0.55 + 1.55$) MHz, 2.6 ($= 1.00 + 1.55$) MHz, and 3.1 ($= 1.55 + 1.55$) MHz, and are assigned to combination frequencies of the fundamental quadrupole frequencies. Such combination frequencies are observed in the three-pulse ESEEM when *more than one* equivalent nucleus is contributing to the modulation pattern. Spectral simulation of the line intensities indicates that two equivalent ^{14}N nuclei contribute to the ESEEM, and, therefore, there are at least two histidine residues at the metal site in this enzyme.[7]

There is a broad peak at 4.1 MHz in Figure 9, which can be assigned to the field orientation dependent $\Delta m_I = 2$ "double quantum" transition. This transition takes place when a single photon of microwave energy induces a transition between the outermost levels 1 and 3 of the lower $m_s = -\frac{1}{2}$ spin manifold of Figure 3 causing the nuclear spin angular momentum to change by $2\hbar$. For randomly oriented samples, the position of the double quantum peak of the lower electron spin manifold is given to a good approximation by Eq. (22),[11]

$$\nu_{13} = 2[(\nu_n + a/2)^2 + K^2(3 + \eta^2)]^{1/2} \tag{22}$$

where a is the isotropic coupling constant of the ^{14}N nucleus for the "powder sample". Equation (22) is valid when $(\nu_n + a/2) \geq K$, a situation that applies here since $(0.95 + 1.9/2) = 1.90 > 0.425$, and when the hyperfine coupling is primarily isotropic. Substitution of the above values of $\nu_n = 0.95$ MHz, $a = 1.9$ MHz, $K = 0.425$ MHz and $\eta = 0.65$ in Eq. (22) gives $\nu_{13} = 4.1$ MHz in agreement with the observed frequency and in accord with its assignment to a double quantum transition. The 4.1-MHz line is broad because there is some

---

[h] In cases where the frequencies of first two lines do not exactly sum to give the frequency of the third line, it is preferable to use the "least-squares" equations $K = (\nu_+ + \nu_-)/6$ and $\eta = (2\nu_0 + \nu_+ - \nu_-)/6K$. See Flanagan, H. L.; Gerfen, G. J.; Singel, D. J., "Multifrequency Electron Spin–Echo Envelope Modulation: The Determination of Nitro Group ^{14}N Hyperfine and Quadrupole Interactions of DPPH in Frozen Solution" *J. Chem. Phys.* **1987**, *88*, 20–24.

anisotropy in the ^{14}N nuclear hyperfine interaction. The transitions ν_{12} and ν_{23} between the inner level and the outer levels of the lower $m_S = -\frac{1}{2}$ manifold (Fig. 3) are highly dependent on the orientation of the magnetic field within the molecular axis frame and are usually very broad and difficult to discern in frozen solution samples.

It should be evident to the student, that in situations where exact cancellation ($\nu_n \approx a/2$) does not occur, the spectrum can be considerably more complicated to analyze. The double quantum peaks then become important sources of information in these cases. Frequently, one must resort to simulations and to making measurements at more than one microwave frequency to interpret the spectrum fully.[10]

2. An $S = \frac{1}{2}$, $I = \frac{1}{2}$ Example

We now turn our attention to the analysis of an $S = \frac{1}{2}$, $I = \frac{1}{2}$ system, namely, the ^{13}C couplings in copper transferrin. Transferrin is the transport protein for iron but also binds a variety of other transition metal ions, including Cu^{2+}. In the case of Fe^{3+}, the complex is six coordinate with a carboxylate group of an aspartate residue, two phenolate groups of two tyrosine residues, an imidazole group from histidine, and a bidentate carbonate anion as ligands.[12] Electron spin echo envelope modulation has been a very useful tool in studies of copper coordination in transferrin.[13]

The Fourier transforms of two-pulse ESEEM patterns of Cu^{2+}–transferrin–CO_3 are illustrated in Figure 10 for a field of 3185 G in the perpendicular region of the CW EPR spectrum. Here "parallel" and "perpendicular" refer to molec-

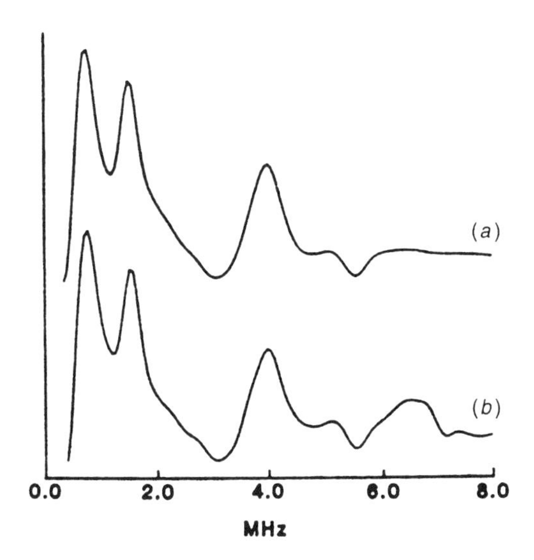

Figure 10
Two-pulse ESEEM frequency spectra for Cu^{2+}–transferrin–CO_3 at 4.2 K. (*a*) $[^{12}C]O_3$ and (*b*) $[^{13}C]O_3$ containing samples. [Reproduced with permission from Eaton et al.[13]]

ular orientations within the frozen solution sample where the principal z axis of the g and Cu nuclear hyperfine matrices are parallel or perpendicular to the magnetic field direction (Chapter 3). In Figure 10, we see a pair of barely resolved lines at 0.74 and 0.86 MHz as well as lines at 1.60 and 4.05 MHz. These lines correspond to the quadrupole lines ν_0, ν_-, and ν_+, and ν_{13} of the histidine ligand, the origin of which has been discussed in detail above for phenylalanine hydroxylase. The parameters obtained from an analysis of the spectra are $a = 1.8$ MHz, $e^2qQ = 1.64$ MHz and $\eta = 0.90$ ($\nu_n = 0.98$ MHz for ^{14}N at 3185 G). The quadrupole coupling constant e^2qQ and the hyperfine coupling a are comparable to those obtained for histidine coordination in phenylalanine hydroxylase, but the quadrupole tensor is more nearly completely rhombic ($\eta = 0.90$ vs. 0.65). In the FT of the three-pulse ESEEM (not shown), no combination bands are observed, hence, the ESEEM spectrum does not reveal the presence of more than a single histidine ligand in the complex.

In addition to the ^{14}N lines, the ^{13}C-carbonate sample displays a line at 6.6 MHz, which is not present in the ^{12}C-carbonate sample [Fig. 10 (a and b)]. This line is assigned to the high-frequency ^{13}C ($I = \frac{1}{2}$) component of the pair of lines predicted by Eq. (6b) to occur at $\nu_n \pm A/2$. At the 3185-G field of the experiment, $\nu_n(^{13}C) = 3.41$ MHz from which a coupling constant of $A = 2(\nu_{ESEEM} - \nu_n) = 2(6.6 \text{ MHz} - 3.4 \text{ MHz}) = 6.4$ MHz is calculated. The other component of the pair is predicted to occur at $\nu_n - A/2 = 0.2$ MHz, which is below the low-frequency cutoff of the spectrum.

The ^{13}C coupling constant A varies in magnitude depending on the field selected in the CW EPR spectrum. For example, at a field of 2615 G corresponding to a Cu^{2+} parallel line in the CW EPR spectrum, a coupling constant of $A = 8.4$ MHz is obtained, which compares with $A = 6.4$ MHz for the perpendicular line (see above). This anisotropy in the ^{13}C coupling constant A can be attributed to an anisotropic magnetic dipole interaction between the electron spin and the ^{13}C nuclear spin, that is, Eq. (17e). In summary, the ESEEM data demonstrate that carbonate and a single histidine ligand are coordinated in Cu^{2+}-transferrin, a finding confirmed by X-ray crystallography.[14]

IV. Electron Nuclear Double Resonance Spectroscopy

In contrast to ESEEM, ENDOR spectroscopy is a more established technique. This spectroscopy has contributed enormously to our understanding of metal coordination in proteins and chelate complexes and of the electronic structure of radicals. Like ESEEM, ENDOR observes nuclear spin flip transitions *within* an electron spin manifold; however, they are achieved in a very different manner. The ENDOR spectrometer essentially consists of a standard CW EPR spectrometer with an NMR radio frequency (rf) source and an rf coil within the microwave cavity. This arrangement enables EPR and NMR experiments to be performed simultaneously on the sample.

The main value of ENDOR over CW EPR spectroscopy comes from enhanced spectral resolution that is of the order of a thousand or more in frequency. The resonance lines are inherently narrower in ENDOR spectroscopy. An additional resolution enhancement comes from the decrease in density of nuclear hyperfine lines that are additive in ENDOR but multiplicative in EPR. This fact is a consequence of the different selection rules for the two spectroscopies, namely, $\Delta m_s = \pm 1$, $\Delta m_I = 0$ for EPR versus $\Delta m_s = 0$, $\Delta m_I = \pm 1$ for ENDOR. For example, an electron spin coupling to three sets of equivalent protons, $n_i = 1$, 2 and 2, produces 18 lines in the EPR spectrum, $\Pi_i(2n_iI_i + 1) = [2(1)\frac{1}{2} + 1][2(2)\frac{1}{2} + 1][2(2)\frac{1}{2} + 1] = 18$, compared to only six lines, $\Sigma_i 4I_i = 4\frac{1}{2} + 4\frac{1}{2} + 4\frac{1}{2} = 6$, in the ENDOR spectrum, two from each set of equivalent protons.

Another advantage of ENDOR is the ability to determine directly the identity of the interacting nucleus via the nuclear g_n factor (also true for ESEEM) and the nuclear quadrupole coupling constant for nuclei with $I > \frac{1}{2}$. It is also possible to determine the relative signs of the nuclear hyperfine coupling constants from the TRIPLE experiment where two radio frequencies are employed.[15] Thus ENDOR can provide a detailed picture of the electron spin density distribution in transition metal–ligand complexes.

Consider again the energy level diagram in Figure 1 for the $S = \frac{1}{2}$, $I = \frac{1}{2}$ case. In the ENDOR experiment, the magnetic field is set on a particular line in the EPR spectrum, say the one corresponding to the EPR transition $2 \rightarrow 4$. The EPR transition is then saturated (or partially saturated) with microwave power, resulting in an equalization of the electron spin populations of the two states, 2 and 4. Consequently, the EPR absorption is no longer observed. While maintaining this saturation, the radio frequency is swept. When the rf frequency matches the NMR transition frequency between levels 3 and 4, nuclear spin flips occur and the populations of levels 2 and 4 will no longer be equal and the EPR absorption will be restored. Thus ENDOR is simply NMR spectroscopy detected on an EPR spectrometer.

The various relaxation rates that govern the observation of ENDOR are discussed in detail elsewhere (see Kevin and Kispert, 1976, in General References). The ENDOR line shapes and amplitudes are affected by complicated relaxation mechanisms that are generally not easily accounted for. The principal information obtained is the ENDOR frequencies, not the homogeneous line shapes or amplitudes. Thus ENDOR cannot be used to quantify the number of *equivalent* nuclei giving rise to the signal. Furthermore, the failure to observe signals from a particular nucleus does not mean that it is not present since relaxation conditions may not favor its observation. This technique requires saturation of both the EPR and NMR lines. The requirements for saturation are

$$\gamma_e^2 B_1^2 T_{1e} T_{2e} \geq 1 \tag{23a}$$

$$\gamma_n^2 B_2^2 T_{1n} T_{2n} \geq 1 \tag{23b}$$

where B_1 is the microwave magnetic field strength and B_2 is the rf magnetic field strength. The quantities T_{1i} and T_{2i} (i = e or n) are the spin–lattice (longitudinal) and spin–spin (transverse) relaxation times, respectively, of the electron and nuclear spins, and γ_i (i = e or n) is the magnetogyric ratio. The relaxation times T_{1e}, T_{2e}, T_{1n}, and T_{2e} increase markedly as the temperature is lowered. In order to achieve sufficiently long relaxation times such that Eqs. (23) are satisfied, ENDOR

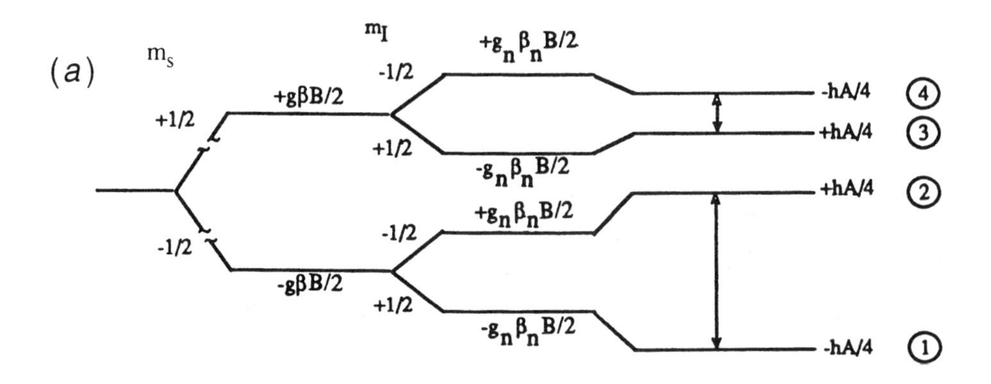

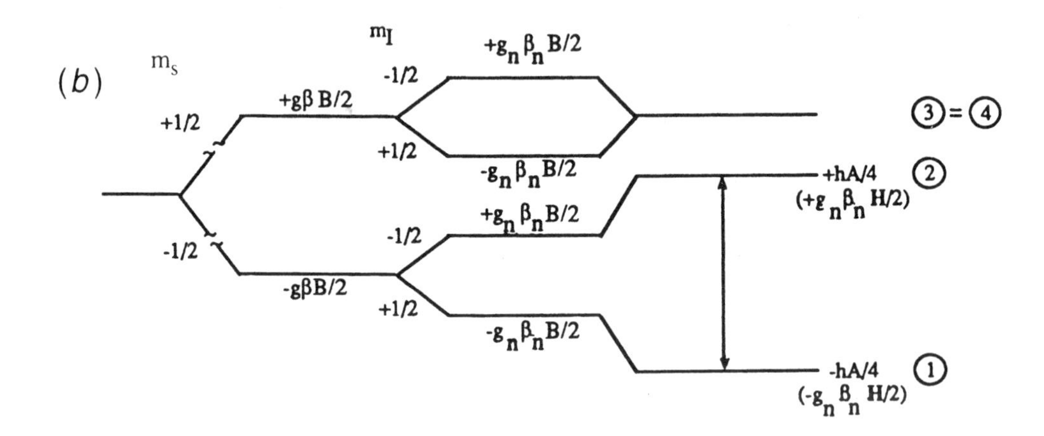

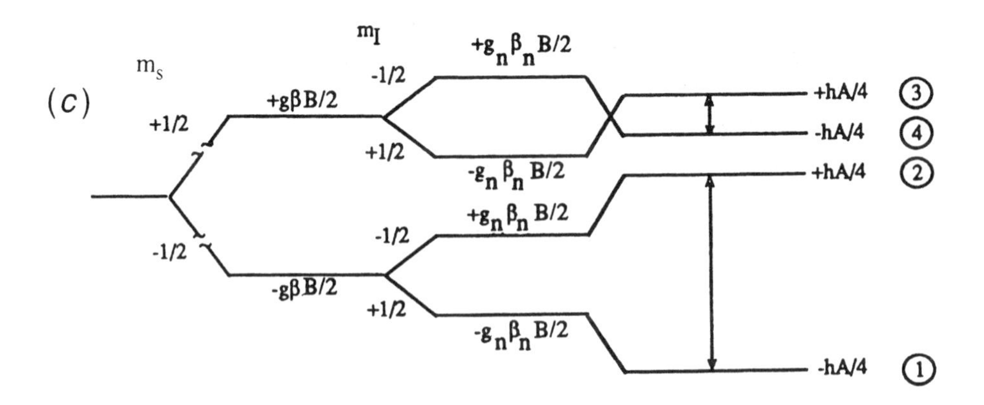

Figure 11
Energy level diagrams for the limiting cases (a) $A/2 < \nu_n$; (b) $A/2 = \nu_n$; (c) $A/2 > \nu_n$. Here it is assumed that the field is along a principal axis of the **A** matrix. [Reproduced from Snetsinger.[1]]

of transition metal complexes is typically run below 30 K. The ENDOR of radicals, however, can often be obtained in solution at room or somewhat lower temperature. Because of the ${\gamma_n}^2$ term, saturation of nuclei with low magnetic moments such as ^2H and ^{14}N can be difficult to achieve and, consequently, their ENDOR signals can be hard to observe except below 4 K.

A. The Limiting Cases

The first-order equations for ENDOR (and ESEEM) for three limiting cases where $S = \frac{1}{2}$, $I = \frac{1}{2}$ are given below.

Case I $\quad \nu_n > A/2 \quad \nu_{\mathrm{ENDOR}} = \nu_{\mathrm{n}} \pm A/2$ (24a)

Case II $\quad \nu_{\mathrm{n}} = A/2 \quad \nu_{\mathrm{ENDOR}} = \nu_{\mathrm{n}} + A/2 = 2\nu_{\mathrm{n}}$ (24b)

Case III $\quad \nu_{\mathrm{n}} < A/2 \quad \nu_{\mathrm{ENDOR}} = A/2 \pm \nu_{\mathrm{n}}$ (24c)

The energy level diagrams for these three cases are shown in Figure 11 and the corresponding stick spectra in Figure 12. Cases I and III are sufficient to describe

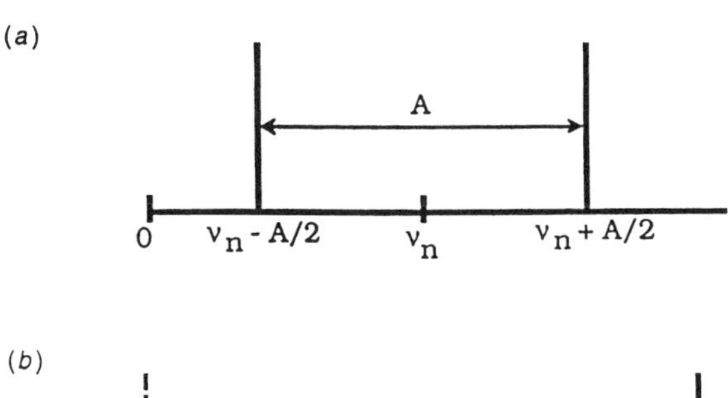

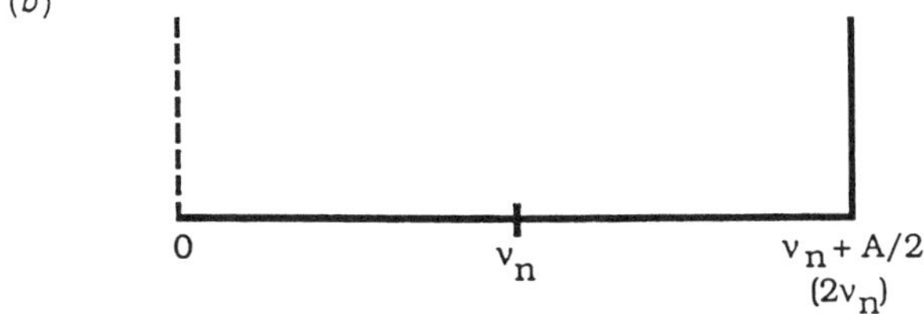

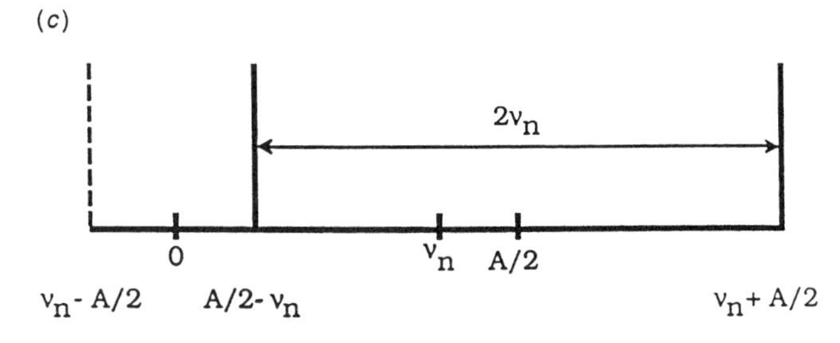

Figure 12
The ENDOR stick spectra for the limiting cases (*a*) $A/2 < \nu_{\mathrm{n}}$; (*b*) $A/2 = \nu_{\mathrm{n}}$; and (*c*) $A/2 > \nu_{\mathrm{n}}$ corresponding to the energy level diagrams in Figure 11. [Reproduced from Snetsinger.[1]]

most ENDOR spectra obtained at fields of 3000–3400 G ($g = 2$ spin systems at X band (9.0–9.5 GHz) microwave frequencies). Case I has already been considered [Fig. 1, Eqs. (6)].

Consider a hypothetical experiment in which $A/2$ is made larger and larger relative to ν_n. Initially, one would observe a two line spectrum with a splitting of A according to Eq. (24a) [Fig. 12(a)]. As $A/2$ is made progressively larger, the lower frequency line moves closer and closer to zero frequency. Eventually, $A/2$ would become equal to ν_n. This situation is known as "exact cancellation" and would result in the energy level diagram shown in Figure 11(*b*) and a single line in the ENDOR [Fig. 12(*b*)]. As $A/2$ is made larger than ν_n, a two peak spectrum centered at ν_n would result with the low-frequency peak being at some negative frequency $\nu_n - A/2$ as shown as a dotted line in Figure 12(*c*). By looking at the energy level diagram in Figure 11(*c*), one can see that the energy levels corresponding to levels 3 and 4 have crossed, thus the "negative" peak is actually observed at the positive frequency $A/2 - \nu_n$. The high-frequency peak will continue to appear at $A/2 + \nu_n$. This situation thus corresponds to the limiting Case III where the pair of lines is centered around $A/2$ as given by Eq. (24c).

Most ENDOR spectra are performed at X-band microwave frequency where protons and other nuclei with weak couplings and relatively large precessional frequencies belong to Case I. The majority of other $I = \frac{1}{2}$ nuclei belong to Case III. In complicated situations with multiple overlapping ENDOR lines, the choice of spectrometer microwave frequency can be exploited to cause the exact cancellation condition of certain nuclei and to shift lines relative to one another, thus facilitating interpretation of the spectrum. Q-band (35 GHz) ENDOR is increasingly being employed in conjunction with ENDOR measurements at X band.[16]

In situations involving $S = \frac{1}{2}$, $I = 1$ systems, one must take into account the quadrupole interaction. In this case, ENDOR frequencies are given by Eq. (11). A maximum of four allowed lines is predicted for the spectrum (Fig. 2). Whether all four lines are observed depends on the relative magnitudes of the nuclear Zeeman frequency, the nuclear hyperfine and quadrupolar coupling constants, and on the various relaxation processes that govern the ENDOR intensities. Powder pattern effects obtained with frozen solution samples can also complicate the spectrum and lead to loss of resolution.

B. Examples of ENDOR Spectral Assignments

Studies of the solvation structure of metal ions can be addressed nicely by ENDOR spectroscopy. Figure 13 illustrates the CW EPR spectrum of the VO^{2+} ion in a 50% water–methanol solvent.[17] To a first approximation, the EPR spectrum consists of two sets of eight overlapping lines that arise from coupling of the unpaired electron with the ^{51}V ($I = \frac{7}{2}$) nucleus ($2I + 1 = 8$ lines). The eight low-intensity lines having the larger hyperfine coupling are due to those VO^{2+} complexes oriented in the frozen solution such that their short V=O bond axis is approximately parallel to the magnetic field vector **B**. The more intense set of eight lines with the smaller hyperfine coupling are from complexes oriented with

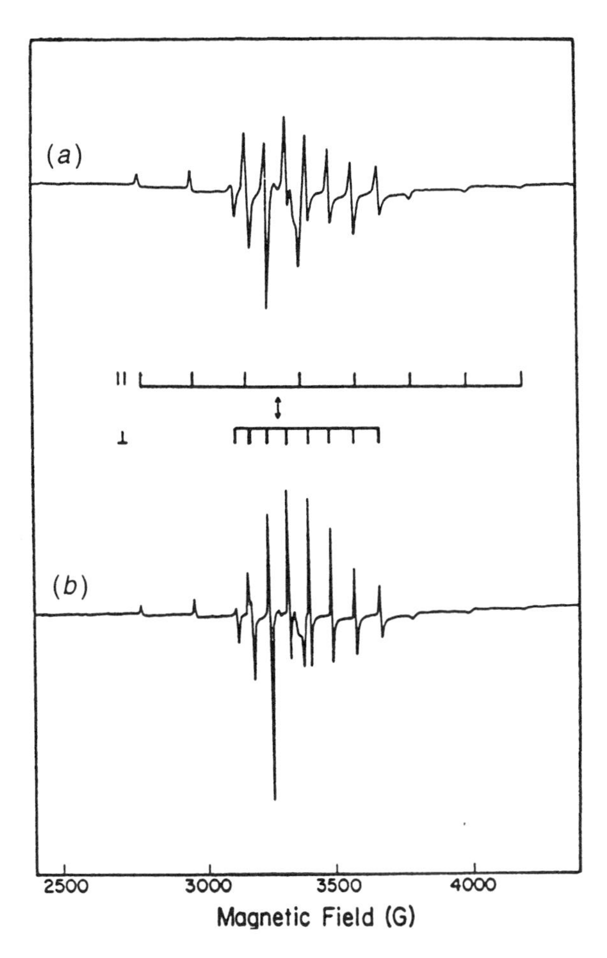

Figure 13
The CW EPR spectrum of VO^{2+} in frozen 50 : 50 (*a*) H_2O/CH_3OH and (*b*) D_2O/CD_3OD solvents. [Reproduced with permission from Mustafi, D.; Makinen, M. W. *Inorg. Chem.* **1988**, *27*, 3360–3368. Copyright © 1988 American Chemical Society.]

the V=O bond perpendicular to **B**. These features correspond to turning points in the "powder pattern" EPR spectrum of frozen solutions (see Chapter 3). (Spectra from other orientations also contribute to the observed spectrum but do not correspond to turning points, and, thus, do not show up in a first derivative display.) Therefore, by positioning the magnetic field on particular EPR lines in the ENDOR experiment, one can selectively excite molecules oriented either parallel or perpendicular to the magnetic field direction. In this way, one can obtain the anisotropic components of the ligand ^{1}H nuclear hyperfine matrices in the coordinate system of the g and ^{51}V hyperfine matrices. As will be illustrated below, the ^{1}H couplings can then be used to estimate vanadium–proton distances in the complex.

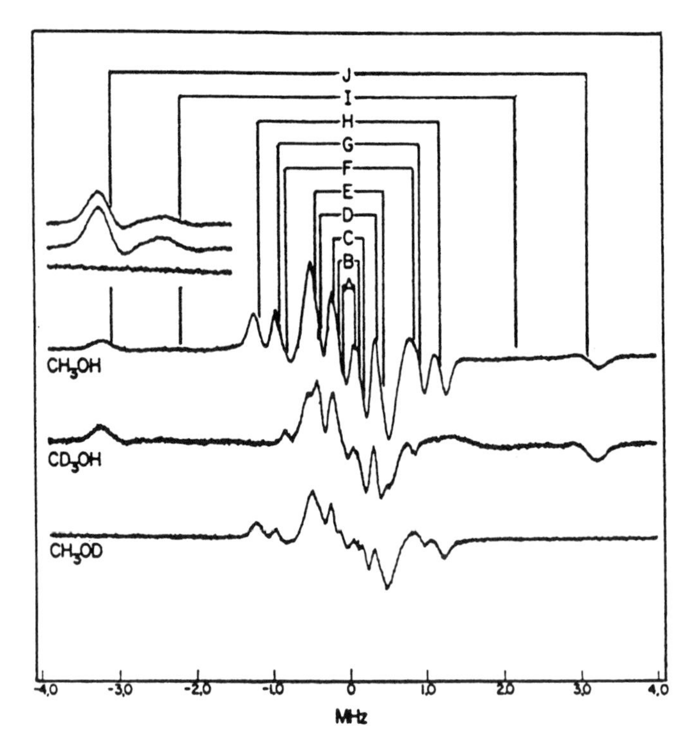

Figure 14
Proton ENDOR spectra of VO^{2+} at 5 K in methanol for B set at the $-7/2$ || EPR line at 2790 G. Couplings *A–G* are indicated. [Reproduced with permission from Mustafi; D.; Makinen, M. W. *Inorg. Chem.* **1988**, *27*, 3360–3368. Copyright © 1988 American Chemical Society.]

Changing from a CH_3OH/H_2O solvent to the deuterated CD_3OD/D_2O solvent causes a reduction in the CW EPR line widths of around 6 G, indicating that unresolved proton couplings are present (cf. Fig. 13). The magnitude of these couplings decrease in proportion to the ratio of the nuclear g factors of deuterium and hydrogen (Table I), $g_D/g_H = 0.857/5.585 = 0.153$, hence, the observed reduction in heterogeneous broadening of the EPR line. While the proton couplings cannot be measured directly from the EPR, they are readily seen in the ENDOR spectrum (Fig. 14).[17]

Figure 14 shows the ^1H ENDOR of VO^{2+} in methanol when the magnetic field is set on the ^{51}V $m_I = -\frac{7}{2}$ parallel line (lowest field line at 2790 G) in the EPR spectrum. In this instance, ^1H couplings are being sampled from molecules oriented parallel to the magnetic field direction. The spectrum corresponds to Case I discussed above with the pairs ENDOR lines centered about the Zeeman frequency of the proton $\nu_n \pm A/2$, in this instance, $\nu_n = 11.88$ MHz at $B = 2790$ G.

Ten different ^1H couplings, *A–J*, are evident in the spectrum. By measuring ENDOR spectra at parallel and perpendicular EPR field positions on CH_3OH, CH_3OD, and CD_3OD samples, the various lines can be assigned to particular protons.[17] The anisotropic coupling constants so derived enable one to estimate the vanadium–proton distances. For example, the coupling constant for the J proton in Figure 14 is 6.12 MHz, which is the separation between the high- and low-frequency ENDOR lines. The 6.12-MHz coupling is assigned to $A_{\parallel}$ of a hydroxyl proton of an axially coordinated methanol positioned opposite the vanadyl V=O oxygen atom. From measurements on the perpendicular $m_I = -\frac{3}{2}$ EPR line, a value of $A_{\perp} = 3.21$ MHz is obtained. By assuming that $A_{\perp}$ is negative [as it would be if the isotropic coupling constant a were small relative to the dipole interaction F in Eq. (17e) with $\theta = 90°$], an isotropic coupling of $a = A_{\parallel}/3 + 2A_{\perp}/3 = 6.12/3 + 2(-3.21)/3 = -0.10$ MHz and a dipolar term $F = a - A_{\perp} = -0.10 - (-3.21) = 3.11$ MHz are calculated. From the value of F, a distance of $r = 2.90$ Å is estimated using Eq. (17h).

The small value of the isotropic coupling constant $a = -0.10$ MHz indicates that there is little electron spin density on the hydrogen nucleus, and that the coupling is largely through space and dipolar in nature. From similar analyses of the other ENDOR lines, various vanadium–proton distances in the range 2.60–5.60 Å can be calculated. The largest distance of 5.60 Å corresponds to the methyl protons of a methanol molecule hydrogen bonded to the vanadyl oxygen, namely, V=O- - -HOC$\underline{H}_3$. The student should consult the Mustafi and Makinen (1988) paper[17] for further details concerning the complete assignment of the ENDOR lines and the elucidation of the structure of the complex by point dipole calculations.

Vanadyl porphyrin complexes are often found in crude oils and are of considerable industrial interest because of their effects on petroleum refining, in particular, the poisoning of cracking catalysts. Figure 15(a) shows the EPR spectrum of a sample of asphaltene, the involatile end fraction of the distillation of crude oil.[18] While the g factors and ^{51}V coupling constants from the EPR spectrum are typical of porphyrin complexes, the EPR spectrum alone provides limited information on the coordination environment of the metal. The ^{14}N ENDOR spectra of Boscan and Arabian asphaltenes [Figs. 15(b and c)] compare favorably with that of vanadyl tetraphenylporphyrin [Fig. 15(d)]. The spectra show the four ENDOR lines expected from Eqs. (10) from which values of $A_z = 8.01$ MHz and $P_z = 0.26$ MHz are obtained for the asphaltene samples and $A_z = 7.89$ MHz and $P_z = 0.25$ MHz for the tetraphenylporphyrin complex. Proton ENDOR spectra not shown provide further information about the structure of the porphyrin in asphaltenes.[18] In addition to VO^{2+} complexes as illustrated here, ENDOR has been performed on many other metal ion complexes, including Cu^{2+}, Mo^{5+}, Ni^{3+}, and Fe^{3+} and a host of metalloprotein systems, two of which are discussed below.[19]

Another example of the application of ENDOR spectroscopy comes from studies of the MoFe cluster in nitrogenase, the enzyme that catalyzes the conversion of dinitrogen to ammonia.[20] The molybdenum–iron (FeMo) protein, one of two components of nitrogenase, contains 2 Mo and approximately 30 Fe atoms arranged in a minimum of six polynuclear metal clusters, two of which contain a single Mo. The two molybdenum–iron cofactor clusters (designated FeMoco) are paramagnetic and each contain three unpaired electrons so the net electron spin is $S_{\text{net}} = \frac{3}{2}$. These clusters have an approximate composition $MoFe_{6-8}S_{9\pm1}$. Because they are a constituent of the active site of the enzyme,

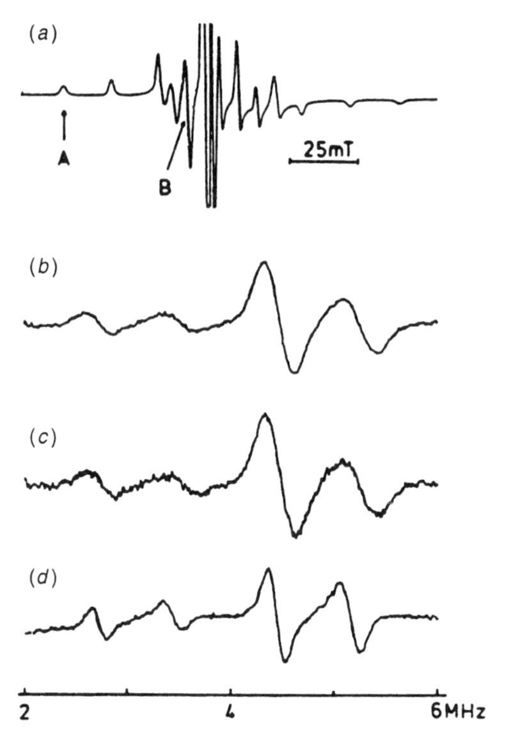

Figure 15
(*a*) An EPR spectrum of Boscan asphaltene. The ENDOR spectra (*b*), (*c*), and (*d*) were obtained by monitoring the feature marked A. (*b*) Boscan heavy asphaltene, (*c*) Arabian heavy asphaltene, and (*c*) vanadyl tetraphenylporphrin diluted in the zinc homolog. The ^{14}N frequencies are $\nu_n = 0.89$ MHz for (*b*) and (*c*) and $\nu_n = 0.88$ MHz for (*d*). 120 K. [ENDOR Spectra of Vanadyl Complexes in Asphaltenes, Atherton, N. M.; Fairhurst, S. A.; Hewson, G. J. Copyright © 1987. Reprinted by permission of John Wiley & Sons, Ltd.]

their structure is of special interest. The ^{1}H, ^{57}Fe, ^{95}Mo, and ^{33}S ENDOR spectroscopies have provided insight into the structural organization of the cluster.[20,21] Here we only consider briefly the ^{57}Fe ENDOR spectrum.

The $S = \frac{3}{2}$ cluster from the bacterium *Azotobacter vinelandii* (AV) exhibits a rhombic EPR signal with g factors $g_1(= g_x) = 4.32$, $g_2(= g_y) = 3.68$, and $g_3(= g_z) = 2.01$ as shown in Figure 16.[21] The EPR spectrum derives from the lower $M_s = \pm\frac{1}{2}$ Kramers doublet of the $S = \frac{3}{2}$ spin system. The ^{57}Fe ENDOR spectra obtained with protein from AV grown on ^{57}Fe enriched media are shown in Figure 17 for field settings corresponding to g_1 and g_3 in the EPR spectrum.[20] Since these g factors are at the extreme ends of the spectrum of the frozen solution sample (Fig. 16), the ENDOR spectra in Figure 17(*a* and *b*) are from molecules oriented with their g_1 and g_3 axes approximately parallel to the applied field, respectively. At the field corresponding to g_2, there are overlapping electron resonances from intermediate molecular orientations so the ENDOR spectrum (not shown) in this case is not a "single-crystal" type spectrum.

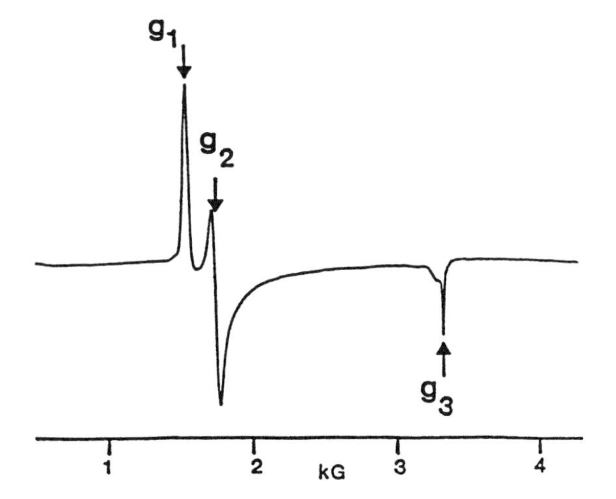

Figure 16
The EPR spectrum of *A. vinelandii* MoFe protein of nitrogenase at 2 K. [Reproduced with permission from Venters, R. A.; Nelson, M. J.; McLean, P. A.; True, A. E.; Levy, M. A.; Hoffman, B. M.; Orme-Johnson, W. H. *J. Am. Chem. Soc.*, **1986**, *108*, 3487–3498. Copyright © 1986 American Chemical Society.]

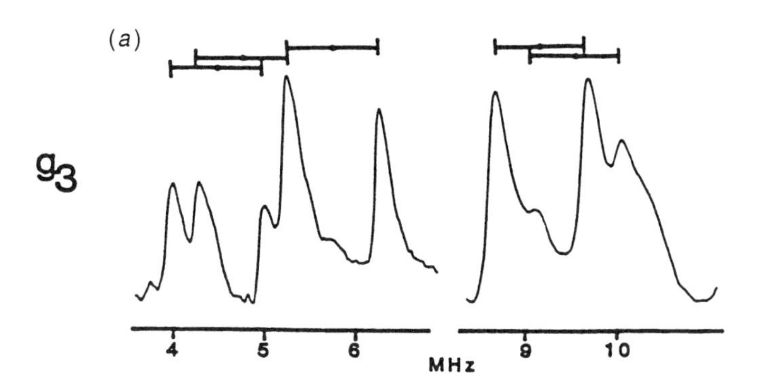

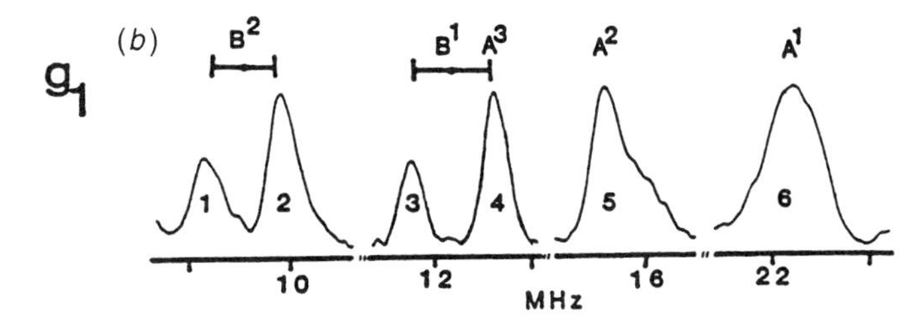

Figure 17
Single-crystal-like ENDOR spectra at 2 K of the FeMoCo protein at (*a*) the high-field g_3 edge and (*b*) the low-field g_1 edge of the EPR spectrum in Figure 16. [Reproduced with permission from True, A. E.; Nelson, M. J.; Venters, R. A.; Orme-Johnson, W. H.; Hoffman, B. M. *J. Am. Chem. Soc.* **1988**, *110*, 1935–1943. Copyright © American Chemical Society.]

The ^{57}Fe is an $I = \frac{1}{2}$ nucleus and belongs to Case III at magnetic fields corresponding to X-band microwave frequency; the ENDOR line positions are given by Eq. (24c), which predicts two lines centered at $A/2$ and separated by $2\nu_n$. Figure 17(*a*) for the field at 3355 G, corresponding to g_3, shows the assignment of the ENDOR spectrum. The fact that pairs of lines must be separated by $2\nu_n$ (= 0.924 MHz at 3355 G) aids considerably in the assignment of the spectrum. The spectra reveal that there must be at least five distinct iron sites in the FeMoco cluster shown by the bars in Figure 17(*a*). Recall that ENDOR cannot count the total number of ^{57}Fe present but can distinguish types of ^{57}Fe by their different $A/2$ values. The five distinguishable Fe sites are designated A^1, A^2, A^3, B^1, and B^2 in Figure 17(*b*). Thus, the ENDOR data limit the number of possible FeMoco structures to only those with at least five inequivalent Fe atoms. The structure of the cofactor subsequently determined by X-ray crystallography[22] is in accord with the ENDOR prediction and shows the presence of five inequivalent Fe atoms in an $MoFe_7$ cluster, namely,

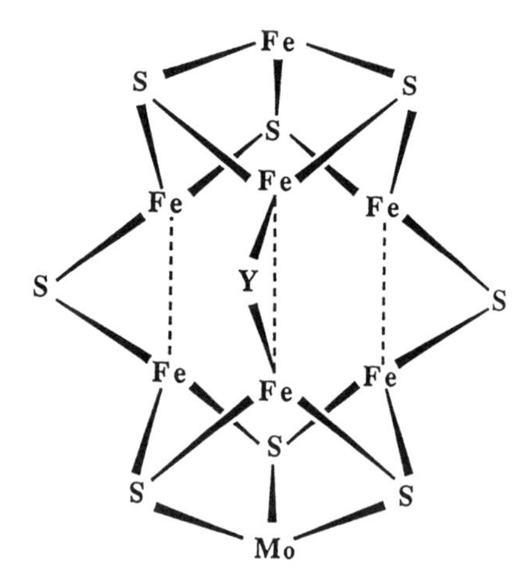

The analysis of the ENDOR spectra of the FeMoco cluster is more complicated than the foregoing suggests due to the pseudonuclear Zeeman effect on the spectrum. The name derives from the fact that for $S > \frac{1}{2}$ spin systems, the *apparent* nuclear g factor, in this case of the ^{57}Fe nucleus, may be larger or smaller than the nuclear g-factor g_n of the free nucleus (Table I) due to the zero-field splitting of the electron spin (see Chapter 3). This effect can give rise to a larger or smaller nuclear Zeeman frequency than predicted from the usual resonance condition $\nu_n = g_n\beta_n B/h$ for the free nucleus. The magnitude and direction of the change in the "observed" nuclear frequency ν_n' depends on the magnitude and sign of the ^{57}Fe hyperfine coupling constant and the energy separation between the $M_s \pm \frac{1}{2}$ and $M_s = \pm\frac{3}{2}$ Kramers' doublets. The parameter ν_n' can differ enormously from ν_n. When the field is along g_z, the pseudonuclear Zeeman effect

[i] A detailed discussion of the pseudo nuclear Zeeman effect is beyond the scope of this chapter. The student should consult Abragam, A.; Bleaney, B. *Electron Paramagnetic Resonance of Transition Ions*; Clarendon Press: Oxford, 1970 and True et al.[20] and Venters et al.[21] further details.

is zero and $\nu_n' = \nu_n$; that is the reason that the spectrum in Figure 17(*a*) can be analyzed using Eq. (24c), which is in fact derived for an $S = \frac{1}{2}$, $I = \frac{1}{2}$ spin system.

The assignment of the spectrum in Figure 17(*b*) is achieved by measuring ENDOR spectra systematically as a function of magnetic field between g_1 and g_3 in Figure 16 and tracking the position of the lines. Through spectral simulations, the principal values of the ^{57}Fe hyperfine matrix and the relative orientation of the principal axis systems for the hyperfine and g matrices have been determined for all five iron sites, a considerable accomplishment!

In another example, the use of ^{15}N isotopic substitution for ^{14}N to facilitate interpretation of ENDOR spectra is illustrated by studies of phthalate dioxygenase (PDO), an enzyme that catalyzes oxidative degradation of aromatic compounds.[23] This enzyme contains a Rieske-type center [2Fe–2S], named after its discoverer. This binuclear iron–sulfur center probably is involved in transfer of electrons to the active site of the enzyme. The reduced form of the center is composed of an Fe^{2+} ($S = 2$) and an Fe^{3+} ($S = \frac{5}{2}$) antiferromagnetically coupled (spin paired) dinuclear cluster to give a net spin $S_{net} = \frac{5}{2} - 2 = \frac{1}{2}$, which is observable by EPR. The $S = \frac{1}{2}$ rhombic EPR spectrum is characterized by the g-factors $g_1 = 2.01$, $g_2 = 1.92$, and $g_3 = 1.76$ in frozen solution. The goal of the ENDOR study was to test the model shown below for the structure of the Rieske-type center.

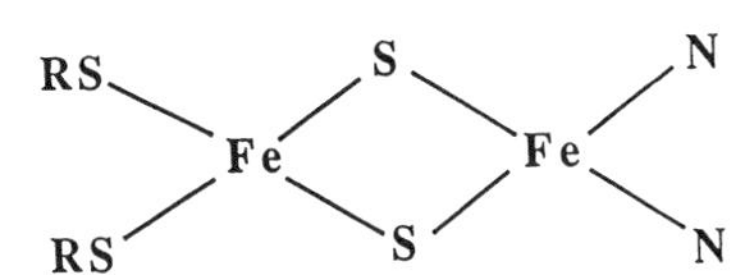

The ENDOR measurements were carried out on phthalate dioxygenase samples from the bacterium *Pseudomonas cepacia* grown in various isotopic nitrogen media (Sample 1: natural abundance ^{14}N; Sample 2: uniformly ^{15}N labeled; Sample 3: [^{15}N]histidine in a ^{14}N background; and Sample 4: [^{14}N]histidine in a ^{15}N background). Here background refers to all other amino acid nitrogens in the protein. A four line ENDOR pattern could arise from either a single ^{14}N nucleus with a resolvable quadrupole interaction of $3P_i$ ($i = x, y$, or z) between pairs of low-frequency and pairs of high-frequency lines as in Figure 2(*c*) or from two ^{14}N nuclei with two different $A_i/2$ values but unresolved quadrupole interactions. The use of ^{15}N ($I = \frac{1}{2}$) eliminates this ambiguity. The ^{15}N ENDOR line positions are given by Eq. (24c), where $A/2 > \nu_n$, that is, Case III.

Figure 18 shows the ENDOR spectra of ^{15}N enriched PDO (Sample 2) and natural abundance ^{14}N (Sample 1) measured at the field position in the EPR spectrum corresponding to g_2.[23] While the ^{14}N spectrum [Fig. 18(*b*)] demonstrates the presence of nitrogen coordination, its interpretation is not straightforward and requires computer simulation using data from the ^{15}N spectra (see Gurbiel et al.[23] for further details). The ^{15}N spectrum [Fig. 18 (*a*)] clearly shows the presence of two different ^{15}N nuclei exhibiting sharp doublets centered at $A/2$ values of 2.55 and 3.35 MHz with a splitting $2\nu_n = 3.1$ MHz. Although the g_2 field position samples a range of molecular orientations, ENDOR measurements at the g_1 field position, which does correspond to a unique molecular orientation,

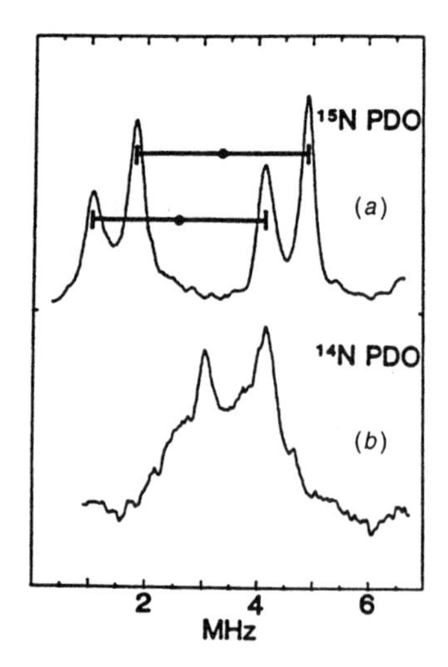

Figure 18
The ENDOR spectra of (*a*) ^{15}N-enriched (Sample 2) and (*b*) ^{14}N natural abundance (Sample 1) phthalate dioxygenase at 2 K. The assignment of $A/2$ (·) and twice the Larmor frequency (⊢⊣) are indicated on the spectra. [Reproduced with permission from Gurbiel, R. J.; Batie, C. J.; Sevaraja, M.; True, A. E.; Fee, J. A.; Hoffman, B. M.; Ballon, D. P.; *Biochemistry* **1989**, *28*, 4861–4871. Copyright © 1989 American Chemical Society.]

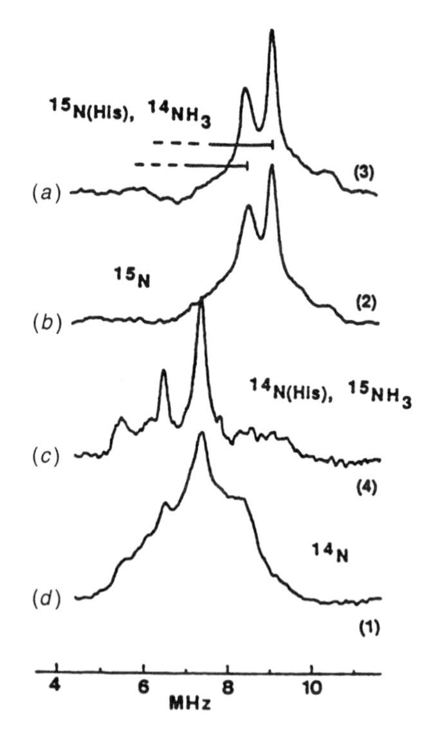

Figure 19
The Q-band ENDOR spectra at $g_2 = 1.92$ of PDO at 2 K. Protein with (*a*) [^{15}N]histidine and background ^{14}N (Sample 3), (*b*) uniformly labeled ^{15}N (Sample 2), (*c*) [^{14}N] histidine in a ^{15}N background (Sample 4), and (*d*) natural abundance nitrogen. [Reproduced with permission from Gurbiel, R. J.; Batie, C. J.; Sevaraja, M.; True, A. E.; Fee, J. A.; Hoffman, B. M.; Ballou, D. P. *Biochemistry* **1989**, *28*, 4861–4871. Copyright © 1989 American Chemical Society.]

also show two ^{15}N doublets (not shown) so the interpretation of the spectrum in Figure 18 is correct.

The ENDOR measurements on [^{15}N]histidine identify the two nitrogenous ligands as histidine. Figure 19 shows the Q-band (35.3-GHz EPR frequency) ENDOR spectrum of the four samples of PDO with the field at g_2 (13137 G). Curve (*a*) shows the spectrum of the [^{15}N]histidine enriched sample with all other nitrogen as ^{14}N. Two prominent ^{15}N peaks at 8.2 and 9.2 MHz are observed, clearly indicating that there are two histidines present. These derive from the peaks at 4.1 and 4.8 MHz in the X-band ENDOR spectrum [Fig. 18(*a*)]. However, at the higher magnetic field of the Q-hand spectrum ^{15}N belongs to Case I (i.e., $\nu_n > A/2$) and the ENDOR line positions are therefore given by Eq. (24a), whereas at X-band microwave frequency ^{15}N belongs to Case III. The two lower frequency partners are absent from spectrum (*a*) in Figure 19 because they are predicted to occur at 2.3 and 3.1 MHz, which are below the low-frequency end of the spectrum. The fact that the ENDOR spectrum of the uniformly ^{15}N labeled Sample 2 [curve (*b*)] is the same as for curve (*a*) further demonstrates that only histidine nitrogens contribute to the spectrum.

Curve (*c*) for the [^{14}N]histidine labeled sample with a ^{15}N background (Sample 4) and curve (*d*) for the naturally abundant ^{14}N sample (Sample 1) both show ^{14}N features at 5.6, 6.5, and 7.4 MHz but the spectrum is better resolved in curve (*c*). The fact that additional ^{15}N resonances are not observed in curve (*c*) in comparison to curve (*d*) indicates that there are no non-histidyl nitrogens contributing to the spectrum. The better resolution in curve (*c*) relative to curve (*d*) may be related to an unidentified nuclear relaxation pathway from bulk ^{14}N in the protein.[23]

Through a combination of measuring ^{14}N and ^{15}N ENDOR spectra at various fields across the EPR spectrum and ENDOR spectral simulations, the authors have been able to determine the principal values of the ^{14}N and ^{15}N hyperfine tensors of the two histidine ligands designated site 1 and site 2 and their orientation in the *g*-matrix principal axis system (Table II). The relative orientation of the axis systems of the *g* and *A* matrix are related by the Eulerian angles α, β, γ given in Table II.[23] Only the *A* values for ^{15}N are listed in the table since the *A* values for ^{14}N can be calculated from the ratio of nuclear g_n factors of the two isotopes (Table I) via Eq. (25).

$$A(^{14}N) = A(^{15}N) g_n(^{14}N)/g_n(^{15}N) = -0.713 \cdot A(^{15}N) \tag{25}$$

We now proceed to use the hyperfine coupling constants to gain insight into the structure of the cluster. The ENDOR simulations indicate that the signs of the coupling constants A_1, A_2, and A_3 for each site must be the same (positive or negative). It therefore follows that the magnitudes of the three values can be averaged to obtain the magnitude of the angular independent isotropic coupling constant, $a = (A_1 + A_2 + A_3)/3$. Values of a (site 1) = 6.0 MHz and a (site 2) = 7.7 MHz are obtained. The isotropic coupling constant comes from unpaired electron density in the 2s orbital on the nitrogen due to some electron delocalization from the metal onto the ligands. Since the two a values are so close, they

Table II
Principal Values of the Hyperfine and Quadrupolar Tensors for the Histidyl Ligands of the [2Fe–2S] Cluster of Phthalate Dioxygenase[a]

	Principal Values (MHz)	
	Site 1	Site 2
A (^{15}N)		
A_1	4.6 (2)	6.4 (1)
A_2	5.4 (1)	7.0 (1)
A_3	8.1 (1)	9.8 (2)
P (^{14}N)		
P_1	0.85 (8)	0.80 (8)
P_2	0.45 (8)	0.35 (8)
P_3	−1.30	−1.15
	Euler Angles	
α	0 (30)	0 (10)
β	35 (5)	50 (5)
γ	0 10)	0 (30)

[a] From Gurbiel et al.[23]

must derive from two different histidine ligands rather than from the two nitrogens of the imidazole ring of a single histidine ligand. (Directly coordinated and remote nitrogens of the same imidazole ring would be expected to have very different isotropic coupling constants.)

The **A** matrix is approximately axial for both sites, that is, $A_1 \approx A_2$ for each, with the larger A_3 $(= A_{\|})$ corresponding to the unique axis in both instances (Table II). It follows that the directions of A_3 for the two sites must approximately correspond to the two Fe–N bond axes of the coordinated imidazole rings since that is the direction of the maximum electron–nuclear dipole interaction. Furthermore, an analysis of the magnitude of the isotropic and dipolar parts of the A_3 coupling constant as detailed in Gurbiel[23] is consistent with nitrogen coordination to the Fe^{2+} rather than to the Fe^{3+} in the dinuclear cluster. (Coordination of the two histidine residues to the two different irons is unlikely since different histidine nitrogen coupling constants would be expected.)

The directions of the principal axes of P_1, P_2, and P_3 of the quadrupole interaction gives the orientation of the imidazole ring, P_2 being the tensor component perpendicular to the ring. Figure 20 shows the most reasonable structure of

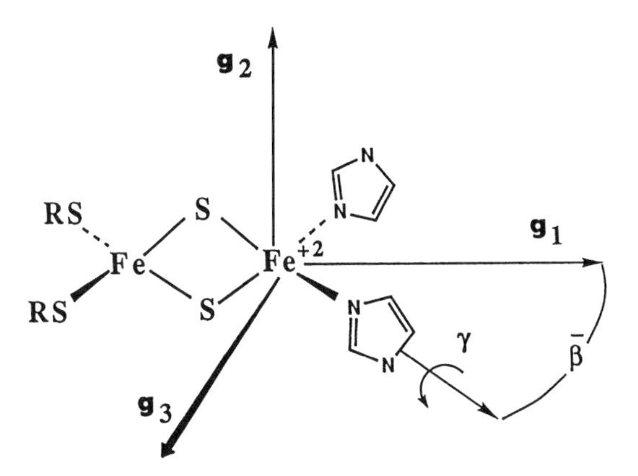

Figure 20
Proposed structure of the Rieske-type [2Fe–2S] cluster of phthalate dioxygenase as determined by ENDOR spectroscopy. The [2Fe–2S] core and the $g_1 - g_2$ plane lie in the paper. The angles γ and β with $\bar{\beta} = \pi/2 - \beta$ are given in Table II. [Reproduced with permission from Gurbiel, R. J.; Batie. C. J.; Sevaraja, M.; True, A. E.; Fee, J. A.; Hoffman, B. M.; Ballou, D. P. *Biochemistry* **1989**, *28*, 4861–4871. Copyright © 1989 American Chemical Society.]

the Rieske-type [2Fe–2S] cluster based on the analysis of the ENDOR data. The imidazole rings lie nearly in the N–Fe–N plane since the torsion angle γ is nearly zero (Table II). The two cysteine protein ligands (RS) and the two imidazole ligands lie in the plane defined by the g_2 and g_3 axes; the Fe–Fe vector is along the g_1 direction. The coordination about the two irons is approximately tetrahedral.

The equations presented in this chapter for ENDOR (and ESEEM as well) apply only to $S = \frac{1}{2}$ electron spin systems. While most ENDOR and ESEEM of biological materials has been performed on $S = \frac{1}{2}$ spin systems, increasingly studies are being done on $S > \frac{1}{2}$ metal centers in metalloproteins.

C. Recent Developments

In recent years, there has been increased emphasis on the application of multi-frequency ESEEM and ENDOR to aid in spectral assignments. By carrying out experiments at various frequencies, the magnitude of the field dependent nuclear Zeeman term relative to the nuclear hyperfine and quadrupole terms can be adjusted to achieve the exact cancellation condition required for pure ^{14}N quadrupole spectra. Similarly, the field dependence of lines in complicated ESEEM spectra obtained at different EPR frequencies can be a tremendous aid in assigning their origins. Performing ENDOR at X-band (9.5 GHz) and Q-band (35 GHz) frequencies has proven valuable in a number of studies where lines from ^{14}N and ^{1}H overlap at the lower frequency, making their assignment difficult.[23] Spectrometers operating at 95 GHz and above are becoming more commonplace and are expecially useful in studies of radicals.[24,25] Pulsed ENDOR spectroscopy has come into increased use, owing in part to the availability of commercial instrumentation.[26–30] Pulsed ENDOR does not have the same stringent relaxation time requirements imposed on CW ENDOR, so in principle, pulsed ENDOR can detect nuclei coupled to the electron spin that might otherwise not be observed in the CW experiment. It has some of the same advantages as pulsed NMR such as a better signal/noise (S/N) ratio and the ability to do spectral editing by two-dimensional methods. Similarly, multidimensional ESEEM experimental techniques have emerged that provide a means of separating and correlating spin interactions, thus facilitating the assignment and analysis of ESEEM spectra for the purposes of structure determination.[31–33]

ACKNOWLEDGMENTS

The helpful comments of Professor David J. Singel on the chapter are gratefully acknowledged. This work was supported in part by grant R37 GM20194 from the National Institute of General Medical Sciences.

References

1. Snetsinger, P. A. *ENDOR and ESEEM Studies of a Cyanide Adduct of Transferrin*, Ph.D. Thesis, University of New Hampshire, Durham, NH, 1990; 200 pp.
2. Gerfen, G. J.; Hanna, P. M.; Chasteen, N. D.; Singel, D. J. "Characterization of the Ligand Environment of Vanadyl Complexes of Apoferritin by Multifrequency Electron Spin–Echo Envelope Modulation" *J. Am. Chem. Soc.* **1991**, *113*, 9513–9519.
3. Magliozzo, R. S.; McCracken, J.; Peisach, J. "Electron–Nuclear Coupling in Nitrosyl Heme Proteins and in Nitrosyl Ferrous and Oxy Cobaltous Tetraphenylporphyrin Complexes" *Biochemistry* **1987**, *26*, 7923–7931.
4. See de Beer, R.; van Ormondt, D. "Resolution Enhancement in Time Domain Analysis of ESEEM" In *Advanced EPR: Applications in Biology and Biochemistry*; Hoff, A. J., Ed.; Elsevier: Amsterdam, the Netherlands, 1989; pp 135–176 and references cited therein.
5. (a) Mims, W. B., "Envelope Modulation in Spin–Echo Experiments", *Phys. Rev.* **1972**, *B5*, 2409–2419; (b) Mims, W. B., "Amplitudes of Superhyperfine Frequencies Displayed in the Electron-Spin-Echo Envelope", *Phys. Rev.* **1972**, *B6*, 3543–3545; (c) Mims, W. B. and Peisach, J. "Electron Spin Echo Spectroscopy and the Study of Metalloproteins" In *Biological Magnetic Resonance*, Berliner, L. J., Reuben, J., Eds.; Plenum P: New York, 1981; Vol. 3; pp 213–263.
6. Mims, W. B.; Peisach, J., "ESEEM and LEFE of Metalloproteins and Model Compounds" In *Advanced EPR: Applications in Biology and Biochemistry*; Hoff, A. J., Ed., Elsevier: Amsterdam, the Netherlands, 1989; pp 1–57.
7. McCracken, J.; Pember, S.; Benkovic, S. J.; Villafranca, J. J.; Miller, R. J.; Peisach, J. "Electron Spin–Echo Studies of the Copper Binding Site in Phenylalanine Hydroxylase from *Chromobacterium violaceum*" *J. Am. Chem. Soc.* **1988**, *110*, 1069–1074.
8. Zweier, J.; Aisen, P.; Peisach, J.; Mims, W. B. "Pulsed Electron Paramagnetic Resonance Studies of the Copper Complexes of Transferrin" *J. Biol. Chem.* **1979**, 254, 3512–3515.
9. McCracken, J.; Peisach, J.; Dooley, D. M., "Cu(II) Coordination Chemistry of Amine Oxidases: Pulsed EPR Studies of Histidine, Imidazole, Water, and Exogenous Ligand Coordination" *J. Am. Chem. Soc.* **1987**, *109*, 4064–4072.
10. Flanagan, H. L.; Singel, D. J. "Analysis of ^{14}N ESEEM Patterns of Randomly Oriented Solids" *J. Chem. Phys.* **1987**, *87*, 5606–5616.
11. Dikanov, S. A.; Astashkin, A. V. "ESEEM of Disordered Systems: Theory and Applications" In *Advanced EPR: Applications in Biology and Biochemistry*; Hoff, A. J., Ed.; Elsevier: Amsterdam, the Netherlands, 1989; pp 59–117 and references cited therein.
12. Baker, E. N.; Lindley, P. F. "New Perspectives on the Structure and Function of Transferrins" *J. Inorg. Biochem.*, **1992**, *47*, 147–160.
13. Eaton, S. S.; Dubach, J.; Eaton, G. R.; Thurman, G.; Ambruso, D. R. "Electron Spin Echo Envelope Modulation Evidence for Carbonate Binding to Fe(III) and Copper(II) Transferrin and Lactoferrin" *J. Biol. Chem.* **1990**, *265*, 7138–7141.
14. Smith, C. A.; Anderson, B. F.; Baker, H. M.; Baker, E. N., "Metal Substitution in Transferrins: The Crystal Structure of Human Copper-Lactoferrin at 2.1-Å Resolution" *Biochemistry* **1992**, *31*, 4527–4533.
15. (a) Biehl, R.; Plato, M.; Möbius, K., "General TRIPLE Resonance on Free Radicals in Solution. Determination of Relative Signs of Isotropic Hyperfine Coupling Constants" *J. Chem. Phys.* **1975**, *63*, 3515–3522; (b) Kurreck, H.; Bock, M.; Bretz, N.; Elsner, M.; Kraus, H.; Lubitz, W.; Müller, F.; Geissler, J.; Kroneck, P. M. H. "Fluid Solution and Solid-State Electron Nuclear Double Resonance Studies of Flavin Model Compounds and Flavoenzymes" *J. Am. Chem. Soc.* **1984**, *106*, 737–746.
16. (a) Werst, M. M.; Davoust, C. E.; Hoffman, B. M. "Ligand Spin Densities in Blue Copper Proteins by Q-band ^{1}H and ^{14}N ENDOR Spectroscopy" *J. Am. Chem. Soc.* **1991**, *113*, 1533–1548; (b) Fan, C.; Teixeira, M.; Moura, J.; Moura, I.; Huynh; B.-H.; Le Gall, J.; Peck, Jr., H. D.; Hoffman, B. M. "Detection and Characterization of Exchangeable Protons Bound to the Hydrogen-Activation Nickel Site of *Desulfovibrio gigas* Hydrogenase: A ^{1}H and ^{2}H Q-band ENDOR Study" *J. Am. Chem. Soc.* **1991**, *113*, 20–24.

17. Mustafi, D.; Makinen, M. W. "ENDOR-Determined Solvation Structure of VO^{2+} in Frozen Solution" *Inorg. Chem.* **1988**, *27*, 3360–3368.
18. Atherton, N. M.; Fairhurst, S. A.; Hewson, G. J. "ENDOR Spectra of Vanadyl Complexes in Asphaltenes" *Magn. Reson. Chem.* **1987**, *25*, 829–830.
19. (a) Hütterman, J.; Kappl, R. "ENDOR: Probing the Coordination Environment in Metalloproteins" In *Metal Ions in Biological Systems*, Vol. 22; Sigel, H., Ed.; Marcel Dekker: New York, 1987; pp 1–80; (b) Hoffman, B. M.; Gurbiel, R. J.; Werst, M. M.; Sivaraja, M. "Electron Nuclear Double Resonance (ENDOR) of Metalloenzymes In *Advanced EPR. Applications in Biology and Biochemistry*; Hoff, A. J., Ed.; Elsevier: Amsterdam, The Netherlands, 1989; pp 541–591.
20. True, A. E.; Nelson, M. J.; Venters, R. A.; Orme-Johnson, W. H.; Hoffman, B. M. "^{57}Fe Hyperfine Coupling Tensors of the FeMo Cluster in *Azotobacter vinelandii* MoFe Protein: Determination by Polycrystalline ENDOR Spectroscopy" *J. Am. Chem. Soc.* **1988**, *110*, 1935–1943 and references cited therein.
21. Venters, P. A.; Nelson, M. J.; McLean, P. A.; True, A. E.; Levy, M. A.; Hoffman, B. M.; Orme-Johnson, W. H. "ENDOR of the Resting State of Nitrogenase Molybdenum–Iron Proteins from *Azotobacter vinelandii, Klebsiella pneumoniae*, and *Clostridium pasteurianum*: ^{1}H, ^{57}Fe, ^{95}Mo, and ^{33}S Studies" *J. Am. Chem. Soc.* **1986**, *108*, 3487–3498.
22. Chan, M. K.; Kim, J.; Rees, D. C. "The Nitrogenase FeMO-Cofactor and P-Cluster Pair: 2.2 Å Resolution Structures" *Science* **1993**, *260*, 792–794.
23. Gurbiel, R. J.; Batie, C. J.; Sivaraja, M.; True, A. E.; Fee, J. A.; Hoffman, B. M.; Ballou, D. P. "Electron–Nuclear Double Resonance Spectroscopy of ^{15}N-Enriched Phthalate Dioxygenase from *Pseudomonas cepacia* Proves That Two Histidines are Coordinated to the [2Fe–2S] Rieske-Type Clusters" *Biochemistry* **1989**, *28*, 4861–4871.
24. Möbius, K.; "High-Field/High Frequency EPR/ENDOR—A Powerful New Tool in Photosynthesis Research" *Appl. Magn. Reson.* **1995**, *9*, 389–407.
25. Gerfen, G. J.; Bellew, B. F.; Griffin, R. G.; Singel, D. J.; Ekber, C. A.; Whittaker, J. W. "High-Frequency Electron Paramagnetic Resonance Spectroscopy of the Apogalactose Oxidase Radical" *J. Phys. Chem.* **1996**, *100*, 16739–16748.
26. Thomann, H.; Bernardo, M.; Adams, W. W. "Pulsed ENDOR and ESEEM Spectroscopic Evidence for Unusual Nitrogen Coordination in the Novel H_2-Activating Fe–S Center of Hydrogenase" *J. Am. Chem. Soc.* **1991**, *113*, 7044–7046.
27. Davoust, C. E.; Doan, P. E.; Hoffman, B. M. "Q-Band Pulsed Electron Spin-Echo Spectrometer and Its Application to ENDOR and ESEEM" *J. Magn. Reson. Ser A* **1996**, *119*, 38–44.
28. Hoffman, B. M.; DeRose, V. J.; Doan, P. E.; Gurbiel, R. J.; Houseman, A. L. P.; Telser, J. "Metalloenzyme Active-Site Structure and Function through Multifrequency CW and Pulsed ENDOR" In *Biological Magnetic Resonance*, Berliner, L. J., Reuben, J., Eds.; Plenum: New York, 1993; *vol. 13*; pp 151–218.
29. Gemperle, C.; Schweiger, A. "Pulsed Electron–Nuclear Double Resonance Methodology" *Chem. Rev.* **1991**, *91*, 1481–1505.
30. Goldfarb, D.; Bernardo, M.; Thomann, H.; Kroneck, P. M. H.; Ullrich, V. "Study of Water Binding to Low-Spin Fe(III) in Cytochrome P450 by Pulsed ENDOR and Four-Pulse ESEEM Spectroscopies" *J. Am. Chem. Soc.* **1996**, *118*, 2686–2693.
31. Höfer, P. "Distortion-Free Electron–Spin-Echo Envelope-Modulation Spectra of Disordered Solids Obtained from Two- and Three-Dimension HYSCORE Experiments" *J. Magn. Reson. Ser. A* **1994**, *111*, 77–86.
32. Song, R.; Zhong, Y. C.; Noble, C. J.; Pilbrow, J. R.; Hutton, D. R. *Chem. Phys. Lett.* **1995**, *237*, 86–90.
33. Dikanov, S. A.; Xun, L.; Karpiel, A. B.; Tyryshkin, A. M.; Bowman, M. K. "Orientationally-Selected Two Dimensional ESEEM Spectroscopy of the Rieske-Type Iron Sulfur Cluster in 2,4,5-Trichlorophenoxyacetate Monooxygenase from *Burkholderia cepacia* AC1100" *J. Am. Chem. Soc.* **1996**, *118*, 8408–8416.

General References

The following treatments of ESEEM and ENDOR vary considerably in their level of sophistication and difficulty in reading. One should start with chapters from the textbooks followed by the more advanced material. Sources are listed in order of sequence of recommended reading.

Pilbrow, J. R. *Transition Ion Electron Paramagnetic Resonance*, Clarendon: Oxford, 1990; Chapter 10, pp 440–480. A basic introduction to the principles of ESEEM and pulsed ENDOR with a few examples of applications.

ESEEM SPECTROSCOPY

Mims, W. B.; Peisach, J. "Electron Spin Echo Spectroscopy and the Study of Metalloproteins" In *Biological Magnetic Resonance*, Vol. 3; Berliner, L. J.; Reuben, J., Eds., Plenum: New York, 1981; Chapter 5, pp 213–261. A good introduction to the subject for the nonexpert.

Hoff, A. J., Ed., *Advanced EPR. Applications in Biology and Biochemistry*; Elsevier: Amsterdam, the Netherlands, 1989. A multiauthored volume with the first six chapters devoted to pulsed EPR methods. An excellent source for more advanced reading and applications.

Dikanov, S. A.; Tsvetkov, Y. D., "Structural Applications of the Electron Spin Echo Method" In *J. Structural Chem.* **1986**, *26*, 766–801. An overview of the theory of ESEEM with applications to polyoriented systems, ions, radicals, triplet state and biological systems. Somewhat advanced reading.

Tsvetkov Y. D.; Dikanov, S. A. "Electron Spin Echo: Applications to Biological Systems" *Metal Ions Biol. Sys.* **1987**, *22*, 207–262. An overview of the applications of ESEEM to biological problems.

Schweiger, A. "Pulsed Electron Spin Resonance Spectroscopy: Basic Principles, Techniques and Examples of Applications" *Angew. Chem. Int. Ed. Engl.* **1991**, *30*, 265–292. A detailed mathematical presentation of the theory and applications of pulse methods at the advanced level.

ENDOR SPECTROSCOPY

Wertz, J. E.; J. R. Bolton, J. B. *Electron Spin Resonance. Elementary Theory and Practical Applications*, McGraw-Hill: New York, 1972; Chapter 13, pp 353–377. A good introduction to the subject.

Kevan, L.; L. D. Kispert, L. D. *Electron Double Resonance Spectroscopy*, Wiley: New York, 1976. An excellent volume devoted solely to ENDOR. A good reference book.

Hoffman, B. M. "Electron Nuclear Double Resonance (ENDOR) of Metalloenzymes" *Acc. Chem. Res.* **1991**, *24*, 164–170. An account of applications of ENDOR to metal ions in proteins with emphasis on work from the author's laboratory.

Hoffman, B. M.; Gurbiel, R. J.; Werst M. M.; Sivaraja, M. "Electron Nuclear Double Resonance (ENDOR) of Metalloproteins" In *Advanced EPR. Applications in Biology and Biochemistry*; Hoff, A. J., Ed.; Elsevier: Amsterdam, The Netherlands, 1989; Chapter 15, pp 541–591. A good review of the scope of ENDOR as applied to metalloproteins.

Hüttermann J.; Kappl, R. "ENDOR: Probing the Coordination Environment in Metalloproteins" In *Metal Ions in Biological Systems*, Vol. 22; Sigel, H., Ed., Marcel Dekker: New York, 1987; Chapter 1, pp 1–80. A good review of the application of ENDOR to metalloproteins.

Eachus, R. S.; Olm, M. T. "Electron Nuclear Double Resonance Spectroscopy" *Science* **1985**, *230*, 268–274. An overview of the applications of ENDOR for the nonspecialist.

Kurreck, H.; Kirste, B.; Lubitz, W. *Electron Nuclear Double Resonance Spectroscopy of Radicals in Solution*, VCH: Weinheim, 1988. A book on ENDOR of radicals with many examples and data on various systems. Limited treatment of the principles.

5

CD and MCD Spectroscopy

MICHAEL K. JOHNSON

Department of Chemistry and Center for Metalloenzyme Studies
University of Georgia
Athens, GA 30602

I. Introduction

This chapter provides an introduction for the nonspecialist to the use of circular dichroism (CD) and magnetic circular dichroism (MCD) for investigating the electronic and magnetic properties of transition metal centers in chemical and biochemical systems. By restricting the focus to metal centers, new and developing areas such as vibrational CD and the use of CD to investigate the secondary structure of proteins will not be covered. The objective is to make the reader aware of the physical basis for both CD and MCD and to show how and why they are used in conjunction with conventional absorption spectroscopy for assigning electronic excited states. In addition, variable temperature MCD (VTMCD) can provide ground-state information complementary to that obtained by electron paramagnetic resonance (EPR), Mössbauer, and magnetic susceptibility studies. Both the methodology and utility of this application are also discussed. A basic understanding of the concepts and use of group theory and symmetry, molecular orbital (MO) theory, ligand field theory, and quantum mechanics is assumed throughout.

At the outset, it is important to emphasize that the origins of CD and MCD are quite distinct and that the resultant spectra are additive. Experimentally, they are very similar in that both involve the measurement of the differential absorption of left and right circularly polarized light as a function of wavelength. The only difference lies in the presence of a magnetic field parallel to the direction of propagation of the light in the case of MCD. Circular dichroism is a manifestation of natural optical activity, and hence requires an inherent structural dissymmetry (chirality) in the molecular structure of the chromophore or its environment. In contrast, the origin of MCD lies in the Faraday effect, that is, optical activity is induced in all matter by applying a magnetic field parallel to the direction of propagation of light. As we shall see, the Faraday effect is in turn a direct consequence of the Zeeman interaction, that is, the effect of the magnetic field on the electronic energy levels of the chromophore. Consequently, only optically active samples exhibit CD, whereas MCD is a universal property of all matter.

Even from this brief introduction, the utility of CD and MCD in resolving and assigning electronic transitions, as well as the differences in the type of information obtained from these two techniques, should already be apparent. Both CD and MCD have different selection rules compared to conventional absorption and consist of negative and positive features leading to greater fine structure than the corresponding absorption spectra. In relation to the second point, CD will be sensitive to asymmetry in the spatial structure of the chromophore or its environment, whereas MCD will be more sensitive to the detailed electronic structure of the chromophore.

II. Circularly Polarized Light

We begin by discussing the concept of circularly polarized light, what it is, and how it can be generated and described mathematically. A circularly polarized light beam is one in which the electric vector rotates uniformly about the direction of propagation by 2π during each cycle, prescribing a helix at any instant of time (see Fig. 1). Right and left circularly polarized light differ in the direction of rotation: clockwise and anticlockwise, respectively, from the viewpoint of an observer facing the oncoming beam. As illustrated in Figure 1, plane-polarized light is the resultant of left and right circularly polarized beams of the same amplitude moving in-phase at the same velocity.

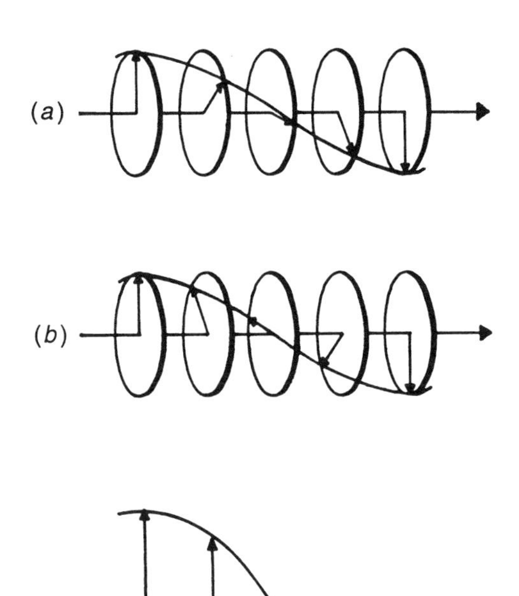

Figure 1
(a) Right circularly polarized light. (b) Left circularly polarized light. (c) The plane-polarized resultant of (a) and (b). Horizontal arrows give the direction of propagation, and arrows perpendicular to this direction denote the instantaneous spatial direction of the electric vector.

Circularly polarized light is conveniently generated by passing plane-polarized light through a quarter-wave ($\lambda/4$) retardation plate. This phenomena is readily understood in terms of the classical theory of electromagnetic radiation (see Fig. 2). Plane-polarized light at 45° to the optic axis of an optically active crystal can be resolved into two perpendicular components: the extraordinary wave, which is parallel to the optic axis, and the ordinary wave, which is perpendicular to the optic axis. In the crystal, the ordinary wave is retarded relative to the extraordinary wave with the magnitude of the retardation dependent on the thickness of the crystal. As illustrated in Figure 2, a $\lambda/2$ retardation plate results in rotation of the plane of polarized light by 90°, whereas a $\lambda/4$ retardation plate generates circularly polarized light.

Armed with this understanding of how circularly polarized light can be generated, it is straightforward to derive the classical expression for a circularly polarized waveform. The electric vector of x- or y-polarized light, $\mathbf{E}_x$ or $\mathbf{E}_y$, propagating along the z direction can be represented as

$$\mathbf{E}_x = \mathbf{i}E_0 \cos[2\pi\nu(t - nz/c)] \quad (1)$$

$$\mathbf{E}_y = \mathbf{j}E_0 \cos[2\pi\nu(t - nz/c)] \quad (2)$$

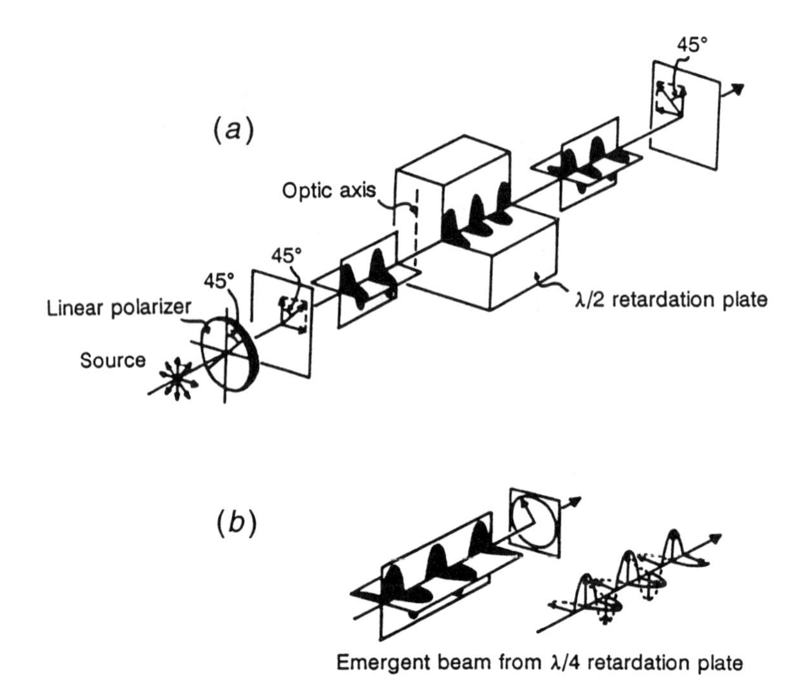

Figure 2
Generation of circularly polarized light. (*a*) Rotation of the plane of plane-polarized light by a half-wave retardation plate. (*b*) Generation of circularly polarized light from plane-polarized light by a quarter-wave retardation plate.

where **i** and **j** are unit vectors along the x and y axes, respectively, v is the frequency, t is time, n is the refractive index of the medium, and c is the velocity of light in vacuo. Introducing a phase difference of $\pi/2$ into the y-polarized component and using simple trigonometrical relationships leads to

$$\mathbf{E}_y = \pm\,\mathbf{j}E_0 \sin[2\pi v(t - nz/c)] \tag{3}$$

and vector addition of Eqs. (1) and (3) gives expressions for the electric vector of left and right circularly polarized light, $\mathbf{E}^+$ and $\mathbf{E}^-$, respectively,

$$\mathbf{E}^{\pm} = E_0\{\mathbf{i}\cos[2\pi v(t - nz/c)] \pm \mathbf{j}\sin[2\pi v(t - nz/c)]\} \tag{4}$$

Alternatively, the electric vectors for plane-polarized light and circularly polarized light can be written as

$$\mathbf{E}_x = \mathbf{i}E_0 \exp[i2\pi v(t - nz/c)] \tag{5a}$$

and

$$\mathbf{E}^{\pm} = (\mathbf{i} \mp i\mathbf{j})E_0 \exp[i2\pi v(t - nz/c)] \tag{5b}$$

where the understanding is that we take only the real part of the right-hand side of Eqs. (5a and b). Note that left circularly polarized light, $\mathbf{E}^+$, corresponds to the + sign in Eq. (4) and the − sign in Eq. (5b), while right circularly polarized light, $\mathbf{E}^-$, refers to the − sign in Eq. (4) and the + sign in Eq. (5b). These signs are of crucial importance when we come to consider the selection rules for the absorption of left and right circularly polarized light.

III. Optical Activity

Optical activity refers to the ability of a substance to rotate the plane of polarized light. It was discovered by Biot in 1812 and quantified in terms of the specific rotation $[\alpha]$,

$$[\alpha] = \alpha/\ell dp = \alpha/\ell b' \tag{6}$$

where α is the observed rotation in degrees, ℓ is the pathlength in decimeters, d is the density of the medium in grams per cubic centimeter ($\mathrm{g\,cm^{-3}}$), ρ is the fraction by weight of the optically active substance, and b' is the concentration in grams per cubic centimeter. The molecular or molar rotation, $[M]$, in units of degree square centimeters per decimole ($\mathrm{deg\,cm^2\,dmol^{-1}}$) is then defined as

$$[M] = M_r[\alpha] \times 10^{-2} = M_r\alpha/(100\ell b') = 100\alpha/lb \tag{7}$$

where M_r is the molecular weight of the optically active compound, and l and b are the pathlength and concentration in more conventional units (cm and $\mathrm{mol\,L^{-1}}$, respectively). [The 10^{-2} factor is a consequence of the historical practice of expressing the concentration in grams per deciliter ($\mathrm{g\,dL^{-1}}$) and the pathlength in decimeters (dm), which leads to an expression analogous to the Beer–Lambert law, namely, $\alpha = [M]\ell b''$, where b'' is the concentration in moles per deciliter ($\mathrm{mol\,dL^{-1}}$). Despite these archaic units of concentration and pathlength, IUPAC recommends retaining the above definition of molecular rotation due to the wealth of information in the literature.]

In 1835, Fresnel attributed the optical rotation to a difference in the refractive indices for left and right circularly polarized light, n_L and n_R, respectively. As illustrated in Figure 1, plane-polarized light can be resolved into right and left circularly polarized components. If $n_L > n_R$, then the left circularly polarized component is delayed relative to the right circularly polarized component in traversing the optically active medium, giving rise to a clockwise, (+) or dextrorotation of plane-polarized light (see Fig. 3). For incident x-polarized radiation, E_y is zero, but both E_y and E_x have finite values for the emergent plane-polarized beam with the ratio

$$E_y/E_x = \tan[\pi\nu l(n_L - n_R)/c] \tag{8}$$

The ratio represents the tangent of the angle of rotation that leads directly to Fresnel's equation for α in radians per unit pathlength, measured in the same units as λ

$$\alpha = (n_L - n_R)\pi/\lambda \tag{9}$$

Polarimetry involves measuring α at a fixed wavelength, the sodium D line at 589 nm unless otherwise indicated, and is often satisfactory of identifying the specific enantiomer or quantifying the ratio of enantiomers for molecules that do not absorb in the visible region. Dextrorotatory corresponds to a positive α and a clockwise rotation, and levorotatory corresponds to a negative α and an anti-clockwise rotation. It should, however, be borne in mind that the nature of the medium can alter the magnitude and even the sign of the rotation measured at a fixed wavelength. Consequently, the measurement of α as a function of wavelength, termed optical rotatory dispersion (ORD), facilitates a more definitive assessment of the nature and extent of optical isomerism.

Thus far the discussion has been confined to optical activity in regions where the sample is optically transparent. However, it is important to remember that refraction and absorption are not independent phenomena. Rather they are interrelated, with the refractive index related to the real part and the absorption index related to the imaginary part of a complex variable, the complex refractive index, $\hat{n} = n - ik$. Pairs of parameters related in this way are related by a Kronig–Kramers transformation, so that it is possible to derive the absorption spectrum from the dispersion of the refractive index and vice versa. Consequently, if n_L is greater than n_R at wavelengths where the optically active medium is transparent, then k_L will be greater than k_R in regions where light is absorbed, where k_L and k_R are the absorption indices for left and right circularly polarized, respectively. Hence, optically active materials exhibit CD, that is, differential absorption of left and right circularly polarized light, in regions where the material absorbs. This effect was first observed in solution by Cotton in 1895 and is often termed the "Cotton effect". It is also apparent in anomalous ORD, which is superimposed on the monotonically changing ORD that is observed in optically transparent regions.

The physical consequence of passing the plane-polarized light, depicted in Figure 3(*a*), through an optically active medium in a region of absorption is shown in Figure 4. The simultaneous existence of a difference between n_L and n_R and between k_L and k_R leads to an emergent, resultant electric vector that prescribes an ellipse, defined by the ellipticity, ψ. The angle between the major axis of the ellipse and the plane of the incident polarized light corresponds to the optical rotation, α, and the major and minor axes of the ellipse correspond to the sum and difference in the amplitudes of the two circularly polarized components on emerging from the optically active medium. It is important to realize that the ellipticity in Figure 4 has been greatly exaggerated for the purpose of clarity, and that the difference in the absorption of left and right circularly polarized light, while measurable, is usually extremely small. Consequently, to a good approximation, it is still appropriate to consider optical activity in terms of rotation of the plane of polarized light.

For historical reasons, CD is generally measured in terms of ellipticity, ψ, rather than ΔA, the differential absorption of left and right circularly polarized

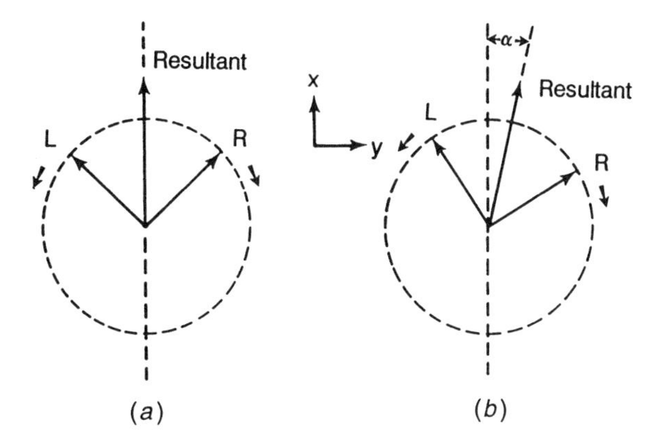

Figure 3
Electric vectors of left and right circularly polarized light, and the plane-polarized resultant (*a*) before and (*b*) after, traversing an optically active, dextrorotatory medium ($n_L > n_R$).

light, $A_L - A_R$. Hence, it is important to determine the relationship between these two quantities. From Figure 4, it is apparent that

$$\tan\psi = (E_R - E_L)/(E_R + E_L) \tag{10}$$

and, for a unit pathlength, the amplitude of the electric vector decreases as the light traverses an absorbing medium according to

$$E = E_0 \exp(-2\pi k/\lambda) \tag{11}$$

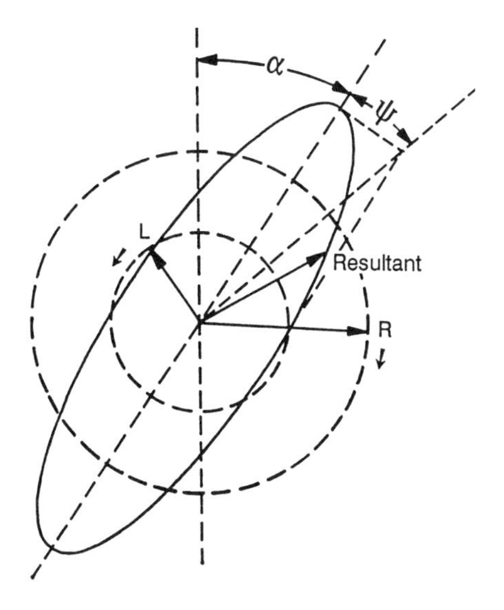

Figure 4
The rotation, α, and ellipticity, ψ of plane-polarized light emerging from a optically active absorbing medium ($n_L > n_R$ and $k_L > k_R$). The ellipticity is greatly exaggerated for the purpose of clarity.

Combining Eqs. (10) and (11), approximating tan ψ by ψ, and taking the first two terms only in the expansion of the exponentials, both good approximations since ψ and $k_L - k_R$ are always very small, leads to

$$\psi = (k_L - k_R)\pi/\lambda \tag{12}$$

with ψ in units of radians per unit pathlength. Note that the expression for ψ is formally analogous to that for α, Eq. (9), with absorption indices in place of refractive indices. Since the intensity of light, I, is proportional to the square of the amplitude of the electric vector, we also have

$$A = \log_{10}(I_0/I) = \log_{10}[\exp(4\pi k/\lambda)] \tag{13}$$

which when combined with Eq. (12) gives

$$\Delta A \approx 3.0 \times 10^{-2}\psi \tag{14}$$

where ψ is in degrees. The approximations used in obtaining Eq. (14) do not justify a more exact conversion factor. By analogy with the molecular rotation, $[M]$, the molecular or molar ellipticity, $[\theta]$, is defined by

$$[\theta] = M_r\psi/(100\ell b') = 100\psi/lb \tag{15}$$

where $[\theta]$ has the same units as $[M]$, namely, degree square centimeters per decimole. Therefore the relationship between molecular ellipticity in degrees and the decadic molar circular dichroism, $\Delta\varepsilon = \Delta A/lb$, where $\Delta\varepsilon = \varepsilon_L - \varepsilon_R$, is given by

$$[\theta] \approx 3300\ \Delta\varepsilon \tag{16}$$

While the trend in the literature is to use $\Delta\varepsilon$ in quantifying both CD and MCD spectra, it is important to remember this conversion factor since $[\theta]$ is still in common use.

The absorption, ORD, and CD for a single electronic transition of the (+) isomer of an optically active molecule are shown in Figure 5. For the (−) isomer, the CD will be negative and the ORD will be the mirror image with the trough at lower energy than the peak. The anomalous ORD in the region of the absorption represents essentially the derivative of the CD curve with respect to frequency. Moreover, from the above discussion it is apparent that the ORD and CD spectra are related by a Kronig–Kramers transformation, and Moscowitz[1] has set out the methods for calculating one type of spectrum from the other. The dispersion–absorption relationships of Kuhn[2], which assume a Gaussian line shape for the CD, provide a more simple semiempirical conversion between ORD and CD and can be summarized as

$$[A] = [M]_{max} - [M]_{min} = 4028\ \Delta\varepsilon_{max} \tag{17}$$

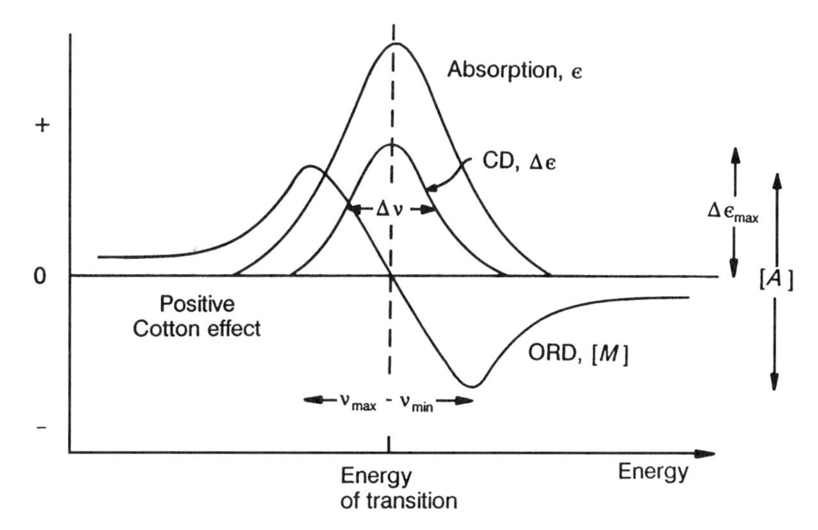

Figure 5
Absorption, CD, and ORD curves for a single electronic transition of the (+) isomer of an optically molecule.

and

$$\Delta\nu = 0.925(\nu_{\max} - \nu_{\min}) \tag{18}$$

where $\Delta\nu$ is the CD bandwidth at half-maximum height and $\nu_{\max}$ and $\nu_{\min}$ are the maximum and minimum of the anomalous ORD.

Clearly, it is unnecessary to measure both the ORD and CD spectra and for both experimental and practical reasons CD has become the method the choice. As discussed below in Section IV, CD can be easily and accurately measured with modern instruments and the simple shape combined with fact that the intensity drops to zero not far from the peak maximum, makes CD more useful in resolving and assigning electronic transitions in optically active samples.

IV. Instrumentation

As is the case with any modern physical technique, an understanding of how and why the spectrometer works is essential to obtain optimal and meaningful results. Commericial CD spectrophotometers (spectropolarimeters) are now available in the range 170–2000 nm and a block diagram of a typical instrument for measuring CD spectra in the ultraviolet–visible (UV/vis) region is shown in Figure 6. The spectrometer consists of a source, monochromator, linear polarizer, photoelastic modulator (PEM), sample holder, and detector with an appropriate signal processing system. In the UV/vis region, the source is usually a xenon-arc lamp and the detector is a photomultiplier (PM) tube, whereas in the near-infrared (IR) region the source is typically a tungsten filament lamp and the detector is a

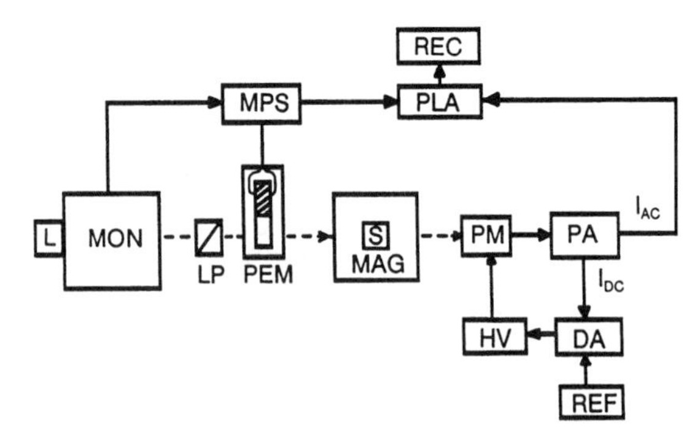

Figure 6
Block diagram of a CD spectrometer. Broken line indicates light path. The abbreviation L = light source; MON = monochromator; LP = linear polarizer; PEM = photoelastic modulator; MPS = modulator power supply; S = sample; MAG = magnet (necessary for MCD studies); PM = photomultiplier tube; HV = high-voltage power supply for PM; PA = preamplifier; DA = differential amplifier; REF = reference signal; PLA = phase-lock amplifier; REC = recorder.

photoconductive device such as a liquid nitrogen cooled InSb detector. Most modern instruments are interfaced to a microcomputer, which facilitates data handling and baseline subtraction, in addition to improvements in signal-to-noise (S/N) via multiple scanning.

The major advance in the sensitivity and reliability of CD instruments came from the development of photoelastic modulators, which convert plane-polarized light into alternating left and right circularly polarized light at a fixed modulation frequency, usually 50 kHz. They consist of a single-crystal of quartz with evaporated gold electrodes on two opposite faces, to which an alternating voltage is applied. The crystal is cut such that this alternating voltage induces oscillation along the z axis (piezoelectric effect), which results in alternating stress-induced birefringence in the attached, resonantly matched fused quartz block. By varying the magnitude of the applied voltage, the fused quartz block becomes a $\lambda/4$ plate at any λ, and incident plane-polarized light at 45° to the z axis is converted into alternating left and right circular polarized light. The applied voltage is programmed to the wavelength drive to maintain optimal circularly polarized output at any wavelength.

After passing through the sample, the CD signal is extracted from the detector output as the alternating current (ac) component of the outgoing signal, which is phase-sensitively detected using a lock-in-amplifier referenced to the modulation frequency. The direct current (dc) component provides a baseline for the measurement of both the absorption and CD signal. Most instruments operate by maintaining a constant dc current by varying the high-tension voltage delivered to the PM using a differential amplifier. The response of the PM to the applied voltage is nearly logarithmic, so that the PM voltage measures the sample absorbance to a good approximation. In this way the small 50 kHz ac signal directly measures ΔA, the differential absorbance of left and right circularly polarized light. However, for historical reasons, most commercial instruments record the signal in terms of ellipticity in units of millidegrees (mdeg). Regular calibration of the ellipticity scale is essential for quantitative work and is carried out with optically active standards such as an aqueous solution of dry 10-camphorsulfonic acid ($[\theta]_{290.5} = +4260$ deg cm^2 dmol^{-1}) or androsterone in dioxane ($[\theta]_{304} = +11{,}170$ deg cm^2 dmol^{-1}).

Magnetic circular dichroism measurements involve having the sample in longitudinal magnetic field and the magnetically induced CD is superimposed on the natural CD "baseline". It is conventional to subtract the natural CD in presenting MCD data and the reader should assume this is the case unless told otherwise. Both electromagnets (0–1.5 T) and superconducting magnets (0–7 T) are used and in the latter case it is usually necessary to shield or spatially separate sensitive spectrometer components. To obtain magnetic ground-state information by monitoring the temperature and magnetic field dependence of discrete bands, it is necessary to use magnetic fields up to 7 T with sample temperatures as low as 1.5 K and commercial superconducting magnets with such capabilities are available. Clearly, for low-temperature studies, the sample must form a glass on freezing and this is usually effected by judicious choice of solvents for inorganic complexes or addition of 50% (v/v) glycerol or ethylene glycol for aqueous solutions. Low-temperature glasses have a variable amount of birefringent strain that tends to depolarize the circularly polarized incident light. The consequences are attenuation of the intensity and baseline shifts in the natural CD. The latter usually makes natural CD studies of frozen solution only useful for qualitative investigations. However, since the natural CD is subtracted from magnetically induced CD, this is not a problem in MCD, and the attenuation of the intensity can be corrected for quantitatively by measuring the natural CD of a standard placed after magnet with the sample in and out of the light path.

V. Rotation Strength and Circular Dichroism Selection Rules

The mechanisms by which electronic absorption can occur, in order of relative importance, are electric dipole, magnetic dipole, and electric quadrupole. Only the first two are of practical importance and the dipole strengths for electric dipole and magnetic dipole intensities, D_e and D_m, respectively, are given by the square of the electric and magnetic dipole transition moments:

$$D_\mathrm{e} = \langle \Psi_i | \hat{\mu}_\mathrm{e} | \Psi_j \rangle^2 \tag{19}$$

$$D_\mathrm{m} = \langle \Psi_i | \hat{\mu}_\mathrm{m} | \Psi_j \rangle^2 \tag{20}$$

where $\hat{\mu}_\mathrm{e}$ and $\hat{\mu}_\mathrm{m}$ are the electric and magnetic dipole moment operators and Ψ_i and Ψ_j are the ground- and excited-state wave functions, respectively. Since D_m is much smaller than D_e, the absorption intensity primarily depends on the latter. The electric dipole moment operator, $\hat{\mu}_\mathrm{e}$, transforms as a translation, since electric dipole intensity arises from a translation of electronic charge, whereas the magnetic dipole moment operator, $\hat{\mu}_m$, transforms as a rotation, since magnetic dipole intensity arises from a rotation of electronic charge. When charge is moved along a helical path, the translational motion produces an electric dipole moment whereas the rotational motion gives a magnetic dipole moment. The two moments are parallel or antiparallel according to the right or left chirality (handedness) of the helical path. Hence, the basis for the quantum mechanical theory of optical activity is that the transition is required to be *both* electric and magnetic dipole allowed. By analogy with absorption, where the area of the band is proportional to the electric dipole strength of the transition, the area of a CD band is proportional to the rotational strength of the transition, R, which represents the scalar product of the electric and magnetic dipole transition moments:

$$R = Im\{\langle\Psi_i|\hat{\mu}_e|\Psi_j\rangle \cdot \langle\Psi_i|\hat{\mu}_m|\Psi_j\rangle\} \tag{21}$$

where Im indicates that the imaginary part is to be taken, since the magnetic dipole operator, $\hat{\mu}_m$ is pure imaginary. For practical purposes, Eq. (21) can be written as

$$R = \mu_e \mu_m \cos\phi \tag{22}$$

where μ_e and μ_m are the magnitudes of the electric and magnetic transition moments and ϕ is the angle between them.

The requirement for optical activity can now be formalized in terms of a nonzero scalar product of electric and magnetic transition moments. Since $g \rightarrow g$ and $u \rightarrow u$ transitions are formally electric dipole forbidden, whereas $g \rightarrow u$ and $u \rightarrow g$ are magnetic dipole forbidden, it is immediately apparent that centrosymmetric molecules are not optically active. More generally, it can readily be demonstrated that a molecule must lack *any* S_n axis (including $S_1 \equiv \sigma$ or $S_2 \equiv i$) to be optically active, which translates to only those molecules with nonsuperposable mirror images. Hence, only molecules belonging to the point groups C_n, D_n, O, T, or I are optically active and discrete transitions must be both electric and magnetic dipole allowed to exhibit CD. Metal centers in biological systems often have low symmetry coordination geometries, and hence exhibit intrinsic optical activity. However, even highly symmetrical chromophores, such as hemes, can exhibit optical activity if they are in an environment of lower symmetry.

From the above discussion, it should be apparent that magnetic dipole allowed transitions will usually dominate the CD spectrum of optically active transition metal complexes. Indeed, this has lead to a criterion for assigning magnetic dipole allowed transitions based on the so-called anisotropy factor, g, which is defined as $\Delta\varepsilon/\varepsilon$. (The anisotropy factor is not related to the g value used in EPR spectroscopy). Since the rotational strength, R, and the dipole strength, D, can be obtained experimentally from the area of the CD and absorption bands, respectively, that is,

$$R = [3hc10^3 \log_e 10/(32\pi^3 N)] \int (\Delta\varepsilon/\nu)\, d\nu \tag{23}$$

and

$$D = [3hc10^3 \log_e 10/(8\pi^3 N)] \int (\varepsilon/\nu)\, d\nu \tag{24}$$

where N, and h are Avogadro's number and Planck's constant, respectively, the anisotropy factor can be approximated as

$$g = \Delta\varepsilon/\varepsilon \approx 4R/D = 4\mu_m \cos\phi/\mu_e \tag{25}$$

By inserting realistic values of μ_e and μ_m, it can be shown that values of $g > 0.01$ can only arise from magnetic dipole allowed transitions. Hence, comparison of absorption and CD intensities provides a convenient method for identifying magnetic dipole allowed transitions of optically active chromophores.

Correlating the magnitude and sign of CD bands with the absolute configuration of optically active chromophores involves estimating the rotational strengths of individual transitions, which in turn entails constructing approximate descriptions of the ground and excited electronic states. Numerous methods of assessing the rotational strengths both of inherently dissymmetric chromophores and symmetric chromophores in chiral environments have been advanced with widely varying levels of success. Detailed discussion and critique of these methods is beyond the scope of this chapter. However, a cursory outline of the currently prevailing methods, with references for further reading, is presented in the next two paragraphs.

To date, the most widely exploited and successful model for inorganic coordination complexes is the so-called "independent-systems" approach.[3–5] A chiral molecule is divided into the chromophore, M, and the surrounding groups or ligands, L, which form the dissymmetric molecular environment. As implied by the method, the M and L subsystems are treated as quasi-independent systems by neglecting overlap between their respective charge distributions. The electronic functions of an enantiomer are expressed as products of the chromophore functions, $|M_m\rangle$, and ligand functions, $|L_l\rangle$, which interact via an electrostatic operator, V. The interaction is handled by perturbation theory, which leads to both "static" and "dynamic" contributions. The static contributions consist of matrix elements of the form $\langle M_m L_0 | V | M_n L_0 \rangle$ that are similar to the traditional crystal field perturbations employed in transition metal complexes. The ligands or surrounding groups play an active role in light absorption in the dynamic contributions that have the general form $\langle M_m L_0 | V | M_n L_k \rangle$. An allowed magnetic dipole transition of the chromophore induces a coupled electric dipole (virtual excitation) in each of the surrounding groups. The resultant induced electric dipole will usually be finite in noncentrosymmetric molecules and will enhance isotropic absorption by interacting with and taking energy from the radiant field.

The extent to which static or dynamic coupling terms dominate and the order of the perturbation employed varies from system to system, but the overall approach has been successful in rationalizing CD spectra of numerous optically active chromophores. The elegance of the approach lies in its generality, since the subsystems need not refer to a transition metal–ligand set. In other words, the source of the chirality can reside in more remote groups rather than the intrinsic dissymmetry of the metal–ligand complex. In contrast, parameterized ligand field approaches are generally only applicable to the latter case. However, recent reports of such methods, particularly those using the cellular ligand field model of Gerloch and co-worker,[6] have been very successful in quantitatively reproducing the signs and magnitudes of rotational strengths of d–d transitions in chiral transition metal complexes.

VI. Applications of Circular Dichroism Spectroscopy

The three principal applications of CD spectroscopy in the study of metal complexes and metal centers in biological systems are summarized below.

1. Identifying optical isomers of chiral metal complexes. The sign of the CD spectrum is related to the absolute configuration of the molecule.
2. Resolving and assigning electronic transitions. The difference in the selection rules for optical activity and electronic absorption can result in bands with weak absorption exhibiting strong CD hands. For example, spin forbidden d–d bands may show up strongly in the CD spectrum. In addition, the biphasic nature of the CD spectrum often permits the resolution of overlapping bands. The requirement of a chiral chromophore greatly limits the utility of CD for investigating the electronic structure of highly symmetrical transition metal complexes. However, the low symmetry environments of biological transition metal centers almost invariably lead to intrinsic CD that is useful in effecting detailed assignments of metal centered electronic transitions.
3. Monitoring changes in the local environment of a chromophoric metal center. For example, the electronic CD originating from metal centers in biological systems is often far more sensitive to conformational changes of the macromolecule than the corresponding absorption spectrum.

The third application is potentially the most useful, but is also the most difficult to interpret in terms of the nature of the conformational change. The first and second applications are illustrated below using a chiral, substitutionally inert coordination complex, $[Co(en)_3]^{3+}$ (en = ethylenediamine), and a simple metalloprotein, reduced rubredoxin.

VII. Circular Dichroism Examples

The solution absorption and CD spectra of Λ-(+)-$[Co(en)_3]Cl_3$[a] are shown in Figure 7. This example provides a good illustration of how magnetic dipole allowed transitions under the parent symmetry usually dominate the optical activity even in complexes of much lower formal symmetry. Based on the intensities of the dominant absorption and CD features, the anisotropy factors for the absorption bands centered around 21,300 and 29,400 cm^{-1} are 0.022 and 0.0033,

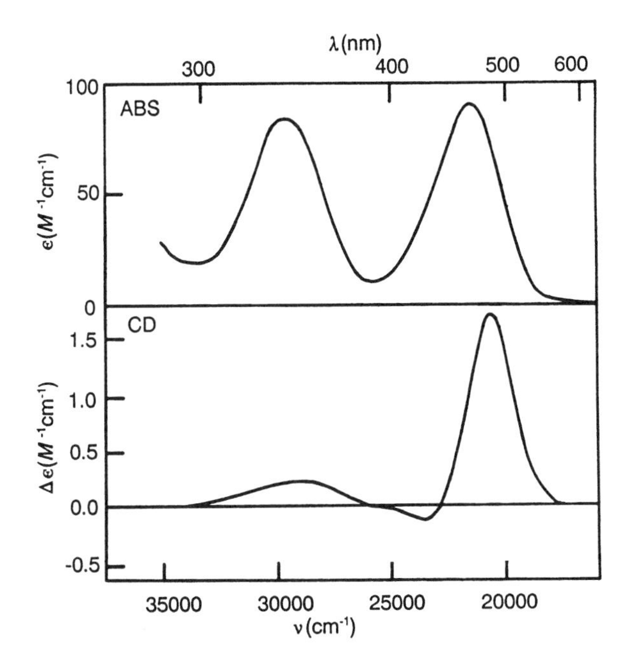

Figure 7
Absorption and CD spectra of Λ-(+)-$[Co(en)_3]Cl_3$.

[a] Following the IUPAC convention (1970), the absolute configurations of the principal types of optically active octahedral chelate complexes are labeled Λ or Δ according to

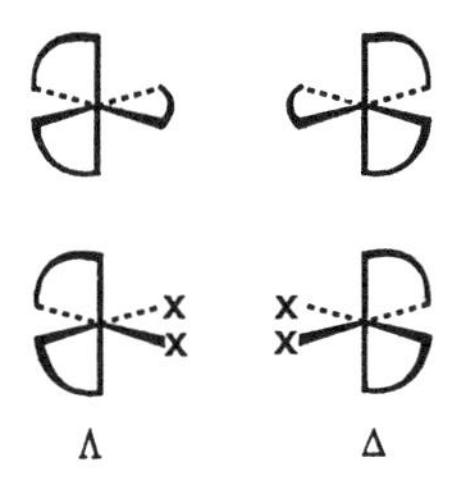

That the (+) isomer of $[Co(en)_3]^{3+}$ has the Λ configuration has been determined by anomalous X-ray scattering. It is important to realize that the sign of the rotation at the sodium D line does not provide a criteria for predicting the absolute configuration. For example, (+)-$[Co(en)_2(NH_3)Cl]^{2+}$ and (−)-$[Co(en)_2(NCS)Cl]^{+}$ have the same configuration.

respectively, suggesting that only the former corresponds to a magnetic dipole allowed transition under the parent octahedral symmetry. This prediction is in accord with theory, since the absorption bands at 21,300 and 29,400 cm^{-1} are assigned to the $^1A_{1g} \rightarrow {}^1T_{1g}$ and $^1A_{1g} \rightarrow {}^1T_{2g}$ d–d transitions, respectively, and the magnetic dipole moment operator transforms as T_{1g} under the parent octahedral symmetry.

More careful examination of the CD spectrum reveals two oppositely signed features with maxima and minima at 20,500 and 23,300 cm^{-1}, respectively, corresponding to the absorption band centered at 21,300 cm^{-1}. These originate from the overlap of the two oppositely signed CD bands expected under D_3 symmetry, that is, $^1A_1 \rightarrow {}^1E$ and $^1A_1 \rightarrow {}^1A_2$, both of which are electric and magnetic dipole allowed. The separation between these components is small and the absorption bands are broad, so they are not resolved in the absorption spectrum. In contrast, only a positive CD band centered close to the absorption maximum is observed for the higher energy band. In this case, both the absorption and CD band originate from the $^1A_1 \rightarrow {}^1E$ component, under D_3 symmetry, which is both electric and magnetic dipole allowed. The $^1A_1 \rightarrow {}^1A_1$ component is electric and magnetic dipole forbidden, and hence is not expected to appear in either the CD or absorption spectra.

Rubredoxins are a group of low molecular weight (M_r = 5,000–8,000 Da) iron–sulfur proteins found in both anaerobic and aerobic bacteria. These proteins are characterized by a single $Fe^{2+,3+}$ ion in a distorted tetrahedral coordination environment of cysteinyl S. The combination of near-IR CD and absorption studies has been effective in identifying ligand field transitions and thereby assessing the nature of the excited state distortion in reduced rubredoxin.[7] In a tetrahedral ligand field, Fe^{2+} (d^6) is expected to exhibit a single spin-allowed d–d band, $^5E \rightarrow {}^5T_2$ in the near-IR region at energy $10Dq$. It is usually difficult to observe weak electronic transitions in the near-IR absorption spectra of metalloproteins due to the intense O–H, N–H, and C–H vibrational overtones. Preparing samples in D_2O partially alleviates this problem, but it is generally desirable to use CD and/or MCD to identify and characterize such transitions. These techniques are particularly useful in this region since vibrational CD and MCD are several orders of magnitude weaker than electronic CD and MCD.

Absorption and CD spectra for reduced rubredoxin from *Clostridium pasteurianum* in D_2O are shown in Figure 8. The sharp absorption bands in the 4,000–9,000-cm^{-1} region are vibrational overtones (O–H, N–H, and C–H) from the protein and some residual HOD, and the region between 5,500 and 4,700 cm^{-1} is not shown due to the intense vibrational overtones of D_2O. Since oxidized rubredoxin has no spin allowed d–d transitions (high-spin Fe^{3+}, $^6A_1$ ground state), the weak vibrational overtones can be removed by subtraction of the oxidized spectrum. The resulting absorption spectrum of reduced rubredoxin consists of a band centered at 6,250 cm^{-1} and the beginning of another band that is estimated to be centered at 3,700 cm^{-1}, assuming a Gaussian shape and comparable bandwidth. The intensities and frequencies of these bands are consistent with assignment to components of the $^5E \rightarrow {}^5T_2$ d–d transition.

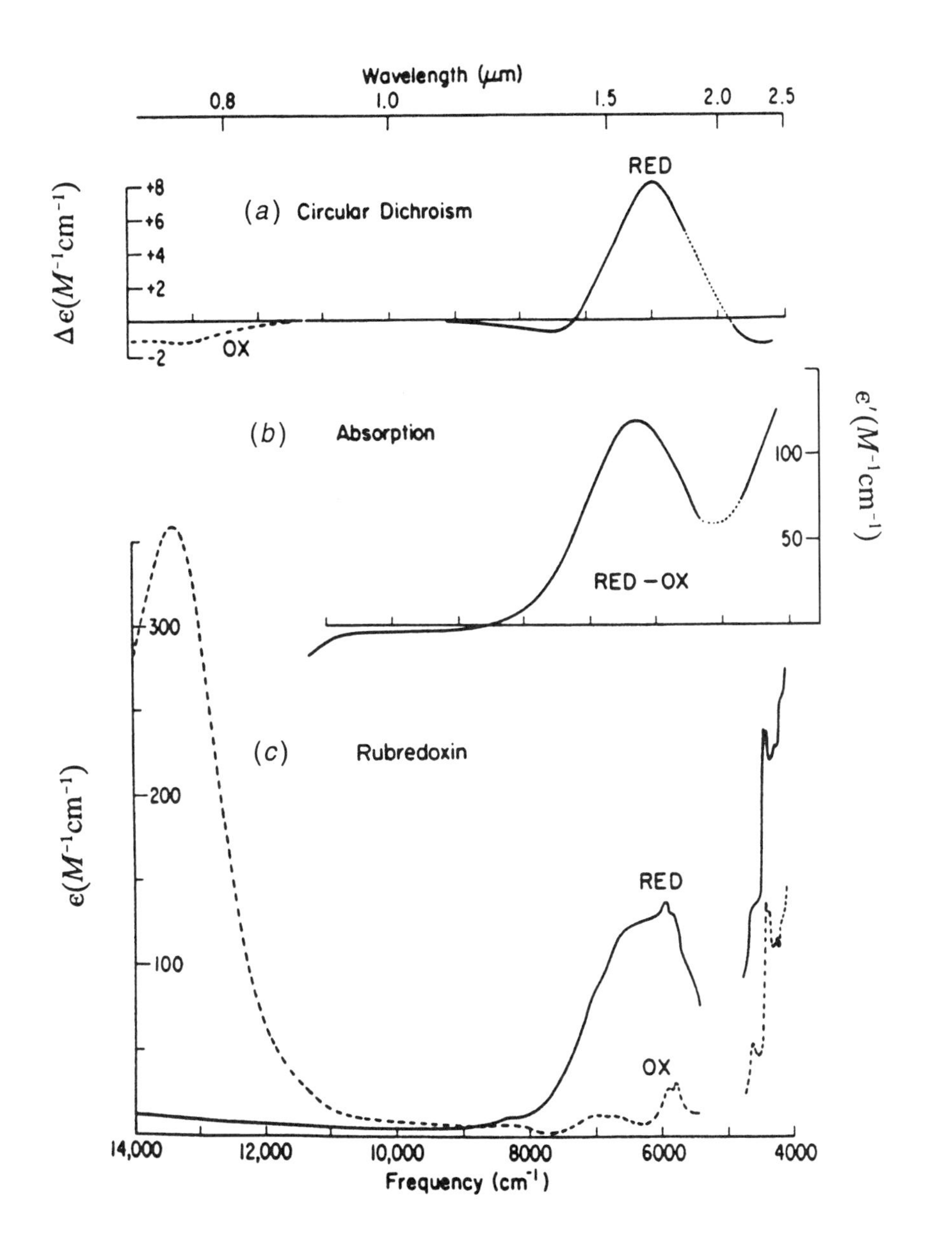

Figure 8
Absorption and CD spectra of oxidized (OX, dashed line) and reduced (RED, solid line) *C. pasteurianum* rubredoxin. Panel (*c*) is the difference absorption spectrum (RED–OX). [Taken from Eaton and Lovenberg[7] with permission.]

This assignment is supported and extended by the CD spectrum. The large anisotropy ratio, $g = 0.07$, for the 6,250-cm^{-1} band is consistent with the $^5E \rightarrow {}^5T_2$ assignment, since this transition is intrinsically magnetic dipole allowed under the parent tetrahedral symmetry. Furthermore, the presence of a negative CD band at 7,400 cm^{-1} and a positive CD band at 6,000 cm^{-1} reveals that the 6,250-cm^{-1} absorption band consists of two components. The $^5T_2$ excited state is therefore split into three components and the splittings are too large to be explained by spin–orbit coupling. Consequently, the effective symmetry for the FeS_4 unit must be D_2 or lower. This finding is consistent with the X-ray crystal structure for oxidized *C. pasteurianum* rubredoxin (1.2-Å resolution), which is best interpreted in terms of effective C_2 symmetry for the FeS_4 unit.

VIII. Electronic Spectroscopy in Magnetic Fields

Before considering the theoretical aspects of MCD that are applicable to the interpretation of spectra obtained for inorganic and bioinorganic systems, we begin by considering the effects of a magnetic field on the polarized electronic spectroscopy of an atomic system. An isolated atomic transition between ^1S and ^1P states will be used for simplicity to illustrate the basic principles. The purpose is to show the effect of a magnetic field on electronic energy levels, to derive the selection rules for the absorption of plane polarized and circularly polarized light in a magnetic field, and to distinguish between and assess the utility of the classical Zeeman experiment and the MCD experiment.

In zeroth order, solutions to Schrödinger's equation are given by

$$\hat{\mathscr{H}}|\Psi_n\rangle = E_n|\Psi_n\rangle \tag{26}$$

where Ψ_n are orthonormal wave functions of energy E_n. If the application of a magnetic field, H, results in a perturbation, $\hat{\mathscr{H}}'$, then the first-order correction, E_n', to the energy E_n is given by

$$E_n' = \langle\Psi_n|\hat{\mathscr{H}}'|\Psi_n\rangle \tag{27}$$

The Zeeman perturbation, $\hat{\mathscr{H}}'$, corresponds to the scalar product of the magnetic moment operator, $\hat{\boldsymbol{\mu}}$, and the applied field, $\mathbf{H}$, which leads to

$$\hat{\mathscr{H}}' = -\hat{\boldsymbol{\mu}} \cdot \mathbf{H} = \beta(\hat{\mathbf{L}} + 2\hat{\mathbf{S}}) \cdot \mathbf{H} \tag{28}$$

where β is the Bohr magneton[b] and $\hat{\mathbf{L}}$ and $\hat{\mathbf{S}}$ are the sum of the individual electron orbital and spin angular momentum operators, respectively. Hence, for a magnetic field along the z axis, the first-order perturbation to the energy is given by

$$E_n' = \beta H\langle\Psi_n|\hat{L}_z + 2\hat{S}_z)|\Psi_n\rangle \tag{29}$$

[b] The Bohr magneton is defined as $\beta = |e|h/(4\pi mc) = 0.4669$ cm^{-1} T^{-1}, where e and m are the charge and mass of an electron, respectively.

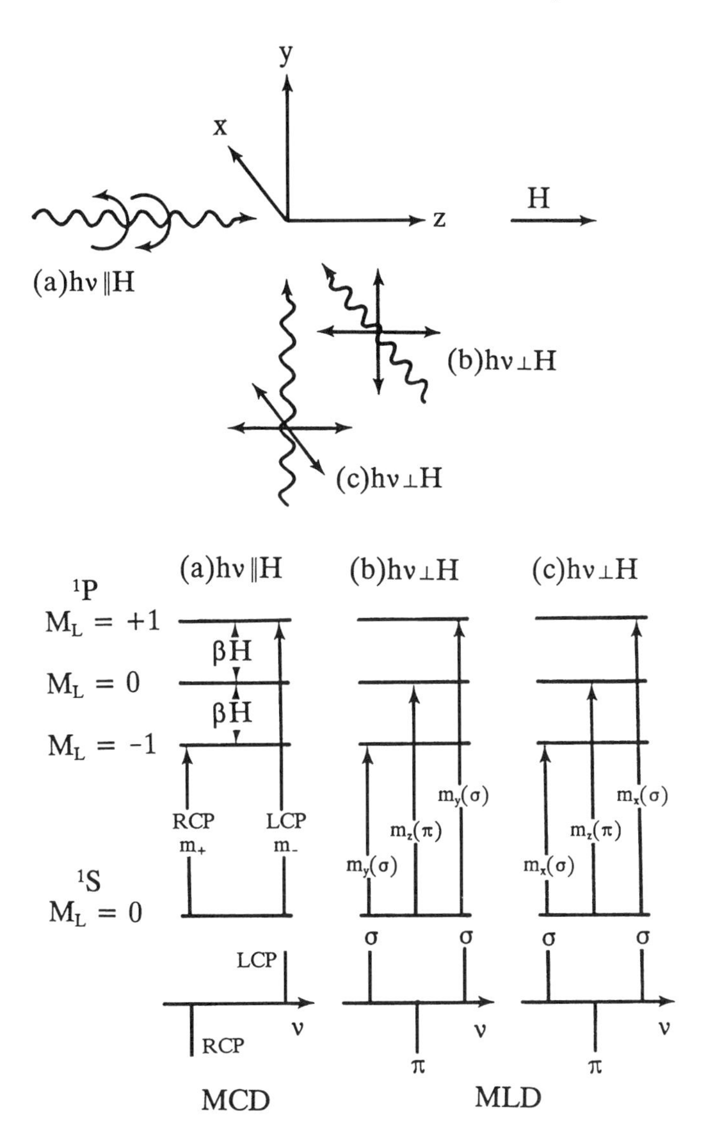

Figure 9
Schematic MCD and MLD spectra for a $^1S \rightarrow {}^1P$ transition. Upper panel: Types of experiment with the magnetic field parallel and perpendicular to the direction of light propagation. Lower panel: Resultant transitions and schematic MCD and MLD spectra for a $^1S \rightarrow {}^1P$ transition.

For example, an orbitally degenerate ^1P state will be split into three components separated in energy by βH by the application of a magnetic field, see Figure 9. Such removal of degeneracy by the application of a magnetic field is often termed a first-order Zeeman effect and the resultant states are termed Zeeman components. In contrast, the energy of an orbitally and spin nondegenerate state, such as a ^1S state, will be unaffected in first order by the application of a magnetic field. Consequently, a magnetic field can, in principle, split a $^1S \rightarrow {}^1P$ transition into three components. The result $(\hat{L}_z + 2\hat{S}_z)|\Psi_n\rangle$ may not be explicitly known for systems that are less theoretically understood than atoms, but the splitting can always be written in terms of the g factor, such that the Zeeman splitting is given by $g\beta H$[c]. The g factor is the same as that used in EPR spectroscopy and provides information about orbital and spin angular momenta.

The selection rules for the absorption of plane-polarized light and circularly polarized light can be readily deduced by considering transitions between hydrogenic atomic orbitals of the form

$$\Psi = R_{nl}(r)P_{lm_l}(\theta)e^{i\phi m_l} \times \text{spin} \tag{30}$$

where $x = r\sin\theta\cos\phi$, $y = r\sin\theta\sin\phi$, $z = r\cos\phi$, and $d\tau = r^2\sin\theta\; dr \cdot d\theta \cdot d\phi$ (neglecting spin).

A. Plane Polarized Light

The electric dipole moment operator for plane-polarized light with the electric vector polarized along the z axis, $\hat{m}_z$, corresponds to ez, which leads to the following expression for the transition moment integral between a ground orbital, Ψ, and an excited state orbital, Ψ'

$$e\int_\phi\int_\theta\int_r R(r)^* P(\theta)^* e^{-i\phi m_l} r\cos\theta R(r)' P(\theta)' e^{i\phi m_l'} r^2 \sin\theta\, dr \cdot d\theta \cdot d\phi \tag{31}$$

The selection rule can be derived by considering integration over ϕ from 0 to 2π since

$$\int_0^{2\pi} e^{i(m_l' - m_l)\phi} d\phi = 0 \quad \text{unless} \quad m_l' - m_l = 0 \tag{32}$$

For x-polarized light, the electric dipole moment operator, $\hat{m}_x$, corresponds to ex and the integration over ϕ is now

$$\int_0^{2\pi} e^{i(m_l' - m_l)\phi}\cos\phi\, d\phi = \frac{1}{2}\int_0^{2\pi} e^{i(m_l' - m_l + 1)\phi}\, d\phi + \frac{1}{2}\int_0^{2\pi} e^{i(m_l' - m_l - 1)\phi}\, d\phi = 0$$
$$\text{unless} \quad m_l' - m_l = \pm 1 \tag{33}$$

A similar result is obtained for y-polarized light. Hence, the selection rules are $\Delta m_l = 0$ for z-polarized light and $\Delta m_l = \pm 1$ for x- and y-polarized light.

[c] For atoms with Russell–Saunders coupling, the g factor can be calculated using

$$g = 1 + \frac{J(J+1) - L(L+1) + S(S+1)}{2J(J+1)}$$

B. Circularly Polarized Light

From Eq. (5b), it is apparent that the electric dipole moment operator for circularly polarized light propagating along the z direction, $\hat{m}_{\pm}$, is given by $(1/\sqrt{2})$ $(\hat{m}_x \pm i\hat{m}_y)$, where $\hat{m}_-$ corresponds to $\boldsymbol{E}^+$, that is, left circularly polarized light, and $\hat{m}_+$ corresponds to $\boldsymbol{E}^-$, that is, right circularly polarized light. Therefore, since

$$x \mp iy = r \sin\theta(\cos\phi \mp i \sin\phi) = r \sin\theta e^{\mp i\phi} \tag{34}$$

the integral over ϕ takes the form

$$\int_0^{2\pi} e^{(m_l' - m_l \mp 1)\phi} d\phi \tag{35}$$

and the selection rules are $\Delta m_l = +1$ for left circularly polarized light and $\Delta m_l = -1$ for right circularly polarized light.

These results are summarized in Figure 9, which shows schematic magnetic linear dichroism (MLD) and MCD spectra for a $^1S \rightarrow {}^1P$ transition. The former is derived from the classical Zeeman experiment and the polarizations are depicted by π (electric vector parallel to magnetic field direction) and σ (electric vector perpendicular to the magnetic field direction). Clearly, the Zeeman experiment can be used to determine magnetic splittings of electronic energy levels provided the bandwidth of the transition is smaller than the Zeeman splitting. However, this situation only arises for atomic transitions, since Zeeman splittings are of the order of a few wavenumbers even at magnetic fields of 5 T. In principle, this difficulty can be overcome for broad-band systems by measuring MLD, which corresponds to the differential absorption of light polarized parallel and perpendicular to a magnetic field that is perpendicular to the direction of light propagation. In practice, MLD has thus far proven to be of limited utility for inorganic and bioinorganic systems.

Broad-band systems are, however, amenable to study by MCD, which measures the differential absorption of left and right circularly polarized light propagating along the magnetic field direction. A transition from a nondegenerate ground state to a degenerate excited state, such as the $^1S \rightarrow {}^1P$ atomic transition depicted in Figure 10, will therefore result in temperature-independent, derivative-shaped MCD dispersion, called an A term, with the cross-over point corresponding to the absorption maximum. If the ground state is degenerate, such as in a $^1P \rightarrow {}^1S$ atomic transition as shown in Figure 10, then the allowed transitions remain the same, but there is a Boltzmann population distribution over the ground-state Zeeman components. The result is greater absorption of left circularly polarized light than right circularly polarized light, which leads to temperature-dependent MCD dispersion called a C term. For broad band systems where the Zeeman splitting is much smaller the bandwidth, the C term will exhibit absorption-shaped dispersion, with the maximum at the same energy as the absorption maximum. Moreover, provided thermal energy is much greater than the Zeeman splitting, the C term intensity will exhibit a simple $1/T$ dependence.

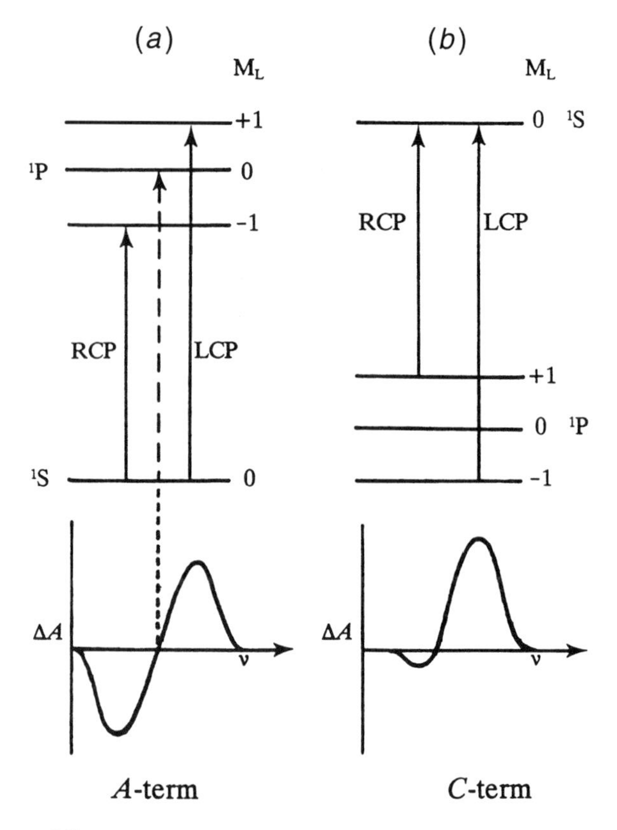

Figure 10
Schematic depiction of MCD A (a) and C terms (b) for transitions between ^{1}S and ^{1}P states ($\Delta A = A_L - A_R$).

Thus far the discussion has only included the origin of MCD A and C terms, which arise from splitting of degenerate states in a magnetic field, that is, a first-order Zeeman effect. However, application of a magnetic field can also result in mixing of electronic states that is often referred to as a second-order Zeeman effect. For a magnetic field along the z direction that mixes states $|\Psi_k\rangle$ into an electronic state $|\Psi_n\rangle$, the resultant electronic state, $|\Psi_{n'}\rangle$ is given by

$$|\Psi_{n'}\rangle = |\Psi_n\rangle - \sum_{k \neq n} \frac{\langle \Psi_k | \hat{\mu}_z | \Psi_n \rangle}{(E_n - E_k)} H_z |\Psi_k\rangle \tag{36}$$

Such field-induced mixing manifests itself by induced CD known as a MCD B term. This phenomena is illustrated by the following example. Consider a special case of a ground-state $|\Psi_a\rangle$ and two excited-states $|\Psi_j\rangle$ and $|\Psi_k\rangle$ such that the $|\Psi_a\rangle \rightarrow |\Psi_j\rangle$ transition is x polarized (i.e., only $\langle \Psi_a | \hat{m}_x | \Psi_j \rangle \neq 0$) and the $|\Psi_a\rangle \rightarrow |\Psi_k\rangle$ transition is y polarized (i.e., only $\langle \Psi_a | \hat{m}_y | \Psi_k \rangle \neq 0$). Such transitions cannot result in MCD A or C terms, which require *two* nonzero components of the electric dipole transition moment, since they depend on the difference in the squares of the transition dipole moments for left and right circularly polarized light (see the more detailed discussion of A and C terms presented below). However, $|\Psi_j\rangle$ and $|\Psi_k\rangle$ will be mixed by H_z since rotation about z takes x into y. After mixing the transition, $|\Psi_a\rangle \rightarrow |\Psi_j\rangle$ becomes $|\Psi_a\rangle \rightarrow |\Psi_j + \lambda\Psi_k\rangle$, where the mixing coefficient, λ, is pure imaginary. The MCD intensity for this transition is proportional to the difference in the transition dipole moments for the absorption of circularly polarized light, that is,

$$\Delta A \propto |\langle\Psi_a|\hat{m}_-|\Psi_j + \lambda\Psi_k\rangle|^2 - |\langle\Psi_a|\hat{m}_+|\Psi_j + \lambda\Psi_k\rangle|^2 \tag{37}$$

and simple algebraic manipulation leads to

$$\Delta A \propto -2i\lambda[\langle\Psi_a|\hat{m}_x|\Psi_j\rangle\langle\Psi_k|\hat{m}_y|\Psi_a\rangle + \langle\Psi_a|\hat{m}_y|\Psi_k\rangle\langle\Psi_j|\hat{m}_x|\Psi_a\rangle] \tag{38}$$

Clearly, the MCD intensity will not be zero and the magnitude will depend on the electric dipole moments for the transitions and the extent of mixing. Furthermore, if only $|\Psi_j\rangle$ and $|\Psi_k\rangle$ are mixed by the magnetic field, then the MCD under $|\Psi_a\rangle \rightarrow |\Psi_j\rangle$ will be equal in magnitude but opposite in sign to that under $|\Psi_a\rangle \rightarrow |\Psi_k\rangle$.

IX. Magnetic Circular Dichroism Theory

A. General Expression for MCD Dispersion

Extending MCD theory to molecules with broad absorption bands composed of numerous overlapping vibronic transitions is clearly necessary for inorganic and bioinorganic systems. The simplest theoretical approach, known as the rigid shift (RS) approach, assumes that the Born–Oppenheimer and Franck–Condon approximations hold and that the overall bandshape does not change as a result of the Zeeman interaction. The reader is referred to Stephens,[8] Schatz and McCaffery,[9] or Piepho and Schatz,[10] for derivation of the MCD dispersion equation using this approach. Here the results are summarized, noting that the basic form of the expression is predicted by the preceding discussion.

For a transition between states A and J, the RS MCD dispersion is given by

$$\Delta A(\mathrm{A} \rightarrow \mathrm{J}) = \gamma\left\{-A_1\left(\frac{\partial f}{\partial E}\right) + \left(B_0 + \frac{C_0}{kT}\right)f\right\}\beta Hbl \tag{39}$$

where γ is a spectroscopic constant, k is Boltzmann's constant; f is a normalized line shape function (e.g., Gaussian) that is dependent on the energy of the transition, $E_{\mathrm{JA}} = E_{\mathrm{J}} - E_{\mathrm{A}}$, and the incident photon energy, E; and A_1, B_0, and C_0 are parameters that depend on the electric dipole selection rules for the absorption of circularly polarized light. They are given explicitly below for oriented or isotropic molecules with light propagating parallel to a magnetic field applied along the z direction,[d]

$$A_1 = +\frac{1}{d_A}\sum_{\alpha,\lambda}[|\langle A_\alpha|\hat{m}_-|J_\lambda\rangle|^2 - |\langle A_\alpha|\hat{m}_+|J_\lambda\rangle|^2] \times [\langle J_\lambda|\hat{L}_z + 2\hat{S}_z|J_\lambda\rangle - \langle A_\alpha|\hat{L}_z + 2\hat{S}_z|A_\alpha\rangle] \quad (40a)$$

$$B_0 = \frac{-2}{d_A}\sum_{\alpha,\lambda} Re\left\{ \sum_{K_\kappa \neq J} [\langle A_\alpha|\hat{m}_-|J_\lambda\rangle\langle K_\kappa|\hat{m}_+|A_\alpha\rangle| - \langle A_\alpha|\hat{m}_+|J_\lambda\rangle\langle K_\kappa|\hat{m}_-|A_\alpha\rangle] \frac{\langle J_\lambda|\hat{L}_z + 2\hat{S}_z|K_\kappa\rangle}{E_K - E_J} + \sum_{K_\kappa \neq A} [\langle A_\alpha|\hat{m}_-|J_\lambda\rangle\langle J_\lambda|\hat{m}_+|K_\kappa\rangle - \langle A_\alpha|\hat{m}_+|J_\lambda\rangle\langle J_\lambda|\hat{m}_-|K_\kappa\rangle] \frac{\langle K_\kappa|\hat{L}_z + 2S_z|A_\alpha\rangle}{E_K - E_A} \right\} \quad (40b)$$

$$C_0 = -\frac{1}{d_A}\sum_{\alpha,\lambda}[|\langle A_\alpha|\hat{m}_-|J_\lambda\rangle|^2 - |\langle A_\alpha|\hat{m}_+|J_\lambda\rangle|^2]\langle A_\alpha|\hat{L}_z + 2\hat{S}_z|A_\alpha\rangle \quad (40c)$$

where A_α and J_λ are the ground and excited electronic states, respectively, with components α and λ, K_κ, is any other state that can mix with A_α or J_λ, when $H \neq 0$, and d_A is the degeneracy of the state A_α. The basis functions for states A and J are chosen to be diagonal in μ_z.

From Eq. (39), it is apparent that MCD dispersion is the sum of three terms, $-A_1(\partial f/\partial E)$, $B_0 f$, and $C_0 f/kT$, which correspond to the A, B, and C terms described phenomenologically in Section VIII. B. The origin of A, B, and C terms can now be rationalized in the more rigorous theoretical framework of Eqs. (40a–c). Since only orbitally or spin degenerate states will have nonzero magnetic dipole matrix elements, inspection of Eqs. (40a and c) reveals that A terms are only possible when the ground or excited states are degenerate and that C terms require ground-state degeneracy. If only the ground state is degenerate then $A_1 = C_0$. Degeneracy in both the ground and excited states leads to $A_1 \neq C_0$, however, the C term usually dominates in broad band systems. In contrast, B terms do not require degeneracy in either the ground or excited state, and result from mixing of the ground and excited state with all other excited states. Since all quantum mechanical states will mix to some extent in a magnetic field, B terms account for the universality of the Faraday effect. Summation over all possible

[d] For nonisotropic randomly oriented molecules, the expressions are more complex involving summations over several terms because of the inequality of the space-oriented x, y, and z axes. Derivation of appropriate expressions for A_1, B_0, and C_0 for randomly oriented molecules is given in Piepho and Schatz.[10]

mixing states is clearly impractical in evaluating *B* terms and it is generally only necessary to include states that are relatively close in energy, that is, small $E_K - E_J$ and/or $E_K - E_A$.

Inspection of Eq. (39) also indicates that *A*, *B*, and *C* term contributions to the MCD spectrum can in principle be identified by their different dispersion and temperature-dependence characteristics. *A* terms are temperature independent and exhibit derivative-shaped dispersion; *B* terms are temperature independent and exhibit absorption-shaped dispersion; *C* terms are inversely dependent on temperature (provided the Zeeman splitting, $g\beta H$, is much less than the thermal energy, kT) and exhibit absorption-shaped dispersion. However, using these characteristics to identify *A*, *B*, and *C* terms in spectra where there are overlapping transitions can be difficult. For example, closely spaced and oppositely signed *B* or *C* terms can result in derivative-shaped features, called "pseudo-*A*-terms". Pseudo-*A*-terms resulting from oppositely signed *C* terms are commonly encountered for transitions to spin–orbit split components of an orbitally degenerate excited state, for example, the Soret band of low-spin ferric hemes. Fortunately, this situation is readily distinguishable from an *A* term or a pseudo-*A*-term resulting from overlapping *B* terms, by virtue of its inverse temperature dependence. A potential complication in identifying *C* terms is that *B* terms can become temperature dependent if the mixing state itself becomes populated over the temperature range of the experiment. For example, field-induced mixing of the closely spaced zero-field components of a ground-state manifold can give rise to temperature-dependent *B* terms. However, temperature-dependent *B* terms can usually be distinguished from *C* terms by their linear dependence on magnetic field under conditions where *C* terms exhibit magnetic saturation. The examples discussed below, illustrate each of these types of MCD band and show how they can be distinguished.

At this juncture, it is important to note that the expressions for both *A* and *C* terms contain the term $|\langle A_\alpha|\hat{m}_-|J_\lambda|^2 - |\langle A_\alpha|\hat{m}_+|J_\lambda\rangle|^2$. First, it is the basis of the requirement for *two* nonzero components of the electric dipole transition moment to observe *A* or *C* terms in the MCD spectrum. A transition that is polarized along a unique axis can, therefore, only exhibit *B*-term MCD. Second, these electric dipole matrix elements are only nonzero if there is orbital degeneracy in either the ground or excited state. This requirement would appear to limit the utility of MCD spectroscopy for investigating metals centers in low symmetry biological environments, since orbital degeneracy requires at least a threefold symmetry axis. However, spin–orbit coupling in transition metal complexes is usually sufficiently large to mix states and thereby provide a mechanism for MCD intensity. Another important consequence of the low-symmetry environments encountered for metal centers in biology is that ground-state degeneracy can usually be equated with spin degeneracy, which leads to the useful generalization that only paramagnetic transition metal centers will exhibit temperature-dependent MCD spectra. Consequently, variable temperature MCD spectroscopy provides a convenient and invaluable method for distinguishing and deconvoluting the electronic transitions originating from diamagnetic and paramagnetic metal chromophores in bioinorganic systems.

Since the contributions of A, B, and C terms to the MCD spectrum are additive, it is useful to consider the factors that govern their relative intensities. From the preceding discussion, it should be apparent that these three components are favored by narrow bandwidths, nearby mixing states, and low temperatures, respectively. Assuming Gaussian line shape, that is,

$$\left(\frac{\partial f}{\partial E}\right)_{\max} \approx \frac{(f)_{\max}}{\Gamma} \tag{41}$$

where Γ is the bandwidth at half-height, the maximum contributions of A, B, and C terms to ΔA can be approximated based on Eq. (39) by

$$A : B : C \approx \frac{1}{\Gamma} : \frac{1}{\Delta E} : \frac{1}{kT} \tag{42}$$

where ΔE is the energy separation to the closest mixing state. For example, at room temperature, with $\Gamma = 1{,}000\ \mathrm{cm}^{-1}$, $\Delta E = 10{,}000\ \mathrm{cm}^{-1}$, and $kT = 200\mathrm{cm}^{-1}$, the relative intensities would be

$$A : B : C \approx 10 : 1 : 50 \tag{43}$$

Such considerations lead to the conclusion that C terms, if present, are likely to dominate the MCD even at room temperature. Moreover, they will be overwhelmingly dominant at liquid helium temperatures, since reducing the temperature to 4.2 K results in an enhancement of up to 70-fold in the C-term contribution. Consequently, low-temperature MCD is a sensitive and selective probe for the optical transitions associated with paramagnetic metal centers. If a sample contains both paramagnetic and diamagnetic metal centers, then the C terms from the paramagnetic component will usually dominate the low-temperature MCD spectrum at liquid helium temperatures. As suggested by Eq. (42), B terms will generally be much smaller than A or C terms for high symmetry chromopores with ground- and/or excited-state degeneracy. However, for low-symmetry chromophores with no spin or orbital degeneracy, the MCD will consist exclusively of B terms. Hence, B terms can account for much of the MCD intensity from diamagnetic metal centers in low-symmetry environments. The B terms can also become comparable in intensity to C terms at low temperatures, if the mixing state is close in energy. Temperature-dependent B terms originating from field-induced mixing of the zero-field split components of a ground-state manifold provide a good illustration of this type of behavior (see examples below).

The complexity of Eqs. (40a–c) makes it extremely difficult to predict values and signs for A_1, B_0, and C_0 for low-symmetry metal chromophores or for multinuclear clusters with ill-defined and complex electronic structures. However, this is not the case for A_1 and C_0, provided the system under investigation consists of isotropic or highly symmetrical oriented molecules. First, they can often be greatly simplified by noting that the zero-field absorption oscillator strength for the transition, D_0, which can be calculated from the integrated area of the absorption band, is given by

$$D_0(\mathrm{A} \to \mathrm{J}) = \frac{1}{2d_{\mathrm{A}}} \sum_{\alpha,\lambda} [|\langle \mathrm{A}_\alpha | \hat{m}_+ | \mathrm{J}_\lambda \rangle|^2 + |\langle \mathrm{A}_\alpha | \hat{m}_- | \mathrm{J}_\lambda \rangle|^2] \tag{44}$$

Comparison with Eqs. (40a and c) shows that electric-dipole matrix elements of the type $\langle \mathrm{A}_\alpha | \hat{m}_\pm | \mathrm{J}_\lambda \rangle$ appear in the expressions for A_1, C_0, and D_0, and these will often cancel out, within a proportionality factor, when the ratios A_1/D_0 and C_0/D_0 are taken. Consequently, the ratios A_1/D_0 and C_0/D_0 lead to estimates of excited- and ground-state g values, respectively, in favorable cases. For example, the interested reader should try verifying that $A_1/D_0 = 2g$ for a ${}^1\mathrm{A}_{1g} \to {}^1\mathrm{T}_{1u}$ transition and that $C_0/D_0 = 2g$ for a ${}^1\mathrm{T}_{1u} \to {}^1\mathrm{A}_{1g}$ transition for systems with octahedral system (see Stephens[8] for the worked solution). Unfortunately, determining the ratio B_0/D_0 is not usually effective for cancelling out difficult-to-evaluate electric-dipole matrix elements. In addition, the use of group theory, albeit at a level generally unfamiliar to most chemists, permits enormous simplification and/or evaluation of the expressions for A_1, C_0, and D_0. For highly symmetrical, oriented systems, group theory shows that many of the matrix elements in the summations are zero and the remaining ones can be simplified or evaluated using the Wigner–Eckart theorem. Piepho and Schatz[10] wrote a comprehensive treatise on the use of group theoretical techniques, including irreducible tensor methods, for the quantitative interpretation of the MCD of highly symmetrical transition metal complexes. The interested reader is referred to this text for detailed discussion of this important aspect of MCD spectroscopy.

B. Determining Electronic Ground-State Properties

In addition to resolving and assigning electronic transitions, variable temperature and magnetic field MCD studies can also provide estimates of the ground-state properties of paramagnetic transition metal chromophores. These properties include spin state, g values, zero-field splitting (zfs) parameters for $S > \frac{1}{2}$ ground states, and magnetic coupling constants for magnetically interacting centers. Two distinct types of experiments are used to obtain this information. The first involves monitoring the saturation properties of discrete MCD bands as a function of decreasing temperature and/or increasing magnetic field. The second involves monitoring the MCD intensity at fixed field as a function of temperature under conditions where the Zeeman splitting is much less than kT, and the MCD intensity is linear in H, that is, the linear or Curie law limit. The utility and theoretical basis for each approach is discussed separately below.

1. Saturation Curves

Inspection of Eq. (39) indicates that the C-terms MCD from a paramagnetic chromophore is linearly dependent on H/T. However, this relationship only holds provided the Zeeman splitting is $\ll kT$, since the differential absorption of left and right circularly polarized light is a direct consequence of the Boltzmann population distribution over the Zeeman-split sublevels of the ground state (see Fig. 10). Eventually, as H increases and/or T decreases, the MCD intensity must become independent of H/T when there is population of only the lowest Zeeman component and the system is fully magnetized or saturated. Explicitly including the Boltzmann populations of the ground-state Zeeman sublevels in deriving the expression for C_0 leads to

$$\Delta A = \gamma\left[\sum_{\alpha,\lambda} \frac{\exp(\langle \mathrm{A}_\alpha|\hat{\mu}_z|\mathrm{A}_\alpha\rangle H/kT)}{\sum_\alpha \exp(\langle \mathrm{A}_\alpha|\hat{\mu}_z|\mathrm{A}_\alpha\rangle H/kT)} \times [|\langle \mathrm{A}_\alpha|\hat{m}_-|\mathrm{J}_\lambda\rangle|^2 - |\langle \mathrm{A}_\alpha|\hat{m}_+|\mathrm{J}_\lambda\rangle|^2]f\right] bl \tag{45}$$

The simplest example is provided by an isotropic $S = \frac{1}{2}$ ground state ($M_s = \pm\frac{1}{2}$), for which

$$\langle \mathrm{A}_\alpha|\hat{\mu}_z|\mathrm{A}_\alpha\rangle = \pm\tfrac{1}{2}g\beta H \tag{46}$$

Combining Eqs. (45 and 46) leads to the following simple expression for $\Delta\varepsilon$ at fixed wavelength

$$\frac{\Delta\varepsilon}{K} = \tanh\left(\frac{g\beta H}{2kT}\right) \tag{47}$$

where K is a proportionality constant that includes the electric dipole matrix elements, which are constant and independent of magnetic field and temperature for a specified $A \rightarrow J$ transition.[e] Hence, for a ground-state doublet with isotropic Zeeman interaction, a plot of $\Delta\varepsilon$ versus H/T or $\beta H/2kT$ (termed a saturation or magnetization curve) is a simple tanh function that depends only on the magnitude of the isotropic g value. This dependence is illustrated in Figure 11, which shows theoretical magnetization curves obtained using Eq. (47) for isotropic g values between 1 and 3. The y axis is expressed as percent magnetization, which corresponds to the MCD intensity as a percentage of the intensity at magnetic saturation. Clearly, MCD magnetization measurements afford a means of assessing ground-state g values that is independent of absorption measurements and depolarization effects. As indicated above, it is in principle possible to assess ground-state g values from C_0/D_0 ratios by integration of the MCD C term and the *corresponding* absorption band area. In practice, this method is fraught with difficulties for real systems due to baseline uncertainties in absorption measurements, and the problems associated with deconvoluting overlapping bands to ascertain absorptions corresponding to individual C terms. For example, in a multicomponent metalloenzyme, a C term from a paramagnetic metal center may dominate the MCD, whereas electronic transitions from diamagnetic components may dominate the absorption spectrum in the same spectral region. Magnetization measurements circumvent these problems and provide a more reliable optical method for assessing ground-state g values. Although not as accurate as g

[e] The parameter K is used as a generic symbol for a proportionality constant in many of the equations that follow. Actual numerical values will differ in each expression.

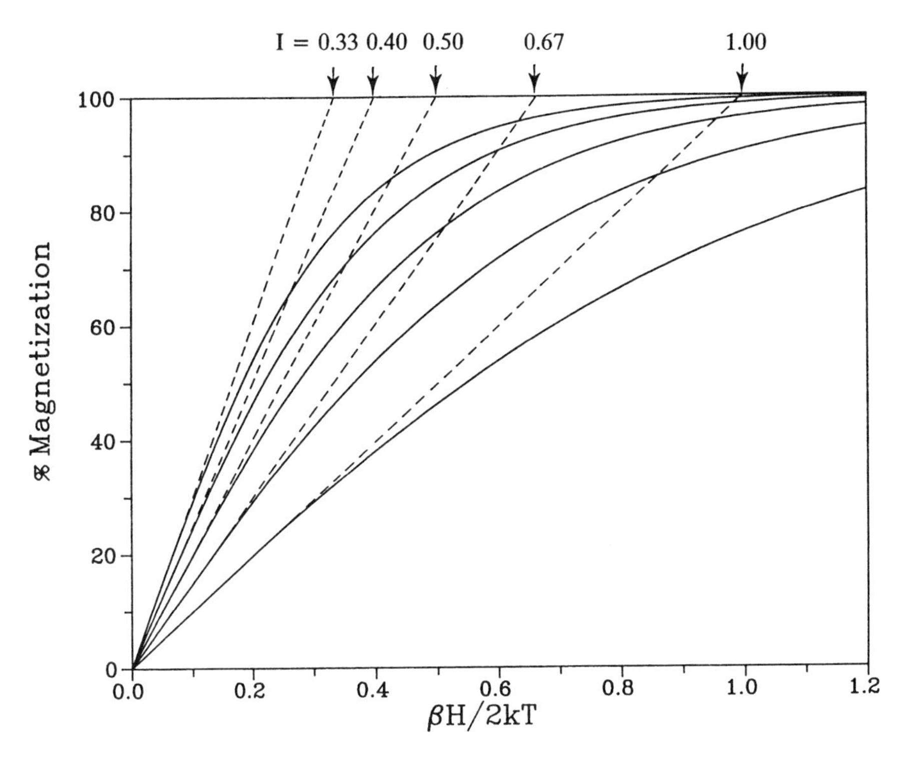

Figure 11
Magnetic circular dichroism magnetization curves for an isotropic $S = \frac{1}{2}$ ground states with $g = 1.0$, 1.5, 2.0, 2.5, and 3.0 (initial slope increasing with g), computed according to Eq. (47). The arrows on the x axis indicate intercept values, $I = 1/g$.

values determined by EPR spectroscopy, they provide a useful correlation between EPR signals and specific electronic transitions. Moreover, in $S > \frac{1}{2}$ systems, effective g values can be assessed for zero-field doublets that do not exhibit readily detectable EPR signals, for example, systems with integer spin ground states such as high spin Ni(II) ($S = 1$), see Section XI.

Thus far, the discussion of MCD saturation has been limited to an $S = \frac{1}{2}$ Kramers doublet with an isotropic Zeeman interaction. To be generally useful for transition metal complexes, it is necessary to develop theoretical expressions for magnetization curves arising from randomly oriented molecules with anisotropic $S = \frac{1}{2}$ ground states and to consider the MCD magnetization behavior expected for Kramers (odd number of unpaired electrons) and non-Kramers (even number of unpaired electrons) $S > \frac{1}{2}$ ground states. Following Schatz et al.,[11] the appropriate expression for an axial $S = \frac{1}{2}$ chromophore is

$$\frac{\Delta\varepsilon}{K} = m_+^2 \left[\int_0^{\pi/2} \frac{\cos^2\theta \sin\theta}{\Gamma} g_{\parallel} \tanh\left(\frac{\Gamma\beta H}{2kT}\right) d\theta - \sqrt{2}\frac{m_z}{m_+} \int_0^{\pi/2} \frac{\sin^3\theta}{\Gamma} g_{\perp} \tanh\left(\frac{\Gamma\beta H}{2kT}\right) d\theta \right] \tag{48}$$

where θ is the angle between the molecular z axis and the applied field, m_z and m_+ are the molecular-fixed electric dipole transition matrix elements for the corresponding operators (i.e., transition dipole moments for the molecular z and xy polarized transitions) and $\Gamma = ({g_\parallel}^2 \cos^2\theta + {g_\perp}^2 \sin^2\theta)^{1/2}$. Orientation averaging is accomplished by numerical integration over θ. Inspection of Eq. (48) shows it to be the sum of terms arising from the xy- and z-polarized components. Consequently, MCD magnetization curves arising from anisotropic doublets are dependent on the polarization of the electronic transition, and hence can be wavelength dependent. The requirement to consider explicitly the transition polarization adds an additional complication to the interpretation of MCD magnetization data compared to that obtained in a magnetic susceptibility experiment. However, MCD magnetization experiments do offer some advantages for obtaining ground-state magnetic information, particularly for multicomponent metalloenzyme systems. First, it is possible to investigate selectively individual components provided each has well-resolved contributions to the MCD spectrum. Second, MCD magnetization studies are not sensitive to nonchromophoric paramagnetic impurities [such as nonspecifically bound Fe(III) ions], which complicate the interpretation of bulk magnetic susceptibility measurements of metalloproteins.

It is worthwhile to consider three special cases for which MCD magnetization data will be independent of the wavelength of measurement. The first is an isolated doublet ground state with isotropic Zeeman interaction, that is, $g_\parallel = g_\perp = g$, in which case Eq. (48) simplifies to Eq. (47). The second is when the electronic spectrum is dominated by xy-polarized transitions, such as in hemoproteins. In this instance, the second term in Eq. (48) can be neglected. Since low-spin Fe(III) hemes usually exhibit rhombic g tensors, it is useful to extend Eq. (48) to include xy-polarized transitions from randomly oriented molecules with a rhombic $S = \frac{1}{2}$ ground state (i.e., $g_z \neq g_y \neq g_x$). The appropriate expression, derived by Thomson and Johnson,[12] is

$$\frac{\Delta\varepsilon}{K} = m_+^2 \int_{\phi=0}^{2\pi} \int_{\theta=0}^{\pi} \frac{\cos^2\theta \sin\theta}{\Gamma} g_z \tanh\left(\frac{\Gamma\beta H}{2kT}\right) d\theta\, d\phi \tag{49}$$

where $\Gamma = ({g_x}^2 \sin^2\theta \sin^2\phi + {g_y}^2 \sin^2\theta \cos^2\phi + {g_z}^2 \cos^2\phi)^{1/2}$. The third is a ground-state doublet with $g_\parallel \neq 0$ and $g_\perp = 0$, a situation that frequently arises for individual doublets in the zero field split manifolds of $S > \frac{1}{2}$ ground states. In this case, Eq. (48) reduces to

$$\frac{\Delta\varepsilon}{K} = m_+^2 \int_0^{\pi/2} \cos\theta \sin\theta \tanh\left(\frac{g_\parallel \beta H \cos\theta}{2kT}\right) d\theta \tag{50}$$

Since $g_\perp = 0$, all temperature-dependent MCD transitions from the ground state are xy polarized and the magnetization data is invariant as a function of wavelength.

The protocol for extracting ground-state g values for systems with isolated doublet ground states therefore involves nonlinear least-squares fitting of plots of $\Delta\varepsilon$ *versus* $\beta H/2kT$ to Eq. (48–50). However, there is a much simpler procedure that facilitates rapid estimates of ground-state g values from MCD magnetization curves for systems with isotropic or axial ($g_{\parallel} \neq 0$, $g_{\perp} = 0$) doublets lowest in energy.[12] By using the appropriate equations, it is simple to demonstrate that the ratio of the asymptotic limit of the saturation curve to the initial slope, which corresponds to the intercept, I, on the abscissa (see Fig. 11), are solely dependent on the ground-state g values. For an isotropic doublet $I = 1/g$, whereas for the axial doublet $I = 3/(2g_{\parallel})$ provided $g_{\perp} = 0$.

Complete analysis of MCD magnetization data originating from $S > \frac{1}{2}$ ground states is a complex undertaking. However, this situation is usually readily identified by one or more of the following observations: (a) "nesting" of magnetization plots, that is, noncoincidence of plots of $\Delta\varepsilon$ versus $\beta H/2kT$ obtained at different temperatures; (b) marked deviation from the theoretical data anticipated for an isotropic $S = \frac{1}{2}$ doublet with $g = 2$; and (c) plots of $\Delta\varepsilon$ versus $1/T$ in the "linear limit" only become linear at high temperatures. Illustrative examples of such behavior are given in Section XI. All three observations are a consequence of having a zero-field split manifold of levels for the ground state. This lifting of ground-state spin degeneracy is a consequence of low-symmetry crystal fields and results from mixing of excited states into the ground state via second-order spin–orbit coupling. It is generally treated using an electronic spin Hamiltonian, $\hat{\mathscr{H}}_{\mathrm{e}}$, of the form

$$\hat{\mathscr{H}}_{\mathrm{e}} = g_0\beta \mathbf{H} \cdot \hat{\mathbf{S}} + D[\hat{S}_z^{\,2} - S(S+1)/3] + E(\hat{S}_x^{\,2} - \hat{S}_y^{\,2}) \qquad (51)$$

where g_0 is an isotropic g tensor, and D and E are the axial and rhombic zfs parameters, respectively (see Chapter 3 on EPR spectroscopy for a more detailed discussion of this spin Hamiltonian). Since Kramers and non-Kramers ground states can exhibit marked differences in their MCD magnetization properties, these two cases are treated separately below.

For Kramers ground states, zfs results in $(2S+1)/2$ doublets in the ground-state manifold. Provided the zero-field energy separation between the doublets is significantly larger than the Zeeman interaction, each doublet can be treated as an effective $S = \frac{1}{2}$ system and effective g values can be deduced from the spin Hamiltonian. The temperature dependent MCD from such a ground state will be the sum of the C terms originating from each doublet, with the individual contributions weighted according to their Boltzmann population. However, there is the additional complication that complete analysis of saturation magnetization data must explicitly include field-induced mixing of zero-field components. Orientationally averaged theoretical expressions can be derived, but they are extremely cumbersome with too many parameters for meaningful fitting (i.e., D, E, effective g values and transition polarizations for *each* zero-field doublet). In the limit where the zfs is much larger than the Zeeman interaction, only the lowest doublet will be significantly populated at temperatures <2 K. Therefore, in favorable cases, analysis of the magnetization data obtained at very low temperatures, using the expressions and/or intercept relationships discussed above for a magnetically isolated doublet, can yield estimates of effective g values for the lowest zero-field doublet of the ground-state manifold, and hence the ground-state spin.

The situation can be even more complex for non-Kramers ground states. In the limit of a completely axial system ($D \neq 0$, $E = 0$), zfs results in levels with $M_S = \pm S, \pm(S-1), \ldots, 0$, with the $\pm S$ doublet lowest in energy for $D < 0$ and the nondegenerate $M_S = 0$ level lowest in energy for $D > 0$. For the $D < 0$ case, the C term from the $M_S = \pm S$ doublet will usually dominate the MCD intensity at very low temperatures. Once again, provided the zfs is much larger than the Zeeman interaction, the saturation magnetization data obtained at very low temperatures can be analyzed using the methods developed for a magnetically isolated doublet state. Since $g_{\parallel} = 4S$ and $g_{\perp} = 0$ for the $M_S = \pm S$ doublet when $D \neq 0$ and $E = 0$, the relationship $I = 3/(2g_{\parallel})$ can be used to obtain $g_{\parallel}$ and hence the ground-state spin, S.

Very different MCD magnetization behavior is anticipated for an axial non-Kramers ground state with $D > 0$, since this results in a nondegenerate level lowest in energy. Temperature-independent MCD that is linearly dependent on the applied field strength will be observed with increasing temperature until thermal energy becomes comparable with the zfs and higher lying zero-field components become populated. As illustrated by some of the examples presented in Sector XI, C terms from higher lying non-Kramers doublets can dominate the MCD at higher temperatures giving rise to anomalous temperature dependence behavior. Alternatively, the MCD intensity at fixed applied fields may appear to saturate quite normally as a function of $1/T$. Such behavior, coupled with linear field dependence at constant low temperature, is uniquely indicative of temperature dependent B terms as the dominant contributor to the low-temperature MCD intensity. Field-induced mixing of closely spaced zero-field components is particularly prevalent for $S > \frac{1}{2}$ ground states, In such cases, B terms can be of comparable intensity to C terms, since the energy separation between the mixing levels is often quite small (i.e., <20 cm^{-1}). In addition, they become temperature dependent, since the mixing state(s) are populated as a function of increasing temperature.

The saturation characteristics anticipated for a temperature-dependent B term are best illustrated by considering a simple system consisting of two nondegenerate levels, separated in energy by ΔE, that are mixed by an applied field. Since the B terms originating from each level will be equal in magnitude, but opposite in sign (see above), it is easily shown that

$$\frac{\Delta\varepsilon}{K} = \tanh\left(\frac{\Delta E}{2kT}\right) \tag{52}$$

Hence, temperature-dependent B terms undergo saturation at low temperatures in the same way as C terms [cf. Eqs. (47 and 52)], except that ΔE corresponds to the energy separation between the mixing states rather than the Zeeman splitting. Moreover, it is clear that analysis of the temperature dependence using Eq. (52), or a similar expression that allows for degeneracy in the upper level, can lead to an estimate of ΔE, and hence the zero-field splitting between levels in the ground-state manifold.

The situation becomes more complicated for non-Kramers ground states with a significant rhombic distortion (i.e., $D \neq 0$ and $E \neq 0$), since this removes all ground-state degeneracy. In general, such ground-state properties result in MCD magnetization data that can appear qualitatively similar irrespective of the sign of D. However, Solomon and co-worker[13] showed that it is possible to distinguish the sign of D through quantitative analysis of data. Detailed discussion of this type of quantitative analysis to yield ground-state zfs, g values and polarizations is beyond the scope of this chapter. The interested reader is referred to an excellent recent review by Pavel and Solomon.[13]

The potential for temperature-dependent B-term contributions to the temperature-dependent MCD exhibited by transition metal centers with $S > \frac{1}{2}$ ground states has been recognized relatively recently. In many instances, they appear to be the dominant contributor of low-temperature MCD intensity for non-Kramers systems with $D > 0$. In some case, they may be partially responsible for the nesting exhibited by other types of $S > \frac{1}{2}$ ground states and the extent of their contribution must be established on a case-by-case basis.

2. Temperature Dependence in the Linear Limit

Estimates of the energies of low-lying levels resulting from zfs or magnetic exchange interaction can be made by monitoring the MCD intensity at fixed wavelength and small applied field as a function of $1/T$.[14] In the linear limit (i.e., $kT \gg g\beta H$ and $\Delta\varepsilon$ linearly dependent on H), deviations from a simple Curie law dependence will occur if low-lying states are populated with increasing temperature. The temperature-dependent MCD intensity can be expressed as the sum of the A, B, and C term contributions from each level, weighted according to their Boltzmann populations

$$\frac{\Delta\varepsilon}{K} = \sum_i \alpha_i \left(A_i + B_i + \frac{C_i}{kT} \right) \tag{53}$$

where α_i is the fractional population of level i, and A_i, B_i, and C_i are the A, B, and C term contributions that originate from this level. The axial zfs parameter, D, or the magnetic coupling constant, J, appear in the numerator of the exponent in the exponential terms that make up α_i.

Before proceeding to derive the appropriate expression for nonlinear least-squares fitting to obtain D or J, it is usually necessary to have some knowledge of the nature of the ground-state manifold for the system under investigation, that is, spin state and sign of D or the nature (antiferromagnetic or ferromagnetic) of magnetic coupling. Such information is usually obtained from MCD magnetization (see above), EPR, magnetic susceptibility, or Mössbauer studies, or some combination thereof. The expression used can then be simplified compared to Eq. (53) to minimize the number of variable parameters. For example, the neglect of A-term contributions is usually a good approximation, and nonzero C terms will only originate from degenerate or near-degenerate levels. In contrast, B terms, particularly those arising from field-induced mixing of the low-lying levels (temperature-dependent B terms) can be comparable in magnitude to C terms, and hence must be explicitly included in the analysis, unless there are reasons to believe that such mixing will be negligible, for example, xy-polarized transitions from an axial $S > \frac{1}{2}$ ground state ($D \neq 0$ and $E = 0$).

In general, the uniqueness of the fit to obtain D or J will be dependent on the number of low-lying levels, since this will govern the number of variable parameters, and the extent of deviation from linearity of the $\Delta\varepsilon$ versus $1/T$ plot. In all cases, it is important to assess the uniqueness of the fit by conducting experiments at several distinct wavelengths. The B and C term contributions will vary at different wavelengths, whereas the energy separation of the low-lying levels will remain constant.

X. Applications of Magnetic Circular Dichroism Spectroscopy

Magnetic circular dichroism spectroscopy has been extensively used to investigate the electronic and magnetic properties of inorganic complexes and metalloproteins. Reviews have appeared on iron–sulfur proteins,[15] porphyrins, and heme proteins,[16,17] copper, rare earth, cobalt and non-heme iron bioinorganic systems in general,[18] and non-heme ferrous enzymes.[19] In contrast to CD and absorption studies, MCD provides *both* excited- and ground-state information, and hence a more powerful probe of the coordination geometry and structure of metal chromophores.

Excited-state information is accessible by room-temperature MCD studies, although variable-temperature studies are obviously desirable in effecting band assignments. In general, MCD is effective in identifying, resolving, and assigning electronic transitions. Compared to absorption, it has the same advantages as CD in that bands have sign and intensity. In addition, the temperature dependence of bands associated with paramagnetic chromophores adds a further dimension to the resolving power and detection limits of MCD. This attribute is particularly useful in identifying electronic transitions arising from paramagnetic centers in multicentered metalloproteins. Moreover, it should be emphasized that the selection rules for absorption, CD, and MCD spectroscopies are quite distinct and use of all three in the same spectral region affords a powerful and complementary approach to assigning electronic excited states.

Ground-state information is accessible only by variable-temperature MCD studies down to liquid helium temperatures, via saturation magnetization and/or temperature-dependence studies in the linear limit. The type of information that can be obtained includes g values, spin state, zfs parameters, and magnetic coupling constants. In providing estimates of ground-state g values, MCD magnetization studies provide an important link between EPR resonances and specific electronic transitions. Knowing which EPR signals correspond to a particular set of electronic transitions, and vice versa, clearly enhances the information content of both techniques. An even more direct link, as well as the potential for more precise optically detected g values, is provided by the MCD detected optical-microwave double resonance experiment. The interested reader is directed to the work of Thomson and co-workers[20] for a description of the theory and applications of this novel experimental approach.

XI. Magnetic Circular Dichroism Examples

The principal applications of variable-temperature MCD spectroscopy are illustrated below by a few specific examples. We start with an inorganic example, the ferricyanide ion, and then proceed to hemoprotein, iron–sulfur protein, and nickel protein bioinorganic examples. The $[Fe(CN)_6]^{3-}$ example demonstrates that detailed electronic assignments based on the predicted sign and magnitude of MCD bands can be made for highly symmetrical complexes. Although detailed band assignments are not always possible for the bioinorganic examples, they do illustrate how variable-temperature MCD studies provide useful electronic and magnetic information from which structural data can be inferred. A few comments on units of MCD intensity are appropriate at the outset. It is customary to use $\Delta\varepsilon$ or $\Delta\varepsilon/H$ rather than $[\theta]$ or $[\theta]/H$ for MCD intensity. The parameter $\Delta\varepsilon$ is usually used for low-temperature and high-magnetic-field measurements where saturation effects may be observed. The parameter $\Delta\varepsilon/H$ is frequently used for measurements in the linear limit where $\Delta\varepsilon \propto H$.

The analysis of the MCD spectrum of $[Fe(CN)_6]^{3-}$ by Stephens[21] was a milestone in development of MCD spectroscopy, since it marked the first time that the technique was used to distinguish between two proposed assignments in a molecular system. The absorption spectrum in the UV–vis region comprises three intense bands centered at 420, 305, and 262 nm [see Figure 12(*a*)] that are assigned to ligand–metal charge-transfer transitions involving transfer of a ligand t_{1u} or t_{2u} electron to the hole in the metal t_{2g} set of d orbitals [see Fig. 12(*b*)]. Hence, the three absorption bands correspond to $^2T_{2g} \rightarrow {}^2T_{2u}$, $^2T_{2g} \rightarrow {}^2T_{1u}(\pi + \sigma)$, and $^2T_{2g} \rightarrow {}^2T_{1u}(\sigma + \pi)$, each with unresolved splittings arising from spin–orbit coupling. However, the assignment of these bands was not known prior to MCD studies. As shown in Figure 12(*a*), each of these transitions gives rise to a temperature-dependent C term; negatively signed for the 305-nm band and positively signed for the 420- and 262-nm bands. Since the relative signs of the C terms are determined solely by the symmetry designations of the transitions, it follows that the $^2T_{2g} \rightarrow {}^2T_{1u}$ transitions must correspond to the positive C terms and that the $^2T_{2g} \rightarrow {}^2T_{2u}$ transition must correspond to the negative C term. To establish if the signs of the C terms agree with the experimental result, it is necessary to evaluate C_0/D_0 values for each type of transition. Neglecting spin–orbit coupling, this can be accomplished using the Wigner–Eckart theorem using symmetry arguments alone, except for the calculation of the ground-state orbital angular momentum reduced matrix element, which can be done quite reliably assuming that the t_{2g} orbital is a pure metal d orbital (see Piepho and Schatz[10] for a detailed discussion). The predicted C_0/D_0 ratios for the 420-, 305-, and 262-nm bands are 1.0, −1.0, and 1.0. Hence, the observed signs for the C terms are as predicted and the intensities relative to their respective absorptions are in reasonable quantitative agreement.

The role of variable temperature MCD as a probe for paramagnetic chromophores and the utility of this approach in assigning specific transitions to individual chromophores in multicomponent metalloenzymes are well illustrated by

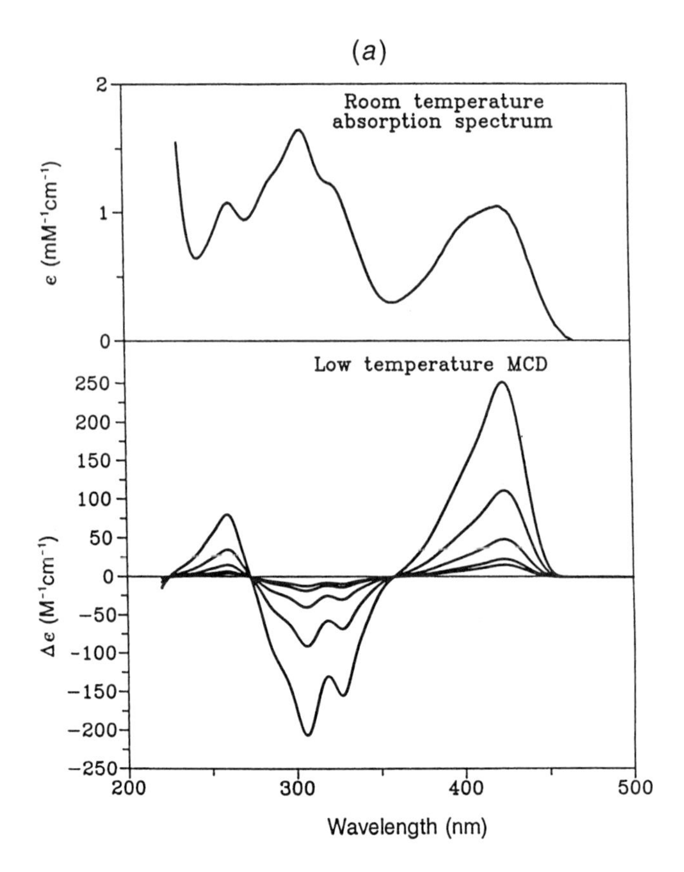

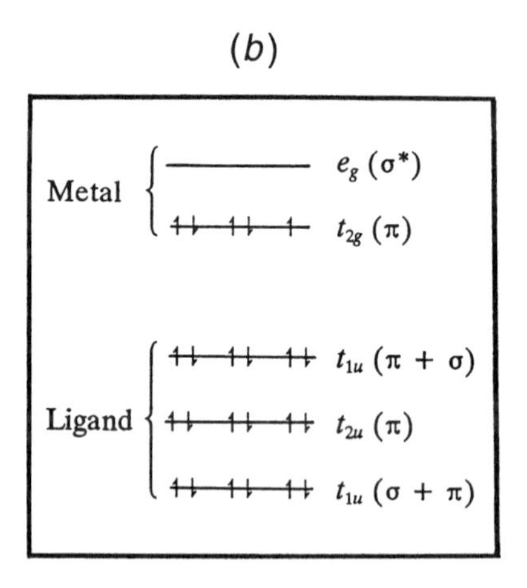

Figure 12
(*a*) Room temperature absorption (top panel) and variable temperature MCD spectra (lower panel) of potassium ferricyanide in 50 : 50 glycerol/water. MCD spectra were recorded with a magnetic field of 0.5 T at 1.82, 4.22, 10.1, 22.1, and 34.5 K. The MCD intensity is proportional to $1/T$, as expected for C terms in the linear limit. (*b*) Partial orbital energy level diagram for $[Fe(CN)_6]^{3-}$.

studies of the fully reduced nitrite reductase from *Pseudomonas aeruginosa*[22] [see Figure 13(*a*)]. This enzyme contains two distinct types of heme group; heme *c* and a chlorin-type heme classified as heme d_1, both of which contribute to the room temperature absorption spectrum. Five coordinate Fe(II) hemes are generally high spin with a paramagnetic $S = 2$ ground state, whereas six-coordinate Fe(II) hemes are frequently low spin and diamagnetic with an $S = 0$ ground state. Although the electronic spectrum in the UV–vis region is dominated by the two $^1A_{1g} \rightarrow {}^1E_u$, $\pi \rightarrow \pi^*$ transitions of the porphyrin (Soret or *B* band in the region 380–500 nm and α or *Q* band in the region 500–700 nm), the magnetic properties of the iron center are manifest in the MCD properties of these transitions as a result of mixing of the iron dπ and porphyrin pπ orbitals. Inspection of the MCD spectra reveals both temperature-dependent and temperature-independent regions suggesting that one of the hemes is low spin and diamagnetic and the other is high spin and paramagnetic.

Comparison with the absorption and MCD spectra of ferrocytochrome c_{551} from *P. aeruginosa* [see Fig. 13(*b*)], shows that the temperature-independent bands between 500 and 550 nm arise from the α band of low-spin reduced heme *c*. The heme *c* Soret band at 420 nm, which dominates the absorption spectrum, is apparent in the MCD spectra as a weak temperature-independent derivative (marked by arrows) superimposed on a broad temperature-dependent band. The MCD features of reduced cytochrome c_{551} are dominated by intense *A* terms that arise from the orbital degeneracy of the two ^1E_u excited states and the relative intensities reflect the much greater orbital moment associated with the α band.[17] The MCD intensity of reduced cytochrome c_{551} does increase by a factor of about 2 between room temperature and 4.2 K. However, this is entirely due to the narrowing of the bandwidths on cooling, since *A*-term intensity is critically dependent on the line width [see Eq. (42)]. The sharpening is largely complete on cooling to 100 K and, as shown in Figure 13(*b*), no increase intensity is observed between 38 and 16 K. Hence, the heme *c* in the nitrite reductase is low spin and diamagnetic with MCD arising from temperature-independent *A* terms. In contrast, heme d_1 is high spin, $S = 2$ ground state, with temperature-dependent MCD bands that arise from *C* terms and/or temperature-dependent *B* terms. The electronic spectrum of reduced heme d_1 therefore comprises a very broad Soret band spanning the region from 390 to 490 nm and a broad α band between 550 and 680 nm. The breadth of the transitions originates from removal of the *x*, *y* degeneracy that results from reduction of one of the pyrrole rings of the chlorin.

Magnetic circular dichroism spectroscopy has been extensively used to investigate the ground- and excited-state properties of heme-containing proteins as well as iron porphyrins.[16,17] Most recently, near-IR MCD has emerged as a reliable method, in conjunction with parallel EPR studies, for assessing the axial ligands of low-spin Fe(III) hemes.[23] Low-spin Fe(III) hemes exhibit charge-transfer transitions arising from the highest filled porphyrin molecular orbitals, $a_{1u}(\pi)$ and $a_{2u}(\pi)$, to the hole in the Fe(III) t_{2g} subshell (see Fig. 14). These transitions occur in the near-IR region and their energies are very sensitive to the nature of the axial ligands. However, they are difficult to detect reliably in the absorption spectra of hemoproteins due to overlap with the intense C–H, O–H,

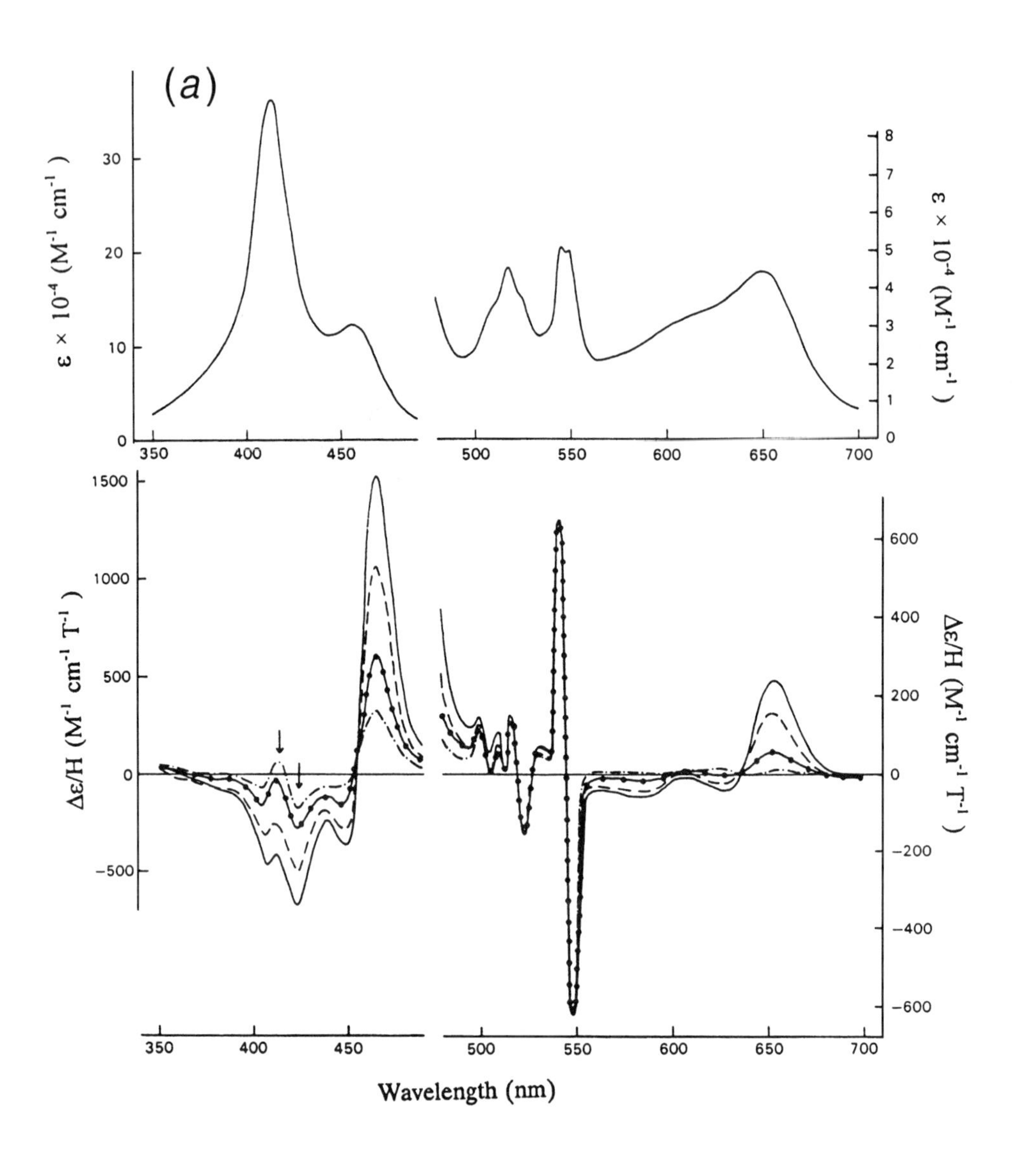

Figure 13

(*a*) Room-temperature absorption (top panel) and variable temperature MCD spectra (lower panel) of reduced *P. aeruginosa* nitrite reductase. The MCD spectra were recorded at 16 K (——), 27 K (- - -), 53 K (-•—•-), and 95 K (-·-·-). [Taken with permission from Walsh et al.[22]]
(*b*) Room-temperature absorption (top panel) and variable temperature MCD spectra (lower panel) of reduced *P. aeruginosa* cytochrome c_{551}. The MCD spectra were recorded at 16 K (——) and 38 K (-•—•-). [Taken with permission from Walsh et al.[22].]

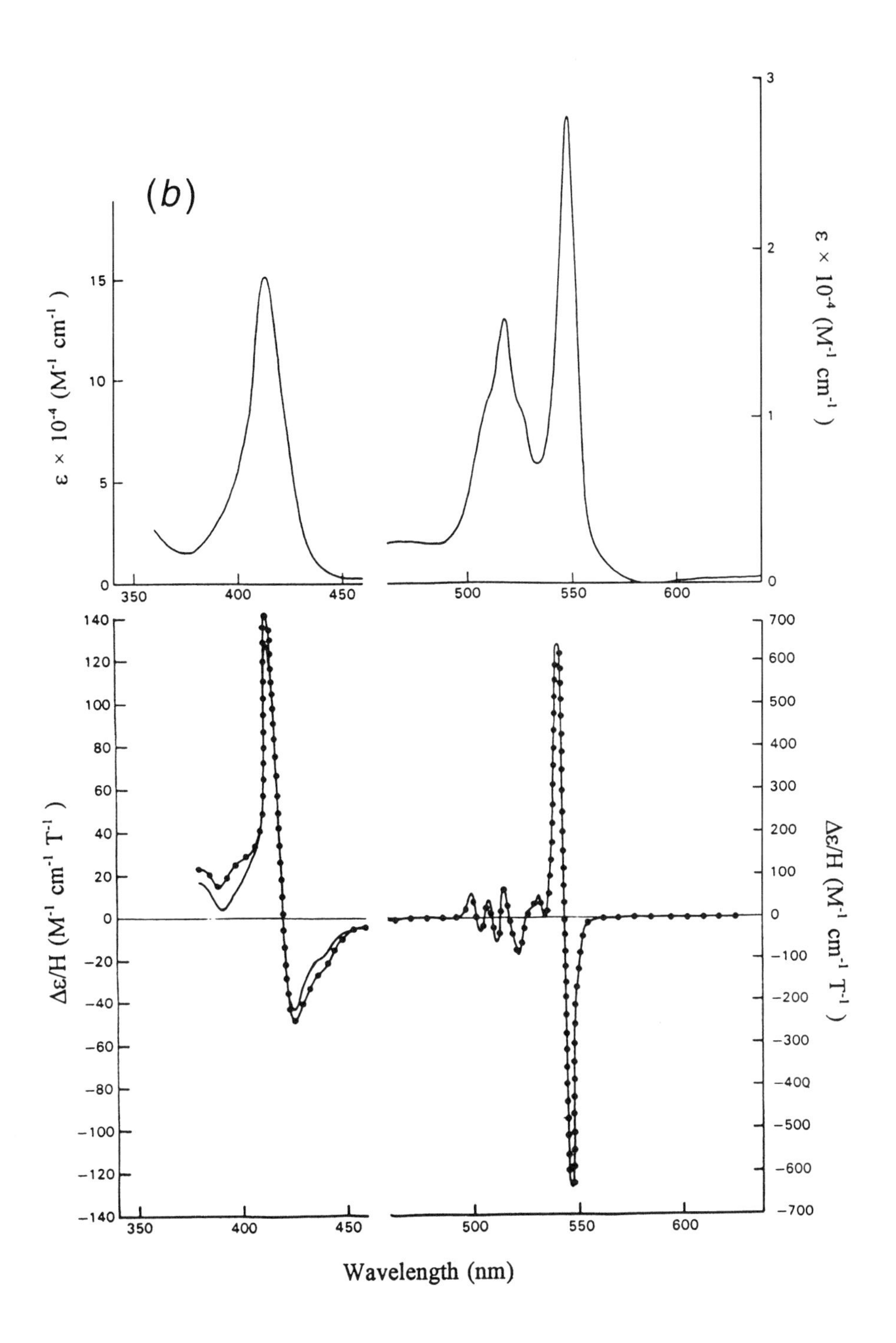
(b)
ε × 10⁻⁴ (M⁻¹ cm⁻¹)
ε × 10⁻⁴ (M⁻¹ cm⁻¹)
Δε/H (M⁻¹ cm⁻¹ T⁻¹)
Δε/H (M⁻¹ cm⁻¹ T⁻¹)
Wavelength (nm)

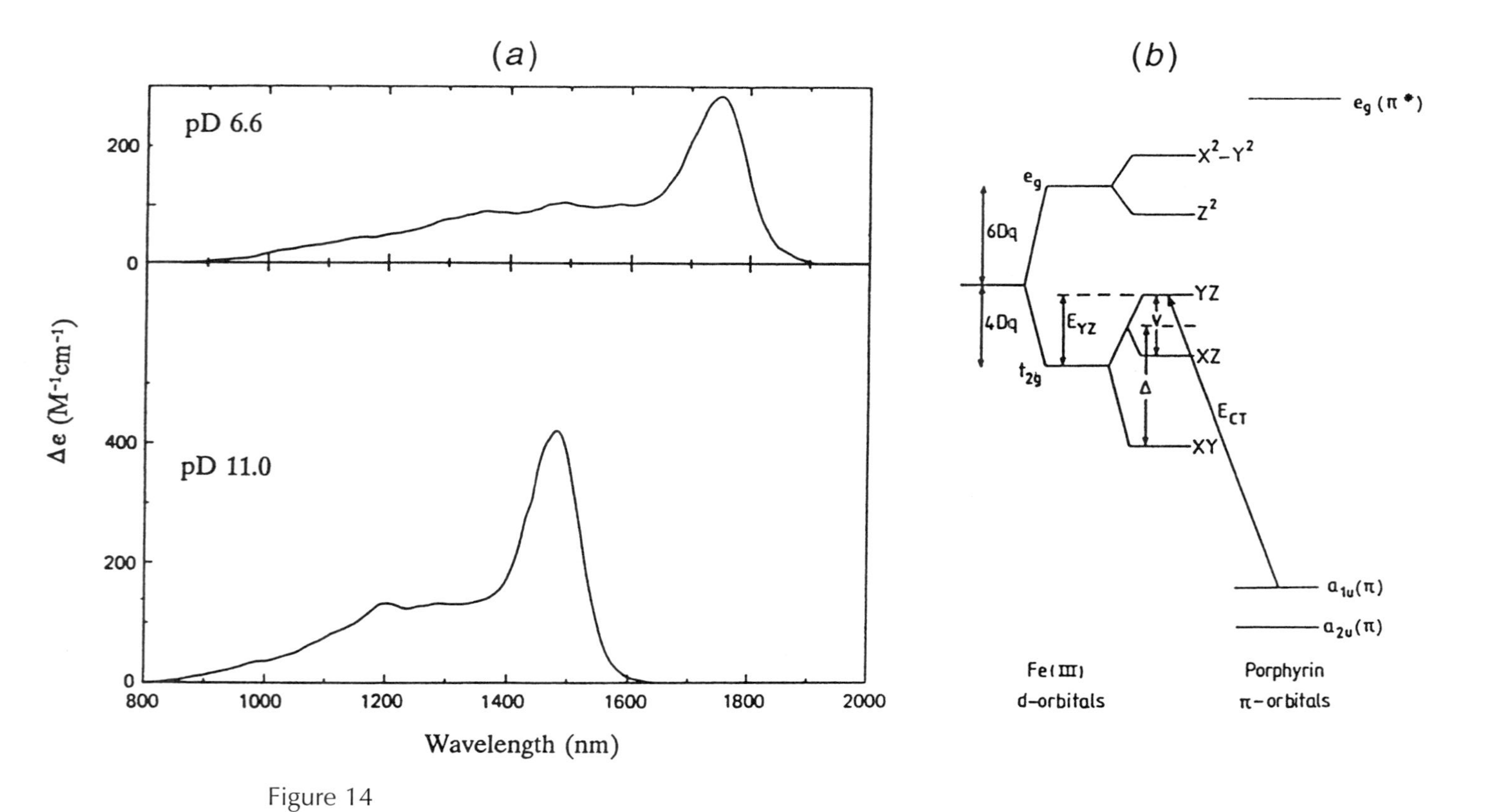

Figure 14
(*a*) Near-IR MCD spectra of horse heart cytochrome *c* at 4.2 K and 5.0 T. [Modified from Gadsby et al.[24] with permission.] (*b*) The d orbital energy levels of low-spin Fe(III) and the π, π^* orbitals of the porphyrin ring showing the near-IR transition between the highest filled porphyrin π levels and the d orbitals. [Reprinted with permission from Gadsby, P. M. A.; Thomson A. J. *J. Am. Chem. Soc.* **1990**, *112*, 5003–5011. Copyright © 1990 American Chemical Society.]

and N–H vibrational overtones of the polypeptide backbone. In contrast, they are readily detected in the MCD spectrum, since vibrational MCD is several orders of magnitude weaker than electronic MCD. Moreover, the MCD intensity arises from C terms, since low-spin Fe(III) hemes are paramagnetic with $S = \frac{1}{2}$ ground states, and is consequently greatly enhanced at low temperatures. The sensitivity of the energy of these transitions to the nature of the axial ligands is illustrated in Figure 14 by low-temperature near-IR MCD spectra for horse heart cytochrome c as a function of pD[f].[24] At neutral pD, cytochrome c is known to have methionine and histidine as the axial ligands, and the change in properties that occurs at alkaline pD values has been interpreted in terms of replacement of axial methionine ligand by an adjacent lysine. However, this interpretation was controversial prior to the MCD studies. The MCD studies reveal that a marked change in the energy of the porphyrin to iron charge-transfer transitions accompanies the alkaline transition. The energy of these transitions, as depicted by the MCD maximum of the lowest energy component, E_{CT}, has been found to correlate with the energy of Fe(III) hole relative to the baricenter of the t_{2g} 3d subshell, E_{YZ}. This correlation is the basis for the assignment of axial ligands for low-spin Fe(III) protoheme IX in proteins: methionine–histidine (E_{CT} 1740–1950 nm); histidine–histidine (or imidazole) (E_{CT} 1500–1630 nm); histidine–histidine (or imidazolate) (E_{CT} 1350–1370 nm); methionine–histidinate (E_{CT} 1550 nm); lysine–histidine (E_{CT} 1480–1550 nm); thiolate–histidine (E_{CT} 1035–1200 nm). Hence, the MCD data are clearly consistent with methionine–histidine and lysine–histidine axial ligation for the neutral ($E_{CT} = 1750$ nm) and alkaline ($E_{CT} = 1480$ nm) forms, respectively.

Iron–sulfur proteins have been another fruitful area for MCD investigations.[15] This class of metalloproteins contains one or more [2Fe–2S], [3Fe–4S], [4Fe–4S], or [8Fe–7S] cluster, with cysteinyl S generally completing tetrahedral coordination about Fe. The Fe–S clusters generally function in electron transport, although structural, regulatory or catalytic roles have been proposed and/or established in some cases. The UV–vis absorption spectra of iron–sulfur proteins are broad and featureless as a result of multiple overlapping S $\rightarrow$ Fe charge-transfer transitions and are usually of little diagnostic use in determining cluster type. In contrast, clusters with paramagnetic ground states [resulting from magnetic coupling of the high spin Fe(II) and Fe(III) ions] invariably exhibit temperature-dependent MCD spectra that are rich in detail and serve to illustrate the complexity of the electronic structure. For the simplest clusters, this has permitted electronic excited-state assignments. In general, the pattern of bands in the low-temperature MCD spectrum, coupled with the ground-state properties as revealed by MCD magnetization studies of discrete bands, provides a secure basis for assigning cluster type and oxidation state.

An example from our own work is provided by spinach glutamate synthase.[25] This monomeric enzyme ($M_r \approx 160,000$) contains one flavin mononucleotide (FMN), 3–4 non-heme iron and 4 acid labile S^{2-} per molecule. The absorption spectrum is dominated by transitions from the FMN and the nature

[f] Samples for near-IR MCD are usually prepared in deuterated buffers to minimize noise in the spectral region associated with intense C–H, O–H, and N–H vibrational overtones, 1400–1500 nm.

of the iron–sulfur cluster(s) was not known prior to MCD measurements. Variable-temperature MCD spectra and magnetization data for the isolated (OX) and dithionite-reduced (RED) forms of spinach glutamate synthase are shown in Figure 15. Clearly, a paramagnetic iron–sulfur cluster is present in both redox states and the saturation of the magnetization data at the lowest temperature investigated shows that the temperature dependence arises exclusively from *C* terms. Comparison with the low-temperature MCD spectra of structurally characterized iron–sulfur proteins, indicates that the intensity and pattern of bands observed for oxidized and reduced glutamate synthase consistent with one $[3Fe–4S]^{1+}$ and one $[3Fe–4S]^{0}$ cluster, respectively.

Further confirmation is obtained by analysis of magnetization data, which affords insight into the ground-state magnetic properties. All $[3Fe–4S]^{1+}$ clusters characterized thus far exhibit $S = \frac{1}{2}$ ground states as a result of antiferromagnetic coupling of three high-spin Fe(III) ions, whereas the one-electron reduced $[3Fe–4S]^{0}$ clusters have $S = 2$ ground states with $D < 0$ leaving a $M_s = \pm 2$ non-Kramers doublet lowest in cnergy (see inset in Fig. 15). The intercept value for the oxidized sample, $I = 0.51$ translates to an isotropic g value of 1.97, and hence is characteristic of a $S = \frac{1}{2}$ ground state. Parallel EPR studies revealed an $S = \frac{1}{2}$ EPR signal, $g_{\parallel} = 2.02$ and $g_{\perp} = 1.94$ and the experimental data are well fit by theoretical magnetization data computed for these g values using Eq. (48). For such small g value anisotropy, the magnetization data is insensitive to the polarization ratio, m_z/m_+, and acceptable fits can be obtained with values in the range -1 to $+1$. The reduced cluster magnetizes with a much steeper initial slope, which translates into a much smaller intercept value, $I = 0.19$. For a non-Kramers doublet with $g_{\parallel} = 4S$ and $g_{\perp} = 0.0$, this intercept value translates to $g_{\parallel} = 8.0$, and hence a ground-state spin $S = 2$. Moreover, the lowest temperature magnetization data is well fit by theoretical data computed using Eq. (50) with $g_{\parallel} = 8.0$ assuming a completely *xy*-polarized transition. The nesting that is observed in the magnetization data at higher temperatures results from population of higher zero-field components ($M_s = \pm 1$ and 0) and/or field-induced mixing of the levels of the ground-state manifold.

Nickel metalloproteins provide good illustrations of how variable-temperature MCD measurements enable estimates of ground-state zfs parameters and magnetic coupling constants. First, we consider Ni(II) substituted rubredoxin[26] (see Fig. 16). The close similarity between the electronic and magnetic properties of Ni(II) substituted rubredoxin and the structurally defined inorganic analog complex $[Ni(SPh)_4]^{2-}$, indicates a tetragonally distorted tetrathiolate coordination for Ni(II), resulting in a $^3A_2$ ground state under the effective D_{2d} symmetry. The intense absorption bands centered at 360 and 450 nm are attributed to S $\rightarrow$ Ni(II) charge-transfer transitions and the much weaker bands at 670 and 720 nm to components of the highest energy d–d band ($^3T_1 \rightarrow {}^3T_1$ under parent tetrahedral symmetry). Corresponding MCD bands with anomalous temperature dependence are found in the MCD spectrum. Below 15 K, the MCD spectrum is independent of temperature and linearly dependent on the magnetic field. Together with the dispersion characteristics, this shows that the MCD intensity arises from *B* terms originating from a nondegenerate level. The temperature

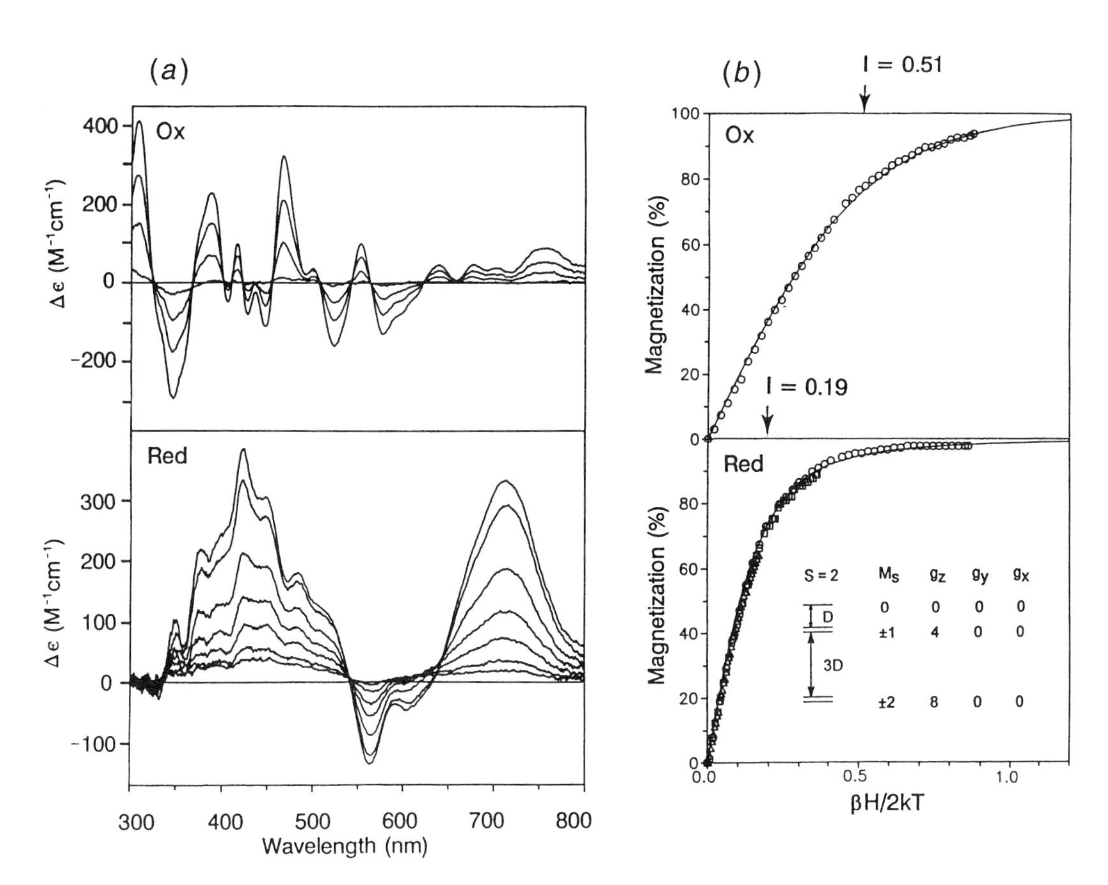

Figure 15
Variable-temperature MCD spectra and magnetization data for as isolated (ox) and dithionite-reduced (red) spinach glutamate synthase. (*a*) Magnetic circular dichroism at 4.5 T and 1.69, 4.22, 9.9, and 53 K for as isolated and 1.68, 4.22, 9.9, 17, 26, 53, and 97 K for dithionite reduced enzyme. All bands arise from *C* terms and increase in intensity with decreasing temperature. (*b*) The MCD magnetization 0–4.5 T; as isolated at 460 nm and 1.74 K, solid line is theoretical magnetization computed using Eq. (48) with $g_{\parallel} = 2.02$, $g_{\perp} = 1.94$, and $m_z/m_+ = -1$; dithionite reduced at 716 nm and 1.77 K (○), 4.22 K (□), 9.0 K (△), solid line is theoretical magnetization computed according to Eq. (50) with $g_{\parallel} = 8.0$. The arrows indicate the intercept values, *I*, for each magnetization plot. [Modified from Knaff et al.[25] with permission.]

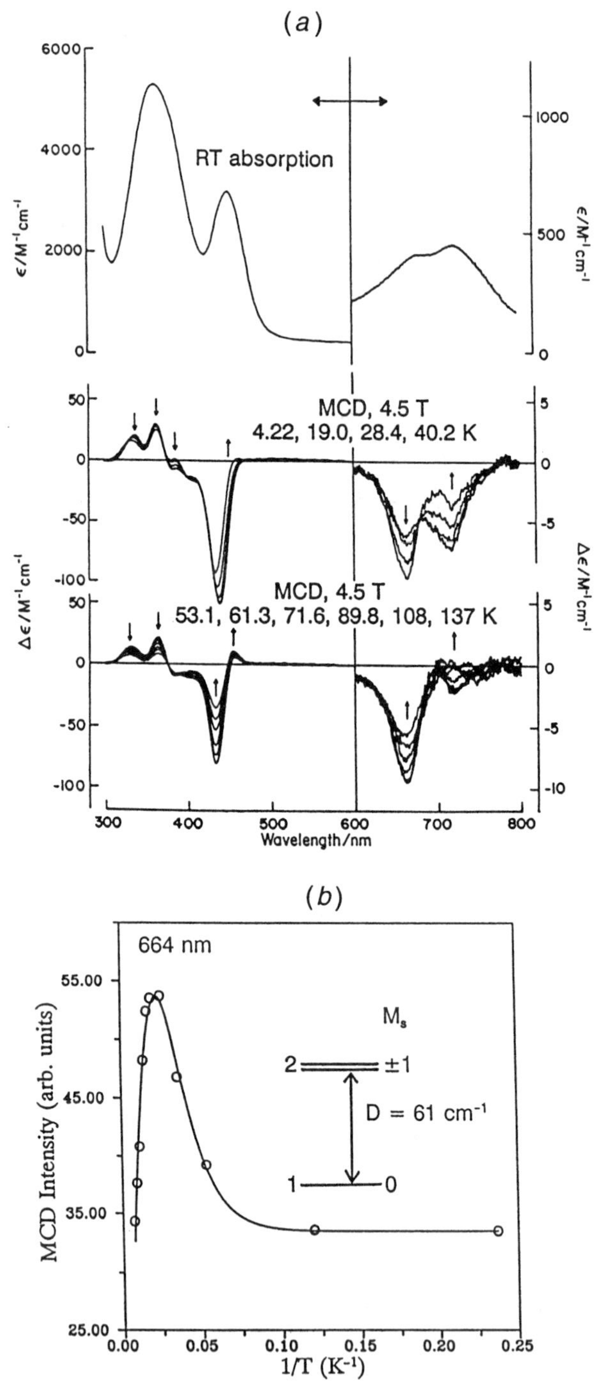

Figure 16
(*a*) Room-temperature absorption and variable-temperature MCD spectra of Ni(II) substituted rubredoxin from *Desulfovibrio gigas*. Arrows indicate the direction of change of MCD intensity with increasing temperature. [Reprinted with permission from Kowal A. T.; Zambrano, I. C.; Moura, I.; Moura; J. J. G.; Le Gall, J.; Johnson, M. K., *Inorg. Chem.* **198**, *27*, 1162–1166. Copyright © American Chemical Society.] (*b*) Temperature dependence of the MCD intensity of Ni(II) substituted rubredoxin at 664 nm. Solid line is best fit to Eq. (54). Best-fit parameters are $B_1 = 33.5$, $B_2 = -13.6$, $C_2 = 6233$ (arbitrary units), and $D = 61\ \text{cm}^{-1}$. The inset shows the zfs and labeling of the levels for the $^3A_2$ ground state (D_{2d} symmetry).

dependence with increasing temperature is strongly wavelength dependent. For some bands, the MCD intensity decreases with increasing temperature, while others, such as the 664-nm band, show an initial increase and subsequent decrease at higher temperatures [see Fig. 16(*b*)]. This type of behavior indicates population of a low-lying degenerate level and the wavelength variability results from differences in the magnitude and signs of the B and C terms from the degenerate and nondegenerate levels.

Large zfs in the $^3A_2$ ground state is anticipated as a result of spin–orbit coupling in the parent $^3T_1$ state. Consequently, the MCD data indicate positive axial zfs leaving the $M_s = 0$ state lowest in energy with the near degenerate $M_s = \pm 1$ doublet separated in energy by D, the axial zfs parameter [see Fig. 16(*b*)]. The value of D is readily determined by nonlinear least-squares fitting of the temperature dependence to

$$\begin{aligned} \Delta\varepsilon(\text{au}) &= \alpha_1 B_1 + \alpha_2(B_2 + C_2/kT) \\ \alpha_1 &= 1/[1 + 2\exp(-D/kT)] \\ \alpha_2 &= 2\exp(-D/kT)/[1 + 2\exp(-D/kT)] \end{aligned} \tag{54}$$

where B_1 is the B term from the nondegenerate $M_s = 0$ level and B_2 and C_2 are the B and C terms, respectively, from the upper $M_s = \pm 1$ "doublet". At 664 nm, the best-fit parameters were $D = 61\ \text{cm}^{-1}$, $B_1 = 33.5$, $B_2 = -13.6$, and $C_2 = 6{,}233$ (au). Hence, the anomalous temperature dependence at this wavelength arises from a large C term contribution that has the same sign as the B-term contribution from the lower level. It is also noteworthy that $B_1 \approx -2B_2$, which suggests that the dominant B-term contribution at this wavelength results from field-induced mixing of the $M_s = 0$ and $M_s = \pm 1$ zero-field components (i.e., a temperature dependent B term). Fits at other wavelengths afforded values of D within $\pm 5\%$ of the value at 664 nm, with markedly different values and relatively signs for the B- and C-term contributions.

In some instances, the temperature-dependent MCD spectra for high-spin Ni(II) centers with positive axial zfs can be almost exclusively dominated by temperature-dependent B terms. A good example is provided by the Ni(II) tetrapyrrole cofactor, F_{430}, in methyl-CoM reductase.[27] Spectroscopic evidence indicates that F_{430} is six coordinate in the intact enzyme. In accord with this, the MCD associated with the $\pi \rightarrow \pi^*$ transitions centered around 430 nm is temperature dependent, indicating that the Ni(II) is high spin, $S = 1$. While the temperature-dependence behavior, that is, saturating as a function of $1/T$ at fixed applied fields [see inset in Fig. 17(*b*)], is characteristic of C terms or temperature-dependent B terms, the almost linear dependence on magnetic field at fixed low temperatures [see Fig. 17(*b*)] shows that a nondegenerate state is lowest in energy. Coupled with the observation of negligible MCD intensity on extrapolation to infinite temperature, this leads to the conclusion that a temperature-dependent B term arising from the mixing of $M_s = 0$ and $M_s = \pm 1$ zero-field components is the dominant contributor to the low-temperature MCD intensity. Analysis of the data to yield an estimate of the axial zfs parameter, D, involves fitting sets of

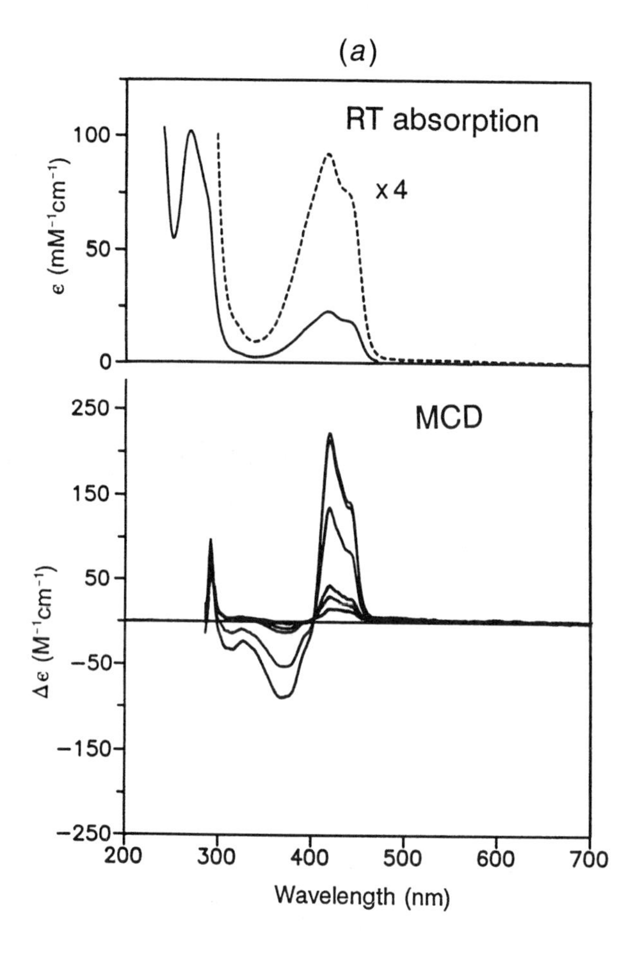

(a)
RT absorption
x 4
ε (mM⁻¹cm⁻¹)
100
50
0
MCD
250
150
50
−50
−150
−250
Δε (M⁻¹cm⁻¹)
200
300
400
500
600
700
Wavelength (nm)

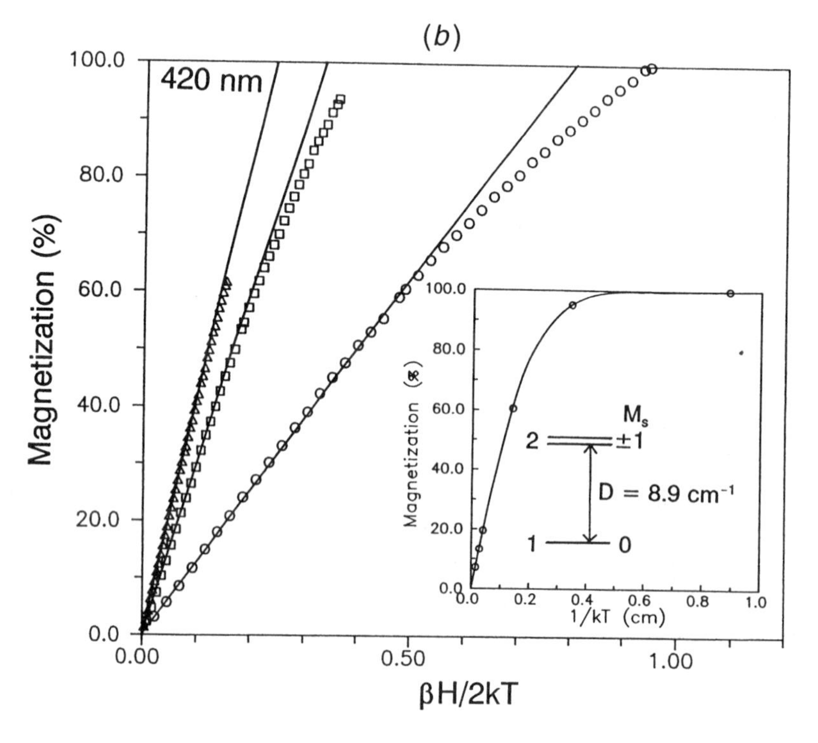

(b)
420 nm
Magnetization (%)
100.0
80.0
60.0
40.0
20.0
0.0
0.00
0.50
1.00
βH/2kT
Magnetization (%)
1/kT (cm)
M_s
2
±1
D = 8.9 cm⁻¹
1
0

temperature-dependence data at fixed applied fields to Eq. (54). A typical set of data obtained at 3 T is shown in the inset of Figure 17(*b*). The best-fit parameters were $D = 8.9\ \mathrm{cm}^{-1}$, $B_1 = 223$, $B_2 = -114$, and $C_2 = 991$ (au), which shows that $B_1 \simeq -2B_2$ as required by a temperature-dependent B term involving the mixing of a nondegenerate state with a low-lying doubly degenerate state. The value of D was independent of the applied field and is consistent with octahedral coordination [D is typically 0–10 cm^{-1} for octahedral Ni(II) complexes]. Using Eq. (54), it is possible to estimate the relative contributions of the B and C terms to the total MCD intensity as a function of temperature. This procedure leads to the conclusion that the C-term contribution is less than 16% of the total MCD at any temperature.

The final example shows how MCD was used to address the nature of the active site Ni(II) center in urease, that is, monomer or dimer. Subsequent crystallographic studies of *Klebsiella aerogenes* urease have confirmed that the active is dimeric and shown that the binickel center is bridged by a carbamylated lysine. At the time of the initial MCD investigations, Jack bean urease, a hexamer of identical 90-kDa subunits each containing one catalytic site and two Ni(II) ions, was the best studied example. Variable-temperature MCD studies of this enzyme provide a means of identifying Ni(II) d–d transitions that are obscured by the light scattering background of this globular protein and probing the ground-state magnetic properties of the active site Ni(II) center.[28] The MCD spectra for the 2-mercaptoethanol inhibited form are shown in Figure 18. By comparison with the low-temperature MCD of the native enzyme and numerous octahedral Ni(II) coordination complexes, the temperature-dependent MCD bands centered at 422 and 750 nm were initially assigned to the two highest energy d–d bands under approximately octahedral symmetry, ${}^3A_{2g} \rightarrow {}^3T_{1g}(P)$ and ${}^3A_{2g} \rightarrow {}^3T_{1g}(F)$, respectively. However, subsequent studies of both the native and in the 2-mercaptoethanol inhibited form in the near-IR region failed to locate an intense negative MCD band corresponding to the lowest energy, ${}^3A_{2g} \rightarrow {}^3T_{2g}$ d–d band, predicted to occur at about 1300 nm based on the ligand field analysis. Such a band has been observed in all octahedral Ni(II) complexes investigated to date. This observation coupled with the anomalously high MCD intensity of native urease in comparison to octahedral Ni(II) complexes has lead us to reassign these

Figure 17 (*facing page*)
(*a*) Room-temperature absorption and variable-temperature MCD spectra of *Methanobacterium thermoautotrophicum* (strain ΔH) methyl-CoM reductase. The MCD spectra were recorded at 1.62, 4.22, 10.2, 36, 51, and 98 K, with a magnetic field of 4.5 T. All MCD bands increase in intensity with decreasing temperature. (*b*) The MCD magnetization data for methyl-CoM reductase. Wavelength, 420 nm; temperatures, 1.61 K (○), 4.22 K (□), 9.9 K (△); magnetic fields, 0–4.5 T. The inset shows a plot of percent magnetization *versus* $1/kT$ at 3.0 T and the zfs and labeling of the $S = 1$ ground state. The solid line is the best fit to Eq. (54). Best-fit parameters are $B_1 = 223$, $B_2 = -114$, $C_2 = 991$ (au), and $D = 8.9\ \mathrm{cm}^{-1}$. [Modified from Hamilton et al.[27] with permission.]

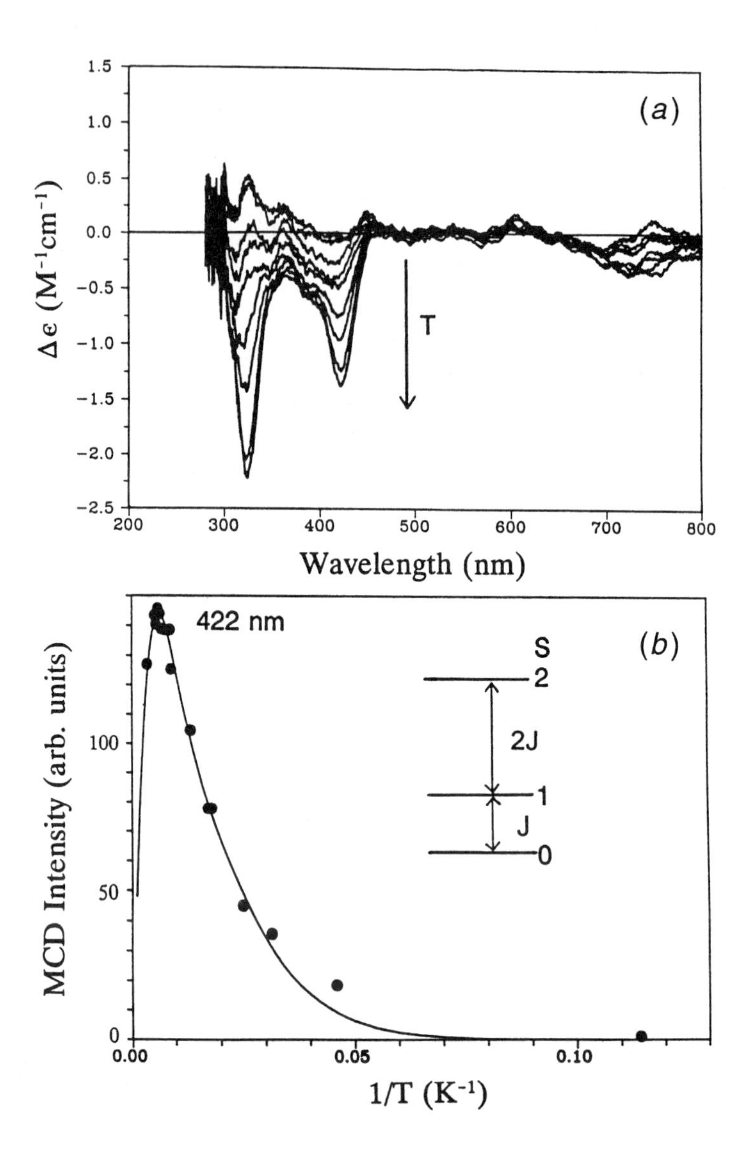

Figure 18
(a) Variable-temperature MCD spectra of 2-mercaptoethanol inhibited Jack bean urease. The MCD spectra were recorded at 4.2, 8.8, 22, 32, 40, 56, 75, 115, and 150 K with a magnetic field of 4.5 T. As indicated by the arrow, the MCD intensity of all bands increases with *increasing* temperature. (b) Temperature dependence of the MCD intensity of 2-mercaptoethanol inhibited Jack bean urease at 422 nm. Solid line is best fit to Eq. (55), using $C_1 = 9084$, $C_2 = 61124$ (au), and $J = 80\ \mathrm{cm}^{-1}$. The inset shows the energy level scheme for the antiferromagnetically coupled binuclear Ni(II) center. [Reprinted with permission from Finnegan, M. G.; Kowal, A. T.; Werth, M. T.; Clark, P. A.; Wilcox, D. E.; Johnson, M. K. *J. Am. Chem. Soc.* **1991**, *113*, 4030–4032. Copyright © 1991 American Chemical Society.]

d–d bands in terms of five-coordinate Ni(II) center.[29] In contrast, the band at 324 nm is not observed in the native enzyme and is assigned to a S $\rightarrow$ Ni(II) charge-transfer band arising from direct thiolate binding to the nickel. By facilitating identification of the ligand field bands, the MCD studies demonstrate that 2-mercaptoethanol binds directly at the nickel with retention of Ni coordination geometry and minimal perturbation of the ligand field.

Of particular interest is the anomalous temperature-dependence behavior of the MCD bands for 2-mercaptoethanol inhibited urease (see Fig. 18). Below 10 K, the sample exhibits negligible MCD intensity; the spectra at 4.22 and 8.8 K are almost superimposed. The observed bands all *increase* in intensity with increasing temperature up to 150 K before decreasing slightly as the temperature approaches 273 K. Such behavior can only be reconciled with a nondegenerate state lowest in energy with degenerate state(s) becoming populated at higher temperatures. Assuming a monomeric Ni(II) center with $D > 0$, the data was fit to Eq. (54). However, the fits were very poor and gave values of D between 60 and 100 cm^{-1}; much greater than the axial zfs expected for a five- or six-coordinate Ni(II) center. The alternative possibility is an antiferromagnetically coupled binuclear Ni(II) center. The resultant energy levels using an isotropic exchange Hamiltonian, $\mathscr{H} = J\hat{S}_1 \cdot \hat{S}_2$, are shown as an inset in Figure 18(*b*). Since the MCD intensity at low temperatures is negligible, B terms can be ignored to a first approximation and the appropriate expression for analyzing the temperature dependence is

$$\Delta\varepsilon(\mathrm{au}) = \alpha_1 C_1/kT + \alpha_2 C_2/kT \tag{55}$$

$$\alpha_1 = 3\exp(-J/kT)/[1 + 3\exp(-J/kT) + 5\exp(-3J/kT)]$$

$$\alpha_2 = 5\exp(-3J/kT)/[1 + 3\exp(-J/kT) + 5\exp(-3J/kT)]$$

where α_1, C_1 and α_2, C_2 are the fractional populations and C terms from the threefold degenerate $S = 1$ and fivefold degenerate $S = 2$ levels, respectively. Nonlinear least-squares fitting to Eq. (55) gave $J = 80 \pm 10$ cm^{-1} and the best fit is shown in Figure 18(*b*). The large value of J justifies the neglect of zfs of the $S = 1$ and 2 levels and indicates magnetic exchange coupling via a bridging ligand. The MCD data therefore provided strong evidence in favor of a binuclear Ni(II) active site in urease and an assessment of the magnitude and type of the magnetic coupling in the 2-mercaptoethanol inhibited derivative.

All of the above example of CD and MCD spectroscopy have concerned elucidation of properties of static molecular structures corresponding to stable, equilibrium species. Clearly, it would useful to apply these techniques in addition to absorption measurements to studying the properties of transient reaction intermediates or nonequilibrium conformations and the kinetics of dynamical behavior. The recent development of CD and MCD instrumentation that is capable of nanosecond time resolution is therefore an important advance. This technique has been applied to studying the photolytic dissociation and rebinding of CO to heme proteins such as hemoglobin, myoglobin, and cytochrome *c* oxidase. The interested reader is referred to articles by Goldbeck and co-workers[30,31] for a description of the instrumentation and a discussion of the applications of time-resolved CD and MCD spectroscopy.

ACKNOWLEDGMENTS

The preparation of this chapter was supported by grants from the National Institutes of Health (GM-33806 and GM-51962) and the National Science Foundation (DMB-8921986, MCB-9419019, and DBI-9413236). Special thanks are due to Michael G. Finnegan for help in preparing figures, Kristi L. Kiick for critical reading of the manuscript, and the collaborators who supplied the samples used for spectroscopic investigations.

References

1. Moscowitz, A. "Theory and Analysis of Optical Rotatory Dispersion Curves" In *Optical Rotatory Dispersion*; Djerassi, C., Ed.; McGraw-Hill: New York, 1960; pp 150–177
2. Kuhn, W. "Optical Rotatory Power" *Annu. Rev. Phys. Chem.* **1958**, *9*, 417–448.
3. Richardson, F. S. "Theory of Optical Activity in the Ligand-Field Transitions of Chiral Transition Metal Complexes" *Chem. Rev.* **1979**, *79*, 17–36.
4. Mason, S. F. *Molecular Optical Activity and the Chiral Discriminations*; Cambridge University Press: Cambridge, UK, 1982; pp 103–137.
5. Schipper, P. E.; Rodger, A. "Generalized Selection Rules for Circular Dichroism: A Symmetry-Adapted Perturbation Model for Magnetic Dipole Allowed Transitions" *Chem. Phys.* **1986**, *109*, 173–193.
6. Fenton, N. D.; Gerloch, M. "Chirality and Bonding in the First Coordination Shells of Three CoN_2Cl_2 Chromophores: Quantitative Reproduction of Rotatory Strengths" *Inorg. Chem.* **1990**, *29*, 3718–3726.
7. Eaton, W. A.; Lovenberg, W. "The Iron–Sulfur Complex in Rubredoxin" In *Iron–Sulfur Proteins, Vol II*; W. Lovenberg, Ed.; Academic: New York, 1973; pp 131–162.
8. Stephens, P. J. "Magnetic Circular Dichroism" *Adv. Chem, Phys.* **1976**, *35*, 197–264.
9. Schatz, P. N.; McCaffery, A. J. "The Faraday Effect" *Q. Rev. Chem. Soc.* **1969**, *23*, 552–584.
10. Piepho, S. B.; Schatz, P. N. *Group Theory in Spectroscopy with Applications to Magnetic Circular Dichroism*; Wiley, New York, 1983.
11. Schatz, P. N.; Mowery, R. L.; Krausz, E. R. "M.C.D./M.C.P.L. Saturation Theory with Applications to Molecules in $D_{\infty h}$ and its Subgroups" *Mol. Phys.* **1978**, *35*, 1537–1557.
12. Thomson, A. J.; Johnson, M. K. "Magnetization Curves of Haemoproteins Measured by Low-Temperature Magnetic-Circular-Dichroism Spectroscopy" *Biochem. J.* **1980**, *191*, 411–420.
13. Pavel, E. G.; Solomon, E. I. "Recent Advances in Magnetic Circular Dichroism Spectroscopy" In *Spectroscopic Methods in Bioinorganic Chemistry*; Solomon, E. I. and Hodgson, K. O., Eds.; American Chemical Society Books Symposium Series, Vol. 692, Washington, DC; 1998; pp 119–135.
14. Browett, W. R.; Fucaloro, A. F.; Morgan, T. V.; Stephens, P. J. "Magnetic Circular Dichroism Determination of Zero field Splitting in Chloro(*meso*-tetraphenylporphinato)-iron(III)" *J. Am. Chem. Soc.* **1983**, *105*, 1868–1872.
15. Johnson, M. K.; Robinson, A. E.; Thomson, A. J. "Low-Temperature Magnetic Circular Dichroism of Iron–Sulfur Proteins" In *Iron–Sulfur Proteins*; Spiro, T. G., Ed.; Wiley, New York, 1982; pp 367–406
16. Dawson, J. H.; Dooley, D. M. "Magnetic Circular Dichroism Spectroscopy of Iron Porphyrins and Heme Proteins" In *Iron Porphyrins, Part 3*; Lever, A. B. P., Gray, H. B., Eds.; VCH Publishers, New York, 1989; pp 1–92.
17. Cheesman, M. R.; Greenwood, C.; Thomson, A. J. "Magnetic Circular Dichroism of Hemoproteins" In *Advances in Inorganic Chemistry, Vol. 36*; Academic, San Diego, CA, 1991; pp 203–255.
18. Dooley, D. M.; Dawson, J. H. "Bioinorganic Applications of Magnetic Circular Dichroism Spectroscopy: Copper, Rare-Earth Ions, Cobalt and Non-Heme Iron Systems" *Coord. Chem. Rev.* **1984**, *60*, 1–66.
19. Solomon, E. I.; Pavel, E. G.; Loeb, K. E.; Campochiaro, C. "Magnetic Circular Dichroism Spectroscopy as a Probe of the Geometric and Electronic Structure of Non-heme Ferrous enzymes" *Coord. Chem. Rev.* **1995**, *144*, 369–460.
20. Barrett, C. P.; Peterson, J.; Greenwood, C.; Thomson, A. J. "Optical Detection of Paramagnetic Resonance by Magnetic Circular Dichroism. Applications to Aqueous Solutions of Metalloproteins" *J. Am. Chem. Soc.* **1986**, *108*, 3170–3177.
21. Stephens, P. J. "The Faraday Rotation of Allowed Transitions: Charge-Transfer Transitions in $K_3Fe(CN)_6$" *Inorg. Chem.* **1965**, *4*, 1690–1692.

22. Walsh, T. A.; Johnson, M. K.; Greenwood, C.; Barber, D.; Springall, J. P.; Thomson, A. J. "Some Magnetic Properties of *Pseudomonas* Cytochrome Oxidase" *Biochem. J.* **1979**, *177*, 29–39.
23. Gadsby, P. M. A.; Thomson, A. J. "Assignment of the Axial Ligands of Ferric Ion in Low-Spin Hemoproteins by Near-Infrared Magnetic Circular Dichroism and Electron Paramagnetic Resonance Spectroscopy" *J. Am. Chem. Soc.* **1990**, *112*, 5003–5011.
24. Gadsby, P. M. A.; Peterson, J.; Foote, N.; Greenwood, C.; Thomson, A. J. "Identification of the Ligand-Exchange Process in the Alkaline Transition of Horse Heart Cytochrome *c*" *Biochem. J.* **1987**, *246*, 43–54.
25. Knaff, D. B.; Hirasawa, M.; Ameyibor, E.; Fu, W.; Johnson, M. K. "Spectroscopic Evidence for a [3Fe–4S] Cluster in Spinach Glutamate Synthase" *J. Biol. Chem.* **1991** *266*, 15080–15084.
26. Kowal, A. T.; Zambrano, I. C.; Moura, I.; Moura, J. J. G.; LeGall, J.; Johnson, M. K. "Electronic and Magnetic Properties of Nickel-Substituted Rubredoxin: A Variable-Temperature Magnetic Circular Dichroism Study" *Inorg. Chem.* **1987**, *27*, 1162–1166.
27. Hamilton, C. L.; Scott, R. A.; Johnson, M. K. "The Magnetic and Electronic Properties of *Methanobacterium thermoautotrophicum* (Strain ΔH) Methyl Coenzyme M Reductase and its Nickel Tetrapyrrole Cofactor F_{430}" *J. Biol. Chem.* **1989**, *264*, 11605–11613.
28. Finnegan, M. G.; Kowal, A. T.; Werth, M. T.; Clark, P. A.; Wilcox, D. E.; Johnson, M. K. "Variable-Temperature Magnetic Circular Dichroism Spectroscopy as a Probe of the Electronic and Magnetic Properties of Nickel in Jack Bean Urease" *J. Am. Chem. Soc.* **1991**, *113*, 4030–4032.
29. Park. I.-S.; Michel, L. O.; Pearson, M. A.; Jabri, E.; Karplus, P. A.; Wang, S.; Dong, J.; Scott, R. A.; Koehler, B. P.; Johnson, M. K.; Hausinger, R. P. "Characterization of the Mononickel Metallocenter in H1324A Mutant Urease" *J. Biol. Chem.* **1996**, *271*, 18632–18637.
30. Goldbeck, R. A.; Kliger, D. S. "Natural and Magnetic Circular Dichroism Spectroscopy on the Nanosecond Timescale" *Spectroscopy* **1992**, *7*, 17–29.
31. Goldbeck, R. A.; Kimshapiro, D. B.; Kliger, D. S. "Fast Natural and Magnetic Circular Dichroism Spectroscopy" *Annu. Rev. Phys. Chem.* **1997**, *48*, 453–479.

General References

CIRCULAR DICHROISM, GENERAL

"Circular Dichroism: Principles and Applications", Nakanishi, K., Berova, N., and Woody, R. W., Eds., VCH Publishers, New York, 1994.

CIRCULAR DICHROISM OF TRANSITION METAL COMPLEXES

Richardson, F. S. "Theory of Optical Activity in the Ligand-Field Transitions of Chiral Transition Metal Complexes" *Chem. Rev.* **1979**, *79*, 17–36.

Mason, S. F. Molecular Optical Activity and the Chiral Discriminations; Cambridge University Press: Cambridge, UK, 1982; pp 103–137.

Schipper, P. E.; Rodger, A. "Generalized Selection Rules for Circular Dichroism: A Symmetry-Adapted Perturbation Model for Magnetic Dipole Allowed Transitions" *Chem. Phys.* **1986**, *109*, 173–193.

ULTRAVIOLET CIRCULAR DICHROISM OF PROTEINS AND POLYPETIDES

Greenfield, N. J. "Methods to Estimate the Conformation of Proteins and Polypeptides from Circular Dichroism Data" *Anal. Biochem.* **1996**, *235*, 1–10.

PRACTICAL ASPECTS OF MAGNETIC CIRCULAR DICHROISM

Johnson, M. K. "Variable Temperature Magnetic Circular Dichroism of Metalloproteins" In Metal Clusters in Proteins, Que, L., Jr., Ed. ACS Symposium Series Vol. 372, American Chemical Society, Washington, DC, 1988; pp 326–342.

Thomson, A. J.; Cheeseman, M. R.; George, S. J. "Variable Temperature Magnetic Circular Dichroism" *Methods Enzymol.* **1993**, *226*, 199–232.

MAGNETIC CIRCULAR DICHROISM THEORY (LISTED IN ORDER OF INCREASING LEVEL OF SOPHISTICATION)

Schatz, P. N.; McCaffery, A. J. "The Faraday Effect" *Q. Rev. Chem. Soc.* **1969**, *23*, 552–584.

Stephens, P. J. "Magnetic Circular Dichroism" *Adv. Chem, Phys.* **1976**, *35*, 197–264.

Piepho, S. B.; Schatz, P. N. Group Theory in Spectroscopy with Applications to Magnetic Circular Dichroism; Wiley, New York, 1983.

MAGNETIC CIRCULAR DICHROISM SATURATION MAGNETIZATION THEORY

Schatz, P. N.; Mowery, R. L.; Krausz, E. R. "M.C.D./M.C.P.L. Saturation Theory with Applications to Molecules in $D_{\infty h}$ and its Subgroups" *Mol. Phys.* **1978**, *35*, 1537–1557.

Thomson, A. J.; Johnson, M. K. "Magnetization Curves of Haemoproteins Measured by Low-Temperature Magnetic-Circular-Dichroism Spectroscopy" *Biochem. J.* **1980**, *191*, 411–420.

Bennett, D. E.; Johnson, M. K. "The Electronic and Magnetic Properties of Rubredoxin: A Low Temeprature Magnetic Circular Dichroism Study" *Biochim. Biophys. Acta* **1987**, *911*, 71–80.

Pavel, E. G.; Solomon, E. I. "Recent Advances in Magnetic Circular Dichroism Spectroscopy" In Spectroscopic Methods in Bioinorganic Chemistry; E. I. Solomon and K. O. Hodgson, Eds.; American Chemical Society Books Symposium Series, Vol. 692, Washington, DC; 1998; pp 119–135.

REVIEWS ON SPECIFIC BIOINORGANIC TOPICS

Johnson, M. K.; Robinson, A. E.; Thomson, A. J. "Low-Temperature Magnetic Circular Dichroism of Iron–Sulfur Proteins" In Iron–Sulfur Proteins; T. G. Spiro, Ed.; Wiley, New York, 1982; pp 367–406.

Dawson, J. H.; Dooley, D. M. "Magnetic Circular Dichroism Spectroscopy of Iron Porphyrins and Heme Proteins" In Iron Porphyrins, Part 3; A. B. P. Lever, H. B. Gray, H. B., Eds.; VCH Publishers, New York, 1989; pp 1–92.

Cheesman, M. R.; Greenwood, C.; Thomson, A. J. "Magnetic Circular Dichroism of Hemoproteins" In Advances in Inorganic Chemistry, Vol. 36; Academic, San Diego, 1991; pp 203–255.

Dooley, D. M.; Dawson, J. H. "Bioinorganic Applications of Magnetic Circular Dichroism Spectroscopy: Copper, Rare-Earth Ions, Cobalt and Non-Heme Iron Systems" *Coord. Chem. Rev.* **1984**, *60*, 1–66.

Solomon, E. I.; Pavel, E. G.; Loeb, K. E.; Campochiaro, C. "Magnetic Circular Dichroism Spectroscopy as a Probe of the Geometric and Electronic Structure of Non-heme Ferrous Enzymes" *Coord. Chem. Rev.* **1995**, *144*, 369–460.

Landrum, G. A.; Ekberg, C. A.; Whiltaker, J. W. "A Ligand-Field Model for MCD Spectra of Biological Cupric Complexes" *Biophys. J.* **1995**, *69*, 674–689.

6

Aspects of ^{57}Fe Mössbauer Spectroscopy

ECKARD MÜNCK

Department of Chemistry
Carnegie Mellon University
Pittsburgh, PA 15513

I. Introduction

Ever since the discovery of recoilless nuclear gamma resonance by Rudolf Mössbauer in 1958, the technique, known as Mössbauer spectroscopy, has developed into a powerful tool to investigate the electronic structure of compounds with high resolution. The Mössbauer effect, which requires a nucleus with low-lying excited states, has been observed for 43 elements. The technique has been employed to study the electronic environment of Mössbauer-active atoms in metals and insulators by probing, for example, magnetism, electronic relaxation, chemical bonding, and redox reactions. By far the most utilized Mössbauer nucleus is ^{57}Fe, and for that reason this chapter will focus entirely on this isotope.

This chapter will focus on those aspects of Mössbauer spectroscopy that have played a decisive role in the characterization of iron-containing compounds in bioinorganic chemistry and metallobiochemistry. These aspects concern the properties of paramagnetic centers as observed for mononuclear sites and exchange-coupled dinuclear clusters by election paramagnetic resonance (EPR) and Mössbauer spectroscopy. Rather than describing and tabulating typical Mössbauer data of high-spin, intermediate-spin, or low-spin Fe(II), Fe(III), or Fe(IV) compounds, we will pick a high-spin ($S = \frac{5}{2}$) iron(III) site and describe typical spectra that have been observed for such a site. While Mössbauer spectroscopy can well stand on its own, its power is substantially increased when it is applied in combination with EPR. After discussing the connections between the two techniques and describing spectra of an Fe(III) site, we will couple an Fe(III) site ferromagnetically and antferromagnetically to an Fe(II) site. Such spin-coupled pairs are presently under intensive study in many laboratories with bioinorganic and biochemical interests. While the choice of material, at first sight, may not seem to be entirely satisfactory to the reader, it is hoped that this chapter will facilitate access to a very active field of the current research literature.

It is important to know that the Mössbauer phenomenon rests on the fact that γ radiation (high-energy photons) can be emitted or absorbed without imparting recoil energy ($E_R = E_\gamma^2/2M = 1.95 \times 10^{-3}$ eV, where $E_\gamma = 14.4$ KeV and M is the mass of the ^{57}Fe nucleus). In a solid, most of the recoil energy is converted into lattice vibration energy. Mössbauer has shown, however, that there is a certain probability, described by the recoil-free fraction f, that γ emission and absorption take place without recoil. In order to observe the Mössbauer effect, the ^{57}Fe nucleus must be placed in a solid or frozen solution matrix. Among the many texts[1–4] that deal with the fundamentals of this technique, the book by Gütlich et al.[1] is highly recommended.

II. Some Background on Basic Interactions

Iron-57 is a stable isotope with 2.2% natural abundance. In ^{57}Fe Mössbauer spectroscopy, we observe transitions between the nuclear ground state of ^{57}Fe (nuclear spin $I_g = \frac{1}{2}$; nuclear g factor, $g_g = 0.181$) and a nuclear excited state at 14.4 KeV ($I_e = \frac{3}{2}, g_e = -0.106$, nuclear quadrupole moment Q) (Fig. 1). For the bare nucleus, the nuclear ground and excited states would exhibit twofold and fourfold degeneracies, respectively, and only a single Mössbauer transition would be observed. If the nucleus is embedded in an electronic environment of symmetry lower than spherical, cubic, or tetrahedral, the degeneracy of the nuclear excited state is partially lifted by the quadrupole interaction between the nuclear quadrupole moment Q and the electric field gradient tensor $\mathbf{V}$ generated by the surrounding charges at the site of the ^{57}Fe nucleus. The electric field at the nucleus is the negative gradient of the potential, $V(x, y, z)$, and the electric field gradient (EFG) tensor is a 3×3 second-rank tensor, with element V_{ij}, formed by taking the spatial derivatives of the electric field. However, only five of the nine components are independent. Mixed derivatives, such as $\partial^2 V/\partial x \partial y = \partial^2 V/\partial y \partial x$, are the same, so the EFG tensor is symmetric. It can be shown that s electrons do not contribute to the EFG and, therefore, Laplace's equation evaluated at the nucleus requires that the EFG tensor is traceless, (i.e., $\sum V_{ii} = 0$). In an

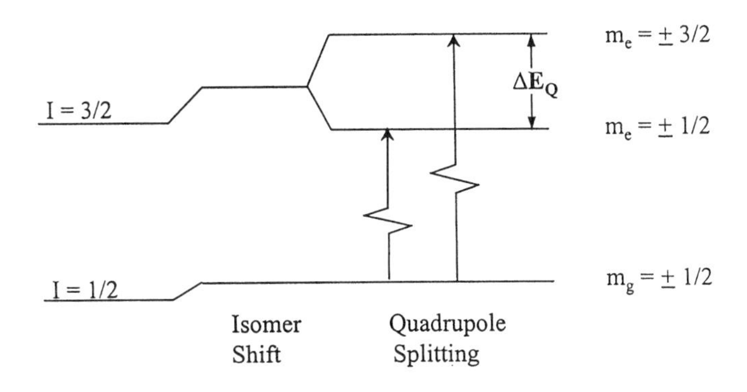

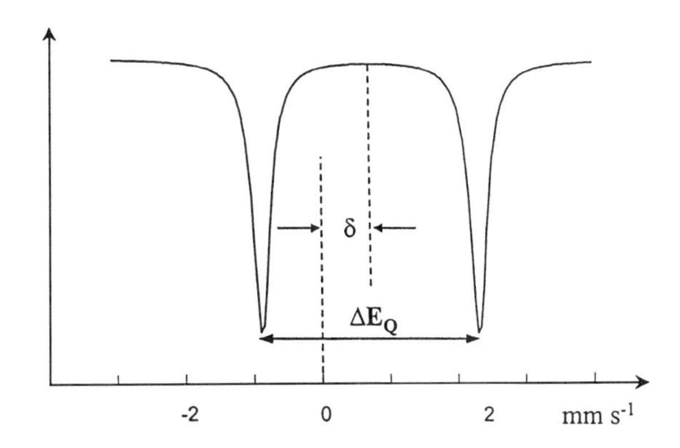

Figure 1
Quadrupole splitting of the ^{57}Fe excited state and shift of the nuclear states by the electric monopole interaction that gives rise to the isomer shift. Also shown is a typical Mössbauer spectrum for a sample containing randomly oriented molecules. In the absence of heterogeneities or relaxation effects the absorption lines have Lorentzian shape. The velocity scale is drawn relative to the centroid of Fe metal.

environment of spherical, octahedral, or tetrahedral symmetry we must have $V_{xx} = V_{yy} = V_{zz}$; thus $V_{ii} = 0$.

In the principal axis system of the EFG tensor, the off-diagonal elements vanish, and since the EFG tensor is traceless, only two independent components need to be specified. By convention, these parameters are V_{zz} and the asymmetry parameter η, defined by $\eta = (V_{xx} - V_{yy})/V_{zz}$. By choosing the coordinate system such that $|V_{zz}| \geq |V_{yy}| \geq |V_{xx}|$, the asymmetry parameter can be restricted to $0 \leq \eta \leq 1$. The conventional Hamiltonian of the quadrupole interaction for the ^{57}Fe nuclear *excited* state is

$$\mathcal{H}_Q = \frac{eQV_{zz}}{12}[3\mathbf{I}_z^2 - I(I+1) + \eta(\mathbf{I}_x^2 - \mathbf{I}_y^2)] \tag{1}$$

where e is the proton charge. One remark may be in order here. It is true that $\eta = 0$ for complexes with tetragonal or trigonal point symmetry. However, the observation of $\eta = 0$ does not necessarily imply such a symmetry. For example, a d_{xy} orbital can be an eigenstate in rhombic symmetry; yet a d_{xy} orbital has a charge distribution with fourfold symmetry which yields $V_{xx} = V_{yy}$ and thus $\eta = 0$.

In the absence of magnetic interactions, $\mathscr{H}_Q$ splits the nuclear excited state into two degenerate doublets. For $\eta = 0$, these doublets are labeled by the magnetic quantum numbers $m_e = \pm\frac{3}{2}$ and $m_e = \pm\frac{1}{2}$; for $V_{zz} > 0$ the $m_e = \pm\frac{3}{2}$ levels have the higher energy. The two doublets are separated in energy by the quadrupole splitting

$$\Delta E_Q = \frac{eQV_{zz}}{2}\sqrt{1 + \eta^2/3} \tag{2}$$

Figure 1 shows the splitting of the nuclear excited state and the resulting Mössbauer spectrum, a quadrupole doublet, for a sample containing randomly oriented molecules such as found in polycrystalline samples or in frozen solution. For this case, the intensities of both lines are equal and the sign of ΔE_Q is thus undetermined. Although one should therefore write $|\Delta E_Q|$ in Eq. (2), it is customary to quote simply ΔE_Q. For diamagnetic (electronic spin $S = 0$) samples, the sign of ΔE_Q and η can be determined by studying the sample in strong applied magnetic fields. For compounds exhibiting magnetic effects, various methods can be used to determine the components of the EFG tensor; we will discuss one method below.

In Figure 1, we have indicated another quantity measured by Mössbauer spectroscopy. Electric monopole interactions between the nuclear charge distributions and the electrons at the nucleus cause a shift of the nuclear ground and excited states without affecting the splittings of the two multiplets. These interactions are described by the parameter δ, which measures the isomer shift. This parameter can be obtained directly from the centroid of the Mössbauer spectrum. Since both the Mössbauer source and the absorber experience an isomer shift, the experimental result depends on the nature of the radiation source used. Therefore, it is customary to quote δ relative to a standard. The standard most commonly used is Fe metal or sodium nitroprusside dihydrate, $Na_2[Fe(CN)_5NO] \cdot 2H_2O$, at 298 K.

The isomer shift is a measure of the s-electron density at the nucleus. This density can be influenced either by altering the s population of a valence shell or, via shielding effects, by increasing or decreasing electron density with p or d character. The parameter δ depends on the oxidation state, the spin state, the coordination environment and the degree of covalency, and thus this parameter conveys important structural and chemical information. If one deals with an iron compound of unknown structure and oxidation state, one can make use of correlation diagrams in which δ is plotted as a function of the oxidation and spin state.[1,2] Unfortunately, among the common states only the high-spin ($S = 2$) iron(II) state has a unique range of δ values; the isomer shifts of other states overlap considerably. However, by taking into account quadrupole splittings and magnetic data, one can generally assign the spin and oxidation state with high

Table I
Values for ΔE_Q and δ for Some Compounds of Biological Interest[a]

Oxidation State	Spin State	Ligand Set	ΔE_Q(mm s^{-1})	δ(mm s^{-1})
Fe(IV)	$S=2$	Fe–(O,N)[b]	0.5–1.0	0.0–0.1
	$S=1$	Hemes	1.0–2.0	0.0–0.1
		Fe–(O,N)	0.5–4.3	−0.20–0.10
Fe(III)	$S=\frac{5}{2}$	Hemes	0.5–1.5	0.35–0.45
		Fe–S	<1.0	0.20–0.35
		Fe–(O,N)	0.5–1.5	0.40–0.60
	$S=\frac{3}{2}$	Hemes	3.0–3.6	0.30–0.40
	$S=\frac{1}{2}$	Hemes	1.5–2.5	0.15–0.25
		Fe–(O,N)	2.0–3.0	0.10–0.25
Fe(II)	$S=2$	Hemes	1.5–3.0	0.85–1.0
		Fe–S	2.0–3.0	0.60–0.70
		Fe–(O,N)	2.0–3.2	1.1–1.3
	$S=0$	Hemes	<1.5	0.30–0.45

[a] The entries give *typical* numbers at 4.2 K. The isomer shifts are quoted relative to the centroid of Fe metal at 298 K, the standard most commonly used. Table I is by no means complete. The classes of compounds chosen are hemes, localized valence states of Fe with tetrahedral sulfur ligation (Fe–S), and hexacoordinate and pentacoordinate Fe sites with oxygen and/or nitrogen ligands [Fe–(O,N)].
[b] Only two examples; see Lee et al.[26] and Dong et al.[29]

confidence. Table I lists typical values for ΔE_Q and δ for three classes of compounds important in bioinorganic chemistry and biochemistry, namely, hemes, iron–sulfur centers containing Fe sites in a distorted tetrahedral sulfur environment, and compounds with a hexa- or penta-coordinate N/O environment.

The nuclear ground and excited states of the ^{57}Fe nucleus have magnetic moments that can interact with a magnetic field **B**. This interaction is described by

$$\mathscr{H} = -\mu \cdot \mathbf{B} = -g_n \beta_n \mathbf{B} \cdot \mathbf{I} \tag{3}$$

where g_n is the nuclear g factor and β_n is the nuclear magneton. In the absence of quadrupole interactions, the Hamiltonian in Eq. (3) splits the nuclear states into equally spaced levels of energy

$$E(m_I) = -g_n \beta_n B m_I \tag{4}$$

where m_I is the nuclear magnetic quantum number. Because $g_e/g_g < 0$, the sequence of m_I values is in reverse order for the excited state. Figure 2 shows schematically the splittings of the nuclear ground and excited states. The allowed γ transitions between the nuclear sublevels follow from the selection rules for

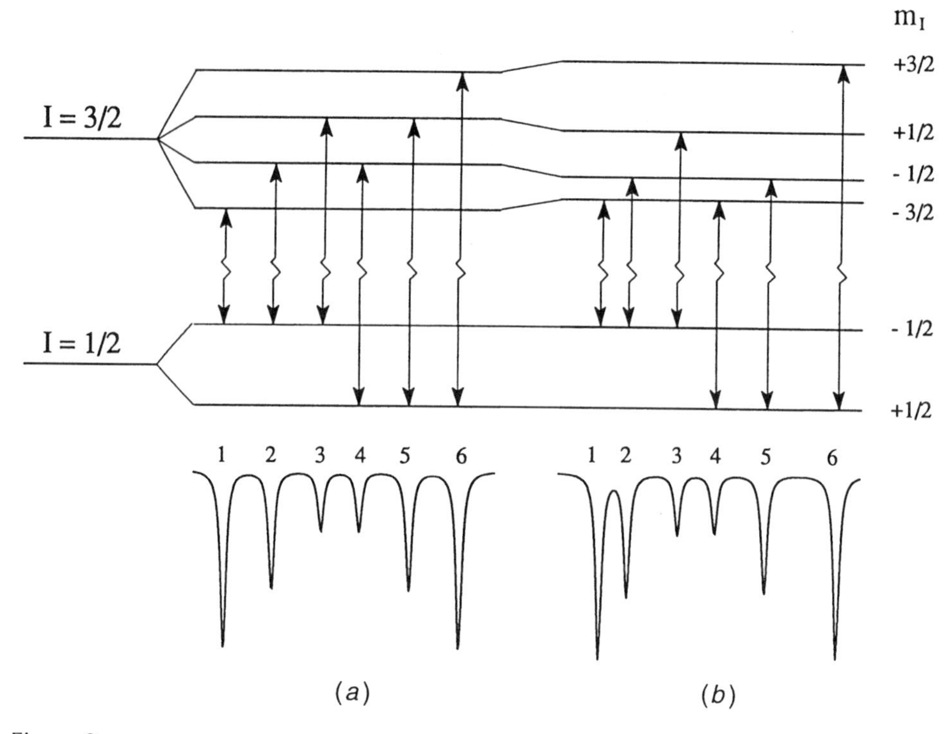

Figure 2
Magnetic hyperfine splittings of the nuclear ground state and the 14.4-keV level according to Eq. (4). The diagram drawn in (*a*) assumes $\Delta E_Q = 0$ while that of (*b*) is drawn for the case where a large magnetic interaction is perturbed by a quadrupole interaction with $eQV_{zz} > 0$ and $\eta = 0$. Note that the energy of the γ transition is 14,400 eV, whereas typical hyperfine splittings are about 3×10^{-7} eV. A typical line width is 1.5×10^{-8} eV.

magnetic dipole transitions: $\Delta I = 1$, $\Delta m_I = 0, \pm 1$. Note that the splitting of the six-line Mössbauer spectrum is a measure of the strength of the magnetic field at the nucleus. Note also that lines 1, 3, 4, and 6 derive from $|\Delta m_I| = 1$ transitions, while lines two and five result from $\Delta m_I = 0$ transitions.

In Eq. (3), the direction of the magnetic field defines the nuclear quantization axis. This field is generally not the applied field but an effective field B_{eff}, which we introduce below. If a γ quantum approaches the nucleus at an angle θ relative to the nuclear quantization axis, the angular dependence of the absorption probability is proportional to $\sin^2 \theta$ for $\Delta m_I = 0$ transitions and proportional to $(1 + \cos^2 \theta)$ for $\Delta m_I = \pm 1$ transitions. Thus, if we study a sample for which the effective magnetic field at the nucleus is *parallel* to the observed γ rays, we have $\theta = 0$ and the intensity of the $\Delta m_I = 0$ lines vanishes [Fig. 3(*a*)]. On the other hand, if the nuclear quantization axis is *perpendicular* to the observed γ rays, the intensity of the $\Delta m_I = 0$ lines is maximized [Fig. 3(*b*)]. Now, for paramagnetic systems it is frequently possible to orient B_{eff} by application of an external magnetic field of moderate strength (10 mT is generally sufficient). The degree of this

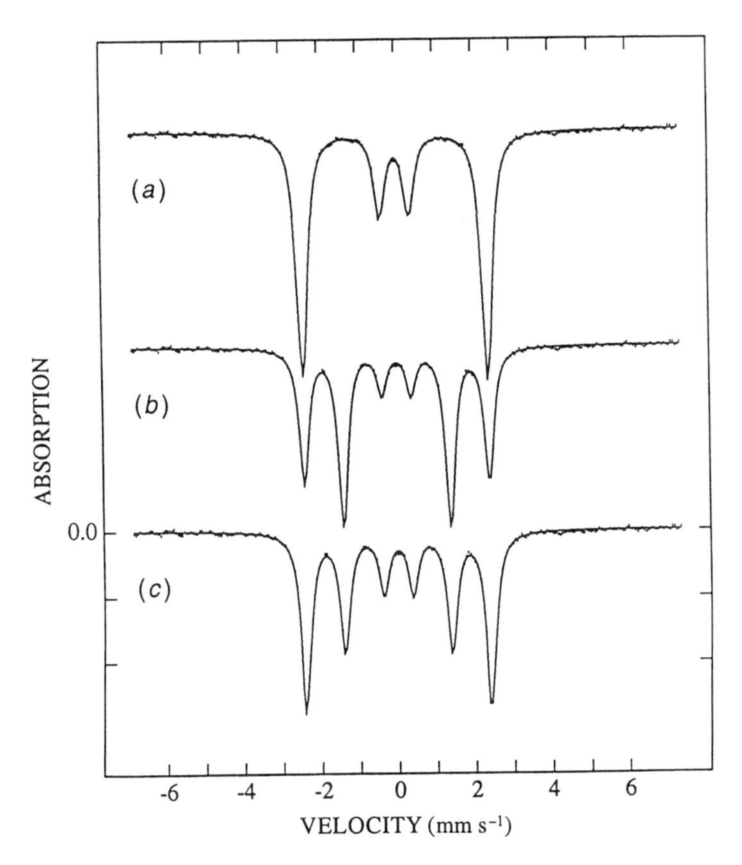

Figure 3
Illustration of Mössbauer spectra of an $S = \frac{1}{2}$ system with *isotropic* (*a* and *b*) and *uniaxial* (*c*) magnetic properties. It is assumed that the samples consist of a collection of randomly oriented molecules as occurs in frozen solution or polycrystalline powders. The spectra shown in (*a*) and (*b*) apply to the isotropic case, Eq. (6), with a weak magnetic field applied either parallel (*a*) or perpendicular (*b*) to the observed γ radiation. The spectrum in (*c*) illustrates the case $g_x = g_y = 0$ and $g_z \neq 0$; the intensities of the absorption lines are independent of the orientation of the (weak) applied field.

orientation depends on the properties of the electronic system, for example, on the electronic Zeeman term and the zero-field splitting (zfs). Since the properties of the latter interactions are sensitively probed by EPR, there must be a connection between the shapes of the Mössbauer and EPR spectra. This connection allows us to draw very powerful conclusions about the type and number of iron sites in a cluster of unknown structure. Moreover, it allows us to identify a paramagnetic species even if most of the iron in the sample belongs to other species or impurities. In the following section, we will elucidate these connections in a simplified but systematic way. For the remainder of this chapter, we will assume that the nuclear quantization axis is solely determined by B_{eff}. This is not strictly true because the nuclear quantization axis is determined by an interplay of magnetic and quadrupole interactions. In many practical cases, however, the magnetic interactions dominate and we can safely ignore quadrupolar effects for the type of considerations we wish to develop here.

It is customary to describe the magnetic properties of the electronic ground manifold of a transition metal ion by a spin Hamiltonian. Let us consider an iron site, such as low-spin Fe(III), with electronic spin $S = \frac{1}{2}$. A suitable spin Hamiltonian for describing the EPR and Mössbauer spectra of such a system can be written as

$$\mathscr{H} = \beta\mathbf{S}\cdot\mathbf{g}\cdot\mathbf{B} + \mathbf{S}\cdot\mathbf{A}\cdot\mathbf{I} - g_{\mathrm{n}}\beta_{\mathrm{n}}\mathbf{B}\cdot\mathbf{I} + \mathscr{H}_Q \tag{5}$$

The first and the third terms describe the electronic and nuclear Zeeman interactions, respectively. The magnetic hyperfine interaction between the electronic system and the nucleus is described by the magnetic hyperfine tensor **A**. This tensor contains spin–dipolar and orbital contributions as well as a contribution due to the polarization of the core electrons by unpaired electrons (Fermi contact term); **A** carries information about covalency and the orbital state of the unpaired d electrons. We start our considerations by assuming that **g** and **A** are *isotropic*, that is, $g_x = g_y = g_z = g$ and $A_x = A_y = A_z = A$, and discuss

$$\mathscr{H} = g\beta\mathbf{S}\cdot\mathbf{B} + \mathbf{AS}\cdot\mathbf{I} - g_{\mathrm{n}}\beta_{\mathrm{n}}\mathbf{B}\cdot\mathbf{I} \tag{6}$$

For the following considerations, the quadrupole interactions are of little consequence and we therefore ignore this term. The A values of iron compounds are $|A| \leq 0.001\ \mathrm{cm}^{-1}$, whereas $g\beta B = 0.3\ \mathrm{cm}^{-1}$ for EPR at X band ($B \approx 0.33$ T at $g = 2$); the nuclear Zeeman term is more than four orders of magnitude smaller than the electronic Zeeman term. Thus, the electronic Zeeman term dominates and it follows that the quantization axis (the z axis) of S and the expectation values $\langle S\rangle = \langle\psi|S|\psi\rangle$ are solely determined by the electronic Zeeman interaction, even in applied fields as small as $H = 10$ mT. For the system of Eq. (6) the electronic spin is quantized along **B** (the magnetic moment $\boldsymbol{\mu} = -g\beta\mathbf{S}$ precesses around the applied field) and $\langle S_z\rangle = M_S = \pm\frac{1}{2}$ and $\langle S_x\rangle = \langle S_y\rangle \approx 0$.

The first term in Eq. (6) depends only on the electronic variable S, whereas the second term couples the electronic with the nuclear system. Since the electronic Zeeman interaction dominates, the $\mathbf{AS}\cdot\mathbf{I}$ term has little influence in determining the expectation value of $\langle\mathbf{S}\rangle$. This fact allows us to replace the operator **S** in Eq. (6) by its expectation value $\langle\mathbf{S}\rangle$ to obtain a nuclear Hamiltonian for each sublevel of the electronic doublet

$$\mathscr{H}_{\mathrm{n}} = \mathbf{A}\langle\mathbf{S}\rangle\cdot\mathbf{I} - g_{\mathrm{n}}\beta_{\mathrm{n}}\mathbf{B}\cdot\mathbf{I} \tag{7}$$

Although the electronic Zeeman term does not explicitly appear in Eq. (7), we have to keep in mind that it determines $\langle\mathbf{S}\rangle$. For the "spin-up" state, we take $\langle S_z\rangle = +\frac{1}{2}$, for the "spin-down" state $\langle S_z\rangle = -\frac{1}{2}$. Since $\langle S_x\rangle = \langle S_y\rangle \approx 0$, the magnetic hyperfine term is reduced to $A\langle S_z\rangle I_z$, that is, only the component of the magnetic hyperfine term along the electronic quantization axis matters. This is a general result from perturbation theory; the term $A(\mathbf{S_xI_x} + \mathbf{S_yI_y})$ yields off-diagonal elements that give second-order contributions of $A(A/g\beta B) \approx A/300$ to the energies for the conditions stated above. We can rewrite Eq. (7) as

$$\mathscr{H}_{\mathrm{n}} = -g_{\mathrm{n}}\beta_{\mathrm{n}}\left(\frac{-A\langle \mathbf{S}\rangle}{g_{\mathrm{n}}\beta_{\mathrm{n}}} + \mathbf{B}\right) \cdot \mathbf{I}$$

$$= -g_{\mathrm{n}}\beta_{\mathrm{n}}(\mathbf{B}_{\mathbf{int}} + \mathbf{B}) \cdot \mathbf{I} = -g_{\mathrm{n}}\beta_{\mathrm{n}}\mathbf{B}_{\mathbf{eff}} \cdot \mathbf{I} \tag{8}$$

where we have defined the internal magnetic field

$$\mathbf{B}_{\mathbf{int}} = -A\langle \mathbf{S}\rangle / g_{\mathrm{n}}\beta_{\mathrm{n}} \tag{9}$$

Equation (8) connects the spin Hamiltonian [Eq. (6)] with Eq. (3); in a paramagnetic system it is $\mathbf{B}_{\mathbf{eff}}$ that acts on the nuclear magnetic moment and causes the splitting of the nuclear levels. Iron compounds exhibit internal magnetic fields up to 60 T and for moderate applied fields ($B < 0.3$ T) the magnetic splitting of the Mössbauer spectrum is determined by $\mathbf{B}_{\mathbf{int}}$.

Assume now that we have an iron site described by Eq. (6). If the sample is either polycrystalline or consists of molecules in frozen solution, the molecules are randomly oriented and spatially fixed. Since $\mathbf{B}_{\mathbf{int}}$ is parallel to $\langle \mathbf{S}\rangle$ and thus parallel to the applied field $\mathbf{B}$, we can align the nuclear quantization axis of every molecule in the sample either parallel or perpendicular to the observed γ radiation by a suitable orientation of the applied field. If the field is applied parallel to the observed γ radiation, we have $\theta = 0$ for the entire population of molecules and a Mössbauer spectrum as shown in Figure 3(a) will result (angle θ has been defined above). Rotating the applied field by 90° will result in a spectrum as shown in Figure 3(b).

Let us now consider the anisotropic case of Eq. (5). Since the product of a tensor with a vector gives another vector, we can write $\mathbf{g} \cdot \mathbf{B} = \mathbf{B}'$ to obtain the Zeeman term $\beta\mathbf{S} \cdot \mathbf{B}'$. Now , $\mathbf{B}'$ gives the quantization direction of the electronic spin, that is, the only nonvanishing component of $\langle \mathbf{S}\rangle$ is along $\mathbf{B}'$. Expressing the components of $\mathbf{B}$ in the principal axis frame of the g tensor we have

$$\begin{aligned} B_{x'} &= g_x B_x = g_x B \sin\beta \cos\alpha \\ B_{y'} &= g_y B_y = g_y B \sin\beta \sin\alpha \\ B_{z'} &= g_z H_z = g_z B \cos\beta \end{aligned} \tag{10}$$

where g_x, g_y, and g_z are the principal components of g and where α and β are polar angles. If the g tensor is anisotropic, $\mathbf{B}'$ is not parallel to $\mathbf{B}$, and thus $\mathbf{B}_{\mathbf{int}} = -\mathbf{A} \cdot \langle \mathbf{S}\rangle / g_{\mathrm{n}}\beta_{\mathrm{n}}$ is not parallel to the applied field. Therefore, the field dependence (parallel to the γ rays vs. perpendicular) of the Mössbauer spectrum is less pronounced as $\mathbf{g}$ becomes more anisotropic. Since the g values are determined by EPR, one can make reasonable predictions (usually by computer simulations) about the distribution of $\mathbf{B}_{\mathbf{int}}$ and thus the shape of the Mössbauer spectrum once the result of an EPR study are known.

Let us now consider the case where one g value is very large compared to the other two (e.g., $g_z \gg g_x, g_y$). This case of extreme anisotropy, called the *uniaxial* case, is frequently encountered. Examples are the low-spin Fe(III) highly anisotropic low-spin (HALS) complexes (see Chapter 3) and subdoublets of multiplets with $S \geq \frac{3}{2}$. As can be seen from Eq. (10), $B_x' \approx B_y' \approx 0$ and $B_z \neq 0$ for almost all orientations of the applied field relative to the molecular axes, that is, $\mathbf{B}$ is directed along the molecular z axis (the z axis of $\mathbf{g}$), and for moderate anisotropies of $\mathbf{A}$ the internal magnetic field, $\mathbf{B_{int}} = -\mathbf{A} \cdot \langle \mathbf{S} \rangle / g_n\beta_n$, will be directed along z as well. (If $\mathbf{A}$, on the other hand, is anisotropic such that $A_x \gg A_z$, the internal magnetic field will acquire a sizable x component if $\langle S_x \rangle \neq 0$. This anisotropy would cause the internal field to tilt away from the z axis.) If the internal magnetic field is locked along the molecular z axis, the intensities of the Mössbauer spectrum do not change when the applied field is changed from parallel to perpendicular. The six-line pattern, averaged over all molecular orientations, will have a 3:2:1:1:2:3 intensity pattern as shown in Figure 3(c). Aasa and Vånngard[5] gave a very useful expression for the EPR transition probabilities observed in the "derivativc" spectrum of samples with randomly oriented molecules. In the uniaxial case, the area under the peak at g_z is proportional to $g_x{}^2 + g_y{}^2$ and thus vanishingly small in the limit $g_x \approx g_y \approx 0$. In EPR jargon, such a system is called "EPR silent". Thus, a field-independent Mössbauer spectrum suggests an EPR silent state, and conversely, a field-dependent Mössbauer spectrum implies that an EPR signal must be observed. Next, we extend our discussion to a slightly more complicated case, namely, the high-spin ($S = \frac{5}{2}$) iron(III)ion.

III. Effective *g*-Values for High-Spin Fe(III)

For describing the magnetic properties of the electronic ground manifold of paramagnetic transition metal ions, it is customary to replace the true Hamiltonian by an effective Hamiltonian, which operates only on spin variables. The spin on which this so-called spin Hamiltonian operates is not necessary the true spin; it may be an effective spin or a fictitious spin that obeys the same rules as the real spin.[6] The spin Hamiltonian approach is extremely useful, because it allows one, without going into the messy details of fundamental calculations, to compare the results obtained by different techniques such as EPR, electron nuclear double resonance, magnetic circular dichroism (MCD), NMR, Mössbauer spectroscopy, and magnetic susceptibility. The spin Hamiltonian parameters such as the zfs parameters D and E, the g tensor and the A tensor of similar compounds fall into narrow ranges.

A suitable spin Hamiltonian for describing the EPR spectra of the high-spin ($S = \frac{5}{2}$) system can be written as

$$\mathcal{H}_e = D(\mathbf{S_z^2} - \tfrac{35}{12}) + E(\mathbf{S_x^2} - \mathbf{S_y^2}) + g_0\beta\mathbf{B} \cdot \mathbf{S} \tag{11}$$

where D and E are the axial and rhombic zfs parameters, respectively. (These parameters are the principal axis components of the traceless zfs tensor, $\mathbf{D}$). For the high-spin iron(III) ion $g_0 = 2.0$ is appropriate. For $B = 0$ the zfs term of

Eq. (11) partially removes the sixfold degeneracy of the spin sextet and three (Kramers) doublets result (see Fig. 11 of Chapter 3).

In order to describe the Mössbauer spectra of high-spin iron(III) complexes, Eq. (11) needs to be augmented by terms describing the hyperfine interactions

$$\mathscr{H}_{\mathrm{hf}} = A_0\mathbf{S}\cdot\mathbf{I} - g_n\beta\mathbf{B}\cdot\mathbf{I} + \mathscr{H}_Q \tag{12}$$

The orbital ground state of the iron(III) ion has a vanishingly small orbital angular momentum, and the magnetic hyperfine interaction is therefore dominated by the isotropic Fermi contact term. The majority of the high-spin iron(III) complexes of interest to the inorganic and bioinorganic chemist has D values larger than 0.5 cm^{-1}, ranging up to about 20 cm^{-1} for certain heme complexes. We can thus study Mössbauer spectra under conditions where $\beta B \ll |D|$, that is, under conditions where the three Kramers doublets are well separated in energy. It is customary in the EPR literature and useful for the consideration of the Mössbauer spectra to consider each doublet separately and describe its magnetic properties with an effective spin $S' = \frac{1}{2}$ Hamiltonian, as illustrated by the following example.

Consider Eq. (11) for $D > 0$ and $E = 0$ under the condition that $\beta B \ll D$. Without loss of generality we assume that the applied magnetic field is in the xz plane with polar angle β relative to the z axis of the zfs tensor. We then wish to solve

$$\mathscr{H} = D(\mathbf{S}_z^2 - \tfrac{35}{12}) + g_0\beta B(\mathbf{S}_x\sin\beta + \mathbf{S}_z\cos\beta) + A_0\mathbf{S}\cdot\mathbf{I} \tag{13}$$

For $B = 0$ the eigenstates of the electronic system are $|S = \frac{5}{2}, M_s\rangle$, which we order in such a way that the two levels that comprise a Kramers doublet stay together, that is, in the sequence $M_s = +\frac{5}{2}, -\frac{5}{2}, +\frac{3}{2}, -\frac{3}{2}, +\frac{1}{2}, -\frac{1}{2}$. In order to solve Eq. (13), we would have to diagonalize the 12×12 matrix $\{(2S+1)(2I_g+1) = 12\}$ involving the nuclear ground state and a 24×24 matrix involving the nuclear excited state. However, since $\beta B \ll D$ we can neglect off-diagonal elements of the Zeeman and magnetic hyperfine interactions that involve states belonging to different Kramers doublets. The 12×12 matrix is then in a block-diagonal form consisting of three 4×4 matrices that can be solved separately, each 4×4 matrix belonging to a specific Kramers doublet. With the nomenclature $|S = \frac{5}{2}, M_s\rangle|I_g = \frac{1}{2}, m_g\rangle \equiv |\phi_{M_S}\rangle|m_g\rangle$ and abbreviating $\Delta = \beta B/2$ we obtain for the matrix of the $M_s = \pm\frac{1}{2}$ ground doublet:

	$\|\phi_{+1/2}\rangle\|+\frac{1}{2}\rangle$	$\|\phi_{-1/2}\rangle\|+\frac{1}{2}\rangle$	$\|\phi_{+1/2}\rangle\|-\frac{1}{2}\rangle$	$\|\phi_{-1/2}\rangle\|-\frac{1}{2}\rangle$
$\|\phi_{+1/2}\rangle\|+\frac{1}{2}\rangle$	$g_0\Delta\cos\beta + A_0/4$	$3g_0\Delta\sin\beta$	0	0
$\|\phi_{-1/2}\rangle\|+\frac{1}{2}\rangle$	$3g_0\Delta\sin\beta$	$-g_0\Delta\cos\beta - A_0/4$	$3A_0/2$	0
$\|\phi_{+1/2}\rangle\|-\frac{1}{2}\rangle$	0	$3A_0/2$	$-g_0\Delta\cos\beta - A_0/4$	$3g_0\Delta\sin\beta$
$\|\phi_{-1/2}\rangle\|-\frac{1}{2}\rangle$	0	0	$3g_0\Delta\sin\beta$	$g_0\Delta\cos\beta + A_0/4$

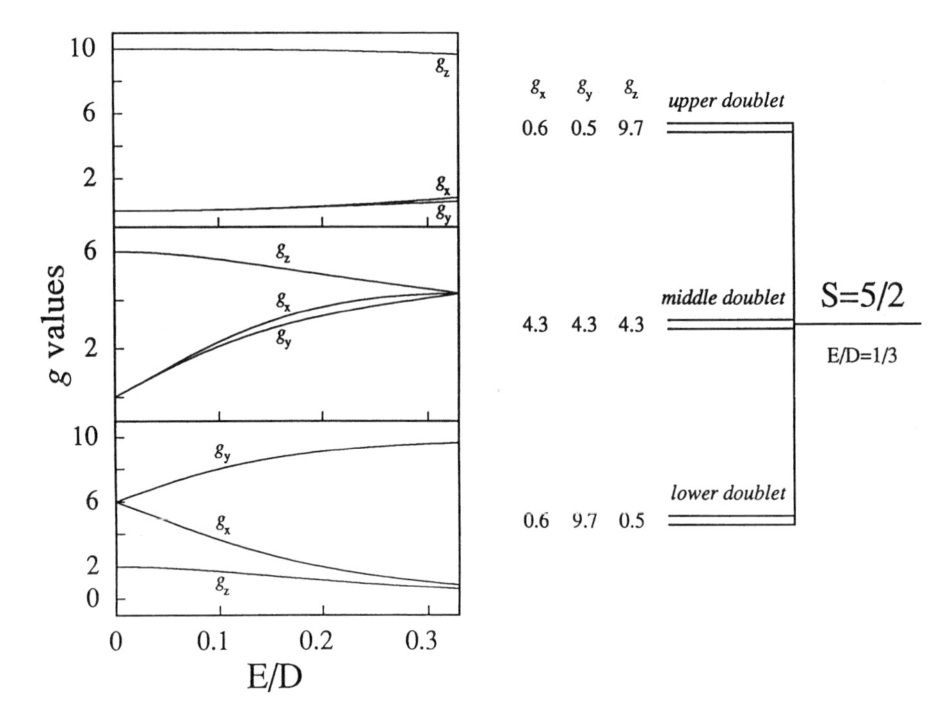

Figure 4
Effective g values for the three Kramers doublets of an $S = \frac{5}{2}$ spin multiplet plotted versus the rhombicity parameter E/D. It is assumed that $g_0 = 2$. The plot covers the whole range of g values. For $E/D > \frac{1}{3}$, one can relabel the coordinate axes and scale E/D back into the range $0 \leq E/D \leq \frac{1}{3}$. Also shown is an energy level diagram for the frequently occurring "rhombic" case $E/D = \frac{1}{3}$.

Of interest here are the factors 3 in the off-diagonal elements. They occur because $S_+|\phi_{-1/2}\rangle = \sqrt{S(S+1)+\frac{1}{4}}|\phi_{+1/2}\rangle = [2S+1)/2]|\phi_{+1/2}\rangle = 3|\phi_{+1/2}\rangle$. Now, identical matrix elements are obtained for the $S' = \frac{1}{2}$ spin Hamiltonian

$$\mathcal{H} = \beta \mathbf{S}' \cdot \mathbf{g} \cdot \mathbf{B} + \mathbf{S}' \cdot \mathbf{A} \cdot \mathbf{I} \tag{14}$$

where $\mathbf{g}$ and $\mathbf{A}$ are tensors with principal axis values $g_x = g_y = 3g_0, g_z = g_0$ and $A_x = A_y = 3A_0, A_z = A_0$, and where $S' = \frac{1}{2}$ is an effective spin obeying the same rules as a real spin $S = \frac{1}{2}$. Thus for $g_0 = 2$, we can treat the ground doublet as a spin $\frac{1}{2}$ system with principal g values $g_x = g_y = 6$ and $g_z = 2$. Similarly, we obtain for the Kramers doublet at energy $2D$ the g values $g_x = g_y = 0$ and $g_z = 6$, and for the upper doublet $g_x = g_y = 0$ and $g_z = 10$.

We can generalize these ideas for $E \neq 0$ and generate plots of the effective g values as a function of E/D. Such a plot, introduced by Wickman,[7] is shown in Figure 4. Similar plots can be generated for other multiplets. Such graphs are extremely useful for identifying the spin state by EPR studies. Note that because of $S_+|S, -\frac{1}{2}\rangle = ((2S+1)/2)|S, +\frac{1}{2}\rangle$, we obtain $g_x = g_y = (2S+1)g_0/2$ for the $M = \pm\frac{1}{2}$ doublet for the axial case $E = 0$. (Thus, $g_x = g_y = 4.0$ implicates a spin $S = \frac{3}{2}$.) Note also that the following relations[7] hold for each Kramers doublet,

$$A_x/g_x = A_y/g_y = A_z/g_z = A_0/g_0 \tag{15}$$

Before presenting specific examples, we wish to discuss briefly the effects of electronic spin relaxation on the Mössbauer spectra. Equation (9) shows that the internal field $\mathbf{B}_{\mathbf{int}}$ is proportional to $\langle\mathbf{S}\rangle$. Thus, if the electronic spin fluctuates rapidly due to spin–lattice or spin–spin relaxation, the internal field will fluctuate. If this fluctuation is fast compared to the nuclear precession frequencies, which are of the order of 10 MHz, determined by $\mathbf{A}$ and $\mathscr{H}_Q$, the nucleus will sample an average field that can be computed by evaluating the thermal average $\langle\mathbf{S}\rangle_{\text{th}}$,

$$\langle S_i\rangle_{\text{th}} = \frac{\sum\limits_n \langle S_i\rangle_n \exp(-\Delta E_n/kT)}{\sum\limits_n \exp(-\Delta E_n/kT)} \tag{16}$$

where $i = x, y, z$ and where n sums over all thermally accessible states. If only one spin multiplet is thermally accessible, $\langle S_i\rangle_{\text{th}}$ is proportional to $g_i B_i S(S+1)/3kT$ for $kT \gg \Delta E_n$ and thus follows the temperature dependence of the magnetic susceptibility.

In the slow fluctuation limit, which prevails for most Fe(III) compounds for $T < 10$ K, $\langle S\rangle$ is computed separately for each thermally populated level and thus each level contributes its own Mössbauer spectrum with an intensity according to the Boltzmann factor that determines the population of the level [Fig. 5(*a–c*)].

At intermediate fluctuation rates, the spectra are quite complex and difficult to analyze unless one knows some of the A values and quadrupolar parameters from studies at very low temperatures. In strong applied magnetic fields, $B > 5$ T, and at temperatures around $T = 1.5$ K, achieved by pumping on the liquid helium bath, one can usually work under conditions where only the lowest level is measurably populated. It follows from Eq. (16) that $\langle S_i\rangle_{\text{th}} = \langle\psi|S_i|\psi\rangle$, where $|\psi\rangle$ designates the ground state.

A. Example of a Mössbauer Spectrum for a High-Spin Iron(III) Center

Many well-analyzed Mössbauer spectra of high-spin iron(III) compounds have been published. Nevertheless, here we will discuss a fictitious compound, because this will allow us to adjust the parameters in such a way that we can see the crucial features of the Mössbauer spectra of high-spin Fe(III) in one example. Our compound of interest exhibits a weak EPR resonance at $g = 9.7$ and an intense derivation-type signal at $g = 4.3$. The diagram of Figure 4 (plus a bit familiarity with the shapes of EPR spectra, see Chapter 3) informs us that these resonances belong to an $S = \frac{5}{2}$ system with rhombicity $E/D = \frac{1}{3}$. Let us refer to the three Kramers doublets as the lower, middle ($g = 4.3$), and upper doublet (see Fig. 4).

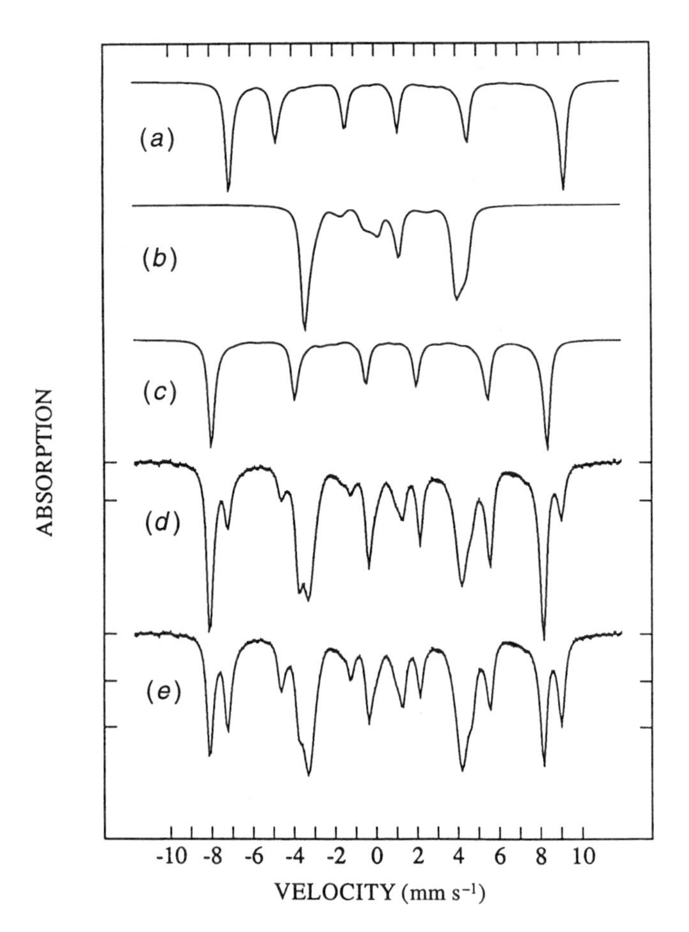

Figure 5
Mössbauer spectra of a high-spin iron(III) system with $E/D = \frac{1}{3}$. The "experimental" spectra shown in (*d*) and (*e*) were computed (and then endowed with random noise) with the following parameter set. $D = 0.5\ \text{cm}^{-1}$, $E/D = \frac{1}{3}$, $g_0 = 2.0$, $A_0 = -29$ MHz, $\Delta E_Q = 1.0\ \text{mm s}^{-1}$, $\eta = 1$, and $\delta = 0.5\ \text{mm s}^{-1}$, parallel field of $H = 60$ mT, $T = 4.2$ K for (*d*) and $T = 15$ K for (*e*). It is assumed that the electronic spin relaxation rate is slow so that each of the three Kramers doublets produces its own Mössbauer spectrum. The solid curves in A–C give the contributions of each doublet.

For $E/D = \frac{1}{3}$ the lower and upper doublets are uniaxial, whereas the middle doublet is magnetically isotropic. For $D > 0$ the lower doublet has an easy axis of magnetization along y ($g_y = 9.7$); the upper doublet, on the other hand, has the easy axis along z. Thus, for the ground doublet B_{int} will be directed along the molecular y axis (y is defined by the zfs). Because of relations (15), we expect the magnetic splittings of the Mössbauer spectra of the upper and lower doublets to be the same. The middle doublet should exhibit a splitting that is reduced by 4.3/9.7 relative that of the ground doublet. Figures 5(*d* and *e*) show computer-generated spectra "measured" at $T = 4.2$ and 15 K, respectively. In order to make them look more like real data, we have added some random noise to the baseline. The spectra in Figures 5(A–C) show the contributions of the lower doublet (c), of the middle doublet in parallel field (b), and of the upper doublet (a).

The Mössbauer spectra of the lower and upper doublets are readily identified by inspection of the data of Figure 5(d). At 4.2 K, the population of the upper doublet is significantly less than that of the ground doublet; the intensity ratio of the spectral components gives the Boltzmann factor and thus, with E/D known, the zfs parameter D. In our example, the spectra of the upper and lower doublets have the same magnetic hyperfine splittings. If one would observe that these splittings differ, and EPR would suggest that E/D is exactly $\frac{1}{3}$, the assumption of an isotropic magnetic hyperfine interaction, A_0, would not be justified. Note that the internal magnetic field for the lower doublet depends on $\langle S_y \rangle A_0$, whereas that of the upper doublet depends on $\langle S_z \rangle A_0$. For $E/D = 1/3$, we have $|\langle S_z \rangle|_{\text{upper doublet}} = |\langle S_y \rangle|_{\text{lower doublet}} = 9.7/4$. Thus, if the magnetic hyperfine splittings differ, it follows that the magnetic hyperfine interaction in Eq. (13) is not isotropic and that $A_{0z} \neq A_{0y}$. Some spectra are so well defined that 2% differences between A_y and A_z are readily determined. Despite the identical magnetic splittings the lines of the spectra of the upper and lower Kramers doublets do not overlap. Rather, the outermost lines of the spectrum of the lower doublet are shifted to the left while those belonging to the upper doublet are shifted in the opposite direction. These shifts are caused by quadrupole interactions. For high-spin iron(III) compounds the magnetic interactions are generally much larger than the quadrupole splittings, and thus only the component of $\mathbf{V}$ along $\mathbf{B}_{\text{int}}$ matters. For our compound, the ground-state spectrum measures V_{yy}; the shift pattern (see also Fig. 2) shows that $V_{yy} < 0$. From the spectrum of the upper doublet, we infer $V_{zz} > 0$. Closer inspection reveals that $eQV_{zz}/2 = -eQV_{yy}/2 = 0.86\ \text{mm s}^{-1}$. Since $\mathbf{V}$ is traceless, it follows that $V_{xx} = 0$ and thus $\Delta E_Q = 1.0\ \text{mm s}^{-1}$ and $\eta = 1$.[a]

The magnitude of A_0, but not its sign, can be determined from the magnetic splittings of the spectra of each of the three doublets. These splittings depends on $|B_{\text{eff}}|$. For the two levels of a Kramers doublet, $\langle S \rangle_{\text{spin up}} = -\langle S \rangle_{\text{spin down}}$, and thus both levels yield the same Mössbauer spectrum in weak applied fields. In strong fields, say for $B > 3.0$ T, only the lowest level is appreciably populated at $T \leq 4.2$ K, and since $\langle S_i \rangle < 0$ for the lowest level the sign of B_{int}, and thus the sign of A_0, can be obtained by comparing $|B_{\text{eff}}| = |B_{\text{int}} + B|$ in strong and weak fields. Finally, by evaluating the centroid of the spectra of Figures 5(d and e) the isomer shift can be obtained. The spectra shown are plotted relative to the centroid of Fe metal, using $\delta = 0.5\ \text{mm s}^{-1}$.

We have also measured a spectrum of our fictitious compound at $T = 250$ K. The spectrum consists of a quadrupole doublet (fast electronic spin relaxation!) with $\Delta E_Q = 1.0\ \text{mm s}^{-1}$. The observation that the same ΔE_Q is observed at $T = 4.2$ and 250 K is in accord with the electronic properties for high-spin iron(III) compounds. The quadrupole splitting would exhibit a temperature dependence if other *orbital* states become thermally populated. But the lowest

[a] The three Kramers doublets are sublevels of a spin multiplet. All six levels of the sextet belong to the same orbital state and thus the three doublets have the same EFG.

excited states of high-spin iron(III) compounds with weak ligand fields are spin quartets located at least 2000 cm^{-1} above the ground multiplet (1 cm^{-1} corresponds to 1.44 K in units of kT); thus the observed temperature-independent ΔE_Q.

The situation just described is frequently encountered in the literature; there are many iron(III) compounds, inorganic, bioinorganic, and biochemical that exhibit a $g = 4.3$ EPR signal. For further examples, we refer the reader to the literature: studies of the protein rubredoxin,[8] the enzyme protocatechuate 3,4-dioxygenase,[9] and an Fe(III) · EDTA, EDTA-ethylenediaminetetraacetic acid, complex.[10]

Given the EPR information, an experienced graduate student with a good background in Mössbauer spectroscopy can analyze spectra such as those of Figure 4 in less than 1 h. Analyses, however, become challenging when one deals with clusters that contain three, four, six, or eight iron sites. We have struggled with such cases in our laboratory and our efforts have been facilitated by the knowledge of some of the principles discussed above.

IV. Exchange Coupled High-Spin Fe(II)–Fe(III) Complexes

A. Background

Two paramagnetic metal ions separated by intervening diamagnetic atoms, or group of atoms, can interact with each other. Although the fundamental interactions that give rise to this coupling involve the orbital coordinates of the electrons, strong restrictions are imposed on the spins of the two metal ions, because the wave functions must be antisymmetric with respect to the exchange of electrons. In the framework of a spin Hamiltonian, the interaction between the two metal sites, a and b, can be described by

$$\mathcal{H}_{ex} = \mathbf{J}\mathbf{S_a} \cdot \mathbf{S_b} \tag{17}$$

where $\mathbf{S_a}$ and $\mathbf{S_b}$ are the spin operators of the two metal sites. The constant J measures the strength of the interaction. It conveys information about the nature of the bridging ligands and the geometry of the coordination enviroments of the metals. In many cases, the exchange interaction is much stronger than all other terms in the spin Hamiltonian and thus it is prudent to determine the energies and eigenstates of $\mathcal{H}_{ex}$ first. The exchange term has the same mathematical structure as the more familiar spin–orbit interaction for atoms, $\lambda\mathbf{L} \cdot \mathbf{S}$, and the problem is solved in the same way, namely, by solving Eq. (17) in the coupled representation obtained by combining the site spins $\mathbf{S_a}$ and $\mathbf{S_b}$ to a system spin $\mathbf{S}, \mathbf{S_a} + \mathbf{S_b} = \mathbf{S}$. Since $(\mathbf{S_a} + \mathbf{S_b})^2 = \mathbf{S_a^2} + \mathbf{S_b^2} + 2\mathbf{S_a} \cdot \mathbf{S_b} = \mathbf{S^2}$ we write

$$\mathcal{H}_{ex} = (J/2)(\mathbf{S^2} - \mathbf{S_a^2} - \mathbf{S_b^2}) \tag{18}$$

which yields for the energies

$$E(S) = (J/2)[S(S+1) - S_a(S_a+1) - S_b(S_b+1)] \tag{19}$$

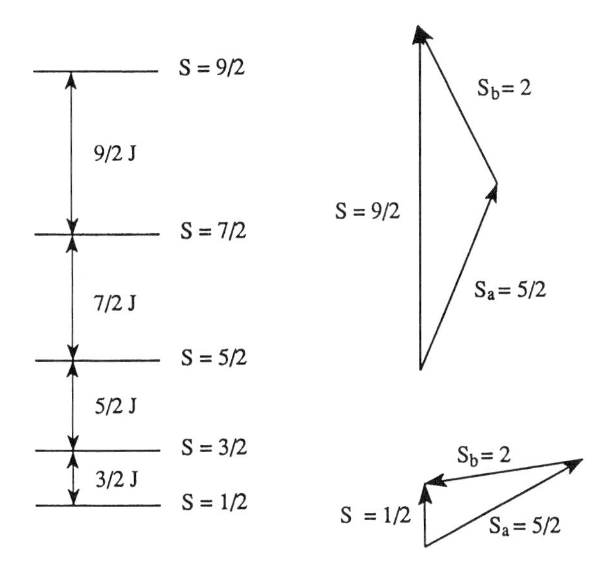

Figure 6
Spin ladder resulting from Eq. (19) for isotropic coupling of two spins $S_a = \frac{5}{2}$ and $S_b = 2$. Also shown are the arrangements of the spin vectors for the $S = \frac{1}{2}$ and $S = \frac{9}{2}$ system states.

The allowed values of S are restricted to $|S_a - S_b| \leq S \leq |S_a + S_b|$. For fixed values of S_a and S_b, we obtain a spin ladder with a level spacing according to the Landé interval rule. For $J > 0$ the ground state has the minimum allowed S and the coupling is called antiferromagnetic; for ferromagnetic coupling ($J < 0$) the level scheme is reversed. Figure 6 shows a level diagram resulting from antiferromagnetic coupling of of two spins $S_a = \frac{5}{2}$ and $S_b = 2$, appropriate for a high-spin iron(III) and high-spin iron(II) ion, respectively. Also shown are the arrangements of the spin vectors for the $S = \frac{1}{2}$ and $S = \frac{9}{2}$ system states. A spin Hamiltonian for describing the Mössbauer spectra of such a dinuclear pair of ions can be written as

$$\mathscr{H} = J\mathbf{S_a} \cdot \mathbf{S_b} + \mathscr{H}_e(\mathrm{a}) + \mathscr{H}_e(\mathrm{b}) + \mathscr{H}_{\mathrm{hf}}(\mathrm{a}) + \mathscr{H}_{\mathrm{hf}}(\mathrm{b}) \tag{20}$$

with

$$\mathscr{H}_e(\mathrm{i}) = \mathbf{S_i} \cdot \mathbf{D_i} \cdot \mathbf{S_i} + \beta \mathbf{S_i} \cdot \mathbf{g_i} \cdot \mathbf{B} \tag{20a}$$

and

$$\mathscr{H}_{\mathrm{hf}}(\mathrm{i}) = \mathbf{S_i} \cdot \mathbf{A_i} \cdot \mathbf{I_i} - g_n\beta_n \mathbf{B} \cdot \mathbf{I_i} + \mathscr{H}_Q(\mathrm{i}) \tag{20b}$$

where i = a, b. For arbitrary J, this spin Hamiltonian is immensely complex. Considering that Eqs. (20) contain eight tensors, the situation seems rather hopeless. However, under certain circumstances the complexity is reduced considerably. First, **g** and **A** of the high-spin iron(III) ion are isotropic. Second, for strong antiferromagnetic coupling ($J \gg$ zfs) we need to consider only the $S = \frac{1}{2}$ manifold. We will discuss this particular case in some detail.

If $\mathscr{H}_{ex}$ is large compared to all other interactions in Eq. (20), we can treat $\mathscr{H}_e(i)$ and $\mathscr{H}_{hf}(i)$ as perturbations, that is, we evaluate their matrix elements within the coupled representation, $|S, M_s\rangle$, ignoring elements connecting different multiplets S. By using the celebrated Wigner–Eckart theorem, we can obtain for the $S = \frac{1}{2}$ manifold a spin Hamiltonian for which the individual spin operators $\mathbf{S_a}$ and $\mathbf{S_b}$ of Eqs. (20) have been replaced by $\mathbf{S}$. The reader will find a detailed account in the book by Bencini and Gatteschi.[11] In short, the following steps are involved. First, since the zfs are described by (spherical) tensors of second rank, their matrix elements vanish within the $S = \frac{1}{2}$ manifold. Second, the matrix elements of the electronic Zeeman interactions can be cast into a different form by using

$$\langle S = \tfrac{1}{2}, M_s|\mathbf{S_a}|S = \tfrac{1}{2}, M_s'\rangle = \tfrac{7}{3}\langle S = \tfrac{1}{2}, M_s|\mathbf{S}|S = \tfrac{1}{2}, M_s'\rangle$$

and

$$\langle S = \tfrac{1}{2}, M_s|\mathbf{S_b}|S = \tfrac{1}{2}, M_s'\rangle = -\tfrac{4}{3}\langle S = \tfrac{1}{2}, M_s|\mathbf{S}|S = \tfrac{1}{2}, M_s'\rangle$$

Thus, we can replace, for calculations within the $S = \frac{1}{2}$ manifold, the electronic Zeeman terms of Eq. (20a) $\beta\mathbf{S_a} \cdot \mathbf{g_a} \cdot \mathbf{B} + \beta\mathbf{S_b} \cdot \mathbf{g_b} \cdot \mathbf{B}$ with

$$\beta\mathbf{S} \cdot (\tfrac{7}{3}\mathbf{g_a} - \tfrac{4}{3}\mathbf{g_b}) \cdot \mathbf{B} = \beta\mathbf{S} \cdot \mathbf{g} \cdot \mathbf{B} \tag{21}$$

where

$$\mathbf{g} = \tfrac{7}{3}\mathbf{g_a} - \tfrac{4}{3}\mathbf{g_b}$$

In frozen solution or polycrystalline samples, the principal components of **g** are determined from EPR experiments. The same equivalence factors $\frac{7}{3}$ and $-\frac{4}{3}$ come into play when $\mathbf{S_a}$ and $\mathbf{S_b}$ of the magnetic hyperfine tensors are replaced by $\mathbf{S}$ (In the semiclassical vector coupling model these factors arise when $\mathbf{S_a}$ and $\mathbf{S_b}$ are projected onto the direction of the system spin $\mathbf{S}$.). Thus, in the spin $S = \frac{1}{2}$ manifold, the spin Hamiltonian of the coupled pair takes the form

$$\begin{aligned} \mathscr{H} = \beta\mathbf{S} \cdot \mathbf{g} \cdot \mathbf{B} &+ [\mathbf{S} \cdot \mathbf{A_a}^{\mathbf{c}} \cdot \mathbf{I_a} - g_n\beta_n\mathbf{B} \cdot \mathbf{I_a} + \mathscr{H}_Q(a)] \\ &+ [\mathbf{S} \cdot \mathbf{A_b}^{\mathbf{c}} \cdot \mathbf{I_b} - g_n\beta_n\mathbf{B} \cdot \mathbf{I_b} + \mathscr{H}_Q(b)] \end{aligned} \tag{22}$$

where we have replaced the A tensors of the uncoupled representation by those of the coupled system.

$$\mathbf{A_a}^{\mathbf{c}} = \tfrac{7}{3}\mathbf{A_a} \qquad \text{and} \qquad \mathbf{A_b}^{\mathbf{c}} = -\tfrac{4}{3}\mathbf{A_b} \tag{23}$$

Viewing Eq. (22) from a standpoint of the Mössbauer nuclei at sites a and b, we see that the spectrum of each site is described by

$$\mathscr{H}(\mathrm{i}) = \beta \mathbf{S} \cdot \mathbf{g} \cdot \mathbf{B} + \mathbf{S} \cdot \mathbf{A}_\mathrm{i}^\mathrm{c} \cdot \mathbf{I}_\mathrm{i} - g_\mathrm{n}\beta_\mathrm{n}\mathbf{B} \cdot \mathbf{I}_\mathrm{i} + \mathscr{H}_Q(\mathrm{i}) \tag{24}$$

Thus *both* iron sites share the same Zeeman term, and *one* spin thus determines the distribution of *two* internal magnetic fields.

B. Example of Antiferromagnetic Coupling: Fe_2S_2 Clusters

As an example of a spin coupled Fe(II)–Fe(III) cluster, we will discuss some studies of a protein that contains an iron–sulfur cluster with an Fe_2S_2 core. This type of cluster has been studied intensively with a variety of spectroscopic techniques. These studies have established that the cluster contains two iron sites bridged by two sulfide ligands, and that each iron is coordinated additionally to two thiolate ligands provided by cysteine residues. The spectroscopic studies have also suggested that each iron is surrounded by a tetrahedral arrangement of ligands. These ideas were confirmed by subsequent X-ray crystallographic studies.[12]

Figures 7–9 show Mössbauer spectra of MMO reductase, a 40-kDa component protein of the methane monooxygenase (MMO) complex of *Methylosinus trichosporium.*[13] Next, we discuss a few features of these spectra,[14] emphasizing

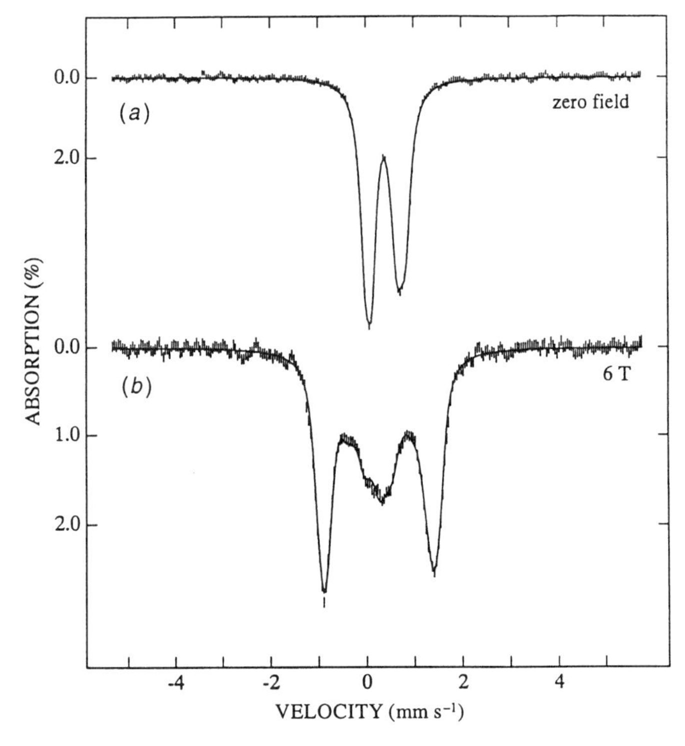

Figure 7
Mössbauer spectra of *oxidized* [Fe^{III}–Fe^{III}] MMO reductase recorded at 4.2 K in zero field (*a*) and in a parallel field of 6.0 T (*b*). Solid lines are theoretical curves calculated with the assumption that $S = 0$. The shape of the 6.0 T spectrum indicates that $\Delta E_Q > 0$ for both sites of the cluster. [Adapted from Fox et al.[14]]

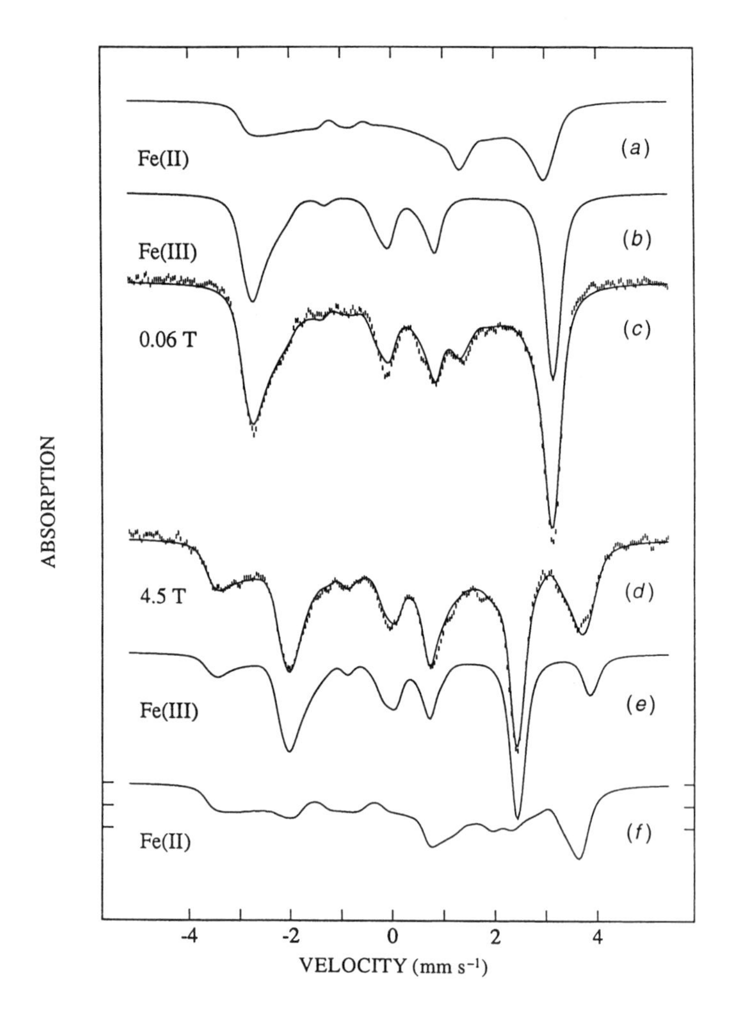

Figure 8
Mössbauer spectra of *reduced* [Fe^{II}–Fe^{III}] MMO reductase recorded at 4.2 K in parallel field of 50 mT (*c*) and 4.5 T (*d*). The solid lines are theoretical spectra generated from Eq. (22) with the following parameters. The Fe(III) site: $\mathbf{A}_{\mathbf{a}}^{\mathbf{c}}$ (MHz) = (−53, −49, −43), $\Delta E_Q = +0.59$ mm s^{-1}, $\eta = 0$, $\delta = 0.31$ mm s^{-1}. The Fe(II) site: $\mathbf{A}_{\mathbf{b}}^{\mathbf{c}}$ (MHz) = (+14, +15, +37), $\Delta E_Q = -3.01$ mm s^{-1}, $\eta = 0.9$, $\delta = 0.65$ mm s^{-1}. The curves in (*a*, *b*, *e*, and *f*) show the spectra of the two sites separately. [Adapted from inconsistent Fox et al.[14]]

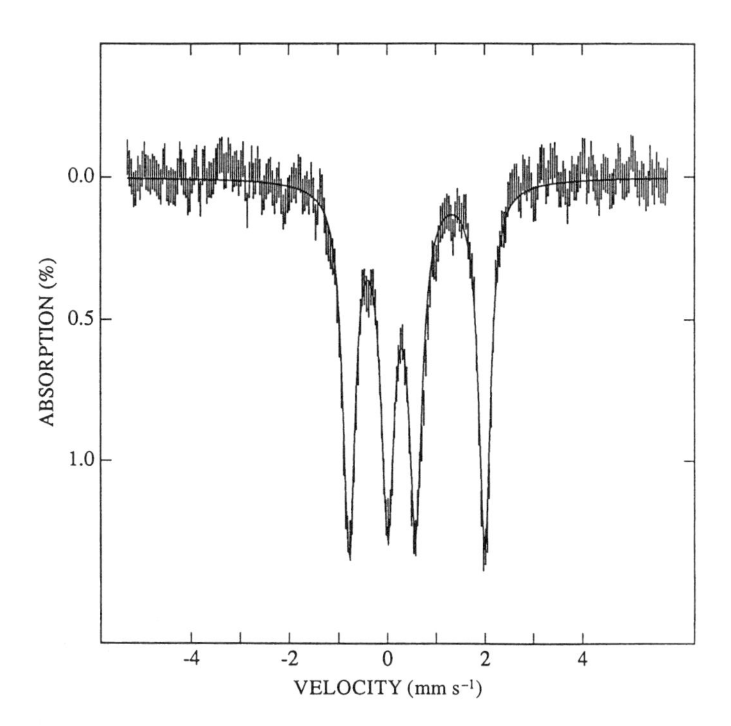

Figure 9
Mössbauer spectra of reduced MMO reductase recorded at $T = 200$ K in zero magnetic field. The theoretical curve is the result of least-squares fitting two quadupole doublets to the data. The two innermost lines belong to the Fe(III) site. [Adapted from Fox et al.[14]]

the utility of Mössbauer spectroscopy for the elucidation of the electronic structure of clusters in general.

1. The Fe_2S_2 cluster of MMO reductase can exist in two oxidation states. Two Mössbauer spectra, recorded at $T = 4.2$ K, of the *oxidized* Fe^{III}–Fe^{III} protein are shown in Figure 7. The spectrum of Figure 7(*a*), recorded in zero magnetic field, consists of a superposition of two very similar quadrupole doublets with ΔE_Q(site 1)= 0.81 mm s^{-1}, $\delta(1) = 0.29$ mm s^{-1}, and $\Delta E_Q(2) = 0.51$ mm s^{-1}, $\delta(2) = 0.25$ mm s^{-1}. Figure 7(*b*) shows a spectrum recorded in a parallel applied field of $B = 6.0$ T; the solid line is a theoretical curve generated by computer simulation under the assumption that the cluster is *diamagnetic*. Recall that $B_{\text{int}} = 0$ for a diamagnetic ($S = 0$) compound. The observed magnetic splittings are entirely attributable to the *applied* field. The values for ΔE_Q and δ are typical for high-spin iron(III) with tetrahedral sulfur coordination. High-spin iron(III) is, however, not diamagnetic. The observed diamagnetism can be explained by assuming antiferromagnetic coupling of two $S = \frac{5}{2}$ spins to a resultant dimer spin $S = 0$.
2. Full reduction of the cluster is achieved by a one-electron reduction. In the *oxidized* form, the cluster is EPR silent. Electron paramagnetic resonance

studies of the *reduced* Fe^{II}–Fe^{III} sample reveal a system with $S = \frac{1}{2}$ and g values at $g_x = 1.86, g_y = 1.94$, and $g_z = 2.06$. Integration of an EPR signal allows one to determine the spin concentration of the sample; the sample discussed here yielded 1.06 ± 0.1 spins per 2 Fe.

3. Consider the 4.2 K Mössbauer spectrum of the *reduced* protein, shown in Figure 8(*c*), and compare the spectrum with that of Figure 7(*a*). It is quite apparent that all the iron of the reduced sample belongs to species exhibiting magnetic hyperfine structure. Thus, the addition of *one* electron has transformed *both* iron atoms from a non-Kramers system (complex with an even number of electrons; S = integer or zero) to a Kramers system (complex with an odd number of electrons; S = half-integer). One conclusion is inescapable: The electron that entered the protein upon reduction must be shared by both iron sites, that is, the two sites must be covalently linked.
4. The Mössbauer spectrum of the reduced reductase of Figure 9 was recorded at $T = 200$ K under conditions where the electronic spin relaxes fast on the Mössbauer time scale. The spectrum consists of two nested quadrupole doublets. The inner doublet has $\Delta E_Q(a) = 0.58\ \mathrm{mm\,s^{-1}}$ and $\delta(a) = 0.26\ \mathrm{mm\,s^{-1}}$, whereas the outer doublet has $\Delta E_Q(\mathrm{b}) = 2.80\ \mathrm{mm\,s^{-1}}$ and $\delta(b) = 0.57\ \mathrm{mm\,s^{-1}}$ (isomer shifts are temperature dependent because of the so-called second-order Doppler shift[1]; for comparison with the shifts of the oxidized protein we use the shifts observed at 4.2 K, $\delta(a) = 0.31\ \mathrm{mm\,s^{-1}}$ and $\delta(\mathrm{b}) = 0.65\ \mathrm{mm\,s^{-1}}$). The parameters of site (b) are typical of those reported for high-spin Fe(II) in a tetrahedral environment of sulfur ligands. The parameters of site (a), on the other hand, are still those of a high-spin iron(III) site. Thus, the reduced cluster consists of a localized Fe(III) and a localized Fe(II) site.
5. From the preceding discussion, it appears plausible that the reduced Fe_2S_2 clusters consists of an antiferromagnetically coupled pair of Fe(II) and Fe(III) ions. Thus, we would anticipate that the 4.2 K Mössbauer spectra need to be evaluated with Eq. (22). Note that the coupling model predicts a sign reversal for the A tensor of the iron(II) site [see Eq. (23)]. It is well established from Mössbauer studies of mononuclear iron(III) and iron(II) compounds with tetrahedral sulfur coordination that the components of the A tensors are negative. Equation (23) predicts negative components for $\mathbf{A_a}^{\mathbf{c}}$ and *positive* components for $\mathbf{A_b}^{\mathbf{c}}$.

A comparison of the features of the spectra of Figure 8(*c* and *d*) confirms this prediction. The essential arguments for determining the sign of the A tensors are as follows. In an applied field of 4.5 T, the Zeeman splitting of the $S = \frac{1}{2}$ doublet is such that at $T = 4.2$ K 81% of the molecules are in the spin-down state for which $\langle S_z \rangle = -\frac{1}{2}$. By measuring the sign of B_{int} the sign of A tensor components can be determined [see Eq. (9)]. As discussed above, the magnetic splitting of a Mössbauer spectrum is determined by $|B_{\mathrm{eff}}| = |B_{\mathrm{int}} + B|$. Thus, depending on the sign of B_{int} the magnetic splitting will either increase or decrease as the applied field is increased. A comparison of the spectra shown in Figure 8(*c* and *d*) shows that one site has $B_{int} > 0$, while the other site has $B_{int} < 0$ (cf. also the theoretical spectra for both sites). Which spectrum belongs to the iron(III) site? This question is answered by computer simulations that show that unacceptable fits result if the iron(III) site has $B_{int} > 0$ [recall that the iron(III) site has a distinctively smaller ΔE_Q and is in this respect quite different fom the iron(II) site].

It turns out that the A tensor of the iron(III) site is slightly anisotropic. [In part, this is so because J is not infinitely large and Eq. (22) is therefore only approximately correct.[14]] If we take the average of the x, y, and z components, $\mathbf{A_a}^{\mathbf{c}}(\mathrm{av}) = -48.3$ MHz we obtain with Eq. (23) $A_a = -20.7$ MHz. This value is in excellent agreement with the value reported for the single Fe(III) site of rubredoxin, $A = -20$ MHz, which has tetrahedral sulfur coordination.[8]

6. Much more can be learned from the parameter set listed in the caption of Figure 8. For instance, consideration of the ligand fields in conjunction with the coupling model explains why g_x and g_y are less than 2; the spin coupling model originally proposed by Gibson et al.[15] has been exceedingly successful. For further details on the pioneering work on the Mössbauer spectra of Fe_2S_2 clusters, the reader is referred to the literature.[16,17] The lessons learned from the studies of Fe_2S_2 clusters have been successfully applied to gain an understanding of the electronic structure of the oxo- and hydroxo-bridged dinuclear clusters of nonheme diiron proteins, such as MMO hydroxylase.[14] The EPR and Mössbauer spectra of these proteins are a bit more complex than those of the Fe_2S_2 clusters, primarily because their J values are substantially smaller. Because the level spacings between the multiplets of the spin coupled system are smaller for iron–oxo proteins, the zfs terms of Eq. (20a) can effectively mix the excited $S = \frac{3}{2}$ multiplet with the $S = \frac{1}{2}$ ground doublet. Consequently, Eqs. (21) and (23) need to be modified by terms that depend on J and the zfs parameters.[14,18,19] This dependence allows one to determine J from an analysis of the EPR and Mössbauer spectra of the ground state. Debrunner and co-workers[18] published a pioneering study of the enzyme purple acid phosphatase.

C. Example of Ferromagnetic Coupling

In this section, we consider briefly the Mössbauer spectra of a ferromagnetically coupled Fe^{II}–Fe^{III} complex. Starting again from Eq. (20), now for $J < 0$, and still using $\beta H \ll |\mathrm{zfs}|$, we obtain a ground multiplet with $S = \frac{9}{2}$. Since $S > \frac{1}{2}$, the zfs terms of Eq. (20a) need to be considered. By using the same methods as descibed above (see also Bencini and Gatteschi[11]), we obtain an effective spin Hamiltonian for the $S = \frac{9}{2}$ multiplet

$$\mathscr{H} = \mathscr{H}_e + \mathscr{H}_{hf} \tag{25}$$

with

$$\mathscr{H}_e = \mathbf{S} \cdot \mathbf{D} \cdot \mathbf{S} + \beta \mathbf{S} \cdot \mathbf{g} \cdot \mathbf{B} \tag{25a}$$

$$\mathscr{H}_{\text{hf}} = [\mathbf{S}\cdot\mathbf{A_a}^{\mathbf{c}}\cdot\mathbf{I_a} - g_n\beta_n\mathscr{H}\cdot\mathbf{I_a} + \mathscr{H}_Q(a)] + [\mathbf{S}\cdot\mathbf{A_b}^{\mathbf{c}}\cdot\mathbf{I_b} - g_n\beta_n\mathbf{B}\cdot\mathbf{I_b} + \mathscr{H}_Q(b)] \tag{25b}$$

$$\mathbf{g} = \tfrac{5}{9}\mathbf{g_a} + \tfrac{4}{9}\mathbf{g_b}, \tag{25c}$$

$$\mathbf{D} = \tfrac{5}{18}\mathbf{D_a} + \tfrac{3}{18}\mathbf{D_b} \tag{25d}$$

$$\mathbf{A_a}^{\mathbf{c}} = \tfrac{5}{9}\mathbf{A_a} \quad \text{and} \quad \mathbf{A_b}^{\mathbf{c}} = \tfrac{4}{9}\mathbf{A_b} \tag{25e}$$

Note that no sign reversal occurs when the A tensor of the iron(II) site, $\mathbf{A_b}$, is referred to the coupled system. For $\beta B \ll |D|$ we can proceed as for the $S = \frac{5}{2}$ multiplet discussed above and consider each Kramers doublet separately.

As an example, we consider the Mössbauer spectra of a dinuclear complex with the trianionic dinucleating ligand salmp [bis(salicylideneamino)-2-methylphenolate(3−)], $(Et_4N)[Fe_2(salmp)_2$[20]; the structure of the compound is indicated in Figure 10. From these spectra we can draw the following conclusions.

1. At $T = 100$ K we observe two quadrupole doublets with a 1 : 1 intensity ratio (data not shown). Doublet (a) has parameters typical of high-spin iron(III) in an octahedral environment of N, O ligands, with $\Delta E_Q(a) = 0.98$ mm s^{-1} and $\delta(a) = 0.60$ mm s^{-1}. Doublet (b), with $\Delta E_Q(b) = 2.29$ mm s^{-1} and $\delta(b) = 1.13$ mm s^{-1}, reflects a high-spin iron(II) site.
2. Figure 11 shows a Mössbauer spectrum recorded at $T = 1.5$ K (for $3\text{ K} \le T \le 50$ K the spectra are poorly resolved due to relaxation effects.) The intensities of this exceptionally well-resolved spectrum are the same whether the 50-mT field is applied parallel or perpendicular to the observed γ radiation. As discussed above, this situation implies that the electronic ground state is a Kramers doublet with uniaxial magnetic

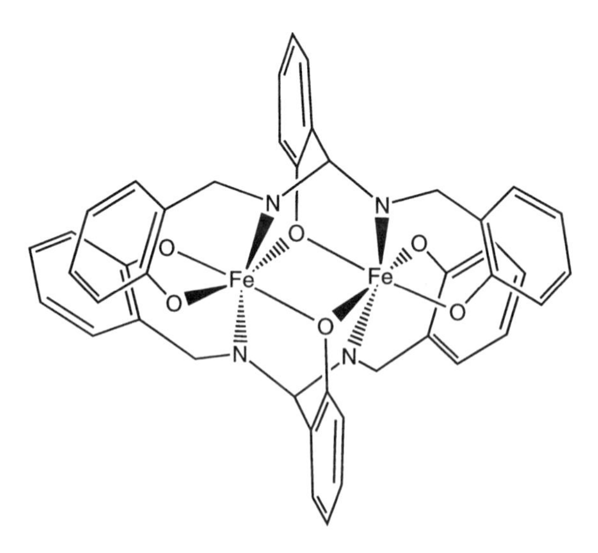

Figure 10
Schematic structure of $[Fe_2(salmp)_2]^-$.

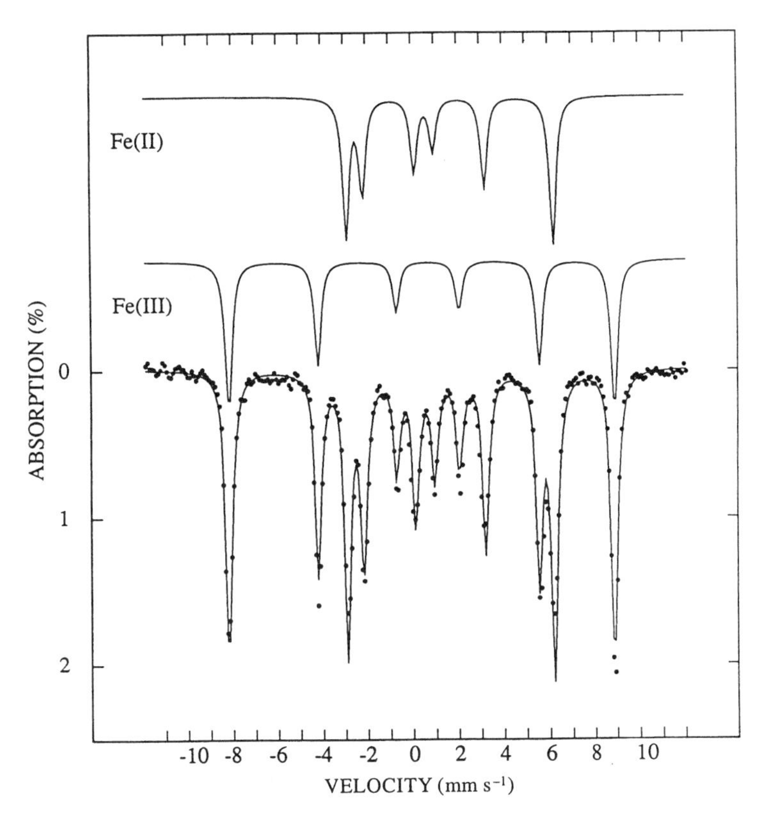

Figure 11
Mössbauer spectrum of polycrystalline $(Et_4N)[Fe_2(salmp)_2]$ recorded at 1.5 K. The spectrum results from the lowest Kramers doublet of an $S = \frac{9}{2}$ multiplet. The theoretical curves were generated from Eq. (25); the parameters used are given in Surerus et al.[21] [Adapted from Surerus et al.[21]]

properties. For a pair of high-spin iron(III) and iron(II) ions, the $S = \frac{1}{2}$ system state cannot have an uniaxial (or highly anisotropic) g tensor. The g tensor of the high-spin Fe(III) is nearly isotropic, $g_0 = 2.0$, and the g values of high-spin Fe(II) are slightly above $g = 2$, with anisotropies smaller than 15%; the system g tensor, computed with Eq. (21), therefore cannot be uniaxial. Thus, we can conclude from the 1.5 K spectrum that the ground multiplet must be the $S = \frac{9}{2}$ manifold that results from ferromagnetic coupling. This conclusion is in accord with magnetic susceptibility studies[21] and subsequent EPR studies, which confirm that the ground multiplet has $S = \frac{9}{2}$. Resonances were observed for the first and second excited-state Kramers doublet of an $S = \frac{9}{2}$ system with $D \approx 2.5\ \text{cm}^{-1}$ and $E/D \approx 0.25$. These parameters produce a ground Kramers doublet with $g_x \approx g_z \approx 0$ and $g_y \approx 17.5$. For $D \approx 2.5\ \text{cm}^{-1}$ and $E/D = 0.25$ the first excited Kramers doublet is at energy $14.9\ \text{cm}^{-1}$ and thus not populated at the temperature the spectrum of Figure 10 was recorded.

3. With the known values for ΔE_Q and δ and the knowledge that B_{int} has to be along the y direction for both iron sites (they share the same electronic system), it is relatively straightforward to decompose the experimental spectrum of Figure 10 into an iron(III) and an iron(II) subspectrum. This decomposition yields for the iron(III) site $\mathbf{A_a}^c = -16.1$ MHz, and using Eq. (25e) we obtain $\mathbf{A_a} = -29$ MHz, which is typical for such a site. Since high-spin iron(II) A tensors are in general quite anisotropic and only the y component of B_{int} is measured here, a comparison with A tensors of high-spin iron(II) sites is not meaningful.

D. Double Exchange, a Coupling Mechanism in Delocalized Mixed-Valence Systems

In the two preceding examples, we have discussed dinuclear clusters with *localized* Fe(III) and Fe(II) sites. Valence localization was evident by the observation of *two* quadrupole doublets with values for ΔE_Q and δ typical for high-spin Fe(III) and Fe(II) sites. In a fully *delocalized* system, one would observe only one quadrupole doublet with values for ΔE_Q and δ, which are averages between those observed for localized sites. Formally, the two sites are in the $Fe^{2.5+}$ oxidation state. Recently, it has been shown that for *delocalized* systems the Heisenberg–Dirac–Van Vleck (HDVV) Hamiltonian $\mathscr{H}_{\mathrm{ex}} = J\mathbf{S_a} \cdot \mathbf{S_b}$ needs to be augmented by a term that takes the delocalization into account. The new term, referred to as double exchange or spin-dependent delocalization, produces splittings that are linear in the system spin S, in contrast to the HDVV term, which yields the characteristic $S(S+1)$ dependence. For a dinuclear cluster with valence delocalization, one has to replace Eq. (19) with

$$E(S) = (J/2)[S(S+1) - S_a(S_a+1) - S_b(S_b+1)] \pm B(S+\tfrac{1}{2}) \qquad (26)$$

where B measures the strength of the double exchange. For sufficiently large B, the ground state can thus have $S = \frac{9}{2}$ even if the HDVV coupling is *antiferromagnetic* ($J > 0$). Thus, without information from a Mössbauer study, one could easily mistake the situation for HDVV ferromagnetism ($J > 0$). Girerd and co-workers have introduced a spin Hamiltonian with a novel transfer operator that produces the correct energies of Eq. (26), and that can describe double exchange in more complicated system[22]; Münck et al.[23] provides a relatively elementary account. To date, only one dinuclear complex exhibiting double exchange has been well characterized; Wieghardt and co-workers[24] synthesized an Fe^{II}–Fe^{III} complex, which has an $S = \frac{9}{2}$ ground state and yields identical Mössbauer spectra for both sites. Inclusion of double exchange is also important for the description of the magnetic properties of clusters with Fe_3S_4 and Fe_4S_4 cores. The interested reader will find an illuminating discussion about the interplay of Heisenberg exchange, double exchange, and vibronic interactions in an article by Blondin et al.[25]

E. An Example for Combined EPR and Mössbauer Studies

Much attention in metalloprotein research and in bioinorganic chemistry is being devoted to the elucidation of the structure and function of diiron proteins, a class that includes the enzymes methane monooxygenase, ribonucleotide reductase, and the purple acid phosphatases as well as the oxygen carrier hemerythrin.[13] For our discussion, it is of interest that an iron(IV) intermediate has been proposed for the catalytic cycle of methane monooxygenase; Mössbauer studies have shown that this oxidation level is indeed attained.[26] The active sites of diiron proteins contain oxo (or hydroxo)-bridged dinuclear clusters, and much effort is under way to synthesize appropriate structural and functional models.

Que and co-workers[27–29] characterized a number of formally Fe^{III}–Fe^{IV} intermediates having the formulation $[Fe_2(O)_2L_2]^{3+}$ derived from the reactions of (μ-oxo)-diiron(III) complexes $[Fe^{III}_2\ O(L)_2OH(H_2O)](ClO_4)_3$ with H_2O_2 (Fig. 12). For L = TPA, 5-Me$_3$-TPA, 5-Et$_3$-TPA [TPA = tris(pyridyl-2-methyl)-amine], the intermediates have an $Fe_2(\mu\text{-}O)_2$ core based on extended X-ray absorption fine structure (EXAFS), resonance Raman, and electrospray ionization

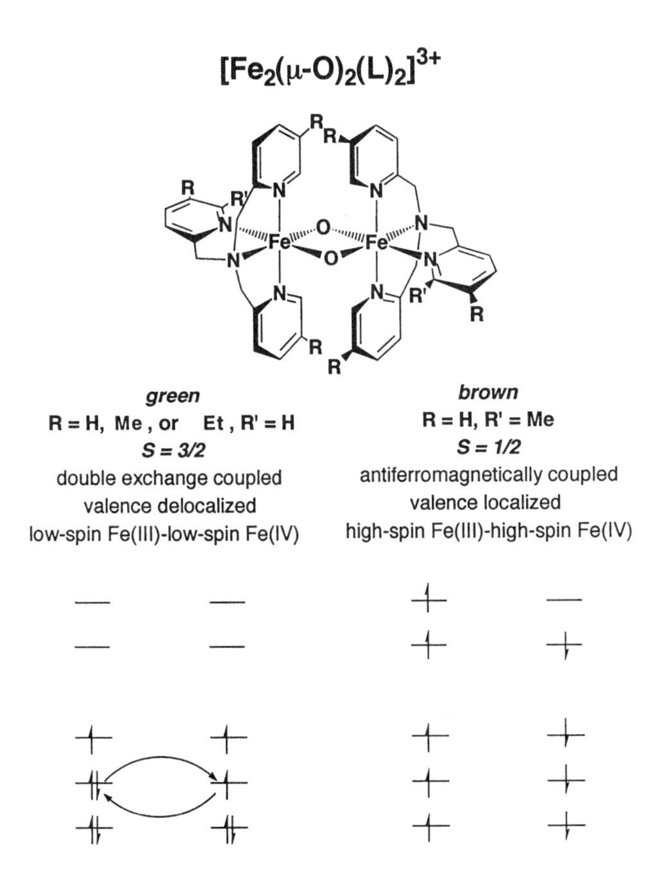

Figure 12
Proposed structures and spin coupling schemes for $[Fe_2O_2L_2]^{3+}$ complexes.

mass spectral data.[27,28] The presence of the 5-alkyl substituents does not significantly affect the electronic properties of this core. The introduction of a 6-methyl substituent, that is, 6-Me-TPA, affords an intermediate with a similar formulation on the basis of its electrospray ionization mass spectrum but with dramatically different electronic structure.[29] Mössbauer spectroscopy has played a key role in elucidating the nature of these novel Fe^{III}–Fe^{IV} species.

For L = TPA, 5-Me_3-TPA, 5-Et_3-TPA, treatment of the diiron(III) starting material $[Fe^{III}{}_2O(L)_2OH(H_2O)](ClO_4)_3$ with H_2O_2 in MeCN affords a green (λ_{max} 614 nm) metastable intermediate at −40 °C. It exhibits a well-resolved EPR spectrum with g values at $g_x = 3.90, g_y = 4.45$, and $g_z = 2.01$. From our discussion of effective g values of Kramers doublets, we recognize these g values as those belonging to the $M = \pm\frac{1}{2}$ doublet of a system with $S = \frac{3}{2}$. The temperature dependence of the EPR spectra revealed that the $M = \pm\frac{1}{2}$ doublet is the ground state (i.e., $D > 0$) and that the $M = \pm\frac{3}{2}$ doublet is at least 40 cm^{-1} above the ground state. Thus, at 4.2 K only the $M = \pm\frac{1}{2}$ doublet is populated.

The Mössbauer spectra of the green intermediate recorded at $T = 140$ and 4.2 K are shown in Figure 13. The spectra are dominated by two unresolved quadrupole doublets with $\Delta E_Q(1) = 1.57$ mm s^{-1} and $\Delta E_Q(2) = 1.03$ mm s^{-1} and $\delta(1) = \delta(2) = 0.42$ mm s^{-1} at 140 K, which arise from the starting diiron-(III) complex; note the major doublet in Figure 13(a) (experimental spectrum minus the contribution of the doublet outlined by the solid line). The starting material is diamagnetic, as expected for an antiferromagnetically coupled pair of high-spin iron(III) sites. Except for a shoulder on the left, the spectrum of the green intermediate is hidden under the major doublet in Fig. 13(b), because the species of interest has small magnetic hyperfine interactions. Since the green intermediate is EPR active, it should have a magnetic component in the $T = 4.2$ K spectrum, whose shape we can describe very well by using the correlations between EPR and Mössbauer spectroscopy as described below.

Consider a diamagnetic compound exhibiting a quadrupole doublet at 4.2 K. Application of a weak field of 50 mT will not cause any measurable changes in the spectrum; the nuclear Zeeman interaction is too small for $B = 50$ mT. Assume that we measure the spectrum in parallel as well as in perpendicular field. The difference spectrum will yield a flat baseline. Contrast this situation with that of a *paramagnetic EPR active* compound. Assume that we obtain spectra similar to those of Figure 3(a and b). Subtracting the perpendicular spectrum from that obtained in parallel field will produce a difference spectrum for which the $\Delta m_I = 0$ lines point upward. Thus, if we have a sample containing an EPR active and a diamagnetic (or integer spin paramagnetic) species, we can cancel the contribution of the latter by taking the difference spectrum. The difference spectrum shown in Figure 13(c) is the contribution of the desired EPR active green intermediate. By computer simulations, one can fit the difference spectrum and determine the magnetic hyperfine interactions, the quadrupole splittings, and the isomer shift. The solid lines in Figure 13(b and c) show the results of such calculations. By matching the simulated spectrum to the left shoulder of the spectrum of Figure 13(b), we infer that 30% of the total Fe in the sample belongs to the green intermediate. Most interestingly, the computer simulations showed

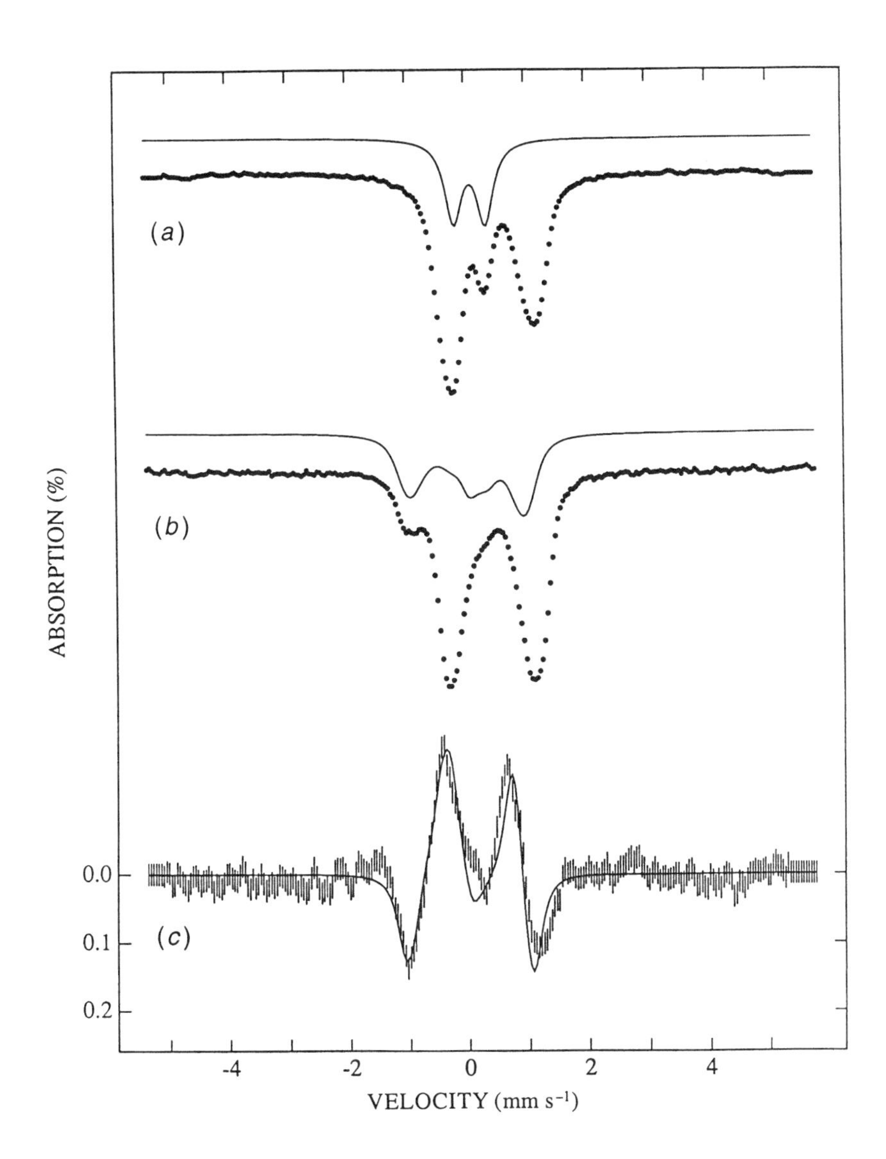

Figure 13
Mössbauer spectra obtained to characterize the green intermediate $[Fe_2O_2(TPA)_2]^{3+}$ generated by reacting $[Fe_2O(TPA)_2(OH)(H_2O)](ClO_4)_3$ at −40 °C in acetonitrile with H_2O_2. The spectra were obtained at $T = 140$ K in zero field (*a*) and at $T = 4.2$ K in a 0.15 T parallel field (*b*). The difference spectrum shown in (*c*) was obtained by subtracting a spectrum obtained in 60 mT perpendicular field from the corresponding parallel field spectrum. The solid lines are theoretical curves computed with an $S = \frac{3}{2}$ spin Hamiltonian. [Adapted from Leising et al.[27]]

that the paramagnetic species consists of only one spectral component, that is, if the $S = \frac{3}{2}$ system represents a diiron center, both iron sites are equivalent.

At 140 K, the electronic spin of the $S = \frac{3}{2}$ species relaxes fast on the time scale of Mössbauer spectroscopy, and the magnetic spectrum observed at 4.2 K collapses into one quadrupole doublet. This doublet again accounts for 30% of the iron in the sample and, when correlated with is EPR signal intensity, corresponds to about 1 spin/2 Fe for the $S = \frac{3}{2}$ species. The Mössbauer parameters for the doublet are $\Delta E_Q = 0.53$ mm s^{-1} and $\delta = 0.07$ mm s^{-1}. The decrease in the isomer shift from 0.42 mm s^{-1} indicates that the diiron center has been oxidized in the intermediate, in agreement with the formal Fe^{III}–Fe^{IV} state deduced from the mass spectral data. However, the fact that the iron sites are equivalent in this species implies a valence delocalized description. The system spin $S = \frac{3}{2}$ is proposed to result from parallel spin alignment of low-spin ($S_1 = \frac{1}{2}$) Fe(III) and low-spin ($S_2 = 1$) Fe(IV) sites, likely coupled via a double exchange mechanism (Fig. 12).[28,30]

The electronic description that is emerging for the green intermediate is corroborated by more recent data obtained on a related "brown intermediate". The brown intermediate $[Fe_2(O)_2(6\text{-Me-TPA})_2]^{3+}$ is generated by the reaction of H_2O_2 with the corresponding (μ-oxo)diiron(III) 6-Me-TPA complex.[29] It exhibits a brown color ($\lambda_{max} = 350$ nm) and an isotropic EPR signal at $g = 2.0$. This signal broadens when either ^{57}Fe or ^{17}O is introduced into the intermediate, showing that the resonance originates not from a radical but from the Fe_2O_2 core. Figure 14 shows a Mössbauer spectrum of the brown intermediate recorded at 1.5 K in the presence of a 2.5 T applied field. Analysis of this spectrum is straightforward. The central part of the spectrum is a quadrupole doublet (53% of the total Fe) with parameters identical to that of the diiron(III) starting material. The remainder (47%) is a paramagnetic component that corresponds to the $S = \frac{1}{2}$ species. Correlation of the Mössbauer and EPR intensities gives a spin concentration of 1 spin per 2Fe for the $S = \frac{1}{2}$ species.

The Mössbauer spectrum of the brown intermediate consists of two subspectra of equal intensity. By studying the sample in parallel and transverse applied fields, it was possible to prove that these two subspectra belong to the $S = \frac{1}{2}$ spin system. One subspectrum, labeled Fe(III) in Figure 14, has parameters typical of a high-spin ($S_1 = \frac{5}{2}$) Fe(III) site; for example, its isomer shift, $\delta = 0.48$ mm s^{-1}, is identical to that observed for the two sites of the diiron(III) starting material. The second site has $\delta = 0.08$ mm s^{-1}, which implies that one of the iron sites is oxidized to the Fe(IV) state.[31] By taking into account that the intermediate has a system spin $S = \frac{1}{2}$ and that one of the iron sites is high-spin Fe(III), it follows that the second site must be high-spin ($S_2 = 2$) Fe(IV). One can use Eqs. (21) and (23) to analyze the properties of the brown intermediate, since an octahedral high-spin d^4 electronic configuration is complementary to a tetrahedral high-spin d^6 configuration. Such an analysis shows that the electronic structure of the brown intermediate is appropriately described as an antiferromagnetically coupled diiron center with localized high-spin Fe(III) and Fe(IV) sites.

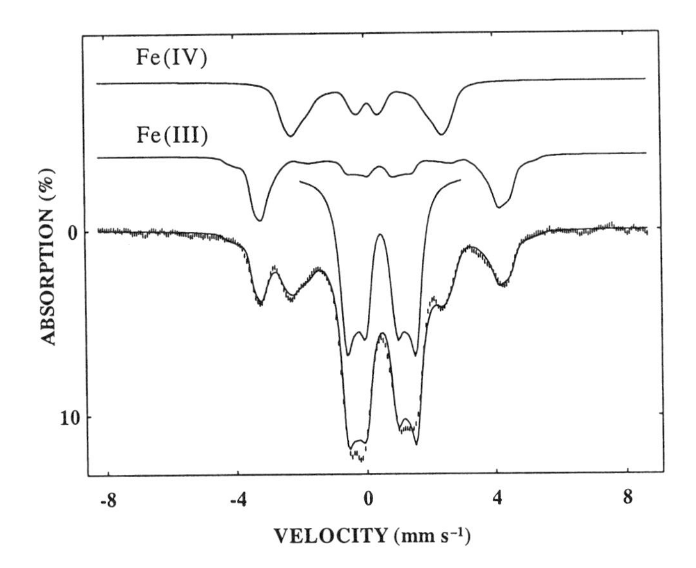

Figure 14
Mössbauer spectrum of the brown intermediate $[Fe_2O_2(6\text{-Me-TPA})_2]^{3+}$ recorded at 1.5 K in a parallel applied field of 2.5 T. The solid lines drawn above the data are theoretical curves describing the Fe(III) site and the Fe(IV) site, as well as a remnant (53%) of the diiron(III) starting material. The solid line drawn through the data is the sum of the three theoretical curves. [Adapted from Dong et al.[29]]

The differing electronic properties of the green and brown intermediates can be understood by considering the effects of the 6-methyl substituent. Based on a comparison of structures of Fe(III) TPA and 6-Me-TPA complexes, introducing the 6-methyl substituent imposes steric constraints that discourages formation of the shorter metal–ligand bonds required by a low-spin configuration. Thus a low-spin configuration for an Fe(TPA) center would likely be converted to a high-spin configuration for the corresponding Fe(6-Me-TPA). For the green intermediate, the low-spin configurations of the individual iron sites give rise to the $S = \frac{3}{2}$ system spin via double exchange, while for the brown intermediate, the high-spin configurations of the individual iron sites give rise to the $S = \frac{1}{2}$ system spin via antiferromagnetic exchange (Fig. 12). These insights into electronic structure could only be obtained by a judicious application of Mössbauer principles in combination with EPR analysis.

ACKNOWLEDGMENTS

Preparation of this chapter was supported by grant DMB-9096231 from the National Science Foundation and grant GM 22701 from the National Institutes of Health.

References

1. Gütlich, P.; Link, R.; Trautwein, A. *Mössbauer Spectroscopy and Transition Metal Chemistry*; Springer-Verlag: Berlin, 1979.
2. Greenwood, N.; and Gibb, T. *Mössbauer Spectroscopy*; Chapman & Hall: London, 1971.
3. May, L., Ed., *An Introduction to Mössbauer Spectroscopy.*; Plenum: New York, 1971.
4. Goldanskii, V. I.; Herber, R., Eds., *Chemical Applications of Mössbauer Spectroscopy*; Academic: New York, 1968.
5. Aasa, R.; Vånngard, T. "EPR Signal Intensity and Powder Shapes: A Reexamination" *J. Magn. Reson.* **1975**, *17*, 308–305.
6. Abragam, A.; Bleaney, B. *Electron Paramagnetic Resonance of Transition Metal Ions*; Dover, New York, 1986.
7. Wickman, H. H.; Klein, P. M.; Shirley, D. A. "Paramagnetic Hyperfine Structure and Relaxation Effects in Mössbauer Spectra: ^{57}Fe in Ferrichrome A" *Phys. Rev.* **1966**, *152*, 345–357.
8. Schulz, C. E.; Debrunner, P. G. "Rubredoxin, a Simple Iron-Sulfur Protein: Its Spin Hamiltonian and Hyperfine Parameters" *J. Phys. (Paris) Colloq.* **1976**, *37*, 154–158.
9. Whittaker, J. W.; Lipscomb, J. D.; Kent, T. A.; Münck, E. "*Brevibacterium fuscum* Protocatechuate 3,4-Dioxygenase, Purification, Crystallization and Characterization" *J. Biol. Chem.* **1984**, *259*, 4466–4475.
10. Lang, G.; Aasa, R.; Garbett, K.; Williams, R. J. P. "Paramagnetic Mössbauer Spectra of Some Rhombic Fe^{3+} Materials: Correlation with ESR" *J. Chem. Phys.* **1971**, *55*, 4539–4578.
11. Bencini, A.; Gatteschi, D. *EPR of Exchange Coupled Systems.*; Springer-Verlag: Berlin, 1990.
12. Fukuyama, K.; Hase, T.; Matsumoto, S.; Tsukihara, T.; Katsube, Y.; Tanaka, N.; Kakudo, M.; Wada, K.; Matsubara, H. "Structure of *S. platensis* [2Fe–2S] Ferredoxin and Evolution of Chloroplast-type Ferredoxins" *Nature* (*London*) **1980**, *286*, 522–524.
13. (a) Que, L., Jr., and True, A. E. "Dinuclear Iron- and Manganese-Oxo Centers In Biology" *Prog. Inorg. Chem.* **1990**, *38*, 97–200. (b) Lipscomb, J. D. "Biochemistry of the Soluble Methane Monooxygenase" *Annu. Rev. Microbiol.* **1994**, *48*, 371–399.
14. Fox, B. G.; Hendrich, M. P.; Surerus, K. K.; Andersson, K. K.; Froland, W. A.; Lipscomb, J. D.; Münck, E., "Mössbauer, EPR, and ENDOR Studies of the Hydroxylase and Reductase Components of Methane Monooxygenase from *Methylosinus trichosporium* OB3b" *J. Am. Chem. Soc.* **1993**, *115*, 3688.
15. Gibson, J. F.; Hall, D. O.; Thornley, J. H. M.; Whatley, F. R. "The Iron Complex in Spinach Ferredoxin" *Proc. Natl. Acad. Sci. USA* **1966**, *56*, 987–990.
16. Münck, E.; Debrunner, P. G.; Tsibris, J. C. M.; Gunsalus, I. C. "Mössbauer Parameters of Putidaredoxin and Its Selenium Analogue" *Biochemistry* **1972**, *11*, 855–863.
17. Sands, R. H.; Dunham, W. R. "Spectroscopic Studies on Two-Iron Ferredoxins" *Q. Rev. Biophys.* **1975**, *7*, 443–504.
18. Sage, J. T.; Xia, Y.-M.; Debrunner, P. G.; Keough, D. T.; de Jersey, J.; Zerner, B. "Mössbauer Analysis of the Binuclear Iron Site in Purple Acid Phosphatase from Pig Allantoic Fluid" *J. Am. Chem. Soc.* **1989**, *111*, 7239–7247.
19. Bertrand, P.; Guigliarelli, B.; More, C. "The Mixed Valence Fe(III)Fe(II) Binuclear Centers of Biological Molecules Viewed Through EPR Spectroscopy" *New J. Chem.* **1991**, *15*, 445–454.
20. Snyder, B. S.; Patterson, G. S.; Abrahamson, A. H.; Holm, R. H. "Binuclear Iron System Ferromagnetic in Three Oxidation States: Synthesis, Structures, and Electronic Aspects of Molecules with a $Fe_2(OR)_2$ Bridge Unit Containing Fe(III,III), Fe(III,II), and Fe(II,II)" *J. Am. Chem. Soc.* **1989**, *111*, 5214–5223.
21. Surerus, K. K.; Münck, E.; Snyder, B. S.; Holm, R. H. "A Binuclear Mixed-Valence Ferromagnetic Iron System with an $S = 9/2$ Ground State and Valence Trapped and Detrapped States" *J. Am. Chem. Soc.* **1989**, *111*, 5501–5502.

22. Papaefthymiou, V.; Girerd, J. J.; Moura, I.; Moura, J. J. G.; Münck, E. "Mössbauer Study of *D. gigas* Ferredoxin II and Spin Coupling Model for Fe_3S_4 Cluster with Valence Delocalization" *J. Am. Chem. Soc.* **1987**, *109*, 4703–4710.
23. Münck, E.; Papaefthymiou, V.; Surerus, K. K.; Girerd, J.-J. "Mössbauer Study of *D. gigas* Ferredoxin II and Spin-Coupling Model for the Fe_3S_4 Cluster with Valence Delocalization" In *Metal Clusters in Proteins*; L. Que, Jr., Ed; ACS Symposium Series 372, American Chemical Society: Washington DC, 1988; Chapter 15.
24. Ding, X.-Q.; Bominaar, E. L.; Bill, E.; Winkler, H.; Trautwein, A. X.; Drüke, S.; Chaudhuri, P.; Wieghardt, K. "Mössbauer and Electron Paramagnetic Resonance Study of the Double-Exchange and Heisenburg-Exchange Interactions in a Novel Binuclear Fe(II/III) Delocalized-Valence Compound" *J. Chem. Phys.* **1989**, *92*, 178–186.
25. Blondin, G.; Borshch, S.; Girerd, J.-J. "When Must Double Exchange Be Used?" *Comments Inorg. Chem.* **1992**, *12*, 315–340.
26. Lee, S.-K.; Fox, B. G.; Froland, W. A.; Lipscomb, J. D. and Münck, E. "A Transient Intermediate of the Methane Monooxygenase Catalytic Cycle Containing an $Fe^{IV}Fe^{IV}$ Cluster" *J. Am. Chem. Soc.* **1993**, *115*, 6450–6451.
27. Leising, R. A.; Brennan, B. A.; Que, L., Jr., Fox, B. G. and Münck, E. "Models for Non-Heme Iron Oxygenases: A High Valent Iron-Oxo Intermediate," *J. Am. Chem. Soc.* **1991**, *113*, 3988–3990.
28. Dong, Y.; Fujii, H.; Hendrich, M. P.; Leising, R. A.; Pan, G.; Randall, C. R.; Wilkinson, E. C.; Zang, Y.; Que, L., Jr.; Fox, B. G.; Kauffmann, K. and Münck, E. "A High-Valent Nonheme Iron Intermediate. Structure and Properties of $[Fe_2(\mu\text{-}O)_2(5\text{-}Me_3\text{–}TPA)_2]\cdot(ClO_4)_3$" *J. Am. Chem. Soc.* **1995**, *117*, 2778–2792.
29. Dong, Y.; Que, L., Jr.; Kauffmann, K. and Münck, E. "An Exchange Coupled Complex with Localized High-Spin Fe^{IV} and Fe^{III} Sites of Relevance to Cluster X of *E. coli* Ribonucleotide Reductase" *J. Am. Chem. Soc.*, **1995**, *117*, 11377–11378.
30. A. Ghosh, J. E. Almlöf, and L. Que, Jr., "Electronic Structure of Non-Heme High-Valent Iron–Oxo Complexes With the Unprecedented $[Fe_2(\mu\text{-}O)_2]^{3+}$ Core", *Angew. Chem. Int. Ed. Engl.* **1966**, *35*, 770–772.
31. Kostka, K. L.; Fox, B. G.; Hendrich, M. P.; Collins, T. J.; Rickard, C. E. F.; Wright, L. J. and Münck, E. "High Valent Transition Metal Chemistry. Mössbauer and EPR Studies of High-Spin ($S = 2$) Iron(IV) and Intermediate spin ($S = \frac{3}{2}$) Iron(III) Complexes with a Macrocyclic Tetraamido-N Ligand" *J. Am. Chem. Soc.*, **1993**, *115*, 6746–6757.

General References

Gütlich, P.; Link, R.; Trautwein, A. *Mössbauer Spectroscopy and Transition Metal Chemistry*. Springer-Verlag, Berlin, 1979.

Kahn, O. *Molecular Magnetism*, VCH, New York, 1993.

7

Molecular Magnetism in Bioinorganic Chemistry

JEAN-JACQUES GIRERD and YVES JOURNAUX

Laboratoire de Chimie Inorganique
UMR CNRS 8613
Université Paris-Sud, France

CONTENTS

I. Introduction

Molecular magnetism has been the subject of much interest in inorganic chemistry for several decades and has provided insights into the electronic structures of both mononuclear and polynuclear metal complexes.[1–3] For the latter in particular, the temperature dependence of their magnetic susceptibilities has revealed interactions between metal centers that relate to their structures, and such magnetostructural correlations have developed from subsequent characterization of the structures of these polynuclear systems by X-ray diffraction. From these efforts, two fields have emerged: one is the application of the acquired knowledge for the design and synthesis of molecular magnetic materials and the other is the study of magnetic properties of metallobiomolecules and models. In this chapter, we discuss the main concepts and methods of magnetochemistry with a special focus on their applications to bioinorganic chemistry.

II. Basic Definitions and Units

Let us consider a sample containing N molecules, each molecule having an electronic spin $S = \frac{1}{2}$. Later in this chapter we will see that in the presence of a magnetic induction $\mathbf{B}$[a] each molecule in this sample has one of the two following energies: $\frac{1}{2}g\beta B$ or $-\frac{1}{2}g\beta B$, g being the Landé factor and β the Bohr magneton. The magnetic moment for a molecule in the lower energy state has a projection on $\mathbf{B}$ equal to $\mu_1 = \frac{1}{2}g\beta$ (parallel to the magnetic induction), while the magnetic moment for a molecule in the higher energy state has a projection $\mu_2 = -\frac{1}{2}g\beta$ (antiparallel to $\mathbf{B}$). The distribution of the N molecules into these two states obeys Boltzmann's law, where N_1 represents the number of molecules at the lower energy and N_2 represents the number of molecules at the higher energy [Eq. (1)].

$$
\begin{aligned}
N_1 &= \frac{N \exp(g\beta B/2kT)}{\exp(g\beta B/2kT) + \exp(-g\beta B/2kT)} \\
N_2 &= \frac{N \exp(-g\beta B/2kT)}{\exp(g\beta B/2kT) + \exp(-g\beta B/2kT)}
\end{aligned}
\tag{1}
$$

where k is the Boltzmann constant.

[a] The unit of B is the tesla (T). The constant $\beta = 9.2740154 \times 10^{-24}$ J T^{-1} is the Bohr magneton.

The magnetization of the sample is obtained by summing the magnetic moments of all the molecules. The contribution of the N_1 molecules, with their magnetic moment aligned along **B**, is $N_1\mu_1$, while that of the N_2 molecules, with their magnetic moment aligned in the opposite direction, is $N_2\mu_2$. Thus

$$M = N_1 g\beta/2 - N_2 g\beta/2 \tag{2}$$

As there are more molecules in the lower energy state than in the higher one ($N_1 > N_2$), we see that M is positive, that is, the sample has a magnetization **M** parallel to the magnetic induction **B**. Such a sample is said to be paramagnetic. If we assume that $g\beta B/kT \ll 1$, we can mathematically develop the exponentials to arrive at a much simpler expression

$$M = Ng^2\beta^2 B/4kT \tag{3}$$

Note that the magnetization is proportional to the magnetic induction. In particular, it is zero in the absence of magnetic induction. It is also inversely proportional to the temperature T so that it doubles each time we divide T by a factor of 2.[b] This inverse dependence of M on T results from the fact that more and more molecules occupy the lower energy state as T decreases.

As far as definitions are concerned, the primary quantity defined in textbooks on electricity and magnetism is the magnetization per unit of volume, that is, we define N in Eq. (3) as the number of molecules per unit of volume. The unit of magnetization per unit volume is joule per tesla per cubic meter or ampere per reciprocal meter. ($\mathrm{J\,T^{-1}\,m^{-3}}$ or $\mathrm{A\,m^{-1}}$). For N equals Avogadro's number (N_A), the units for molar magnetization M_M are joule per tesla per mole or ampere square meters per mole ($\mathrm{J\,T^{-1}\,mol^{-1}}$ or $\mathrm{A\,m^2\,mol^{-1}}$). Another important quantity is the magnetic susceptibility per unit of volume defined as[c]

$$\chi = \mu_0 M/B \tag{4}$$

where $\mu_0 = 4\pi \times 10^{-7}\ \mathrm{A^{-1}\,mT}$. With this definition χ is dimensionless. For a sample containing molecules with $S = \frac{1}{2}$, we obtain with the use of Eq. (3)

$$\chi = \mu_0 Ng^2\beta^2/4kT \tag{5}$$

This equation is known as the Curie law for magnetic susceptibility. We see that $\chi > 0$ for a paramagnetic sample. In practice, the susceptibility is expressed as χ_g (gram susceptibility) or χ_M (molar susceptibility) where

---

[b] When the temperature T becomes very small, the magnetization does not become infinite but saturates instead. The same happens when B becomes very large (see later in the chapter).

[c] It turns out that the conventional definition is $\chi = M/H$, where **H** is the magnetic field related to the magnetization **M** and the magnetic induction **B** through the fundamental formula, $\mathbf{B} = \mu_0(\mathbf{H} + \mathbf{M})$, where **H** is in units of ampere per reciprocal meter ($\mathrm{Am^{-1}}$). This definition leads to a quantity that, from a strictly mathematical point of view, does not follow the Curie law. With most samples one can make the assumption $M \ll H$; under this assumption, both definitions lead in fact to the same numbers.

$$\chi_g = \chi/\rho \qquad \text{(in units of m}^3\,\text{kg}^{-1}) \tag{6}$$

$$\chi_M = \chi M_0/\rho \qquad \text{(in units of m}^3\,\text{mol}^{-1}) \tag{7}$$

ρ and M_0 being the density of the sample in kilograms per cubic meter ($kg\,m^{-3}$) and the molar mass, respectively. Since the cgs system is still often used for the molar magnetic susceptibility, it is useful to note that χ_M (SI, $m^3\,mol^{-1}$) = $(4\pi \times 10^{-6})\chi_M$ (cgs, $cm^3\,mol^{-1}$). In this chapter we will keep apparent the conversion factor $4\pi \times 10^{-6}$ in order to facilitate the passage from one system to the other.

For molecules that have no electronic spin, **M** is antiparallel to **B** and $\chi < 0$; such samples are referred to as being diamagnetic. Diamagnetism corresponds to the effect of magnetic induction on the motions of paired electrons. It is always present, even in paramagnetic samples. In this chapter, we will focus on electronic paramagnetism. Nuclear paramagnetism is very weak and is usually masked by electronic diamagnetism.

The magnetization represented by Eq. (3) is in fact only the paramagnetic contribution to the magnetization of a sample made of molecules with an electronic spin $S = \frac{1}{2}$. Such a sample will also have a diamagnetic contribution to magnetization. The intensity of electronic paramagnetism is evaluated experimentally after correction for diamagnetism:

$$M_M{}^{\text{para}} = M_M{}^{\text{meas}} - M_M{}^{\text{dia}} \tag{8}$$

$$\chi_M{}^{\text{para}} = \chi_M{}^{\text{meas}} - \chi_M{}^{\text{dia}} \tag{9}$$

$\chi_M{}^{\text{dia}}$ can be computed from tables of diamagnetic molar susceptibilities by adding the contribution of each atom or groups of atoms. A selection of values is given in Table I. In general, paramagnetism is much stronger than diamagnetism,

Table I
Examples of Atomic and Molecular Molar Diamagnetic Susceptibilities[a]

Atom or Molecule	Diamagnetic Susceptibility ($4\pi \times 10^{-12}$ $m^3\,mol^{-1}$)	Atom or Molecule	Diamagnetic Susceptibility ($4\pi \times 10^{-12}$ $m^3\,mol^{-1}$)
H	−2.9	F	−6.3
C	−6.0	Cl	−20.1
N (open chain)	−5.6	Br	−30.6
O (ether or alcohol)	−4.6	I	−44.6
O (carbonyl)	−1.7	S	−15.0
P	−26.3	Se	−23.0
H_2O	−13	Pyridine	−49
NH_3	−18	o-Phenanthroline	−128

and a small error in the evaluation of the diamagnetic correction does not significantly affect the values. When paramagnetism is weak (e.g., biological samples, where the diamagnetic components greatly outnumber the paramagnetic centers), it is appropriate to measure the diamagnetic susceptibility of a control sample in the absence of the paramagnetic center(s).

Diamagnetism is independent of temperature, while paramagnetism is linearly proportional to $1/T$ as stated by the Curie law [Eq. (6)]. Therefore measurements are routinely carried out at liquid helium temperatures (4.2 K) to enhance the accuracy. A number of molecules, however, exhibit non-Curie behavior; in which some deviation from the Curie law is observed. This behavior can occur in systems with one metal per molecule but also in cases where there is more than one paramagnetic center in the molecule. Analysis of such magnetic properties can provide insight into interactions between the paramagnetic centers and is the subject of much present study.

Electronic paramagnetism is related to electron spin, which is why magnetic measurements contribute to the understanding of the electronic properties of matter. The aim of this chapter is to give some examples of the findings in this area with a special focus on cases in bioinorganic chemistry.

III. Measurements

We will consider only two types of magnetic measurements in this chapter, the Faraday method and the superconducting quantum interference device (SQUID) magnetometry; O'Connor[4] can be consulted for a more extensive treatment of methods.[5]

A. Measurement of Susceptibility: The Faraday Method

This method is based on the experimental fact that in an inhomogeneous field, a paramagnetic sample is attracted into the zone with the largest field and a diamagnetic one is repulsed by this zone. A sample of volume dv experiences a force equal to

$$d\mathbf{F} = \tfrac{1}{2}\mu_0(\chi - \chi_{\text{med}})\ \mathbf{grad}(\mathbf{H}^2)\ dv \tag{10}$$

where χ_{med} is the volumic susceptibility of the medium in which the experiment is done. The magnet of the apparatus is built in such a way that $\mathbf{H}^2$ has a nonzero gradient only along the vertical axis (let us call it z) and that this gradient is maximum and constant in a volume comparable to that of the sample (v). The sample is brought into this zone and Eq. (10) then simplifies to

$$F_z = \mu_0(\chi - \chi_{\text{med}})v\mathbf{H}(d\mathbf{H}/dz) \tag{11}$$

The force can be measured with a balance; if the medium is air, the presence of dioxygen makes $\chi_{\text{med}} > 0$ and this diminishes χ so that it is better to make this measurement under vacuum or He. A more important reason to avoid dioxygen is that it can condense on the sample at low temperature (at a temperature depending on pressure) and give erroneous results. The value of $\mathbf{H}(d\mathbf{H}/dz)$ is determined with a standard, often $HgCo(SCN)_4$, for which $\chi_g{}^0$ and mass m^0 are known. So one has

$$\begin{aligned} F_z &= \mu_0 \chi_g m \mathbf{H}(d\mathbf{H}/dz) + F' \\ F_z{}^0 &= \mu_0 \chi_g{}^0 m^0 \mathbf{H}(d\mathbf{H}/dz) + F' \end{aligned} \tag{12}$$

where F' is the force measured for the sample holder. Therefore

$$\chi_M = \chi_g{}^0 [m^0/(F_z{}^0 - F')][(F_z - F')/m] M_0 \tag{13}$$

If the sample is paramagnetic, one can be sure that it is in the right zone of measurement by adjusting the sample position to obtain a maximum value for F. With a diamagnetic sample, only a check of the geometry can ensure that the sample is in the right zone. Using a balance sensitive to the microgram scale allows one to measure samples of a few milligrams. However, some problems can arise with the sample holder and/or the sample. It is very difficult to find a sample holder that is free of paramagnetic impurities. It is thus prudent to measure the magnetic susceptibility of the sample holder in the absence of sample over the entire temperature range and subtract this from the experimental data. Another common problem is the presence of a ferromagnetic impurity in the sample, which can arise from either the metal salts used in the synthesis or the use of a metal spatula. Such an impurity, however, is easy to detect. It corresponds to an essentially field independent magnetization (M_s)

$$M^{\text{meas}} = M^{\text{sample}} + M_s \tag{14}$$

so that

$$\chi^{\text{meas}} = \chi^{\text{sample}} + M_s/H \tag{15}$$

In the presence of such an impurity, χ^{meas}, instead of staying constant, increases when the field is decreased. The only way to avoid such an error is to use ultrapure chemicals and a nonmetallic spatula!

B. Measurement of Magnetization: The SQUID Magnetometer

A SQUID magnetometer uses quite a different principle from the Faraday method and gives much more information, since it measures the magnetization, rather than the susceptibility as the Faraday balance does. In a SQUID magnetometer, the sample is in a uniform field and acquires some magnetization. When it is moved back and forth through two superconducting pick-up coils (Fig. 1), the total flux that threads through the coils is disturbed, and a current, which contains the information on the value of M, is induced to maintain the total flux constant. This phenomenon occurs because the total flux threading a closed superconducting circuit cannot change and moreover is quantized. The quantum

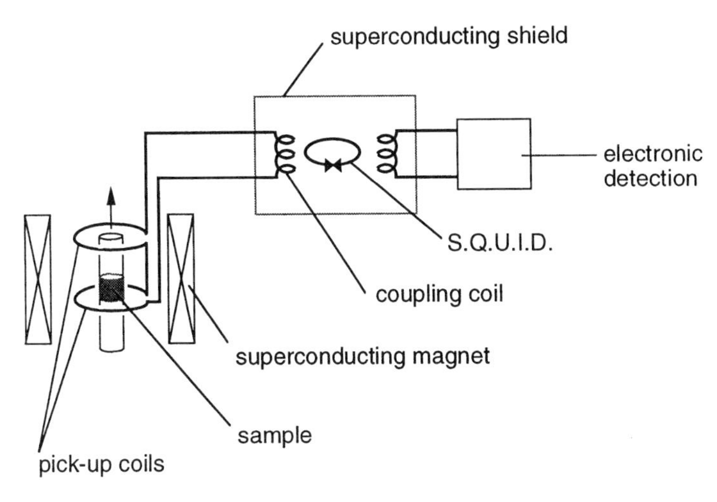

Figure 1
Scheme for a SQUID magnetometer.

unit of flux is called a fluxon and is equal to $h/2e$. The sample must be thermally isolated from the pick-up coils, since the temperature of the sample has to be varied and that of the coils has to be maintained below the critical temperature for superconductivity. The induced current generates a local variation of flux in the coupling coil, which in turn induces an alternating current in the SQUID; the number of periods of the ac is equal to the number of fluxons. Due to the small value of the fluxon, very weak magnetizations can be measured, as needed for biological samples. The increasing availability of these magnetometers has provided a strong boost for this area of research.

In their pioneering SQUID work on metalloproteins, Day and co-workers showed[6,7] that D_2O has to be used as a solvent, because the nuclear magnetic moments of the solvent protons contribute significantly to the magnetization of the sample. The principal problem is that the solvent protons relax slowly, a process that may take several hours to achieve equilibrium after a change in sample temperature. This slow relaxation introduces irreproducibility in the measurements for low concentrations of metalloproteins and increases the time required for a complete temperature-dependence study. The use of D_2O as solvent provides faster relaxing deuterons (due to quadrupolar relaxation of the $I = 1$ nuclei) and allows the sample to achieve its equilibrium magnetization value in a shorter period of time.

IV. The Relationship between Magnetic and Electronic Properties

Once a reliable function $M^{\text{para}} = f(T, B)$ or $\chi^{\text{para}} = f(T)$ has been measured for the molecule in question, we will want to relate its magnetic properties with its molecular properties to gain insight into its electronic structure.

A. Magnetization and Energy Levels

Let us denote E_n the energy of the quantum state $|n\rangle$ of a molecule. The change of E_n in response to a variation of the magnetic induction **B** is determined by the magnetic moment μ_n associated with the level $|n\rangle$:

$$\mu_{n\alpha} = -\frac{\partial E_n}{\partial B_\alpha} \tag{16}$$

The component of the magnetic induction **B** along the axis α is noted B_α. The molar magnetization of the system is defined as the Boltzmann average of the magnetic moments of the populated levels for N_A molecules (N_A is Avogadro's number):

$$M_{M\alpha} = \frac{N_A \sum_n \mu_{n\alpha} \exp(-E_n/kT)}{\sum_n \exp(-E_n/kT)} \tag{17}$$

k is the Boltzmann constant ($k = 1.380658 \times 10^{-23}$ J K^{-1} = 0.69503877 cm^{-1} K^{-1}).

B. Susceptibility and Energy Levels

From these components of the molar magnetization, a second-rank tensor of molar zero-field susceptibility $[\chi_{M\alpha\beta}]$ can be defined, the components of which are

$$\chi_{M\alpha\beta} = \mu_0(\partial M_{M\alpha}/\partial B_\beta)_{\mathbf{B}=0} \tag{18}$$

If the system is isotropic, $[\chi_{M\alpha\beta}]$ reduces to a scalar. When **B** is oriented along one of the principal axes $i = 1, 2, 3$ of $[\chi_{M\alpha\beta}]$, $\mathbf{M}_M$ is colinear to **B** and one has, for a weak enough field, $\chi_{Mi} = \mu_0 M_{Mi}/B$ [see Eq. (4)]. Next, we will consider isotropic systems or anisotropic systems in which the field is oriented along one of the principal axes.

Van Vleck showed that χ_M can be computed from the dependence of the energy versus the magnetic induction. Equation (19) describes the dependence of E_n as a second-order function of B

$$E_n = E_n^{(0)} + E_n^{(1)}B + E_n^{(2)}B^2 \tag{19}$$

By using Eqs. (16) and (17), the molar magnetization is calculated to be

$$M_M = \frac{N_A \sum_n [-E_n^{(1)} - 2E_n^{(2)}B] \exp(-E_n/kT)}{\sum_n \exp(-E_n/kT)} \tag{20}$$

which, using the approximation $\exp(-E_n^{(1)}B/kT) = 1 - E_n^{(1)}B/kT$, can be re-written as

$$M_M = \frac{N_A \sum_n [-E_n^{(1)} - 2E_n^{(2)}B] \exp[-E_n^{(0)}/kT][1 - E_n^{(1)}B/kT]}{\sum_n \exp[-E_n^{(0)}/kT][1 - E_n^{(1)}B/kT]} \quad (21)$$

By using the condition $M_M(B = 0) = 0$, we find

$$\sum_n E_n^{(1)} \exp(-E_n^{(0)}/kT) = 0 \quad (22)$$

so that

$$M_M = \frac{BN_A \sum_n \{-2E_n^{(2)} + [E_n^{(1)}]^2/kT\} \exp[-E_n^{(0)}/kT]}{\sum_n \exp(-E_n^{(0)}/kT)} \quad (23)$$

From which the well-known Van Vleck equation is obtained

$$\chi_M = \frac{\mu_0 N_A \sum_n \{-2E_n^{(2)} + [E_n^{(1)}]^2/kT\} \exp[-E_n^{(0)}/kT]}{\sum_n \exp[-E_n^{(0)}/kT]} \quad (24)$$

Perturbation theory can be used to compute the energy as a function of the magnetic induction. The Hamiltonian in zero magnetic induction, $\mathscr{H}_0$, is assumed to be completely solved (with eigenvalues and eigenfunctions). In the low magnetic induction limit, one can consider the magnetic induction dependent Hamiltonian as a perturbation $\mathscr{V}$. The terms $E_n^{(1)}B$ and $E_n^{(2)}B^2$ are the eigenvalues of the perturbation in first- and second-order approximations, respectively. We are now ready to venture through the dense forest of magnetochemistry.

V. Mononuclear Systems with Metal Ions of the First Transition Series

First, we will study the case of systems with one transition metal ion. We will assume that the effect of electron–nuclei and electron–electron interactions for the compounds of interest is completely known. We thus assume to have a good knowledge of the electronic ground and excited states. Each state ${}^{2S+1}\Gamma$ is labeled by an electron spin value S and a symmetry label Γ for the wave function. For example, these states are precisely those studied in ultraviolet–visible (UV–vis) spectroscopy. The value of $2S + 1$ corresponds to the spin degeneracy. The label

Γ contains the information on the orbital degeneracy. The A or B, E, and T designations reflect the orbital degeneracy (1, 2, and 3, respectively).

In magnetochemistry, consideration of spin–orbit (SO) coupling and Zeeman effect is essential. Their mathematical expression is

$$\mathscr{H}_{\mathrm{SO}} + \mathscr{H}_{\mathrm{Zeeman}} = \zeta \Sigma_i \mathbf{l}_i \cdot \mathbf{s}_i + \beta \Sigma_i (\mathbf{l}_i + 2\mathbf{s}_i) \cdot \mathbf{B} \tag{25}$$

where ζ, $\mathbf{l}_i$ and $\mathbf{s}_i$ are the spin–orbit coupling constant and the orbital and spin momenta of electron i, respectively. The following discussion will be divided into two parts: (A) the case wherein the ground state is well separated from the excited states before application of $\mathscr{H}_{\mathrm{SO}} + \mathscr{H}_{\mathrm{Zeeman}}$ and has an orbitally nondegenerate wave function; (B) the case in which the ground state is orbitally degenerate or has excited states close to it in energy.

A. Systems with an Orbitally Nondegenerate Ground State, Well Separated from Excited States

Which systems have such a ground state? Let us consider a transition metal ion in a strong octahedral (O_h) ligand field. The five 3d orbitals are split into three t_{2g} and two e_g orbitals, the latter two being at higher energy than the former three (Fig. 2). They are separated by the energy Δ, which reflects the strength of the interaction of the metal ion with the ligands. In tetrahedral symmetry (T_d), the ordering of the orbitals is reversed (Fig. 2). Table II summarizes the orbital and spin designations of the ground states for the various d electron configurations in cubic symmetry. We see in Table II that d^3, d^5 (high spin), d^6 (low spin), d^8 ions in octahedral symmetry and d^2, d^5, d^7 ions in tetrahedral symmetry have an orbitally nondegenerate ground state. For example, the state $^3A_{2g}$ for d^8 corresponds to an orbitally nondegenerate state and a spin triplet. If the ligand field is strong enough, this ground state is well separated from excited states.

The other ions, in cubic symmetry, have an orbitally degenerate ground state and will a priori have magnetic properties that cannot be understood within the

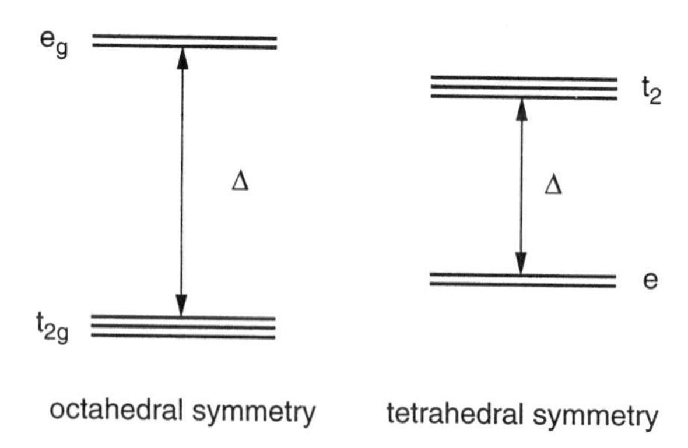

Figure 2
Splitting of the d orbitals in cubic symmetry.

Table II
Ground States in Cubic Symmetry

	O_h				T_d	
d^1	t_{2g}^1	$^2T_{2g}$			e^1	^2E
d^2	t_{2g}^2	$^3T_{1g}$			e^2	$^3A_2$
d^3	t_{2g}^3	$^4A_{2g}$			$e^2t_2^1$	$^4T_1$
d^4	$t_{2g}^3e_g^1$	^5E_g	t_{2g}^4	$^3T_{1g}$	$e^2t_2^2$	$^5T_2$
d^5	$t_{2g}^3e_g^2$	$^6A_{1g}$	t_{2g}^5	$^2T_{2g}$	$e^2t_2^3$	$^6A_1$
d^6	$t_{2g}^4e_g^2$	$^5T_{2g}$	t_{2g}^6	$^1A_{1g}$	$e^3t_2^3$	^5E
d^7	$t_{2g}^5e_g^2$	$^4T_{1g}$	$t_{2g}^6e_g^1$	^2E_g	$e^4t_2^3$	$^4A_2$
d^8	$t_{2g}^6e_g^2$	$^3A_{2g}$			$e^4t_2^4$	$^3T_1$
d^9	$t_{2g}^6e_g^3$	^2E_g			$e^4t_2^5$	$^2T_2$

limits of the following theory. Fortunately, the symmetry of the environment of a transition metal ion is frequently lower than cubic, especially in biological systems. This lowering of symmetry will transform orbitally degenerate ground states in cubic symmetry into nondegenerate ground states. For example, a Cu(II) ion, which would have a ^2E_g ground state in octahedral symmetry, will become in tetragonal symmetry either $^2A_{1g}$ (z axis elongated) or $^2B_{1g}$ (*z*-axis compressed). Similarly, an Mn(III) ion, which has a ^5E_g ground state in octahedral symmetry, will become in tetragonal symmetry either $^5A_{1g}$ (*z*-axis compressed) or $^5B_{1g}$ (*z*-axis elongated). We see that, by taking low symmetry into account, a large number of systems will have an orbitally nondegenerate ground state. Nevertheless it is good to remember that the ions that have an orbitally degenerate ground state in cubic symmetry (Table II) can give rise to difficulties if the lowering of symmetry is not strong enough. It is not an all-or-nothing problem.

It can be shown quite generally that, for systems with an orbitally nondegenerate ground state well separated from the excited states, the magnetic properties can be calculated with the help of the following "effective Hamiltonian" (also called a spin Hamiltonian since it acts on an "effective" spin), which acts only on the ground state (see Appendix I):

$$\mathscr{H}_S = \beta\mathbf{B}[\mathbf{g}]\mathbf{S} + D[S_z^2 - S(S+1)/3] + E[S_x^2 - S_y^2] \qquad (26)$$

The first part describes the effect of the magnetic induction **B** and the second part is called the zero-field splitting (zfs) Hamiltonian, since it persists even when the magnetic induction is zero. Three parameters appear: [**g**] is a tensor and *D* and *E* have the dimensions of energy. Note that the orbital momentum and the spin–orbit coupling are hidden in this formalism. In fact, [**g**], *D*, and *E* are related to

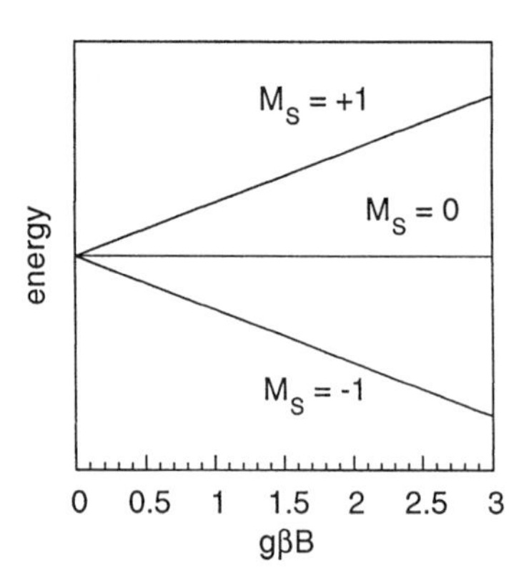

Figure 3
Energy levels for a spin $S = 1$ in a magnetic induction. Energy of the states and the energy $g\beta B$ are in the same arbitrary unit.

fundamental parameters: energy of the states ${}^{2S+1}\Gamma$, orbital momentum, and spin–orbit coupling. The corresponding mathematical expressions arc obtained through the building process of the spin Hamiltonian (see Appendix I).

1. Magnetic Properties of Cubic Systems

In cubic symmetry, one has $D = E = 0$, so there is no contribution from the zfs Hamiltonian (see Appendix I). Moreover, for cubic symmetry, the three axes x, y, and z are equivalent and the tensor $[\mathbf{g}]$ becomes a scalar. Since there is no unique axis, we can decide to take the direction of the field as the z axis; hence, the simpler formulation in Eq. (27).

$$\mathscr{H}_S = g\beta \mathbf{B} \cdot \mathbf{S} = g\beta B S_z \tag{27}$$

For a system with electronic spin S, there are $2S + 1$ quantum states $|M_S\rangle$ with $M_S = -S, -(S-1), \ldots, (S-1), S$. The eigenvalues of Eq. (27) are

$$E = g\beta B M_S \tag{28}$$

This relationship, illustrated in Figure 3 for the case $S = 1$, yields straight lines with slope M_S for the energies of the three spin states. The ground state is the state $|-S\rangle$. By using Eqs. (16) and (17), the molar magnetization is thus:

$$M_{\mathrm{M}} = \frac{-N_{\mathrm{A}} g\beta \sum_{M_S} [M_S \exp(-xM_S)]}{\sum_{M_S} \exp(-xM_S)} \tag{29}$$

with $x = g\beta B/kT$. One can show that

$$\sum_{M_S} \exp(-xM_S) = \frac{sh[(S+\frac{1}{2})x]}{sh(x/2)} \tag{30}$$

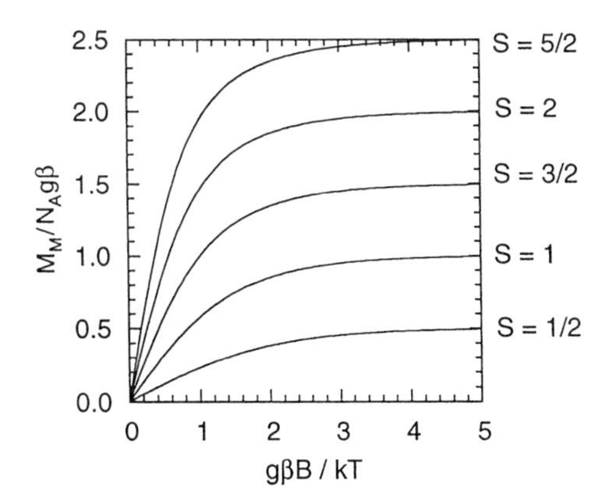

Figure 4
Magnetizations for spin values $S = \frac{1}{2}$, 1, $\frac{3}{2}$, 2, and $\frac{5}{2}$, as functions of the ratio $g\beta B/kT$. Note the linear regime for low values of $g\beta B/kT$. The saturation magnetization is equal to $M_{\mathrm{M}} = N_{\mathrm{A}}g\beta\mathrm{S}$.

Taking the first derivative relative to x affords

$$\frac{d}{dx}\left[\sum_{M_S}\exp(-xM_S)\right] = -\sum_{M_S} M_S\exp(-xM_S) \tag{31}$$

from which one can conclude

$$M_{\mathrm{M}} = N_{\mathrm{A}}g\beta SB_S(x) \tag{32}$$

where

$$B_S(x) = \{(S+\tfrac{1}{2})\coth[x(S+\tfrac{1}{2})] - \tfrac{1}{2}\coth(x/2)\}/S \tag{33}$$

is the Brillouin function. Figure 4 shows the Brillouin curves for $S = \frac{1}{2}$, 1, $\frac{3}{2}$, 2, and $\frac{5}{2}$. Interestingly, in this case, the magnetization depends only on the ratio of the magnetic induction over the temperature. We will see that for lower symmetry $S > \frac{1}{2}$ systems, this is no longer true and the corresponding analysis will be very informative. At infinite x, the magnetization approaches the limit $M_{\mathrm{M}} = N_{\mathrm{A}}Sg\beta$, which is called the saturation magnetization. Physically, saturation corresponds to a situation where the only populated state of Figure 3 is the groundstate $|-S\rangle$; thus the magnetic moment for 1 mol of complex is $N_{\mathrm{A}}g\beta S$.

At the other limit, when $x \ll 1$ (Fig. 4), the magnetization becomes linear in x. By using the approximations $\coth x = 1/x + x/3$, one gets the Curie law [see also Eq. (4)]

$$\chi_{\mathrm{M}} = \mu_0 N_{\mathrm{A}}g^2\beta^2 S(S+1)/3kT \tag{34}$$

This result can also be deduced from the Van Vleck equation. Curie behavior can be represented in a number of ways: as a hyperbola in a χ_M versus T plot [Fig. 5(*a*)], as a line parallel to the T axis in a $\chi_M T$ versus T plot [Fig. 5(*b*)], where the ordinate gives the Curie constant C, or as a straight line passing through the origin in either a $1/\chi_M$ versus T plot whose slope gives 1/C [Fig. 5(*c*)] or a χ_M versus $1/T$, where the slope gives C [Fig. 5(*d*)].

The Curie constant

$$\mathrm{C} = \chi_M T = \frac{(\mu_0 N_A g^2 \beta^2) S(S+1)}{3k}$$

can be used to characterize the magnetic properties of the system. In SI units, the quantity $\mu_0 N_A \beta^2/3k$ is close to $4\pi\frac{1}{8}10^{-6}$ so that

$$8\chi_M T/4\pi \times 10^{-6} \approx g^2 S(S+1) \tag{35}$$

Others prefer instead to use the magnetic moment μ to characterize the system,

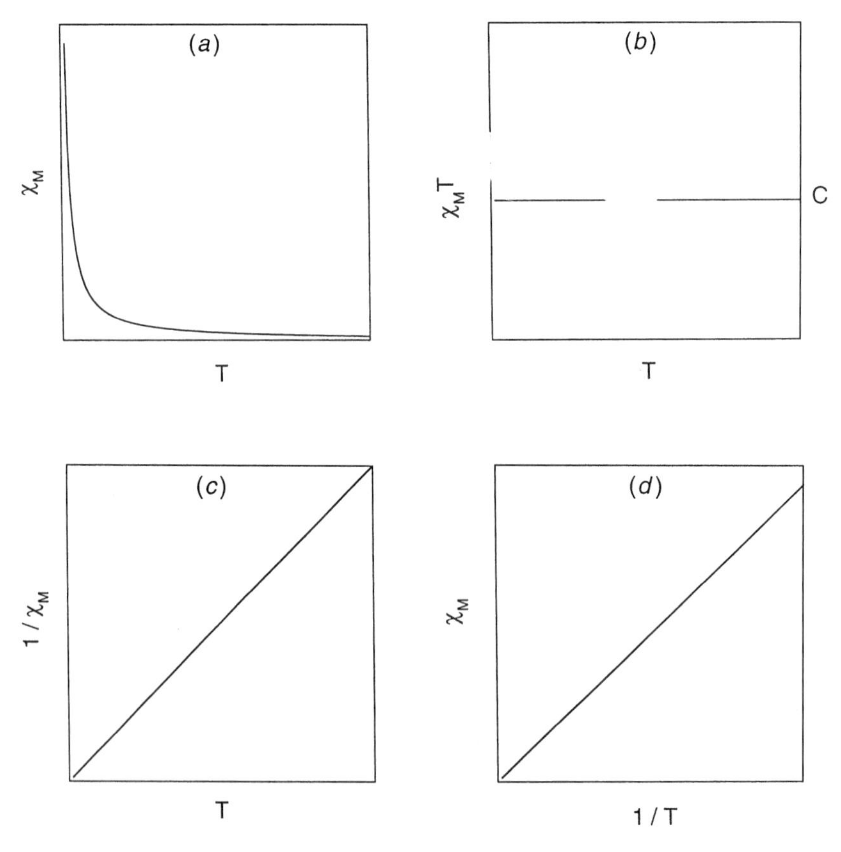

Figure 5
Representations of the Curie law as (*a*) $\chi_M = f(T)$; (*b*) $\chi_M T = f(T)$; (*c*) $1/\chi_M = f(T)$; (*d*) $\chi_M = f(1/T)$ (see text).

Table III
Effective Moment μ and Curie Constant $\chi_M T$ for Different Values of the Electronic Spin S

S	μ/β	$\chi_M T/(4\pi \times 10^{-6}$ m^3 mol^{-1} K)
$\frac{1}{2}$	1.732	0.375
1	2.828	1
$\frac{3}{2}$	3.873	1.876
2	4.899	3.001
$\frac{5}{2}$	5.916	4.377

where

$$\mu^2 = \frac{3k\chi_M T}{\mu_0 N_A \beta^2} = g^2 S(S+1) \tag{36}$$

$g^2S(S+1)$ further simplifies into $n(n+2)$ for $g = 2.0$, where n = number of unpaired electrons. Table III summarizes the values of μ and $\chi_M T$ for different values of S with $g = 2.0$.

Table IV gives a selection of g factors for metal ions. Note that the g factors are not always close to 2.0, with significant deviations for octahedral Ni(II)

Table IV
Representative g Values for A and E Ions in Cubic Symmetry

	O_h	
d^3	$^4A_{2g}$	V^{II} 1.98, Cr^{III} 1.98, Mn^{IV} 1.98
d^4	^5E_g	Cr^{II} 1.99, Mn^{III} 1.98
d^5	$^6A_{1g}$	Mn^{II} 2.00, Fe^{III} 2.00
d^7	^2E_g	Co^{II} 2.00
d^8	$^3A_{2g}$	Ni^{II} 2.25
d^9	^2E_g	Cu^{II} 2.11
	T_d	
d^1	^2E	V^{IV} 1.87
d^2	$^3A_2$	
d^5	$^6A_1$	Mn^{II} 2.00, Fe^{III} 2.00
d^6	^5E	Fe^{II} 2.21
d^7	$^4A_2$	Co^{II} 2.37–2.46

($g = 2.25$) and tetrahedral Co(II) complexes ($g \sim 2.4$) due to spin–orbit coupling [see Eq. (A.I.6)].

2. Lower Symmetry

When the symmetry is lower than cubic, the system becomes anisotropic and the magnetic properties depend on the orientation of the field relative to the molecular axes.

a. An Anisotropic $S = \frac{1}{2}$ System: A Cu(II) ion In the case of an $S = \frac{1}{2}$ species, the zfs term in Eq. (26) vanishes and we only need the term $\mathcal{H}_S = \beta\mathbf{B}[\mathbf{g}]\mathbf{S}$. The expression of the [**g**] tensor is given in Appendix I. It depends on the nature of the ground state and on the spin–orbit coupling between that ground and excited states. For example, Cu(II) in tetragonally elongated environments has a $d_{x^2-y^2}$ ground state and [**g**] is axial with typical values $g_{//} = 2.22$, $g_{\perp} = 2.05$. See also Chapter 3, the EPR chapter, for other examples of systems with anisotropic g values.

b. Anisotropic $S > \frac{1}{2}$ Systems In this case, the full Hamiltonian of Eq. (26) has to be used. Consider the case of an $S = 1$ ground state in axial symmetry that is the simplest case we can think of. There are now two principal directions: z and that perpendicular to it. First, let us consider the magnetic induction parallel to z. In this case, the Hamiltonian can be written as

$$\mathcal{H}_S = \beta g_{//} B S_z + D[S_{z^2} - S(S+1)/3] \tag{37}$$

The eigenvalues are easily found to be

$$E_S = \beta g_{//} B M_S + D[(M_S)^2 - \tfrac{2}{3}] \tag{38}$$

These energies are represented in Figure 6(a) as functions of $\beta g_{//} B/D$ for $D > 0$. For $B = 0$, the lowest state has $M_S = 0$ and is at $E_S = -\frac{2}{3}D$ (or 0 if we define the energy of this state as the zero of energy); at $E_S = D/3$ (or D if we define the energy of the $M_S = 0$ state as the zero of energy), there are two degenerate states $|M_S = \pm 1\rangle$. When B increases, the lowest state is $|M_S = 0\rangle$ till the crossing point $\beta g_{//} B/D = 1$, where $|M_S = -1\rangle$ becomes the ground state. Figure 6(b) represents the energy level diagram for $D < 0$; in this case the $|M_S = -1\rangle$ is always the ground state.

The magnetization is along the z axis since $\langle S_z\rangle \neq 0$ and $\langle S_x\rangle = \langle S_y\rangle = 0$[d]; the parallel magnetization is easily computed as

$$M_{M//} = \frac{N_A\left\{-\beta g_{//}\exp\left[\dfrac{-(\beta g_{//}B + D)}{kT}\right] + \beta g_{//}\exp\left[\dfrac{-(-\beta g_{//}B + D)}{kT}\right]\right\}}{1 + \exp\left[\dfrac{-(\beta g_{//}B + D)}{kT}\right] + \exp\left[\dfrac{-(-\beta g_{//}B + D)}{kT}\right]} \tag{39}$$

[d]The eigenstates of the problem are the $|M_S\rangle$ states. Thus the eigenvalues of S_z are known. Both S_x and S_y cannot then be simultaneously known and their average for each $|M_S\rangle$ state is 0.

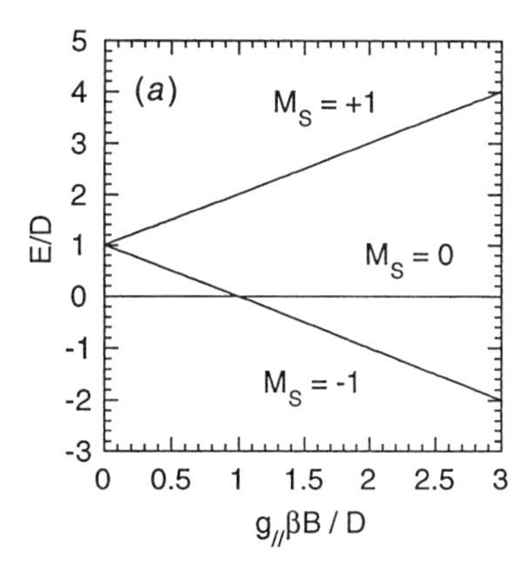

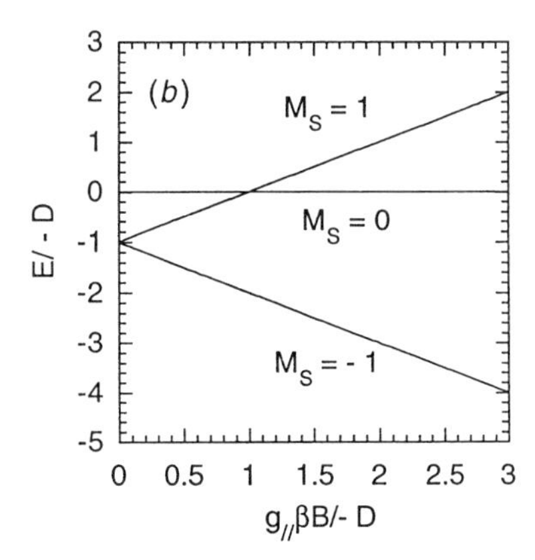

Figure 6
Energy levels as functions of $g_{//}\beta B/D$ for a spin $S = 1$ with an axial zfs and a magnetic field parallel to the axis of distortion. (*a*) $D > 0$; (*b*) $D < 0$.

This expression is distinct from the Brillouin equation. Let us recall that, following the Brillouin equation, the magnetization depends only on the ratio B/T. Here it also clearly depends on the ratio B/D. The variation of $M_{M//}$ as a function of $g_{//}\beta B/kT$ is represented in Figure 7(*a*) for $D > 0$, for different ratios $\beta g_{//}B/D$. When $\beta g_{//}B/D$ is infinite, the curve is the Brillouin one. When $\beta g_{//}B/D$ decreases, saturation is more difficult to obtain and eventually cannot be obtained $(\beta g_{//}B/D = \frac{1}{2})$ since in this last case the $\beta g_{//}B/D$ value is below the crossing value $\beta g_{//}B/D = 1$ in Figure 6(*a*). Then, the ground state is the $|M_S = 0\rangle$ one. When D is negative, the magnetization is higher than the Brillouin value [Fig. 7(*b*)] but tends toward the same limit. We will see below the experimental importance of the difference between Eq. (39) and the Brillouin expression for magnetization.

When the magnetic induction is perpendicular to z, let us say parallel to x,

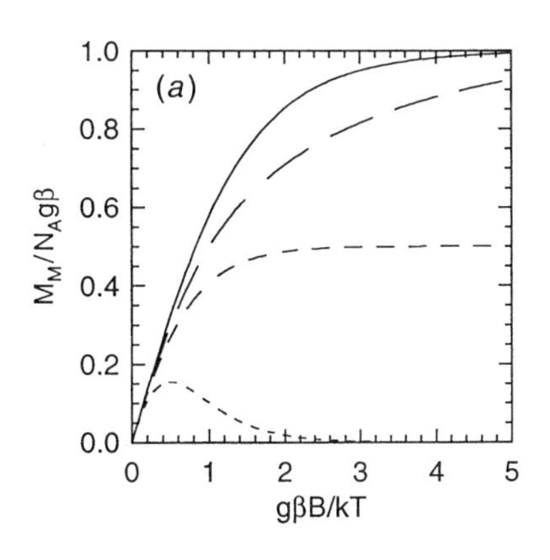

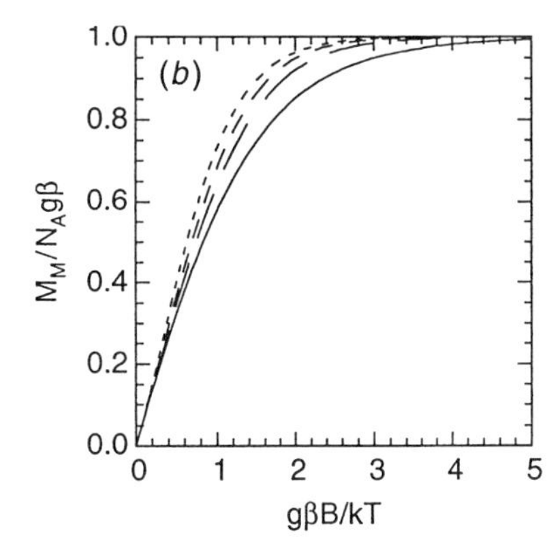

Figure 7
Isofield parallel magnetization as a function of $g_{//}\beta B/kT$ for a spin $S = 1$ with an axial zfs for $g_{//}\beta B/D = 0$ (---), 0.5 (_ _ _), 1 (__ __), 2 (___). (*a*) $D > 0$; (*b*) $D < 0$. $g_{//} = g$.

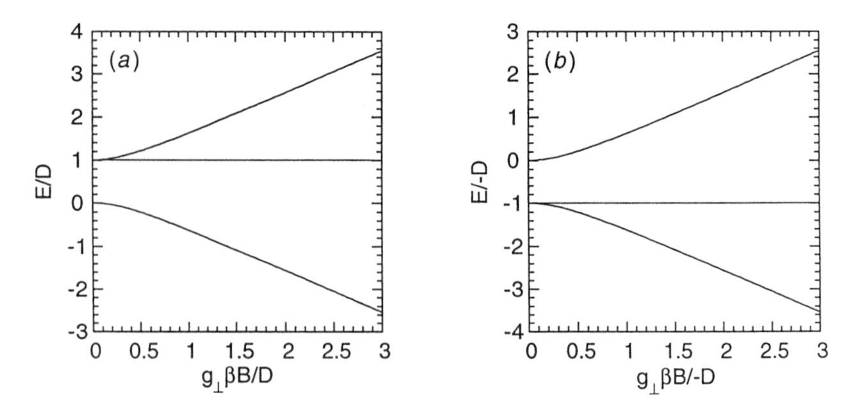

Figure 8
Energy levels as functions of $g_\perp \beta B/D$ for a spin $S = 1$ with an axial zfs effect and a magnetic field perpendicular to the axis of distorsion. (*a*) $D > 0$: (*b*) $D < 0$.

the Hamiltonian becomes

$$\mathscr{H}_S = \beta g_\perp B S_x + D[S_{z^2} - S(S+1)/3] \tag{40}$$

Now the eigenvalues are slightly more complicated to evaluate. We have to build the matrix of $\mathscr{H}_S$ in the $|M_S\rangle$ basis. What is the action of S_x on the $|M_S\rangle$ states? One uses the $S_+ = S_x + iS_y$ and $S_- = S_x - iS_y$ operators since $S_x = (S_+ + S_-)/2$. The effect of S_+ and S_- on $|M_S\rangle$ is

$$\begin{aligned} S_+|M_S\rangle &= [S(S+1) - M_S(M_S+1)]^{1/2}|M_S+1\rangle \\ S_-|M_S\rangle &= [S(S+1) - M_S(M_S-1)]^{1/2}|M_S-1\rangle \end{aligned} \tag{41}$$

For our $S = 1$ problem, the following matrix is obtained

$$\begin{bmatrix} 0 & g_\perp \beta \mathrm{B}\sqrt{2}/2 & g_\perp \beta B\sqrt{2}/2 \\ g_\perp \beta B\sqrt{2}/2 & D & 0 \\ g_\perp \beta B\sqrt{2}/2 & 0 & D \end{bmatrix}$$

The eigenvalues are

$$\begin{aligned} E_S &= D \\ E_S &= (D \pm (D^2 + 4{g_\perp}^2\beta^2 B^2)^{1/2})/2 \end{aligned} \tag{42}$$

These exact solutions are represented in Figure 8(*a*) for $D > 0$ and in Figure 8(*b*) for $D < 0$. The magnetization is now parallel to x since $\langle S_x \rangle \neq 0$ and

---

[e]The eigenstates of the problem are given by diagonalization of the previous matrix. One can check that the eigenvalues of S_x are known. Both S_z and S_y cannot then be simultaneously known and their average on each eigenstate is 0.

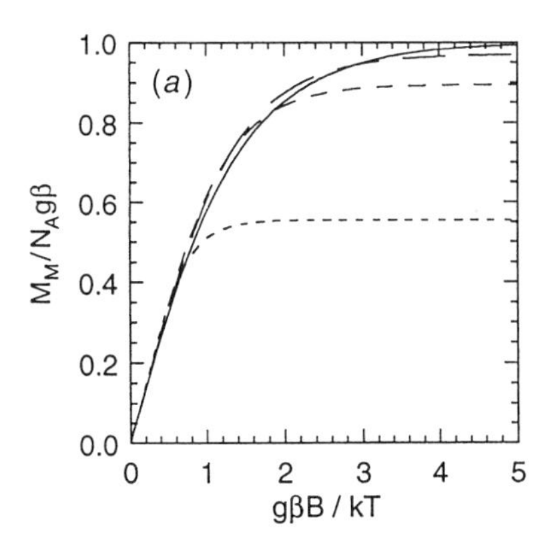

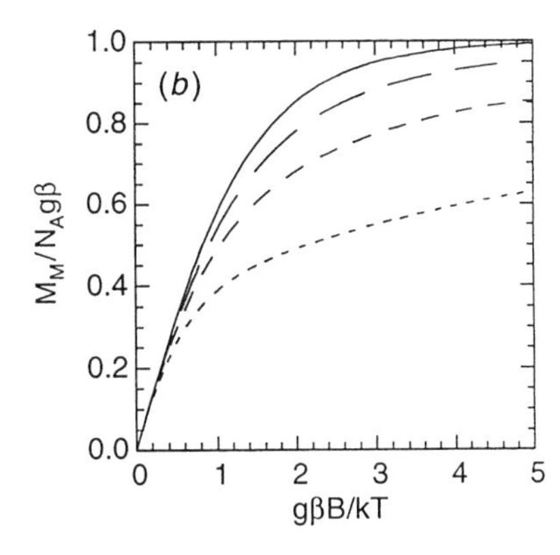

Figure 9
Isofield perpendicular magnetization as a function of $g_{\perp}\beta B/kT$ for a spin $S = 1$ in the presence of an axial zfs for $g_{\perp}\beta B/D = 0$ (---), 0.5 (_ _ _), 1 (__ __), 2 (___). (*a*) $D > 0$; (*b*) $D < 0$. $g_{\perp} = g$.

$\langle S_z\rangle = \langle S_y\rangle = 0^e$ (perpendicular magnetization) and is represented in Figure 9(*a* and *b*) for different ratios $\beta g_{\perp}B/D$ for $D > 0$ and $D < 0$, respectively. When $g_{\perp}\beta B/D$ decreases, the magnetization at 0 K, which is the derivative of the lowest energy curve of Figure 8 versus the magnetic induction, decreases.

The Van Vleck equation may be used to calculate the susceptibility. When the field is parallel to the z axis, one gets

$$\chi_{M//} = \frac{(\mu_0 N_A \beta^2 g_{//}{}^2/kT)2[\exp(-D/kT)]}{[1 + 2\exp(-D/kT)]} \tag{43}$$

This expression can also be obtained by calculating the magnetization for low magnetic induction values.

For the more difficult case of the field perpendicular to the z axis, the coefficients for the development of the energy as a function of the magnetic induction are

n	$E_n^{(0)}$	$E_n^{(1)}$	$E_n^{(2)}$
1	0	0	$-g_{\perp}{}^2\beta^2/D$
2	D	0	$g_{\perp}{}^2\beta^2/D$
3	D	0	0

We finally arrive at

$$\chi_{M\perp} = \frac{\mu_0 N_A[(g_{\perp}\beta)^2/D][2 - 2\exp(-D/kT)]}{[1 + 2\exp(-D/kT)]} \tag{44}$$

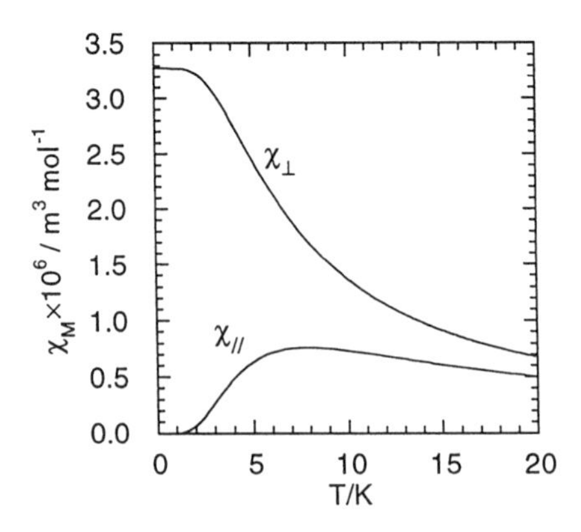

Figure 10
Parallel and perpendicular molar magnetic susceptibilities for a spin $S = 1$ with an axial zfs $D = +8$ cm^{-1} and $g = 2$.

These results are represented in Figure 10 for $D = +8\,\text{cm}^{-1}$ and in Figure 11 for $D = -8\,\text{cm}^{-1}$. The first remark is that, despite a rather large value of D, the effects are noticeable only at low temperature. For $D = +8$ cm^{-1}, the parallel susceptibility tends to 0 at low T. The reason is that the $M_S = 0$ state is the only one occupied at low T [Fig. 6(a)]. As for the perpendicular susceptibility, it goes to a nonzero value since the second derivative of the energy at 0 K is nonzero [Fig. 8(a)]. For $D = -8$ cm^{-1}, the parallel susceptibility diverges since it corresponds to occupation of the $M_S = -1$ state [Fig. 6(b)].

These theoretical calculations have been wonderfully experimentally verified by Carlin et al.[8] on single crystals of $Cs_3VCl_6 \cdot 3H_2O$, which contains V(III) (d^2) in the *trans*-$[VCl_4(H_2O)_2]^-$ anion (axial symmetry). Figure 12 gives $\chi_{M\perp}$ and $\chi_{M//}$ as functions of temperature. The lines correspond to the theoretical expressions with $D = +8.05$ cm^{-1} and $g_{//} = 1.93$ and $g_\perp = 1.74$. The agreement is perfect.

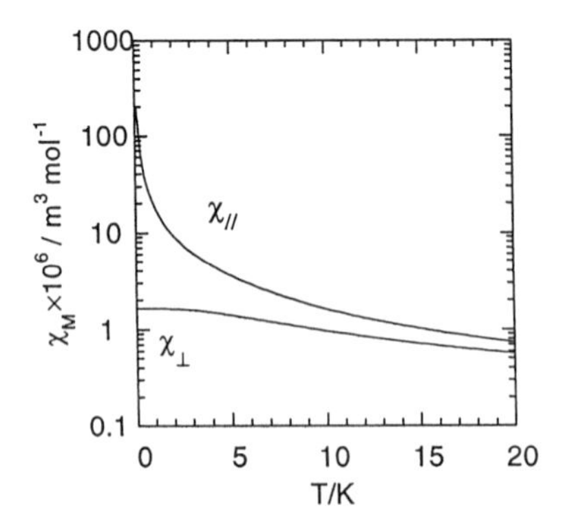

Figure 11
Parallel and perpendicular molar magnetic susceptibilities for a spin $S = 1$ with an axial zfs $D = -8$ cm^{-1} and $g = 2$. The ordinate is logarithmic.

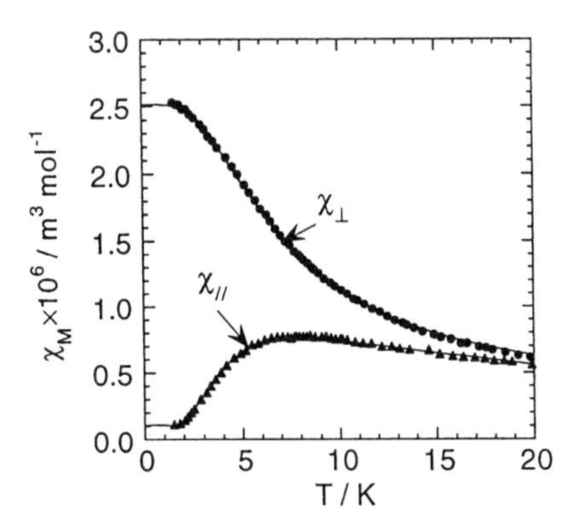

Figure 12
Parallel and perpendicular molar magnetic susceptibilities for the axially distorted $S = 1$ system $Cs_3VCl_6 \cdot 3H_2O$. The theoretical curves correspond to $D = +8.05$ cm^{-1} and $g_{//}$ 1.93 and $g_\perp = 1.74$. [Adapted from Carlin et al.[8] with permission from the American Chemical Society].

The magnetization for a powder sample can be computed as

$$M_M = \frac{M_{M//} + 2M_{M\perp}}{3} \tag{45}$$

The result is represented in Figure 13(a and b) for $D > 0$ and $D < 0$, respectively. We see that, as said above, the magnetization depends not only on B/T but also on B/D. Experimentally each of these curves can be determined by measurement on a powder or a frozen solution of the magnetization as a function of T for a given B value (isofield curve). The very fact that different curves are for different B values is proof that the sample contains an anisotropic $S > \frac{1}{2}$ system. For an isotropic system, one would get the universal Brillouin curve. One sees

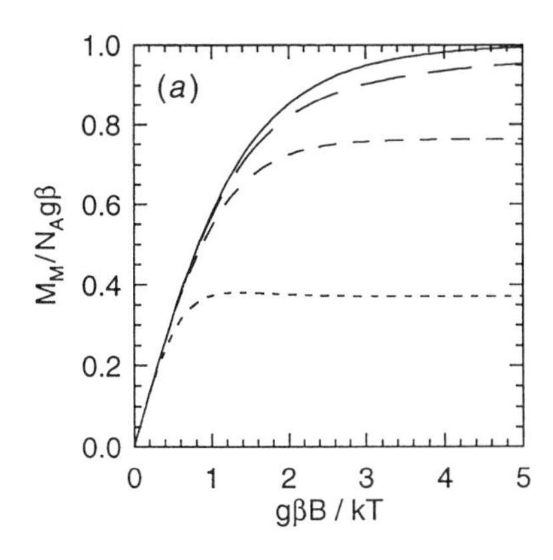

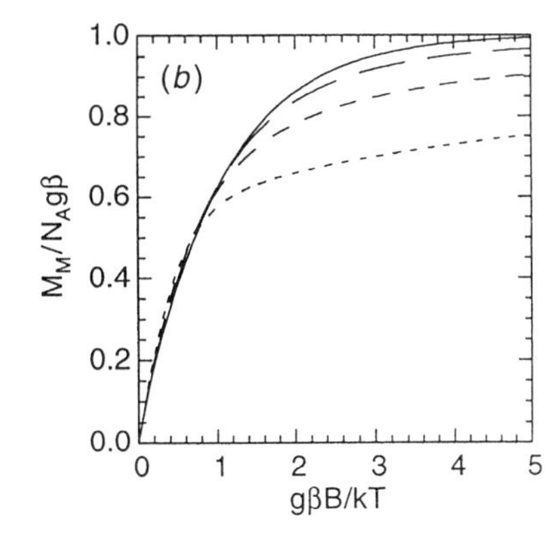

Figure 13
Isofield powder magnetization as a function of $g\beta B/kT$ for a spin $S = 1$ in the presence of an axial zfs effect for $g\beta B/D = 0$ (---), 0.5 (_ _), 1 (__ __), 2 (___). (a) $D > 0$; (b) $D < 0$.

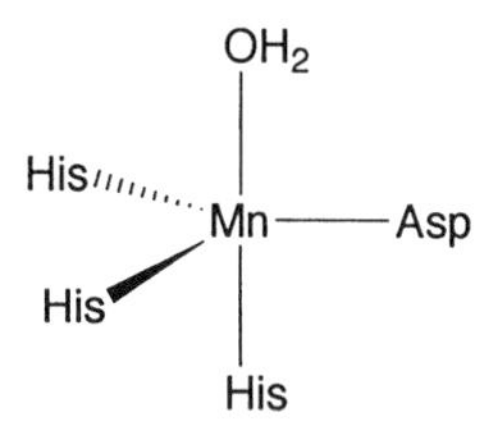

Figure 14
Schematic representation of the structure of the active site of Mn superoxide dismutase.

that such a family of curves allows the very precise determination of the value of D (and E). This method has been introduced in the study of metalloproteins by Day.[10]

As an example, let us consider manganese superoxide dismutase (SOD) of *Thermus thermophilus* (MnSOD),[9] which catalyzes the disproportionation of the superoxide anion. The enzyme exists in two oxidation states, Mn(III) or Mn(II). The structure of the active site has been determined to be trigonal bipyramidal[10] as shown in Figure 14. Figure 15 represents the isofield magnetization of Mn^{III}SOD as a function of $\beta B/kT$ for different magnetic inductions. Note that there are different curves for different fields, which must arise from the presence of zfs. Analysis of the data shows that Mn(III) is in the high spin state $S = 2$ and the zfs has been found to be axial with $D = +2.44\ cm^{-1}$ and $g = 2.0$. Confirmation by UV–vis and magnetic circular dichroism (MCD) spectroscopy of this positive value of D has been made by Whittaker and Whittaker,[11] along with an analysis in terms of the geometry of the site.

B. Other Systems

We now comment briefly on systems that do not meet the requirement of an orbitally nondegenerate ground state well separated from excited states; such systems complicate the interpretation of magnetic susceptibility data.

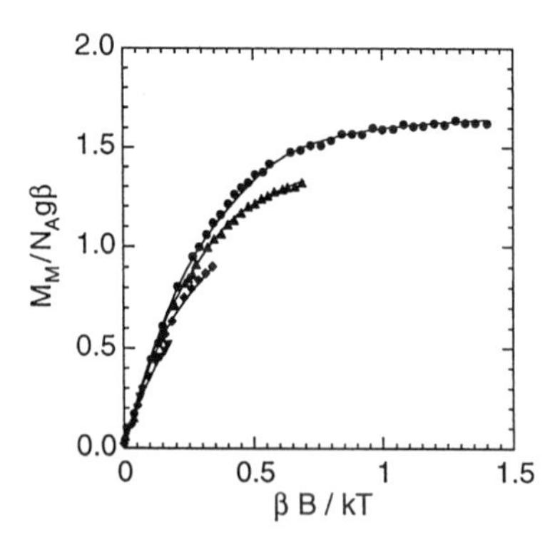

Figure 15
Isofield magnetization of a D_2O solution of Mn(III) SOD of *T. thermophilus* as a function of $\beta B/kT$. Data are at applied fields of 4.0 (•), 2.0 (▲), 1.0 (♦), and 0.5 (▼) tesla. The solid lines were calculated from a spin $S = 2$ spin Hamiltonian with $D = +2.44\ cm^{-1}$, $E/D = 0$, and $g = 2$. [Adapted from Peterson et al.[9] with permission from Elsevier Science].

1. An Orbitally Degenerate Ground State

One can see from Table II that a low-spin Fe(III) in a strong octahedral field has a $^2T_{2g}$ ground state. The same situation is encountered for a d^1 ion in octahedral geometry. The magnetic susceptibility for such situations is treated in Appendix II, and, as expected, it does not follow the Curie law. The low-spin Fe(III) center in porphyrin complexes is close to such a situation.[12,13] In fact, it has a lower symmetry that can be in the first approximation considered as tetragonal wherein the $^2T_{2g}$ state splits into one ^2E_g (lower energy) and one $^2B_{2g}$ (higher energy) state. The susceptibility, taking the tetragonal symmetry into account, has been computed.[14] In reality, the symmetry is still lower and EPR becomes the best method to study this problem.[b]

In Table II, one sees that Fe(II) (high spin) in O_h symmetry has a $^5T_{2g}$ ground state. In a tetragonal environment, this state splits into ^5E_g and $^5B_{2g}$. If the axial perturbation is such that the ground state is $^5B_{2g}$, we go back to the case of an orbitally nondegenerate state for which a zfs approach can be applied. Nevertheless, due to the small gap between the two components of the $^5T_{2g}$, which are coupled by spin–orbit coupling, the zfs parameter D will be quite large. The case where the ground state is the orbitally degenerate ^5E_g state cannot be treated by the zfs model and requires a more complex approach developed by Whittaker and Solomon.[15] High-spin Co(II) in an O_h environment represents a well-documented case of an orbitally degenerate ground state. The reader is referred to more advanced textbooks, for an in-depth discussion.[1–3]

2. Ground and Excited States Close in Energy: Spin transitions, Quantum Spin Mixing

For the d^4, d^5, d^6, and d^7 ions in O_h symmetry, there are two alternative ground states: low spin or high spin depending on how the d orbitals are filled (Table II). When the ligand field is weak, the high-spin situation is favored as Δ is less than the energy required to pair-up electrons; in strong field, Δ is greater than the pairing energy and the low-spin state becomes the ground state. For some metal complexes, the two states are close in energy and a spin equilibrium is obtained. Typically, the low-spin state is slightly lower in energy than the high-spin state so that a change from high to low spin is observed upon cooling the sample from ambient temperature.

An interesting situation has been found in Fe(III) porphyrins. The magnetic properties of high-spin Fe(III) porphyrin complexes can be adequately interpreted with the zfs approach. It has been shown that the excited states that efficiently contribute to the zfs are, in D_{4h}, the $^4A_{2g}$ and ^4E_g states derived from the $^4T_{1g}$ state. Some other Fe(III) porphyrin complexes have a definite $^4A_{2g}$ ground state. Still in some others, the $^4A_{2g}$ and the $^6A_{1g}$ states are so close in energy that the spin–orbit coupling efficiently mixes them to give a "quantum mechanically admixed intermediate-spin state".[16–18] The difference between a spin admixed state and a spin equilibrium is that only one type of species can be spectroscopically detected in the first case, while two species in equilibrium are detected in the second case. Further discussion of these interesting systems can be found in more specialized books.[1–3]

VI. Polynuclear Complexes of First-Row Transition Metal Ions

A. Dinuclear Complexes

The presence of more than one paramagnetic metal center in a number of metalloprotein active sites has stimulated investigations of their magnetic properties and those of related synthetic model complexes. What distinguishes these sites from the mononuclear species just discussed is essentially the interaction between the electronic spins on each ion. This interaction creates several new energy levels for the molecule that are usually accessible at normal temperatures and can thus give rise to a non-Curie behavior for the magnetic properties of the molecule. Heisenberg, Dirac and Van Vleck (HDVV) demonstrated that the following Hamiltonian adequately describes the effect of electronic spin coupling between two neigbouring ions:

$$\mathcal{H}_{\mathrm{exch}} = J\mathbf{S_A} \cdot \mathbf{S_B} \tag{46}$$

where J is the exchange coupling parameter and has the dimension of energy. This Hamiltonian belongs to the category of effective Hamiltonians like the spin Hamiltonian discussed in the preceding section (also see Appendix III). While there are other forms of this Hamiltonian, namely, $-J\mathbf{S_A} \cdot \mathbf{S_B}$, $-2J\mathbf{S_A} \cdot \mathbf{S_B}$, or $2J\mathbf{S_A} \cdot \mathbf{S_B}$, we will use the convention shown in Eq. (46); it is thus important to check the form of the HDVV Hamiltonian used in a particular study before comparing J values from different studies to make sure that the signs and magnitudes of the J values being compared follow the same convention.

A good example of a complex with metal–metal interactions is copper(II) acetate. Its magnetic susceptibility varies with temperature as shown in Figure 16; the temperature dependence does not follow the Curie law and exhibits a maxi-

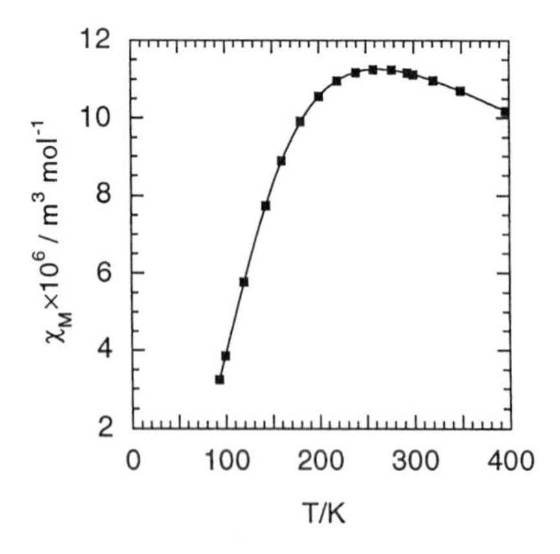

Figure 16
Molar magnetic susceptibility of $Cu(MeCO_2)_2 \cdot H_2O$ as a function of T. The curve corresponds to the theoretical expression given in the text with an exchange parameter $J = 293.7\ \mathrm{cm}^{-1}$.

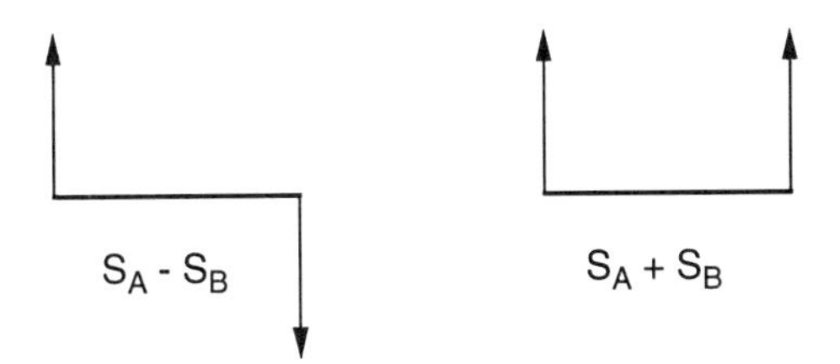

Figure 17
Antiferromagnetic and ferromagnetic ordering of two spins.

mum at 250 K. The very small values of the magnetic susceptibility and the presence of a maximum could not be explained by considering a mononuclear unit. This magnetic behavior can be satisfactorily explained by assuming the presence of two interacting copper(II) ions, each with an unpaired electron (d^9 configuration). The two copper(II) ions interact with each other to afford a diamagnetic singlet state ($S = 0$) and a paramagnetic triplet state ($S = 1$) by the combination of angular momenta such that $|S_A - S_B| \leq S \leq S_A + S_B$ (Fig. 17). When the ground state has $S = 0$, the copper ions are said to be antiferromagnetically coupled. On the other hand, ferromagnetic coupling of the copper ions would result in a ground $S = 1$ state.

The energies of the two states can easily be found using Eq. (47), which is obtained from Eq. (46) by substituting $\mathbf{S_A} \cdot \mathbf{S_B}$ with the expression derived from rearranging the identity $\mathbf{S}^2 = (\mathbf{S_A} + \mathbf{S_B})^2 = \mathbf{S_A}^2 + 2\mathbf{S_A} \cdot \mathbf{S_B} + \mathbf{S_B}^2$, that is,

$$\mathscr{H}_{\text{exch}} = J[\mathbf{S}^2 - \mathbf{S_A}^2 - \mathbf{S_B}^2]/2 \tag{47}$$

The wave functions of a pair can be written as $|S_A, M_{SA}, S_B, M_{SB}\rangle$. These functions are eigenfunctions of $\mathbf{S_A}^2$ and $\mathbf{S_B}^2$ but not of $\mathbf{S}^2$

$$\mathbf{S_i}^2|S_A, M_{SA}, S_B, M_{SB}\rangle = S_i(S_i + 1)|S_A, M_{SA}, S_B, M_{SB}\rangle \qquad i = \text{A or B} \tag{48}$$

Nevertheless it is possible to make linear combinations of these wavefunctions to get new ones which are eigenfunctions of $\mathbf{S_A}^2$, $\mathbf{S_B}^2$, and $\mathbf{S}^2$. These coupled spin functions are noted $|S_A, S_B, S, M_S\rangle$. They are eigenfunctions of Eq. (46) with the eigenvalues

$$E(S) = J[S(S+1) - S_A(S_A+1) - S_B(S_B+1)]/2 \tag{49}$$

For $S_A = S_B = \frac{1}{2}$,

$$\begin{aligned} E(0) &= -3J/4 \\ E(1) &= J/4 \end{aligned} \tag{50}$$

To derive the magnetic susceptibility expression, we need to obtain the dependence of the energy levels as a function of magnetic induction. The applicable Hamiltonian is

$$\mathscr{H}_{\text{total}} = \mathscr{H}_{\text{S}} + \mathscr{H}_{\text{exch}} \tag{51}$$

where

$$\mathscr{H}_{\text{S}} = \beta\mathbf{B}[\mathbf{g}_{\text{A}}]\mathbf{S}_{\text{A}} + \beta\mathbf{B}[\mathbf{g}_{\text{B}}]\mathbf{S}_{\text{B}} \tag{52}$$

For simplicity, we will assume $[\mathbf{g}_{\text{A}}]$ and $[\mathbf{g}_{\text{B}}]$ to be isotropic and have a common g value, so that

$$\mathscr{H}_{\text{S}} = g\beta\mathbf{B}(\mathbf{S}_{\text{A}} + \mathbf{S}_{\text{B}}) = g\beta BS_z \tag{53}$$

Application of Eq. (51) to thc copper acetate problem affords the following dependence of the energy levels $E(S, M_S)$ on the magnetic field:

$$\begin{aligned} E(0,0) &= -3J/4 \\ E(1,1) &= J/4 + g\beta B \\ E(1,0) &= J/4 \\ E(1,1) &= J/4 - g\beta B \end{aligned} \tag{54}$$

The J value represents the energy that separates the ground $S = 0$ state from the excited $S = 1$ state. For the four states, we thus obtain

n	$E_n^{(0)}$	$E_n^{(1)}$
1	$-3J/4$	0
2	$J/4$	$-g\beta$
3	$J/4$	0
4	$J/4$	$g\beta$

from which the magnetic susceptibility may be derived as

$$\chi_{\text{M}} = \frac{2\mu_0 N_{\text{A}} g^2 \beta^2}{kT[3 + \exp(J/kT)]} \tag{55}$$

A very good fit to the experimental data is obtained using a J value of 294 cm^{-1}. The near zero value of the magnetic susceptibility at low temperatures reflects the population of only the $S = 0$ state at these temperatures; as the temperature

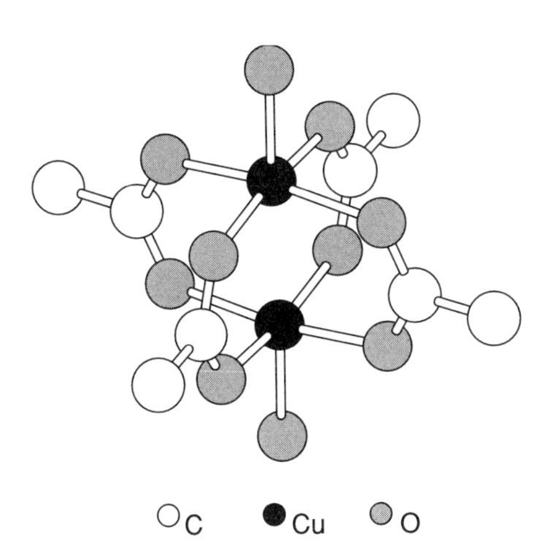

Figure 18
Structure of $Cu(MeCO_2)_2 \cdot H_2O$.

increases, the $S = 1$ state becomes populated and the magnetic susceptibility increases. The susceptibility reaches a maximum at a temperature where the increased susceptibility due to the increased population of the $S = 1$ state is counterbalanced by the decreasing susceptibility of the $S = 1$ state as the temperature increases (Curie law).

The dimeric structure of copper(II) acetate dihydrate was subsequently confirmed by X-ray crystallography. It has the molecular arrangement shown in Figure 18, with the two copper(II) ions in a square pyramidal environment. The basal plane consists of the four carboxylato oxygen atoms, and the apical position is occupied by the oxygen atom of a water molecule. The copper–copper internuclear distance is 2.64 Å.

The presence of antiferromagnetic or ferromagnetic coupling interactions can be identified from plots similar to those drawn in Figure 5. Figure 19 illustrates plots for two interacting $S = \frac{1}{2}$ centers. In the χ_M versus T plot of Figure 19(*a*), antiferromagnetic coupling ($J > 0$) is characterized by the presence of a maximum, as discussed in Section V.B.2. For ferromagnetically coupled molecules ($J < 0$), χ_M follows the profile of the Curie law curve for an uncoupled system but with $\chi_M > C/T$ at all temperatures. Such a plot is thus a more useful diagnostic for the presence of antiferromagnetic coupling.

Figure 19(*b*) plots $\chi_M T$ versus T. At infinite temperature, the Curie constant approaches the uncoupled value $4\pi \frac{1}{8} 10^{-6} \times g^2[S_A(S_A + 1) + S_B(S_B + 1)]$. As the temperature decreases ($T \to 0$ K), $\chi_M T \to 0$ for antiferromagnetic coupling, while for ferromagnetic coupling $\chi_M T$ approaches the Curie constant associated with the $S_A + S_B$ state.

Equation (55) becomes at the high-temperature limit

$$\chi_M = \frac{2\mu_0 N_A g^2 \beta^2}{4k(T + J/4k)} \tag{56}$$

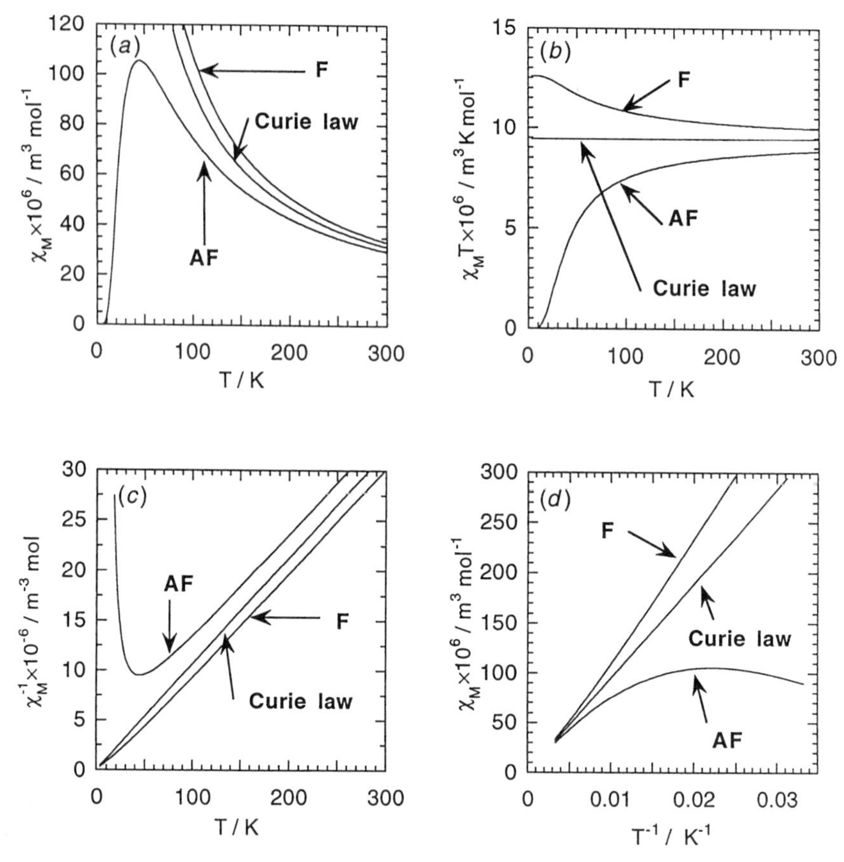

Figure 19
Representation of the magnetic behavior of a dimer of spins $S_A = S_B = \frac{1}{2}$ for $J = 0$ (Curie law); $J > 0$ (antiferromagnetic coupling); $J < 0$ (ferromagnetic coupling). (*a*) $\chi_M = f(T)$; (*b*) $\chi_M T = f(T)$; (*c*) $1/\chi_M = f(T)$; (*d*) $\chi_M = f(1/T)$ (see text).

and takes the form

$$\chi_M = \frac{C}{(T + \theta)} \tag{57}$$

with $\theta = J/4k$. This equation defines the Curie–Weiss law that applies only at high temperature. As illustrated by the graphs $\chi_M{}^{-1}$ versus T in Figure 19(*c*), the shift of the straight line relative to the Curie law to higher values indicates antiferromagnetic coupling, while the opposite shift indicates ferromagnetic coupling. Extrapolation of these lines to $T = 0$ can in principle provide a crude estimate of J. Some authors prefer to represent the data in a χ_M versus T^{-1} plot as in Figure 19(*d*).

In the absence of complicating factors, it is easy to assemble a compact formula for the susceptibility of two ions of spin S_A and S_B by using the energies given in Eq. (49) and the Boltzmann law, that is,

$$\chi_M = \frac{(\mu_0 N_A g^2 \beta^2 / 3kT) \sum_S (2S+1) S(S+1) \exp[-JS(S+1)/2kT]}{\sum_S (2S+1) \exp[-JS(S+1)/2kT]} \tag{58}$$

with $|S_A - S_B| \leq S \leq S_A + S_B$. Magnetic measurements from 300 to 4.2 K or lower allow the characterization of the nature of the ground state and how the excited-state levels are ordered. Thus an accurate value for J can be obtained from fitting the temperature dependence of the susceptibility. Though not strictly a kind of "spectroscopy", these measurements provide information about the magnetic levels that is otherwise very difficult to determine.

Examples of coupled dinuclear sites are found in copper[19] and iron proteins.[20] Hemocyanin, the reversible oxygen carrier protein in arthropods and molluscs, has an active site that consists of two copper ions. In oxyhemocyanin, the metal ions are in the copper(II) oxidation state, but are essentially diamagnetic due to strong antiferromagnetic coupling. Thus the ground state is $S = S_A - S_B = 0$. The inability to populate the $S = S_A + S_B = 1$ state at ambient temperature allows a lower limit to be placed on J at $>600\ \text{cm}^{-1}$. The large value for J arises from the presence of bridging ligands that efficiently mediate the metal–metal interaction. In oxyhemocyanin, the bound dioxygen has been shown crystallographically to bridge the copper centers in a μ-$\eta^2 : \eta^2$ fashion (Fig. 20),[21] a structure that rationalizes the very strong antiferromagnetic interaction observed.

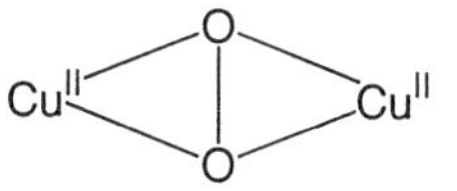

Figure 20
Structure of the Cu_2O_2 unit in oxyhemocyanin.

The energy levels are more complicated for a dinuclear iron(III) complex in which each Fe(III) ion has $S = \frac{5}{2}$. When the two ions are antiferromagnetically coupled ($J > 0$), the possible values for the system spin are $S = 0$–5, as represented in Figure 21. Examples of exchange-coupled diiron(III) centers in metal-

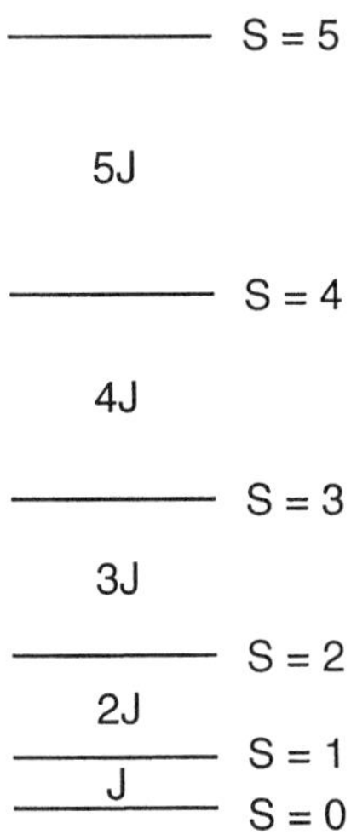

Figure 21
Ordering of spin levels derived from the coupling of two spins $S_A = S_B = \frac{5}{2}$.

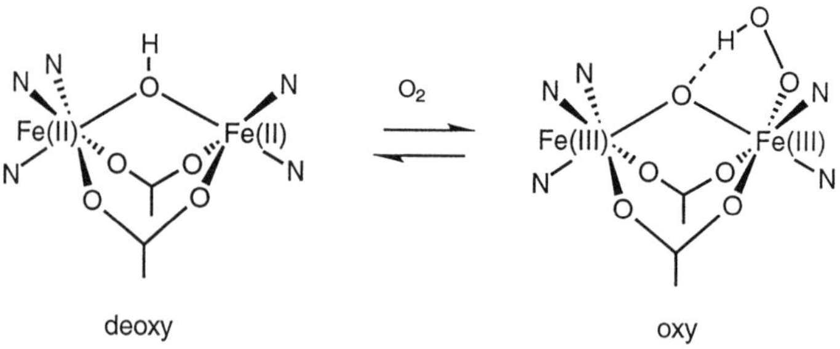

Figure 22
Reversible binding of O_2 by hemerythrin.

loenzymes include 2-Fe ferredoxins and hemerythrin. The 2Fe ferredoxins participate in electron-transfer functions and contain Fe_2S_2 clusters. Small molecule analogues of such sites have been synthesized and extensively characterized.[22] The iron centers exhibit strong antiferromagnetic coupling due to the presence of two bridging sulfides and a short metal–metal separation (2.7 Å). From fits of the temperature dependent magnetic properties of these synthetic complexes, J is found to be about $300\,cm^{-1}$.

Hemerythrin, the respiratory protein for some marine organisms, also contains a dinuclear Fe active site; crystal structures of several forms have been determined.[23] The iron centers are bridged by an oxygen atom and two carboxylates with five histidines serving as terminal ligands (Fig. 22). There is an open site on one iron for dioxygen to bind. Upon binding to the diiron(II) form, dioxygen is reduced to peroxide and the iron centers are oxidized to the iron(III) state.

The protonation state of the single oxygen atom bridge relates to the observed magnetic properties. Removing the hydroperoxo group from oxyhemerythrin gives rise to methemerythrin, another diiron(III) form that has been crystallographically characterized. Its temperature-dependent magnetic suceptibility has been measured and shown to exhibit strong antiferromagnetic coupling with $J = 268\ cm^{-1}$ (Fig. 23).[24] Such strong coupling derives from the presence of

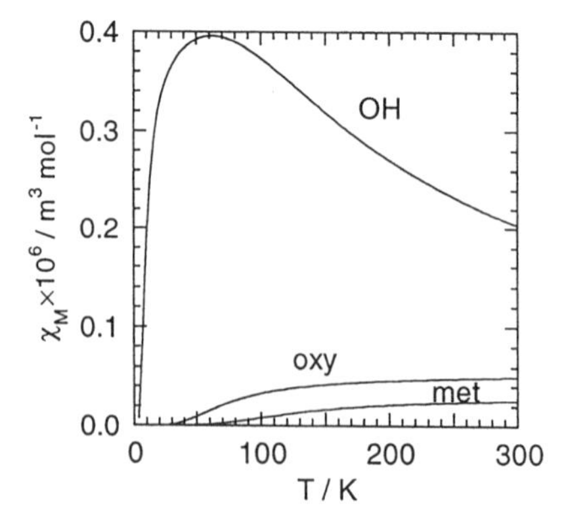

Figure 23
Molar magnetic susceptibilities for $[Fe_2(\mu\text{-}OH)(\mu\text{-}O_2CMe)_2(HBpz_3)_2]^+$ (OH), oxyhemerythrin (oxy), and methemerythrin (met).

the oxo bridge and is corroborated by the large number of (μ-oxo)diiron(III) complexes reported in the literature.[25] Oxyhemerythrin also contains a (μ-oxo)-diiron(III) center but exhibits a somewhat lower J value of 154 cm^{-1} (Fig. 23)[24] This weaker coupling interaction results from the hydrogen-bonding interaction between the bound hydroperoxide proton and the oxo bridge, which has been established by resonance Raman spectroscopy.[26] Indeed the protonation of the oxo bridge in the hemerythrin model complex $[\{HB(pz)_3\}_2Fe_2(\mu\text{-}O)(\mu\text{-}MeCO_2)_2]$ to $[\{HB(pz)_3\}_2Fe_2(\mu\text{-}OH)(\mu\text{-}MeCO_2)_2]^+$ results in the decrease of J from 242 to 34 cm^{-1}.[27]

We have just discussed examples wherein equivalent spins $S_A = S_B$ are coupled to afford ground states of $S = |S_A - S_B| = 0$ or $S = 2S_A$. Inequivalent spins may also couple. One prominent example in bioinorganic chemistry is the coupled dinuclear site in resting cytochrome oxidase that consists of a high-spin iron(III) heme ($S = \frac{5}{2}$) and a copper(II) ($S = \frac{1}{2}$) center (Fig. 24) and may afford

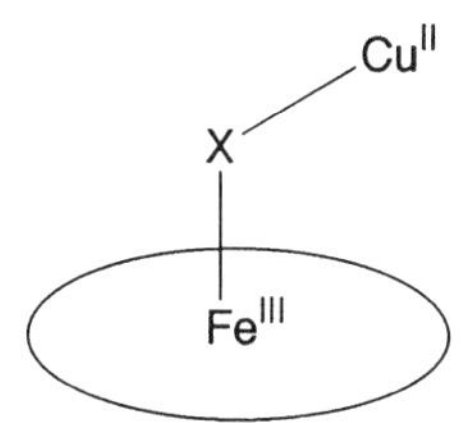

Figure 24
The Cu^{II}–Fe^{III} pair in cytochrome *c* oxidase.

either $S = 2$ (antiferromagnetic, $S_A - S_B$) or 3 (ferromagnetic, $S_A + S_B$) ground states. In the resting oxidase, the dinuclear cluster appears to be strongly coupled antiferromagnetically,[28] while the addition of cyanide makes the coupling weakly ferromagnetic. This magnetic behavior is modeled in synthetic complexes. When an oxo group bridges the heme iron and the copper center, strong antiferromagnetic coupling is observed,[29] while the substitution of the oxo bridge by cyanide leads to a weak ferromagnetic coupling.[30] The various factors that lead to the different coupling interactions are not yet fully understood; in particular, the possibility that the low-spin Fe(III) oxidation state may be involved adds complications, in which angular momentum and spin–orbit coupling effects may need to be considered.[31]

As we have seen for mononuclear ions, zfs effects can come into play at lower temperatures. Consider the following example. The synthetic $[(TACN)_2Mn^{III}{}_2O(MeCO_2)_2]^{2+}$, where TACN = 1,4,7-triazacyclononane, has a temperature-dependent magnetic susceptibility represented in Figure 25 under the form $\chi_M T$ as a function of T.[32] The product $\chi_M T$ increases from 6.69 $(4\pi \times 10^{-6})$ m^3 mol^{-1} K at 281.4 K to 9.47 $(4\pi \times 10^{-6})$ m^3 mol^{-1} K at 26.6 K and then decreases to 8.38 $(4\pi \times 10^{-6})$ m^3 mol^{-1} K at 2.9 K, which was the lowest temperature measured. In the absence of a coupling interaction, each Mn ion in an S = 2 state has a $\chi_M T$ value close to 3 $(4\pi \times 10^{-6})$ m^3 mol^{-1} K from Table III. At room temperature the experimental value for the dimanganese complex is

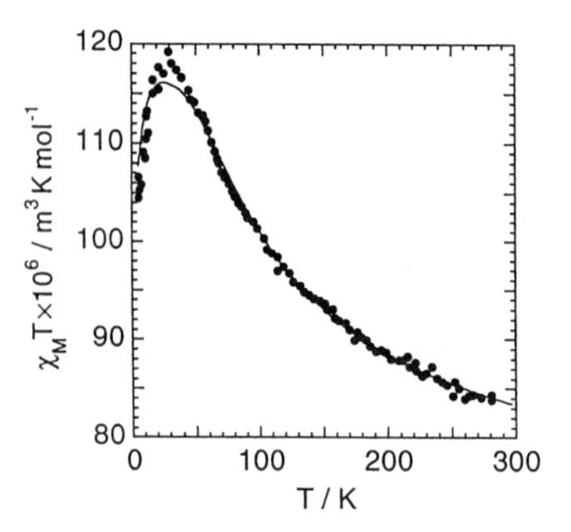

Figure 25
Molar magnetic susceptibility of $[(TACN)_2Mn^{III}{}_2O(MeCO_2)_2]^{2+}$ as $\chi_M T = f(T)$. The theoretical curve corresponds to the spin Hamiltonian given in the text with $J = -18\ \text{cm}^{-1}$ and $D_A = D_B = +3\ \text{cm}^{-1}$. [Reprinted from Wieghardt et al.[32] with permission from the American Chemical Society].

already larger than $2 \times 3\ (4\pi \times 10^{-6})\ \text{m}^3\,\text{mol}^{-1}\,\text{K}$. Moreover $\chi_M T$ increases as the temperature decreases, which from Figure 19(*b*) is the signature of a ferromagnetically coupled case. The ferromagnetic coupling between $S_A = 2$ and $S_B = 2$ gives $S = 4$ as the ground state; so for $S = 4$ with $g = 2$, $\chi_M T$ is expected to be equal to $g^2S(S+1)/8 = 10\ (4\pi \times 10^{-6})\ \text{m}^3\,\text{mol}^{-1}\,\text{K}$. This value is not reached at any temperature experimentally attainable. The decrease of $\chi_M T$ at low temperatures can be attributed to the zfs of the individual ions. The data have been quantitatively analyzed using the Hamiltonian

$$\mathcal{H}_{\text{total}} = J\mathbf{S_A} \cdot \mathbf{S_B} + \sum_{i=A,B} \{D_i[S_{iz}{}^2 - S_i(S_i+1)/3] + E_i[S_{ix}{}^2 - S_{iy}{}^2] + \beta\mathbf{B}[\mathbf{g_i}]\mathbf{S_i}\} \quad (59)$$

in which it is assumed that the principal axes of the zfs tensors at the two sites are colinear and E was set to zero. The results of the best fit of the experimental data using the Hamiltonian of Eq. (59) gives $J = -18\ \text{cm}^{-1}$, $D_A = D_B = 3\ \text{cm}^{-1}$, $g_{A//} = g_{B//} = 2.0$ and $g_{A\perp} = g_{B\perp} = 1.925$. The positive sign of the D parameter was not completely ascertained by this type of experiment but is in agreement with the sign predicted by theory for a high spin $Mn^{(III)}$ ion in a compressed octahedron[33] the geometry determined by X-ray crystallography.[32]

B. Trinuclear Complexes

Metalloproteins with trinuclear metal sites are also known. One particularly interesting example is aconitase, an enzyme involved in the tricarboxylic acid cycle responsible for the conversion of citrate to isocitrate.[34] Active aconitase contains an $[Fe_4S_4]^{2+}$ cluster of the cubane type; however, the enzyme is usually isolated in

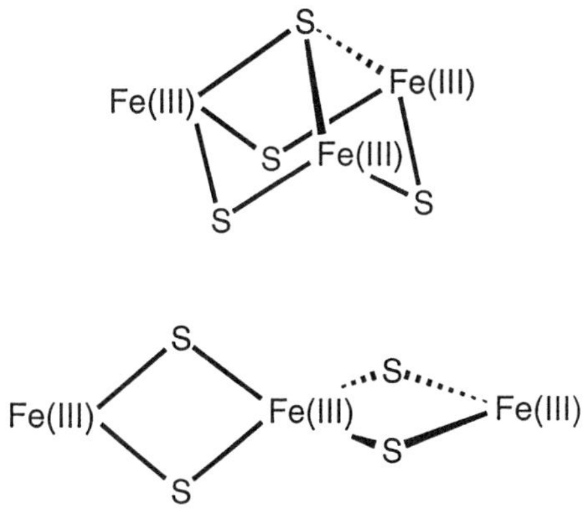

Figure 26
Schematic structures of the $[Fe_3S_4]^+$ cluster in the equilateral and linear configurations.

an inactive form that contains a C_3-symmetric $[Fe_3S_4]^+$ cluster, similar to that observed in ferredoxin II of *D. gigas* (Fig. 26). At high pH, the inactive aconitase cluster is converted into a different form which has properties very similar to that of yet another $[Fe_3S_4]^+$ cluster with a linear geometry (Fig. 26). The magnetic properties of the C_3 symmetric $[Fe_3S_4]^+$ cluster presented in Figure 27 follow the Curie law for $S = \frac{1}{2}$ very closely.[35] The magnetic properties of a synthetic linear

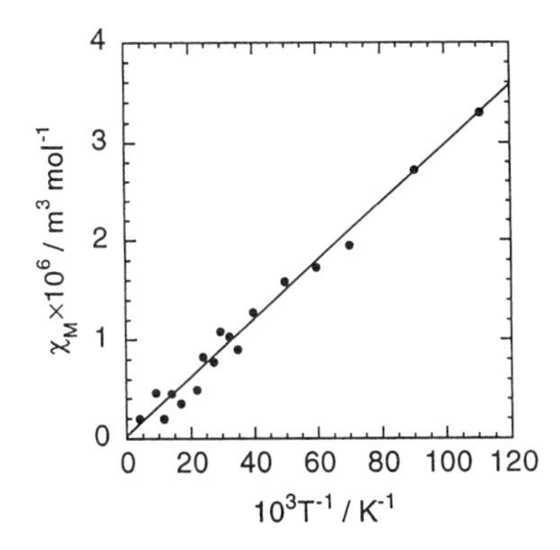

Figure 27
Molar susceptibility of the equilateral $[Fe_3S_4]^+$ cluster in *D. gigas* as function of $1/T$. [Adapted from Day et al.[35] with permission from the American Society for Biochemistry and Molecular Biology].

$[Fe_3S_4]^+$ cluster (Fig. 28) also follow the Curie law but with a slope indicative of $S = \frac{5}{2}$.[36] How can we understand this difference in spin for the ground states of these clusters?

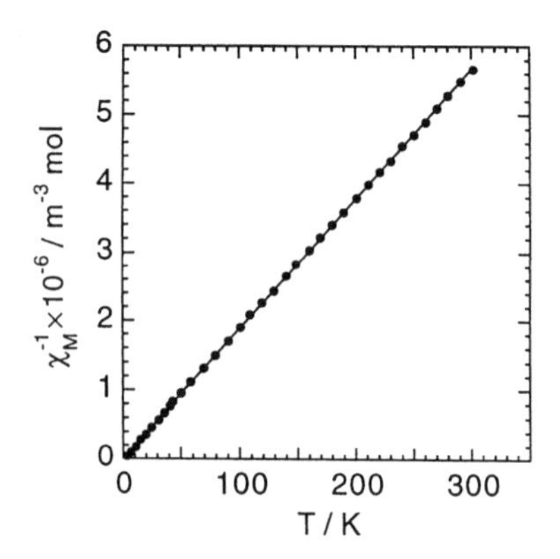

Figure 28
Inverse of molar susceptibility of the linear $[Fe_3S_4]^+$ cluster in a model complex as a function of temperature. [Adapted from Girerd et al.[36] with permission from the American Chemical Society].

Let us consider an isosceles spin triangle (Fig. 29). By extension from the dinuclear systems, the HDVV Hamiltonian will be

$$\mathcal{H}_{\text{exch}} = J[\mathbf{S_A} \cdot \mathbf{S_C} + \mathbf{S_B} \cdot \mathbf{S_C}] + J'\mathbf{S_A} \cdot \mathbf{S_B} \tag{60}$$

or

$$\mathcal{H}_{\text{exch}} = J\mathbf{S_C} \cdot [\mathbf{S_A} + \mathbf{S_B}] + J'\mathbf{S_A} \cdot \mathbf{S_B} \tag{61}$$

which suggests that the pair $\mathbf{S_{AB}} = \mathbf{S_A} + \mathbf{S_B}$ may play a special role. Using the total spin moment $\mathbf{S} = \mathbf{S_A} + \mathbf{S_B} + \mathbf{S_C}$, we write

$$\mathcal{H}_{\text{exch}} = J[\mathbf{S}^2 - \mathbf{S_{AB}}^2 - \mathbf{S_C}^2]/2 + J'[\mathbf{S_{AB}}^2 - \mathbf{S_A}^2 - \mathbf{S_B}^2]/2 \tag{62}$$

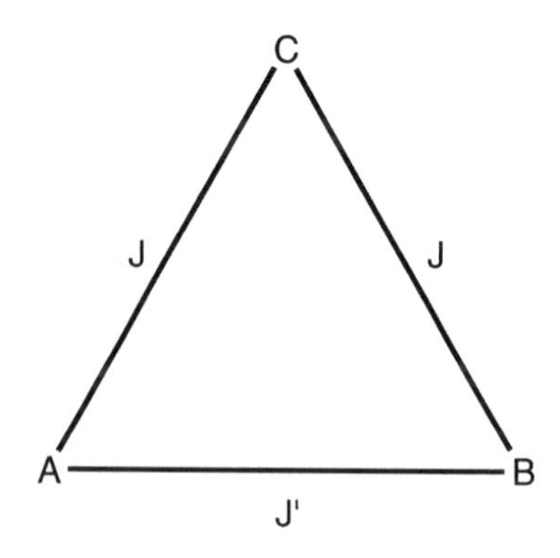

Figure 29
An isosceles spin triangle.

Table V
Possible Spin Values for a Triad of Fe(III) High-Spin Ions

S_{AB}	S_C	S
0	$\frac{5}{2}$	$\frac{5}{2}$
1		$\frac{3}{2}, \frac{5}{2}, \frac{7}{2}$
2		$\frac{1}{2}, \frac{3}{2}, \frac{5}{2}, \frac{7}{2}, \frac{9}{2}$
3		$\frac{1}{2}, \frac{3}{2}, \frac{5}{2}, \frac{7}{2}, \frac{9}{2}, \frac{11}{2}$
4		$\frac{3}{2}, \frac{5}{2}, \frac{7}{2}, \frac{9}{2}, \frac{11}{2}, \frac{13}{2}$
5		$\frac{5}{2}, \frac{7}{2}, \frac{9}{2}, \frac{11}{2}, \frac{13}{2}, \frac{15}{2}$

In these clusters, the Fe(III) ion is in an environment close to T_d and the spin of each ion is $S_A = S_B = S_C = \frac{5}{2}$. To couple three angular momenta, we have to combine any two first to get a coupled pair and add the third one to the coupled pair. It is convenient, considering Eq. (62), to couple $\mathbf{S_A}$ and $\mathbf{S_B}$ first. The possible values of the spin of the coupled pair and the total spin are indicated in Table V, obtained using the relations $|S_A - S_B| \leq S_{AB} \leq S_A + S_B$ and $|S_{AB} - S_C| \leq S \leq S_{AB} + S_C$. What differs from the dinuclear situation is that almost all spin states appear several times. The only one to appear just once is the $S = \frac{15}{2}$ state that corresponds to the unique arrangement of three local spins parallel. There are two $S = \frac{1}{2}$ ($S_{AB} = 2$ or 3) and six $S = \frac{5}{2}$ (all values of S_{AB}) combinations. The wave functions can be written as $|(S_A S_B)S_{AB}, S_C, S, M_S\rangle$ and are eigenfunctions of Eq. (62) with the eigenvalues

$$E(S_{AB}, S) = \frac{J[S(S+1) - S_{AB}(S_{AB}+1) - S_C(S_C+1)]}{2} + \frac{J'[S_{AB}(S_{AB}+1) - S_A(S_A+1) - S_B(S_B+1)]}{2} \qquad (63)$$

The energies possible for the triad are represented in Figure 30 as functions of J'/J. The C_3-symmetric situation corresponds to $J' = J$ so that

$$E(S_{AB}, S) = \frac{J[S(S+1) - S_A(S_A+1) - S_B(S_B+1) - S_C(S_C+1)]}{2} \qquad (64)$$

In this case, all states with the same spin value have the same energy (Fig. 30). If $J > 0$, the ground state will be a spin doublet $|(S_A S_B)S_{AB} = 2$ or 3, S_C, $S = \frac{1}{2}$, $M_S >$. In the linear situation, $|J'| \ll |J|$, so a good approximation is $J' = 0$,

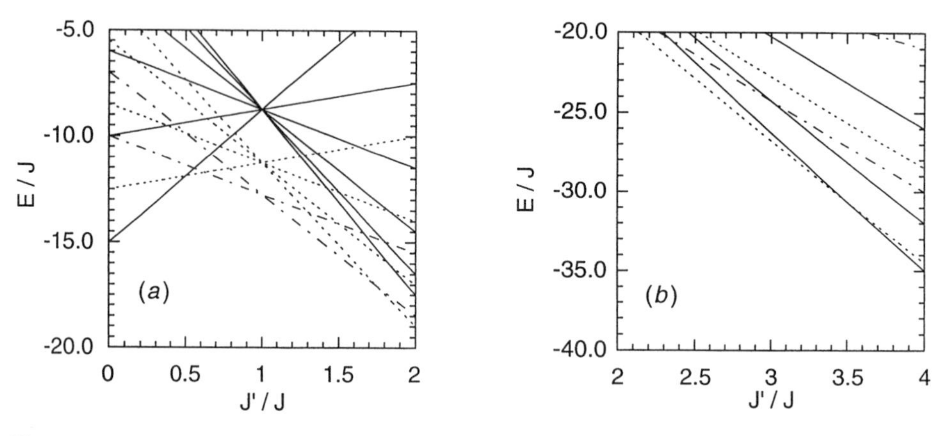

Figure 30
Energy levels for an Fe(III) isosceles triad in unit of $J > 0$ as functions of J'/J. (-.-.-) $S = \frac{1}{2}$; (...) $S = \frac{3}{2}$; (___) $S = \frac{5}{2}$. The other levels are not represented. (*a*) The $0 \leq J'/J \leq 2$; the spin of the ground-state changes from $S = \frac{5}{2}$, $S = \frac{3}{2}$, $S = \frac{1}{2}$ and then becomes $S = \frac{3}{2}$. (*b*) The $2 \leq J'/J \leq 4$; the spin of the ground-state changes from $\frac{3}{2}$ to $\frac{5}{2}$.

which leads to

$$E(S_{AB}, S) = \frac{J[S(S+1) - S_{AB}(S_{AB}+1) - S_C(S_C+1)]}{2} \tag{65}$$

The ground state is a spin sextet $|(S_A S_B)S_{AB} = 5,\ S_C,\ S = \frac{5}{2},\ M_S\rangle$, which intuitively is the only way to arrange three spins in a row with antiferromagnetic coupling between nearest neighbor pairs (Fig. 31). The two terminal spins are

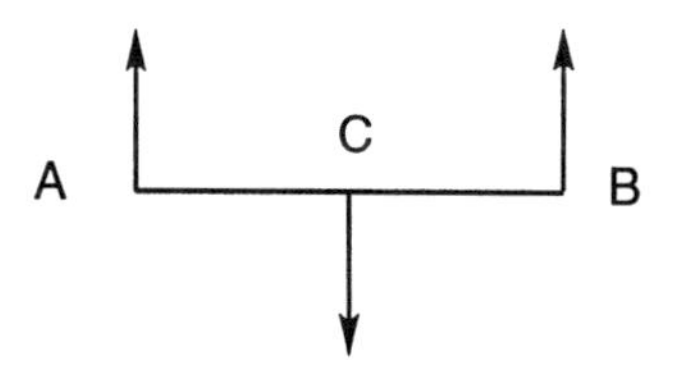

Figure 31
Spin ordering for a linear antiferromagnetic triad.

parallel to give $S_{AB} = 5$ and are aligned antiparallel with respect to S_C to give $S = \frac{5}{2}$. The metal centers in these clusters interact strongly, so J is quite large ($300\,cm^{-1}$) affording a large gap between the ground state and the first excited state. Only the ground state is populated between 300 and 4.2 K, the temperature range of the magnetic measurements.[36]

Trinuclear clusters with three different J values have more complex magnetic properties; other methods need to be employed to solve the HDVV Hamiltonian. A review by Griffith can be consulted.[37]

C. Interaction Mechanism

The HDVV Hamiltonian describes the interaction of two spins in a molecule. Though these interactions may also reflect dipolar interactions between the magnetic moments of the localized spins, such interactions are too small to explain the observed phenomena. It is useful to localize the unpaired electrons on specific orbitals called magnetic orbitals, that is, orbitals containing unpaired spin density. In many cases, they are essentially the metal d atomic orbitals partially delocalized on the ligands. In other words, they are the highest singly occupied orbitals representing the antibonding combination of the metal d atomic orbitals with the symmetry adapted combination of s and p orbitals of the ligands. The magnetic orbitals are illustrated in Figure 32 in the case of V(IV) (d^1) and Cu(II) (d^9) ions in their more common geometries.

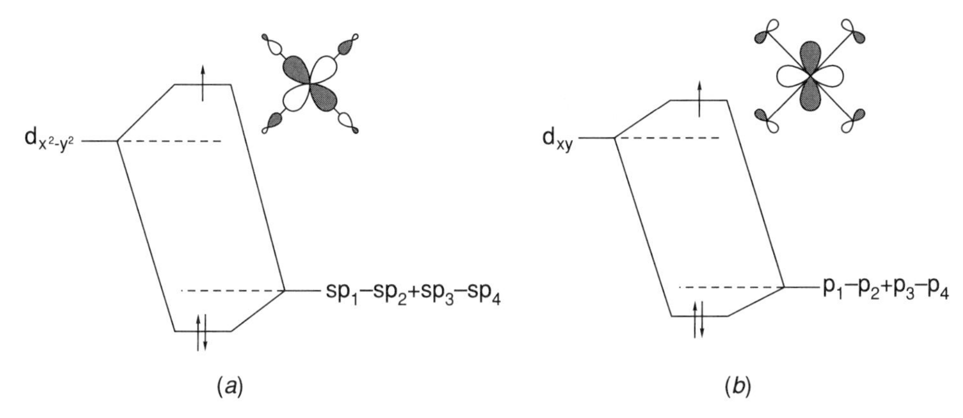

Figure 32
Magnetic orbitals for a planar Cu(II) ion (*a*) and a VO^{2+} ion (*b*).

What can happen between two orbitals, each containing one unpaired electron, on two adjacent centers? A chemical bond may form, the strength of which depends on the overlap between the two orbitals, and the two electrons will become spin paired. The same type of spin pairing in fact occurs in antiferromagnetic coupling, but the stabilization of the spin-paired state relative to that of the spin-free state is significantly smaller and corresponds to that of a very weak chemical bond, its strength being quantified by the magnitude of J.

These interactions are usually mediated by bridging atoms. Thus in the case of hemerythrin and its models (e.g., $[\{HB(pz)_3\}_2Fe_2(\mu\text{-}O)(\mu\text{-}MeCO_2)_2]$; see Fig. 23), the iron(III) centers are bridged by an oxo group, which gives rise to short Fe–μ-O_{oxo} (1.8 Å) bonds and strong antiferromagnetic coupling ($J > 200\,cm^{-1}$).[20] When the oxo bridge is protonated as in the case of $[\{HB(pz)_3\}_2Fe_2(\mu\text{-}OH)\cdot(\mu\text{-}MeCO_2)_2]^+$, the Fe–$\mu$-O bond increases to 1.95 Å and J weakens to $34\,cm^{-1}$.[27] Protonation of the oxo bridge disrupts the Fedπ–Opπ interaction, thus diminishing the pathways by which the two iron centers can communicate and decreasing the overlap between the magnetic orbitals of the iron centers.

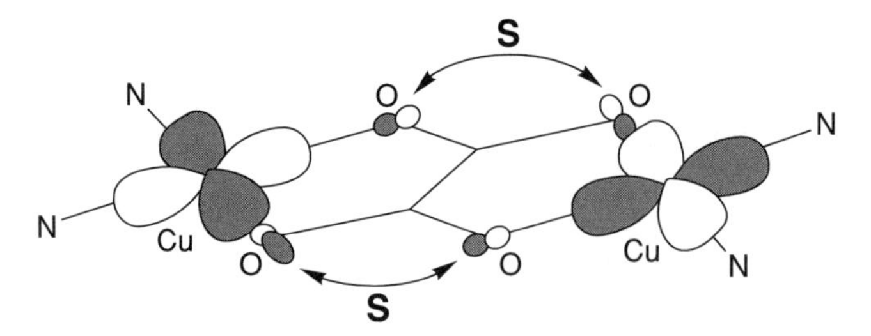

Figure 33
Structure and magnetic orbitals for $[\text{tmen}(H_2O)Cu(\mu\text{-}C_2O_4)Cu(H_2O)\text{tmen}]^{2+}$.

If antiferromagnetic coupling corresponds to spin pairing, to what does ferromagnetic coupling correspond? Let us consider the situation wherein the magnetic orbitals are not far apart but, due to symmetry, have a zero overlap. The orbitals can spatially "interact" but the overlap integral is 0 over the entire space; the magnetic obitals are said to be orthogonal and no antiferromagnetic coupling is possible. This situation is where ferromagnetic coupling arises. A consideration of the symmetry properties of the magnetic orbitals has led to relationships often referred to as Goodenough–Kanamori rules, presented here in a slightly modified formulation by Anderson[38]:

1. When the two ions have lobes of localized orbitals pointing toward each other in such a way that the orbitals would have a reasonably large overlap integral, the exchange is antiferromagnetic.
2. When the orbitals are arranged in such a way that they are expected to be in contact but have a zero overlap integral, the exchange is ferromagnetic (but usually weaker than the corresponding antiferromagnetic exchange).

1. Tunable Exchange in Dinuclear Copper(II) Complexes

The importance of the overlap between magnetic orbitals is illustrated by two (μ-oxalato)dicopper(II) complexes. The molecular structure of $[\text{tmen}(H_2O)Cu(C_2O_4)Cu(H_2O)\text{tmen}]^{2+}$, where tmen = N,N,N',N'-tetramethyl-1,2-diaminoethane (Fig. 33) shows a centrosymmetric dinuclear unit with an intramolecular Cu–Cu separation of 5.15 Å.[39] The individual copper environments do not deviate much from an ideal square pyramid, and the two basal planes are coplanar with the $(C_2O_4)^{2-}$ bridge. The Cu(II) ion is a d^9 ion; consequently, its magnetic orbital would be the one pointing toward the closest ligands, that is, that in the basal plane of the square pyramid. For this complex, the two magnetic orbitals, illustrated in Figure 33, are in the same plane and have a reasonably large overlap via the oxalato bridge. As expected by the first Goodenough–Kanamori rule, the observed coupling is antiferromagnetic. The stabilization of the singlet state with regard to the triplet state is equal to 385 cm^{-1}.

In contrast, the coupling interaction in $[(\text{terpy})Cu(C_2O_4)Cu(\text{terpy})]^{2+}$ where terpy = $2,2':6',2''$-terpyridine is extremely weak with $J \approx 0\,\text{cm}^{-1}$.[40] Its molecular structure (Fig. 34) also shows a centrosymmetric dinuclear unit with a

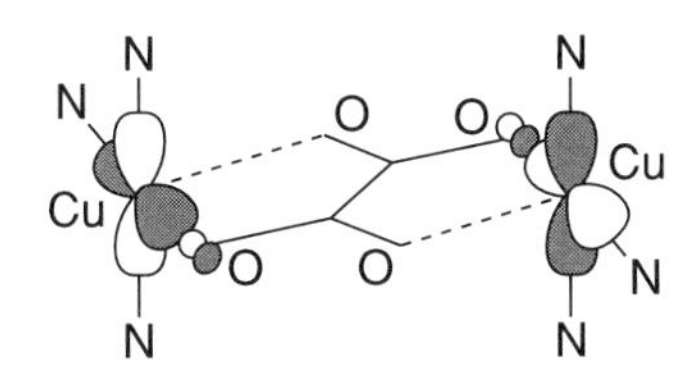

Figure 34
Structure and magnetic orbitals for [terpyCu(μ-C_2O_4)Cu(H_2O)terpy]$^{2+}$.

Cu–Cu separation of 5.47 Å, and the individual copper environments do not deviate much from ideal square pyramid. However, the basal planes in this complex are not coplanar with $C_2O_4{}^{2-}$ but are instead perpendicular to it. The magnetic orbitals are unfavorably oriented (illustrated in Fig. 34) for any significant overlap; therefore the negligible coupling observed experimentally can easily be rationalized.

2. Orthogonality of Magnetic Orbitals and Ferromagnetic Coupling

The second Goodenough–Kanamori rule can be illustrated by the interaction of a $d_{x^2-y^2}$ and a d_{xy} magnetic orbital as illustrated in Figure 35. The former is antisymmetric with respect to the plane containing the two metal ions and perpendicular to the plane of the molecule, while the latter is symmetric. The two magnetic orbitals have different symmetries and consequently are orthogonal; a ferromagnetically coupled complex would be expected. This situation is nicely illustrated by the dinuclear complex [CuVO(fsa)$_2$en(H_2O) · MeOH] with (fsa)$_2$en = N,N'-(2-hydroxy-3-carboxybenzylidene)-1,2-diaminoethane.[41] The

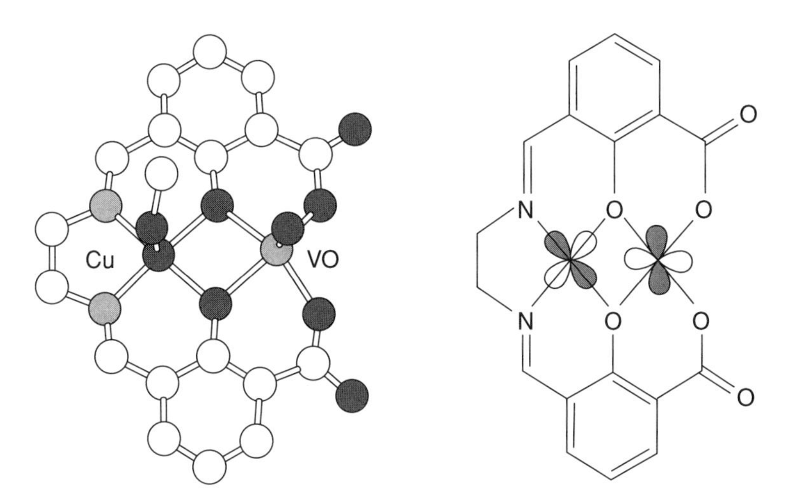

Figure 35
Structure and magnetic orbitals for [CuVO(fsa)$_2$en(H_2O)(MeOH)].

molecular structure of this compound depicted in Figure 35 has close to C_s symmetry. The oxygen atoms of both the vanadyl group and the methanol ligand, the copper(II) and vanadium(IV) ions all belong to a mirror plane that bisects both CuN_2O_3 and VO_5 square pyramids. The magnetic orbital of the Cu(II) ion (d^9 ion) is the $d_{x^2-y^2}$ orbital and transforms as the a'' irreducible representation of the C_s point group; it is antisymmetric with respect to the CuOVO mirror plane. The magnetic orbital of the V(IV)O (d^1 ion) ion, constructed from the metal d_{xy} orbital, transforms as a' and is symmetric with respect to the same mirror plane. The relative orientations of the magnetic orbitals are represented in Figure 35, and the overlap integral $\langle Cu|VO \rangle = 0$. Ferromagnetic coupling between the two ions is predicted and experimentally observed ($J = -118$ cm^{-1}).

3. Metal–Radical Interactions

Interactions between metal centers and ligand radicals is another recurrent theme in bioinorganic chemistry. For example, the porphyrin radical plays a prominent role in the storage of oxidizing equivalents in heme peroxidases. Indeed the formally Fe(V) oxidation state ascribed to Compound I of horseradish peroxidase is best described as an Fe(IV) complex with a porphyrin ligand which has been oxidized by one electron.[42] The two oxidizing equivalents above Fe(III) are thus stored on the iron center and on the porphyrin ligand.

The participation of porphyrin radicals in heme oxidation chemistry has spurred the synthesis of a number of iron(porphyrin radical) complexes.[43] For example, the six-coordinate complex $[Fe^{III}(TPP^{\bullet})(ClO_4)_2]$ [$TPP^{\bullet}$ = *meso*-tetraphenylporphyrin radical monoanion] has an $S = 3$ ground state derived from the ferromagnetic coupling of the $S = \frac{5}{2}$ Fe(III) center to the porphyrin ligand radical. The magnetic orbital for meso-substituted tetraarylporphyrin radicals has a_{2u} character in D_{4h} symmetry, but the magnetic orbitals of the high-spin iron(III) ion all have gerade character and so are orthogonal to the magnetic orbital of the porphyrin, which is ungerade. The gap between the $S = 3$ ground state and the $S = 2$ excited state is $|3J| = 240$ cm^{-1} such that $J = -80$ cm^{-1}.

In contrast, the five-coordinate $[Fe(TPP^{\bullet})Cl]^+$ complex has an $S = 2$ ground state with J estimated to be $> +1000$ cm^{-1}. In this complex, the porphyrin ring is ruffled, thereby lowering the molecular symmetry to C_{2v}. Under this point group, the porphyrin magnetic orbital symmetry reduces from a_{2u} to a_1, while the metal d orbitals have a_1, a_1, b_1, b_2, and a_2 symmetries. Thus overlap between the ligand orbital and two of the metal d orbitals becomes symmetry allowed and antiferromagnetic coupling is observed.

The complex $[Fe^{IV}(O)Cl(TMP^{\bullet})]$ serves as a model for peroxidase Compounds I and has been shown to have an $S = \frac{3}{2}$ ground state with $-J >$ 60 cm^{-1}.[44] Such a ground state derives from the ferromagnetic coupling of an $S = 1$ Fe(IV) center with an $S = \frac{1}{2}$ ligand radical. Such a ferromagnetic interaction may be easily rationalized by considering the magnetic orbitals of the components. The $S = 1$ Fe(IV) center in C_{2v} symmetry would have magnetic (d_{xz} and d_{yz}) orbitals of b_1 and b_2 character, respectively, which are clearly orthogonal to the porphyrin magnetic orbital, which has a_1 symmetry. An interesting question

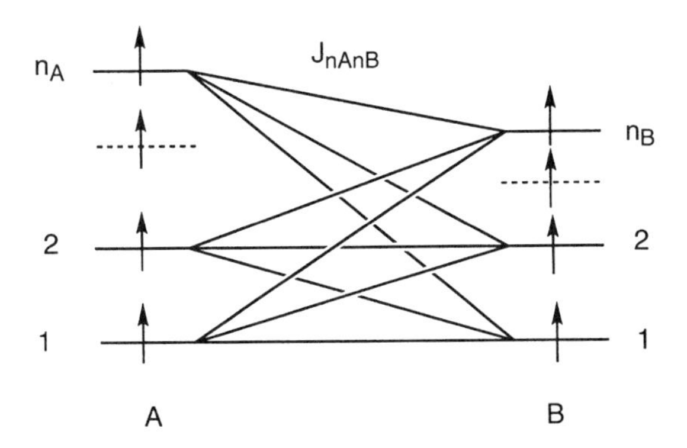

Figure 36
Individual exchange interactions between two metal ions.

that remains is why the Compounds I of horseradish peroxidase and chloroperoxidase exhibit much weaker coupling ($J \sim 2$ and $36\,\mathrm{cm}^{-1}$, respectively) and why the observed coupling is opposite in sign to that of the model complex. The coupling interaction could be weakened by low unpaired spin density at the pyrrole nitrogens in the porphyrin magnetic orbital. Furthermore, structural distortions in the proteins may lower the porphyrin symmetry and favor orbital overlap with the metal magnetic orbitals thus promoting antiferromagnetic coupling.

As explained in Appendix III, the exchange interaction J between centers A and B can be related to exchange interactions between unpaired electrons at the two sites (Fig. 36). The relation is

$$J = (1/n_A n_B) \sum_{i,j} J_{ij} \tag{66}$$

where n is the number of unpaired electrons on site A or B. Each J_{ij} is made of several contributions, some positive, some negative. A reasonable degree of understanding of these different contributions in term of orbital interactions has been achieved.[3] For example, a rationalization of the coupling constants for M^{III}–O–M^{III} units has been proposed.[45]

When only one exchange pathway is really efficient, say J_{lk}, J from Eq. (66) is $J = (1/n_A n_B) J_{lk}$, all the other J_{ij} terms being neglected. The metal–porphyrin radical systems discussed above could correspond to such a situation. The complex $[\mathrm{Cu^{II}(TPP^{\bullet})}]^{+}$ exhibits $J = J_{dx^2-y^2-a_{2u}} = -400\ \mathrm{cm}^{-1}$ [n_A for the radical is 1 and n_B for Cu(II) is 1].[43] For the analogous Fe(III) complex, one may suggest $J = (\frac{1}{5})\ J_{dx^2-y^2-a_{2u}}$ (here $n_A = 1$ and $n_B = 5$), which would lead to $J = -80\ \mathrm{cm}^{-1}$ in agreement with the observed value.[43]

D. Mixed-Valence Diiron Complexes and Double Exchange Coupling

Mixed-valence systems are particularly important in bioinorganic chemistry. They are frequently implied in electron-transfer proteins. One recent example is

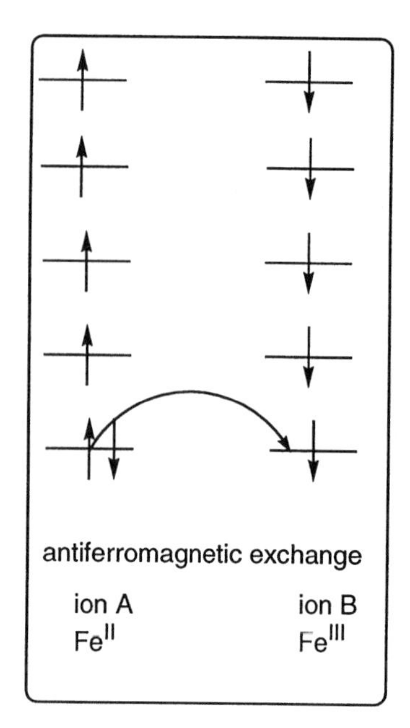

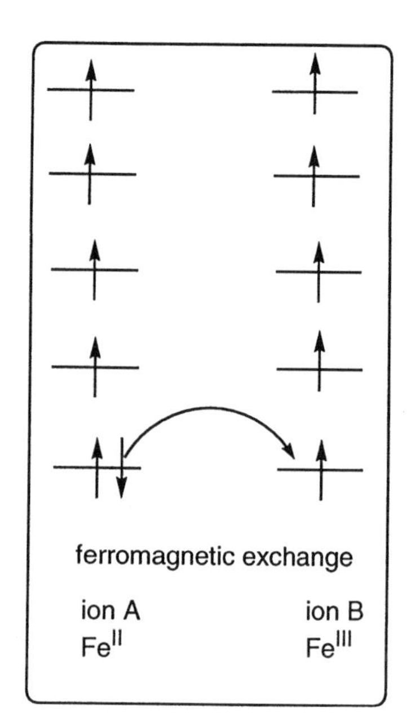

Figure 37
Double exchange between Fe(II) and Fe(III).

the Cu^{II}–Cu^{I} pair of the Cu_A site in cytochrome *c* oxidase.[46] These systems raise interesting problems concerning intra- and intercluster electron transfer. In a Cu^{II}–Cu^{I} pair, the Cu(I) is $S = 0$ and the Cu(II) is $S = \frac{1}{2}$. Then there is no spin coupling between these two ions. The only question is the extent of delocalization of the extra electron between both metal centers.

In an Fe^{III}–Fe^{II} pair, both electron delocalization and the coupling of the electronic spins must be considered. It is easy to show qualitatively that exchange coupling must affect electron delocalization. Let us consider Figure 37 and consider first the case of ferromagnetic exchange. Of the two electrons in the lowest d orbital, the only electron able to shift to the other metal center is the down spin one, following the Pauli exclusion principle. After the transfer, A will be Fe(III) with all the individual spins aligned; Hund's rule is not violated. So electron transfer can occur between two spin aligned centers without impediment. It is possible to show that the energy of delocalization in this case is β, where β is the resonance integral between the orbitals involved in the transfer (not to be confused with the Bohr magneton that also uses the same symbol!). It is the same quantity as the resonance integral used in Hückel molecular orbital (MO) theory. The larger the overlap between the orbitals is, the larger $|\beta|$. In the case of antiferromagnetic coupling of the spins at centers A and B, the delocalized electron has up spin. After its transfer, center A becomes Fe(III) but with a spin projection $M_{S_A} = +\frac{3}{2}$. This state is a mixture of spin quintet and sextet. The effect is that the delocalization energy will be greatly reduced relative to the first case. One can show that the delocalization energy in this case is $\beta/5$. So delocalization favors

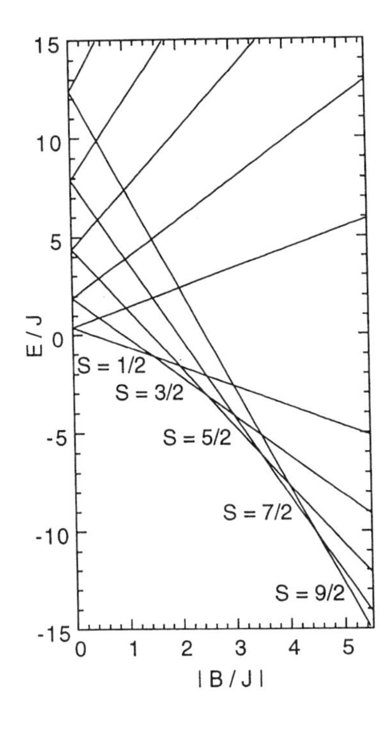

Figure 38
Energy levels for an $Fe^{II}Fe^{III}$ pair.

ferromagnetic coupling of the spins. In this reasoning, we consider that the local exchange energy is large versus the delocalization energy, which seems correct for the type of mixed-valence clusters studied. In other words, we consider Hund's rule to be obeyed.[47] This phenomenon could be called "spin-dependent electron transfer" and is a modified version of the "double exchange" theory of Anderson and Hasegawa,[48] which covers the more general case of any delocalization and local exchange energies. It is possible to show that the energy of an $Fe^{II}Fe^{III}$ pair in the "spin-dependent electron transfer" theory is given by

$$E(S) = (J/2)S(S+1) \pm \frac{\beta(S+0.5)}{(2S_0+1)} \tag{67}$$

where all symbols have been already defined except S_0, which is the core spin (the larger between S_A and S_B minus $\frac{1}{2}$; for example, in $Fe^{II}Fe^{III}$, $S_0 = 2$). The nature of the ground state will depend on the ratio $[\beta/(2S_0+1)]/J$, which is illustrated for the $J > 0$ case in Figure 38, which represents $E(S)$ as a function of $|B/J|$, where $B = \beta/(2S_0+1)$. For $\beta = 0$, the usual Heisenberg ordering is found.

The spin-dependent electron-transfer theory is supported experimentally by studies on the mixed-valence complex $[(Me_3TACN)Fe^{II}(\mu\text{-}OH)_3Fe^{III}(Me_3TACN)]^{2+}$.[49] The Fe ions are found to be identical by Mössbauer spectroscopy and an intense band is observed at 13,000 cm^{-1}. The complex is an example of a class III mixed-valence system, where the valences are fully delocalized (Fe oxidation state = 2.5). At room temperature its magnetic moment is 9.95 μ_B, corresponding to $S = \frac{9}{2}$! Thus even at room temperature the spins are aligned, which implies a very large ferromagnetic coupling. For $[(Me_3TACN)Fe^{II}(\mu\text{-}OH)_3Fe^{III}(Me_3TACN)]^{2+}$, the Fe–Fe distance is 2.5 Å and the odd electron is delocalized between two d_{z^2} orbitals that point towards each other (the z axis being the Fe–Fe axis); this leads to a large delocalization energy ($|\beta| = 6500\ cm^{-1}$), which can be determined by optical spectroscopy since the previously mentioned near-IR band corresponds to the transition between the $\sigma_g(d_{z^2})$ bonding MO and the antibonding $\sigma_u(d_{z^2})$ one ($2|\beta| = 13{,}000\ cm^{-1}$).

Other mixed-valence dinuclear units are found in the reduced forms of iron–sulfur (e.g., 2Fe ferredoxins)[50] and iron–oxo proteins (e.g., hemerythrin, methane monooxygenase, purple acid phosphatases, and ribonucleotide reductase).[51] In most of these cases, the two iron centers are antiferromagnetically coupled, and the ground state is $S = \frac{1}{2}$; an EPR spectrum with $g_{ave} < 2$ is observed (see the EPR chapter, Chapter 3). The valences are observed to be localized on the Mössbauer time scale. These systems could correspond to a smaller $[\beta/(2S_0 + 1)]/J$ ratio. The consequence is a reduction of the delocalization energy, which makes the system susceptible to vibronic or static trapping.

VII. Conclusion

In this chapter, we have discussed the relationship between magnetic and electronic properties. In the past, molecular magnetochemistry has played a major role in the understanding of electronic states of monometallic sites. We have tried here to provide a concise picture of these studies, focusing on systems that have no orbital degeneracy in their ground state, that state moreover being well separated from excited states. We could only briefly mention other systems with very interesting properties (e.g., spin transitions), and interested readers are directed to more specialized books. The appendices provide an idea of how spin Hamiltonians are built; we have tried to give the reader the feeling that these Hamiltonians are valid under some assumptions that may not apply in every situation. Recently, interest in polymetallic systems has led to a true "Renaissance" in magnetochemistry. We have approached this problem both from the phenomenological point of view and from an electronic structure perspective to provide insight into how interactions between metal centers affect the magnetic properties of the multimetal unit. The notion of "spin-dependent electron transfer" has been introduced to rationalize the high-spin states found for some mixed-valence metal clusters. The study of magnetic properties of polynuclear complexes has led to interesting applications not only in bioinorganic chemistry but also in the study of molecular ferromagnets; this illustrates the richness of a field which, together with other physical methods discussed in this book, helps break down the barriers between chemistry, physics, and biology.

The Spin Hamiltonian for Mononuclear Systems

In this appendix, we will give only the main ideas for building a spin Hamiltonian [Eq. (26)]. Our goal is to make the reader aware of the approximations used and the building procedure. The Hamiltonian describing the system will be separated into two components:

$$\mathscr{H}_{\text{total}} = \mathscr{H}_0 + \mathscr{V} \tag{A.I.1}$$

$\mathscr{H}_0$ contains all the electron–electron and electron–nuclei interactions. The solutions to $\mathscr{H}_0$ are assumed to be known. The energies are $E_n{}^0$. The corresponding states are represented as ${}^{2S_n+1}\Gamma_n$. The term $\mathscr{V}$ contains the spin–orbit coupling and the Zeeman interactions, that is,

$$\mathscr{V} = \zeta \sum_i \mathbf{l_i} \cdot \mathbf{s_i} + \beta \sum_i (\mathbf{l_i} + 2\mathbf{s_i}) \cdot \mathbf{B} \tag{A.I.2}$$

If the ground state is well separated from the excited states, the effect of these excited states on the ground state may be evaluated using perturbation theory. Let us assume that the ground state ${}^{2S_0+1}\Gamma_0$ is orbitally nondegenerate; the $2S_0 + 1$ spin degeneracy is retained. The $\mathscr{V}$ operator connects this degenerate ground state with the excited state and has the effect of removing the $2S_0 + 1$ degeneracy. Information about this splitting may be obtained experimentally from the magnetic properties of a metal complex.

Now comes a big trick! One can build a Hamiltonian, called an effective Hamiltonian, which acts only in the ground subspace (dimension $2S_0 + 1$) but reproduces the splitting pattern due to the excited staes as calculated using perturbation theory. An essential difference between the Hamiltonian of Eq. (A.I.1) and the effective Hamiltonian is related to the spaces within which these Hamiltonians act.

Let us introduce the projection operators P_{n} on the subspaces of energy $E_n{}^0$. One can show that an effective Hamiltonian correct to the second order is

$$\mathscr{H}_{\text{eff}} = P_0 \mathscr{V} P_0 + \sum_n \frac{P_0 \mathscr{V} P_n \mathscr{V} P_0}{(E_0{}^0 - E_n{}^0)} \tag{A.I.3}$$

When this procedure is applied using $\mathscr{V}$ of Eq. (A.I.2), the result is the spin Hamiltonian of Eq. (26). When working with an effective Hamiltonian, one must not forget that it is based on a perturbational approach and that it may not apply in some situations. In the building procedure, one obtains the expressions for D,

E, and [**g**] in terms of the parameters appearing in the basic Hamiltonian. With the restrictive assumption that we mix in the ground state only states derived from the ground term (L,S) for the free ion, the following simplified version of Eq. (A.I.2) may be used.

$$\mathcal{V} = \lambda \mathbf{L} \cdot \mathbf{S} + \beta(\mathbf{L} + 2\mathbf{S}) \cdot \mathbf{B} \tag{A.I.4}$$

To evaluate the second-order term in Eq. (A.I.3), one uses

$$\Lambda_{\mu\nu} = \sum_n \frac{\langle \Psi_0 | \mathbf{L}_\mu | \Psi_n \rangle \langle \Psi_n | \mathbf{L}_\nu | \Psi_0 \rangle}{(E_n{}^0 - E_0{}^0)} \tag{A.I.5}$$

where Ψ_0 is the orbital part of the wave function of the ground state $^{2S+1}\Gamma_0$ (energy E_0) and Ψ_n is the orbital part of the wave function of the excited state $^{2S+1}\Gamma_n$ (energy E_n). The x, y, and z axes are the principal axes of the tensor $[\Lambda_{\mu\nu}]$. Then Eq. (A.I.6) and (A.I.7) are obtained.

$$g_{\mu\nu} = 2(\delta_{\mu\nu} - \lambda \Lambda_{\mu\nu}) \tag{A.I.6}$$

$$\begin{aligned} D &= -\lambda^2[\Lambda_{zz} - (\Lambda_{xx} + \Lambda_{yy})/2] \\ E &= -\lambda^2[\Lambda_{xx} - \Lambda_{yy}]/2 \end{aligned} \tag{A.I.7}$$

where $\delta_{\mu\nu} = 1$ if $\mu = \nu$, and 0 otherwise.

If the symmetry is cubic, $\Lambda_{xx} = \Lambda_{yy} = \Lambda_{zz} = \Lambda$; therefore $D = E = 0$ and $g = 2(1 - \lambda\Lambda)$. The zfs Hamiltonian is operative only if there is a distortion from cubic symmetry. One sees that D and E are inversely proportional to the gaps between the ground state and the relevant excited states. In case of an axial distortion, which will correspond to the z axis, $D \neq 0$, but $E = 0$.

As previously mentioned, the preceding equations are valid only for the contributions to the zfs coming from excited states derived from the same atomic term. Only under this assumption can the spin–orbit coupling term be written as $\lambda\mathbf{L} \cdot \mathbf{S}$. The origin of zfs for a high-spin Fe(III) cannot be rationalized on this basis, since there is no excited state that has the same spin ($S = \frac{5}{2}$) as the ground state. Fortunately, Griffith[52] showed that the spin Hamiltonian of Eq. (26) stays valid even when more general interactions are included. Only the expression of the parameters are modified. A very clear derivation of the zfs parameters for Fe(III) is given in Trautwein et al.[16]

APPENDIX II

Magnetic Susceptibility of a $^2T_{2g}$ State

Here, we compute the magnetic susceptibility of a $^2T_{2g}$ state. Six functions form a basis for this state $|xy\rangle, |xz\rangle, |yz\rangle, |xy\beta\rangle, |xz\beta\rangle, |yz\beta\rangle$. The β notation stands for the $-\frac{1}{2}$ spin projection. The Hamiltonian to be considered is that for a one-electron problem:

$$\mathscr{H}_{\mathrm{SO+Zeeman}} = \lambda \mathbf{l} \cdot \mathbf{s} + \beta(l_z + 2s_z) \cdot \mathbf{B} \tag{A.II.1}$$

The axes x, y, z have been chosen along the metal–ligand directions. The x, y, and z axes are equivalent. The field is oriented along the z axis, but the susceptibility would be the same when the field is oriented along the x or y axes. The effect of applying l_z on the wave functions of $^2T_{2g}$ is listed below[53]:

$$\begin{aligned} l_z|xy\rangle &= -2i|x^2 - y^2\rangle \\ l_z|xz\rangle &= i|yz\rangle \\ l_z|yz\rangle &= -i|xz\rangle \end{aligned} \tag{A.II.2}$$

The eigenvalues and eigenvectors of l_z in the $^2T_{2g}$ basis are

$$\begin{aligned} &|xy\rangle && m_l = 0 \\ &|\eta\rangle = (|xz\rangle + i|yz\rangle)/\sqrt{2} && m_l = 1 \\ &|\eta\rangle^* = (|xz\rangle - i|yz\rangle)/\sqrt{2} && m_l = -1 \end{aligned} \tag{A.II.3}$$

We can thus consider an orbital momentum $l = 1$ to be associated with the subspace $^2T_{2g}$.

The action of $\lambda \mathbf{l} \cdot \mathbf{s}$ on the basis set $|xy\rangle, |xy\beta\rangle, |\eta\rangle, |\eta\rangle^*, |\eta\beta\rangle, |\eta\beta\rangle^*$ is easily found from the table in Balhausen.[53] A state with a fourfold degeneracy is obtained at $-\lambda/2$ with the following eigenvectors

$$|\eta\rangle^*, |\eta\beta\rangle, |\phi_1\rangle = \tfrac{2}{3}^{1/2}[|xy\beta\rangle - (i/\sqrt{2})|\eta\rangle], \quad |\phi_2\rangle = \tfrac{2}{3}^{1/2}[|xy\rangle - (i/\sqrt{2})|\eta\beta\rangle^*] \tag{A.II.4}$$

and one doubly degenerate at λ:

$$|\phi_3\rangle = \tfrac{1}{3}^{1/2}\,[|xy\beta\rangle + i\sqrt{2}|\eta\rangle]$$
$$|\phi_4\rangle = \tfrac{1}{3}^{1/2}\,[|xy\rangle + i\sqrt{2}|\eta\beta\rangle^*] \tag{A.II.5}$$

It is easy to check that the action of $\mathscr{H}_{\text{SO+Zeeman}}$ on the preceding basis set gives the following matrix:

$$\begin{bmatrix} -\lambda/2 & 0 & 0 & 0 & 0 & 0 \\ 0 & -\lambda/2 & 0 & 0 & 0 & 0 \\ 0 & 0 & -\lambda/2 & 0 & -\beta B\sqrt{2} & 0 \\ 0 & 0 & 0 & -\lambda/2 & 0 & \beta B\sqrt{2} \\ 0 & 0 & -\beta B\sqrt{2} & 0 & \lambda + \beta B & 0 \\ 0 & 0 & 0 & \beta B\sqrt{2} & 0 & \lambda - \beta B \end{bmatrix}$$

There are six eigenstates. Two of them, $|\eta\beta\rangle$ and $|\eta\rangle^*$ have the invariant energy $-\lambda/2$. The four others have their energies given by the diagonalization of the following 2×2 matrices:

$$\begin{bmatrix} -\lambda/2 & -\beta B\sqrt{2} \\ -\beta B\sqrt{2} & \lambda + \beta B \end{bmatrix}$$

and

$$\begin{bmatrix} -\lambda/2 & \beta B\sqrt{2} \\ \beta B\sqrt{2} & \lambda - \beta B \end{bmatrix}$$

It is possible to find the exact solutions, but we will use perturbation theory to compute the first- and second-order corrections. The energies of these states have a second order dependence on field

$$E(|\phi_1\rangle) = -\lambda/2 - \tfrac{4}{3}\beta^2 B^2/\lambda$$
$$E(|\phi_3\rangle) = \lambda + \beta B + \tfrac{4}{3}\beta^2 B^2/\lambda$$
$$E(|\phi_2\rangle) = -\lambda/2 - \tfrac{4}{3}\beta^2 B^2/\lambda \tag{A.II.6}$$
$$E(|\phi_4\rangle) = \lambda - \beta B + \tfrac{4}{3}\beta^2 B^2/\lambda$$

By using the Van Vleck formula, one gets

$$\chi_{\mathrm{M}} = \frac{(\mu_0 N_{\mathrm{A}}\beta^2/3kT)[8 + (3x - 8)\exp(-3x/2)]}{x[2 + \exp(-3x/2)]} \tag{A.II.7}$$

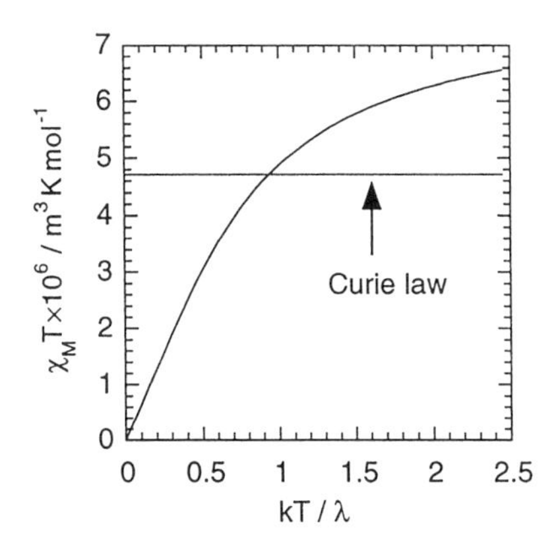

Figure A.II.1
Curie constant $\chi_M T = f(kT/\lambda)$ and effective magnetic moment for the ${}^2T_{2g}$ level of a d^1 system in O_h. The Curie law for $S = \frac{1}{2}$ with $g = 2$ is shown for comparison (see text).

with $x = \lambda/kT$. It is represented in Figure (A.II.1) as $\chi_M T$ as a function of kT/λ. When T goes to infinity, $\chi_M T \rightarrow 0.525\ 4\pi \times 10^{-6}\ \mathrm{m^3\,mol^{-1}\,K}$, which is larger than the spin $S = \frac{1}{2}$ value, that is, there is an orbital contribution to the magnetic moment. When $T \rightarrow 0$, $\chi_M T \rightarrow 0$: the orbital momentum cancels the spin contribution. This behavior is quite different from that expected from the Curie law.

APPENDIX III

The Heisenberg–Dirac–Van Vleck Hamiltonian

The HDVV Hamiltonian belongs to the class of effective Hamiltonians. The basic Hamiltonian is written as

$$\mathscr{H}_{\text{total}} = \mathscr{H}_0 + \mathscr{V} \tag{A.III.1}$$

where $\mathscr{H}_0$ describes the individual metal ions and $\mathscr{V}$ contains all terms corresponding to the metal–metal interaction. Both $\mathscr{H}_0$ and $\mathscr{V}$ contain only electron–electron and electron–nuclei interactions. When $\mathscr{V}$ is small, a perturbation treatment can be applied. The ground state of $\mathscr{H}_0$ corresponds to both ions in their respective ground states: ${}^{2S_{A0}+1}\Gamma_{A0}$ and ${}^{2S_{B0}+1}\Gamma_{B0}$; when restricted to ions with orbitally nondegenerate ground states, the degeneracy of the ground state of $\mathscr{H}_0$ is $(2S_{A0}+1)(2S_{B0}+1)$. Using Eq. (A.I.3), one can build the effective Hamiltonian

$$\mathscr{H}_{\text{exch}} = \sum_{i,j} J_{ij}\mathbf{s}_{Ai} \cdot \mathbf{s}_{Bj} \tag{A.III.2}$$

where i and j refer to the magnetic orbitals on centers A and B, respectively. This Hamiltonian acts only on the ground-state subspace. As in the demonstration of the zfs Hamiltonian, expressions for the J_{ij} parameters as functions of fundamental quantities are found by using this procedure. These expressions depend on the models used to describe the metal–metal interaction. A common feature is that each J_{ij} can be written as a sum of positive ($J_{ij}{}^{AF}$) and negative ($J_{ij}{}^{F}$) terms, $J_{ij} = J_{ij}{}^{AF} + J_{ij}{}^{F}$, where $J_{ij}{}^{AF}$ is zero if the overlap between the orbitals a_i and b_j is zero.

When both ions A and B are in their ground states, the spin is maximum (Hund's rule), and one has $\langle \mathbf{s}_{Ai} \rangle = (1/n_A)\langle \mathbf{S}_A \rangle$ and $\langle \mathbf{s}_{Bj} \rangle = (1/n_B)\langle \mathbf{S}_B \rangle$ for any value of i and j. In this case, Eq. (A.III.2) simplifies to

$$\mathscr{H}_{\text{exch}} = J\mathbf{S}_A \cdot \mathbf{S}_B \tag{A.III.3}$$

with

$$J = (1/n_A n_B) \sum_{i,j} J_{ij} \tag{A.III.4}$$

In fact, Eq. (A.III.2) is valid (but not A.III.3!) also for situations involving ions in excited states originating from the same electronic configuration as the ground state, for example, Cr(III) ion in its ground-state ($t_{2g}{}^3$, ${}^4A_{2g}$) coupled to another Cr(III) ion in the excited-state ($t_{2g}{}^3$, ${}^2E_g$). The validity of Eq. (A.III.2) for those excited states has been used in the spectroscopy of metal pairs[54] to get information on the individual exchange interactions J_{ij}.

References

1. Boudreaux, E. A.; Mulay, L. N. *Theory and Applications of Molecular Paramagnetism*; Wiley: New York, 1976.
2. Carlin, R. L. *Magnetochemistry*, Springer-Verlag: New York, 1986.
3. Kahn, O. *Molecular Magnetism*; VCH: New York, 1993.
4. O'Connor, C. J., "Magnetochemistry-Theory and Experimentation" *Prog. Inorg. Chem.* **1982**, *29*, 203.
5. For the NMR method for measuring magnetic susceptibility, see Evans, D. F. "The Determination of the Paramagnetic Susceptibility of Substances in Solution by Nuclear Magnetic Resonance" *J. Chem. Soc.* **1959**, 2003 and Sur, S. K. "Measurement of Magnetic Susceptibility and Magnetic Moment of Paramagnetic Molecules in Solution by High-Field Fourier Transform NMR Spectroscopy" *J. Mag. Reson.* **1989**, 82, 169–173. For a pedagogical example see Crawford, T. H.; Swanson, J. "Temperature Dependent Magnetic Measurements and Structural Equilibria in Solution" *J. Chem. Educ.* **1971**, 48, 382–386.
6. Day, E. P.; Kent, T. A.; Lindahl, P. A.; Münck, E.; Orme-Johnson, W. H.; Roder, H.; Roy, A. "SQUID Measurements of Metalloprotein Magnetization. New Methods Applied to the Nitrogenase Proteins." *Biophys. J.* **1987**, *52*, 837–853.
7. Day, E. P. "Multifield Saturation Magnetization of Metalloproteins" In *Methods in Enzymology*; Riordan, J. F.; Vallee, B. L. Eds.; **1993**, *227*, 437–463.
8. Carlin, R. L.; O'Connor, C. J.; Bhatia, S. N. "Large Zero-Field Splitting in $Cs_3VCl_6 \cdot 3H_2O$." *Inorg. Chem.* **1976**, *15*, 985.
9. Peterson, J.; Fee, J. A.; Day, E. P. "Magnetization of Manganese Superoxide Dismutase from *Thermus Thermophilus*" *Biochim. Biophys. Acta* **1991**, *1079*, 161.
10. Ludwig, M. L.; Metzger, A. L.; Pattridge, K. A.; Stallings, W. C. "Manganese Superoxide Dismutase from *Thermus thermophilus*. A Structural Model Refined at 1.8 Å Resolution" *J. Mol. Biol.* **1991**, *219*, 335–358.
11. Whittaker, J. W.; Whittaker, M. M. "Active Site Spectral Studies on Manganese Superoxide Dismutase" *J. Am. Chem. Soc.* **1991**, *113*, 5528–5540.
12. Scheidt, W. R.; Reed, C. A. "Spin-State/Stereochemical Relationships in Iron Porphyrins: Implications for the Hemoproteins" *Chem. Rev.* **1981**, *81*, 543–555.
13. Adams, K. A.; Rasmussen, P. G.; Scheidt, W. R.; Hatano, K. "Structure and Properties of an Unsymmetrically Substituted Six-Coordinate Iron(III) Porphyrin" *Inorg. Chem.* **1979**, *18*, 1892–1899.
14. Figgis, B. N. "The Magnetic Properties of Transition Metal Ions in Asymmetric Ligand Fields" *Trans. Faraday Soc.* **1961**, *57*, 198.
15. Whittaker, J. W.; Solomon, E. I. "Spectroscopic Studies on Ferrous Non-Heme Iron Active Sites: Magnetic Circular Dichroism of Mononuclear Fe Sites in Superoxide Dismutase and Lipoxygenase" *J. Am. Chem. Soc.* **1988**, *110*, 5329–5339.
16. Trautwein, A. X.; Bill, E.; Bominaar, E. L.; Winkler, H. "Iron-Containing Proteins and Related Analogs-Complementary Mössbauer, EPR and Magnetic Susceptibility Studies" *Struct. Bonding (Berlin)*, **1991**, *78*, 1–95.
17. Maltempo, M. M.; Moss, T. H. "The Spin 3/2 State and Quantum Spin Mixtures in Haem Proteins" *Q. Rev. Biophys.* **1976**, *9*, 181–215.
18. Kintner, E. T.; Dawson, J. H. "Spectroscopic Studies of Ferric Porphyrins with Quantum Mechanically Admixed Intermediate-Spin States: Models for Cytochrome c" *Inorg. Chem.* **1991**, *30*, 4892–4897.
19. Solomon, E. I.; Lowery, M. D. "Electronic Structure Contributions to Function in Bioinorganic Chemistry" *Science*, **1993**, *259*, 1575–1581.
20. Lippard, S. J. "Oxo-Bridged Polyiron Centers in Biology and Chemistry" *Angew. Chem. Int. Ed. Engl.* **1988**, *27*, 344–361.
21. Magnus, K. A.; Ton-That, H.; Carpenter, J. E. "Recent Structural Work on the Oxygen Transport Protein Hemocyanin" *Chem. Rev.*, **1994**, *94*, 727–735.

22. Holm, R. H. "Synthetic Approaches to the Active Sites of Iron–Sulfur Proteins" *Acc. Chem. Res.* **1977**, *10*, 427.
23. Stenkamp, R. E. "Dioxygen and Hemerythrin" *Chem. Rev.* **1994**, *94*, 715–726.
24. Dawson, J. W.; Gray, H. B.; Hoenig, H. E.; Rossman, G. R.; Schredder, J. M.; Wang, R.-H. "A Magnetic Susceptibility Study of Hemerythrin Using an Ultrasensitive Magnetometer" *Biochemistry*, **1972**, *11*, 461–465.
25. Kurtz, D. M., Jr., "Oxo- and Hydroxo-Bridged Diiron Complexes: A Chemical Perspective on a Biological Unit" *Chem. Rev.* **1990**, *90*, 585–606.
26. Shiemke, A. K.; Loehr, T. M.; Sanders-Loehr, J. "Resonance Raman Study of Oxyhemerythrin and Hydroxomethemerythrin. Evidence for Hydrogen Bonding of Ligands to the Fe–O–Fe Center" *J. Am. Chem. Soc.* **1986**, *108*, 2437–2443.
27. Armstrong, W. H.; Lippard, S. J. "Reversible Protonation of the Oxo Bridge in a Hemerythrin Model Compound. Synthesis, Structure, and Properties of (μ-Hydroxo)bis(μ-acetato)-bis[hydrotris(1-pyrazolyl)borato]diiron(III), $[\{HB(pz)_3\}_2Fe_2(\mu\text{-}OH)(\mu\text{-}MeCO_2)_2]^+$" *J. Am. Chem. Soc.* **1984**, *106*, 4632–4633.
28. Day, E. P.; Peterson, J.; Sendova, M. S.; Schoonover, J.; Palmer, G. "Magnetization of Fast and Slow Oxidized Cytochrome c Oxidase" *Biochemistry*, **1993**, *32*, 7855–7960.
29. Lee, S. C.; Holm, R. H. "Synthesis and Characterization of an Asymmetric Bridged Assembly Containing the Unsupported $[Fe^{III}\text{–}O\text{–}Cu^{II}]$ Bridge: An Analogue of the Binuclear Site in Oxidized Cytochrome c Oxidase" *J. Am. Chem. Soc.* **1993**, *115*, 11789–11798.
30. Lee, S. C.; Scott, M. J.; Kauffmann, K.; Münck, E.; Holm, R. H. "Cyanide Poisoning: An Analogue to the Binuclear Site of Oxidized Cyanide-Inhibited Cytochrome c Oxidase" *J. Am. Chem. Soc.* **1994**, *116*, 401–402.
31. Bencini, A.; Gatteschi, D. *EPR of Exchange Coupled Systems*; Springer-Verlag: Berlin, 1990.
32. Wieghardt, K.; Bossek, U.; Nuber, B.; Weiss, J.; Bonvoisin, J.; Corbella, M.; Vitols, S. E.; Girerd, J.-J. "Synthesis, Crystal Structures, Reactivity, and Magnetochemistry of a Series of Binuclear Complexes of Manganese(II), -(III), and -(IV) of Biological Relevance. The Crystal Structure of $[L'Mn^{IV}(\mu\text{-}O)_3Mn^{IV}L'](PF_6)_2 \cdot H_2O$ Containing an Unprecedented Short Mn–Mn Distance of 2.296 Å" *J. Am. Chem. Soc.* **1988**, *110*, 7398–7411.
33. Gerritsen, H. J.; Sabisky, E. S. "Paramagnetic Resonance of Trivalent Manganese in Rutile" *Phys. Rev.* **1963**, *132*, 1507–1512.
34. Kennedy, M. C.; Stout, C. D. "Aconitase: An Iron–Sulfur Enzyme" *Adv. Inorg. Chem.* **1992**, *38*, 323–339.
35. Day, E. P.; Peterson, J.; Bonvoisin, J. J.; Moura, I.; Moura J. J. G. "Magnetization of the Oxidized and Reduced Three-Iron Cluster of *Desulfovibrio gigas* Ferredoxin II" *J. Biol. Chem.* **1988**, *263*, 3684–3689.
36. Girerd, J.-J.; Papaefthymiou, G. C.; Watson, A. D.; Gamp, E.; Hagen, K. S.; Edelstein, N.; Frankel, R. B.; Holm, R. H. "Electronic Properties of the Linear Antiferromagnetically Coupled Clusters $[Fe_3S_4(SR)_4]^{3-}$, Structural Isomers of the $[Fe_3S_4]^+$ Unit in Iron–Sulfur Proteins" *J. Am. Chem. Soc.* **1984**, *106*, 5941–5947.
37. Griffith, J. S. "On the General Theory of Magnetic Susceptibilities of Polynuclear Transition-Metal Compounds" *Struct. Bonding (Berlin)* **1971**, *10*, 87–126.
38. Anderson, P. W. "Exchange in Insulators: Superexchange, Direct Exchange, and Double Exchange" In "Magnetism," Rado, G. T.; Suhl, H., Eds; Vol. I, Academic: New York, 1963.
39. Julve, M.; Verdaguer, Y.; Gleizes, A.; Philoche-Levisalles, M.; Kahn, O. "Design of μ-Oxalato Copper(II) Binuclear Complexes Exhibiting Expected Magnetic Properties" *Inorg. Chem.* **1984**, *23*, 3808–3818.
40. Castro, I.; Faus, J.; Julve, M.; Gleizes, A. "Complex Formation between Oxalate and $(2,2':6',2''$-Terpyridyl)-copper(II) in Dimethyl Sulphoxide Solution. Synthesis and Crystal Structures of Mono- and Di-nuclear Complexes" *J. Chem. Soc. Dalton Trans.* **1991**, 1937–1944.

41. Kahn, O.; Galy, J.; Journaux, Y.; Jaud, J.; Morgenstern-Badarau, I. "Synthesis, Crystal Structure and Molecular Conformations, and Magnetic Properties of a Cu^{II}–VO^{II} Heterobinuclear Complex: Interaction between Orthogonal Magnetic Orbitals" *J. Am. Chem. Soc.* **1982**, *104*, 2165–2176.
42. Schulz, C. E.; Rutter, R.; Sage, J. T.; Debrunner, P. G.; Hagre, L. P. "Mössbauer and Electronic Paramagnetic Resonance Studies of Horseradish Peroxidase and its Catalytic Intermediates" *Biochemistry*, **1984**, *23*, 4743–4754.
43. Reed, C. A.; Orosz, R. D. "Spin Coupling Concepts in Bioinorganic Chemistry" In *Research Frontiers in Magnetochemistry*; O'Connor, C. J., Ed.; World Scientific: London, 1993, pp 351–393.
44. Boso, B.; Lang, G.; McMurry, T. J., Groves, J. T. "Mössbauer Effect Study of Tight Spin Coupling in Oxidized Chloro-5,10,15,20-tetra(mesityl)prophyrinatoiron(III)" *J. Chem. Phys.* **1983**, *79*, 1122.
45. Hotzelmann, R.; Wieghardt, K.; Flörke, U.; Haupt, H.-J.; Weatherburn, D. C.; Bonvoisin, J.; Blondin, G., Girerd, J.-J. "Spin Exchange Coupling in Asymmetric Heterodinuclear Complexes Containing the μ-Oxo-bis(μ-acetato)dimetal Core" *J. Am. Chem. Soc.* **1992**, *114*, 1681–1696.
46. Iwata, S.; Ostermeier, C.; Ludwig, B.; Michel, H. "Structure at 2.8 Å Resolution of Cytochrome *c* Oxidase from *Paracoccus denitrificans*" *Nature (London)* **1995**, *376*, 660–669 and references cited therein.
47. Blondin, G.; Borshch, S.; Girerd, J.-J. "When Must Double Exchange Be Used?" *Comments Inorg. Chem.* **1992**, *12*, 315–340.
48. Anderson, P. W.; Hasegawa, H. "Considerations on Double Exchange" *Phys. Rev.* **1955**, *100*, 675–681.
49. Ding, X.-Q.; Bominaar, E. L.; Bill, E.; Winkler, H.; Trautwein, A. X.; Drüeke, S.; Chaudhuri, P.; Wieghardt, K. "Mössbauer and EPR Study of the Double-Exchange and Heisenberg-Exchange Interactions in a Novel Binuclear Fe(II/III) Valence-Delocalized Complex" *J. Chem. Phys.* **1990**, *92*, 178–186.
50. Cammack, R. "Iron–Sulfur Clusters In Enzymes: Themes and Variations" *Adv. Inorg. Chem.* **1992**, *38*, 281–322.
51. Que, L., Jr.; True, A. E. "Dinuclear Iron- and Manganese–Oxo Sites in Biology" *Prog. Inorg. Chem.* **1990**, *38*, 97–200.
52. Griffith, J. S. *The Theory of Transition-Metal Ions*; Cambridge University Press: Cambridge, UK, 1971.
53. Balhausen, C. J. *Introduction to Ligand Field Theory*, McGraw-Hill: New York, 1962.
54. McCarthy, P. J.; Güdel, H. U. "Optical Spectroscopy of Exchange-Coupled Transition Metal Complexes" *Coord. Chem. Rev.* **1988**, *88*, 69–131.

General References

BASIC TREATMENTS

Drago, R. S., Physical Methods for Chemists, 2nd ed.; Saunders: Fort Worth, 1992. Contains a simple and clear chapter on magnetism.

O'Connor, C. J., "Magnetochemistry—Theory and Experimentation" *Prog. Inorg. Chem.* **1982**, *29*, 203. A concise and broadscope review with a good description of a SQUID magnetometer.

MORE DIFFICULT

Balhausen, C. J., Introduction to Ligand Field Theory, McGraw-Hill: New York, 1962. Consists mainly of a very clear desciption of electronic properties of mononuclear metal complexes including a part on magnetic susceptibility. Also contains a short presentation of magnetic properties of dinuclear systems.

Kahn, O., Molecular Magnetism, VCH: New York, 1993. A rather complete and up to date description of magnetic properties of molecules containing transition metal ions, emphasizing exchange interactions in metal clusters.

Craik, D. "Magnetism. Principles and Applications" Wiley: New York, 1995. A classical text such as this by a physicist may be particularly useful to study the different systems of units in electromagnetism.

8

Nuclear Magnetic Resonance of Paramagnetic Metal Centers in Proteins and Synthetic Complexes

LI-JUNE MING

Department of Chemistry and
Institute for Biomolecular Science
University of South Florida
Tampa, FL 33620

CONTENTS

I. Basic Nuclear Magnetic Resonance Principles for Paramagnetic Molecules

Nuclear magnetic resonance (NMR), a versatile tool for the study of molecular structure and dynamics, has been used to solve problems in bioinorganic chemistry, despite the fact that many systems of interest contain a paramagnetic metal center. The basic principles of NMR in diamagnetic species can be applied, in principle, to the understanding of the NMR properties of paramagnetic species by taking into account the significant influence of electron magnetic moment, which is 658 times larger than that of the proton. The unpaired electrons in these paramagnetic molecules present some challenges and require some adaptation of approaches used for NMR studies of diamagnetic molecules. Nevertheless useful information can be extracted from the NMR spectra of paramagnetic metal complexes and metalloproteins. This chapter is intended to serve as an illustration of how NMR can be applied to the investigation of paramagnetic systems.

A. The NMR Transition

A spinning nucleus in a magnetic field $\mathbf{B}_0$ produces a nuclear magnetic moment μ_N that is proportional to the magnitude of the angular momentum ($\hbar\mathbf{I}$) of the nucleus and gives rise to the NMR Hamiltonian ($\mathscr{H}$) shown in Eq. (1),

$$\mathscr{H} = -\mu_N \cdot \mathbf{B}_0 = -g_N\beta_N B_0 I_z = -\hbar\gamma_N B_0 I_z \tag{1}$$

where $\gamma_N (= \mu_N/\hbar I_z = g_N\beta_N/\hbar)$ is the nuclear gyromagnetic ratio, g_N the nuclear g factor, β_N the nuclear magneton, and I the nuclear spin quantum number. A transition between the two energy levels $I = \frac{1}{2}$ and $I = -\frac{1}{2}$ of the ^{1}H nucleus requires an energy $\Delta E (= h\nu = -\hbar\gamma_N B_0(-\frac{1}{2} - \frac{1}{2}) = \hbar\gamma_N B_0$, Fig. 1), which corresponds to an electromagnetic wave in the radio frequency range (e.g., 500.28 MHz at 11.75 T). Equation (1) can be further modified to include shielding and de-

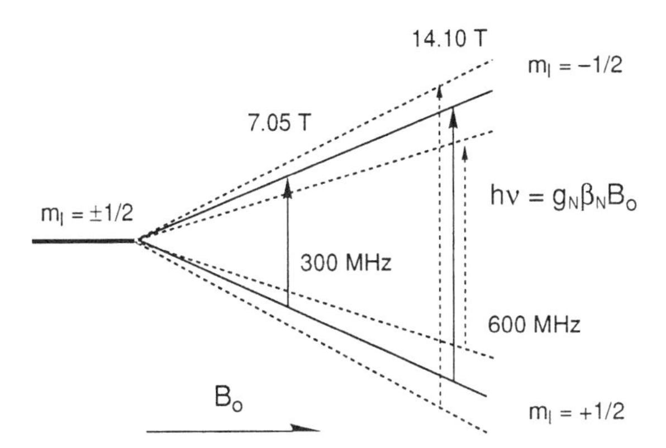

Figure 1
The NMR transitions described in Eqs. (1) and (2) at two different fields to afford proton resonances at 300 and 600 MHz (arrows). The influences caused by the deshielding effect (i.e., the outermost transition causing downfield shifts or higher frequencies) and shielding effect (the transition between the two inner levels affording upfield shifts or lower frequencies) are indicated by dashed arrows at 600 MHz. These arrows are drawn separately for clarity.

shielding effects (Fig. 1) to afford the following equation commonly encountered in the discussion of organic compounds,

$$\mathscr{H} = -\mu_N \cdot \mathbf{B}_0(1 - \sigma) \tag{2}$$

where σ is the shielding constant. This term modulates the magnetic field experienced by the nucleus depending on its environment and is the cause of the "chemical shift" in ^1H NMR spectra of diamagnetic organic and organometallic compounds. For example, the "downfield" shifted signals of aromatic compounds result from a large deshielding effect, and the "upfield" shifted organometallic hydride signals <0 ppm are attributable to a significant shielding effect. The signals shift to lower and higher field, respectively, at fixed frequency or to higher and lower frequency, respectively, at fixed magnetic field (Fig. 1). The readers may find that the equations used for the description of NMR transitions are similar to those for electron paramagnetic resonance (EPR) transitions, but NMR transitions have a much smaller energy gap ΔE [radio frequency (rf) for NMR vs. microwave/infrared (IR) for EPR].

The extremely small energy gap between the two nuclear energy levels (3.315×10^{-25} J for ^1H at 11.75 T) results in only a very small difference in spin population between the two levels at room temperature as quantified by the Boltzmann distribution [Eq. (3)]. This small difference makes NMR spectroscopy one of the least sensitive spectroscopic methods.

$$\frac{P(m_I = -\frac{1}{2})}{P(m_I = \frac{1}{2})} = \exp(-\Delta E/kT) \cong 1 - \Delta E/kT \qquad \text{when } \Delta E \ll kT \tag{3}$$

Thus, the study of a dilute biological sample by means of modern multidimensional NMR techniques may be time consuming and sometimes requires hours of data accumulation. This becomes even more serious for paramagnetic molecules that give rise to broad and low-intensity NMR features with line widths of hundreds or even thousands of hertz.

B. Chemical Shift

Chemical shift is always the first property we examine for signal assignment in order to determine molecular structures of diamagnetic compounds by means of NMR spectroscopy. The chemical shift provides valuable information about the local environment of a nucleus in a diamagnetic molecule, reflecting subtle changes of the paired electron clouds on the nucleus due to perturbations by nearby atoms [Eq. (2)]. Thus, the downfield shifted ^{1}H NMR signals of pyridine (a common ligand moiety used in metalloprotein model compounds) can be ascribed to the deshielding effect of the aromatic ring current. On the other hand, the upfield shifted proton signals of organometallic hydrides are due to a shielding effect of the electron cloud on the hydride (Fig. 1).

The NMR studies of diamagnetic species show that chemical shift values very consistently reflect the electronic environment of the resonating nuclei. While this concept remains valid in paramagnetic molecules, that is, the electron spin density on the resonating nuclei still controls the chemical shift, the large magnetic moment of the unpaired electrons in paramagnetic molecules dramatically alters the chemical shifts of resonating nuclei. For example, instead of having chemical shift values of less than 10 ppm for pyridine protons, values of greater than 100 ppm are observed for some pyridyl ring protons in several paramagnetic metal complexes (Fig. 2).[1] This large chemical shift may be the result of two interactions: a Fermi contact interaction in which the unpaired spin interacts with the nuclei via delocalization onto the nuclei through chemical bonds, and a through-space dipolar interaction between the magnetic moments of the nucleus and the unpaired electron. Typical ^{1}H NMR spectra of a few paramagnetic Fe(II) complexes with coordinated pyridyl ligands are shown in Figure 2 with broad signals observed over a large spectral window. Unlike the chemical shift anisotropy effects commonly found in ^{31}P and ^{195}Pt NMR spectra,[2] the chemical shifts of these features are not affected by the orientation of the molecules with respect to the magnetic field and are therefore called "isotropically shifted" NMR signals.

1. Fermi Contact Interaction

At this point, it should be clear that NMR spectra of paramagnetic species are dramatically different from those of diamagnetic organic and organometallic compounds. The large magnetic moment of the unpaired electrons in a paramagnetic complex very likely overwhelms all other factors and makes the dominant contribution to the large chemical shift of such NMR signals. This large contribution derives from the electron–nuclear hyperfine interaction (with a

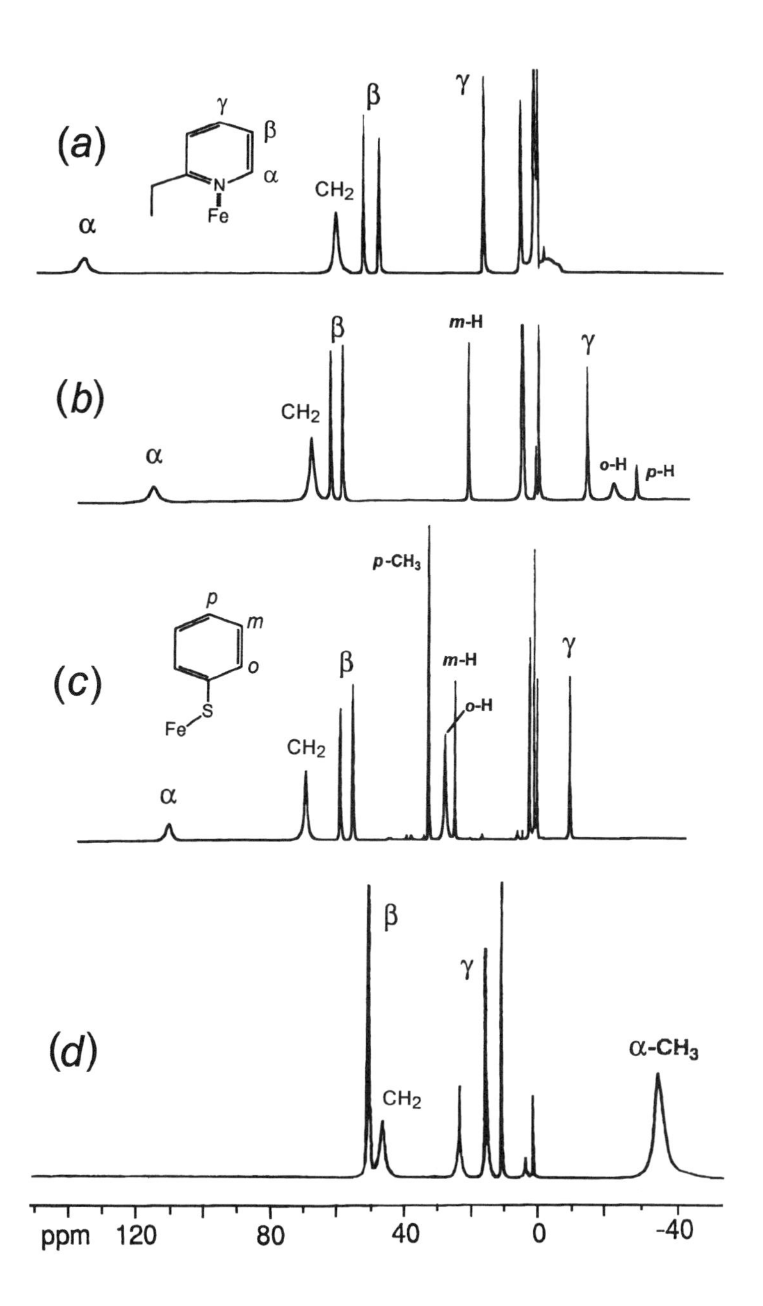

Figure 2
The ¹H NMR spectra of the paramagnetic complexes (*a*) $[Fe^{II}(TPA)Cl_2]^{2+}$, (*b*) $[Fe^{II}(TPA)SC_6H_5]^+$, (*c*) $[Fe^{II}(TPA)SC_6H_2\text{-}2,4,6\text{-}(CH_3)]^+$, and (*d*) $\{Fe^{II}[6\text{-}(CH_3)_3\text{-}TPA]$ (benzoylformate)$\}^+$, where the ligand TPA is tris-2-pyridylmethylamine and 6-$(CH_3)_3$-TPA is tris-[6-methyl-2-pyridyl)methyl]amine. [Adapted from the papers cited in Zang et al.[1]]

hyperfine coupling constant A_c) between a resonating nucleus and unpaired electrons. To get a global picture of the energy level diagram of a system in the presence of both nuclear and electron spin transitions, the readers are urged to study the sections about electron–nuclear hyperfine interactions in the chapters on EPR and ENDOR spectroscopies (Chapter 3 and 4 in this book, respectively).

For simplicity, let us consider a system possessing an $I = \frac{1}{2}$ nucleus and unpaired electrons with an isotropic g tensor. In this system, only the through-bond or contact interaction can be observed; the through-space or dipolar interaction vanishes (see Section I.B.2). The Hamiltonian for this system can be written as in Eq. (4) to include the electronic and unclear Zeeman terms (the first two terms) for isolated electron spin and nuclear spin transitions, respectively, and the Fermi contact term (the third term) for the electron–nuclear hyperfine interaction,

$$\mathscr{H} = g_e\beta_e B_0\mathbf{S} - g_N\beta_N B_0\mathbf{I} + A_c\mathbf{S}\cdot\mathbf{I} \tag{4}$$

where **S** and **I** arc electron and nuclear spin angular momentum operators, respectively, and A_c is the isotropic hyperfine coupling constant and defined as $\frac{8}{3}\pi g_e\beta_e g_N\beta_N|\Psi(0)|^2$ with $|\Psi(0)|$ being the amplitude of the electron wave function at the nucleus. On the basis of Eq. (4), it is appropriate to say that the contact shift in a paramagnetic molecule is caused by the unpaired electron density at the nucleus, which results from a direct delocalization or a spin polarization mechanism and is reflected by the electron–nuclear hyperfine coupling constant A_c. Since the hyperfine coupling constant is independent of the orientation of the molecule with respect to the magnetic field, the Fermi contact term thus affords an "isotropic shift". The contact shift term $(\Delta\nu/\nu)_{\text{contact}}$ of an $\mathbf{I} = \frac{1}{2}$ nucleus coupled with an electronic spin S with a hyperfine coupling constant A_c is shown in Eq. (5).

$$\left(\frac{\Delta B}{B_0}\right)_{\text{contact}} = -\left(\frac{\Delta\nu}{\nu}\right)_{\text{contact}} = \frac{-g\beta S(S+1)}{3\hbar\gamma_N kT}A_c \tag{5}$$

From Eq. (5), it is clear that a positive hyperfine coupling constant gives rise to a downfield shifted signal and that the Fermi contact shift is dependent on the inverse of the absolute temperature. Thus, a plot of chemical shift versus $1/T$ should be linear with an intercept (as $T \to \infty$) in the diamagnetic region if the contact term is the only shift mechanism. The complete energy diagram that includes the three terms in Eq. (4), and a step-by-step derivation of the contact shift term [Eq. (5)] can be found in Appendix I.

Let us now discuss the pathways for the generation of these hyperfine-shifted signals, and the ways for assigning these signals. The unpaired spins of a paramagnetic center can delocalize onto the coordinated atoms and neighboring nuclei through chemical bonds by two different pathways: (a) a direct spin delocalization and (b) a spin polarization of the neighboring covalent bonds. The direct delocalization pathway seems to be the predominant mechanism within the first coordination sphere of paramagnetic molecules of bioinorganic interest. A parallel spin (a ↑ spin) is assigned to the coordinated atom (i.e., the coordinated

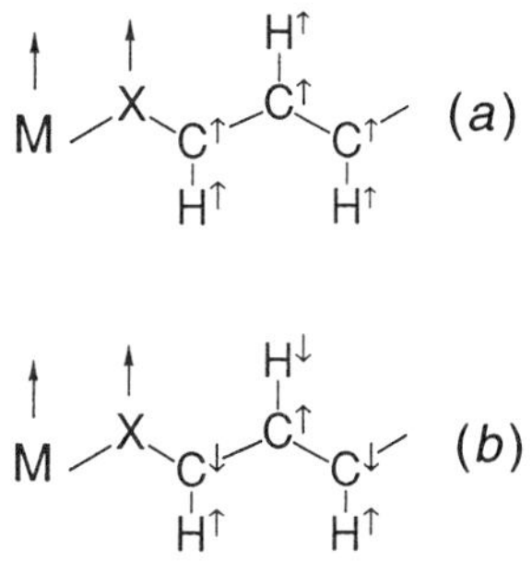

Figure 3
(*a*) Spin delocalization and (*b*) spin polarization patterns of an M–X–CH–CH–CH moiety (M: a paramagnetic metal ion; X: the coordinated atom that is commonly found as O, N, or S in biomolecules; and the rest represents a hydrocarbon moiety.

X atom in Fig. 3) as the starting point. The direct spin delocalization mechanism, as the term reflects, allows the spin, which is aligned with the magnetic field $\mathbf{B}_0$, thus a parallel ↑ spin, on the paramagnetic center to propagate through chemical bonds and remain parallel on the destination nucleus as shown in Figure 3(*a*). The unpaired spin density becomes attenuated as the number of intervening bonds increases. This parallel spin density affords a magnetic moment that generates a parallel "induced field", which gives rise to downfield shifted signals for the resonating nuclei. Consequently, the nuclei closer to the paramagnetic metal center shift farther downfield than those nuclei more distant from the metal center. Examples of this direct delocalization pathway in paramagnetic molecules can be found in Figure 2. The ^{1}H NMR spectrum of the complex $[Fe^{II}(TPA)Cl_2]^{2+}$ in Figure 2(*a*) shows a monotonic attenuation of the chemical shift from the α to the γ pyridyl proton. This shift pattern is very common among pyridine-containing paramagnetic metal complexes.

The spin polarization pathway, on the other hand, results in an alternating pattern of parallel and antiparallel spins on adjacent nuclei along chemical bonds, affording alternating downfield- and upfield-shifted ^{1}H NMR signals, respectively, for parallel (↑) and antiparallel (↓) spins on adjacent CH–CH protons as shown in Figure 3(*b*). Consequently, a direct spin delocalization pathway can be distinguished from a spin polarization pathway on the basis of the shift pattern of the hyperfine-shifted signals of the nuclei on the same ligand. For example, the observation of an alternating upfield and downfield shift pattern for the ring protons of the thiophenol ring [Fig. 2(*b*)] indicates the presence of a predominant spin polarization mechanism on the thiophenol moiety. A similar chemical shift pattern has also been observed in other metal complexes with a phenolate functionality, such as Fe^{III}(salen) complexes [salen = 1,2-bis(salicylideneamine) ethane (2−)].[3]

The presence of a spin polarization mechanism can be corroborated by the substitution of protons with methyl groups. In this case, the methyl-substituted

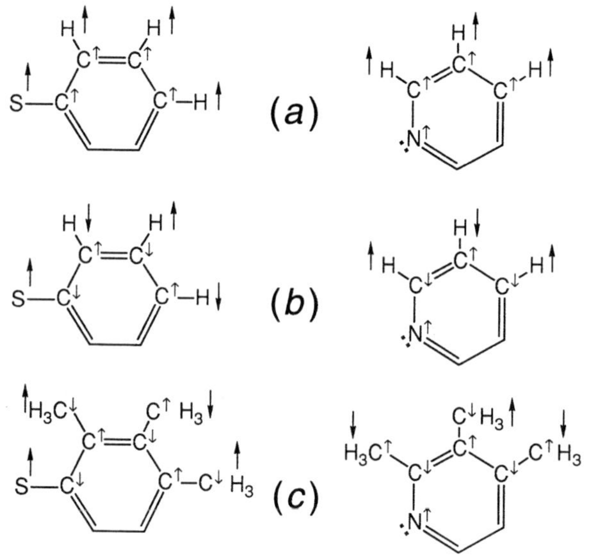

Figure 4
Spin delocalization (*a*) and polarization (*b* and *c*) patterns of thiophenol and pyridine and their methyl derivatives.

derivative would afford a signal shifted to the opposite direction as illustrated by the ortho- and para-CH_3 signals for $\{Fe^{II}TPA[SC_6H_3\text{-}2,4,6\text{-}(CH_3)_3\}^+$ [Fig. 2(*c*)] versus the ortho- and para-H signals for $[Fe^{II}TPA(SC_6H_5)]^+$ [Fig. 2(*b*)]. Unpaired spin density can be delocalized onto the methyl protons by a hyperconjugation interaction between the aromatic p_π orbital and the CH_3 group, engendering the opposite shift. The spin patterns of the thiophenol moiety in these two complexes, illustrated schematically in Figure 4(*b*,*c*), represent simple and clear examples of the spin polarization pathway.

Spectral interpretation can, however, be complicated by the presence of both of these spin delocalization mechanisms operating to different extents. For example, while the α-CH protons in the TPA complexes are downfield shifted [Fig. 2(*a*)], the α-CH_3 proton signal in $[Fe^{II}(6\text{-}Me_3\text{–}TPA)(\text{benzoylformate})]^+$ is upfield shifted [Fig. 2(*d*)]. This finding is indicative of the presence of a spin polarization mechanism, in addition to the direct spin delocalization mechanism, on the pyridine ring in the 6-Me_3–TPA complex. The spin patterns of pyridine and thiophenol ligands and their methyl derivatives are illustrated schematically in Figure 4.

Paramagnetically shifted signals in metalloproteins can arise from amino acid side chains that function as metal ligands. Such amino acids include His, Tyr, Asp, Glu, Cys, and Met (Fig. 5), and their side chains exhibit characteristic chemical shift patterns when bound to a paramagnetic metal center. The isotropic shift patterns for Tyr, Cys, and His are discussed below.

The resonances of the coordinated Tyr in the purple acid phosphatase utero-

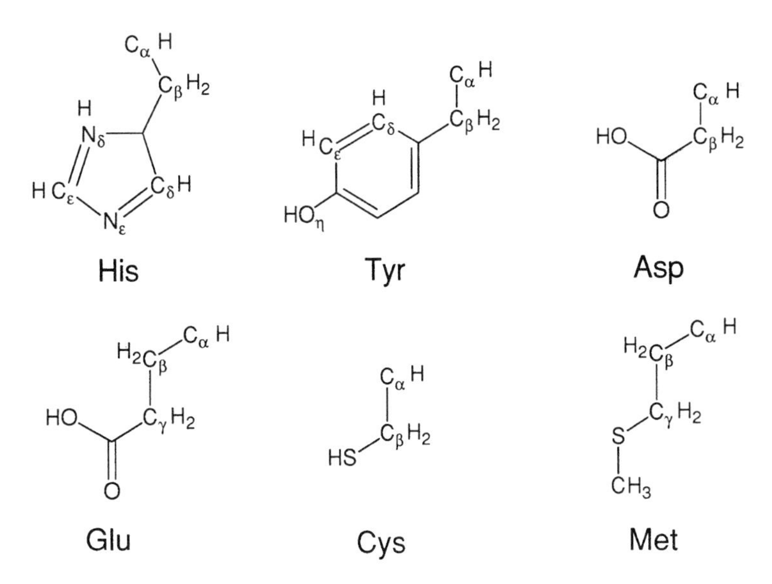

Figure 5
Side-chain structures of His, Tyr, Asp, Glu, Cys, and Met that are often involved in metal binding in metalloproteins.

ferrin [Fig. 6(*a*)] provide a good example of the spin polarization mechanism occurring in a metalloprotein.[4] This enzyme has an Fe^{III}–μ-oxo–Fe^{II} dinuclear active site with the Fe^{III} coordinated tyrosine side chain being responsible for the intense purple color of the enzyme. One Tyr $C_\varepsilon H$ proton (corresponding to an ortho proton of a phenol) is upfield shifted at −70 ppm, while the $C_\delta H$ protons (the meta protons of a phenol) are shifted downfield at 63 and 70 ppm at 30 °C. This shift pattern is consistent with a predominant spin polarization mechanism [Fig. 4(*b*)], as observed in the thiophenol-containing complexes [Figs. 2(*b* and *c*)]. In this context, the $C_\gamma H$ proton, if there were one, would be upfield shifted. Instead, this proton is replaced by the $C_\beta H_2$ group in Tyr, and these protons, as expected, are shifted downfield. However, the two $C_\beta H$ protons have very different shifts, 90 and 12 ppm. This large shift difference is due to restricted rotation of the Tyr C_β–C_γ bond, which makes one of the protons pseudoaxial (H_a) and the other pseudoequatorial (H_e) with respect to the plane of the aromatic ring [Fig. 7(*a*)]. The C–H_a bond would be approximately parallel to the C_γ p_π orbital, and the H_a proton would acquire a larger shift owing to a better hyperconjugative interaction. In contrast, the C–H_e bond would be approximately perpendicular to the C_γ p_π orbital, and the H_e proton would experience a significantly smaller paramagnetic shift [Fig. 7(*a*)]. The observed contact shifts of the $C_\beta H_2$ protons can be related to the dihedral angle θ_H between the p_π orbital and the protons by Eq. (6).

$$\delta_{\mathrm{contact}} = \kappa \rho_\pi \langle \cos^2 \theta_H \rangle \tag{6}$$

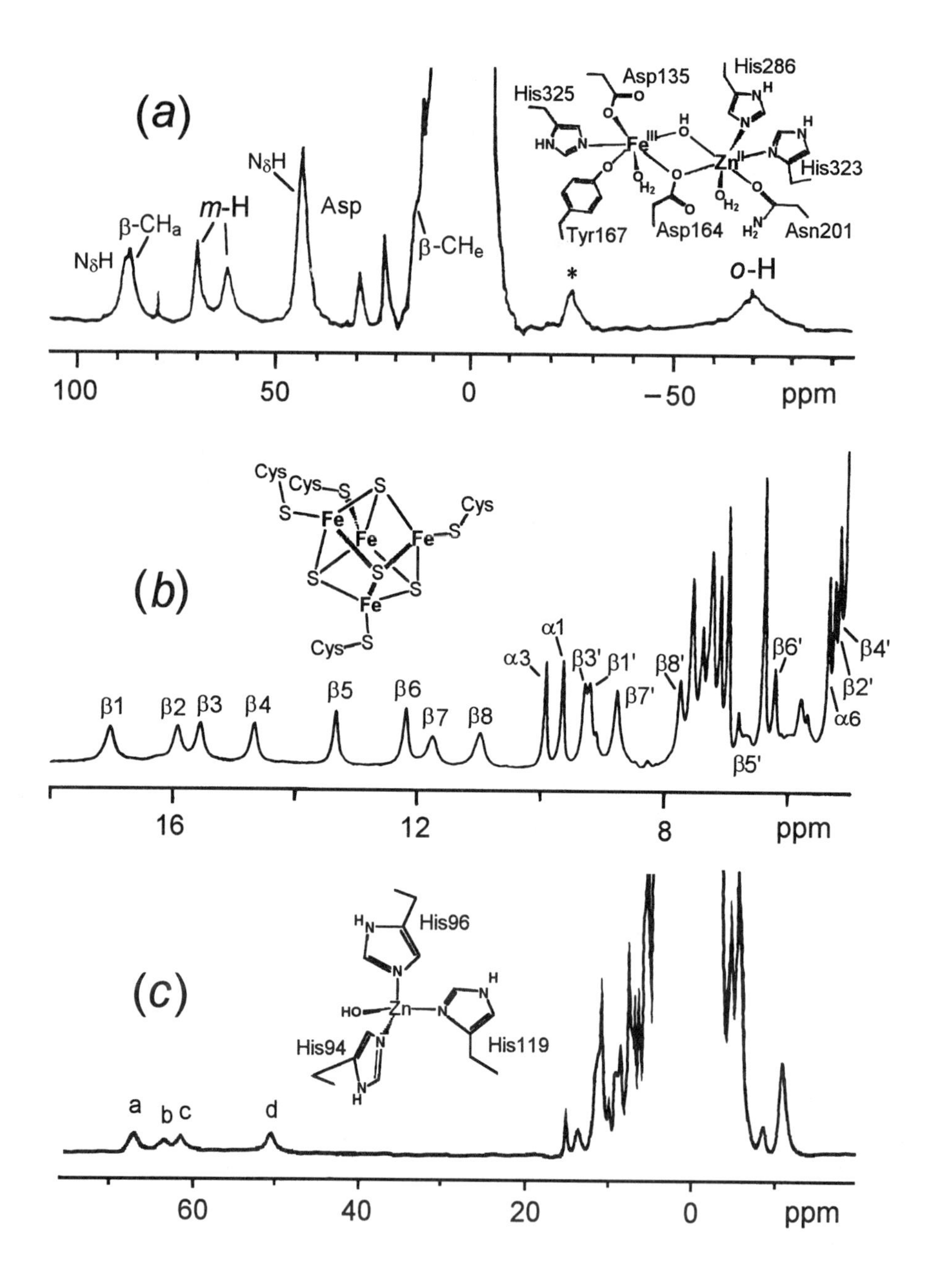

Figure 6
The ^{1}H NMR spectra of (*a*) uteroferrin (300 MHz at 30 °C), (*b*) oxidized *Clostridium pasteurianum* ferredoxin (600 MHz, 300 K), and (*c*) ClO_4^- adduct of bovine Co^{II}–carbonic anhydrase (300 MHz at 300 K). Signals a–d in spectrum (*c*) are assigned to $N_{\delta}H$ protons of His94 and His96 (a and d) and $N_{\varepsilon}H$ (b) and $C_{\delta}H$ (c) of His 119. [Adapted from Lauffer et al.,[4] Bertini et al.,[8] and Banci et al.[9]]

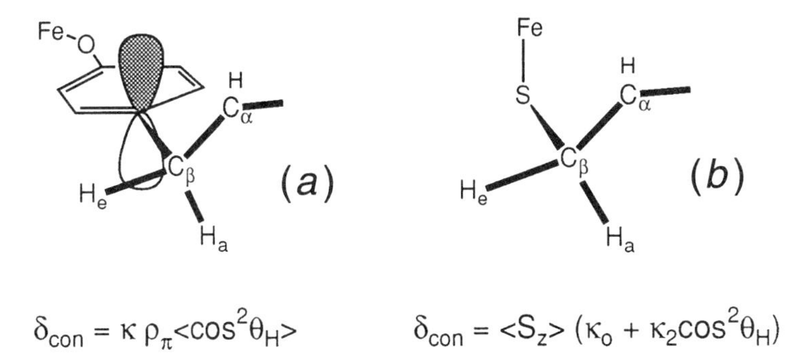

Figure 7
Dihedral angle-dependent hyperconjugation interactions in (*a*) phenolate moiety [e.g., coordinated Tyr in uteroferrin, Fig. 6(*a*)] and (*b*) M–S–C–H systems [e.g., Fe–Cys in ferredoxin, Fig. 6(*b*)].

with κ being the proportionality constant and ρ_π being the delocalized spin on the p_π orbital.[5] An angle of 42° for the axial $C_\beta H$ can be estimated based on its chemical shift. This value is consistent with the values of 46.5°, 53.1°, and 47° for the dihedral angle $\angle C_\alpha–C_\beta–C_\gamma–C_\delta$ estimated from the three available crystal structures of the related $Fe^{III}Zn^{II}$–purple acid phosphatase from kidney bean.[6] This $\pi_p–C_\beta H$ interaction can also be observed by means of EPR in protein-bound tyrosyl radicals, such as in the *Escherechia coli* ribonucleotide reductase, where a large hyperfine interaction was observed between the unpaired electron and the nearly axial $C_\beta H$ proton ($\theta = 0–10°$).[7]

The resonances of the coordinated cysteine residues of ferredoxins illustrate the direct spin delocalization mechanism. The spectrum of *C. pasteurianum* ferredoxin, which contains two Fe_4S_4 clusters, is shown in Figure 6(*b*). Signals $\beta 1/\beta 1'–\beta 8/\beta 8'$ are assigned to the β-CH_2 protons, which are three bonds away from the bound Fe; they are shifted farther downfield than the α-CH protons, which are four bonds away from the bound Fe, signals $\alpha 1–\alpha 8$, as expected for a spin delocalization mechanism. The Cys $C_\beta H_2$ protons also exhibit hyperfine interactions that are dependent on the Fe–S–C_β–H dihedral angle, due to restricted rotation about the $C_\alpha–C_\beta$ bonds of the coordinated cysteines [Fig. 7(*b*)], as illustrated by the range of shifts attributed to the $C_\beta H_2$ protons in Figure 6(*b*).[8] As in the case of the coordinated Tyr residue in uteroferrin [Eq. (6)], the chemical shifts of the $C_\beta H_2$ protons can be correlated to the dihedral angle by a similar equation [Eq. (7)],

$$\delta_{\text{contact}} = \langle S_z \rangle(\kappa_0 + \kappa_2 \cos^2 \theta_H) \tag{7}$$

where $\langle S_z \rangle$ is the spin magnetization on the metal and the average for all the spin states (i.e., $\langle S_z \rangle = -g_e \mu_B B_0 S(S+1)/3kT$), κ_0 and κ_2 are positive constants ($\kappa_2 > \kappa_0$), and θ_H is the Fe–S–C_β–H dihedral angle. The second term is attrib-

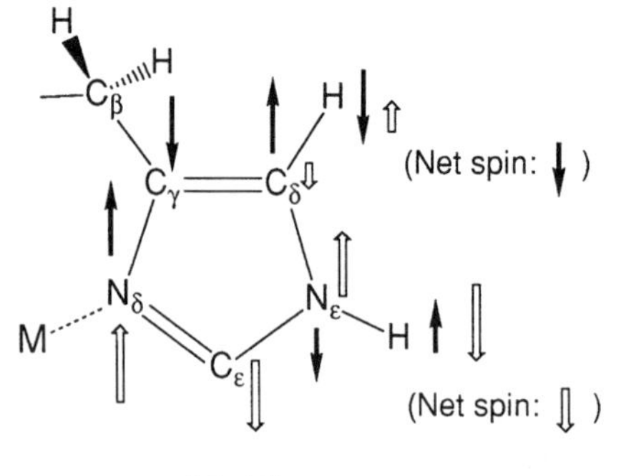

M–His

Figure 8
A "hypothetical" spin polarization mechanism for the imidazole ring of His. The odd-numbered ring would afford opposite spin densities on a ring proton such as the $N_{\varepsilon}H$ proton via the two opposite polarization routes, that is, the predominant four-bond $M{-}N_{\delta}\uparrow{-}C_{\varepsilon}\downarrow{-}N_{\varepsilon}\uparrow{-}H\downarrow$ pathway (hollow arrows) and the five-bond $N_{\delta}\uparrow{-}C_{\gamma}\downarrow{-}C_{\delta}\uparrow{-}N_{\varepsilon}\downarrow{-}H\uparrow$ pathway (filled arrows). A net $\downarrow$ spin is produce on the $N_{\varepsilon}H$ proton to exhibit an upfield shifted signal. A net $\downarrow$ spin is produced on the $C_{\delta}H$ proton in the same way. This model explains that a spin delocalization is the predominant mechanism on the imidazole ring of coordinated His in paramagnetic metalloproteins, since all the ring signals are detected in the downfield region.

utable to the hyperconjugation mechanism similar to that in uteroferrin described above with $\langle S_z \rangle$ substituting for ρ_{π}. The first term is the contribution from spin delocalization as described by Eq. (5), which is smaller than the hyperconjugation term in this case, and becomes negligible in the case of uteroferrin discussed earlier since the β-CH_2 protons are seven bonds away from the metal center.

The His residue is found in many transition metal containing metalloproteins and thus deserves some discussion. The ring NH signal of a coordinated His (usually downfield shifted >30 ppm) can be quite easily identified because of its solvent exchangeability, that is, the N–H resonance disappears when the protein is in D_2O solution. The protons on carbons adjacent to the coordinated nitrogen are often difficult to observe and found only in very few instances, for example, in Cu,Ni-superoxide dismutase (SOD) with the Ni(II) in the Zn site (cf. Section II.C.2). However, when the metal is bound to a His through the N_{δ} nitrogen, the ring $C_{\delta}H$ proton can be detected in the downfield region with a chemical shift comparable to that of the N_{ε}–H proton [Fig. 6(*c*)]. Because these adjacent protons exhibit large shifts of the same sign, they cannot be rationalized by a spin polarization mechanism. Such a mechanism would be expected to give rise to small upfield shifted signals (i.e., with net $\downarrow$ spins), since the odd-numbered ring would give rise to two opposing polarization routes (i.e., the clockwise route illustrated by filled arrows versus the counterclockwise route represented by hollow arrows in Fig. 8) that result in a small net spin density. These observed chemical shifts are more easily rationalized by a predominant spin delocalization mechanism. Figure 6(*c*) shows the NMR spectrum of the ClO_4^- adduct of Co(II) substituted carbonic anhydrase.[9] Consistent with the active site structure, there are three downfield shifted NH signals arising from the three His ligands and one $C_\delta H$ signal attributable to the sole N_δ coordinated His 119 residue.

2. Dipolar Shift and Molecular Structure

In Section I.B.1 we discuss how the paramagnetic effect of the unpaired electron(s) on the metal center can be transmitted to the ligand nuclei through chemical bonds, resulting in the appearance of signals outside the normal diamagnetic region. Besides these through-bond interactions, nuclei can also acquire large shifts via a through-space or dipolar interaction between magnetic moments of the nuclei and the unpaired electrons. Such dipolar shifts arise when the metal center exhibits significant magnetic anisotropy (i.e., $\chi_\parallel \neq \chi_\perp$), the sign and magnitude of which is determined by the geometric relationship between the nuclear and electronic moments.

The dipolar shift is most clearly illustrated by lanthanide complexes. Unlike *d* orbitals, the metal *f* orbitals are shielded and cannot participate in direct chemical bonding with ligand orbitals. Thus, there are no mechanisms for through-bond interactions between the unpaired *f* electrons and the ligand nuclei. Nevertheless, the presence of paramagnetic lanthanide centers can still induce paramagnetic shifts as described below. The addition of $Ln(fod)_3$ complexes (Ln = Pr, Eu, Dy, and Yb; fod = 6,6,7,7,8,8,8-heptafluoro-2,2-dimethyl-3,5-octanedionate) to a $CDCl_3$ solution of 1-hexanol spreads the NMR spectrum of the alcohol over a large spectral window but in significantly different ways (Fig. 9). While the protons of 1-hexanol alone are found at less than 5 ppm and are poorly resolved, the addition of the lanthanide complexes results in well-resolved isotropically shifted signals. In all cases, the most shifted signal arises from the 1-CH_2 protons, and the paramagnetic shift attenuates with increasing distance from the 1-OH oxygen that interacts with the lanthanide center. Since ligand-binding properties of all lanthanide ions are presumably the same, the different shifts exhibited by the different lanthanide complexes must result from different magnetic properties of the lanthanide ions. Thus, the very large shifts of the alcohol signals in the presence of $Dy(fod)_3$ and $Yb(fod)_3$ are attributable to the large magnetic moments of Dy^{3+} and Yb^{3+} ions (10.5 and 4.4 BM, respectively), and the opposite shifts observed in the presence of different complexes are due to the opposite sign of the magnetic susceptibility tensor of the complexes (see below).

The paramagnetic lanthanide(III) ions themselves, such as Yb^{3+}, can also serve as shift reagents for the study of metal-binding sites in biomolecules. For example, the binding of Yb^{3+} to the β-ketophenolate site of the antitumor antibiotic daunomycin in methanol results in upfield shifts for all the protons [Fig. 10(*a*)].[10] The protons on the sugar moiety of the drug are greater than or equal to eight bonds away from the metal, and there are no through-bond mechanisms that can rationalize this chemical shift pattern. Similar to the examples described above, these upfield-shifted signals also result from the dipolar shift mechanism. Finally, the substitution of one of the two Ca^{2+} sites (the EF site) of pike parvalbumin by an Yb^{3+} ion affords the ^1H NMR spectrum shown in Figure

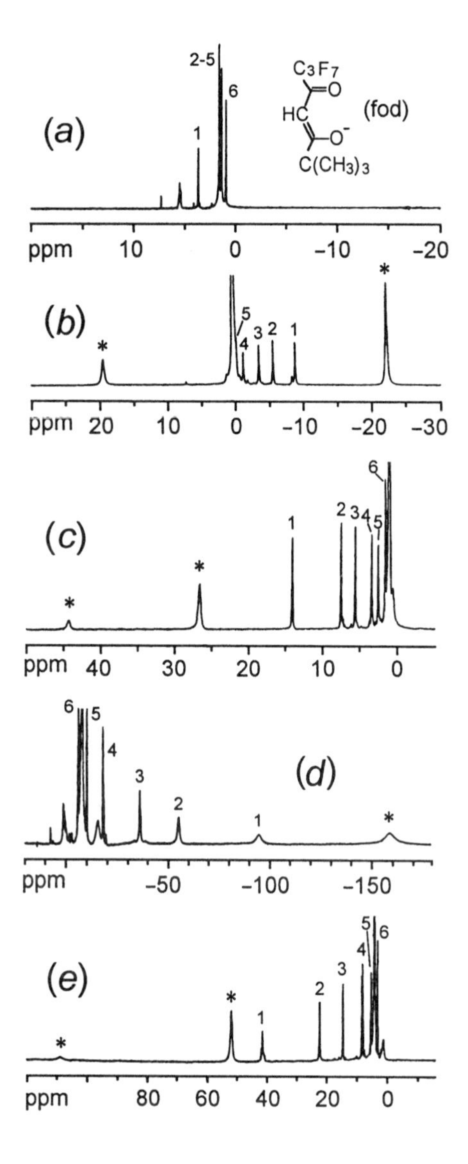

Figure 9
The ^1H NMR spectra at 360 MHz and 298 K of (*a*) 1-hexanol in $CDCl_3$ and the alcohol solution in the presence of, (*b*) $Pr(fod)_3$, (*c*) $Eu(fod)_3$, (*d*) $Dy(fod)_3$, and (*e*) $Yb(fod)_3$. The inset shows the structure of fod (6,6,7,7,8,8,8-heptafluoro-2,2-dimethyl-3,5-octanedionate). The assignment of the alcohol signals are as labeled. The signals marked with asterisks are from the fod ligand of the lanthanide complexes.

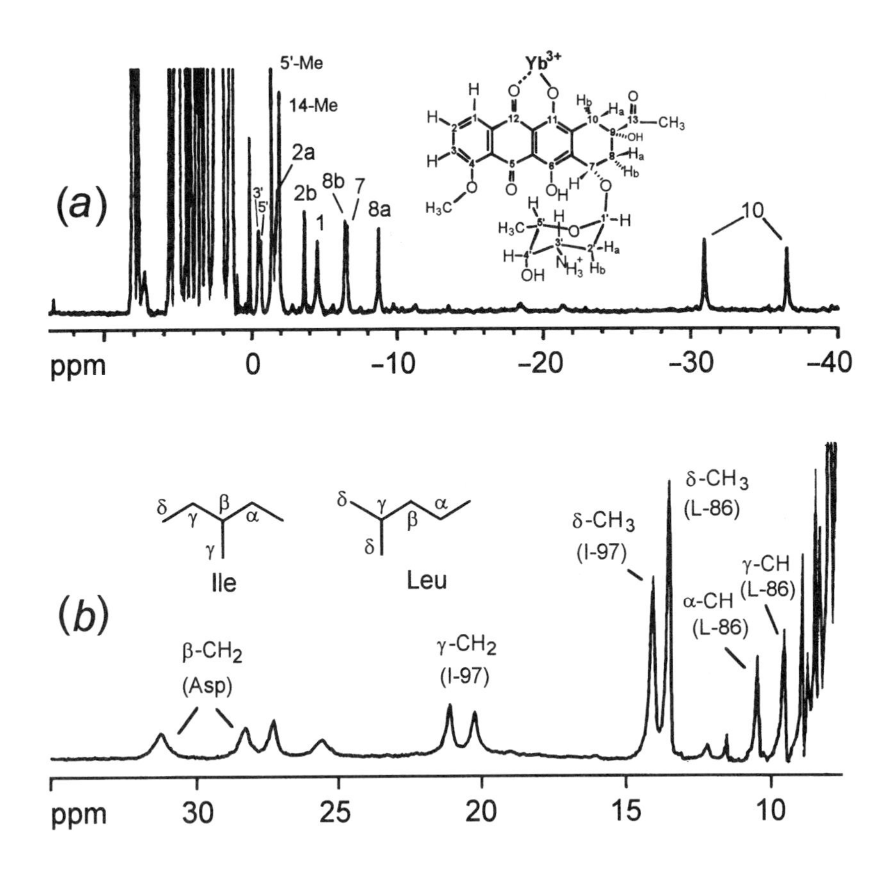

Figure 10
The ^{1}H NMR spectra of (*a*) Yb^{3+} daunomycin (with free daunomycin still visible in the diamagnetic region) in methanol (360 MHz, 298 K) and (*b*) pike parvalbumin with Yb^{3+} bound in the EF site in TRIS buffer at pH 7.8 (360 MHz, 303 K). [Adapted from Ming and Wei[10] and Ming.[11]]

10(*b*).[11] The downfield-shifted signals at 21.1 and 20.2 ppm are assigned to the geminal $C_{\gamma}H_2$ protons of Ile 97, and the two methyl signals at 14.1 and 13.5 ppm can be assigned to the $C_{\delta}H_3$ methyl protons of Ile 97 and Leu 86 (by 2D NMR methods; Section II.C). The crystal structure of parvalbumin shows the C_{γ} carbon of Ile 97 is 5.69 Å away from the metal and separated from the metal by seven bonds, it is unlikely that any through-bond mechanism would give rise to such large paramagnetic shifts for the $C_{\gamma}H_2$ protons. Moreover, Leu 86 is even farther from the metal center (i.e., its C_{γ} and one C_{δ} carbons are 6.02 and 6.60 Å, respectively, away from the EF site metal and separated from the metal by four amino acids), and is not possible to have any contact interaction with the metal to gain isotropic shift. In all these cases, the shifts observed arise from a dipolar mechanism. The presence of the dipolar shift mechanism allows the study of the region more remote (>5 Å) from the metal center. These examples illustrate the utility of lanthanide ions as NMR shift probes for metal-binding sites in biomolecules.

The dipolar shift becomes an important shift mechanism in paramagnetic molecules when magnetic anisotropy is present (i.e., $\chi_\perp \neq \chi_\parallel$). Magnetically anisotropic metal centers encountered in bioinorganic chemistry include many lanthanide ions, high-spin Co^{2+} and Fe^{2+}, and low-spin Fe^{3+}. The dipolar interaction between nuclear and electronic moments is described by the following Hamiltonian [Eq. (8)],

$$\mathscr{H} = g_N \beta_N g_e \beta_e \left[\frac{3(\mathbf{r} \cdot \mathbf{S})(\mathbf{r} \cdot \mathbf{I})}{r^5} - \frac{(\mathbf{S} \cdot \mathbf{I})}{r^3} + \frac{(\ell \cdot \mathbf{I})}{r^3} \right] \tag{8}$$

where $\mathbf{r}$ is electron–nuclear distance vector and ℓ is the net electron orbital angular momentum. All the three terms show an r^{-3} dependence. The dipolar shift can be deduced from Eq. (8) to include the magnetic susceptibility tensor as shown in Eq. (9).

$$\left(\frac{\Delta\nu}{\nu}\right)_{\text{dipolar}} = \frac{1}{3N}\left[\chi_z - \frac{1}{2}(\chi_x + \chi_y)\right]\left(\frac{3\cos^2\theta - 1}{r^3}\right) + \frac{1}{2N}(\chi_x - \chi_y)\left(\frac{\sin^2\theta\cos 2\Omega}{r^3}\right) \tag{9}$$

where N is Avogadro's number, r is the nucleus–metal distance, θ is the angle between $\mathbf{r}$ and the z axis, and Ω is the angle between the x axis and the projection of $\mathbf{r}$ on the xy plane.

Equation (9) can be converted to a more familiar form in terms of the g tensor and the effective spin S, with the identity $\chi_{ii} = g_{ii}{}^2[\beta^2 S(S+1)/3kT]$ for first-row transition metal ions and $\chi = [Ng_J{}^2\beta^2 J(J+1)]/3kT$ for lanthanide ions (J resulting $\mathbf{L} \cdot \mathbf{S}$ coupling in lanthanides). Equation (7) shows that there is no dipolar shift when the paramagnetic center is isotropic (i.e., when $\chi_x = \chi_y = \chi_z$), as both terms vanish. When the magnetic susceptibility tensor is axially symmetric (i.e., $\chi_x = \chi_y$), the second term vanishes in Eq. (7). It can also be seen that the dipolar term vanishes when the term $3\cos^2\theta - 1$ is zero in an axially symmetric system (Fig. 11) ($\theta = 54.74°$, which is often referred to as the "magic angle"), or when at $\theta = 54.74°$ and $\Omega = 45°$ in a rhombic system. The equation also shows that the magnitude of the shift at $\theta = 0°$ and 180° is twice as large as that at $\theta = 90°$ and 270° assuming the same distance from the metal in an axially symmetric system (Fig. 11).

Equation (9) thus implies that appropriate interpretation of the dipolar shifts of a molecule can provide insight into the geometric relationship between the metal center and the shifted nucleus. This geometric factor divides the spherical space about the metal center into regions where the $3\cos^2\theta - 1$ term can be positive or negative (Fig. 11). Thus, the sign of the dipolar shift for a particular nucleus depends on which region it is in and the relative magnitudes of the components of the magnetic susceptibility tensor. For example, for an axial system with $\chi_\parallel > \chi_\perp$, a nucleus at $\theta < 54.74°$ would be predicted to exhibit a downfield-

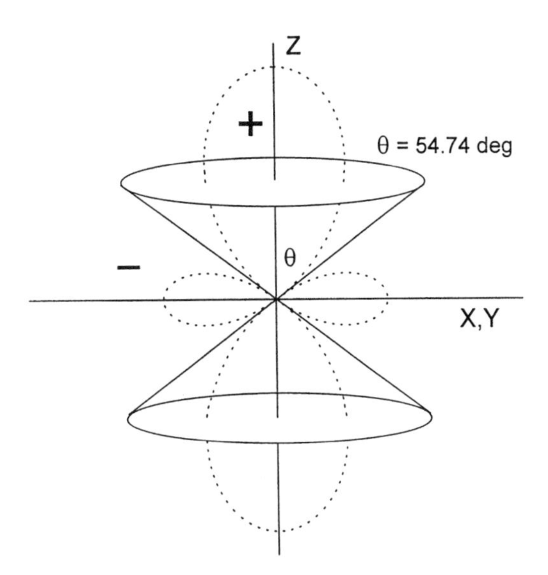

Figure 11
Plot of the term $(3\cos^2\theta - 1)/r^3$ with a fixed distance on the $x(y)$–z plane to show the change of chemical shift (dotted trace) with respect to θ. It also illustrates that chemical shift reaches 0 on the cone formed by precession of the magic angle at $\pm 54.74°$ along the z axis. The sign of chemical shift depends on the relative magnitude of $\chi_\parallel$ and $\chi_\perp$ values in axial systems.

shifted signal (i.e., a positive $\Delta\nu/\nu$ as shown in Fig. 11). Conversely, the analogous nucleus on a metal center with $\chi_\parallel < \chi_\perp$ would be upfield shifted. These factors rationalize why all the alcohol protons described above (Fig. 9) are upfield shifted in the presence of $Pr(fod)_3$ and $Dy(fod)_3$, and downfield shifted in the presence of $Eu(fod)_3$ and $Yb(fod)_3$. The ability to interpret dipolar shifts, however, depends on two things: (1) knowing the principal components of the magnetic susceptibility tensor (χ_{ii} values) of the metal center, and (2) factoring out the dipolar shift contribution from that of the contact interaction. These two pieces of information are not always available in practice.

The identification of the distal residue in the O_2-binding heme protein, myoglobin, illustrates the utility of dipolar shifts. There are two His residues in the active site of myoglobin, the proximal His, which is bound to the iron center, and the distal His, which is not coordinated but assists in the binding of O_2 (see structure in Fig. 12). Two paramagnetically shifted solvent-exchangeable protons are observed in the downfield region of the ^{1}H NMR spectrum of cyano–metmyoglobin (Fig. 12). The broader signal at 24 ppm is attributable to the ring NH of the proximal His due to its proximity to the low-spin Fe(III)-heme center [cf. Eq. (14)], while the signal at 21 ppm has been assigned to the ring NH proton of the distal His. Since the distal His residue is not a ligand to the heme, the observed shift must arise from a dipolar interaction with the low-spin Fe^{3+} center. A few globins from different sources, such as *Aplysia* myoblobin and *Glycera* hemoglobin, lack a solvent-exchangeable signal attributable to a distal His ring proton; it is thus surmised that these globins do not have a distal His residue.[12]

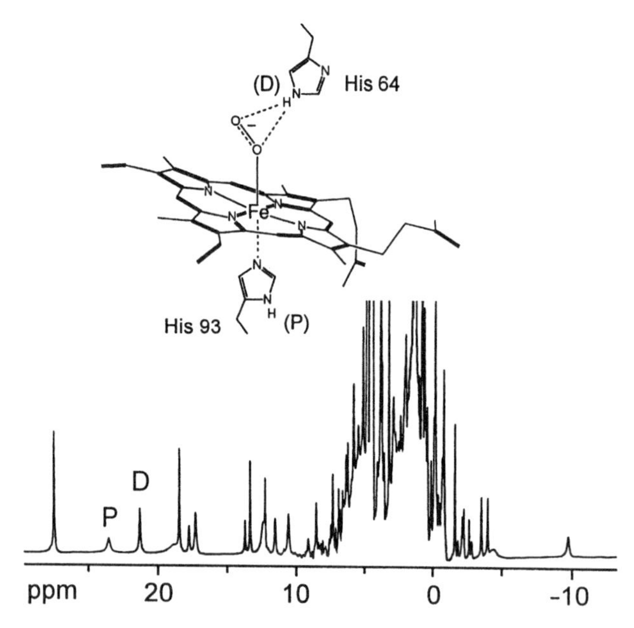

Figure 12
The ^{1}H NMR spectrum at 360 MHz of cyanometmyoglobin at pH 8 and 298 K. The His ring NH signals can be identified by comparing the spectrum with that of a sample in D_2O. This distal His–$N_\varepsilon H$ proton is marked with "D", and the proximal His–$N_\delta H$, "P". The structure is the iron protoporphyrin IX cofactor in myoglobin with a bound O_2 and the proximal and distal histidines in myoglobin.

In the previous sections, we discussed two shift mechanisms, contact and dipolar, ignoring zero-field splitting (zfs) effects. This assumption may not be valid for many high-spin transition metal complexes with multiple unpaired electrons, such as Fe^{3+} ($S = \frac{5}{2}$), Co^{2+} ($S = \frac{3}{2}$), and Ni^{2+} ($S = 1$). Zero-field splitting results in the splitting of the $2S + 1$ degenerate states ($m_s = S, S - 1, \ldots, -S$) into several energy levels that can affect the magnetic properties of the system. The inclusion of zfs introduces a second temperature-dependent term into both contact and dipolar shift equations in the form $\Delta\nu/\nu = C(1/T)[1 + C(D/T)]$. It is thus possible to have a significant T^{-2} dependent term in addition to the T^{-1} term when D/kT is large. A detailed treatment of the zfs term can be found in the chapters on EPR (Chapter 3) and magnetism (Chapter 7).

C. Nuclear Relaxation in the Presence of Unpaired Electrons

As with chemical shifts, the much larger magnetic moment of the unpaired electron relative to the nuclear magnetic moment can dramatically affect nuclear relaxation rates. The NMR relaxation occurs from the interactions of the nuclear spins with fluctuating local magnetic field generated by different motion factors, such as molecular tumbling, chemical exchange, and electronic relaxation. The individual magnetic moments μ_N of all the resonating nuclei in a sample [Eq. (1)] combine to give a net magnetization **M**, which is the vector that represents the macroscopic nuclear magnetic moment. After application of a radio frequency pulse, **M** can follow two pathways to return to its equilibrium value $\mathbf{M}_0$: (1) a longitudinal (spin–lattice) relaxation of the z component of **M** ($\mathbf{M}_z$) and (2) a transverse (spin–spin) relaxation of the x–y components of **M** ($\mathbf{M}_{xy}$).

The longitudinal relaxation of an isolated spin ensemble (i.e., in the absence of cross-relaxation) can be described by a first-order kinetic process with a time constant T_1 as shown in Eq. (10),

$$dM_z/dt = -(M_z - M_0)/T_1 \tag{10}$$

This equation can be integrated to give the familiar relaxation function,

$$M_z(\tau) = M_0[1 - a\exp(-\tau/T_1)] \tag{11}$$

where $a = 1$ when relaxation begins from zero magnetization, that is, in a saturation-recovery experiment with a pulse sequence $D1$-90°-τ-90°, where **M** lies on the y axis after the first 90° pulse, and $a = 2$ when the magnetization relaxes from the state of $-M_0$, that is, in an inversion-recovery experiment with a pulse sequence $D1$-180°-τ-90°, where **M** lies on the $-z$ axis after the 180° pulse (Fig. 13). In principle, a single experiment with a fixed τ value can provide the relaxation time by applying Eq. (11). However, this single-point measurement is prone to error due to imprecise flip angles and an inaccurate value for the starting magnetization $\mathbf{M}_0$. This error can be minimized by fitting Eq. (11) with results of experiments using a series of τ delays according to (Fig. 13).

The transverse (spin–spin) relaxation of an isolated spin is described similarly

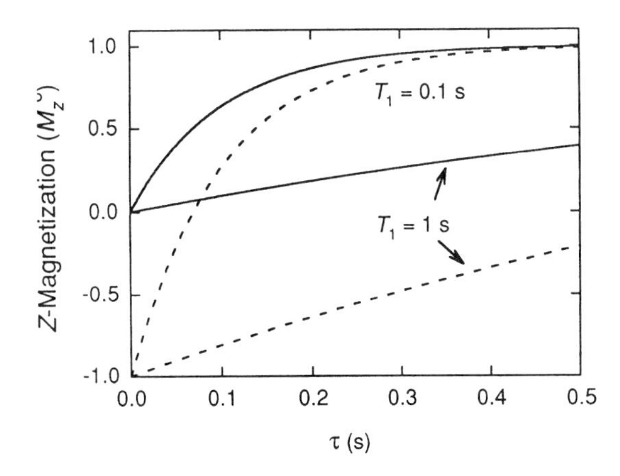

Figure 13
Comparison of longitudinal relaxation behavior (M_z vs. time) of diamagnetic ($T_1 = 1$ s) and paramagnetic ($T_1 = 0.1$ s) species in a saturation-recovery experiment (solid lines) and an inversion-recovery experiment (dotted lines).

as longitudinal relaxation but involves the M_x and M_y components and a different first-order time constant T_2 [Eq. (12)].

$$dM_{x,y}/dt = \pm\omega_0 M_{x,y} - M_{x,y}/T_2 \tag{12}$$

The spin–spin relaxation time T_2 can be related to the half-height line width of the signal of interest, as $T_2 = (\pi\Delta\nu_{1/2})^{-1}$, since solution NMR lineshape is almost always determined by the Lorentzian function, that is,

$$L(\omega) = \frac{1}{\pi}\left[\frac{1/\tau}{(1/\tau)^2 + (\omega - \omega_0)^2}\right]$$

with ω_0 being the center of the peak, τ a relaxation time, and the half-height line width $\Delta\omega_{1/2} = 2/\tau$. The relaxation rates discussed here are the total relaxation rates derived from a number of relaxation mechanisms, such as interactions with nuclear magnetic dipole and quadrupole moments, chemical exchange, and paramagnetic effects.

As with chemical shifts, the effects of an unpaired electron on nuclear relaxation rates can be manifested via a through-bond contact hyperfine coupling and a distance-dependent through-space dipolar interaction between the electron and nuclear magnetic moments. In addition, the coupling of the nuclear magnetic moment with the induced static magnetic moment of all the S spin levels can cause significant nuclear relaxation in a slow-tumbling paramagnetic species like a large metalloprotein; this process is called Curie relaxation. Thus, nuclear relaxation can provide information about nuclear–electron interactions, their distances and molecular dynamics. A more in-depth treatment of nuclear relaxation can be found in specific monographs and books cited in the General Reference section.

1. Dipolar Relaxation

The most common mechanism for nuclear relaxation is a through-space dipolar interaction between the nuclear and electron magnetic moments. This relaxation mechanism is present even in the absence of a covalent link between the paramagnetic center and the resonating nuclei. In the simplest approximation, the rate of dipolar relaxation is directly proportional to τ_e, the electronic relaxation time of the paramgnetic center, and inversely proportional to $r_{M-H}{}^6$, the sixth power of the distance between the paramagnetic center and the resonating nucleus as shown in Eq. (13).

$$\begin{gathered} T_M{}^{-1}(\text{dipolar}) \propto \mathrm{r_{M-H}}{}^{-6}\tau_e \\ T_{1M}{}^{-1}(\text{dipolar}) = T_{2M}{}^{-1}(\text{dipolar}) = \frac{4}{3}\frac{\gamma_N{}^2 g_e{}^2 \beta_e{}^2 S(S+1)}{r_{M-H}{}^6}\tau_e \end{gathered} \tag{13}$$

The individual nuclear relaxation times of the various nuclei can thus be useful for the assignment of isotropically shifted signals when the structure of the

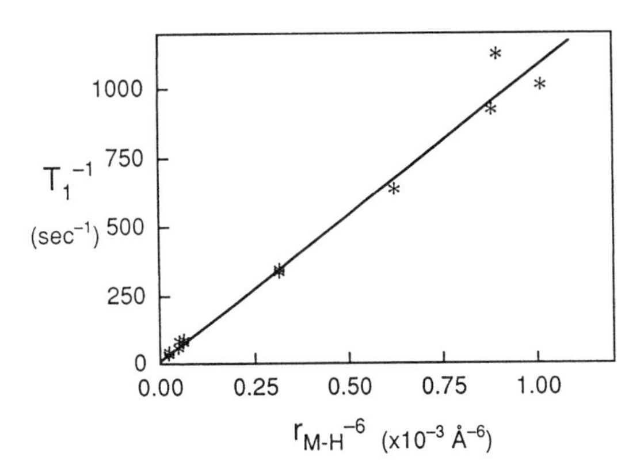

Figure 14
A T_1^{-1} versus r_{M-H}^{-6} plot for the $\{Fe^{II}_2(BPMP)[O_2P(OC_6H_5)_2]_2\}^+$ complex to show their linear relationship as shown in Eq. (13). A generic structure for a dinuclear BPMP complex is shown in Figure 28. [Data are obtained from Ming et al.[13]]

molecule is known, or for the determination of the metal–nuclear distance when the nucleus is assigned but the structure is not known. Owing to their r_{M-H}^{-6} dependence, a 50% error in the measurement of the relaxation times would cause only less than 10% off in metal–nuclear distances. Figure 14 illustrates the linear correlation between the T_1^{-1} values of the individual protons and the inverse sixth power of the metal–nuclear distance derived from the crystal structure of $\{Fe_2(BPMP)[O_2P(OC_6H_5)_2]_2\}^+$, which support the signal assignments.[13] (The spectrum and signal assignments for 2,6-bis{[bis(2-pyridylmethyl)amino] amino methyl}-4-methylphenol (BPMP) complexes will be discussed in detail in Section II.C.)

An important question to consider in understanding the NMR behavior of paramagnetic complexes is which metal centers give rise to relatively sharp paramagnetically shifted NMR features. The relaxation behavior of nuclei on a paramagnetic complex is principally controlled by τ_e, the electronic relaxation time of the paramagnetic center. The shorter the electronic relaxation time τ_e of the unpaired electrons, the slower the nuclear relaxation rate, which affords sharper NMR signals. Thus, metal ions such as high-spin Mn^{2+}, high-spin Fe^{3+}, Cu^{2+}, and Gd^{3+} with τ_e values of 10^{-8}–10^{-11} s afford ^{1}H NMR signals that are often broadened beyond detection for protons close to the metal center. These metal ions are thus designated "relaxation agents". On the other hand, metal ions with τ_e values in the range of 10^{-11}–10^{-13} s give rise to relatively sharp ^{1}H NMR features. Such ions are designated as "shift agents" and include high-spin Fe^{2+}, Co^{2+}, and Ni^{2+}, low-spin Fe^{III}–hemes, and lanthanides such as Pr^{3+}, Eu^{3+}, Dy^{3+}, and Yb^{3+} (see Chapter 3 for a discussion of electronic relaxation times.) In bioinorganic systems, NMR spectroscopy has been useful for the study of a number of iron proteins as well as metalloproteins substituted with Co^{2+}, Ni^{2+}, or lanthanides.

The complete expressions for dipolar relaxation are shown in Eqs. (14) and (15), where ω_S and ω_I are electron and nuclear Larmor frequencies (thus $\omega_S \gg \omega_I$ at a fixed magnetic field). The correlation time τ_c is determined by the relative magnitudes of τ_e, the electronic spin relaxation time (assuming $\tau_{c1} = \tau_{c2}$), τ_r, the rotational correlation time of the molecule, which can be estimated by the Stokes–Einstein equation $\tau_r = 4\pi\eta a^3/3kT$, where η is the viscosity and a is the radius of the molecule, and τ_M, the chemical exchange lifetime as $\tau_c^{-1} = \tau_e^{-1} + \tau_r^{-1} + \tau_M^{-1}$. As mentioned above, τ_e is about 10^{-11}–10^{-13} s for metal ions that act as "shift agents". The time τ_r, being related to molecular mass, is on the order of 10^{-10} s for molecules with a mass of 500 Da, 10^{-8} s for a small protein-like cytochrome *c* (12 kDa), and 7×10^{-6} s for a large protein like myeloperoxidase (155 kDa). Thus, when the chemical exchange rate τ_M^{-1} is negligible, τ_c is controlled by τ_e^{-1} and τ_r^{-1} for kinetically inert metal complexes and by τ_e^{-1} alone in metalloproteins with large τ_r values.

$$T_{1M}^{-1}(\text{dipolar}) = K_1 f_1(\omega, \tau_c) r_{M-H}^{-6}$$

$$= \frac{2}{15} \frac{\gamma_N^2 g_e^2 \beta_e^2 S(S+1)}{r_{M-H}^6} \times \left[\frac{3\tau_c}{1 + \omega_I^2 \tau_c^2} + \frac{\tau_c}{1 + (\omega_S - \omega_I)^2 \tau_c^2} + \frac{6\tau_c}{1 + (\omega_S + \omega_I)^2 \tau_c^2} \right] \qquad (14)$$

$$T_{2M}^{-1}(\text{dipolar}) = K_2 f_2(\omega, \tau_c) r_{M-H}^{-6}$$

$$= \frac{1}{15} \frac{\gamma_N^2 g_e^2 \beta_e^2 S(S+1)}{r_{M-H}^6} \left[4\tau_c + \frac{3\tau_c}{1 + \omega_I^2 \tau_c^2} + \frac{6\tau_c}{1 + \omega_S^2 \tau_c^2} + \frac{\tau_c}{1 + (\omega_S - \omega_I)^2 \tau_c^2} + \frac{6\tau_c}{1 + (\omega_S + \omega_I)^2 \tau_c^2} \right] \qquad (15)$$

In the low-field limit (small ω values) or in the fast-motion regime (short τ_c), Eqs. (14) and (15) can be greatly simplified with the approximations $\omega_{I(S)}^2 \tau_c^2 \ll 1$ and $\omega_S \pm \omega_I \sim \omega_S$, and a predominant contribution of τ_e^{-1} to τ_c^{-1} (e.g., $\tau_e < 10^{-10}$ s) to give Eq. (13). Thus, the shorter the electronic relaxation time τ_e the slower the nuclear relaxation rates $T_{1,2M}^{-1}$ and the sharper the isotropically shifted signals.

Equations (14) and (15) must be considered, however, to obtain a clear picture of nuclear relaxation with respect to its dependence on field ($\omega_{S,I}$), molecular size (τ_r) and chemical exchange (τ_M). The field dependence of the dipolar relaxation rates is illustrated in Figure 15, which plots the correlation functions $f_{1,2}(\omega, \tau_c)$ versus the applied field for different τ_e values. For $\tau_e = 1$ ns, a value representative of relaxation agents, there are two inflection points, when $\omega_S \tau_c = 1$ (at lower field) and $\omega_I \tau_c = 1$ (at higher field). For $\tau_e = 5$ ps, a value representative of shift reagents, only one inflection point is shown in the plot (at $\omega_S \tau_c = 1$); the inflection point associated with $\omega_I \tau_c = 1$ occurs at much higher field beyond the plot range. Irrespective of the relaxation rate of the electron, an inflection

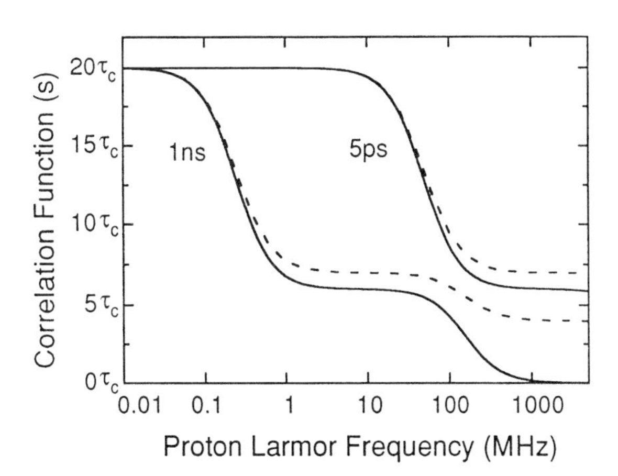

Figure 15
Plots of correlation functions $2f_1$ (solid lines) and f_2 (dotted lines) in Eqs. (14) and (15) versus magnetic field for two systems with $\tau_c = 1$ ns and 5 ps. The correlation function values are expressed in unit of τ_c; thus the relaxation rate of the 1-ns system is 200 times faster than the 5-ps system at 0.01 MHz.

point is observed in the frequency range 40–800 MHz, which is determined by ω_I for a relaxation agent and by ω_S for a shift agent. For the latter with a τ_c of 5 ps, the plot shows that the dipolar relaxation rates of the isotropically shifted NMR signals remain largely unchanged in the range about 200 MHz to 10 GHz, but increase significantly (to afford broad NMR features) below 100 MHz as ω_S approaches τ_c and decrease above 10 GHz as ω_I approaches τ_c (to give sharper signals). The above discussion suggests that detecting isotropically shifted signals may be more easily accomplished at greater than 200 MHz as far as dipolar relaxation is concerned (ignoring Curie relaxation effects; see Section I.C.2). On the other hand, the plot shows that the nuclear relaxation rates undergo very dramatic change in the range 100–1,000 MHz for a relaxation agent with τ_c of 1 ns, with the relaxation rates enhanced more significantly at lower frequency. Thus, the effect of a paramagnetic center such as Cu^{2+} or Mn^{2+} in a metalloprotein on the dipolar relaxation rates of a ligand will be more pronounced at 100 than at 800 MHz (cf. Section I.C.4).

It is also instructive to consider nuclear relaxation rates as a function of the correlation time τ_c, as shown in Figure 16. A correct evaluation of the correlation time can provide insight into the different motion factors that contribute to relaxation. Since these motion factors extend over the range from 10^{-7} to 10^{-13} s, a clear picture about the effect of the correlation function on nuclear relaxation cannot be provided by the plots in Figure 15 with only two τ_c values. However, this influence can be clearly visualized in Figure 16. The plots in this figure show immediately that the dependences of spin–spin and spin–lattice relaxation rates with respect to the correlation time differ significantly. While T_2 relaxation (f_2) increases nearly proportionally with increasing correlation time [due to the pres-

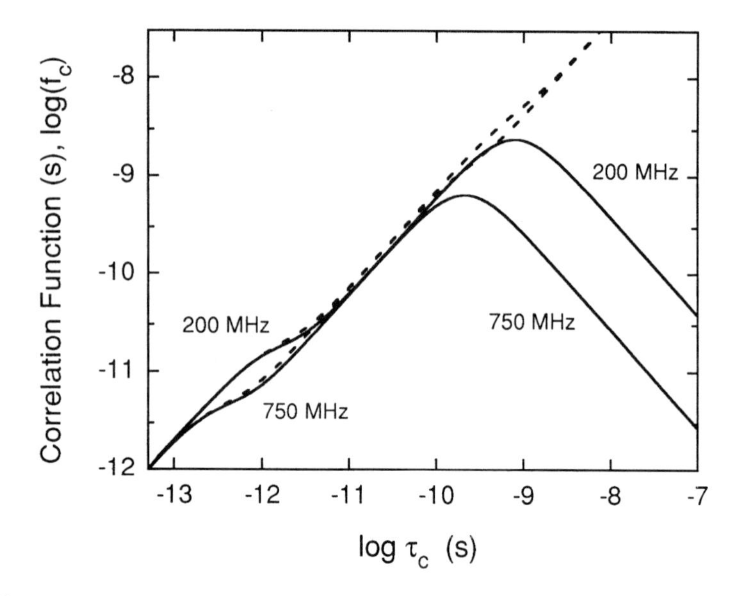

Figure 16
Plots of correlation functions $2f_1$ (solid lines) and f_2 (dotted lines) in Eqs. (14) and (15) versus correlation time at two fixed fields.

ence of the field-independent $4\tau_c$ term in Eq. (15)], T_1 relaxation ($2f_1$) reaches a maximum at $\tau_c = \omega_I^{-1}$ followed by a proportional decrease. Thus the familiar concept that the larger the molecule the broader (the larger the T_2 relaxation) its NMR signals is clearly revealed in this figure; however, the change in T_1 relaxation with respect to τ_c is not obvious from Eqs. (14) and (15) without these plots.

Although nuclear–metal distances can be estimated by the use of Eqs. (13)–(15), the structure of the metal site cannot always be uniquely defined unless the spatial and bond correlations among the nuclei in the surroundings of the paramagnetic metal site can be precisely determined. A combination of the geometric and distance information, obtained from the dipolar relaxation and the dipolar shift [Eq. (9)], may assist signal assignment and provide more insight into the structural features of a paramagnetic center. Drawing a more precise picture of a paramagnetic center will be facilitated if some critical NMR signals can be unambiguously assigned and then used as a starting point for structural elucidation. (Spectral acquisition and signal assignment strategies will be discussed in Section II.)

2. Curie Relaxation

Curie relaxation is another mechanism that can significantly contribute to the line widths (T_2 relaxation) of paramagnetic species of large molecular mass. This mechanism derives from the dipolar relaxation mechanism described above and is thus $r_{\text{M-H}}^{-6}$ dependent, but is proportional to the square of the applied magnetic

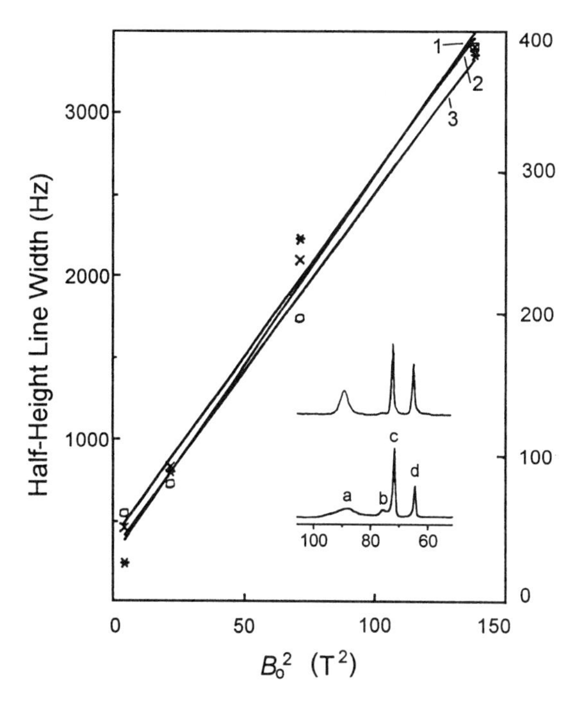

Figure 17
Least-squares fits of the half-height line widths of isotropically shifted ^{1}H NMR signals in Ag,Ni–SOD versus $B_0{}^2$. Signal a (*, left scale, line 1); signal c (×, right scale, line 2); signal d (o, right scale, line 3). The insets show part of the ^{1}H NMR spectra of Ag,Ni–SOD at 90 (top) and 300 (bottom) MHz. [Plots were adapted from Ming.[14a]]

field ($B_0{}^2$). This latter effect is illustrated in Figure 17 by a few isotropically shifted signals of Cu,Zn-SOD wherein the metal centers are respectively substituted with Ag^+ and Ni^{2+} (Ag,Ni–SOD).[14] At higher field, the signals are significantly broader due to Curie relaxation, which counteracts the better sensitivity and higher resolution expected with the higher field. Thus, higher field is not necessarily better when using NMR to study large paramagnetic metalloproteins.

Curie relaxation is the result of the contribution of the magnetic moment of the Curie spin to dipolar relaxation. In the discussion of dipolar relaxation in Section I.C.1, we made a simple assumption that nuclear relaxation is caused by the relaxation of electron spins that are in equally populated spin levels, so there is no static magnetic moment in the system. In reality, there is a finite magnetic moment parallel to the applied magnetic field, because each electron spin level is not equally populated but governed by the Boltzmann distribution that slightly increases the populations of the lower energy levels. The thermal average of all the spin levels, the "Curie spin" $\langle S_z \rangle$, is given as Eq. (16).

$$\langle S_z \rangle = S(S+1) g_e \beta_e B_0 / 3kT \tag{16}$$

To take this effect into account, Eqs. (14) and (15) must be modified to give Eqs.

(17) and (18),

$$T_{1M}^{-1} = \frac{2}{5}\frac{\gamma_N^2 g_e^2 \beta_e^2}{r_{M-H}^6}\left\{\langle S_z\rangle^2 \frac{3\tau_D}{1+\omega_I^2\tau_D^2} + \left[\frac{1}{3}S(S+1) - \langle S_z\rangle^2\right] f_1(\omega,\tau_c)\right\} \tag{17}$$

$$T_{2M}^{-1} = \frac{1}{5}\frac{\gamma_N^2 g_e^2 \beta_e^2}{r_{M-H}^6}\left\{\langle S_z\rangle^2 \left(4\tau_D + \frac{3\tau_D}{1+\omega_I^2\tau_D^2}\right) + \left[\frac{1}{3}S(S+1) - \langle S_z\rangle^2\right] f_2(\omega,\tau_c)\right\} \tag{18}$$

where the correlation times have the relationships, $\tau_c^{-1} = \tau_D^{-1} + \tau_e^{-1}$ and $\tau_D^{-1} = \tau_r^{-1} + \tau_M^{-1}$ with τ_e being the electronic relaxation time, τ_r being the rotation correlation time of the molecule, and τ_M being the lifetime of the nucleus sensing the Curie spin in a system under equilibrium. The first terms in Eqs. (17) and (18) are due to the relaxation caused by the Curie spin, thus designated "Curie relaxation". Curie relaxation is modulated only by molecular motion (τ_D^{-1}), which is molecular rotation in the absence of chemical exchange. The second terms in the equations are modulated by both spin relaxation [the $S(S+1)$ term] and molecular motion (the $\langle S_z\rangle$ term). Under the conditions that $\omega_I\tau_e \ll 1 \ll \omega_I\tau_D$ and $S(S+1) \gg \langle S_z\rangle$, the correlation functions in the second terms can be reduced to $10\tau_c$ and $20\tau_c$, respectively, affording the same formulations as in the simplified dipolar relaxations in Eq. (13).

Curie relaxation can be sizable when τ_r is large. In this case, the first terms in Eqs. (15) and (16) become the predominant terms, such as that in large paramagnetic metalloproteins. Moreover, the $\langle S_z\rangle^2$ term implies that Curie relaxation effects increase as the square of the applied magnetic field and may well exceed the dipolar contribution to the line width (i.e., T_2 relaxation) for a 40-kDa paramagnetic metalloprotein at 11.75 T. Plots of Curie relaxation rates as a function of the square of the magnetic field according to Eqs. (17) and (18) are shown in Figure 18 and illustrated by the example shown in Figure 17.

3. Contact Relaxation

The delocalization of unpaired electrons onto nuclei through chemical bonds (with the coupling constant A_c) can also enhance nuclear relaxation rates when coupled with the motion factors according to the Bloembergen formulation. This nuclear relaxation via unpaired electron delocalization through chemical bonds is the contact relaxation mechanism. There are three factors that contribute to the contact relaxation rates, the electron spin S, the hyperfine coupling constant A_c, and the correlation function $f''(\omega,\tau_c)$ with the assumption of a single electronic relaxation time τ_s, as shown in Eqs. (19) and (20),

$$T_{1M}^{-1}(\text{contact}) = \frac{1}{3}S(S+1)\left(\frac{A_c}{\hbar}\right)^2 f_1''(\omega,\tau_c)$$

$$= \frac{1}{3}S(S+1)\left(\frac{A_c}{\hbar}\right)^2 \frac{2\tau_c}{1+(\omega_S-\omega_I)^2\tau_c^2} \tag{19}$$

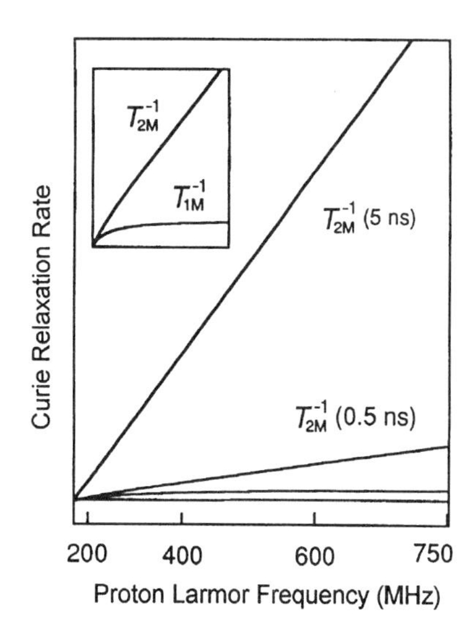

Figure 18
Plot of the Curie relaxation rate versus the square of the static field for a protein of about 15 kDa with $\tau_r = 5$ ns, and a small molecule with $\tau_r = 0.5$ ns, where an 8.6 times increase of the Curie relaxation $T_{2M}{}^{-1}$ can be observed for the protein molecule versus the small molecule at 600 MHz; and a nine times increase of the $T_{2M}{}^{-1}$ relaxation at 600 MHz as compared to that at 200 MHz. The bottom two lines are $T_{1M}{}^{-1}$ relaxations, which exhibit little field dependent as shown in the inset plot at low-field range.

$$T_{2M}{}^{-1}(\text{contact}) = \frac{1}{3}S(S+1)\left(\frac{A_c}{\hbar}\right)^2 f_2''(\omega, \tau_c)$$
$$= \frac{1}{3}S(S+1)\left(\frac{A_c}{\hbar}\right)^2 \left(\tau_c + \frac{\tau_c}{1 + (\omega_S - \omega_I)^2 \tau_c{}^2}\right) \qquad (20)$$

where all the parameters are as defined in previous sections. The rotational correlation time τ_r introduced in the Curie relaxation section [Eqs. (17) and (18)] does not contribute to the contact relaxation because the orientation of a molecule does not affect a through-bond electron–nuclear contact interaction. In the fast-motion regime, that is, when $\omega^2\tau^2 \ll 1$, the two relaxation rates become equal with $f_1''(\omega, \tau_c) = f_2''(\omega, \tau_c) = 2\tau_c$.

Dipolar relaxation is usually assumed to be the predominant mechanism in most studies in the literature that involve relaxation in a paramagnetic molecule. Whether or not this is a valid assumption cannot be easily judged by looking at and comparing equations for contact and dipolar relaxation [Eqs. (14), (15), (19), and (20)]. Moreover, since both mechanisms are functions of the correlation times and the resonance frequencies of an electron and a nucleus, their contributions to the total relaxation are closely dependent on the correlation time of the system and the magnetic field strength at which the measurements are conducted.

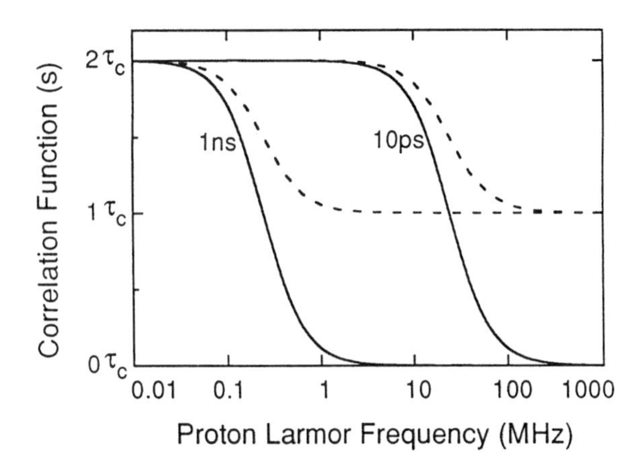

Figure 19
Plots of correlation functions f_1' (solid traces) and f_2' (dashed traces) in Eqs. (19) and (20) versus magnetic field for contact relaxations for two systems with $\tau_c = 1$ ns and 10 ps. The correlation function values are expressed in unit of τ_c; thus the relaxation rate of the 1-ns system is 100 times faster than the 10-ps system at 0.01 MHz.

As in the dipolar relaxation section, the influences of ω and τ_c on the contact relaxation rates can be clearly shown by plotting out the correlation functions with respect to these two terms. A plot of the correlation functions against proton Larmor frequency ω_I is shown in Figure 19, with two correlation times, 1 ns and 10 ps. There is only one inflection point revealed in the plot at $\omega_S = \tau_c^{-1}$, as the nuclear frequency ω_I does not contribute to the correlation function in contact relaxation [$\omega_I \ll \omega_S$ in Eqs. (19) and (20)]. The plots also show clearly that, while contact spin–spin relaxation ($T_{2\text{contact}}^{-1}$) decreases only to one-half of the magnitude in higher magnetic fields, contact spin–lattice relaxation ($T_{1\text{contact}}^{-1}$) vanishes completely. Moreover, for a paramagnetic species with $\tau_e = 1$ ns and $\tau_r > 1$ ns, the contribution of contact spin–lattice relaxation to the total relaxation becomes negligible in a field greater than 10 MHz for ^{1}H; thus, the assumption that dipolar relaxation is predominant becomes valid. The plots also show that $T_{1\text{contact}}^{-1}$ can contribute to the total relaxation at magnetic fields with ^{1}H frequencies greater than 100 MHz when there is significant covalency (i.e., large hyperfine coupling constant A_c) in a paramagnetic system with $\tau_e < 10$ ps and a large S value such as in high-spin Co^{2+} complexes and proteins.

Contact relaxation rates are functions of the correlation time τ_c, and should be significantly affected by changing τ_c. However, as in the case of dipolar relaxation, it is not obvious in Eqs. (19) and (20) how nuclear relaxation rates are affected by the correlation time at a fixed magnetic field. The relationship between nuclear relaxation rates and the correlation time in Eqs. (19) and (20) can be clearly viewed by plotting the correlation functions $f_{1,2}''(\omega, \tau_c)$ against the correlation time τ_c (Fig. 20). Since ω_I is not involved in contact relaxation, there is only one inflection point in the plots at $\tau_c = \omega_S^{-1}$. It is clearly shown in the plots that both spin–spin and spin–lattice nuclear relaxation rates increase with

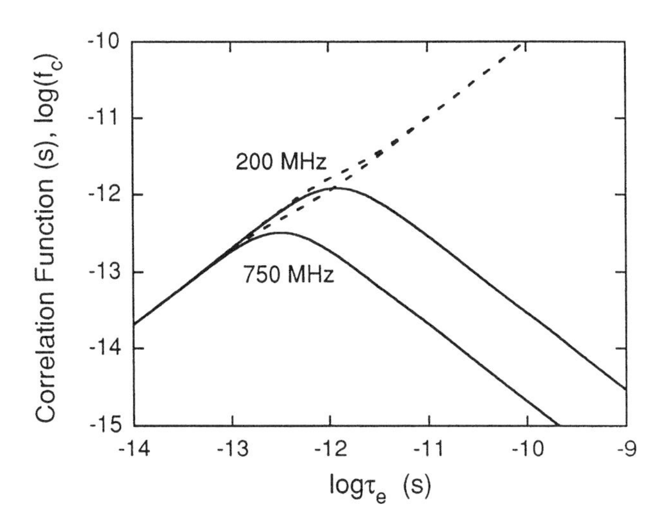

Figure 20
Plot of the correlation functions in Eqs. (19) and (20) against correlation time at two different magnetic fields affording proton Larmor frequencies of 200 and 750 MHz.

correlation time when $\tau_c < \omega_S{}^{-1}$ (e.g., $<10^{-12}$ s at 200 MHz), where the electronic relaxation time τ_e is the predominant term. In this fast-motion limit, Eqs. (19) and (20) can be simplified to Eq. (21) similar to the case of dipolar relaxation. As with dipolar relaxation, $T_{2\text{contact}}{}^{-1}$ has a pseudolinear relationship with the correlation function attributable to the field independent τ_c term [Eq. (20)], whereas $T_{1\text{contact}}{}^{-1}$ relaxation reaches the maximum at $\tau_c = \omega_S{}^{-1}$ then decreases proportionally with τ_c when $\tau_c > \omega_s{}^{-1}$, where Eq. (21) becomes invalid.

$$T_{1M}{}^{-1}(\text{contact}) = T_{2M}{}^{-1}(\text{contact}) = \frac{2}{3}S(S+1)\left(\frac{A_c}{\hbar}\right)^2 \tau_c \tag{21}$$

Contact relaxation can become an important factor when there is significant metal–ligand covalency. If it is present but ignored, the dipolar relaxation rates would be too large, providing estimates of metal–ligand distances that are too short. Such a situation was found in the case of phosphate-bound Co,Zn–, and Co,Co–SOD. Assuming a pure dipolar relaxation pathway, a maximum $Co^{2+}{}_{(\text{Cu site})}$–$^{31}P$ distance of about 2.5 Å was estimated, compared with a value greater than 3 Å for Co–(*O*)–P distance.[15,16] Such underestimated metal–nuclear distances, however, can be taken as evidence for the presence of a direct metal–ligand bond.

4. Applications of Nuclear Magnetic Relaxation

a. Ligand Binding The resonances of a ligand (such as a water molecule or an enzyme inhibitor) bound to a paramagnetic center can often be difficult to observe because of line broadening due to a short metal–nuclear distance and/or unfavorable τ_e values ($>10^{-11}$ s). The properties of the bound ligand can be monitored if there is an equilibrium between free and bound forms. In such a case, the paramagnetic contribution to the nuclear relaxation of the ligand can be described as in Eq. (22),

$$T_p^{-1} = fq(T_M + \tau_M)^{-1} + T_{out}^{-1} \tag{22}$$

with f, q, T_M, and τ_M being the mole fraction, which can be obtained from the affinity constant, stoichiometry, relaxation time, and lifetime, respectively, of the bound ligand, and T_p^{-1} is the observed relaxation rate of the free ligand and T_{out}^{-1} is the outer-sphere paramagnetic contribution, which is usually smaller than the first term in Eq. (22). Under fast exchange conditions, $T_M > \tau_M$, the nuclear relaxation properties of the resonance is an average of all the molecules present in solution and the nuclear relaxation properties of the bound state can be readily extracted, that is, $T_p^{-1} = fqT_M^{-1}$. Thus, particularly for mononuclear Cu^{2+} and Mn^{2+} systems, which have long τ_e values, a very small fraction of a bound ligand can significantly contribute to the nuclear relaxation rates of the bulk ligand under fast-exchange conditions. These systems can thus be studied by means of NMR relaxation techniques to provide structural and mechanistic information as described below.

One example is found in the study of phosphate (P_i) binding with Co^{2+} substituted derivatives of Cu,Zn–SOD (Co,Co– and Co,Zn–SOD).[16] The binding of phosphate to the Co^{2+} in the Cu site can significantly enhance the relaxation rate of its ^{31}P NMR signal. Details of the binding of phosphate to these two proteins can be revealed by a pH-dependent profile of ^{31}P relaxation (Fig. 21). The pH-dependent changes of ^{31}P relaxivity upon phosphate binding to the protein can be described by the equilibrium below [Eq. (23)] with an equilibrium constant K_{a1}/K_f,

$$\text{Co,M–SOD–P}_i \rightleftharpoons \text{Co,M–SOD}^- + \text{H}^+ + \text{P}_i \tag{23}$$

where K_f is the formation constant (300 M^{-1} from optical titrations) for Co,M–SOD–P_i complex (M = Co^{2+} or Zn^{2+}) and K_{a1} is a deprotonation constant of Co,M–SOD. This equilibrium is a combination of the formation of Co,M–SOD–P_i and the deprotonation of Co,M–SOD. The observed ^{31}P relaxivity can be fitted to this equilibrium (Fig. 21), and a best fit is obtained when HPO_4^{2-} is treated as the major species that binds to the enzyme with all the three ionization constants of phosphoric acid taken into consideration in equilibrium (23).

b. Nuclear Magnetic Resonance Dispersion The measurement of nuclear relaxation as a function of Larmor frequency (Figs. 16 and 20) affords the opportunity to gain further insight into the different factors that contribute to the relaxation, such as electronic relaxation, molecular tumbling, and electron–nuclear interactions. The term "nuclear magnetic relaxation dispersion" (NMRD)[17] is used to describe the dispersions in the correlation of nuclear relaxation and

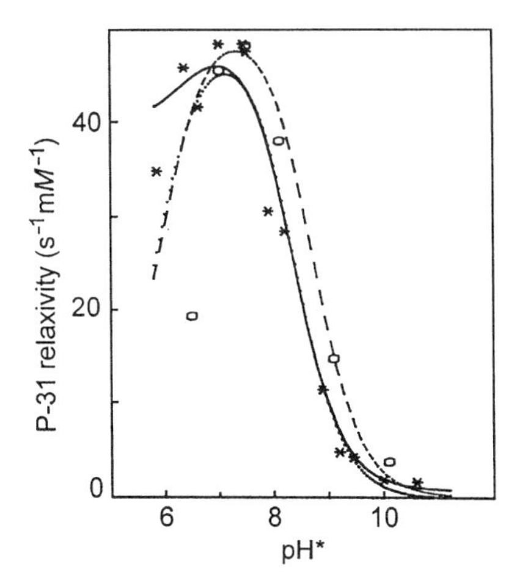

Figure 21
Plots of ^{31}P relaxation (50-mM phosphate) versus pH* at 202.5 MHz in the presence of Co,Co–SOD (∗) and Co,Zn–SOD (○). [Adapted from Ming and Valentine.[16]] The solid line is the fit to equilibrium (23) with affinity constants of 50 and 0.3 M^{-1} respectively for $H_2PO_4^-$ binding to Co,Co–SOD and HPO_4^{2-} binding to the deprotonated form. The dotted line (for Co,Co–SOD) and the dashed line (for Co,Zn–SOD) are the fittings obtained by assuming that the HPO_4^{2-} is the only phosphate interaction for the proteins. Two pK_a values of 7.0 and 7.4 respectively can be obtained for a deprotonation of Co,Co–SOD and Co,Zn–SOD.

magnetic field. In order to obtain a full NMRD profile, measurements of nuclear relaxation in a large range of fields is necessary to afford the inflection points at $\omega_{S,I} = \tau_c^{-1}$. For a paramagnetic species with a relatively long electronic relaxation time of about 1 ns, the entire relaxation dispersion profile can be revealed in the range of about 0.01–500 MHz. The measurement of relaxation times in this range is achieved by the use of a "relaxometer" to cover the lower field region and commercially available NMR spectrometers for the measurements at high field. Due to the low sensitivity of NMR at low magnetic fields, only the water molecule has been studied by means of NMRD techniques. Since water is involved in many metalloenzymatic processes, the study of the status of the water in resting enzymes and during catalysis and inhibition by means of NMRD techniques using paramagnetic metal ions as probes can provide significant mechanistic information about these enzymes. In addition, some detailed physical properties can also be obtained by the use of NMRD techniques as discussed below.

One simple example is the NMRD study of $[Mn(H_2O)_6]^{2+}$.[18] The relaxation rate of water in Mn^{2+} solution acquired in the field range 0.01–100 MHz reveals a dispersion due to ω_s at about 10 MHz. However, this data cannot be fitted well to Eqs. (14) and (15) with an assumption of a predominant dipolar relaxation mechanism but requires the inclusion of a contact interaction with $A_c/\hbar =$ 0.9 MHz (Fig. 22). The contact contribution to relaxation could not have been extracted from a relaxation time measurement at a small field range. In addition,

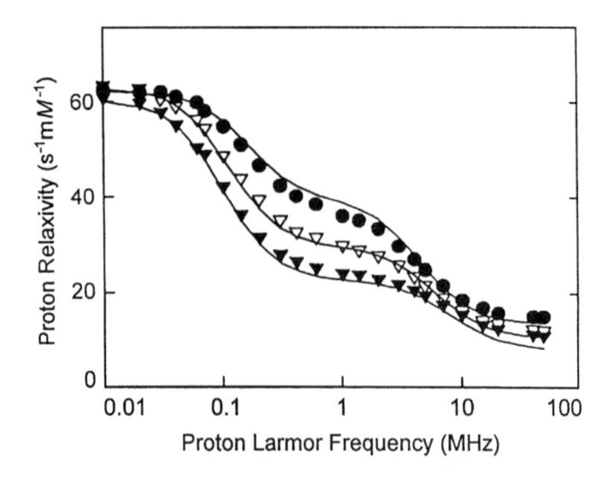

Figure 22
Water ^{1}H NMRD profiles of Mn^{2+} solution at 5 (●), 15 (▽), and 25 (▼) °C, and theoretical fittings considering both contact and dipolar relaxation that afford $A_c/\hbar = 0.9$ MHz, $\tau_r = 55$, 40, and 30 ps, and $\tau_s = 1.5$, 2.2, and 2.5 ns at 5, 15, and 25 °C, respectively. [Plots are adapted from Bertini et al.[18]]

the rotational correlation time τ_r and the electronic relaxation time τ_e are also obtained from the fit. As Mn^{2+} is required in kinase activity as discussed in the example below, a thorough understanding of the relaxation properties of Mn^{2+}–water system is needed to provide more information about the role of the metal in the action of this enzyme family if NMR relaxation is utilized for the study.

The NMRD has also been applied to the study of Co^{2+} substituted human carbonic anhydrase I (CA).[19] This enzyme shows a dispersion with an inflection point at about 30 MHz attributable to ω_S (Fig. 23), which can only be poorly fitted with a pure dipolar relaxation mechanism. However, a significant im-

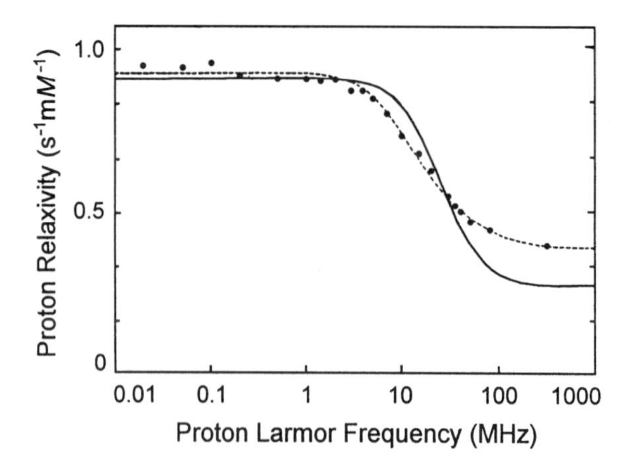

Figure 23
Water ^{1}H NMRD profile of Co^{2+}–CA at pH 9.9, and the fittings with (dashed line) and without (solid line) the consideration of the ZFS. (Adapted from Bertini et al.[19])

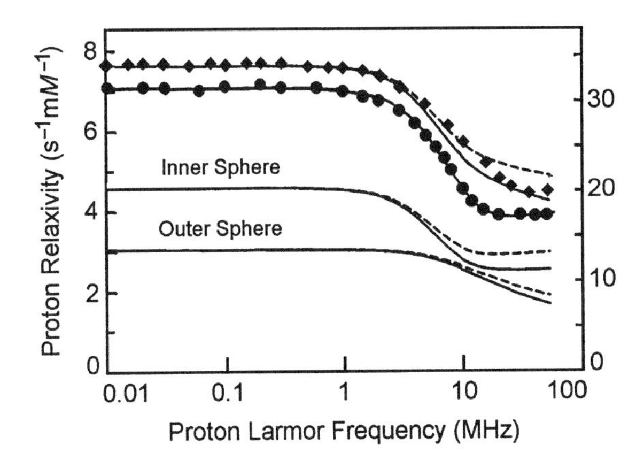

Figure 24
Proton relaxivity profiles (T_1, solid; expected T_2, dashed) of water in the presence of $[Gd(DTPA)(H_2O)]^{2-}$ (♦, left scale, 25 °C) and $[Gd(H_2O)_9]^{3+}$ (●, right scale, 10 °C). The data are fitted with the assumption of nine coordinated water molecules in the latter complex and one in the former. The deconvoluted inner and outer-sphere contribution in the former complex is also shown. [Plots are adapted from Koenig et al.[20a]]

provement is observed when the ZFS of the $S = \frac{3}{2}$ state (into $S = \pm\frac{3}{2}$ and $\pm\frac{1}{2}$ sublevels) of the high-spin Co^{2+} ion is taken into account. The correlation time can be obtained as 33 ps, and a metal–H(water) distance can be estimated to be 2.7 Å. This information is not obtainable at a single field and cannot easily be acquired by other means.

In recent years, magnetic resonance imaging (MRI) has become an indispensable tool in clinical diagnosis. Paramagnetic complexes are utilized to accelarate the relaxation rates of the solvent water to enhance the contrast in the MRI images.[20] Studies of the field-dependent relaxation of paramagnetic metal complexes have provided valuable strategies for designing such "contrast agents". Since the accesibility of water in different tissues or cell compartments to the contrast agents can differ, the water molecules is these compartments can exhibit different relaxation rates. This difference affords different image intensities, thus greatly enhances the contrast of the images. Of all the paramagnetic metal complexes, Gd^{3+} complexes have been found to be the most useful due to their highly paramagnetic nature and long electronic relaxation times.[20] The NMRD profile of an aqueous solution of Gd^{3+} can be fitted with $q = 9$, suggesting that it can be formulated as $[Gd(H_2O)_9]^{3+}$ (Fig. 24). In practice, Gd^{3+} is complexed by the octadentate ligand diethylenetriaminepentaacetic acid (DTPA) to form a complex that is widely used as a contrast agent in clinical MRI diagnosis. The water relaxivity decreases in the presence of this complex relative to the presence of the bare ion, due to the smaller number of coordinated water molecules in the DTPA complex. The NMRD data for this complex can be fitted with $q = 1$, suggesting the formation of the complex $[Gd(DTPA)(H_2O)]^{2-}$ in solution with only one bound water molecule. This result is consistent with the crystal structure of the analogous complex $[Nd(DTPA)(H_2O)]^{2-}$.

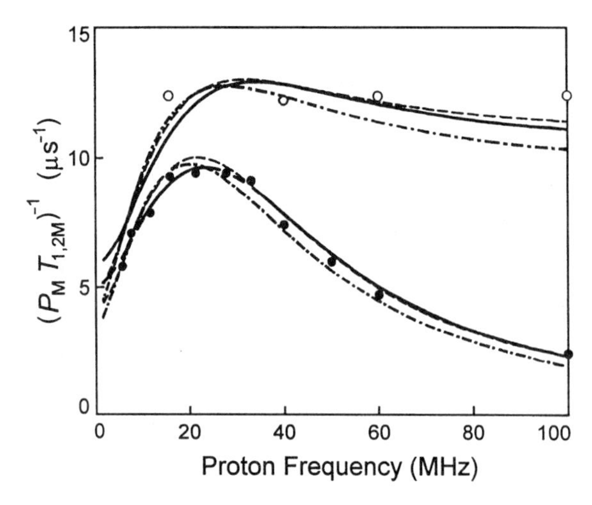

Figure 25
The normalized ^{1}H relaxation rate of water, $(P_M T_{1,2M})^{-1}$ with P_M = [paramagnetic species]/[H_2O], in Mn^{2+}–pyruvate kinase as a function of proton Larmor frequency. [Adapted from Burton et al.[21]]. ●, $(P_M T_{1M})^{-1}$; ○, $(P_M T_{2M})^{-1}$ and the fits with a fixed τ_r (– –), a fixed $q = 2$ (–·–), and the "best-fit" without restrictions on the values of τ_r and q (——).

c. An Example in Enzyme Systems: Pyruvate Kinase Nuclear magnetic resonance dispersion is also useful for the study of the status of bound water in more complicated enzyme systems, such as pyruvate kinase. This enzyme catalyzes the transfer of a phosphate group from phosphoenolpyruvate to adensosine diphosphate (ADP) in the glycolytic pathway [Eq. (24)], and requires Mg^{2+} (or Mn^{2+}) and K^+ for catalysis.

$$^{-}OOC\text{–}C(OPO_3{}^{2-})\text{=}CH_2 + ADP \rightleftharpoons {}^{-}OOC\text{–}C(\text{=}O)\text{–}CH_3 + ATP \quad (24)$$

The NMR relaxation has been useful for obtaining the values of q, τ_M, and τ_r for the enzyme from rabbit muscle (a tetramer of 237 kDa) with Mn^{2+} as the paramagnetic probe. The relaxation rates of water in the Mn^{2+}–kinase solution were measured from 6 to 100 MHz, and fitted by different approaches (Fig. 25)[21]: (a) using a fixed $\tau_r = 1.0 \times 10^{-7}$ s according to the Stokes–Einstein equation (Section I.C.1); (b) using a fixed $q = 2$; and (c) a "best-fit" (with a smallest percent error) without restriction on τ_r and q values. In all the cases, the fits indicate a value of 2–2.5 for q, a metal–proton distance of about 2.8 Å, and τ_M in the range $0.4\text{–}3.3 \times 10^{-8}$ s. The τ_r value obtained in the "best fit" is 20 times shorter than estimated by the Stokes–Einstein equation, which may be accounted for by a possible contribution from a fast local motion. It is also worth noting that a fit with q fixed at 4 affords a large standard deviation (22%). However, a recent crystal structure study of this enzyme revealed that Mn^{2+} is bound to the enzyme through two ligands (Glu 271 and Asp 295), which should leave four coordination sites for water binding.[22] The smaller value of q (2–2.5) derived from the NMRD experiments may reflect the fact that only some of the water molecules are in rapid exchange with bulk solvent, others being constrained due to hydrogen bonding in the active site. On the other hand, since the two parameters q and r are not determined independently in Eq. (22) (in the form of $q/r^6 \propto q/T_M{}^{-1}$) under fast exchange conditions, an erroneous estimate of r can cause a significant variation on the value of q. For example, a change in r from 2.8 to 3 Å increases q from 2.5 to about 3.8. A combination of different methods, such as NMR relaxation, X-ray diffraction, and fluorescence spectroscopy, is thus necessary to draw a more accurate conclusion on the status of coordinated water molecules in complicated metalloprotein systems.

D. NMR Properties of Multinuclear Paramagnetic Metal Centers

In the previous sections we have discussed the NMR chemical shift and nuclear relaxation properties of molecules with isolated paramagnetic centers. However, there are a large number of molecules in chemical and biological systems which contain paramagnetic centers that are not isolated from each other, but rather are magnetically coupled with each other via bridging ligands. As discussed in Chapter 7, the magnetic coupling between the two spins S_1 and S_2 can be described by the Hamiltonian shown in Eq. (25),

$$\mathscr{H} = JS_1 \cdot S_2 \tag{25}$$

where J is the coupling constant between the two spins. The total spin quantum number S' of the coupled system is defined by the individual spin quantum numbers S_1 and S_2, that is, $|S_1 - S_2| \leq S' \leq S_1 + S_2$. When the magnetic coupling results in a spin-paired $S = |S_1 - S_2|$ ground state, the complex is antiferromagnetically coupled and J is positive. When the magnetic interaction affords a spin-parallel $S = S_1 + S_2$ ground state, the complex is ferromagnetically coupled and J is negative. Since magnetic coupling affects the magnetic moment of the paramagnetic center, it can have a dramatic effect on its NMR shift and relaxation properties.

1. Effect on Chemical Shift

The concept of chemical shift in magnetically isolated paramagnetic systems discussed in Section I.B can be extended to the description of a magnetically coupled system when the consequences of magnetic coupling are taken into account. Hence, the spin delocalization and spin polarization pathways are still the mechanisms for unpaired spin density to propagate through chemical bonds. The contact shift described by Eqs. (4) and (5) for systems with isolated paramagnetic centers is valid for a nucleus in a magnetically coupled system when the $\langle S_z \rangle$ value [Eq. (16)] is evaluated over all the magnetically coupled spin states (i.e., $|S_1 - S_2| \leq S_i \leq S_1 + S_2$) after taking multiplicity and the Boltzmann distribution into account (cf. the van Vleck equation in Chapter 7). The contact shift of the nuclei in a magnetically coupled system can therefore be described as in

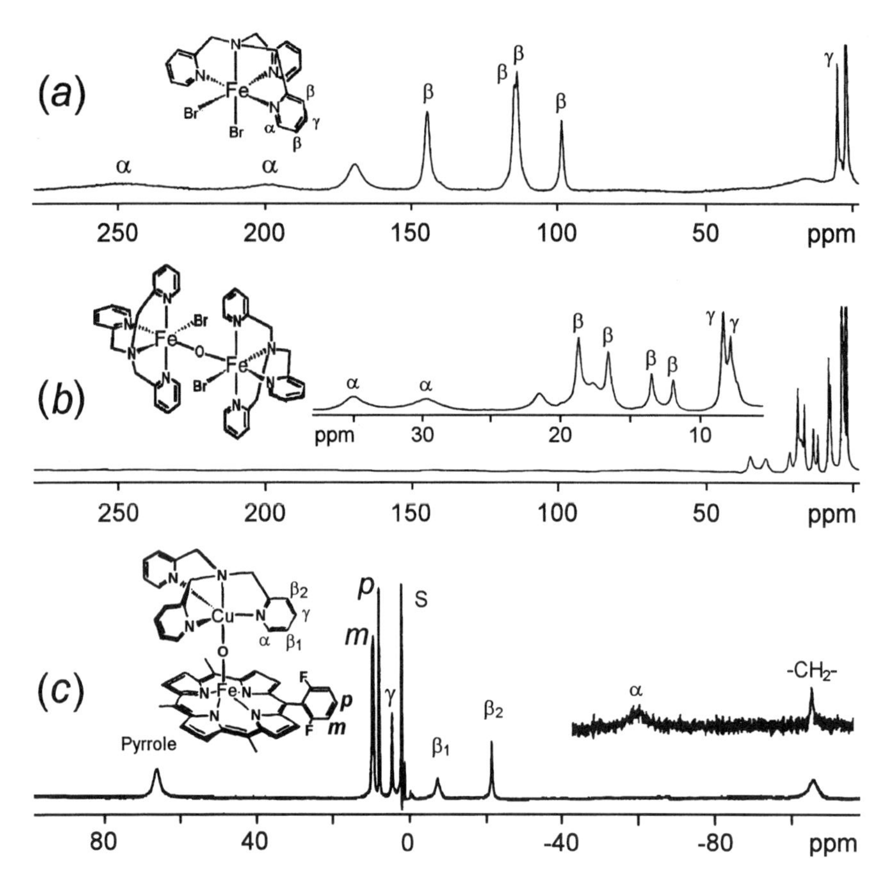

Figure 26
The ^{1}H NMR spectra (300 MHz and ambient temperature) of (*a*) $[Fe^{III}(TPA)Br_2]^+$ and (*b*) $[Fe^{III}_2O(TPA)_2Br_2]^{2+}$. [Adapted from Kojima et al.[23]] The latter, detected in the range 0–40 ppm, can be generated by reacting (*a*) with *t*-butylhydroperoxide for 30 min or synthesized independently. (*c*) The ^{1}H NMR spectra (300 MHz, 295 K) of $[F_8$–TPP)Fe–O–Cu(TPA)](ClO_4) with structure shown. [Adapted from Nanthakumar et al.[32]] The inset in (*c*) is deuterium NMR spectrum of selectively deuterated complexes to confirm proton signal assignment. The signal assignments of the complexes are as labeled.

Eq. (26), with A_i being the hyperfine coupling constant for the interaction of the nucleus with the *i*th magnetic state of the system.

$$\left(\frac{\Delta B}{B_0}\right)_{\text{contact}} = \frac{-1}{\hbar\gamma_{\text{N}} B_0}\Sigma A_i\langle S_{iz}\rangle \tag{26}$$

The consequences of Eq. (26) are illustrated in Figure 26, which compares the ^{1}H NMR spectra of a mononuclear high-spin Fe^{3+} complex $[Fe(TPA)Br_2]^+$ and its oxo-bridged dinuclear derivative $[Fe_2O(TPA)_2Br_2]^{2+}$.[23] The pyridine β protons of the mononuclear complex are observed at 100–150 ppm, while those of the dinuclear complex are found at 10–20ppm. The large difference in chemical shift arises from the fact that the mononuclear complex has an $S = \frac{5}{2}$ ground state while the oxo-bridged complex is strongly antiferromagnetically coupled ($J \sim 200\ \mathrm{cm}^{-1}$) such that only the $S' = 0$ and $S' = 1$ states are highly populated at room temperature. Thus the $\langle S_{iz} \rangle$ values for the dinuclear complex are much smaller and reflected in the smaller shifts. This reduction in $\langle S_{iz} \rangle$ values rationalizes the observed chemical shift of 17–25 ppm for the N_δ–H protons of the histidines bound to an oxo-bridged diiron(III) center of oxyhemerythrin ($J = 154\ \mathrm{cm}^{-1}$), relative to the 90–100 ppm value for imidazole N–H protons on mononuclear Fe^{3+} complexes.[24–26] Similarly, the strong antiferromagnetic coupling between the Fe^{2+} and the Fe^{3+} ions ($J \sim 100\ \mathrm{cm}^{-1}$)[27] of the ferredoxin shown in Figure 6(*b*) accounts for the small shifts observed for the cysteine β–CH_2 protons (~20 ppm) relative to the values found for mononuclear Fe–S centers (~150–250 ppm).[28]

The interactions of a nucleus with the coupled spin pair can be more conveniently discussed by the use of the hyperfine coupling constants $A_i(1)$ or $A_i(2)$, which are the fractional contributions (with a coefficient C_i) of the hyperfine coupling between the nucleus and S_1 or S_2 [$A(1)$ and $A(2)$, respectively] in isolated spin systems as Eq. (27),[29]

$$A_i(1,2) = C_i(1,2)A(1,2) \tag{27}$$

with $C_i(1,2)$ shown in Eq. (28).

$$C_i(1,2) = \frac{S_i(S_i+1) \pm [S_1(S_1+1) - S_2(S_2+1)]}{2S_i(S_i+1)} \tag{28}$$

The contact shift in a magnetically coupled system can thus be obtained as in Eq. (29) where the interactions of the nucleus in question with each of the metal centers are described by the hyperfine coupling constant $A(1)$ or $A(2)$.

$$\left(\frac{\Delta B}{B_0}\right)_{\mathrm{contact}} = \frac{-g\beta}{3\hbar\gamma_N kT} \sum C_i(1,2)A(1,2)S_i(S_i+1)\frac{(2S_i+1)\exp(-E_i/kT)}{\sum(2S_i+1)\exp(-E_i/kT)} \tag{29}$$

When $|J_{12}| < kT$, the chemical shift of a nucleus in question is simply the sum of the effects of each individual paramagnetic center as described in Eq. (29). As J_{12} becomes larger and approaches kT, the excited S' states become less populated and contribute less to the chemical shift. Particularly for antiferromagnetically coupled systems with an $S = 0$ ground state, the chemical shift approaches the diamagnetic value as J_{12} becomes much larger than kT. The NMR can thus be a very useful tool for the study of such systems and allows J_{12} to be estimated.

It is clear that magnetic coupling affects the chemical shifts observed for a paramagnetic complex; it would thus be possible to learn about the strength of the magnetic interaction from the chemical shift properties of a dinuclear com-

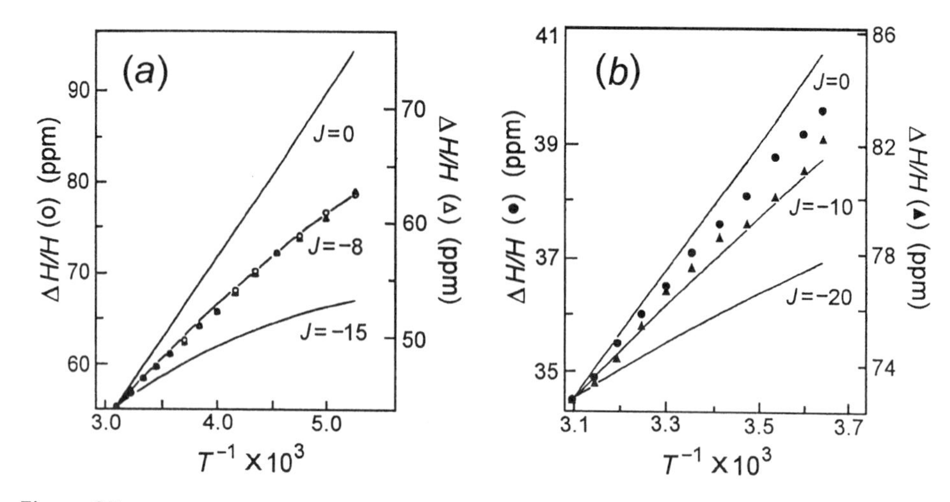

Figure 27
Temperature dependence of isotropic shift in (*a*) $Fe_2(sal_3trien)(OCH_3)Cl_2$ in acetone-d_6 and (*b*) $Fe^{II}Fe^{III}$–uteroferrin. The solid lines are calculations of temperature-dependent isotropic shifts with different J values. The J here is equivalent to $-J/2$ in Eq. (25). [Reprinted from Lauffer et al.[31]]

plex. When the coupling is strong, particularly in a homobimetallic complex, the significant decrease in the chemical shift relative to the uncoupled case can be used to estimate the value of J, assuming that A_i is relatively spin-state independent (i.e., $A_i \approx A$).[30] For example, an antiferromagnetically coupled diiron(III) complex with a J of 200 cm^{-1} would be predicted to have a cumulative $\langle S_{iz} \rangle$ value that is about 10% of that of a mononuclear high-spin Fe(III) complex. This finding is borne out in the comparison of the ^1H NMR chemical shifts of $[Fe_2O(TPA)_2Br_2]^{2+}$ versus $[Fe(TPA)Br_2]^+$ (Fig. 26) as well as those of methemerythrin and its mononuclear counterpart discussed above. For more weakly coupled systems, where the decrease in the shift is much less significant, the temperature dependence of the shift as described by Eq. (29) can be used to assess the strength of the magnetic coupling. Such an approach has been used to estimate the J values for the diiron centers in $Fe_2(sal_3trien)(\mu\text{-}OCH_3)Cl_2$, where sal$_3$trien = tresalicylidenetriethylenetetraamine, and uteroferrin (Fig. 27).[31] However, caution has to be taken when A_i is dependent upon the spin state; for J to be correctly determined, the A_i values for all the thermally accesible spin states must be determined.

Another dinuclear complex, $[Fe^{III}(F_8\text{–}TPP)\text{–}O\text{–}Cu^{II}(TPA)]$,[32] further illustrates the interesting effects of antiferromagnetic coupling on NMR properties. This complex serves as a model for the dinuclear oxygen binding center of cytochrome c oxidase. The complex has an $S' = 2$ ground state derived from an antiferromagnetic interaction of the $S = \frac{5}{2}$ high-spin Fe^{3+}–heme center with the $S = \frac{1}{2}$ Cu^{2+} center, which is separated from the $S' = 3$ excited state by the coupling constant $J = 174$ cm^{-1}. As described in Eq. (27), the hyperfine coupling constant of the nuclei sensing only the Cu^{2+} center in a magnetically coupled system is $C_{i\mathrm{Cu}}A_{\mathrm{Cu}}$. For an antiferromagnetically coupled $S = \frac{5}{2}$ and $S = \frac{1}{2}$ system, the coefficient associated with the Cu ion in the $S' = 2$ ground state is $-\frac{1}{6}$, whereas it is $\frac{1}{6}$ for the $S' = 3$ excited state. Thus for [Fe^{III}(F_8–TPP)–O–Cu^{II}(TPA)], in which the excited state is not fully populated at room temperature due to strong antiferromagnetic coupling, the $C_{i\mathrm{Cu}}A_{\mathrm{Cu}}$ term would reflect the greater contribution of the ground $S' = 2$ state and have a value opposite in sign to A_{Cu}. As the pyridyl proton signals are downfield shifted in a mononuclear Cu^{II}(TPA) complex, antiferromagnetic coupling with an $S = \frac{5}{2}$ Fe^{III}–heme center results in a reversal in sign and the pyridyl protons are upfield-shifted in the [spectrum of Fe^{III}(F_8–TPP)–O–Cu^{II}(TPA)] [Fig. 26(*c*)]. Moreover, the magnitudes of the chemical shifts of the pyridyl protons follow the same order as in the mononuclear complex albeit with opposite sign, that is, 6-H > 3, 5-H > 4-H. This observation suggests the retention of a spin delocalization mechanism, however, initiated by an "antiparallel spin" opposite to the spin delocalization pattern shown in Figure 4.

2. Effect on Relaxation Properties

The coupling of two paramagnetic centers allows a fast relaxing paramagnetic center to increase the relaxation rate of a nearby more slowly relaxing paramagnetic center such that the coupled system essentially adopts the relaxation properties of the faster relaxing ion. This dramatic effect is illustrated in Figures 28 and 29 for a number of magnetically coupled metal complexes[13,33] and metalloproteins.[14] Isotropically shifted ^{1}H NMR signals from ligands bound to high-spin Fe^{3+}, Mn^{2+}, and Cu^{2+} centers, often broadened beyond detection in mononuclear complexes, become relatively sharp in the presence of a faster relaxing paramagnetic center such as high-spin Fe^{2+}, Co^{2+}, or Ni^{2+}.

The effect of magnetic coupling on electronic relaxation can be developed quantitatively along the lines of Eqs. (14) and (15), as described in Eq. (30),

$$\tau_{s1}{}^{-1}(J) = \tau_{s1}{}^{-1}(0) + \frac{2J_{12}{}^2 S_2(S_2+1)}{3\hbar}\left[\frac{\tau_{s2}}{1+(\omega_{s1}-\omega_{s2})^2\tau_{s2}{}^2}\right] \tag{30}$$

where $\tau_{s1}{}^{-1}(J)$ and $\tau_{s1}{}^{-1}(0)$ are the electronic relaxation rates of the paramagnetic center in the presence and absence, respectively, of magnetic coupling with another paramagnetic center having an electronic relaxation time τ_{s2}. Equation (30) indicates that a J_{12} of even 0.5 cm^{-1} is sufficient to significantly accelerate the electronic relaxation rate of the slowly relaxing metal ion. As a result of the magnetic coupling, the shortened electronic relaxation times become the predominant term in the correlation functions as shown in Eq. (13); nuclear relaxation rates then become smaller and the NMR signals become sharper.

Equations (13)–(21) for the treatment of nuclear relaxation in isolated spin systems are still valid for the description of nuclear relaxation in magnetically coupled systems when a fractional contribution of the hyperfine coupling constant is taken into account [Eq. (27)]. Thus, the contact relaxation $T_{1\mathrm{M(contact)}}{}^{-1}$

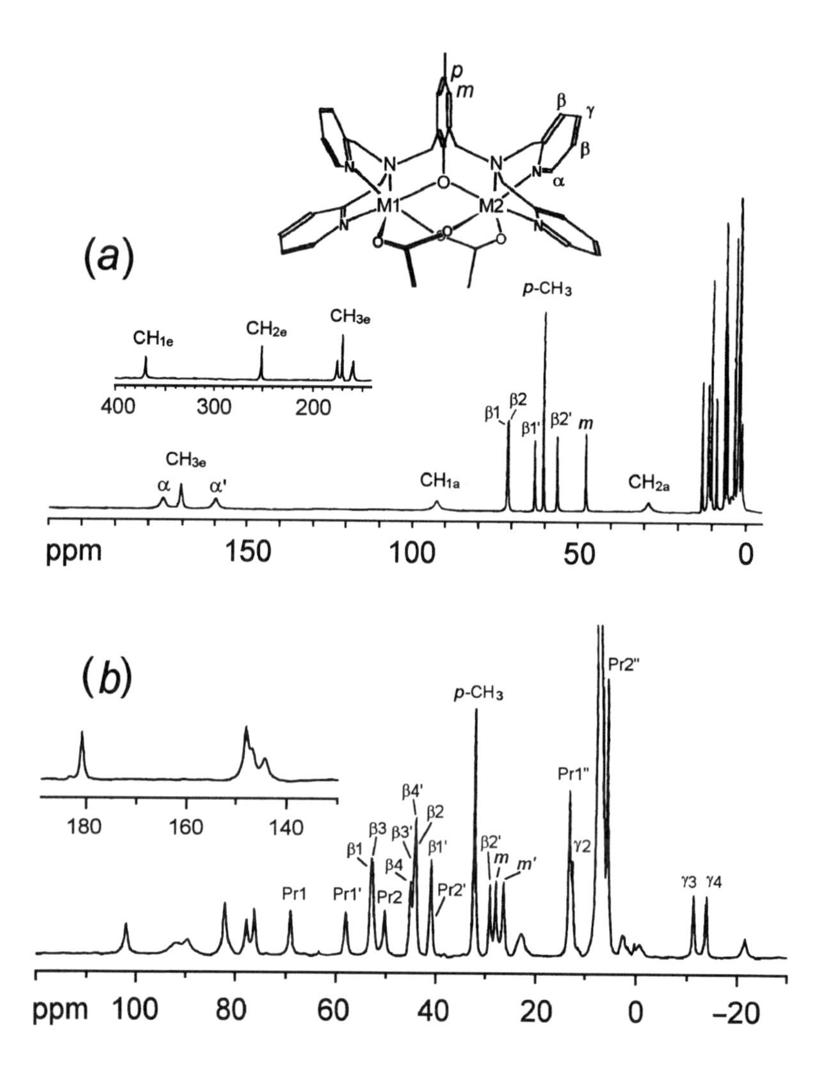

Figure 28
The ^{1}H NMR spectra of magnetically coupled dinuclear metal complexes: (*a*) $\{Fe^{II}Fe^{III}BPMP[O_2P(OC_6H_5)_2]\}^{2+}$ and (*b*) $[Fe^{II}Mn^{II}BPMP(O_2CCH_2CH_3)_2]^{+}$. [Adapted from Ming et al.[13] and Wang et al.[33]] The assignment of the signals by the use of two dimensional (2D) techniques are discussed in Section II.C (cf. Fig. 39).

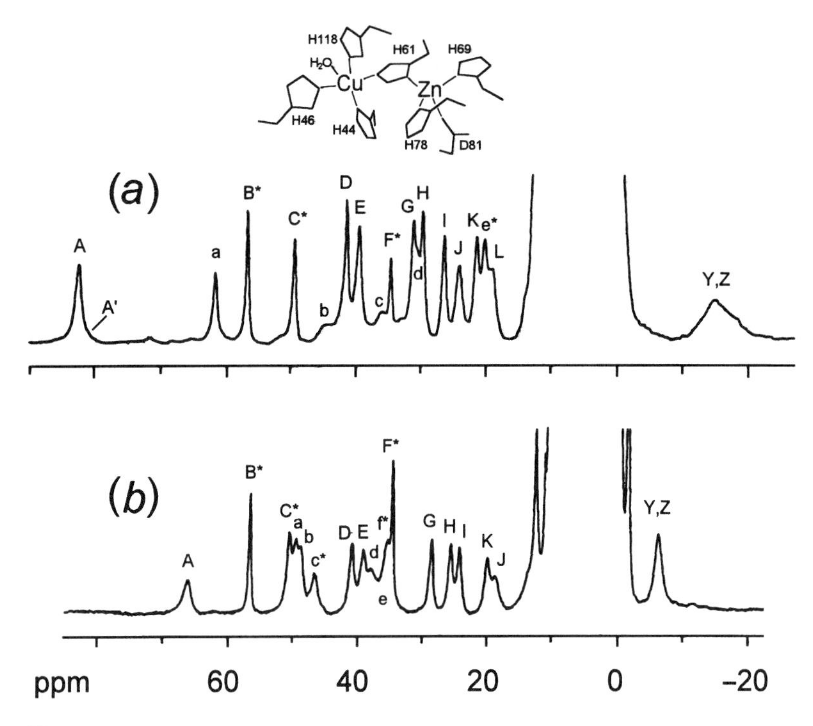

Figure 29
The ^{1}H NMR spectra of magnetically coupled dinuclear metalloproteins (*a*) $Cu^{II}Ni^{II}$SOD and (*b*) $Cu^{II}Co^{II}$SOD. [Adapted from Ming et al.[14]] The signals A–L are attributable to His ring protons on the Cu^{2+} site, the signals a–e are due to His ring protons on the Ni^{2+} or Co^{2+} site, and the asterisked signals are solvent exchangeable His ring NH protons. Assignments of the isotropically shifted signals can be achieved by means of anion titrations and nuclear Overhauser enhancement spectroscopy (NOESY) [Figs. 45 and 48(*a*)].

can be modified to give Eq. (31) for a nucleus that interacts with either metal ion 1 or 2 in a paramagnetic dinuclear metal center with a hyperfine coupling constant $A(1)$ or $A(2)$, where the electronic relaxation in the correlation functions in obtained from Eq. (19). The dipolar relaxation $T_{1M(\text{dipolar})}{}^{-1}$ of a nucleus caused by a paramagnetically coupled dinuclear center can also be obtained [Eq. (32)] via a similar modification of Eq. (14), with the introduction of the z component of the Curie spin $\langle S_{iz}\rangle$. A similar treatment can give $T_{2M}{}^{-1}$ in a magnetically coupled system, with the use of appropriate coefficients and correlation functions from Eqs. (15) and (20).

$$
\begin{aligned}
T_{1M(\text{contact})}{}^{-1} &= \frac{2}{3\hbar}A(1,2)^{2}\Sigma_{i}C_{i}(1,2)^{2}\langle S_{iz}\rangle f_{1}(\tau_{c},\omega_{S,I}) \\
&= X_{1}T_{1M_{1}(\text{uncoupled})}{}^{-1} \qquad (31)
\end{aligned}
$$

$$T_{1\mathrm{M(dipolar)}}^{-1} = \frac{2}{15}\frac{\gamma_\mathrm{N}^2 g_\mathrm{e}^2 \beta_\mathrm{e}^2}{r_{\mathrm{M-H}}^6}\Sigma_i C_i(1,2)^2 \langle S_{iz}\rangle f_1{}'(\tau_\mathrm{c}, \omega_{\mathrm{I,S}})$$
$$= X_1 T_{1\mathrm{M}_1(\mathrm{uncoupled})}^{-1} \quad (32)$$

When the nucleus interacts with both metal centers, a sum of the effects caused by each paramagnetic center is used to describe nuclear relaxation [Eq. (33)]. The description of nuclear relaxation in a magnetically coupled system can be further simplified in the high-temperature limit, where $J \ll kT$ and the reduction coefficients X_n ($n = 1$ or 2) are introduced [Eq. (34)] for the nucleus that interacts with the paramagnetic metal center 1 and 2, respectively, in a coupled system.[34] Thus, an antiferromagnetically coupled Cu^{2+}–Ni^{2+} pair [with $S = \frac{1}{2}$ and 1 as in Cu,Ni–SOD, Fig. 29(*a*)] affords two spin levels $S' = \frac{1}{2}$ and $\frac{3}{2}$, and the two reduction coefficients $X_{\mathrm{Cu}} = \frac{11}{27}$ and $X_{\mathrm{Ni}} = \frac{7}{9}$, which can be utilized in relaxation studies of such a system.

$$T_{1\mathrm{M}}^{-1} = T_{1\mathrm{M}(1)}^{-1} + T_{1\mathrm{M}(2)}^{-1}$$
$$= X_1 T_{1\mathrm{M}_1(\mathrm{uncoupled})}^{-1} + X_2 T_{1\mathrm{M}_2(\mathrm{uncoupled})}^{-1} \quad (33)$$

$$X_n = \frac{1}{S_n(S_n+1)}\frac{\Sigma_i C_{in}^2 S'_i(S'_i+1)(2S'_i+1)}{\Sigma_i(2S'_i+1)} \quad (34)$$

The spectra of the dinuclear centers in Figures 28 and 29 clearly show that relaxation agents such as Cu^{2+} and Mn^{2+} actually give rise to isotropically shifted signals that are sharp enough to be observed. These magnetically coupled dinuclear metal centers can thus serve as useful probes for structural and mechanistic studies of dinuclear metalloproteins. Some additional examples of the application of NMR techniques to the study of magnetically coupled paramagnetic systems can be found in Section II.C.

II. Practical Aspects: Acquiring and Assigning Proton Nuclear Magnetic Resonance Spectra for Paramagnetic Molecules

The very first step for any NMR study is to obtain a spectrum. The presence of a paramagnetic center will increase the spectral width significantly and broaden the resonances dramatically due to a shortening of the relaxation time, thereby decreasing the sensitivity of the paramagnetically shifted signals. The basic principles for obtaining NMR spectra of diamagnetic compounds can nevertheless be applied to the acquisition of informative NMR spectra of paramagnetic species, but very different parameters for data acquisition and processing must be used to take the effect of the paramagnetic center into account. These practical aspects will be discussed in this section.

A. I Just Want a Spectrum!

The acquisition of a one-dimensional (1D) NMR spectrum typically entails a pulse sequence consisting of a relaxation delay, followed by a strong rf pulse for excitation of nuclei, and then the acquisition of the free induction decay (FID). All the parameters in this sequence need to be appropriately adjusted to acquire a useful nice-looking NMR spectrum of a paramagnetic molecule. Spectrum (*a*) in Figure 30 is the ^{1}H NMR spectrum of 5-m*M* horse heart cytochrome *c* in D_2O with optimum adjustment of the instrument, temperature control, and acquisition and processing parameters, namely, a wide spectral window to allow observation

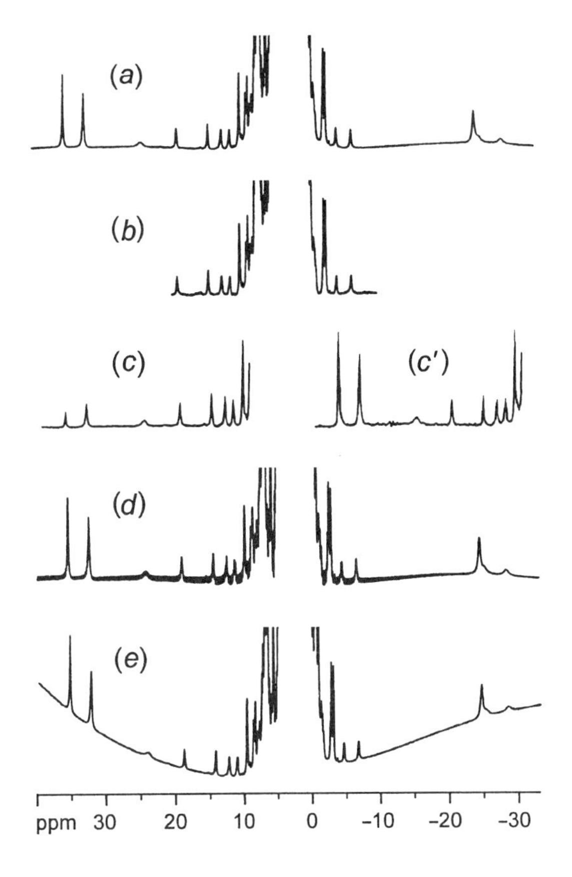

Figure 30
The ^{1}H NMR spectra of horse heart cytochrome *c* in D_2O (*a*) with optimum adjustment of the instrument as described in the text; (*b*) with too narrow a spectral width, which results in signal foldover; (*c*) by the use of a long 90° excitation pulse, which does not excite all the features in a wide spectral window as demonstrated by the lower intensities of the farthest shifted methyl signals at about 34 ppm (0.58 and 1.17 relative to the single-proton signal at 18.8 ppm) and (*c*′) with a short 10° excitation pulse using the same weak rf as in (*c*), which affords a spectrum in which all paramagnetically shifted peaks reflect their correct intensities (note the relative intensities of the 34 ppm methyl resonances are about threefold more intense than the single-proton 18.8 ppm signal); (*d*) with an acquisition time (17.42 ms) that is too short, thereby causing FID truncation; and (*e*) before baseline correction to give spectrum (*a*).

of all isotropically shifted features (>70 ppm), a short 90° pulse (7 μs) to provide a large excitation window, an acquisition time (139.4 ms with 8 K data points) that avoids significant FID truncation of the more slowly relaxation signals, a relaxation delay of 800 ms (usually >5 times the T_1 values of the isotropically shifted signals), appropriate FID apodization (exponential multiplication with a line-broadening factor of 10 Hz) to increase the signal-to-noise (S/N) ratio, followed by a fifth-order polynomial baseline correction.

1. Spectral Width—the Window in the Frequency Domain

Signals with large paramagnetic shifts (especially broad ones) can be easily overlooked if too small a spectral window is set for the acquisition. Figure 30(*b*) shows the ^{1}H NMR spectrum of cytochrome *c*, which was acquired with too narrow a spectral width resulting in a severe signal foldover problem. This problem can be easily corrected by expanding the spectral width. When a large spectral window is required, it may be advantageous to use a spectrometer of lower field, since a given frequency window will correspond to a larger spectral width in terms of ppm at lower field. Thus, a 200-MHz spectrometer would require a dwell time of 10 μs (digitization rate of 100 kHz, readily accessible with modern digitizers) to cover a spectral window of 50,000 Hz or 250 ppm, while a 600-MHz instrument would require the use of a three-times faster digitization rate to detect all the signals. For the purpose of studying paramagnetic compounds, a high-field NMR should be equipped with a fast analogue-to-digital converter that can provide short dwell times to cover a large spectral range in terms of hertz. On NMR spectrometers with digital filters, the use of a spectral window that is too narrow can completely wipe out signals found outside of the window.

2. Radio Frequency Pulse and Relaxation Delay

On a Fourier transform (FT) NMR spectrometer, a high-power rf pulse is applied to simultaneously excite all nuclei of the same type with different frequencies, which may exhibit a very large range in paramagneitc molecules, and then the FID is digitized, weighted with a window function, and Fourier transformed to give a familiar spectrum in unit of parts per million. While a 10-μs rf field can easily cover the excitation window required for diamagnetic species (exhibiting ^{1}H NMR spectra of ~15 ppm and ^{13}C NMR spectra of ~200 ppm), it is not sufficient for the excitation of signals from a paramagnetic molecule with a spectral width of over 200 ppm (>120 kHz at 11.75 T). Such a spectral width would require a 90° pulse of much less than 10 μs to provide effective simultaneous excitation of all the nuclei in the large spectral region. The reason is that the Fourier transformation of a rectangular rf pulse with a duration time τ gives the frequency domain function in the form $\sin(\omega\tau)/\omega + i[1 - \cos(\omega\tau)/\omega]$ with an amplitude of τ at ω_0 and the first null at $\omega_0 \pm 1/\tau$. Consequently, the shorter the rf pulse, the larger the excitation range. This consequence is exaggerated in Figure 30, where spectrum (*c*) was acquired with a weak 90° pulse of 80 μs centered at the water resonace. The inefficient excitation of the methyl signals at about 34 ppm is clearly demonstrated by their decreased intensities relative to the single-proton 18.8 ppm signal. Spectrum (c') acquired with the *same* weak rf pulse but with a much shorter pulse (a small 10° flip angle of 8.9 μs) affords the correct intensities for the methyl signals as a result of the wider range of excitation covered by the shorter pulse.

It is important for NMR studies of paramagnetic species to manipulate the pulse sequence parameters to give a maximum S/N ratio. The optimum pulse angle with a fixed repetition time is given by the Ernst correlation [Eq. (35)][35]

$$\cos \alpha_E = \exp(-T_r/T_1) \tag{35}$$

with α_E being the optimum pulse angle (the Ernst angle) and T_r the repetition time. For example, the optimum pulse angle is 87° for a signal of $T_1 = 10$ ms when a repetition time of 30 ms is used. For more slowly relaxing signals, either the pulse angle or the repetition time needs to be adjusted to achieve optimum S/N ratios. However, in experiments where a 90° pulse is required (such as in 2D studies), the repetition time chosen is a balance of factors such as S/N ratios of the faster relaxing signals versus accurate intensities for the more slowly relaxing signals. The recovery of the magnetization after a 90° excitation pulse is given in Eq. (10), and the S/N ratio of a signal is proportional to the square root of the number of FIDs accumulated, that is, $(T_{total}/T_r)^{1/2}$ with T_{total} the total experimental duration and T_r the repetition time. The relative signal intensity for a partially recovered signal [i.e., M_z at $\tau = T_r$ in Eq. (10)] can be estimated as $M_z'(T_r) = M_z(T_r)(T_{total}/T_r)^{1/2}$, which gives roughly an optimal value with the ratio $T_r/T_1 \sim 1.3$ (i.e., when $dM_z'(T_r)/dT_r = 0$) after the 90° pulse with a fixed experimental time T_{total}. The above discussion provides an alternate criterion for data acquisition when maximum sensitivity rather than the accurate integration of signals is required. A more detailed discussion of optimum parameters for increasing S/N can be found in the literature.[36]

The ^{1}H NMR spectrum of cytochrome *c* shown in Figure 30 was obtained with a 7-μs 90° pulse and a repetition time of 800 ms. These conditions are sufficient to provide quantitative signal intensities for the shifted signals with relaxation times in the range less than 150 ms. However, the sharper features near the diamagnetic region are clearly still suppressed because the relaxation delay is still not long enough for these slowly relaxing signals. This signal suppression is often not a problem as the focus is usually on the resonances most affected by the paramagnetic center.

3. Free Induction Decay: The "Spectrum" in the Time Domain

The familiar frequency domain NMR spectrum is not the direct output from an FT NMR spectrometer but the FT of the time domain spectrum—the free induction decay, FID, acquired after the rf excitation pulse. Thus, the quality of the frequency domain spectrum is dependent upon the quality of the FID. Paramagnetically shifted signals have relatively short relaxation times. The FID of a signal with a 100-Hz line width ($T_2 \cong 3$ ms) will be nearly zero 18 ms after the rf pulse, so an acquisition time of 50 ms is sufficient to collect this data. However,

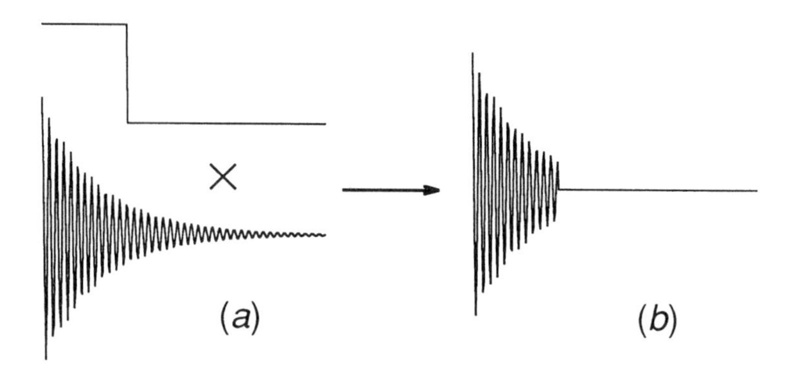

Figure 31
A normal FID (*a*) and an FID with the "tail" truncated (*b*) due to the use of a too short acquisition time, where the "square wave" associated with the truncated FID is responsible of the sine wave after FT of the truncated FID.

such an acquisition time is too short to obtain the entire FID of a 1-Hz-wide signal ($T_2 \cong 300$ ms), and thus the "tail" of the FID would be truncated (Fig. 31). Fourier transformation of the truncated FID results in the appearance of sine wave edges on the sharp signal [Fig. 30(*d*)]. This problem is serious when there is an intense diamagnetic signal in the spectrum, such as the water signal or the diamagnetic envelope of a protein, in which the appearance of sine wave edges on this signal may obscure broad hyperfine-shifted signals nearby. This truncation can be a potential problem when acquiring spectra with large spectral windows, that is, with short dwell times and acquisition times. This problem may be alleviated by using a larger acquisition time or by an appropriate choice of window function (see Section II.A.4).

4. Choice of Window Functions

The application of a window function to FIDs is a versatile strategy for the manipulation of FT NMR spectra. As demonstrated for diamagnetic compounds, the S/N ratio and the resolution of the spectra can be selectively improved with the application of appropriate window functions that respectively emphasize the early and later sections of the FID. The shorter relaxation times associated with paramagnetic compounds require window function parameters that are very different. For example, the introduction of a 0.2-Hz line broadening by an exponential multiplication of the FID of a diamagnetic compound is sufficient to improve the S/N ratio of a ^{1}H NMR spectrum but is not effective for signals with line widths of hundreds of hertz; typically line-broadening factors of 5 Hz or larger are used. A 100-Hz line-broadening factor would not be too large (and is actually quite appropriate) to improve the S/N ratio of a signal of 2000 Hz! For paramagnetic complexes where most ^{1}H NMR signals are well resolved (Figs. 2

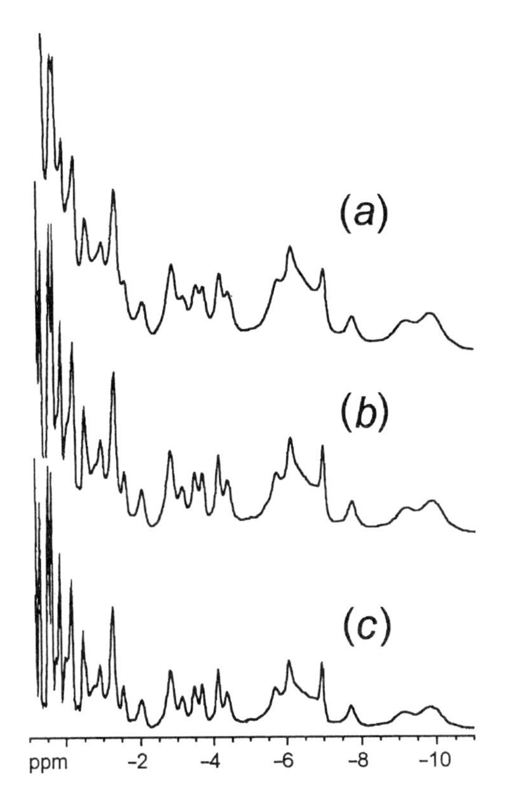

Figure 32
Upfield region of several ^{1}H NMR spectra (360 MHz, 298 K) of Yb^{3+} substituted pike parvalbumin obtained with the application of different window functions prior to FT: (*a*) exponential multiplication with a line-broadening factor of 10 Hz; (*b*) a 30° shifted sine squared bell, and (*c*) a 40% shifted Gaussian bell with a line-broadening factor of −30 Hz.

and 28), the use of an exponential function is sufficient to improve the S/N ratio without loss of signal resolution. However, for systems where signals can seriously overlap, such as metalloproteins, better spectral resolution can only be achieved with appropriate choice of the window functions for the FIDs. Figure 32 shows the upfield region of the ^{1}H NMR spectra of Yb^{3+} substituted pike parvalbumin obtained with the application of different window functions prior to FT. Spectra (*b*) and (*c*) demonstrate that the application of bell-shaped window functions such as sine and Gaussian functions can be more effective than a simple exponential line-broadening function [Spectrum (*a*)] in obtaining NMR spectra suitable for presentation. These functions emphasize different sections of the FID, and the resolution and line-broadening factors used in the window functions need to be adjusted to afford the desired spectra.

The FID truncation can cause a sine wave to appear near the baseline of large signals [Figs. 30(*d*) and 31]. This problem can be eliminated by using appropriate parameters for a window function, such as exponential multiplication with a large line broadening factor, a Gaussian window function with a more negative line broadening factor, or a sine bell function. The choice of an appropriate window function is important for processing 2D NMR spectra since the FIDs in all the 2D experiments are highly truncated owing to the use of limited sizes in both dimensions, usually with 1024×512 or fewer data points. This latter problem can also be alleviated on those spectrometers equipped with a program for extrapolation of a truncated FID.

5. Baseline Correction

Large spectral windows with short rf-pulse delay times are usually needed for the acquisition of NMR spectra of paramagnetic species. The use of these short time parameters results in the appearance of a rolling baseline, that is exacerbated by the presence of an intense signal such as the diamagnetic envelope and/or the residual solvent H_2O signal in paramagnetic protein samples. Often the distorted baseline has to be corrected to give better looking spectra and, more importantly, to reveal useful quantitative data. Spectrum (*e*) in Figure 30 exhibits a distorted baseline, making signal integration difficult. The distorted baseline also makes the measurement of T_1 values more difficult; larger standard deviations for the T_1 values are often observed than those obtained after baseline correction. Baseline distortion can seriously obscure the cross peaks in 2D NMR spectra, and needs to be corrected to obtain a clean 2D contour map.

6. Solvent Signal Suppression

As alluded to earlier, a common problem in obtaining NMR spectra of paramagnetic samples is the presence of resonances with a large range of relaxation times and signal intensities. The large intensity of the diamagnetic envelope, particularly from the solvent signal(s), can obscure the much less intense and broader paramagnetically shifted signals due to dynamic range considerations, that is, the ratio of the most intense signal to the weakest signal. For example, the signal intensities can differ by no more than 2000-fold for a 12-bit digitizer, that is, 2^{12-1} if 1 bit is used for the $\pm$ sign, or around 33,000 for a 16-bit digitizer, that is, 2^{15}. When the more slowly relaxing signals are unimportant, their intensity can be reduced by the use of a fast repetition rate or by the application of a presaturation pulse.[37] Such a strategy is often useful for the suppression of the solvent water signal in biological samples in H_2O solution [Fig. 33(*a*)]. Other useful water suppression techniques include the **W**ater **E**limination **FT** (WEFT) and the modified **D**riven **E**quilibrium **FT** (DEFT) multipulse sequences, selective excitation hard-pulse sequences, and the use of pulse field gradients discussed below.

While the simple presaturation technique is widely used for the suppression of the solvent signal, nearby resonances may also be partially saturated under these conditions. The WEFT pulse sequence uses the simple "inversion-recovery" $D1$-180°-τ-90°-FID sequence [Eq. (11) and Fig. 13] with the delay time τ chosen to give a null signal for the solvent. The signals affected by the paramagnetic center have much shorter relaxation times than the solvent and are thus not sup-

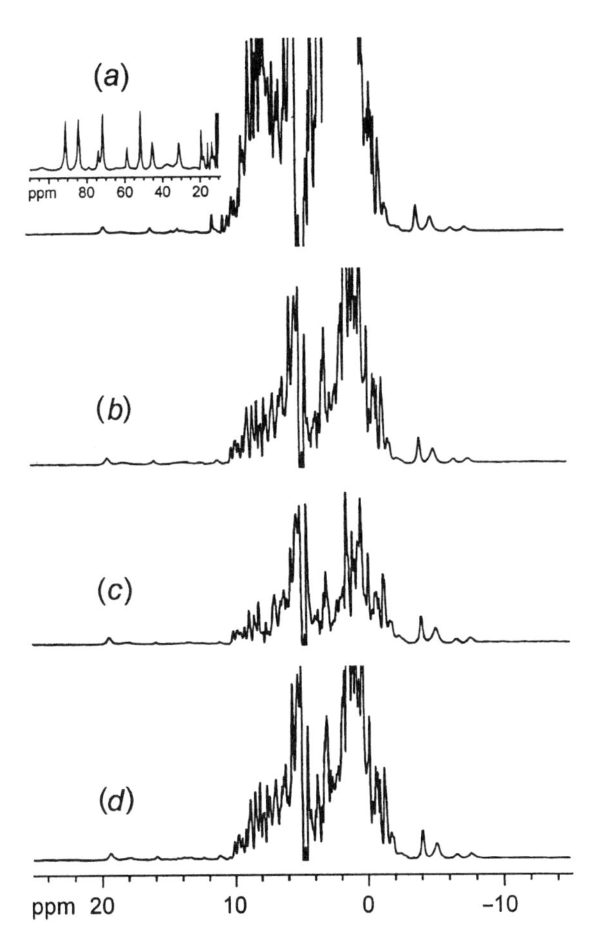

Figure 33
Isotropically shifted ^{1}H NMR spectra at 360 MHz of met-myoglobin in H_2O acquired by the use of different water suppression methods: (*a*) presaturation, (*b*) WEFT pulse sequence with $D1 = 80$ ms and $\tau = 90$ ms, (*c*) WEFT with $D1 = 10$ ms and $\tau = 40$ ms, (*d*) modified-DEFT pulse sequence with $D1 = 80$ ms and $\tau = 80$ ms, The acquisition time in all of the experiments is 40 ms. The inset in (a) is the isotropically shifted signals in the downfield region which are not affected by the pulse sequences.

pressed by the WEFT pulse sequence [Fig. 33(b)]. In the study of paramagnetic metalloproteins, the delay times $D1$ and τ in the WEFT pulse sequence can be significantly shortened (a "super-WEFT" sequence[38]) to give a much shorter repetition time of about $5T_1$ of the fast-relaxing signals. Thus, the signals with longer relaxation times in the diamagnetic envelope are also suppressed along with the water signal [Fig. 33(*c*)]. The resolution of the faster relaxing signals near the edge of the diamagnetic envelope can be significantly improved owing to the significant suppression of the slowly relaxing signals. Nuclear relaxation times can be obtained according to Eq. (36) when a super-WEFT sequence is applied

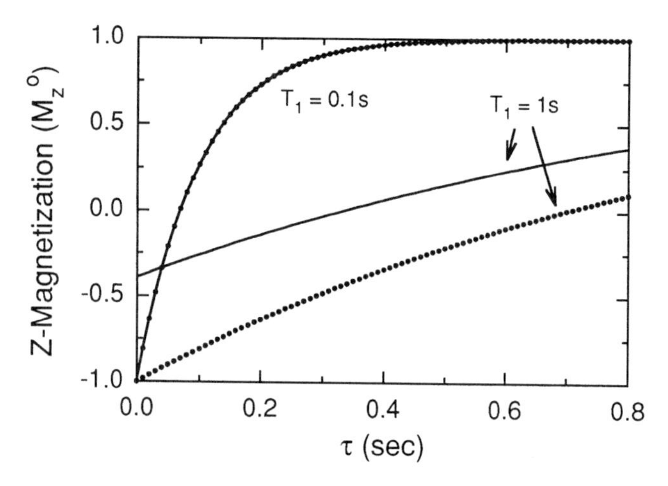

Figure 34
Recovery of isotropically shifted signals with T_1 of 0.1 s, and signals with a long relaxation time of 1 s in the WEFT (dotted lines) and "super-WEFT" (solid lines) pulse sequences according to Eqs. (11) and (36). The $D_1 = 0.5$ s applied here dramatically affects the slowly relaxing signals, but not the fast relaxing isotropically shifted signals with $T_1 = 0.1$ s.

for solvent suppression,

$$M_z(\tau) = M_z{}^{\circ}[1 - (2 - \exp(-D1/T_1))\exp(-\tau/T_1)] \tag{36}$$

where $D1$ is the relaxation delay time after the acquisition pulse and τ is the delay after the 180° pulse. Equation (36) becomes the same as that of Eq. (11) when a relaxation delay $D1 > 5T_1$ is applied, which is clearly shown in Figure 34, where the use of a 0.5-s delay affords a new equilibrium in which the null point for the 1-s signal (representing the solvent signal or/and diamagnetic signals) is shifted from about 0.7 s to about 0.33 s, whereas this short delay does not show any noticeable effect on the faster relaxing signal with $T_1 = 0.1$ s. One advantage of the super-WEFT pulse is that the time needed for the NMR experiment can become much shorter because of the use of shorter delay times, that is, $(0.5 + 0.33)$ s versus $(5 + 0.7)$ s for the suppression of signals with relaxation times of about 1 s.

The modified DEFT pulse sequence, $D1$-90°(θ)-τ-180°(2θ)-τ-90°(θ)-FID,[39] is designed to "drive" both M_z and $M_{x,y}$ magnetizations back to their equilibrium state after data acquisition and in the mean time to suppress the signals with long relaxation times. Again, the more slowly relaxing signals are highly suppressed when short τ values are used (when set to $>5T_1$ of the faster relaxing isotropically shifted signals). This pulse sequence is particularly useful for the observation of very broad isotropically shifted signals by using short delay times, as shown in Figures 33(d).

The recovery of signals with respect to the delay time τ in the modified DEFT

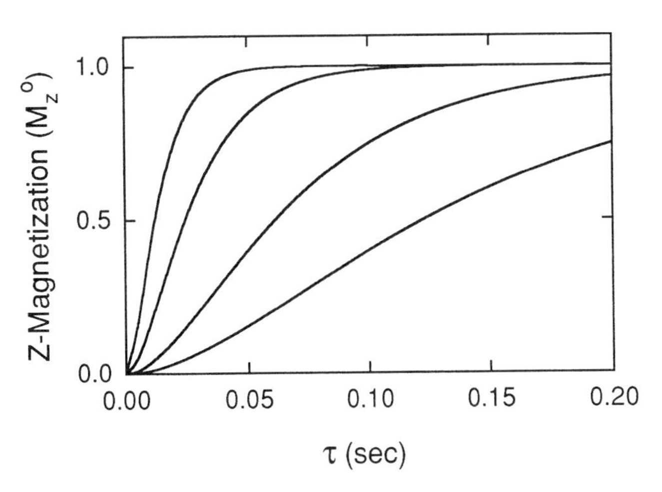

Figure 35
A plot of $M_z(\tau)$ against τ in the modified DEFT pulse sequence for isotropically shifted signals with relaxation times of 10, 20, and 100 ms and a slowly relaxing signal with $T_1 = 1$ s (from top to bottom).

pulse sequence is described by Eq. (37),

$$M_z(\tau) = M_z(\infty)[1 - 2e^{-\tau/T_1} + (1 - \cos\theta)e^{-2\tau/T_1}] \tag{37}$$

where a θ-degree (2θ) pulse is used as the $90°(180°)$ pulse. This equation can be better described as in Figure 35 for a few signals with different relaxation times and the use of $\theta = 90°$ pulses. When a 50-ms delay time τ is chosen, the signal with relaxation time of 10 ms recovers to 99% of its magnetization, the signal of 20 ms gains 84% of its intensity, but the 100-ms signal reaches 15% of its magnetization. A proton farther away from the paramagnetic center with a 1 s relaxation time recovers to 2.4% of its magnetization, thus the large diamagnetic envelope composed of slowly relaxing signals can be significantly suppressed and the isotropically shifted signals near the edge of the diamagnetic envelope can be more clearly revealed as shown in Figure 33(*d*).

Because the three water suppression methods discussed above are based on saturation of the water signal, saturation transfer from the water signal to solvent exchangeable signals may occur. Thus a signal that exchanges rapidly with solvent may also be suppressed along with the saturated solvent. If such a signal is important, a different approach for solvent suppression is required. One efficient way is to selectively excite only the signals of interest, leaving the solvent signal at its equilibrium state. Several binomial hard pulse sequences[40] have been developed for this purpose and consist of pulse trains separated by very short delays to give a cumulative acquisition pulse of less than or equal to 90°. Examples of such pulse sequence are 1-τ-1, 1-τ-2-τ-1, and 1-τ-3-τ-3-τ-1, where τ is set to be $1/2\Delta\nu$

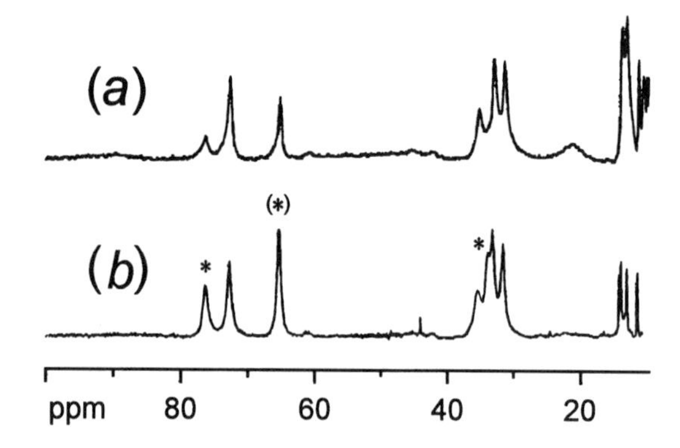

Figure 36
The use of a modified DEFT pulse sequence (*a*) and 1-3̲-3-1̲ pulse sequence (*b*) for the detection of fast solvent exchangeable isotropically shifted ^{1}H NMR signals in Ag^INi^{II}–SOD. [Adapted from Ming et al.[14]] The solvent exchangeable signals are labeled with asterisks, with the third one likely overlapped with the signal about 65 ppm. Note that the solvent exchangeable signal at 33.8 ppm in (*b*) is not observed in spectrum (*a*) by the use of a modified DEFT pulse sequence for sulvent suppression.

with $\Delta\nu$ being the frequency difference between the maximum excitation and the solvent; the numbers represent the relative lengths of pulses that add up to give a θ° acquisition pulse. To illustrate, the 1-τ-3-τ-3-τ-1 sequence consists of the following: $(\theta/8)^\circ$-τ-$(3\theta/8)^\circ$-τ-$(3\theta/8)^\circ$-τ-$(\theta/8)^\circ$. The carrier frequency is set to the position with maximum excitation and the null is adjusted to the solvent signal in the above pulse sequences. In practice, it is more convenient to set the carrier frequency on the solvent and alternate the phases of the pulses by 180°; for example, the 1-τ-$\underline{3}$-τ-3-τ-$\underline{1}$ pulse sequence (underlined pulse indicates a 180° phase shift) consists of $(\theta/8)^\circ$(x)-τ-$(3\theta/8)^\circ$(−x)-τ-$(3\theta/8)^\circ$(x)-τ-$(\theta/8)^\circ$(−x).

These pulse sequences are easy to use, and can be very useful for the observation of solvent exchangeable isotropically shifted signals when the signals are in fast exchange with the water. However, the narrow windows of even excitation of these pulse sequences prevent their being used for quantitative measurement of isotropically shifted features in a large range. Solvent exchangeable isotropically shifted signals in a spectral window of about 20 ppm can be studied quantitatively. Figure 36 shows an example of the successful application of a 1-3̲-3-1̲ pulse sequence for the detection of fast solvent exchangeable signals of a metal-substituted superoxide dismutase in H_2O, an important step for identifying ligands in metalloproteins of unknown structure.

Another useful solvent suppression method is the use of the pulsed-field gradient (PFG) techniques.[41] For high-resolution NMR studies, the presence of a field gradient implies some loss of resolution; that is, the term $\mathbf{B_0}$ in Eqs. (1) and (2) is replaced by $\Delta\mathbf{B_0}$, which generates a variation in the Larmor frequency of the nuclei across the sample. In other words, the presence of field inhomogeneity causes the transverse magnetization $M_{x,y}$ [Eq. (12)] to dephase faster, resulting in signal broadening. This dephasing effect can be reversed by a 180° refocusing pulse along the y axis after an evolution period τ following the initial $90°_x$ pulse. A spin echo is thus generated at time τ after the $180°_y$ pulse, that is, after the sequence $90°_x - \tau - 180°_y - \tau -$ "echo". The very different diffusion rates between macromolecular metalloproteins and water molecules ensures an efficient suppression of the intense water signal by using the PFG technique.

7. Temperature Control

The chemical shift of an isotropically shifted signal is inversely proportional to the temperature, and would extrapolate to its diamagnetic origin in most cases at infinite temperature as shown in Eqs. (5), (A.I.4), and (A.I.5), and Figure 27. Signals with large isotropic shifts would consequently by quite temperature dependent. It is thus important that the sample temperature be well controlled to avoid unnecessary broadening of the signal. This temperature dependent shift is particularly important in experiments where a large amount of power is applied to the probe, such as for a total correlation spectroscopy (TOCSY) experiment with an MLEV-17 spin-lock pulse sequence (cf. Section II.C.1).

B. Nuclear Overhauser Effect

The nuclear Overhauser effect (NOE) is useful for gaining insight into interactions between nuclei that are in close proximity through space but not necessarily through bonds. The effect is manifested as a change in intensity on one nucleus when a nearby nucleus is irradiated (see Appendix II for a more detailed discussion of the fundamental basis for NOE). The magnitude of the NOE on nucleus i upon saturation of nucleus j for a time period t can be calculated from Eq. (38), assuming that j is saturated at $t = 0$, that is, under the initial conditions of $i_z = i_z{}^0$ and $j_z = 0$ in Eq. (10) [or Eq. (A.II.7)],

$$\mathrm{NOE}(i) = (\sigma_{ij}/\rho_i)[1 - \exp(-\rho_i t)] \tag{38}$$

where τ_c is the rotational correlation time, ρ_i is the intrinsic relaxation rate of nucleus i, and σ_{ij} is the cross relaxation term

$$W_2 - W_0 = (\hbar^2\gamma^4/10r_{ij}{}^6)\left[\frac{6\tau_c}{1 + 4\omega^2\tau_c{}^2} - \tau_c\right]$$

where W_2 and W_0 represent relaxation pathways that involve double and zero quantum transitions, respectively, that is, double spin flips [cf. Eqs. (A.II.1)–(A.II.4)], which can be simplified to $-\hbar^2\gamma^4\tau_c/10r_{ij}{}^6$ when $\omega\tau_c \gg 1$ for macromolecules. Internuclear distances can thus be obtained by the measurement of the NOE-buildup NOE(i) with respect to the saturation duration time t, where τ_c can be estimated from the Stokes–Einstein correlation [Eq. (A.II.5)]. Steady-state NOE is reached when the saturation time is long ($t \gg 1/\rho_i$). In this case, NOE is independent of t and Eq. (38) becomes Eq. (39). On the other hand, Eq. (33) can be simplified to Eq. (40) when the saturation duration time is short ($t \ll 1/\rho_i$),

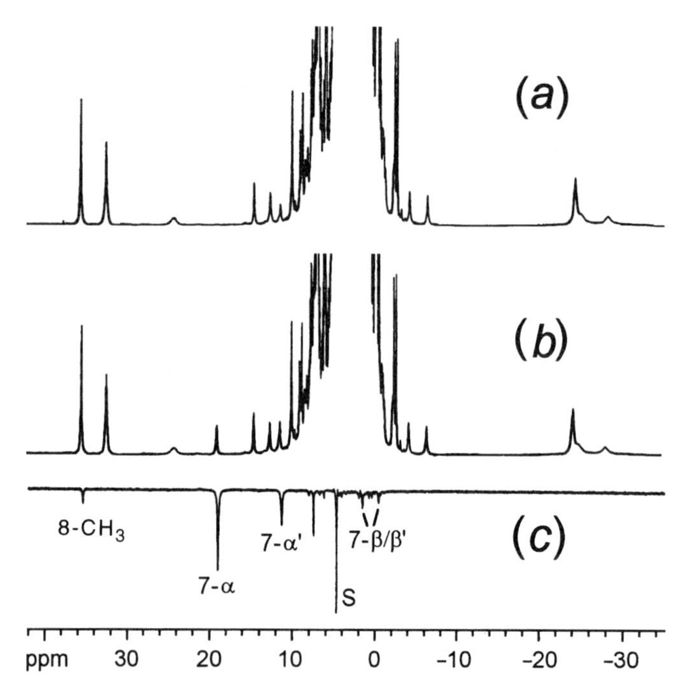

Figure 37
The ^{1}H NMR spectra (360 MHz, 298 K) of bovine Fe^{III}–cytochrome *c* in D_2O with irradiation set on (*a*) and off (*b*) the signal of interest, here one 7-$C_\alpha H_2$ proton, and the NOE difference spectrum (*c*) to show NOEs with 8-CH_3, another 7-$C_\alpha H_2$, and 7-$C_\beta H_2$ protons.

and there is a linear relationship between the NOE observed and t. The slope in this linear correlation [Eq. (40)] is independent of the paramagnetic relaxation rate (ρ_i) of the system, and thus the NOE is only proportional to $r_{ij}{}^{-6}$.

$$\mathrm{NOE}(i) = \sigma_{ij}/\rho_i \tag{39}$$

$$\mathrm{NOE}(i) = \sigma_{ij}t \tag{40}$$

The NOE can be clearly revealed by a difference spectrum obtained by subtraction of a control spectrum with irradiation at a reference position from the spectrum with the signal of interest irradiated (Fig. 37). The NOE difference spectrum can be obtained by the acquisition of a difference FID with the decoupler set alternately on (subtracted) and off (additive) resonance of the signal of interest for a fixed number of transients until a difference spectrum of sufficient S/N ratio is obtained (pulse sequence shown below),

$$[(D1\text{-}D2(\mathrm{off})\text{-}90^\circ - \mathrm{FID})_m - (D1\text{-}D2(\mathrm{on})\text{-}90^\circ\text{-}\mathrm{FID})_m]_n \tag{41}$$

where $D1$ is the relaxation delay time, and $D2$ is irradiation duration time. The choice of a small m value and a large recycling number n have taken into account small variations in the sample over a period of time and ensure good signal averaging. The acquisition of a difference FID is particularly useful for the detection of a very weak NOE or for when the dynamic range of the digitizer is a concern, as is usually the case for the detection of the NOEs in metalloproteins in H_2O solutions.

When a steady-state NOE detection is performed in H_2O, the nonselective WEFT sequence can be used for the suppression of the large H_2O signal (cf. Section II.A.6). The WEFT pulse sequence has been very useful for the detection of NOE of paramagnetic metalloproteins in H_2O (pulse sequence 42), where a saturation pulse on the signal of interest (subtracted) and at a reference position (additive) are alternately turned on for a certain period of time $D2$ during the delay $(\tau + D2)$ after the 180° inversion pulse.

$$[(D1\text{-}180^\circ\text{-}\tau\text{-}D2(\text{off})\text{-}90^\circ\text{-FID})_m - (D1\text{-}180^\circ\text{-}\tau\text{-}D2(\text{on})\text{-}90^\circ\text{-FID})_m]_n \qquad (42)$$

In this experiment, the use of the WEFT sequence with a small m value can overcome the computer dynamic range problem. Alternatively, the polynomial solvent-supression hard pulse sequences, such as $1\text{-}\underline{1}$, $1\text{-}\underline{3}\text{-}1$, and $1\text{-}\underline{3}\text{-}3\text{-}\underline{1}$ pulse sequences (cf. Section II.A.6), can replace the 90° observation pulse in the pulse sequence (41) for the detection of the NOE in H_2O solutions with the excitation maximum set around the signals that exhibit the NOEs with the irradiated signal.

Internuclear distances can be estimated from NOE intensities. However, because NOE is also a function of the correlation time τ_c [Eqs. (38) and (A.II.6), and Fig. A.II.2], an erroneous estimate of τ_c can result in inaccurate NOE based distance measurements. Such errors can arise when coagulation occurs (affording larger τ_c values) or there is local motion for the nuclei of interest (giving smaller τ_c values). In practice, the correlation time can be approximated by the Stokes–Einstein equation [Eq. (A.II.5)] if coagulation is not a concern, or can be derived from the NOE build-up of an easily recognized proton pair with a fixed distance such as a CH_2 group (~1.9 Å) on a ligand (Fig. 5). An example of the latter is shown in Figure 38 for Yb^{3+} substituted bovine α-lactalbumin, where the fits of the NOE buildups with respect to the saturation duration time [Eq. (38)] using a correlation time of 6 ns for this 14-kDa protein affords distances of 1.58 and 1.94 Å for the two pairs. These values are too short to be assigned to any interproton distance in this protein. Alternatively, the use of 12 ns for τ_c, which corresponds to a molecular size double that of a monomeric α-lactalbumin of 14 kDa, gives values of 1.78 and 2.17 Å, which are close to those of a geminal and a vicinal pair, respectively.[42] Coagulation of this protein at the concentrations used for the NMR experiments is thus suggested, in agreement with the results observed by other methods. An NOE build-up experiment can also provide information about local motion of a moiety in paramagnetic metalloproteins. For example, the cross relaxation of −6.1 Hz between the 2-H_β(cis) and 2-H_β(trans) protons of the 2-vinyl moiety in myoglobin obtained from truncated NOE studies is significantly smaller than the value expected ($-14\ s^{-1}$) for a protein with a rotational correlation time in the range of about 8–10 ns deter-

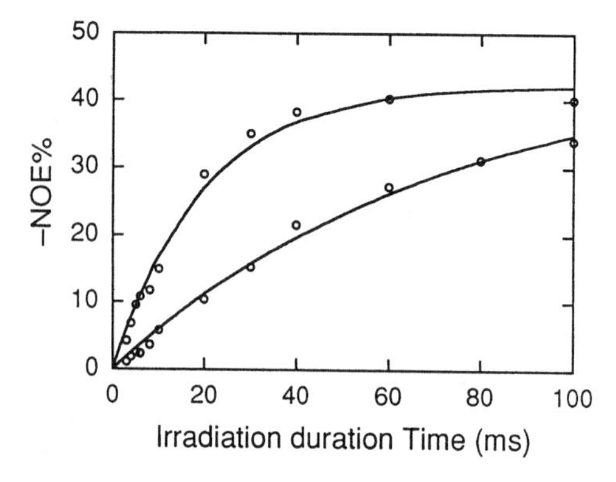

Figure 38
The NOE build-up of proton pairs in Yb^{3+} substituted α-lactalbumin acquired at 360 MHz and 298 K. [Adapted from Ming[42]]. Two groups of proton pairs, geminal and vicinal, are shown here which exhibit two sets of curves with different initial slopes [cf. Eq. (40)]. The data can be resonably fitted only by the use of a correlation time (12 ns) that is double than the estimated value for this 14-kDa protein, suggesting the formation of dimers under the experimental conditions.

mined by several different methods. This observation indicates that the vinyl group in question is significantly more mobile than the protein as a whole.[43]

The fast nuclear relaxation rates ρ_i in paramagnetic species may render the NOE extremely small, and in many cases not detectable. This finding is particularly true for a small metal complex with short molecular rotational correlation time τ_c (affording $\omega\tau_c \ll 1$). However, the detection of the NOE becomes possible when a molecule tumbles slowly, affording a larger τ_c and thus a more negative cross-relaxation term [Eq. (38)]. In practice, increasing the solution viscosity of a sample may increase τ_c to allow observation of the NOE. For example, a typical NOE of about −7% observed for the geminal pair 6-CH_2 in the 16-kDa high-spin met–myoglobin can be enhanced to about −14% when the measurement is conducted in a viscous 30% ethylene glycol solution where the rotaional correlation time of the protein is doubled.[44]

C. 2D NMR Studies of Paramagnetic Metal Complexes and Metalloproteins

The application of 2D NMR techniques to paramagnetic samples extends the utility of NMR for investigating such systems. This section will illustrate some of these approaches.

1. Through-Bond Correlation

a. COrelated SpectroscopY (COSY) As noted earlier, chemical shift information from paramagnetic samples can be difficult to interpret. Signal assignment of paramagnetic species is facilitated by the measurement of distance-dependent relaxation times, distance and geometry-dependent dipolar shifts, and/or the use of isotope labeling or atom-substitution experiments that can be quite time consuming. Since signal line broadening, due to the proximity of the nucleus to a paramagnetic center, usually obliterates the possibility of observing multiplets due to spin–spin coupling, through-bond connectivities are not likely to be established in paramagnetic species based on 1D spectra. Conversely, 2D shift correlation experiments such as COSY are particularly useful for establishing through-bond connectivities via the detection of cross peaks, even in the case where spin multiplets are not observable as in paramagnetic molecules. The simple COSY pulse sequence consisting of $D1$-90°-t_1-90°-AQ, where $D1$ is the relaxation delay and t_1 is a series of incremental time delays to build up the second dimension, can be applied to paramagnetic molecules with suitable adjustment of the parameters to compensate for the short T_1 values of the nuclei. The detection of a pair of COSY cross peaks in the 2D contour map indicates the presence of scalar coupling between the two nuclei related by the cross peaks, and is equivalent to the observation of spin-spin coupling in a 1D spin-decoupling experiment. The intensity of the cross-peak (I_c) can be described as in Eq. (43),

$$I_c = \sin(\pi J_{ab} t_1) \exp(-t_1/T_2) \quad (43)$$

where J_{ab} is the through-bond scalar coupling constant, T_2 is the spin–spin relaxation time, and t_1 is the evolution time for the second dimension. The full development of the coherence transfer occurs when the sine term in Eq. (43) is maximum or when $t_1 = (2J_{ab})^{-1}$ to give $\sin(\pi J_{ab} t_1) = 1$. Thus longer t_1 values are required to observe cross peaks derived from weaker coupling interactions; and the larger the T_2 values (the broader the signals), the weaker the cross peaks. As a result, the chances of observing COSY cross peaks are higher for more strongly coupled nuclei with larger J_{ab} values and for sharper signals with larger T_2 values. Typically the choice of a short acquisition time in the COSY experiment can compensate for the short transverse relaxation time T_2 of the target nuclei and allows the detection of coherence transfer cross peaks in paramagnetic species. In the same time, the use of short repetition times of about 1–2 times the average T_2 of the isotropically shifted features allows the accumulation of a larger number of transients to improve the S/N ratio, which is crucial for the observation of weaker cross peaks.

Figure 39 shows the COSY spectrum of the complex $[Fe^{II}Mn^{II}(BPMP)(CH_3CH_2CO_2)_2]^+$, which illustrates the utility of the COSY experiment.[33] Note the presence of two sets of cross-peaks connecting each methyl signal with two inequivalent methylene signals (Pr1s and Pr2s). These are assigned to the protons of the two bridging propionates, which are distinct from each other because one propionate coordinates more strongly to the Fe^{2+} ion and the other to the Mn^{2+} ion. There are also cross peaks arising from the β, γ, and β' protons of the four distinct pyridine rings on the complex, labeled as Py(Fe or Mn) in the spectrum, in which the γ proton exhibits cross peaks with both β and β' protons. Cross peaks originating from the pyridine α protons are not observed due to their short T_2 values. In general, vicinal proton pairs with signal widths of less than 150 Hz and scalar coupling constants larger than a few hertz usually provide clear cross

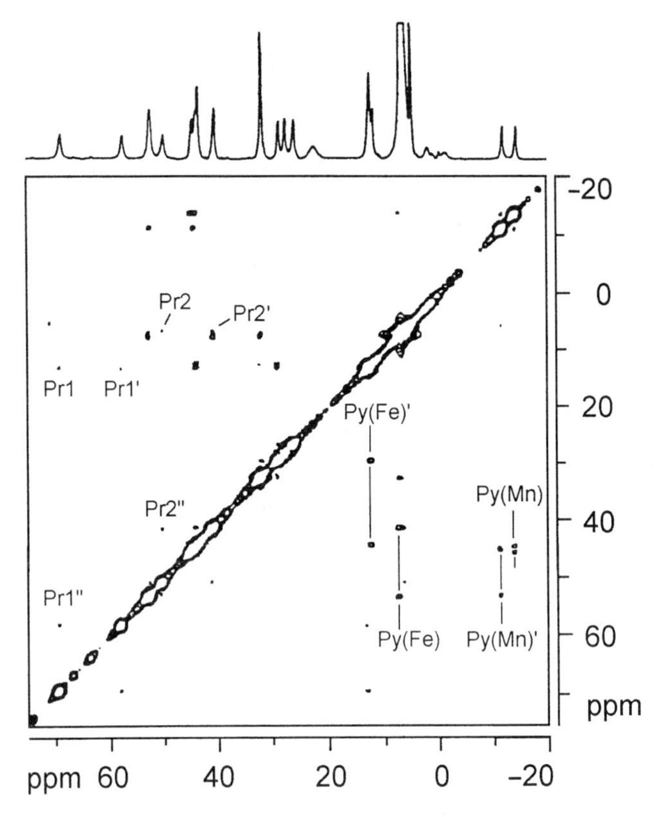

Figure 39
Proton COSY spectrum of $[Fe^{II}Mn^{II}(BPMP)(CH_3CH_2CO_2)_2]B(C_6H_5)_4$ in CD_2Cl_2 (300 MHz, 1024 × 256 data points). [Adapted from Wang et al.[33]] The cross peaks in the two bridging propionate spin systems are labeled as Pr1(2) and Pr1(2)′ for CH_3–$C_\beta H_2$ interactions and Pr1(2)″ for $C_\beta H$–$C_\beta H'$ interactions, and the four pyridine spin systems are labeled as Py(Fe) and Py(Mn).

peaks. The assignment of the pyridine rings to each of the two metal sites can be achieved by comparison with the spectra and relaxation properties of other dinuclear $Fe^{II}M$ complexes.

b. TOtal Correlation SpectroscopY (TOCSY) This spectroscopy is the rotating frame version of COSY and allows the detection of coherence transfer among nuclei within one interconnected spin system. The TOCSY pulse sequence consists of $D1$-90°-t_1-SL-AQ, where $D1$ is the relaxation delay, SL a spin-lock pulse sequence that serves as the "mixing time", and t_1 the evolution time. This technique is particularly useful for grouping together resonances that belong to a particular amino acid residue in protein NMR studies. For example, TOCSY allows the observation of a cross peak between the glutamate $C_\alpha H$ and $C_\gamma H$ protons that are only weakly coupled to each other but are both strongly coupled

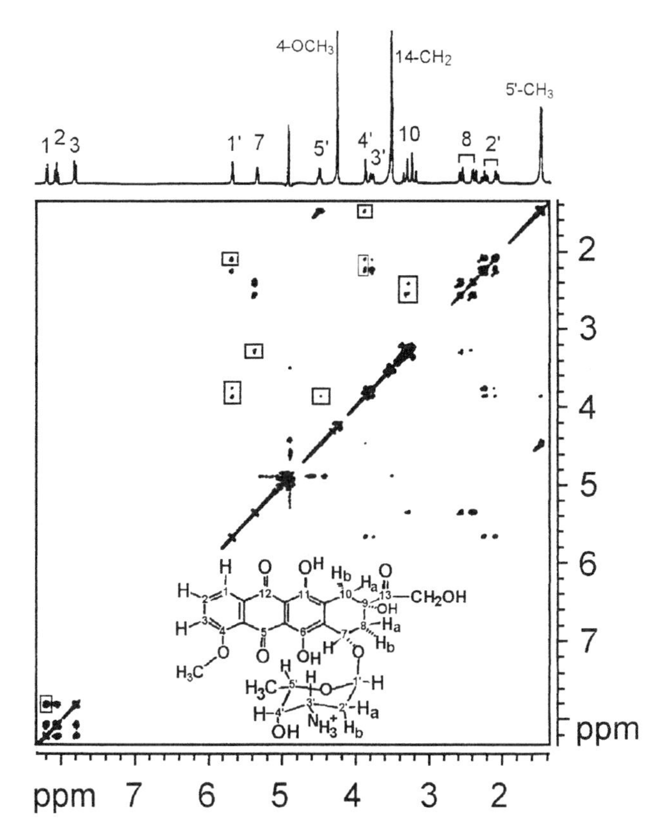

Figure 40
The TOCSY spectrum (360 MHz) of the antitumor drug adriamycin in methanol-d_4. The structure of the drug is similar to that of daunomycin in Figure 10, with the C(O)–CH_3 group replaced by C(O)–CH_2OH. The boxed cross signals are not detected in a simple COSY spectrum.

to the $C_\beta H$ proton. The spectrum of the diamagnetic sugar-containing anthracycline antitumor drug adriamycin (Fig. 40) better illustrates the advantages of TOCSY. This spectrum shows correlations not observable in a COSY experiment between the 1′-H and 3′-H/4′-H protons of the sugar moiety and between the 7-CH and 8-CH_2/10-CH_2 protons (boxed signals). With appropriate adjustment of acquisition parameters, such as a shorter relaxation delay and spin-lock times, the TOCSY technique can also be applied to the study of paramagnetic molecules (see below). However, owing to the broadness of the signals in paramagnetic species, some weak interactions may not be revealed in the TOCSY spectra.

The TOCSY experiments can also exhibit cross peaks that are due to chemical exchange; hence, care has to be taken to distinguish between features due to through-bond correlation and those due to chemical exchange. This differentiation can be achieved by comparing TOCSY spectra with COSY and EXchange

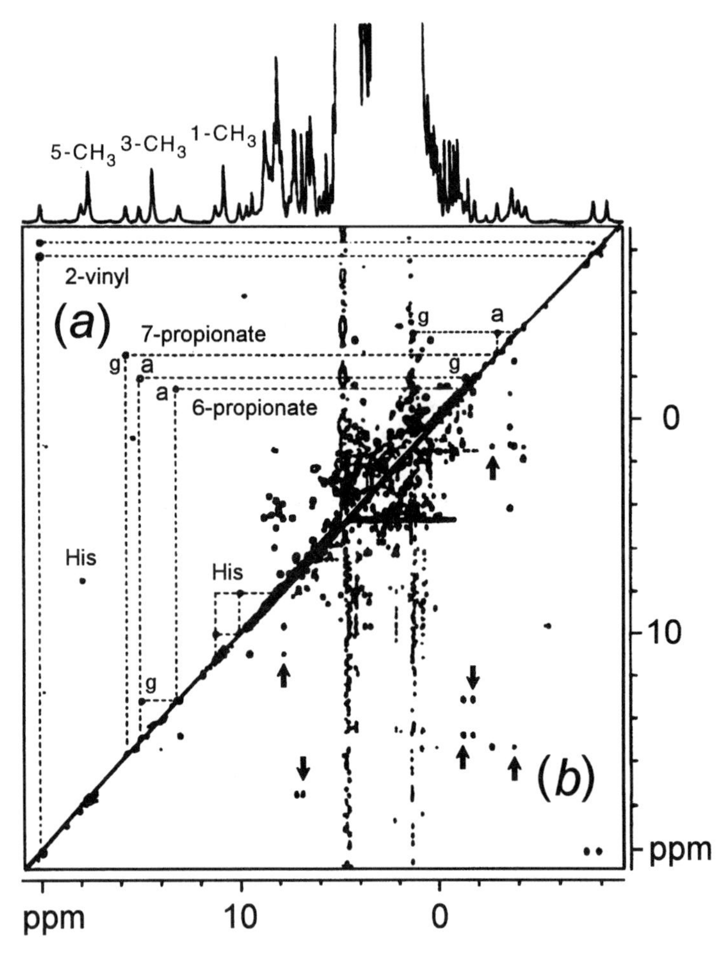

Figure 41
The 360 MHz ^{1}H COSY at 298 K (*a*) and TOCSY at 306 K (*b*) spectra of an overexpressed truncated heme domain (ferricytochrome *b*) of *Chlorella* nitrate reductase. [Adapted from Wei et al.[45]] The two propionate proups on the heme are labeled. The cross peaks marked with "g" originate from geminal pairs, while those marked with "a" are attributable to antivicinal pairs. The vicinal anti-pairs have relatively large scalar coupling constants and are clearly revealed in the COSY spectrum, whereas the weakly coupled vicinal syn-pairs are only visible in the TOCSY spectrum (marked with arrows).

SpectroscopY (EXSY) (see later) spectra; only through-bond correlation is revealed in the COSY spectra and both chemical exchange and through-space interactions are revealed in the latter. Often there is a lack of sufficient power to apply a spin-lock field for a large spectral window, so this technique is not very useful for the study of paramagnetic species with large spectral windows.

Both COSY and TOCSY can also be applied to metalloproteins as demonstrated in Figure 41 for the heme domain of nitrate reductase.[45] Since the isotropically shifted signals due to the protons on the heme, which is bound to the protein via two axial His residues (cf. structure in Fig. 12), are relatively sharp, the use of an acquisition time of about 30 ms at 360 MHz in a magnitude COSY experiment and a 25-ms spin-lock time in the TOCSY experiment can provide spectra of reasonably good quality. In the COSY spectrum, only strongly coupled pairs such as the geminal protons and the vicinal protons with anti configurations on the propionate side chains exhibit detectable cross peaks, whereas cross peaks are detected among all the protons on these groups in the TOCSY spectrum. The configuration of the propionate groups can be determined on the basis of the intensity of the cross peaks in the COSY spectrum. In this case, the heme and the carboxylate group are anti to each other for one propionate (with 2 anti vicinal pairs to give 2 cross peaks labeled "a"), but are syn to each other in the other propionate (affording only 1 anti vicinal pair). The signals at 20.2, −8.3, and −7.6 ppm of a three-spin system can be assigned to a vinyl group on the heme with the first two being the trans pair that exhibit larger scalar coupling, thus more intense cross peaks. The full assignment of the methyl, vinyl, and propionate groups to a specific position on the heme can be achieved by the use of the NOESY and heteronuclear multiple quantum correlation (HMQC) techniques discussed in Sections II.C.1.c and II.C.2.a.

c. Heteronuclear Multiple-Quantum Correlation (HMQC) and Other Methods Besides COSY and TOCSY, there are many other pulse sequences for the observation of coherence transfer, such as phase-sensitive COSY, DQF–COSY (Double-Quantum-Filtered **COSY**) and multiquantum versions like HMQC and HMBC (Heternuclear Multiple-Bond Correlation), that may be useful for signal assignment. Phase-sensitive COSY and DQF-COSY can be used to provide better spectral resolution in the region near the edge of the diamagnetic envelope. There are features with small paramagnetic shifts in this region of the spectra of low-spin Fe^{III}–hemeproteins (Figs. 12 and 33), Fe–S proteins (Fig. 6), and paramagnetic Ln^{3+}-substituted proteins (Fig. 32). In most cases, overlapped isotropically shifted signals may be resolved by performing the experiments at different temperatures due to their different temperature dependences. The DQF–COSY can also suppress diagonal signals (due to their antiphase characteristics) to afford better resolution for the cross peaks near the diagonal signals, and further reduction of the t_1 noise along the second dimension in the spectra. In general, these techniques cannot be applied to broad isotropically shifted signals but are useful for establishing coherence transfer among isotropically shifted signals with less than 100 Hz line width.

Inverse detection techniques (such as the $^{1}H\{X\}$-HMQC[46] and hetero-TOCSY[47] techniques) allow less sensitive heteronuclei to be studied. One bioinorganic application of such techniques is the study of Zn sites in proteins by the use of ^{113}Cd as a probe.[48] Since only the protons scalarly coupled to the ^{113}Cd can be detected, the identity of the coordinated ligands can be determined by means of $^{1}H\{^{113}Cd\}$-HMQC.

For paramagnetic metalloproteins, $^{1}H\{^{13}C\}$-HMQC can be used to identify proton signals that are buried in the diamagnetic region by taking advantage

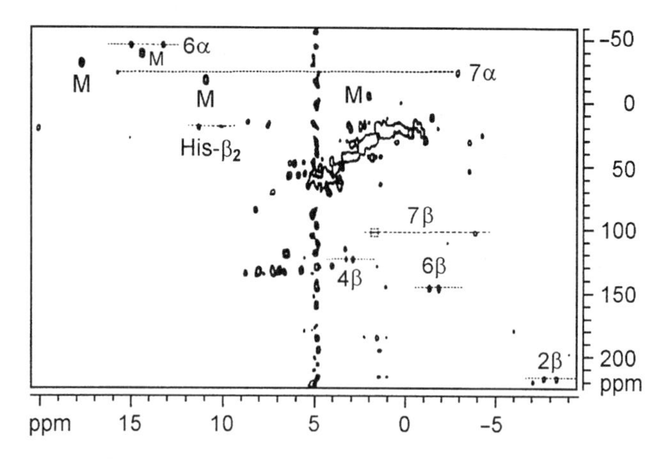

Figure 42
The $^1H\{^{13}C\}$-HMQC spectrum (360 MHz for ^1H, 298 K) of the truncated heme domain of nitrate reductase in D_2O. [Adapted from Wei et al.[45]] The 4 heme–methyl signals, including the 8-CH_3 peak buried in the diamagnetic region, are clearly assigned based on their intensity (labeled with "M"). Seven geminal pairs attributable to the two vinyl, two propionate, and one His–$C_\beta H_2$ groups can also be clearly recognized, since one ^{13}C signal is correlated with two ^1H signals as shown by the dotted lines in the spectrum. The dotted box represents a signal for 7β with a lower intensity that is not revealed at this contour level.

of the large spectral window of ^{13}C NMR to reveal these buried signals, as illustrated in Figure 42 for the heme domain of nitrate reductase. The 8-CH_3 and 4-vinyl heme protons of this protein, which are buried in the diamagnetic envelope (Fig. 41), can be clearly detected in the $^1H\{^{13}C\}$ HMQC spectrum. Unambiguously identified are the methyl signals with their rather intense cross peaks and the geminal pairs which are connected to a common carbon atom. Thus inverse-detection techniques should also be considered as indispensable tools for detailed NMR investigation of paramagnetic molecules.

The HMBC is used for the detection of long-range ^1H–^{13}C couplings (e.g., $^2J_{CH}$ and $^3J_{CH}$) in diamagnetic compounds, but this can be a time consuming task, owing to the low natural abundance of ^{13}C and the weakness of these couplings. The low intensities of isotropically shifted signals in paramagnetic molecules can further impede the detection of such features. The inclusion of PFG in 2D experiments allows the elimination of phase cycling used to subtract out unwanted signals. Consequently, t_1 noise is 2D maps can be significantly reduced, which increases the signal detection limit. The PFG–HMBC method has been applied to the study of a low-spin Fe^{III}–porphyrin complex (Fig. 43).[49] The HMBC map shown in Figure 43(a) acquired in a short time period of 3.5 h clearly shows the expected $^2J_{CH}$ cross-peaks between the pyrrole protons and the α carbons, as well as two $^3J_{CH}$ cross-peaks (boxed). The results obtained by the use of PFG are suprior to those obtained by the conventional method [spectra (b) and (c)] and clearly demonstrate that the benefits of PFG for the detection of

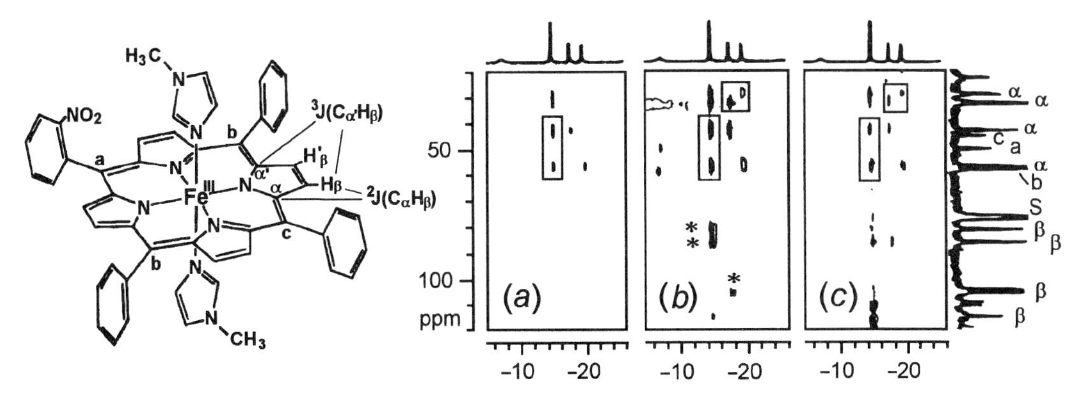

Figure 43
The HMBC of the pyrrole signal region of a 25-mM low-spin Fe^{III}–porphyrin complex [bis(*N*-methylimidazole 5-(*o*-nitrophenyl)-10,15,20-triphenyl porphyrinato)iron(III), structure shown] with PFG (*a*) 1024 transients per FID (3.5 h), (*b*) 8192 transients per FID (29.5 h), and (*c*) with phase cycling and 8192 transients per FID (28.9 h) obtained at 299.92 MHz for ^{1}H. [Adapted from Skidmore and Simonis.[49]] The boxed signals are attributable to the ^3J couplings between the pyrrole protons and the "distal" pyrrole α carbons, and the other cross peaks are due to the ^2J coupling between pyrrole protons and the "proximal" pyrrole α carbons as indicated on the structure. Due to the twofold symmetry of the molecule, there are four pairs of magnetically inequivalent pyrrole protons (with two overlapped signals). The asterisked signals in (*b*) are due to unsuppressed ^1J coupling between the pyrrole proton and the attached carbon.

long-range interactions. This study suggests that PFG–HMBC can be a useful tool for conclusive signal assignment of paramagnetic molecules of high concentrations with relatively sharp features.

2. Through-Space Correlation

a. Two-Dimensional Nuclear Overhauser and Exchange SpectroscopY (NOESY) These experiments are useful for the study of macromolecules to observe cross relaxations among all proton pairs in close proximity. As shown in Eq. (38), the large τ_c values of protein molecules due to slow tumbling can give rise to significant negative NOEs in paramagnetic metalloproteins (cf. Fig. 37). Thus, with the proper choice of acquisition parameters, cross-signals can be detected in NOESY spectra of paramagnetic metalloproteins with relaxation times as short as a few milliseconds, such as in high-spin heme proteins and Co^{2+} substituted proteins. A more detailed discussion on the background of NOE and NOESY can be found in Appendix II. Since the proton relaxation times of isotropically shifted signals in paramagnetic metalloproteins may span a wide range from a few tenths of a millisecond to several hundred milliseconds, the choice of one optimum mixing time for the detection of cross peaks associated with all the isotropically shifted signals is difficult in practice. In most cases, the window for the optimum mixing time is narrow in the NOESY experiments of paramagnetic metalloproteins (Appendix II.B, Fig. A.II.5), and the acquisition of a few spectra with different mixing times becomes necessary to cover signals with different relaxation times. In addition, cross peaks due to chemical exchange can also be detected (see Section II.C.3).

The pulse sequence for the NOESY experiments is comprised of $D1$-90°-t_1-90°-τ_{mix}-AQ(FID), where the delay τ_{mix} is the mixing time to allow the NOE to build up. The typical 300-ms mixing time for the observation of long-range NOEs in a NOESY experiment of a diamagnetic protein cannot be used for the study of isotropically shifted signals with relaxation times of about 10 ms, since the cross-relaxation for NOE buildup cannot successfully compete with the fast paramagnetic relaxation rate! The readers are referred to Eqs. (A.II.7)–(A.II.10) and Figures A.II.5 and A.II.6 for a detailed discussion and illustration of this competition. In practice, a mixing time is chosen that is not much greater than the average relaxation time (T_{1av}) of the nuclei being studied, and a short recycle time of about $2T_{1av}$ is chosen to acquire enough transients to provide a sufficient S/N ratio for broad signals. A NOESY spectrum of cytochrome *c* is shown in Figure 44. Owing to the relatively long relaxation times (e.g., ~70–80 ms for 8- and 3-CH_3 protons) of some of the isotropically shifted signals in this protein, a number of cross signals can still be observed in the NOESY spectrum with a mixing time of 60 ms. Through-space interactions between the 3-CH_3 and 4-$C_\beta H_2$ protons and between 7-propionate and 8-CH_3 protons are observed in this NOESY spectrum as indicated [spectrum (*a*)]. A complete assignment of the signals to specific groups on the heme is thus possible.

The NOESY can also be used for the assignment of broad signals with relatively short T_1 values for which the coherence transfer techniques discussed in Section II.C.1 are ineffective. An example is shown for the Ni^{2+} substituted derivative of Cu,Zn–superoxide dismutase (Cu,Ni–SOD, T_1 ~4–30 ms for protons on the coordinated ligands).[50] The magneitc interaction (Section I.D) between Ni^{2+} and Cu^{2+} in this protein results in well-defined isotropically shifted signals attributable to the His ligands on the Cu^{2+} site, where all the ring protons can be observed clearly (Fig. 29). The NOESY spectrum of Cu,Ni–SOD reveals a few cross peaks originating from the broad His ring protons (Fig. 45), and the coordination modes of the His residues on the Cu site can also be unambiguously recognized (see structure in Fig. 45).[50] For example, the $N_\varepsilon H$ proton of the N_δ-coordinated His 44 shows cross-relaxation with both adjacent $C_\delta H$ and $C_\varepsilon H$ protons (cross peaks 3 and 4), whereas the $N_\delta H$ protons on the N_ε coordinated His residues (His 118 and His 46) show cross peaks only with the adjacent $C_\varepsilon H$ protons (cross peaks 2 and 5) but not with the more distant $C_\delta H$ protons. The NOESY techniques have also been successfully applied to several other paramagnetic metalloproteins that have isotropically shifted signals with T_1 values in the range of only a few milliseconds to greater than 100 ms. These include Fe^{III}–heme proteins, Co^{2+} and Ni^{2+} substituted proteins, and several Ln(III) substituted Ca proteins, thereby demonstrating the broad applicability of NOESY techniques for the study of paramagnetic metalloproteins.

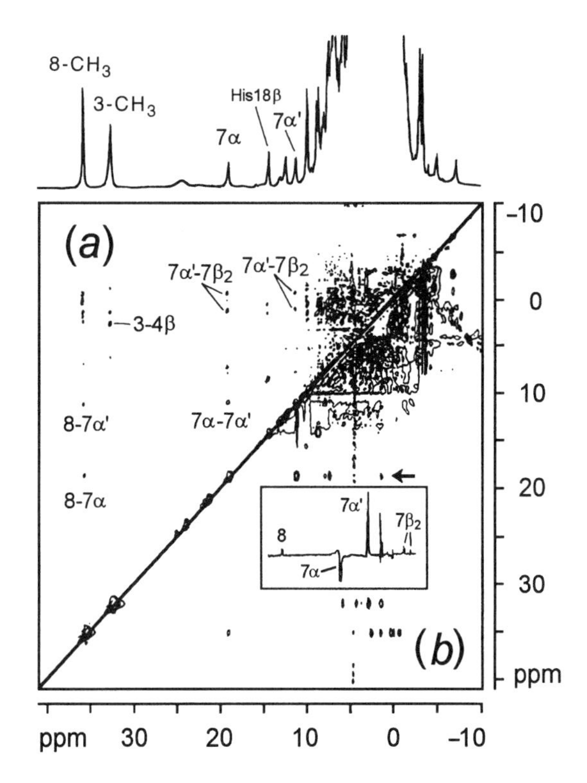

Figure 44
(*a*) The NOESY spectrum (360 MHz at 298 K, and a mixing time of 60 ms), and (*b*) Rotating Frame NOE SpectroscopY (ROESY) spectrum (300 MHz at 298 K, with a spin-lock time of 30 ms) of horse heart cytochrome *c*. The inset in (*b*) is a trace from the ROESY spectrum (marked with an arrow), which shows that the cross peaks (i.e., ROEs) in the ROESY spectrum are positive, whereas the cross peaks (NOEs) in the NOESY spectrum are negative in this protein as shown in Figure 37(*c*).

b. Two-Dimensional Rotating Frame NOE SpectroscopY (ROESY) Another 2D NMR technique is ROESY, which is essentially the NOESY experiment performed in the rotating frame (cf. Appendix II.C for a more detailed discussion on the rotating-frame NOE, the ROE.). One advantage of ROE over NOE is that cross relaxation in the former is always positive for all molecules [Eq. (A.II.21)] and Figure A.II.7 in Appendix II.C). Thus there is no null point as in the case of NOE when $(\omega\tau_c)^2 = 1.25$ [Eq. (A.II.6)]. Moreover, cross peaks due to chemical exchange (see Section II.C.3) are always negative regardless of the value of $\omega\tau_c$ and can be readily distinguished from the ROE cross peaks. Figure 44 compares the NOESY and ROESY spectra of oxidized horse heart cytochrome *c* ($\sim$12 kDa).

The ROESY pulse sequence consists of 90°-t_1-SL-FID. The SL or spin-lock segment is equivalent to the mixing time (τ_{mix}) in the NOESY pulse sequence and

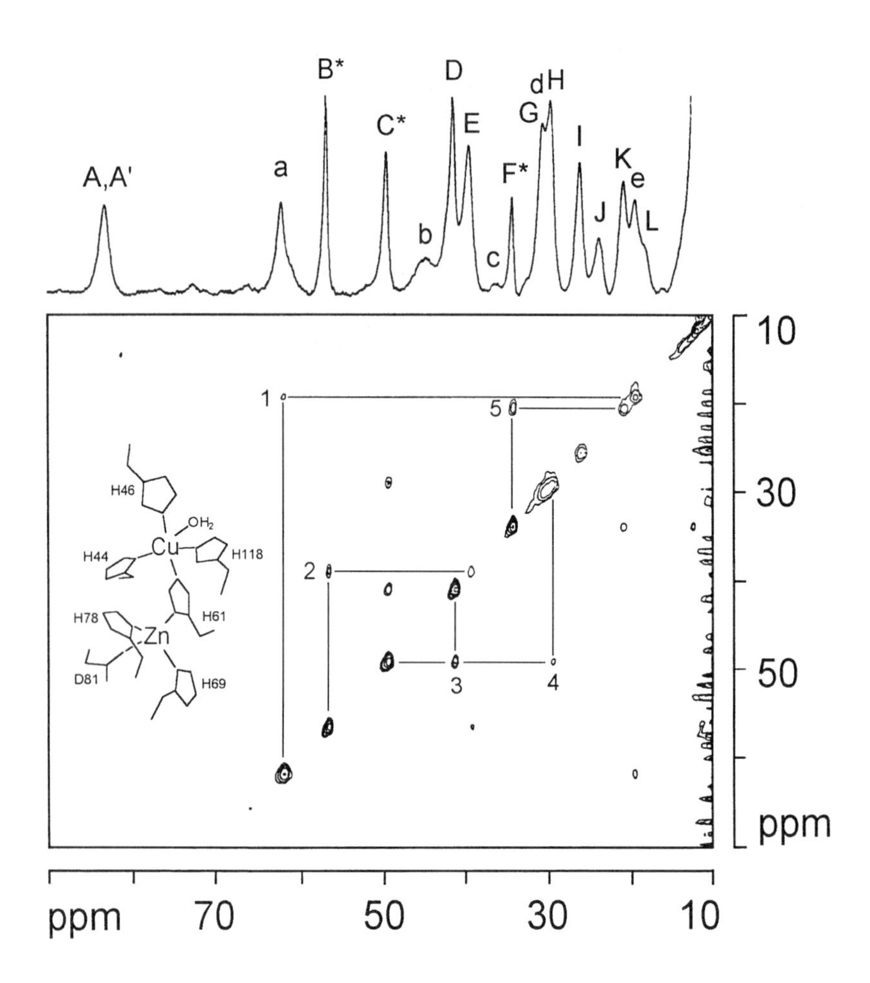

Figure 45
The NOESY spectrum (360 MHz at 298 K) of Cu,Ni–SOD in 50 m*M* phosphate/H_2O buffer at pH 6.5.[50] The signals A–L are attributable to His ring protons on the Cu^{2+} site, and the signals a–e are due to His ring protons on the Ni^{2+} site. The most downfield shifted overlapped signals are assigned to the bridging His 61 ring protons. The asterisked signals are solvent exchangeable His ring NH protons. The solvent exchangeable signal C shows cross relaxations with two ring protons, thus can be unambiguously assigned to the only N_δ coordinated His 44 in the Cu^{2+} site.

can be a weak continuous wave pulse or a pulse train such as MLEV-17 used in TOCSY experiments. The SL power has to be lowered significantly in ROESY experiments in order to suppress the homonuclear Hartmann–Hahn coherence transfer that generates the TOCSY cross peaks with an opposite phase to that in a ROESY spectrum. Thus, the presence of coherence transfer can be a serious artifact that may decrease the intensity of ROESY cross peaks. Although a low SL power in the ROESY experiment works well for the detection of cross-relaxation in diamagnetic molecues, it may pose a problem for the larger spectral windows used for paramagnetic species. Therefore one should only choose to perform a ROESY experiment on a paramagnetic species when sharp isotropically shifted signals are observed in a small spectral window [such as in low-spin heme proteins as shown in Fig. 44(*b*)], and/or when chemical exchange is present in the system.

3. Chemical Exchange

Chemical exchange is an important phenomenon that can be explored by the use of NMR techniques. In the presence of the following equilibrium,

$$\mathrm{A} \underset{k_\mathrm{B}}{\overset{k_\mathrm{A}}{\rightleftharpoons}} \mathrm{B} \tag{44}$$

the change of the Z component of the magnetization of A upon excitation (in the absence of cross relaxation to simplify the case) is further controlled by the two rate constants k_A and k_B [Eq. (45)],

$$dM_\mathrm{A}/dt = (M_{0\mathrm{A}} - M_\mathrm{A})/T_{1\mathrm{A}} - k_\mathrm{A}M_\mathrm{A} + k_\mathrm{B}M_\mathrm{B} \tag{45}$$

where the first term is the change in the magnetization due to the longitudinal relaxation [Eq. (10)], and the latter two terms are due to chemical exchange. The third term vanishes if B is assumed to be saturated instantaneously (i.e., at $t = 0$). Under the conditions of slow exchange (i.e., $k_{\mathrm{A,B}} \ll \Delta\omega$), well-separated signals can be observed for A and B. For the case where the exchange rate is much greater than the frequency difference between the exchange pair (i.e., $k_{\mathrm{A,B}} \gg \Delta\omega$), a complete collapse of the signals will be observed. Let us first consider the situation of slow exchange. When B is saturated for a sufficiently long time, the steady-state magnetization of A ($dM_\mathrm{A}/dt = 0$) can be obtained as $M_{0\mathrm{A}}/(1 + k_\mathrm{A}T_{1\mathrm{A}})$ with the effective observable relaxation rate $T_{1\mathrm{eff}}{}^{-1} = T_{1\mathrm{A}}{}^{-1} + k_\mathrm{A}$. An equation similar to Eq. (45) can be obtained for dM_B/dt. A ratio of $T_{1\mathrm{A}}/T_{1\mathrm{B}}$ can thus be calculated accordingly to provide information on the M–H distance.

Although the measurement of relaxation times has been standardized on all the modern FT NMR spectrometers, caution must be applied in the measurement of relaxation times of paramagnetic species since cross relaxation or chemical exchange can complicate the interpretation of the data. Errors in relaxation times may be generated because the B_1 field may not be sufficient to excite all the isotropically shifted signals in the large spectral windows required for some paramagnetic species. In this case, a "pseudoselective" excitation phenomenon may result in erroneous relaxation time measurements. This phenomenon is illustrated by a study of Fe–amide complexes, where the C–N partial double bond brings about the detection of two widely separated interchanging NH_2 signals ($\Delta\delta$ may reach 80 ppm!).[51] Simultaneous excitation of these two NH signals is not possible when a long 90° pulse of greater than 10 μs is applied. Therefore the measurement of nonselective relaxation times for these signals is not possible. Nevertheless, an estimate of the relative relaxation times of this chemical-exchange pair can be achieved if the exchange kinetics is taken into account as discussed above.

Many chemical exchange systems in bioinorganic chemistry have been studied by the use of NMR techniques. As examples, three commonly encountered chemical exchange phenomena are discussed here: (1) slow electron transfer, (2) metal–ligand binding under slow exchange, and (3) ligand binding to metalloproteins under fast exchange.

a. Slow Electron Transfer Electron transfer is a vital process in living systems, and key roles are played by many metalloproteins, such as cytochromes, Fe–S proteins, and blue Cu proteins. The magnetic properties of metal centers can change dramatically with oxidation state, resulting in a change in isotropic shifts. As in the case of the NOE [Eq. (A.II.7)–(A.II.10) and Fig. A.II.5)], saturation transfer in chemical exchange systems can be observed when the rate of exchange is comparable to the relaxation of the system as shown in Eqs. (A.II.14), (A.II.15), (A.III.1), and (A.III.2) and in Figure A.III.1. Under the appropriate conditions, that is, relatively long relaxation times and relatively fast exchange, one may use the same pulse sequences for 1D NOE difference (Section II.A) or 2D NOESY (dubbed EXSY to differentiate it from NOESY) experiments to establish the correlation between the exchange partners.

Let us now consider an electron-transfer equilibrium between the $Fe^{II}Fe^{II}$ and $Fe^{II}Fe^{III}$ forms of the dinuclear complex $Fe_2(L)(L')_2$ with L being the ligand BPMP and L′ a bridging diphenyl phosphate (Fig. 28).

$$\begin{aligned} &Fe^{II}Fe^{II}(L)(L')_2{}^{+} + Fe'^{II}Fe'^{III}(L)(L')_2{}^{2+} \\ &\quad \rightleftharpoons Fe^{II}Fe^{III}(L)(L')_2{}^{2+} + Fe'^{II}Fe'^{II}(L)(L')_2{}^{+} \end{aligned} \tag{46}$$

Electron transfer between the two oxidation states [Eq. (46)] is slow enough so that isotropically shifted features are clearly resolved and can be associated with each complex. This pair is a perfect system to illustrate the suitability of EXSY for the study of paramagnetic species in the presence of chemical exchange. Figure 46 shows the EXSY spectrum obtained with a very short 3-ms mixing time, which is comparable to the relaxation times of the fast relaxing broad signals. All the cross peaks that connect the exchange partners are clearly detected in this spectrum. Moreover, it is clear that the peaks with relatively long relaxation times of greater than 10 ms show exchange-correlated cross peaks much larger than those associated with the signals with shorter relaxation times. This spectrum also demonstrates that a very wide spectral window ($\sim$160 kHz at 360 MHz) and very short nuclear relaxation times (a few milliseconds) do not always prevent one from performing detailed NMR studies.

b. Metal–Ligand Binding under Slow Exchange The assignment of NMR spectra of diamagnetic compounds is always easier than that of their paramagnetic counterparts. So signal assignment of a paramagnetic metal complex may be achieved via correlation of the paramagnetically shifted features with

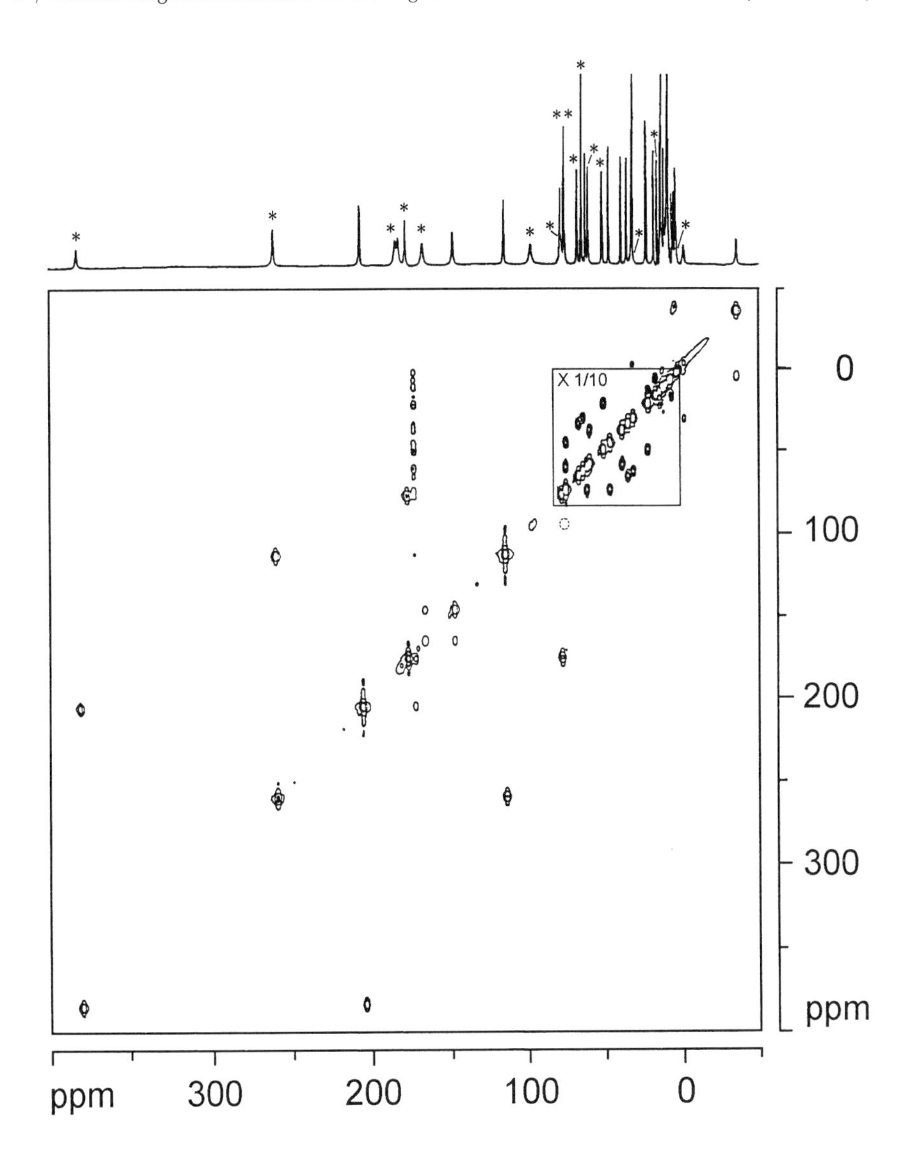

Figure 46
Proton EXSY (360 MHz, 303 K, and a mixing time of 3 ms) spectrum of a sample with a 1 : 1 ratio of the two paramagnetic species, $[Fe^{II}Fe^{II}(BPMP)\{O_2P(OPh)_2\}_2]^+$ and $[Fe^{II}Fe^{III}(BPMP)\{O_2P(OPh)_2\}_2]^{2+}$, which undergo chemical exchange due to electron transfer as shown in equlibrium (46).[13] The structure of the ligand BPMP is shown in Figure 28. The signals attributable to the $Fe^{II}Fe^{III}$ complex are labeled with asterisks. The dotted circle represents a cross peak of low intensity. The signals with longer relaxation times exhibit much more intense cross peaks and are shown with $\frac{1}{10}$ of the signal intensity.

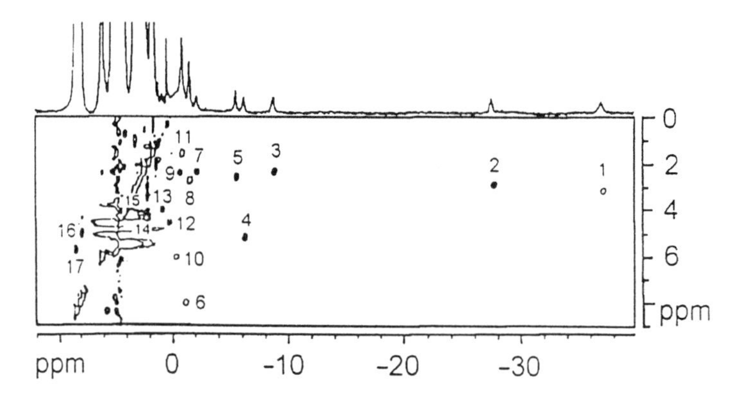

Figure 47
The EXSY spectrum (360 MHz, 298 K, 30-ms mixing time) of a 1 : 1 Yb^{3+}–daumomycin complex in D_2O. The structure of the drug is shown in Figure 10. The cross peaks 1–17 are originated from the protons, respectively, $10H_a$, $10H_b$, $8H_a$, 7H, $8H_b$, 1-H, $2'H_a$, $COCH_3$, $2'H_b$, 1′H, $5'CH_3$, 5′H, 3′H, OCH_3, 4′H, 3H, and 2H.

their diamagnetic ligand counterparts, provided that there is slow enough exchange between free and bound ligand to observe the two types of signals. The example given here is of the anthracyclines, a group of antitumor antibiotics (see Fig. 10 for structure) that bind tightly to DNA by intercalation and hydrogen bond interactions. Metal binding to these antibiotics occurs with high affinity and significantly influences their antibiotic activity. Consequently, the metal binding properties of these antibiotics have been widely studied. A 1:1 Yb^{3+}–daunomycin complex can be prepared in methanol or aqueous solution,[10] where the metal is under exchange between its bound and free forms and thus can be studied by means of saturation transfer techniques. The ^1H NMR spectrum of the free drug can be unambiguously assigned with COSY and TOCSY techniques (Fig. 40). All 17 chemical exchange correlated cross-peaks associated with the isotropically shifted signals and their diamagnetic exchange partners can be found in an EXSY spectrum with a 30-ms mixing time (Fig. 47). The presence of chemical exchange allows the isotropically shifted signals of the complex to be conclusively assigned. A selective metal binding at the 11/12 α-ketophenolate site can be established based on both signal assignments and relaxation time measurements. These data indicate Yb^{3+}–$C(10)H_2$ distances much shorter than the Yb^{3+}–C(7)H distance.

c. Ligand Binding to Metalloproteins Under Fast Exchange In the two examples discussed above, chemical exchange rates are relatively slow, so well-resolved signals for the exchange partners are observed. However, in the case of fast exchange (i.e., $\tau_{ex}{}^{-1} \gg \Delta\omega$), the exchange partners can completely collapse

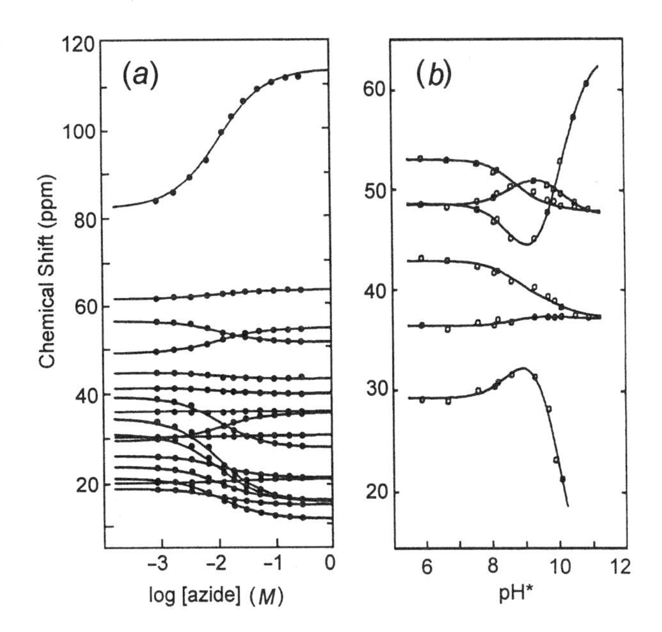

Figure 48
(a) Plot of the chemical shifts of the isotropically shifted ^{1}H NMR signals (300 MHz, 30 °C) of Cu,Ni–SOD against the concentration of azide in 50-mM phosphate at pH 6.5. The significant effect on some of the signals allows the assignment of these signals to protons in the N_3^- binding Cu^{2+} site. [Adapted from Ming et al.[14]] (b) Plot of the chemical shifts of the isotropically shifted ^{1}H NMR signals (200 MHz, 25 °C) of Co,Co–SOD against pH* in 50-mM phosphate. The solid lines in the two plots are numerical fittings according to fast-exchange multiequilibria [Eqs. (47) and (48), and proton associations of phosphate]. [Adapted from Ming and Valentine[16].]

into a single signal with a chemical shift according to Eq. (47),

$$\delta_{\text{ex}} = \sum(\delta_i f_{Mi}) \tag{47}$$

where δ_i and f_{Mi} are the chemical shift and mole fraction, respectively, of each exchange partner. Thus increasing the amount of a ligand (L) added to a metalloprotein (P) sample causes a change in chemical shift owing to the changing mole fractions of the free and bound forms of the protein according to the following equilibrium (Eq. (48)].

$$\text{P} + \text{L} \overset{K_f}{\rightleftharpoons} \text{P} - \text{L} \tag{48}$$

One example is illustrated in Figure 48(a) for the enzyme Cu,Ni–SOD (cf. Figs. 29 and 45) upon azide binding.[14] The addition of azide causes a gradual change in the chemical shifts of the isotropically shifted signals of the protein due to a gradual increase of the mole fraction of the azide-bound form. A simultaneous fitting of all the signals to Eq. (47) according to the equilibrium given in Eq. (48) (with P = Cu,Ni–SOD and L = N_3^-) gives an affinity constant of $111 \pm 5\ M^{-1}$. Several signals are more significantly affected by azide binding, indicating that they belong to ligands at the azide-binding Cu^{2+} site.

A pH dependent chemical shift is another very frequently encountered example of fast chemical exchange phenomena. In metalloprotein systems, a pH titration can be monitored by means of NMR. The deprotonation of a coordinated water molecule is often involved in the action of metalloenzymes, so an accurate measurement of the deprotonation constant of groups near the active site is essential to gain further insight into the mechanism of enzyme action. Acid–base equilibria as shown in Eq. (49) can be monitored by the gradual change of the chemical shifts of the protons in proximity of the deprotonation site, and the K_a for the system can be obtained from the plot of chemical shift versus pH as illustrated in Figure 48(b) for the phosphate-bound Co,Co–SOD.

$$PH^+ \overset{K_a}{\rightleftharpoons} P + H^+ \tag{49}$$

Simultaneous fitting of the chemical shifts of the six isotropically shifted signals in the range of about 30–55 ppm with respect to pH* (in D_2O) can be accomplished when the three ionization constants of phosphate and two ionization constants of Co,Co–SOD–phosphate are taken into account, with the assumption that binding of the dibasic phosphate ion to the protonated protein is the major interaction [cf. Eq. (23)].[16] This fitting gives two pK_a values of 7.5 and 10.1, consistent with results obtained by means of relaxation measurements discussed in Section I.C.4.

III. Perspectives

The NMR studies of diamagnetic proteins have been limited by the size of the proteins owing to excessive signal overlap and significant line broadening in the relatively small approximate 15 ppm spectral window. With the availability of higher magnetic fields and the development of multidimensional NMR techniques, spectral resolution can be dramatically improved. Nevertheless, mechanistic and structural studies of large diamagnetic proteins with molecular sizes close to those of the paramagnetic non-heme iron catechol 1,2-dioxygenase (65 kDa),[52] the di-iron ribonucleotide reductase R2 protein (90 kDa),[53] or the very large hemerythrin octamer (~108 kDa)[54] or myeloperoxidase tetramer (150 kDa)[55], which have been studied with NMR, still pose a significant challenge for the present NMR technology.

In this chapter, we have illustrated the versatile nature of NMR for the study of paramagnetic molecules that are normally not considered "NMR friendly" to most chemists. Although NMR can be a very useful and sophisticated tool for the study of paramagnetic metal centers in small molecules and metalloprotein active sites, it has limitations. For example, relatively high concentrations (millimolar range) of samples are required, 1D and 2D NMR experiments are non-routine, and the prospect of observing all resonances is limited by the relaxation properties of the paramagnetic center. Nevertheless, paramagnetic complexes and metalloproteins have been studied successfully by the use of different NMR techniques to provide information on the metal binding environment and ligand binding properties. With the application of 2D NMR techniques for signal assignment, the use of NMR for the study of paramagnetic metalloproteins has increasingly become an important research tool.

ACKNOWLEDGMENTS

The preparation of this chapter and some 2D NMR studies presented here were partially supported by funds from the University of South Florida (start-up funds, Research and Scholorship Award, and PYF award), the Petroleum Research Fund administered by the American Chemical Society (ACS-PRF 26483-G3), and the American Cancer Society-Florida Division (Edward L. Cole Research Grant, F94USF-3). The Bruker AMX360 NMR spectrometer used for the acquisition of several spectra in this chapter was purchased with the funds from the University of South Florida and from NSF. My co-workers who contribute to the advances of NMR in paramagnetic biomolecules are acknowledged. Jandel Scientic is acknowledged for providing the SigmaPlot® software that was used for the simulation and plotting of several equations in this chapter. Professor Lawrence Que, Jr., is also acknowledged for his many stimulating discussions and suggestions.

APPENDIX I

Fermi Contact Shift

A complete energy level diagram for the NMR transitions of an $I = \frac{1}{2}$ nucleus in the presence of an unpaired electron ($S = \frac{1}{2}$) is shown in Figure A.I.1. In Eq. (4),

$$\mathscr{H} = g_e \beta_e B_0 \mathbf{S} - g_N \beta_N B_0 \mathbf{I} + A_c \mathbf{S} \cdot \mathbf{I} \tag{4}$$

the first two terms give the Zeeman energy levels $\pm\frac{1}{2}(g_e\beta_e \mp g_N\beta_N)B_0$, while the third term gives the splitting $(\pm\frac{1}{4}A_c)$ due to the hyperfine or Fermi contact interaction. Consequently, there are two EPR and two NMR transitions among the energy levels $|\pm m_I, \pm m_S\rangle$, as shown in Figure A.I.1.

While Figure A.I.1 shows that an unpaired electron can have a significant influence on the NMR energy levels, the physical basis for the large paramagnetic chemical shift may still not be obvious at this point. The two NMR transitions shown in Figure A.I.1 are not observed under normal conditions because of relatively fast electronic relaxation. These relaxation rates, typically in the range of 10^9–10^{13} s^{-1}, are larger than the reciprocal of the frequency difference of the NMR transitions. As a result of the fast electronic relaxation between the $S = \pm\frac{1}{2}$ levels, the two NMR transitions will be averaged, and the line width of

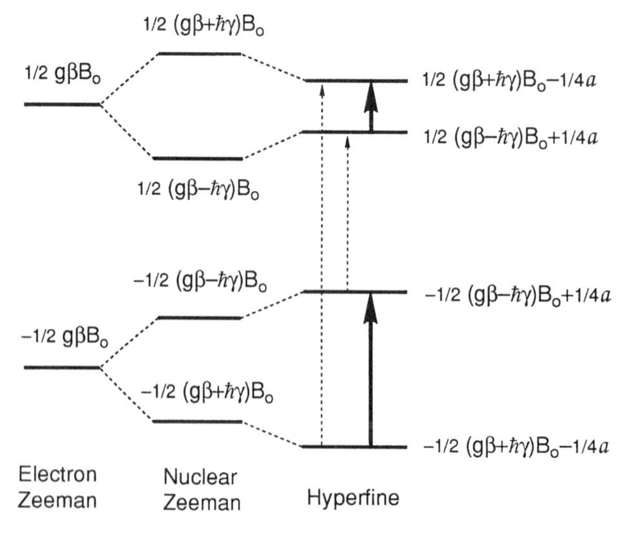

Figure A.I.1
Complete energy diagram of a system with interacting $S = \frac{1}{2}$ and $I = \frac{1}{2}$ spins. The two EPR transitions are indicated by dotted arrows, and the two NMR transitions by solid arrows.

the averaged peak is inversely proportional to the electronic relaxation rate. The position of this averaged transition reflects the fractional Boltzmann populations of the $S = \frac{1}{2}$ and $S = -\frac{1}{2}$ levels [Eq. (A.I.1)],

$$\Delta E_{\text{av}} = P_\alpha(g_N\beta_N B_0 - \tfrac{1}{2}A_c) + P_\beta(g_N\beta_N B_0 + \tfrac{1}{2}A_c) \tag{A.I.1}$$

where $P_{\alpha,\beta}$ are the fractional populations of the $S = \frac{1}{2}$ and $S = -\frac{1}{2}$ levels. According to the Boltzmann distribution, $\text{P} = [1 + \exp(\pm g_e\beta_e H_0/kT)]^{-1}$ or simplified to $(2 \pm g_e\beta_e H_0/kT)^{-1}$ by its Taylor expansion, so the averaged nuclear transition can be now expressed as in Eq. (A.I.2).

$$\Delta E_{\text{av}} = hv = \frac{4g_N\beta_N B_0 + A_c g_e\beta_e B_0/kT}{4 - (g_e\beta_e B_0/kT)^2} \tag{A.I.2}$$

Since $kT \gg g_e\beta_e B_0$, Eq. (A.I.2) can be further simplified to Eq. (A.I.3).

$$\Delta E_{\text{av}} = g_N\beta_N B_0 + \frac{A_c g_e\beta_e B_0}{4kT} \tag{A.I.3}$$

So the influence of the unpaired spin on the averaged NMR transition is $A_c g_e\beta_e B_0/4kT$. For this simple $S = \frac{1}{2}$ and $I = \frac{1}{2}$ system, the Fermi contact shift term $\Delta B/B_0(-\Delta v/v)$ can thus be expressed as in Eq. (A.I.4), which can be further generalized to any paramagnetic center as shown in Eq. (A.I.5),

$$\frac{\Delta B}{B_0} = -\frac{A_c g_e\beta_e}{4g_N\beta_N kT} \quad \left(\text{or} \quad -\frac{A_c g_e\beta_e}{4\hbar\gamma_N kT}\right) \tag{A.I.4}$$

$$\frac{\Delta B}{B_0} = \frac{-\Delta v}{v} = \frac{-g_e\beta_e S(S+1)}{3\hbar\gamma_N kT} A_c = \frac{A_c}{\hbar\gamma_N B_0}\langle S_z\rangle \tag{A.I.5}$$

where $\langle S_z\rangle$ is the average value of S_z over all the spin levels, and is given as in Eq. (A.I.6).

$$\langle S_z\rangle = -S(S+1)g_e\beta_e B_0/3kT \tag{A.I.6}$$

In the case of paramagnetic Ln^{3+} complexes, $\langle S_z\rangle$ is written as in Eq. (A.I.7) to include spin–orbit coupling,

$$\langle S_z\rangle = \frac{-g_J(g_J - 1)J(J+1)\beta_e B_0}{3kT} = -\left(\frac{g_J - 1}{g_J}\right)\frac{\chi B_0}{N\beta_e} \tag{A.I.7}$$

where the magnetic susceptibility can also be obtained according to $\chi = N{\mu_{\text{eff}}}^2/3kT$ with μ_{eff} being $g_J\beta[J(J+1)]^{1/2}$ similar to that for paramagnetic transition metal complexes; and g_J is the Landé factor [Eq. (A.I.8)].

$$g_J = 1 + \frac{J(J+1) - L(L+1) + S(S+1)}{2J(J+1)} \tag{A.I.8}$$

APPENDIX II

Nuclear Overhauser Effect

A. Steady-State NOE

In addition to through-bond scalar coupling ($J_{\mathrm{H-H}}$) among nuclei in a molecule, which gives rise to characteristic splitting patterns in diamagnetic compounds, there is also a through-space dipolar interaction for nuclei in close proximity, which gives rise to the NOE. The NOE provides a convenient way for estimating internuclear distances and is very useful for the structural determination of macrobiomolecules including paramagnetic metalloproteins.

The energy level diagram for two spins magnetically coupled with each other through-space is shown in Figure A.II.1. The NOE (η) is the change in intensity

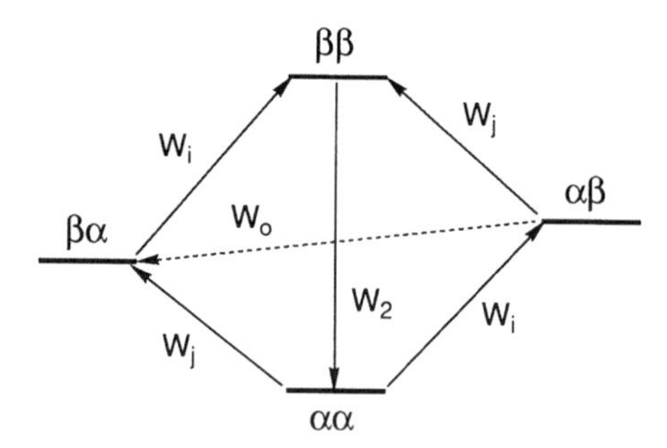

Figure A.II.1
Energy levels for two $I = \frac{1}{2}$ nuclei i and j in close spatial proximity. The NMR transitions of these two nuclei are W_i and W_j, and the transitions W_0 and W_2 are zero-quantum and double-quantum transitions, respectively.

of the transitions of one spin (i or j) when the transitions of the other are saturated by use of a decoupler (called a steady-state NOE). In other words, $\eta_i(j) = (I - I_0)/I_0$ for the enhancement of spin i transitions when spin j is saturated (with respective intensities of I and I_0 in the presence and absence of the perturbation). The energy level diagram in Figure A.II.1 shows that saturating the W_j transitions (i.e., equalizing the populations of the energy levels connected by transitions W_j) can affect the intensities of transitions W_i. This influence in the transitions is particularly true when the W_0 or W_2 relaxation pathway is the dominant mechanism of relaxation. When the system has reached a new equilibrium state, the change of the intensity of spin i upon the perturbation of spin j,

$\eta_i(j)$, can be written as in Eq. (A.II.1).

$$\eta_i(j) = \frac{W_2 - W_0}{2\,W_i + W_2 + W_0} = \frac{\sigma_{ij}}{\rho_i} \quad \text{(A.II.1)}$$

The numerator in this equation is often called the cross relaxation rate with the symbol σ_{ij}, and the denominator is the intrinsic dipolar longitudinal relaxation often designated as ρ_i, which can be obtained with selective relaxation time measurements. Equation (A.II.1) is applicable to two nuclei with same γ values. This equation can be also used for the description of the NOE on a heteronucleus upon proton saturation by incorporating the ratio γ_H/γ_X into Eq. (A.II.1). For example, since $\gamma_H \sim 4\gamma_C$, a significant enhancement of NOE on ^{13}C NMR signals can be observed when protons are saturated.

By taking the results from the dipolar Hamiltonian for the different transition probabilities, the following correlations can be obtained [Eqs. (A.II.2)–(A.II.4)],

$$W_{i(j)} = \frac{3}{20}\hbar^2\gamma_i^2\gamma_j^2 r^{-6}\frac{\tau_c}{1 + \omega_{i(j)}^2\tau_c^2} \quad \text{(A.II.2)}$$

$$W_0 = \frac{2}{20}\hbar^2\gamma_i^2\gamma_j^2 r^{-6}\frac{\tau_c}{1 + (\omega_i - \omega_j)^2\tau_c^2} \quad \text{(A.II.3)}$$

$$W_2 = \frac{12}{20}\hbar^2\gamma_i^2\gamma_j^2 r^{-6}\frac{\tau_c}{1 + (\omega_i + \omega_j)^2\tau_c^2} \quad \text{(A.II.4)}$$

where all the terms are as defined in the chapter, r is the distance between the two nuclei i and j, and τ_c is the rotaional correlation time that can be estimated by the Stokes–Einstein equation (A.II.5), with ζ being the viscosity and a the radius of the molecule.

$$\tau_c = 4\pi\zeta a^3/3kT \quad \text{(A.II.5)}$$

Equations (A.II.2)–(A.II.4) also reveal that a ratio for $W_0 : W_1 : W_2$ of 2 : 3 : 12 can be obtained for small molecules with small τ_c (i.e., in the extreme narrowing limit $\omega\tau_c \ll 1$). A substitution of Eqs. (A.II.2)–(A.II.4) into Eq. (A.II.1) gives the NOE enhancement in terms of nuclear frequencies ω and molecular tumbling time τ_c. In homonuclear systems, further simplification can be achieved by considering $\omega_i \cong \omega_j$ to give Eq. (A.II.6), affording the familiar sigmoidal τ_c dependent NOE as shown in Figure A.II.2.

$$\eta_i(j) = \frac{5 + \omega^2\tau_c^2 - 4\omega^4\tau_c^4}{10 + 23\omega^2\tau_c^2 + 4\omega^4\tau_c^4} \quad \text{(A.II.6)}$$

From Eq. (A.II.6), a maximum of 50% NOE enhancement can be obtained for fast tumbling systems (i.e., with small τ_c values to afford $\eta_i(j) = \frac{5}{10}$) in the absence of unpaired electrons, and a maximum of -100% enhancement can

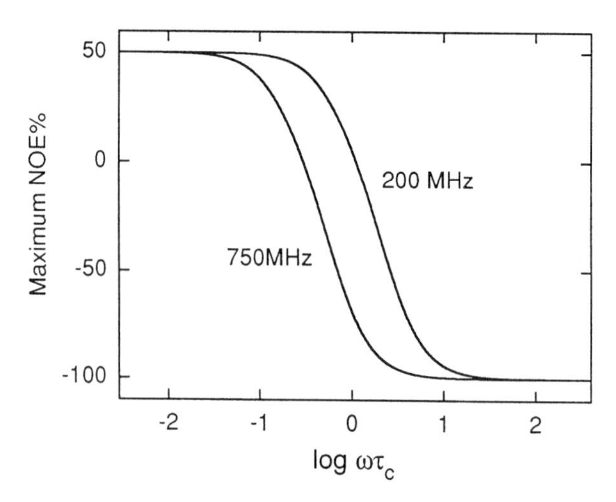

Figure A.II.2
Maximum NOE observed as a function of the rotational correlation time of the molecule. As shown, a small molecule with short τ_c gives rise to positive NOE, whereas a large protein molecule with a long τ_c affords a negative NOE. The NOE vanishes when $(\omega\tau_c)^2 = 1.25$ according to Eq. (A.II.6).

be seen for macromolecules with longer τ_c times (i.e., to afford $\eta_i(j) = -4\omega^4\tau^4/4\omega^4\tau^4$). Moreover, the NOE may not be detectable when the numerator reaches zero [i.e., $(\omega\tau)^2 = 1.25$] for molecules with molecular weights of several kilodaltons. (This problem can be overcome by the use of rotating frame NOE techniques discussed in Section A.II.C.) Figure A.II.3 illustrates the influence of magnetic field on NOE. In a field of 7.05 T (^{1}H frequency of 300 MHz) a small protein with a $\tau_c = 5$ ns exhibits nearly -100% NOE enhancement, whereas a peptide with a $\tau_c = 0.5$ ns gives almost no NOE.

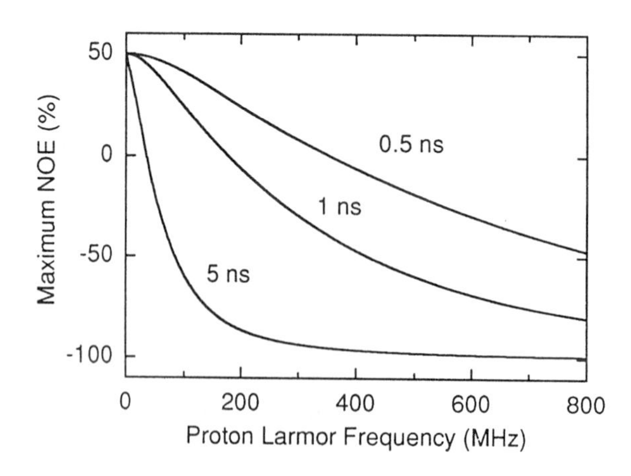

Figure A.II.3
Plot of maximum NOE with respect to the applied magnetic field for a protein with $\tau_c = 5$ ns and a small peptide with $\tau = 0.5$ ns.

B. Transient NOE and NOESY

An NOE can also be observed when one nucleus (j) in a two-spin system (i and j) is selectively inverted by an excitation pulse. When the system is allowed to relax for a time period t prior to FID acquisition, a transient NOE appears on signal i during the course of the recovery of signal j. The cross relaxation between i and j (σ_{ij}) combines with their individual relaxation pathways (ρ) to give a biexponential function for the recovery of their magnetization I_z and J_z as shown in Eqs. (A.II.9) and (A.II.10), which are derived via the solution of the Eqs. (A.II.7) and (A.II.8) under the initial conditions of $I_z = I_z{}^0$ and $J_z = -J_z{}^0$ upon selective inversion of nucleus j. These equations are valid for two-spin systems in the absence of external relaxation mechanisms, that is, when $\rho_i \cong \rho_j$.

$$dI_z/dt = -(I_z - I_z{}^0)\rho_i - (J_z - J_z{}^0)\sigma_{ij} \tag{A.II.7}$$

$$dJ_z/dt = -(J_z - J_z{}^0)\rho_j - (I_z - I_z{}^0)\sigma_{ij} \tag{A.II.8}$$

$$(I_z - I_z{}^0)/I_z{}^0 = \eta_{\text{tr}}(i) = e^{-(\rho-\sigma)t} - e^{-(\rho+\sigma)t} \tag{A.II.9}$$

$$(J_z - J_z{}^0)/J_z{}^0 = -e^{-(\rho-\sigma)t} - e^{-(\rho+\sigma)t} \tag{A.II.10}$$

The longitudinal relaxation is significantly affected by cross relaxation in spin-coupled pairs. Figure A.II.4 shows the effect on $J_z{}^0$ by cross relaxation in a two-spin system. If one uses the "null method" to estimate T_1 (i.e., $T_1 = \tau_{\text{null}}/\ln 2$), the presence of a cross relaxation pathway would result in an overestimate of T_1 for signal j during the course of the recovery of its magnetization as shown in Eq. (A.II.10). In the absence of cross relaxation (i.e., $\sigma = 0$), Eq. (A.II.10) becomes the same as Eq. (11) and follows simple "inversion-recovery" behavior.

The competition of the cross relaxation (σ) and the relaxation (ρ) in the

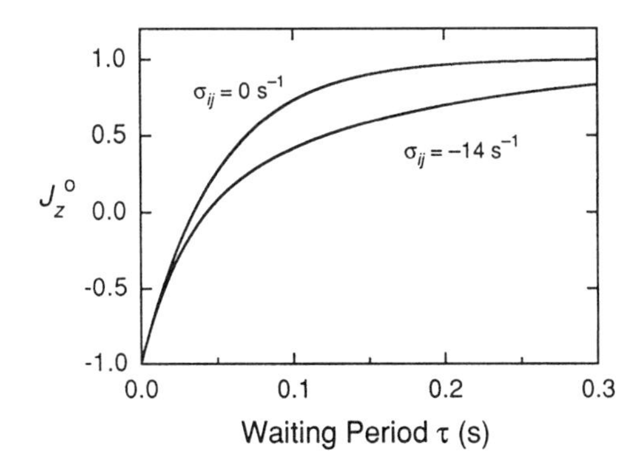

Figure A.II.4
The recovery of the magnetization of a signal with $\rho_j = 20\ \text{s}^{-1}$ followed by selective inversion is affected by cross relaxation ($\sigma_{ij} = -14\ \text{s}^{-1}$). Note that a retardation of the recovery is obsevered in the presence of cross relaxation, which may cause an erroneous estimate of the relaxation time.

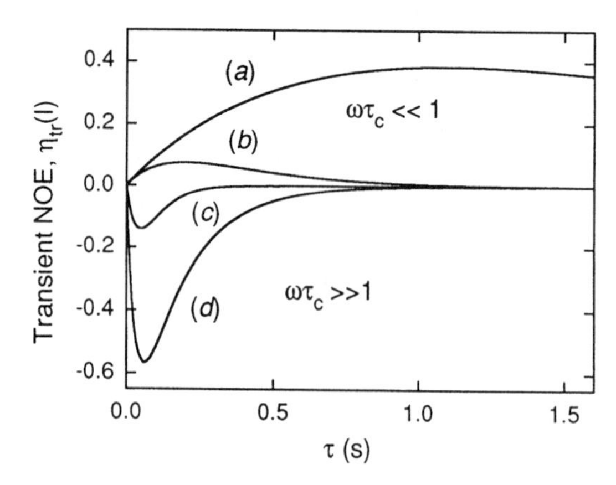

Figure A.II.5
The buildup of transient NOE on signal i followed by selective inversion of its partner j. A positive NOE is observed in the extreme narrowing limit regeme [trace (a), $\rho = 1\ s^{-1}$ and $\sigma = 0.5\ s^{-1}$; trace (b), $\rho = 5\ s^{-1}$ and $\sigma = 0.5\ s^{-1}$] while negative NOE is observed in macromolecular systems [trace (c), $\rho = 20\ s^{-1}$ and $\sigma = -3.8\ s^{-1}$; trace (d), $\rho = 20\ s^{-1}$ and $\sigma = -14\ s^{-1}$]. Traces (c) and (d) are simulated to mimic pairs of protons with a relaxation time of 50 ms in a paramagnetic metalloprotein of about 17 kDa.

buildup of transient NOE can be clearly shown with a plot of Eq. (A.II.9) (Fig. A.II.5). The maximum transient NOE of 38.5% can be achieved [trace(a)] in a system without external relaxation at the extreme narrowing limit of $\rho = 1\ s^{-1}$ and $\sigma = \frac{1}{2}\rho$ [Eq. (A.II.6)]; whereas it becomes negative in macromolecular systems [traces (c) and (d)]. In the presence of external relaxation mechanisms, such as paramagnetic relaxation, the intensity of the NOE becomes smaller and shorter delay times are required for the detection of appreciable NOE [traces (b–d)].

The time t_{max} that gives the maximum value of the transient NOE can be obtained by solving $d\eta_{tr}(i)/dt = 0$, as shown in Eq. (A.II.11). Substitution of $t = t_{max}$ into Eq. (A.II.9) gives the maximum NOE [$\eta_{tr}(\max)$] shown in Eq. (A.II.12) and Figure A.II.6.

$$t_{max} = (1/2\sigma)\ln\left(\frac{\rho+\sigma}{\rho-\sigma}\right) \quad \text{(A.II.11)}$$

$$\eta_{tr}(\max) = -\left(\frac{\rho+\sigma}{\rho-\sigma}\right)^{-(\rho+\sigma)/2\sigma} + \left(\frac{\rho+\sigma}{\rho-\sigma}\right)^{-(\rho-\sigma)/2\sigma} \quad \text{(A.II.12)}$$

In reality, the two spins can have different relaxation times. Thus, Eq. (A.II.9) has to be modified to Eq. (A.II.13), with $D = \frac{1}{2}[(\rho_i - \rho_j)^2 + \sigma^2]^{1/2}$ and $R = \frac{1}{2}(\rho_i + \rho_j)$.

$$(I_z - I_z{}^0)/I_z{}^0 = \eta_{tr}(i) = -(\sigma/D)[e^{-(R+D)t} - e^{-(R-D)t}] \quad \text{(A.II.13)}$$

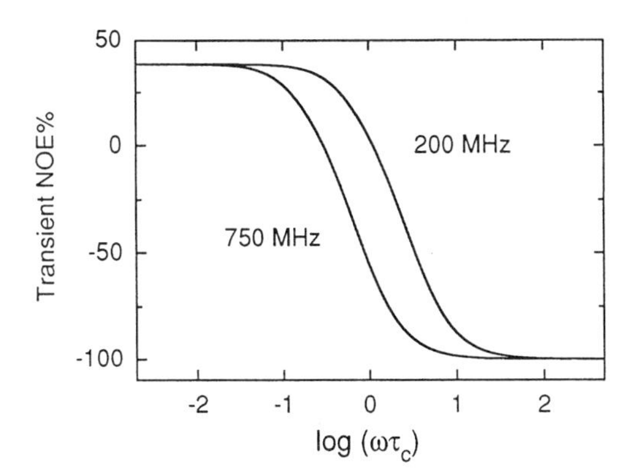

Figure A.II.6
Plot of maximum transient NOE, $\eta_{tr}(\max)$, versus $\log(\omega\tau_c)$ at two different magnetic field strengths.

It is also noteworthy that the initial NOE buildup rate in the transient experiments (i.e., $dI_z/dt(t \to 0) = 2\sigma_{ij}J_z{}^0$ when $I_z = I_z{}^0$ and $J_z = -J_z{}^0$) is twice that in corresponding steady-state NOE experiments ($dI_z/dt(t \to 0) = \sigma_{ij}J_z{}^0$ when $I_z = I_z{}^0$ and $J_z = 0$). However, when the nuclear relaxation times of the two spins are different, the NOE is larger in the steady-state experiments than in the transient experiments for the signal with the longer relaxation time. The best method for the measurement of the NOE depends on the magnetic properties of the system. For example, while the initial rate of NOE buildup in the transient NOE method is larger than in the steady-state experiment, the narrow range of the time for the observation of appreciable NOE and the difficulty of selective inversion of fast relaxing isotropically shifted signals diminishes its utility for paramagnetic systems.

However, the 2D version of the transient NOE method, NOESY, has become a useful technique for the study of paramagnetic metalloproteins. The NOESY experiment consists of the following pulse sequence, $D1$-90°-t_1-90°-τ_m-90°-AQ, that is, delay followed by a 90° pulse, the evolution time for the second dimension, another 90° pulse, then followed by the mixing time τ_m, which is equivalent to the "waiting period" in the transient NOE experiment [t in Eq. (A.II.9)], then data acquisition. The equations for the description of the intensities of the cross signals I_c and the diagonal signals I_d [Eqs. (A.II.14) and (A.II.15)] with respect to the mixing time τ_m in the NOESY experiment are similar to the equations for the transient NOE [Eqs. (A.II.9) and (A.II.10)]

$$I_c = \frac{M_0}{2}\left[e^{-(\rho-\sigma)\tau_m} - e^{-(\rho+\sigma)\tau_m}\right] \tag{A.II.14}$$

$$I_d = \frac{M_0}{2}\left[e^{-(\rho-\sigma)\tau_m} + e^{-(\rho+\sigma)\tau_m}\right] \tag{A.II.15}$$

where M_0 is the intensity of the diagonal peak when $\tau_m = 0$. It is noteworty that the intensity of the cross-signals I_c [Eq. (A.II.14)] is only one-half that of the transient NOE [Eq. (A.II.9)] when the two time-parameters $\tau_{\rm m}$ and t are set the same. The significant influence of fast nuclear relaxation in paramagnetic molecules on the intensity of the cross peaks is similar to that of the transient NOE shown in Figure A.II.5.

C. Rotating Frame NOE

The NOE enhancement vanishes when the numerator in Eqs. (A.II.1) and (A.II.6) reaches zero, that is, when the value of $(\omega\tau_{\rm c})^2$ is equal to 1.25, as in the case of molecules with molecular weights of several kilodaltons. Conventional NOE techniques are thus useless for these systems. The problem can by solved by the observation of transient NOE enhancement in the rotating frame (ROE). For a two spin system i and j that are spin locked along the y axis (in the rotating frame), the change of the transverse magnetizations of the two spins can be written as in Eq. (A.II.16), similar to Eqs. (A.II.7) and (A.II.8).

$$dI_2/dt = -\rho_2 I_2 - \sigma_2 J_2 \quad \text{(A.II.16)}$$

where I_2 and J_2 are the transverse magnetizations, ρ_2 is the intrinsic relaxation rate, and σ_2 is the cross relaxation rate in the rotating frame, which are approximated by Eqs. (A.II.7) and (A.II.8).

$$\rho_2 = \frac{\gamma^4\hbar^2}{20r^6}\left[5\tau_{\rm c} + \frac{9\tau_{\rm c}}{1+\omega^2\tau_{\rm c}{}^2} + \frac{6\tau_{\rm c}}{1+4\omega^2\tau_{\rm c}{}^2}\right] \quad \text{(A.II.17)}$$

$$\sigma_2 = \frac{\gamma^4\hbar^2}{20r^6}\left[4\tau_{\rm c} + \frac{6\tau_{\rm c}}{1+\omega^2\tau_{\rm c}{}^2}\right] \quad \text{(A.II.18)}$$

The initial conditions for the reference spectrum (R) and with signal j selectively inverted (S) for the two-spin system spin-locked in the rotating frame are $I_2 = J_2 = 1$ and $I_2 = -J_2 = 1$, respectively, which give the solutions $e^{-(\rho+\sigma)t}$ and $e^{-(\rho-\sigma)t}$ to Eqn. (A.II.16). The ROE can thus be obtained by subtraction of spectrum S from spectrum R as shown in Eq. (A.II.19). This equation is similar to Eq. (A.II.9) for the description of the transient NOE described earlier.

$$\text{ROE} = e^{-(\rho_2-\sigma_2)t} - e^{-(\rho_2+\sigma_2)t} \quad \text{(A.II.19)}$$

The time $t_{\rm max}$ that gives the maximum value of the ROE can thus be obtained by solving $d(\text{ROE})/dt = 0$, affording Eq. (A.II.20) similar to Eq. (A.II.11) (with ρ and σ replaced by ρ_2 and σ_2). Substitution of $t_{\rm max}$ for t in Eq. (A.II.19) gives the maximum ROE [$\text{ROE}_{\rm max}$, Eq. (A.II.21)], which is similar to Eq. (A.II.12). Note that in the extreme narrowing limit of $\omega\tau \ll 1$, $\text{ROE}_{\rm max}$ becomes virtually

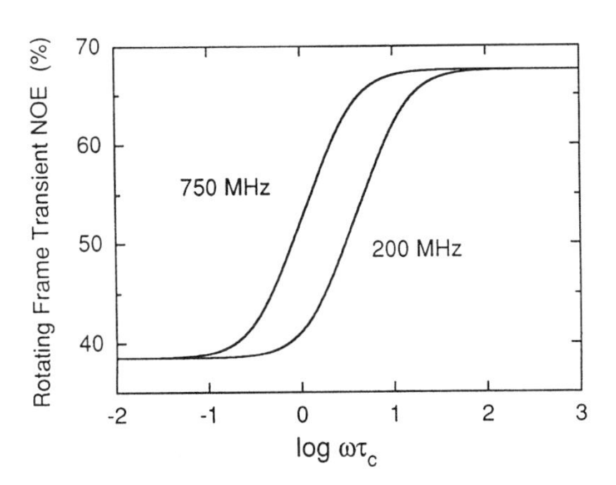

Figure A.II.7
A plot of the ROE_{max} as a function of $\log(\omega\tau_c)$, in which a minimum ROE_{max} value of 38.5% can be obtained when $\omega\tau_c \ll 1$, and a maximum value of 67.5% when $\omega\tau_c \gg 1$; whereas $\eta_{tr}(\max)$ is position when $\omega\tau_c \ll 1$ and negative when $\omega\tau_c \gg 1$.

the same as the maximum transient NOE, $\eta_{tr}(\max)$, with $\rho = \gamma^4\hbar\tau_c/2r^6$ and $\sigma = \gamma^4\hbar\tau_c/r^6$.

$$t_{max} = (1/2\sigma)\ln\left(\frac{\rho_2 + \sigma_2}{\rho_2 - \sigma_2}\right) \tag{A.II.20}$$

$$ROE_{max} = -\left(\frac{\rho_2 + \sigma_2}{\rho_2 - \sigma_2}\right)^{-(\rho_2+\sigma_2)/2\sigma_2} + \left(\frac{\rho_2 + \sigma_2}{\rho_2 - \sigma_2}\right)^{-(\rho_2-\sigma_2)/2\sigma} \tag{A.II.21}$$

A plot of ROE_{max} as a function of $\log(\omega\tau_c)$ is shown in Figure A.II.7; where a minimum ROE_{max} value of 38.5% can be obtained when $\omega\tau_c \ll 1$, and a maximum value of 67.5% when $\omega\tau_c \gg 1$. Thus ROE techniques are very useful for the detection of through-space interactions in molecules in the several kilodalton range. However, paramagnetic molecules can readily lose transverse magnetization during spin lock owing to their fast transverse relaxation times (broad signals), a situation that may limit the use of ROE techniques for the study of such molecules (Section II.B).

APPENDIX III

Chemical Exchange

In the presence of chemical exchange, a selective saturation or excitation of one signal may result in the transfer of magnetization to its exchange partner, showing an intensity change similar to that in the presence of the NOE. Although the change of the magnetization ($dM(t)/dt$) due to chemical exchange has a different mechanism from NOE, it cannot be clearly differentiated from NOE in the $\omega\tau_c \gg 1$ regime, where both phenomena result in the decrease of the magnetization of the spin upon perturbation of its partners (Fig. A.II.2)

Owing to this similarity, the 1D and 2D NMR techniques utilized for the detection of the NOE can also be applied to the study of chemically exchanging systems. The NOESY pulse sequence can be used for the detection of chemical exchange and has been given the acronym EXSY when used for this application to differentiate it from NOESY. The algebraic expressions for the intensity of the cross signals (I_c) and the diagonal signals (I_d) in a simple system of symmetric two-site exchange [i.e., $k_A = k_B = k_{ex}$ in Eq. (44)] and with the two spins possessing the same relaxation rates are shown in Eqs. (A.III.1) and (A.III.2) and correspond to Eqs. (A.II.14) and (A.II.15) for NOESY. A plot of I_c and I_d with respect to the mixing time τ_m is shown in Figure A.III.1

$$I_c = \frac{M_0}{2}[e^{-\rho\tau_m} - e^{-(\rho+2k_{ex})\tau_m}] \qquad \text{(A.III.1)}$$

$$I_d = \frac{M_0}{2}[e^{-\rho\tau_m} + e^{-(\rho+2k_{ex})\tau_m}] \qquad \text{(A.III.2)}$$

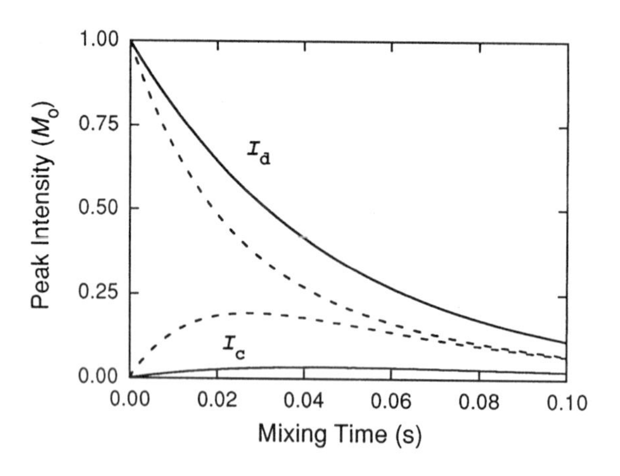

Figure A.III.1

Plot of I_c and I_d versus τ_m in a chemical exchange system with $\rho = 20\ s^{-1}$ at two different exchange rates, $k_{ex} = 2$ (solid) and $20\ s^{-1}$ (dashed).

The mixing time that gives the most intense exchange-correlated cross signals can be derived by solving $dI_c/dt = 0$ [Eq. (A.III.3)], and the exchange rate in this simplified system can be estimated by the use of Eq. (A.III.4).

$$\tau_{max} = \ln(2k_{ex}T_1 + 1)/2k_{ex} \tag{A.III.3}$$

$$\frac{I_d}{I_c} = \frac{1 + \exp(-2k_{ex}\tau_m)}{1 - \exp(-2k_{ex}\tau_m)} \cong \frac{1 - k_{ex}\tau_m}{k_{ex}\tau_m} \tag{A.III.4}$$

References

1. (a) Zang, Y.; Jang, H. G.; Chiou, Y.-M.; Hendrich, M. P.; Que, L., Jr., "Structure and properties of ferromagnetically coupled bis(μ-halo)diiron(II) complexes" *Inorg. Chim. Acta* **1993**, *213*, 41–48. (b) Zang, Y.; Que, L., Jr., "Structure and reactivity of Fe(II)–SAr complexes: relevance to the active site of isopenicillin N synthase" *Inorg. Chem.* **1995**, *34*, 1030–1035. (c) Chiou, Y.-M.; Que, L., Jr., "Models for α-keto acid-dependent non-heme iron enzymes: structure and reactivity of $[Fe^{II}(L)(O_2CCOPh)](ClO_4)$ complexes" *J. Am. Chem. Soc.* **1995**, *117*, 3999–4013.
2. (a) Gorenstein, D. G. *Phosphorus-31 NMR, Principles and Applications*; Academic: New York, 1984. (b) Webb, G. A. "Factors contributing to the observed chemical shifts of heavy nuclei" In Laszlo, P. Ed. *NMR of Newly Accesible Nuclei, Vol. 1*; Academic: New York, 1983.
3. Heistand, R. H., II; Lauffer, R. B.; Fikrig, R.; Que, L., Jr., "Catecholate and phenolate iron complexes as models for the dioxygenases" *J. Am. Chem. Soc.* **1982**, *104*, 2789–2796.
4. Lauffer, R. B.; Antanaitis, B. C.; Aisen, P.; Que, L., Jr., "^{1}H NMR studies of porcine uteroferrin" *J. Biol. Chem.* **1983**, *258*, 14212–14218.
5. Scarrow, R. C.; Pyrz, J. W.; Que, L., Jr., "NMR studies of the dinuclear iron site in reduced uteroferrin and its oxoanion complexes" *J. Am. Chem. Soc.* **1990**, *112*, 657–665.
6. (a) Sträter, N.; Klabunde, T.; Tucker, P.; Witzel, H.; Krebs, B. "Crystal structure of a purple acid phoshatase containing a dinuclear Fe(III)–Zn(II) active site" *Science* **1995**, *268*, 1489. (b) Klabunde, T.; Sträter, N.; Frohlich, R.; Witzel, H.; Krebs, B. "Mechanism of Fe(III)–Zn(II) purple acid phosphatase based on crystal structure" *J. Mol. Biol.* **1996**, *259*, 737.
7. Sjöberg, B.-M.; Gräslund, A. "Ribonucleotide reductase" *Adv. Inorg. Biochem.* **1983**, *5*, 87–110.
8. (a) Bertini, I.; Briganti, F.; Luchinat, C.; Messori, L.; Monnanni, R.; Scozzafava, A.; Vallini, G. "2D ^{1}H NMR studies of oxidized $2(Fe_4S_4)$ ferredoxin from *Clostridium pasteurianum*" *FEBS Lett.* **1991**, *289*, 253–256. (b) Busse, S. C.; La Mar, G. N.; Howard, J. B. "Two dimensional NMR investigation of iron–sufur clusters. Electronic and molecular structure of oxidized *Clostridium pasteurianum* ferredoxin" *J. Biol. Chem.* **1991**, *266*, 23714.
9. Banci, L.; Dugad, L. B.; La Mar, G. N.; Keating, K. A.; Luchinat, C.; Pierattelli, R. "^{1}H nuclear magnetic resonance investigation of cobalt(II) substituted carbonic anhydrase" *Biophys. J.* **1992**, *63*, 530–543.
10. Ming, L.-J.; Wei, X. "An Ytterbium(III) Complex of Daunomycin, a Model Metal Complex of Anthracycline Antibiotics" *Inorg. Chem.* **1994**, *33*, 4617–4618.
11. Ming, L.-J. "Paramagnetic lanthanide(III) ions as NMR probes for biomolecular structure and function" In La Mar, G. N.; Ed. *Nuclear Magnetic Resonance of Paramagnetic Molecules*, NATO-ASI, Kluwer: Dordrecht, The Netherlands, **1995**; pp 245–264.
12. (a) Qin, J.; La Mar, G. N. "Complete sequence specific ^{1}H NMR resonance assignment of hyperfine-shifted residues in the active site of a paramagnetic protein: application to *Aplysia* cyano-metmyoglobin" *J. Biomol. NMR* **1992**, *2*, 597–618. (b) Alam, S. L.; Satterlee, J. D. "Complete heme proton hyperfine resonance assignment of the *Glycera dibranchiata* component IV metcyano monomer hemoglobin" *Biochemistry* **1994**, *33*, 4008–4018.
13. Ming, L.-J.; Jang, H. G.; Que, L., Jr., "2D NMR studies of paramagnetic diiron complexes" *Inorg. Chem.* **1992**, *31*, 359–364.
14. (a) Ming, L.-J. Ph.D. Thesis, "NMR Studies of Nickel(II)- and Cobalt(II)-Substituted Derivatives of Bovine Copper–Zinc Superoxide Dismutase," University of California at Los Angeles, Los Angeles, CA, 1988. (b) Ming, L.-J.; Banci, L.; Luchinat, C.; Bertini, I.; Valentine, J. S. "Characterization of copper–nickel and silver–nickel bovine superoxide dismutase by ^{1}H NMR spectroscopy" *Inorg. Chem.* **1988**, *27*, 4458–4463.

15. Banci, L.; Bertini, I.; Luchinat, C.; Monnanni, R.; Scozzafava, A. "Characterization of the cobalt(II)-substituted superoxide dismutase–phosphate systems" *Inorg. Chem.* **1987**, *26*, 153–156.
16. Ming, L.-J.; Valentine, J. S. "NMR studies of cobalt(II)-substituted derivatives of bovine copper–zinc superoxide dismutase. Effects of pH, phosphate, and metal migration" *J. Am. Chem. Soc.* **1990**, *112*, 4256–4264.
17. Luchinat, C.; Xia, Z. "Paramagnetism and dynamic properties of electrons and nuclei" *Coord. Chem. Rev.* **1992**, *120*, 281–307.
18. Bertini, I.; Briganti, F.; Luchinat, C.; Xia, Z. "Nuclear magnetic relaxation dispersion studies of hexaaquo Mn(II) ions in water–glycerol mixtures" *J. Magn. Reson.* **1992**, *101A*, 198–201.
19. Bertini, I.; Luchinat, C.; Mancini, M.; Spina, G. "The electron-nucleus dipolar coupling. The effect of zero-field splitting of an $S = 3/2$ manifold" *J. Magn. Reson.* **1984**, *59*, 213–222.
20. (a) Koenig, S. H.; Brown, R. D., III "Field-cycling relaxometry of protein solution and tissue: implications for MRI" *Prog. NMR Spect.* **1990**, *22*, 487–567. (b) Lauffer, R. B. "Paramagnetic metal complexes as water proton relaxation agents for NMR imaging: Theory and design" *Chem. Rev.* **1987**, *87*, 901–927.
21. Burton, D. R.; Forsen, S.; Karlstrom, G.; Dwek, R. A. "Proton relaxation enhancement (PRE) in biochemistry: a critical survey" *Prog. NMR Spectrosc.* **1979**, *13*, 1–45.
22. Larsen, T. M.; Laughlin, L. T.; Holden, H. M.; Rayment, I.; Reed, G. H. "Structure of rabbit pyruvate kinase complexes with Mn^{2+}, K^+, and pyruvate" *Biochemistry* **1994**, *33*, 6301–6309.
23. Kojima, T.; Leising, R. A.; Yan, S.; Que, L., Jr., "Alkane functionalization at nonheme iron centers. Stoichiometric transfer of metal-bound ligands to alkane" *J. Am. Chem. Soc.* **1993**, *115*, 11328–11335.
24. Holmes, M. A.; Le Trong, I.; Turley, S.; Sieker, L. C.; Stenkamp, R. E. "Structure of deoxy and oxy hemerythrin at 2.0 Å resolution" *J. Mol. Biol.* **1991**, *218*, 583–593.
25. Dawson, J. W.; Gray, H. B.; Hoenig, H. E.; Rossman, G. R.; Schredder, J. M.; Wang, R.-H., "A magnetic susceptibility study of hemerythrin using an ultrasensitive magnetometer" *Biochemistry* **1972**, *11*, 461–465.
26. Maroney, M. J.; Kurtz, D. M., Jr.; Nocek, J. M.; Pearce, L. L.; Que, L., Jr., "^{1}H NMR probes of the binuclear iron cluster in hemerythrin" *J. Am. Chem. Soc.* **1986**, *108*, 6871–6879.
27. (a) Bertini, I.; Briganti, F.; Luchinat, C.; Scozzafava, A. "^{1}H NMR studies of the oxidized and partially reduced 2(4Fe–4S) ferredoxin from *Clostridium pasteurianum*" *Inorg. Chem.* **1990**, *29*, 1874–1880. (b) Banci, L.; Bertini, I.; Luchinat, C. "The ^{1}H NMR parameters of magnetically coupled dimers—The Fe_2S_2 proteins as an example" *Struct. Bonding (Berlin)* **1990**, *72*, 113–136. (c) Dunham, W. R.; Palmer, G.; Sands, R. H.; Bearden, A. J. "On the structure of the iron-sulfur complex in the two-iron ferredoxins" *Biochim. Biophys. Acta* **1971**, *253*, 373–384. (d) Benini, S.; Ciurli, S.; Luchinat, C. "Oxidized and reduced $[Fe_2Q_2]$ (Q = S, Se) cores of spinach ferredoxin: a comparative study using ^{1}H NMR spectroscopy " *Inorg. Chem.* **1995**, *34*, 417–420.
28. (a) Hagen, K. S.; Holm, R. H. "Synthesis and stereochemistry of metal(II) thiolates of the types $[M(SR)_4]^{2-}$, $[M_2(SR)_6]^{2-}$, and $[M_4(SR)_{10}]^{2-}$ (M = Fe(II), Co(II))" *Inorg. Chem.* **1984**, *23*, 418–427. (b) Werth, M. T.; Kurtz, D. M., Jr.; Moura, I.; LeGall, J. "^{1}H NMR spectra of rubredoxins: new resonances assignable to α-CH and β-CH_2 hydrogens of cysteinate ligands to Fe(II)" *J. Am. Chem. Soc.* **1987**, *109*, 273–275.
29. A general review can be found in Luchinat, C.; Ciurli, S. "NMR of polymetallic systems in proteins" *Biol. Magn. Reson.* **1993**, *12*, 357–420; in *NMR of Paramagnetic Molecules*, Berliner, L. J.; Reuben, J. Eds.; Plenum: New York, 1993.
30. La Mar, G. N.; Eaton, G. R.; Holm, R. H.; Walker, F. A. "Proton magnetic resonance investigation of antiferromagnetic oxo-bridged ferric dimers and related high-spin monomeric ferric complexes" *J. Am. Chem. Soc.* **1973**, *95*, 63–75.

31. Lauffer, R. B.; Antanaitis, B. C.; Aisen, P.; Que, L., Jr., "^{1}H NMR studies of porcine uteroferrin, magnetic interactions and active site structure" *J. Biol. Chem.* **1983**, *258*, 14212–14218.
32. Nanthakumar, A.; Fox, S.; Murthy, N. N.; Karlin, K. D. "Inferences from the ^{1}H-NMR spectroscopic study of an antiferromagnetically coupled heterobinuclear Fe(III)–(X)–Cu(II) S = 2 spin system (X = O^{2-}, OH^-)" *J. Am. Chem. Soc.* **1997**, *119*, 3898–3906.
33. Wang, Z.; Holman, T. R.; Que, L., Jr., "Two-dimensional ^{1}H NMR studies of paramagnetic bimetallic mixed-metal complexes" *Magn. Reson. Chem.* **1993**, *31*, S78–S84.
34. (a) Owens, C.; Drago, R. S.; Bertini, I.; Luchinat, C.; Banci, L. "NMR proton relaxation in bimetallic complexes of Zn(II), Ni(II), and Cu(II)" *J. Am. Chem. Soc.* **1986**, *108*, 3298–3303. (b) Bertini, I.; Owens, C.; Luchinat, C.; Drago, R. S. "Nuclear magnetic resonance proton relaxation in bimetallic complexes containing cobalt(II)" *J. Am. Chem. Soc.* **1987**, *109*, 5208–5212.
35. Ernst, R. R. "Sensitivity enhancement in magnetic resonance" *Adv. Magn. Reson.* **1966**, *2*, 1–135.
36. Becker, E. D.; Ferretti, J. A.; Gambhir, P. N. "Selection of optimum parameter for pulse Fourier transform nuclear magnetic resonance" *Anal. Chem.* **1979**, *51*, 1413–1420.
37. (a) Hore, P. J. "Solvent suppression" *Methods Enzymol.* **1989**, *176*, 64–77. (b) Meier, J. E.; Marshall, A. G. "Methods for suppression of the H_2O signal in proton FT/NMR spectroscopy, A review" *Biol. Magn. Reson.* **1990**, *9*, 199–240.
38. Inubushi, T.; Becker, E. D. "Efficient detection of paramagnetically shifted NMR resonances by optimizing the WEFT pulse sequence" *J. Magn. Reson.* **1983**, *51*, 128–133.
39. Hochmann, J.; Kellerhals, H. "Proton NMR on deoxyhemoglobin: Use of a modified DEFT technique" *J. Magn. Reson.* **1980**, *38*, 23–39.
40. Hore, P. J. "Solvent suppression in Fourier transform nuclear magnetic resonance" *J. Magn. Reson.* **1983**, *55*, 283–300.
41. Keeler, J.; Clower, R. T.; Davis, A. L.; Laue, E. D. "Pulsed-field gradients: Theory and practice" *Methods Enzymol.* **1994**, *239*, 145–207.
42. Ming, L.-J. "Two-dimensional ^{1}H NMR studies of Ca(II)-binding sites in proteins using paramagnetic lanthanides(III) as probes and Yb(III)-substituted bovine α-lactalbumin as an example" *Magn. Reson. Chem.* **1993**, *31*, S104–S109.
43. Ramaprasad, S.; Johnson, R. D.; La Mar, G. N. "Vinyl mobility in myoglobin as studied by time-dependent nuclear Overhauser effect measurements" *J. Am. Chem. Soc.* **1984**, *106*, 3632–3635.
44. Dugad, L. B.; La Mar, G. N.; Unger, S. W. "Influence of molecular correlation time on the homonuclear Overhauser effect in paramagnetic proteins" *J. Am. Chem. Soc.* **1990**, *112*, 1386–1392.
45. (a) Wei, X.; Ming, L.-J.; Cannons, A. C.; Solomonson, L. P. "^{1}H and ^{13}C NMR studies of a truncated heme domain from *Chlorella vulgaris* nitrate reductase: signal assignment of the heme domain" *Biochim. Biophys. Acta* **1998**, *1382*, 129–136. (b) Wei, X. Ph.D. Thesis, "Two-Dimensional NMR Studies of Paramagnetic Metallo-Biomolecules, Metal-antibiotic Drug Complexes and Protein Structure Determination", University of South Florida, Tampa, FL, 1996.
46. Griffey, R. H.; Redfield, A. G. "Proton-detected heteronuclear edited and correlated nuclear magnetic resonance and nuclear Overhauser effect in solution" *Q. Rev. Biophys.* **1987**, *19*, 51–82.
47. (a) Gardner, K. H.; Coleman, J. E. "^{113}Cd–^{1}H heteroTOCSY: A method for determining metal–protein connectivities" *J. Biomol. NMR* **1994**, *4*, 761–774. (b) Schweitzer, B. I.; Gardner, K. H.; Tucker-Kellogg, G. "HeteroTOCSY-based experiments for measuring herteronuclear relaxation in nucleic acids and proteins" *J. Biomol. NMR* **1995**, *6*, 180–188.
48. (a) Otvos, J. D.; Engeseth, H. R.; Wehrli, S. "Multiple-quantum ^{113}Cd–^{1}H correlation spectroscopy as a probe of metal coordination environments in metalloproteins" *J. Magn. Reson.* **1985**, *61*, 579–584. (b) Frey, M. H.; Wagner, G.; Vašák, M.; Sørensen, O. W.;

Neuhaus, D.; Wörgøotter, E.; Kägi, J. H. R.; Ernst, R. R.; Wüthrich, K. "Polypeptide-metal cluster connectivities in metallothionein 2 by novel ^{1}H–^{113}Cd heternuclear two-dimensional NMR experiments" *J. Am. Chem. Soc.* **1985**, *107*, 6847–6851.
49. Skidmore, K. Simonis, U. "Novel strategy for assigning hyperfine shifts using pulse-field gradient heteronuclear multiple-bond correlation spectroscopy" *Inorg. Chem.* **1996**, *35*, 7470–7471.
50. Bertini, I.; Luchinat, C.; Ming, L.-J.; Piccioli, M.; Sola, M., Valentine, J. S. "Two-dimensional ^{1}H NMR studies of the paramagnetic metalloenzyme copper-nickel superoxide dismutase" *Inorg. Chem.* **1992**, *28*, 4433–4435.
51. Ming, L.-J.; Lauffer, R. B.; Que, L., Jr., "Proton nuclear magnetic resonance studies of iron(II/III)-amide complexes, spectroscopic models for non-heme iron proteins" *Inorg. Chem.* **1990**, *29*, 3060–3064.
52. Que, L. Jr.; Lauffer, R. B.; Lynch, J. B.; Murch, B. P.; Pyrz, J. W. "Elucidation of the coordination chemistry of the enzyme–substrate complex of catechol 1,2-dioxygenase by NMR spectroscopy" *J. Am. Chem. Soc.* **1987**, *109*, 5381–5385.
53. Elgren, T. E.; Ming, L.-J.; Que, L. Jr. "Spectroscopic studies of Co(II)-reconstituted ribonucleotide reductase R2 from *E. coli*" *Inorg. Chem.* **1994**, *33*, 891–894.
54. Maroney, M. J.; Kurtz, D. M., Jr.; Nocek, J. M.; Pearce, L. L.; Que, L., Jr., "^{1}H NMR probes of the binuclear iron cluster in hemerythrin" *J. Am. Chem. Soc.* **1986**, *108*, 6871–6879.
55. Dugad, L. B.; La Mar, G. N.; Lee, H. C.; Ikeda-Saito, M.; Booth, K. S.; Caughey, W. S. "A nuclear Overhauser effect study of the active site of myeloperoxidase" *J. Biol. Chem.* **1990**, *265*, 7173–7179.

General References

Banci, L.; Bertini, I.; Luchinat, C. *Nuclear and Electron Relaxation*, VCH: New York, 1991. As its title refelcts, this book covers both nuclear and electron relaxation phenomena and serves as a good reference for better understanding of relaxation phenomenon in paramagnetic species.

Berliner, L. J.; Reuben, J., Eds. *NMR of Paramagnetic Molecules (Biol. Magn. Reson. Vol. 12)*, Plenum: New York, 1993. This is a multiauthored book that provides some valuable experimental approaches and detailed theoretical treatments for proteins with heme, Fe–S, and magnetically coupled centers.

Bertini, I.; Luchinat, C. *NMR of Paramagnetic Molecules in Biological Systems*; Benjamin/Cummings: Menlo Park, CA, 1986. This book is widely accepted as the general reference for this field since its publication. It provides the fundamentals for understanding of paramagnetic systems. An expanded version of this book was published later (Bertini, I.; Luchinat, C. "NMR of Paramagnetic Substances" *Coord. Chem. Rev.* **1996**, *150*, 1–296.) that includes 2D NMR applications.

Bertini, I.; Turano, P.; Vila, A. J. "Nuclear Magnetic Resonance of Paramagnetic Metalloproteins" *Chem. Rev.* **1993**, *93*, 2833–2932. This review is a comprehensive collection of NMR studies on metalloproteins prior to 1993.

Croasmun, W. R.; Carlson, R. M. K. *Two-Dimensional NMR Spectroscopy*; VCH: New York, 1987. This book discusses 2D NMR applications of organic compounds.

Ernst, R. R.; Bodenhausen, G.; Workaun, A. *Principles of Nuclear Magnetic Resonance in One and Two Dimensions*; Clarendon: Oxford, 1987. This book provides more detailed theoretical background about 2D NMR spectroscopy.

La Mar, G. N.; Horrocks, W. D., Jr.; Holm, R. H., Eds. *NMR of Paramagnetic Molecules*; Academic: NY, 1973. This classic book provides good background material on chemical shift, electron delocalization pathways, and nuclear relaxation and early applications on paramagnetic systems.

La Mar, G. N.; Ed. *Nuclear Magnetic Resonance of Paramagnetic Macroolecules*, NATO-ASI, Kluwer: Dordrecht, the Netherlands, 1995. This book collects articles on all aspects of NMR of paramagnetic macromolecules based on presentations at an NATO Advanced Research Workshop in Sintra, Portugal, 1994.

Neuhaus, D.; Williamson, M. P. *The Nuclear Overhauser Effect in Structural and Conformational Analysis*; VCH: NY, 1989. This most recent book has a comprehensive coverage of the title subject.

Sigel, H.; Ed. *Metal Ions in Biological Systems*, Vol 21; Dekker: New York, 1987. This volume on "*Applications of Nuclear Magnetic Resonance of Paramagnetic Species*" contains review chapter covering NMR studies of various paramagnetic systems by a number of authors.

9

X-Ray Absorption Spectroscopy

ROBERT A. SCOTT

Department of Chemistry
University of Georgia
Athens, GA 30602-2556

CONTENTS

I. Introduction

Since X-rays have wavelengths on the order of atomic dimensions, these highly energetic photons can be used to sample the molecular structure of materials. One example of this use is X-ray diffraction from crystalline samples resulting in a complete three-dimensional (3D) crystal structure. X-ray absorption spectroscopy (XAS) can also yield limited molecular structural information on noncrystalline (amorphous) samples. Since an X-ray absorption spectrum is a mea-

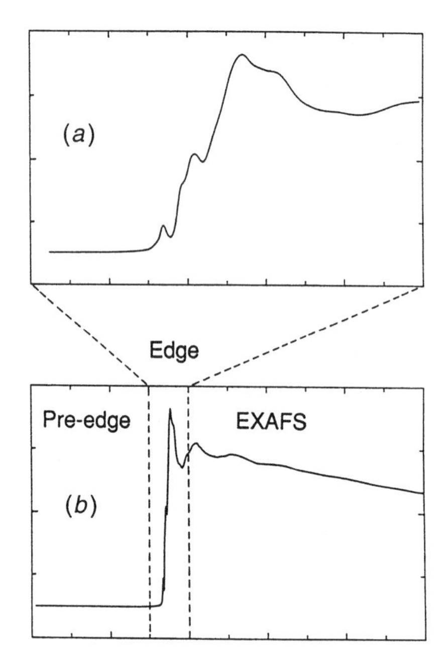

Figure 1
An X-ray absorption spectrum plotted as absorption coefficient versus photon energy. (*a*) Expanded view of the edge region. (*b*) Full spectrum showing the pre-edge, edge, and extended X-ray absorption fine structure (EXAFS) regions.

surement of the energy-dependent absorption coefficient of a material, a tunable X-ray source is required. The advent of synchrotron radiation as a high-intensity, tunable source of X-rays has been responsible for the emergence of XAS within the last three decades as an important new structural and spectroscopic technique.

A typical X-ray absorption spectrum is shown in Figure 1. At a well-defined X-ray photon energy, a sharp rise in absorption coefficient is observed. This rise is called an X-ray absorption *edge* and is due to electron dissociation from a core level of one type of atom (the absorbing atom) in the sample (see below). Spectral features in the edge region are sensitive to the electronic structure of the absorbing atom and can often be used to identify the geometric arrangement of atoms around the absorbing atom. Above the edge, the quasiperiodic modulations in the X-ray absorption coefficient are referred to as EXAFS. The EXAFS may be analyzed to give direct structural information about the local environment around the absorbing atom.

In the literature, a bewildering array of (often redundant) acronyms have been used to describe techniques for extracting the structural and spectroscopic information from the X-ray absorption spectrum. X-ray absorption spectroscopy (XAS) is the generic term that refers to all the techniques. Recently, X-ray absorption fine structure (XAFS) has gained some popularity as another generic reference to analysis of "fine structure" in both edge and EXAFS regions of the spectrum (Fig. 1). The spectral features in the edge region (and their analysis) are most often given the acronym XANES (X-ray absorption near-edge structure), but are sometimes referred to as NEXAFS (near-edge X-ray absorption fine structure). The acronym EXAFS refers to both the modulations in the absorption coefficient above the edge (Fig. 1) and also to the techniques used to analyze them in terms of molecular structure. Other acronyms are sometimes used to describe EXAFS techniques applied to particular types of samples. For example, surface extended X-ray absorption fine structure (SEXAFS) is a technique used to measure the EXAFS of surface species. To avoid confusion, only XAS and EXAFS will be used in this chapter. Any discussion of analysis of the spectral features in the X-ray absorption edge region will simply refer to the *edge* region.

As will be discussed in some detail below, the utility of XAS as a structural technique stems from its ability to give direct structural information about a *local region* around *specific elements* in *amorphous* samples. The technique is element specific since each element in the periodic table exhibits an X-ray absorption edge at a different energy (see Fig. 3). Thus, in a sample with several different elements, the structure around each element may be probed independently. The EXAFS technique yields radial structural information within a 4–5-Å radius around the absorbing atom. Specifically, the question to be answered is *How many* of *what type* of atom are at *what distance* from the absorbing atom? Since EXAFS only gives *radial* distance information, orientation of the absorbing atom sites within a sample is not important. Amorphous samples (e.g., powders, solutions, frozen solutions, and gases) are amenable to study by this technique. The chemical environment of the absorbing element has only a small (yet measurable) effect on the energy of the edge, causing the EXAFS from all occurrences of that element in the sample to overlap. A heterogeneous population of site structures therefore gives rise to an *average* structural environment making EXAFS a poor technique for identifying such structural heterogeneity. The technique is best suited for samples in which all sites of the absorbing atom are structurally identical. In contrast to the simple radial structural information available from EXAFS, the spectral features in the edge region can yield geometrical structural information (e.g., distinguishing tetrahedral from octahedral coordination), making these two spectral regions complementary for structural determinations.

II. Theory of X-Ray Absorption Spectroscopy

In this section, a brief look at the theoretical basis of the X-ray absorption spectroscopic technique will be given. For a more in-depth look at the basic principles, data collection techniques, and applications of XAS, two books are helpful (see Koningsber and Prins[1] and Teo[2].)

A. Edges

The X-ray absorption spectrum of a given sample will exhibit an edge at a photon energy equal to the ionization potential of a bound electron in the constituent atoms of the sample. As illustrated in Figure 2, scanning the photon energy

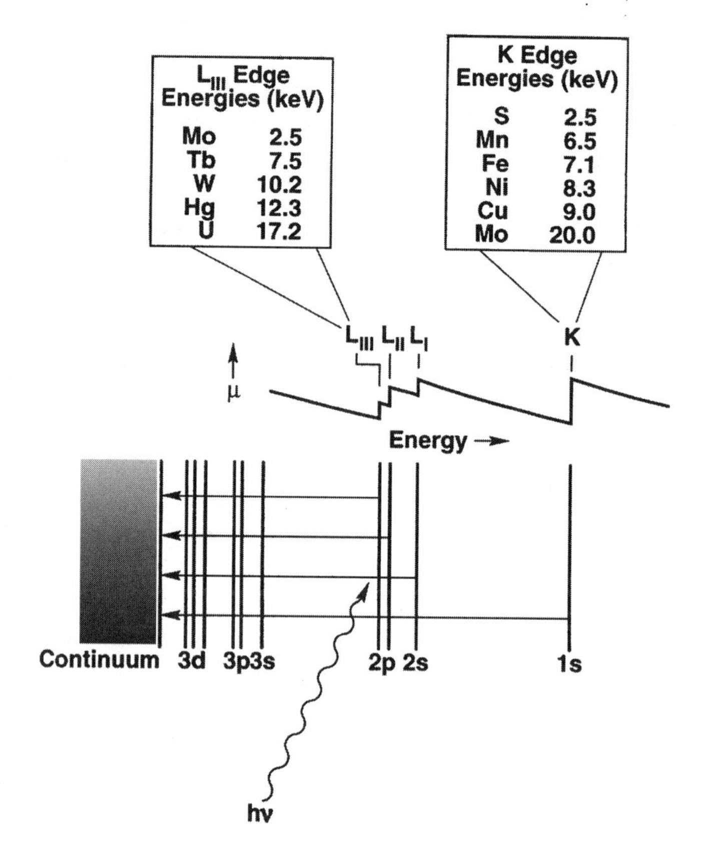

Figure 2
Schematic diagram relating the X-ray absorption spectrum to the atomic energy level diagram. X-ray photon absorption is indicated by the arrows in the energy level diagram, each giving rise to an absorption edge in the top spectrum. Some L_{III} and K X-ray absorption edge energies are tabulated above the diagram.

causes every atom in the sample to give rise to several absorption edges as the photon energy matches the ionization potential of each bound electron $(1s, 2s, 2p_{1/2}, 2p_{3/2}, \ldots)$. These X-ray absorption edges are named for the shells of the Bohr atom (K edges for $n = 1$, L edges for $n = 2$, M edges for $n = 3$, etc.) as indicated in Figure 2. Current synchrotron sources can deliver high-intensity X-ray beams with photon energies ($h\nu$) up to about 30 keV. The low-energy limit is dictated by window material and atmospheric absorption. Thus, unless samples can be handled in ultrahigh vacuum on a windowless beamline, it is difficult to use photon energies below about 2 keV. The approximate K and L_{III} edge energies of a few representative elements are listed in Figure 2. These edge energies are a monotonic function of atomic number (Fig. 3), so the accessible X-ray photon energy region ($\sim$2–30 keV) defines the range of accessible K edges to extend from approximately P to Sn. However, since the same type of structural infor-

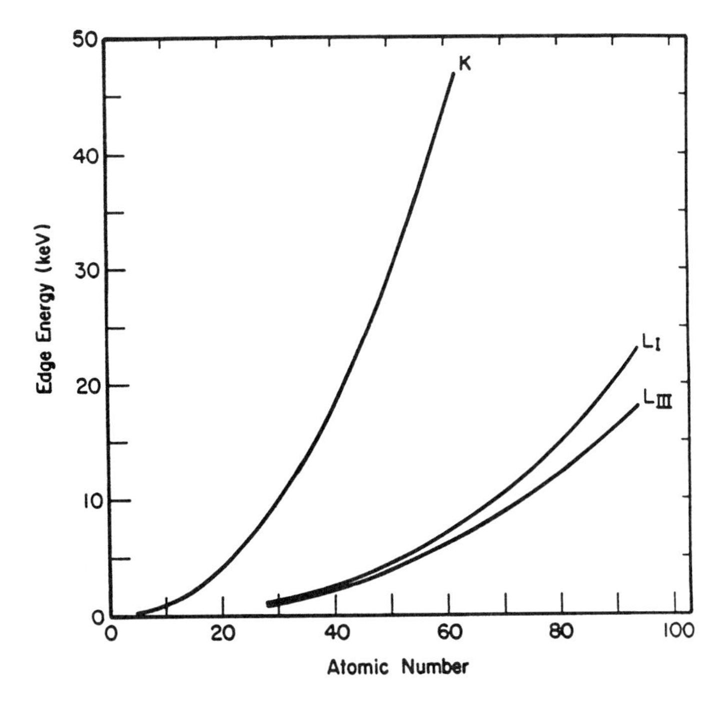

Figure 3
The K, L_I, and L_{III} X-ray absorption edge energies for elements as a function of their atomic numbers. Elements with K-edge energies above the available source photon energies can be studied by examining their L edges.

mation may be obtained from L edges, the rest of the elements in the periodic table are also amenable to the XAS technique. (Accessible L_{III} edges range from ~2.1 keV for Y to ~22 keV for Lr.)

X-ray absorption edges are not simple discontinuities in the absorption coefficient as suggested by the atomic energy level diagram in Figure 2. For example, the spectrum shown in Figure 1 exhibits spectral features at or just before the edge and understanding the origin of these features can yield information about the electronic structure of the absorbing atom site in the sample. The peaks (shoulders) just before the edge arise from electronic transitions from the core level (1s for K edges) to valence levels just below the continuum. Using appropriate selection rules, the intensities of these transitions may be related to the symmetry of the absorbing atom site. Also, the exact energy of the edge is dependent on the charge density at the absorbing atom, which is influenced by the chemical environment (e.g., valence or oxidation state). In general, the higher the oxidation state of the absorbing atom, the higher the energy of the X-ray absorption edge. A simple electrostatic explanation is that it is more difficult to dissociate an electron from an atom with higher positive charge. Some examples of these effects will be discussed in detail later.

B. EXAFS

The information content of the EXAFS technique arises from the physical interaction of the photodissociated core electron (the photoelectron) with electron density surrounding neighboring atoms. It is the influence of this photoelectron–electron scattering on the absorption coefficient for the X-ray photon that gives rise to the EXAFS modulations (see Fig. 5). For this reason, structural information about the neighboring electron density responsible for photoelectron scattering is "encoded" into these absorption coefficient modulations. This section will describe the quantitative details of this photoelectron scattering that allow this structural information to be "decoded" from the EXAFS data.

In the EXAFS region, the X-ray photons absorbed have a higher energy than that necessary to ionize the absorbing atom. Since energy must be conserved, all the energy delivered by the X-ray photon must be accounted for. Defining the X-ray absorption edge energy as E_0 (the threshold energy, equivalent to the ionization potential) and assuming a photon energy of E ($E > E_0$), where did the extra energy ($E - E_0$) delivered by the X-ray photon go? It is transferred into kinetic energy of the photoelectron, which can be considered as a wave. The wavelength of the photoelectron wave (referred to as the de Broglie wavelength) is dependent on its kinetic energy, so that as the photon energy is scanned throughout the EXAFS region, the wavelength of this photoelectron is also "scanned". A photoelectron wave with kinetic energy ($E - E_0$) can be considered to be propagating through-space from the absorbing atom origin at a velocity, v, where $(E - E_0) = m_e v^2/2$ (m_e = electron mass). The de Broglie wavelength of the photoelectron is inversely proportional to its momentum $(m_e v) : \lambda = h/m_e v$. When discussing EXAFS data, it is convenient to use the photoelectron wave vector, k, as the independent variable, rather than the energy, E. The parameter k is proportional to momentum.

$$k = \frac{2\pi m_e v}{h} = \frac{2\pi}{\lambda} = \left[\frac{8\pi^2 m_e}{h^2}(E - E_0)\right]^{1/2} = [0.262449(E - E_0)]^{1/2} \qquad (1)$$

The final numerical expression holds for E and E_0 expressed in units of electron-volts and k in units of reciprocal angstroms.

One early observation that aided understanding of the origin of the EXAFS modulations was the complete absence of these modulations in X-ray absorption spectra of monoatomic gases such as Kr (Fig. 4). In contrast, EXAFS is always observed in condensed media (liquids and solids) or polyatomic gases in which an

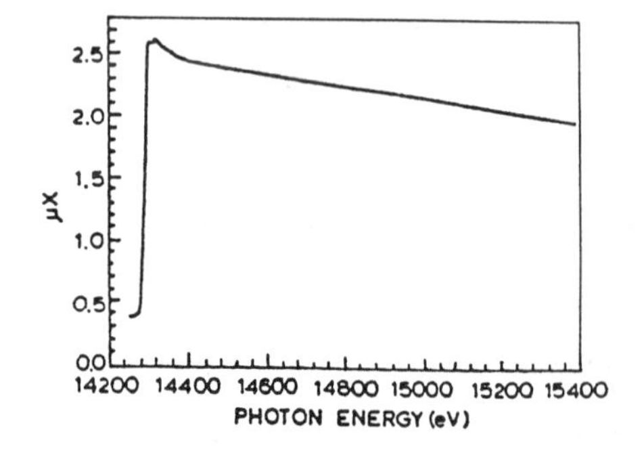

Figure 4
The K X-ray absorption edge spectrum for krypton gas. No EXAFS modulations are visible as expected for a monoatomic gas. [This figure was adapted from Powers.[3]]

absorbing atom is surrounded by other atoms in a regular arrangement. To understand the origin of EXAFS, we must consider the scattering of the photoelectron wave from electron density surrounding neighboring atoms. In this context, the neighboring atoms are often referred to as "scattering atoms" or "scatterers". Consider the simplest possible arrangement of atoms, a diatomic molecule with an absorbing atom *a* and a scattering atom *s*. Figure 5(*a*) shows the photoelectron wave propagating from atom *a* (solid arcs) and scattering from the electron density around atom *s* (dashed arcs). In fact, the photoelectron scatters from *s* in all directions; only the backscattered wave is shown in Figure 5. Mathematically, we

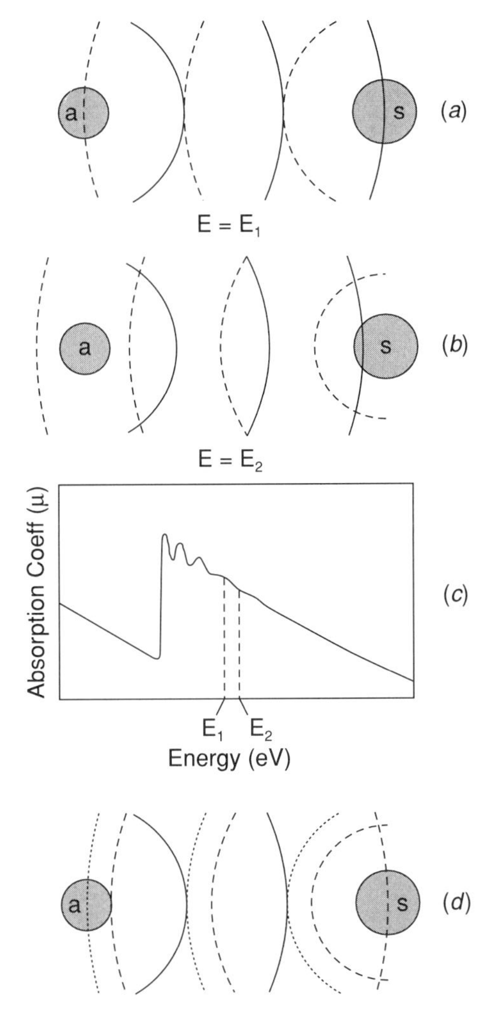

Figure 5
Diagram of photoelectron waves generated by X-ray photon absorption at two different energies. Parts (*a*) and (*b*) indicate the photoelectron scattering for energies E_1 and E_2, respectively, as indicated in the spectrum in (*c*). In (*a*), the scattered photoelectron has a maximum amplitude at the absorbing atom a and in (*b*), the scattered photoelectron has a minimum amplitude at atom a. This interference leads to a maximum and minimum, respectively, in the EXAFS shown in (*c*). In (*d*), the scattering observed at energy E_1 is displayed with a more realistic phase shift (dashed arcs) compared to the simplified diagram of (*a*) (dotted arcs). As the photoelectron wave "passes through" atom *s*, the scattered wave is phase shifted with respect to the outgoing wave (solid arcs).

can write the scattering amplitude as a function of both θ, the angle of scattering, and $k : f_s(\theta, k)$. Then, the backscattering ($\theta = \pi$) amplitude is $f_s(\pi, k)$.

To relate the photoelectron scattering event, which we normally cannot directly observe, to the X-ray absorption coefficient, μ, which we can observe, we must consider the electric dipole transition moment that is equal to the square of the integral $\langle f|\hat{\mathbf{e}}|i\rangle$, where f and i represent wave functions for the final state and the initial state, respectively, and $\hat{\mathbf{e}}$ is the electric dipole operator. For the purposes of this discussion, it suffices to consider only the *overlap* of initial and final state wave functions. For K edges, the initial state is a 1s wave function centered on atom a. (For other edges, the initial state wave function is still centered on atom a.) For an X-ray photon in the EXAFS region, the final state wave function is a combination of the electronic wave function of the relaxed ionized absorbing atom a and the photoelectron wave (both "outgoing" and back-scattered). Since it is only the energy (k) dependence of μ in which we are interested (the EXAFS is just the energy-dependent modulation of μ above the edge), we can ignore the contribution from the ionized atom a and concentrate on the photoelectron wave. Given the localization of the initial state wave function on atom a, only the amplitude of the backscattered photoelectron wave *at the absorbing atom* contributes to the transition moment. Figure 5(*a*) shows that, for energy E_1, this amplitude is a maximum (illustrated by the dashed arc through atom a), thus giving rise to a maximum in the modulation of μ [Fig. 5(*c*)].

Increasing the photon energy to E_2 generates a photoelectron with a shorter wavelength (larger kinetic energy), the scattering of which is illustrated in Figure 5(*b*). At this energy, the backscattered photoelectron wave has a *minimum* amplitude at atom a, generating a minimum in the μ modulation [Fig. 5(*c*)]. Extending this analysis to energies in the rest of the EXAFS region generates the periodic modulation of μ as the photoelectron wavelength is scanned and the backscattered photoelectron wave periodically goes in and out of phase with the outgoing photoelectron wave (an interference effect). Thus, the μ modulation (i.e., the EXAFS) generated by a single scattering atom resembles a sine wave. [The damping of these sinusoidal components that is evident in Figure 5(*c*) is caused by static and dynamic variations in absorber–scatterer distances. This damping is termed the Debye–Waller effect and will be discussed in detail later.]

This description implies a dependence of the EXAFS modulation on the nature and location of the scattering atom s. Thus, proper analysis of the EXAFS can extract this information, elucidating the local structure around the absorbing atom a. In particular, EXAFS can determine *how many* of *what type* of atom are at *what distance* from atom a. Before detailing the mathematical analysis of the EXAFS data to extract this information, a less precise physical picture is useful. Each scattering atom contributes a (damped) sine wave to the overall EXAFS spectrum of a typical coordination site (e.g., in a discrete molecule or a solid lattice). Each of these sine waves can be described by three measurable quantities: frequency, amplitude, and phase. Each of these observables contains structural information about the nature and location of the scattering atom(s) giving rise to that EXAFS component sine wave. As the chart below summarizes, the frequency of the sine wave is a measure of the distance between atoms a and s, the amplitude of the sine wave is a measure of the number of atoms (of that type) at that distance, and the phase of the sine wave helps define the element doing the scattering (i.e., the identity of atom s).

Observable		Information
Frequency	⟶	Distance
Amplitude	⟶	(Coordination) number
Phase	⟶	Atom type

(A diagonal arrow also links Amplitude to Atom type.)

The frequency–distance relationship can be seen by examining Figure 5(*a*) again. At a given photon energy (i.e., a given photoelectron wavelength), a longer a–s distance implies that more periods of the photoelectron wave are required to cover the distance from a to s and back. (The reader should note that the concept of the photoelectron wave *traveling* from a to s and back is a classical description and has limitations.) Thus, for a longer distance, it takes less of a change in photoelectron wavelength (less of a change in photon energy) to go through one period of the interference. The absorption coefficient is therefore modulated at a higher frequency in E (or k).

The relationship of EXAFS sine wave amplitude to the number of atoms s is straightforward. In a coordination site with two identical atoms s (e.g., two oxygen or two sulfur atoms) at the same distance from a, each will contribute identical sine waves to the EXAFS that simply add together to give a sine wave with twice the amplitude. One inherent limitation of the EXAFS technique is that it yields only radial (no angular) structural information. Thus, it does not matter how the two s atoms are arranged around atom a, only that they are at the same distance. This radial dependence gives rise to the concept of "shells" of atoms, which are defined as a collection of atoms (of the same type) all residing at the same distance from the absorbing atom. One shell of atoms gives rise to one (damped) sine wave in the EXAFS.

The phase/atom type relationship is somewhat more difficult to visualize. A subtle effect of the photoelectron scattering illustrated in Figure 5 is the phase shift introduced into the photoelectron wave during the scattering process. For clarity, Figure 5(*a*) was drawn with no phase shift of the photoelectron wave upon backscattering from atom s. Figure 5(*d*) illustrates a more realistic phase shift; the backscattered photoelectron wavelength and EXAFS sine wave frequency remain unaffected. Since the size of this phase shift depends on the electron density at atom s (and at atom a), for a given absorbing atom, different scattering atom types introduce different phase shifts, yielding EXAFS components with different phases. The cross relationship (diagonal arrow on the chart) between amplitude and atom type refers to the fact that heavier scattering atoms (elements with higher atomic number) are better scatterers, yielding EXAFS components with larger amplitudes (for a given number of atoms).

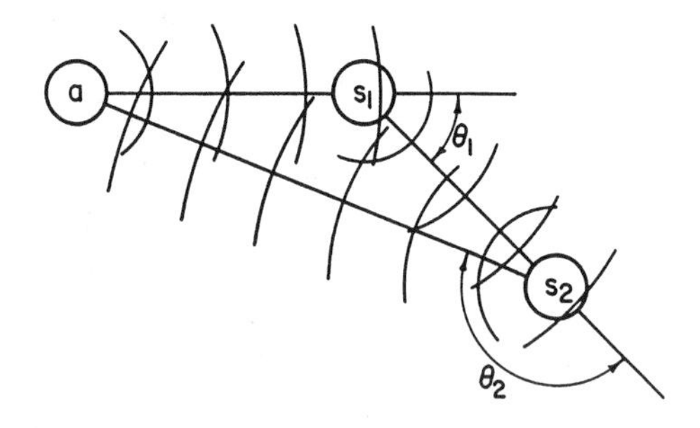

Figure 6
One possible multiple-scattering pathway for a photoelectron generated by X-ray absorption by atom a in the presence of two scattering atoms, s_1 and s_2. The scattering pathway indicated is a–s_1–s_2–a. The multiple-scattering contribution to the EXAFS depends on the distances involved *and* on the angles θ_1 and θ_2.

So far, only single-scattering EXAFS theory has been discussed. In other words, we have only considered the photoelectron backscattering to atom a directly from a single atom s. This treatment ignores the possibility that the photoelectron might encounter two (or more) scattering atoms in its "round trip" back to atom a (Fig. 6), a possibility that is now known to contribute to the EXAFS in certain cases. Although this multiple-scattering phenomenon has been thoroughly treated,[4,5] it is beyond the scope of this text to include the detailed theoretical treatment here. The EXAFS expression to be discussed next only treats single scattering.

The quantitative analysis of EXAFS data in terms of absorber–scatterer (a–s) distances, scatterer numbers, and scatterer types, requires a mathematical expression for the EXAFS quantity, χ, as a function of k. The parameter χ is defined to refer only to the quasiperiodic portion (the modulations) of the EXAFS region of the X-ray absorption coefficient, μ.

$$\chi \equiv \frac{\mu - \mu_0}{\mu_0} \tag{2}$$

In Eq. (2), μ_0 is the energy-dependent X-ray absorption coefficient that would have been observed if the sample contained only the absorbing atom (at the same concentration) without any neighboring scattering atoms. The parameter μ_0 is referred to as the free-atom absorption coefficient. In theory, this expression isolates the quasiperiodic EXAFS modulations from the smooth background free-atom absorption and normalizes the EXAFS to a "per absorbing atom" basis. Operationally, μ_0 proves impossible to measure and difficult to simulate so that Eq. (2) must be modified as discussed later.

The theoretical single-scattering expression for χ is given by Eq. (3):

$$\chi(k) = \sum_{s} \frac{N_s|f_s(\pi, k)|}{kR_{as}{}^2} \exp(-R_{as}/\lambda_f) \exp(-2\sigma_{as}{}^2k^2) \sin[2kR_{as} + \alpha_{as}(k)] \tag{3}$$

Symbol	Units	Definition
N_s		The number of scattering atoms in this shell
$\|f_s(\pi, k)\|$		The inherent backscattering amplitude for this type of scattering atom
R_{as}	Å	The distance between the absorbing atom and the scattering atoms in this shell
λ_f	Å	The mean free path for inelastic scattering of the photoelectron
${\sigma_{as}}^2$	$Å^2$	The rms deviation of R_{as} ($\exp(-2{\sigma_{as}}^2 k^2)$ is referred to as the Debye–Waller factor)
$\alpha_{as}(k)$		The inherent backscattering phase shift for this absorbing atom/scattering atom combination

The summation is over shells of scattering atoms, s, and each term within the summation consists of an amplitude term, an exponential damping term (the Debye–Waller factor), and a sine function to describe the (quasi-)periodic behavior of the EXAFS. As noted previously, the EXAFS amplitude is directly proportional to the number of scattering atoms (N_s) and the k-dependent shape of the EXAFS amplitude (the amplitude envelope) is defined by the backscattering amplitude $|f_s(\pi, k)|$, which is different for every scattering atom type. The amplitude also shows a k^{-1} dependence but this is usually compensated (as is the Debye–Waller damping) by working with the quantity $k^n\chi(k)$ ($n = 3$, usually) (see below). The ${R_{as}}^{-2}$ dependence makes the EXAFS of long-distance shells much weaker than that from nearby atoms. Thus, only atoms within a radius of about 4–5 Å of the absorbing atom contribute significant scattering to the EXAFS. This fall-off of EXAFS amplitude at high R_{as} also has a contribution from inelastic losses of the photoelectron, which are more serious for longer distances. These inelastic losses are usually treated by defining a mean free path for the photoelectron, λ_f, and incorporating the $\exp(-R_{as}/\lambda_f)$ term in Eq. (3).

Within a shell of scattering atoms, there is some variation in R_{as}, which may be static (a spread in the a–s distances from structural distortion or site heterogeneity) or dynamic (e.g., due to a stretching vibration in the a–s bond). This variation leads to a damping of the EXAFS oscillations, which is physically described by ${\sigma_{as}}^2$, a root-mean-square (rms) deviation in the distance, R_{as}. The vibrational portion of ${\sigma_{as}}^2$ has a characteristic temperature dependence.

The quasiperiodic behavior of the EXAFS is described by the sine function of Eq. (3), the argument of which has a frequency, $2R_{as}$, and a phase shift, $\alpha_{as}(k)$. Thus, the measured frequency of each shell's EXAFS contribution is directly proportional to the a–s distance. The phase shift is dependent on k and also dependent on the nature of both the absorbing and scattering atom types. There will be more to say about both the backscattering amplitude, $|f_s(\pi, k)|$, and backscattering phase shift, $\alpha_{as}(k)$, in the discussion of data analysis.

Equation (3) describes the EXAFS, $\chi(k)$, as a sum of damped sine waves in k space, suggesting that a Fourier transform (FT) of $\chi(k)$ might be a useful analysis technique. (Fourier transformation of a sum of sine waves yields peaks in the appropriate frequency space centered at the frequency of each sine wave.) For EXAFS, the frequency of each component is $2R_{as}$, so that a proper FT will yield a set of peaks in distance space, yielding a direct visualization of the shells of scattering atoms at different distances from the absorbing atom.

III. X-Ray Absorption Spectroscopy Data Collection

The spectrometer required to measure an X-ray absorption spectrum contains all the basic components found in a typical ultraviolet–visible (UV–vis) spectrophotometer: a source, a monochromator, and detectors (Fig. 7). Each of these components is specifically adapted for the X-ray region of the spectrum. Two types of data collection will be discussed. Transmission XAS consists of measuring the X-ray photon intensity (I_0) incident on the sample and an intensity (I) transmitted through the sample (Fig. 7). Then $\ln(I_0/I)$ is proportional to the absorption coefficient. The second mode of data collection utilizes fluorescence excitation techniques in which fluorescent X-ray photons are counted (using a fluorescence detector, Fig. 7) as the photon energy is scanned, generating a signal proportional to the absorption coefficient (F/I_0). The fluorescence technique will be discussed in more detail below after each component of the XAS spectrometer is described.

A. Source

The EXAFS technique is inherently insensitive. At $k \approx 10$ Å^{-1}, the EXAFS modulations are always less than about 1% of the size of the edge. Since such a small signal must be extracted from a large background, the X-ray source must be powerful. Four main types of X-ray sources are available: fixed-anode sources, offering monochromatic X-rays typically used for X-ray diffraction; rotating

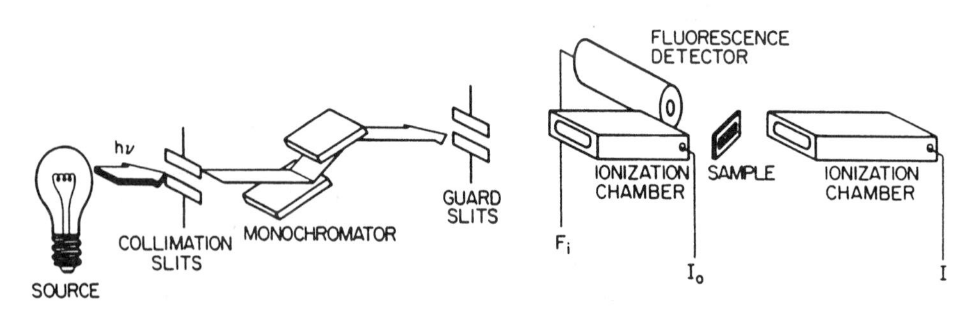

Figure 7
Schematic illustration of a typical experimental setup for X-ray absorption (see text for a description of each component).

anodes, emitting a broader range of X-ray energies and more intensity than fixed-anode sources; plasma sources, offering the possibility of time-resolved measurements[6]; and electron (or positron) storage rings, yielding synchrotron radiation, a broad-band, high-intensity source. Since an X-ray absorption spectrum requires scanning the X-ray photon energy, it is obviously important to have a source with high intensity throughout a range of energies. Fixed-anode sources are thus not useful for XAS. Rotating anode sources are used for XAS, usually as components of laboratory XAS spectrometers, which will be discussed later. Synchrotron radiation first became available for spectroscopic uses in the early 1970s and revolutionized the field of X-ray absorption spectroscopy, since the high-brightness (high intensity in small spot size), tunable synchrotron radiation made XAS feasible for a large range of dilute samples for which other sources were inadequate. For the purposes of this discussion, we can just consider the synchrotron source as providing a high-intensity, small cross section (typically $\sim 2 \times 20$ mm or less), highly collimated beam of X-rays with a spectral range covering the approximate 2–30-keV region necessary for XAS of most elements.

B. Monochromator

Since X-rays penetrate most optical materials, dispersive optical elements (gratings and prisms) are not available for the hard X-ray region. For this reason, X-ray monochromators utilize Bragg reflection from single crystals for monochromatization. The wavelength (λ) of the diffracted beam depends on the angle of incidence (θ) with a Bragg lattice plane of the crystal and upon the d spacing of the crystal (the separation between the lattice planes).

$$n\lambda = 2d \sin\theta \tag{4}$$

For the fundamental reflection, $n = 1$; higher energy harmonics that satisfy this Bragg relationship for $n > 1$ are also reflected from the crystal. Usually, two parallel single crystals (both cut parallel to the same lattice planes) are utilized in a double-crystal monochromator as illustrated in Figure 7. Simply rotating both crystals to vary θ allows tuning of λ [Eq. (4)] and thus energy [Eq. (1)] through a range covering the spectral region of interest. The single-crystal slabs used as monochromator elements are often several centimeters in each dimension (and ~ 1 cm thick) so that silicon is usually (but not exclusively) the material of choice (the semiconductor industry has developed methods for growing silicon crystals large enough for this application) and the Si(111), Si(220), or Si(400) are the most frequently used reflections. As shown in Figure 7, vertical entrance and exit slits are usually present for energy resolution and scattered radiation rejection.

C. Detectors

For quantitative measurement of X-ray absorption spectra, detectors are required to measure the flux (photons s^{-1}) of the X-ray beam. Since hard X-rays can ionize gases, a simple detector can be built to measure the amount of ionized gas per unit time (in a defined volume) which is proportional to X-ray flux. These ionization chambers consist of thin X-ray transparent windows on each end of a chamber holding an inert gas (He, N_2, Ne, or Ar, depending on photon energy range) between two charged plates. Ionized gas molecules (atoms) migrate to the cathodic plate and the electrons migrate to the anodic plate, creating a current proportional to the X-ray flux. Ionization chambers work well as transmission detectors since a significant fraction of the X-ray photons are transmitted through the detector (e.g., to the sample and other detectors; see Fig. 7). For concentrated samples (i.e., ones with a high mole ratio of absorbing atoms), the transmission technique works well. However, in cases for which the sample consists mainly of atom types other than the one being investigated (spectroscopically dilute samples, such as metalloenzymes, highly dispersed supported catalysts, ultrathin films, etc.) fluorescence excitation techniques are much more sensitive.

For K X-ray absorption edges, at photon energies above the edge, a fraction of the ionized absorbing atoms relax by emission of a fluorescent X-ray photon [Fig. 8(*a*)]. This usually occurs by transition of an electron from the $n = 2$ levels to the 1s level, giving rise to Kα fluorescence. Since the $n = 2$ levels lie well below the continuum, the energy of these Kα photons is significantly lower than the K-edge energy. (For example, for Cu with a K edge of 8.98 keV, the Kα fluorescence occurs at 8.05 keV.) The fluorescence excitation XAS technique consists of monitoring the Kα emission (by a photon-counting fluorescence detector, Fig. 7) while the incident X-ray photon energy is scanned through the K edge and EXAFS regions. Since the number of emitted Kα photons is directly proportional to the number of photons absorbed (i.e., the absorption coefficient of the absorbing atom), the fluorescence excitation technique gives a spectrum that mimics the X-ray absorption spectrum of the absorbing atom. The fluorescence excitation technique has a distinct advantage for dilute samples. A transmission spectrum of a dilute sample displays background absorption by all other atoms in the sample [Fig. 9(*a*)], whereas the fluorescence excitation spectrum displays only the absorption by the atom of interest (the absorbing atom, whose Kα emission is being monitored) [Fig. 9(*b*)]. Since the background absorption in the transmission spectrum contributes no signal (only noise), the fluorescence excitation spectrum (with no background) yields a better signal-to-noise (S/N) ratio.

This ideal situation holds only for fluorescence detectors with very high energy resolution. Fluorescence excitation spectra often have a (small) background contribution from scattered photons [Fig. 9(*b*)]. As shown in Figure 8(*b*), most of these scattered photons have energies equal to the energy of the incident X-rays (i.e., they are elastically scattered), so that fluorescence detectors with high-energy resolution, monitoring the Kα fluorescence, will not detect them (this is referred to as scatter rejection). A small amount of Compton scattering may be detected as background near the Kα energy [Fig. 8(*b*)]. Occasionally, low-energy resolution fluorescence detectors (e.g., scintillation detectors or ionization chambers) are used for fluorescence excitation XAS. These detectors cannot resolve the Kα fluorescence from the scattered photons and can only provide improved sensitivity over the transmission technique if another means of scatter rejection is employed. This scatter rejection is usually accomplished by placing between the sample and the fluorescence detector a low-pass filter with a cut-off energy be-

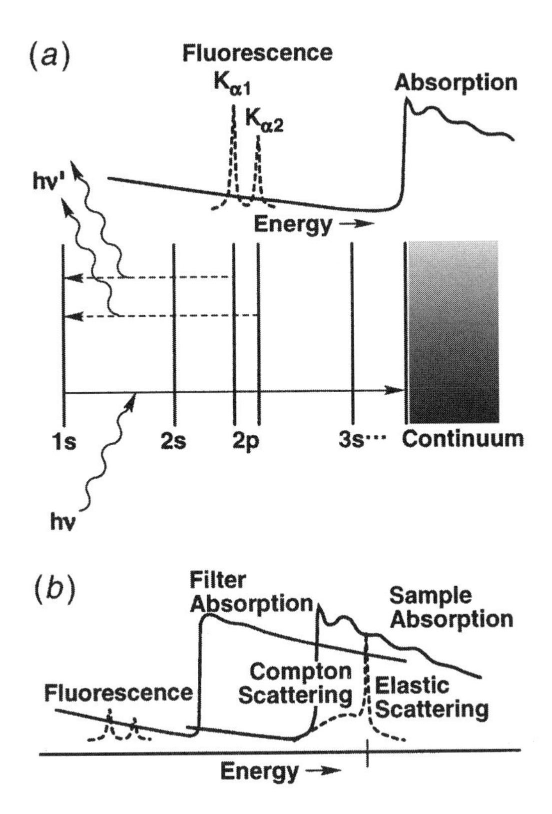

Figure 8
(a) The relationship between the X-ray absorption edge energy and the energies of the $K\alpha$ fluorescence emission lines used for fluorescence excitation detection of X-ray absorption spectra. The $K\alpha$ emission lines result from relaxation of the atom by transition of an electron from an $n = 2$ level to the 1s level. (b) The use of a low-pass filter to reject the background scatter in a fluorescence excitation experiment. The $K\alpha$ fluorescent photons are not absorbed appreciably by the filter but the elastic- and Compton-scattered photons are.

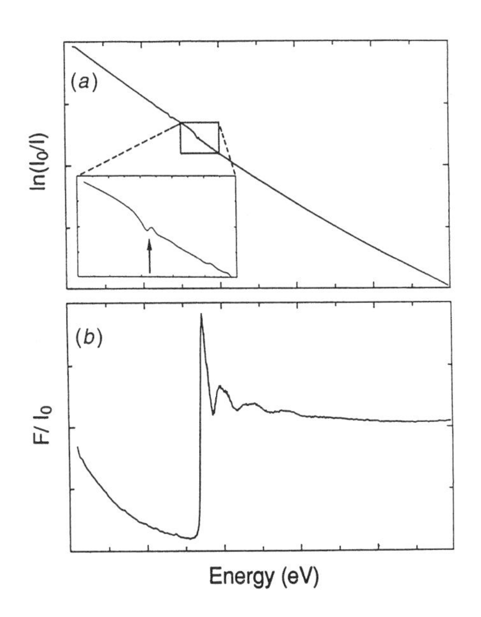

Figure 9
Comparison of (a) transmission and (b) fluorescence excitation X-ray absorption spectra for a dilute sample. These spectra were collected on a 2-mM aqueous sample of a nickel enzyme. The presence of the Ni edge in the transmission spectrum is indicated by the arrow in the inset of (a). Scatter rejection in (b) was achieved by the use of electronic discrimination with a high-energy resolution solid-state detector.

tween the absorption edge and the Kα energies [Fig. 8(*b*)]. This filter preferentially absorbs background scatter photons, allowing the Kα photons through to the detector. A convenient material for such a filter is a thin layer of the element with atomic number (Z) one less than the element being studied (a "Z-minus-one" filter). The development of this filtering strategy with low-energy resolution detectors and the availability of high-count rate high-energy resolution detectors (e.g., solid-state detector arrays) has made fluorescence excitation XAS a standard tool for examining spectroscopically dilute samples.

D. Laboratory Spectrometers

For X-ray absorption spectroscopic investigations of a concentrated sample (e.g., solids and concentrated solutions), the high brightness of synchrotron sources is often unnecessary. Often a laboratory can acquire a rotating anode source, then purchase one of several commercially available X-ray spectrometers designed for XAS. The most common design of such a laboratory XAS spectrometer is shown in Figure 10. Since the source point is much closer to the monochromator than for synchrotron radiation, this "bent crystal" monochromator proves most efficient for collection of a significant amount of the source radiation. (That is, it allows the collection of a large angular spread of source radiation without excessive degradation of spectral resolution.) The spectrometer arrangement shown in Figure 10 is known as the Johannson geometry, with a single crystal bent to a radius of $2R$ and polished to a radius of R. At a given arrangement, all the source (so) X-rays impinging on the polished crystal surface have the same incident angle with the lattice planes (stripes on the crystal cross section in Fig. 10) and all the diffracted rays have the same energy and are focused through the sample (sa) onto the detector (d). To move to the next energy in a scan, the crystal is turned through a small angle θ, while the sample and detector are pivoted around the same point by 2θ.

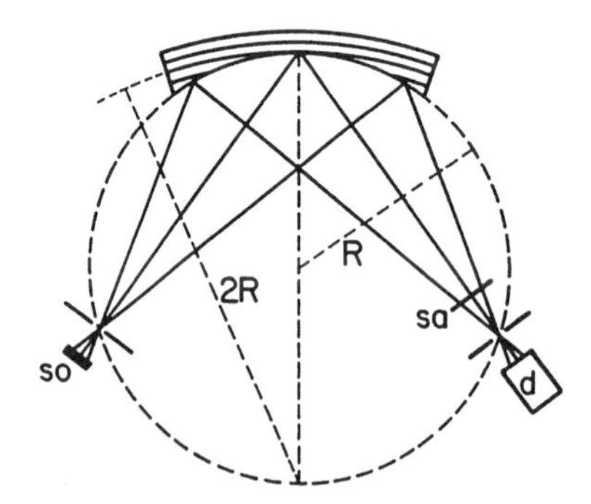

Figure 10
Typical geometry for a laboratory curved-crystal X-ray absorption spectrometer. The curved crystal at the top is shown in cross section with its Bragg planes indicated by the arcs of a circle of radius $2R$. (so = source; sa = sample; d = detector).

IV. Data Reduction and Analysis

The X-ray absorption data collected as discussed in Section III consist of a set of discrete data points: detector counts as a function of monochromator angular position. Depending on the data collection method, the output of I_0 and I detectors (transmission) or F and I_0 detectors (fluorescence) or both were recorded (see Fig. 7). Usually, the angular position of the monochromator crystals [defining the Bragg angle, Eq. (4)] is recorded as a motor position (m_i) of the motor controlling the crystal carriage. This position has been crudely calibrated to the correct energy at data collection, but usually must be precisely calibrated during data reduction. The most common calibration procedure is to use an absorption peak or inflection point from the spectrum of a known standard (often a thin foil of the element being examined) that has been recorded under identical conditions as the sample. The "internal" calibration method involves placing a third ionization chamber (I') behind the two in Figure 7 and placing a standard sample between I and I'. Then, with every sample spectrum, a transmission spectrum of the standard [$\ln(I/I')$] is also recorded. This standard spectrum is analyzed to yield the monochromator angle (m_c) at the defined calibration point, which has a known energy (E_c) from previous measurements. Given this single-point calibration, and knowing the relationship between the motor position measurements and angle (Δ, in units of motor position $\deg^{-1}$), each data point's motor position (m_i) can be converted to a Bragg angle (θ_i) and an absolute energy (E_i):

$$\theta_c = \sin^{-1}(hc/2dE_c) \tag{5}$$

$$\theta_i = \theta_c + [m_i - m_c]/\Delta \tag{6}$$

$$E_i = hc/2d \sin \theta_i \tag{7}$$

The absorption coefficient proportional data (δ_i) to be analyzed are generated by $\delta_i = \ln[I_0(i)/I(i)]$ for transmission or $\delta_i = F(i)/I_0(i)$ for fluorescence detection, yielding the function $\delta(E)$ [e.g., the solid line in Fig. 11(a)]. To calculate the EXAFS quantity χ, we need to know μ [see Eq. (2)], which is the X-ray absorption coefficient for just the absorbing atom site of interest. But the data of Figure 11(a) also contain a background contribution (μ_{back}) from either absorption by other (lower atomic number) atoms in the sample (in transmission) or scatter from the sample (in fluorescence):

$$\delta = \mu + \mu_{back} \tag{8}$$

It is usually very difficult to measure or simulate μ_{back}, so that polynomial or half-Gaussian fitting procedures are used to mimic it as well as possible. This results in an approximate background function, $\mu_{back}(E)$ [dashed line in Fig. 11(a)], which does not completely remove the background:

$$\delta_1 \equiv \delta - \delta_{back} = \mu + \Delta\mu \tag{9}$$

$$\Delta\mu = \mu_{\text{back}} - \delta_{\text{back}} \tag{10}$$

(Here, $\Delta\mu$ is the remaining background absorption.)

If this background-subtraction procedure had worked perfectly ($\Delta\mu = 0$), then we would only have needed to subtract the free-atom X-ray absorption coefficient (μ_0) from δ_1 and normalize to get χ [Eq. (2)]. Unfortunately, one cannot determine the value of $\Delta\mu$ and subtraction of μ_0 generally does not work. Since μ_0 would have been the smooth curve underneath the quasiperiodic EXAFS oscillations [Fig. 11(*b*)], we need a corrected smooth curve [μ_s, dashed line in Fig. 11(*b*)] that does track the absorption coefficient above the edge. This curve is usually calculated by a (cubic or quartic) spline fitting procedure, which generates

$$\mu_s = \mu_0 + \Delta\mu \tag{11}$$

The true free-atom absorption coefficient (μ_0) can be calculated from the Victoreen formula[7] and must be used for normalization (this normalization serves to correct for any pathlength or concentration differences among samples and yields χ data normalized to a "per absorbing atom" basis.)

$$\chi = \frac{\delta_1 - \mu_s}{\mu_0} \tag{12}$$

The resulting EXAFS data (χ) are shown in Figure 11(*c*) for our example.

The EXAFS data, χ, are normally treated not as a function of energy, E, but as a function of photoelectron wave vector, k. To convert from E to k, the threshold energy (E_0) is needed [see Eq. (1)]. The threshold energy is equivalent to the ionization potential for the 1s electron (for K edges), but it is difficult to measure independently. Since there are often features just before the edge corresponding to bound-state transitions, using an inflection point on the observed edge as E_0 is not quite correct. Assuming E_0 is the same for every occurrence of a particular element is not correct either, since E_0 depends slightly on the chemical environment of the absorbing atom. Either method can estimate E_0 to within a few electron volts (eV) and both are used. For our example, assuming $E_0 = 9000$ eV yields the k scale shown at the top of Figure 11(*c*). The EXAFS data are replotted as $\chi(k)$ in Figure 11(*d*). Although the EXAFS oscillations extend to high k values, they are "masked" at high k due to the severe damping from the Debye–Waller term in Eq. (3). For this reason, it is standard practice to work with k^n weighted $\chi(k)$ data ($n = 3$ usually); $k^3\chi(k)$ is plotted in Figure 11(*e*).

Analysis of the EXAFS data in terms of structural information makes use of Eq. (3) to simulate the EXAFS pattern expected for a given hypothetical arrangement of atoms around the absorbing atom. As discussed above, each damped sine wave in the $\chi(k)$ data arises from one shell of atoms at a given distance from the absorbing atom. One effective means of visualizing these shells of atoms is to perform a FT on the $k^3\chi(k)$ data. Fourier transformation effectively does a frequency analysis of the EXAFS data, yielding a peak at the frequency of each sine wave component in the EXAFS. Since the sine wave frequency is directly proportional to the absorber-scatterer separation distance, R_{as} [Eq. (3)],

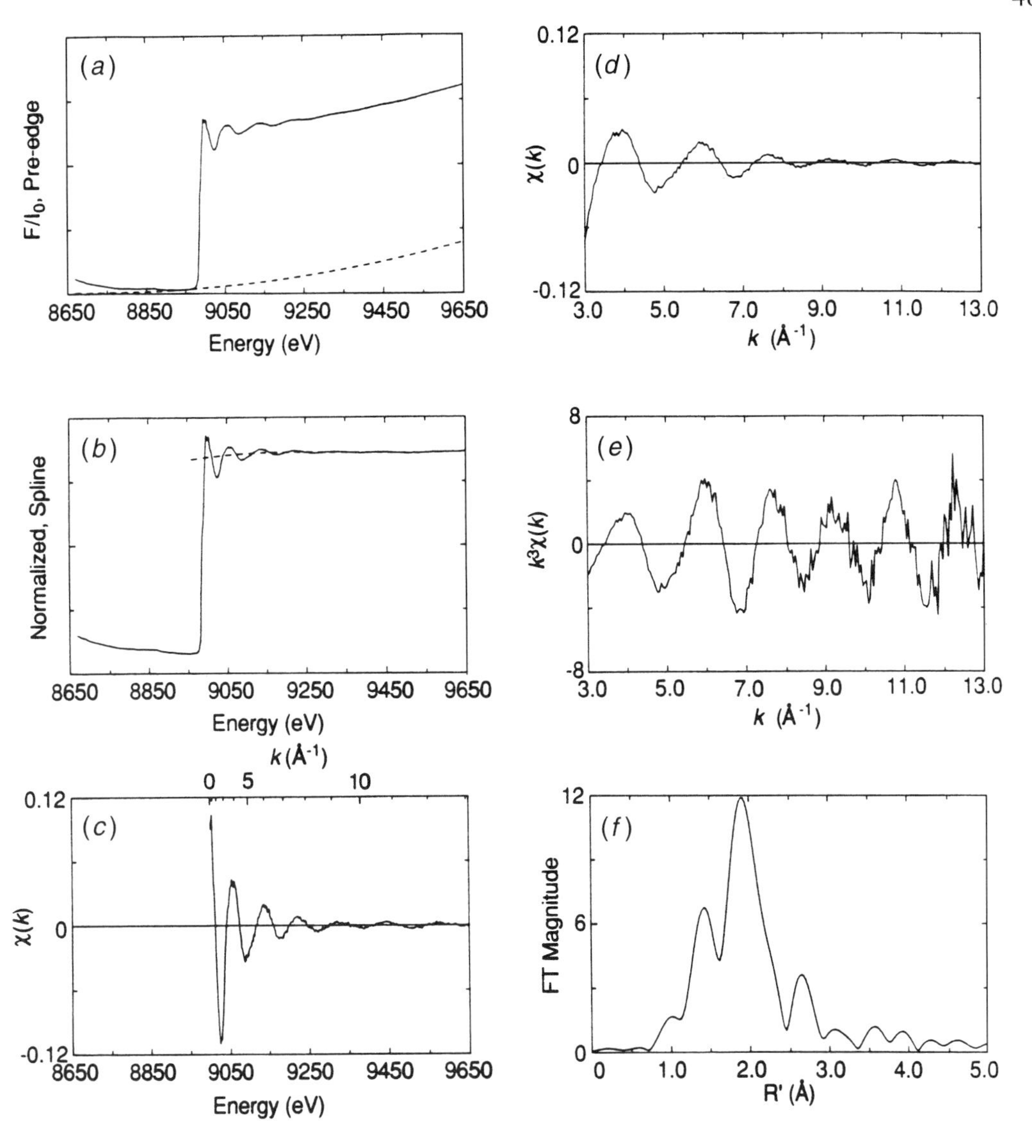

Figure 11
Example of data reduction for EXAFS analysis. (*a*) Fluorescence excitation XAS data (solid line) are corrected by subtraction of a pre-edge background (dashed line); (*b*) a smooth curve (dashed line) is fit to the pre-edge-subtracted data (solid line) by a cubic spline procedure to yield (*c*) the EXAFS data. (*d*) The EXAFS data displayed as a function of photoelectron wave vector, k, and then (*e*) weighted by k^3. (*f*) The magnitude of the FT of the $k^3\chi(k)$ data in (*e*) displayed as a function of the phase-shifted distance, R'.

the FT generates a set of peaks in radial distance space, each peak representing a particular shell of atoms [Fig. 11(*f*)]. [These peaks do not occur at the precise distances owing to contributions from the phase shift, $\alpha_{as}(k)$, in Eq. (3). Therefore a phase-shifted distance, R', is used as the FT abscissa in Fig. 11(*f*). Often $R_{as} \approx R' + 0.4$ Å.] Although Fourier transformation is a useful means for visualizing the radial distribution of scattering atoms around the absorbing atom, one must exercise extreme caution in assigning small peaks in the FT to shells of atoms. Small artifactual FT peaks are generated by truncation of the $k^3\chi(k)$ data, by the k^3 weighting function itself, which acts as an artificial resolution enhancement function in the FT, and by inadequate background subtraction (usually due to poor choice of the spline function used to generate μ_s).

Since the FT of the EXAFS displays a peak for each shell of atoms, it provides a method for separating out contributions from individual shells. The technique of Fourier filtering consists of constructing a (filter) window [dotted lines in Fig. 12(*b*)] around the FT peak of interest, which when multiplied by the FT data yields only that FT peak [solid lines in Fig. 12(*b*)]. Back-transformation of these filtered FT data yields the extracted EXAFS arising solely from that shell of atoms [Fig. 12(*c* and *d*)]. [Adding these two "shells" of EXAFS together generates Fig. 12(*e*), showing that these shells were the main sine wave components of the orginal EXAFS data in Figure 12(*a*).] The Fourier filtering technique is a good way of simplifying the initial curve-fitting simulation of individual shells, which is discussed next.

Although Fourier transformation can yield a qualitative visualization of the radial structure of a site, quantitative structural information (How many of what type of atoms are at what distance from the absorbing atom?) can only be obtained by curve-fitting simulation of the $k^3\chi(k)$ data. In general, a hypothetical set of atom shells is generated and Eq. (3) (or some more sophisticated extension of it) is used to calculate the expected EXAFS data. The calculated EXAFS data are compared to the observed data and (a subset of) the parameters in each "shell" of Eq. (3) (N_s, R_{as}, σ_{as}) are optimized in a least-squares sense. If the fit to the observed data is judged inadequate, additional shells of atoms may be added, or the identity of the atoms in the shells may be changed, and the curve-fitting optimization repeated. Some subjective judgment is required in deciding what constitutes a "good fit". Mathematically, the more shells [terms in the summation of Eq. (3)] that are added, the better the fit should be, but this does not necessarily mean that the added shells are *required*. The art of EXAFS analysis is in deciding which shells are uniquely required to adequately fit the data. Chemical intuition is also helpful in discarding shells with unreasonable distances or coordination numbers. Depending on previous characterization, other pieces of independent information about the site being studied may be available to guide the choice of acceptable hypothetical structures. If nothing is known, one must rely heavily on statistical analysis of the EXAFS data and chemically reasonable bond distances. The more information that is available, the more detailed the structural questions can be.

In the above discussion, it was assumed that once an a–s (absorbing atom–scattering atom) pair was chosen for a shell of the hypothetical structure, it would be straightforward to generate that shell's expected EXAFS using Eq. (3). This assumption implies that one knows the scattering characteristics of the s atoms, that is, that one knows the backscattering amplitude function $|f_s(\pi, k)|$ and the backscattering phase shift function $\alpha_{as}(k)$ for this a–s pair. Two basic approaches have evolved for obtaining these functions. The empirical approach involves

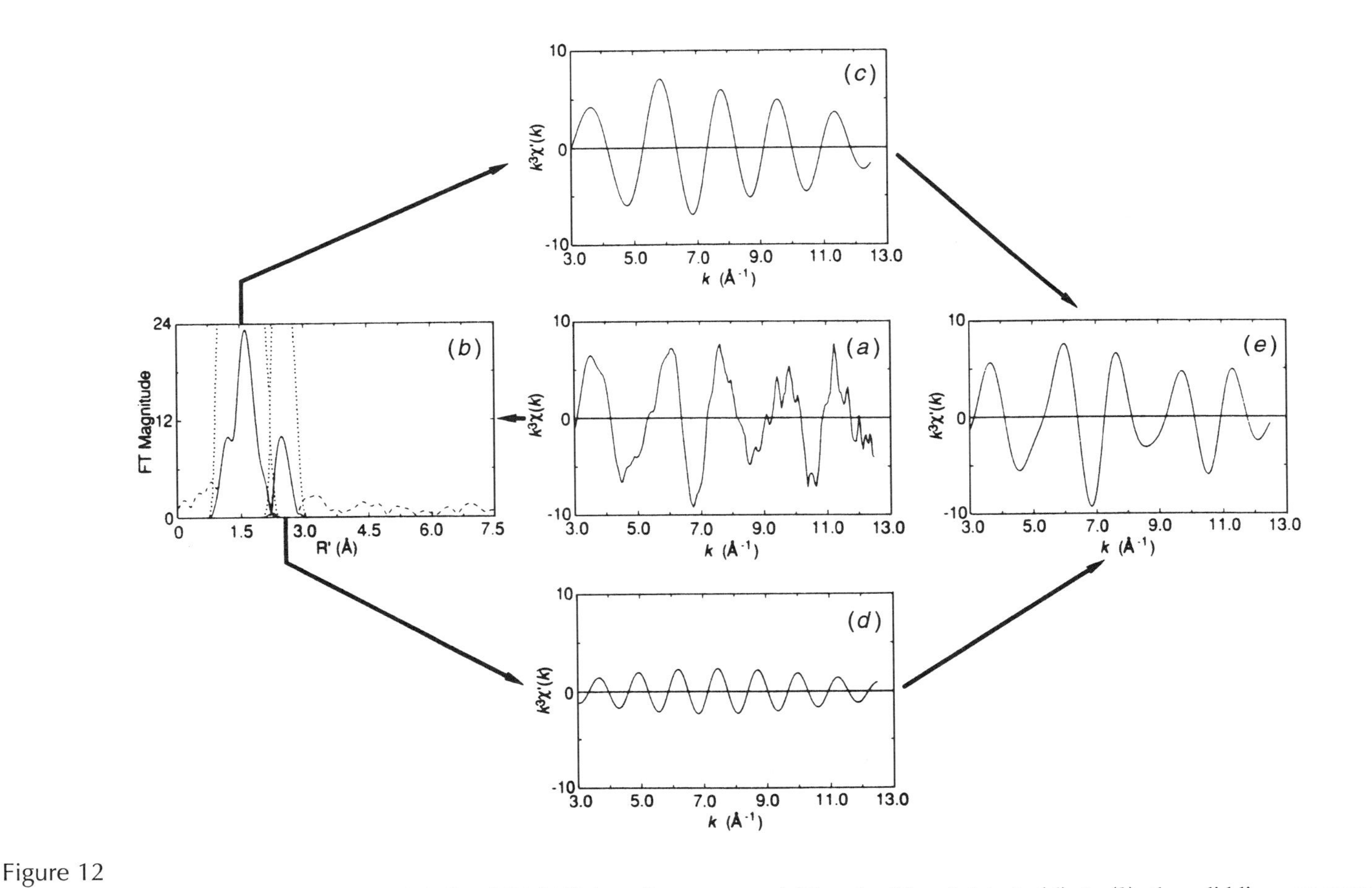

Figure 12
Example of Fourier filtering of individual shells of EXAFS data. Raw EXAFS data (*a*) are Fourier transformed to yield (*b*). In (*b*), the first-shell FT peak is filtered using the window (dotted line) centered at $R' \approx 1.5$ Å giving rise to the filtered $[k^3\chi'(k)]$ data in (*c*). The second-shell FT peak in (*b*) is filtered using the window (dotted line) centered at $R' \approx 2.6$ Å yielding the filtered data in (*d*). In (*b*), the solid lines represent the product of the FT magnitude and the filter windows in each case. The sum of the two filtered EXAFS data sets in (*c*) and (*d*) is displayed in (*e*) and looks very similar to the raw EXAFS data in (*a*), which indicates that these two shells are the main components of the EXAFS.

measuring EXAFS data for structurally characterized (model) compounds with a symmetric shell of one type of atom. The FT of these EXAFS data will yield a single peak that can be Fourier filtered [e.g., see Fig. 12(*c*)] to yield $|f_s(\pi, k)|$ and $\alpha_{as}(k)$ directly. Unfortunately, the empirical method depends on the availability of model compounds with the appropriate a–s combination in a shell that can be separated from any other shells by Fourier filtering. The theoretical approach involves ab initio computation of the scattering behavior of any given a–s pair and depends on having good wave functions for the atoms involved. These theoretical computations continue to improve and the resulting scattering functions now rival those obtained by the empirical method in accuracy.

Analysis of EXAFS data has advanced to the point that reliable radial structural information can be obtained with relatively simple computational tools. Although the spectral structure in the near-edge region (Fig. 1) is rich and varied, the analysis of this region is more complicated and therefore not as well developed as EXAFS analysis. Conceptually, the near-edge energy region can be divided into the region below and the region above E_0 (the threshold energy). In the region below the threshold, bound-state transitions from the 1s level (for K edges) to empty or partially filled valence levels just below the continuum may give rise to peaks or shoulders in the absorption coefficient. For example, for K edges of first-row transition metals, 1s $\rightarrow$ 3d, 1s $\rightarrow$ 4s, and 1s $\rightarrow$ 4p transitions may be observed. Since these are electronic transitions, electric-dipole selection rules are usually sufficient to predict site symmetries from the relative intensities of these transitions. In the region just above E_0, the extra photon energy converted to photoelectron kinetic energy is small and low-energy photoelectrons are more prone to multiple scattering. The predominance of multiple-scattering contributions in this low-k region makes it impossible to extend reliably the single-scattering EXAFS analysis to below about 30 eV above E_0. Multiple-scattering effects also make the X-ray absorption coefficient just above the edge more sensitive to geometry than the higher k EXAFS region. Thus, the information available from the edge region (molecular symmetry and geometry, as well as electronic structure) is complementary to that from EXAFS (radial molecular structure). Although several attempts at ab initio computation of X-ray absorption edge spectra have been made, they have met with limited success, and analyses of the edge region are often qualitative.

V. Applications of X-Ray Absorption Spectroscopy

With the complementary electronic and molecular structural information available from analysis of X-ray absorption edges and EXAFS, it is not surprising that the technique has been applied to structural investigations of a wide variety of materials. XAS techniques have made a significant impact in fields as diverse as condensed-matter physics, materials science, geology, chemistry, biology, and medicine. In any of these fields, successful application of the technique requires a knowledge of both the strengths and limitations of XAS.

In this section, a selection of applications of XAS are presented as examples of the use of the technique in answering structural questions of interest to inorganic chemists. The XAS technique is most useful when applied to samples with a homogeneous population of sites containing the metal being studied. When used to investigate samples with more than one structural type of site (containing the same metal), EXAFS is inherently incapable of distinguishing whether a shell of ligands belongs to the coordination sphere of one site or the other. Only an average metal environment can be deduced from the EXAFS data.

A. What Is the Molecular Symmetry of the Site?

Extended X-ray absorption fine structure is not capable of giving direct information about the geometric arrangement of ligand atoms around a metal (the molecular symmetry) in amorphous samples. (The technique may yield indirect information in the few cases for which the metal–ligand distance is known to be well correlated with molecular symmetry. For samples with long-range order (e.g., single crystals), polarized EXAFS can be used to sense the arrangement of distinguishable ligand atoms.[8]) On the other hand, the X-ray absorption *edge can* yield direct information about the molecular symmetry of metal sites in amorphous samples and has been very useful in this regard. In general, the edge is most sensitive to the *local* molecular symmetry, defined by the atoms directly coordinated to the metal.

Nickel(II) compounds provide an illustrative example of this type of application, which relies on assignment of pre-edge bound-state electronic transitions and a knowledge of the symmetry properties of selection rules for these transitions. Other examples can be found in the literature for Ni,[9] Fe,[10] and Cu.[11] For K X-ray aborption edges of first-row transition metals, bound-state transitions at energies just below the edge have been variously identified as 1s → 3d, 1s → 4s, and 1s → 4p transitions. Only the 1s → 4p transitions are symmetry allowed by electric-dipole selection rules. The others can gain some intensity by orbital mixing or through higher order (e.g., electric quadrupole) transition moments.[12]

A number of representative Ni(II) K X-ray absorption edge spectra are compared in Figure 13. The small pre-edge peak at about 8332 eV in the spectra of approximately tetrahedral (T_d) Ni(II) sites [Fig. 13(*a*)] is assignable to the 1s → 3d transition. This transition is most prominent for tetrahedral sites due to the increase in 4p orbital character of the 3d final state (p–d mixing is allowed in noncentrosymmetric point groups such as T_d), which lends some symmetry allowedness to this transition. The 1s → 3d transition is much less prominent in centrosymmetric sites [e.g., octahedral sites, Fig. 13(*c*), or square planar sites, Fig. 13(*d*)]. Thus, observation of a relatively intense 1s → 3d transition signals the absence of a center of symmetry in the site. This intensity relationship has been quantified for Ni[9] and Fe.[10]

The Ni(II) complexes with square planar (D_{4h}) symmetry give rise to K edge spectra with a large characteristic pre-edge peak about 5 eV below the main edge [Fig. 13(*d*); ~8338 eV for NiN_4 sites, ~8336 eV for NiE_4 sites, E = S or Se]. On

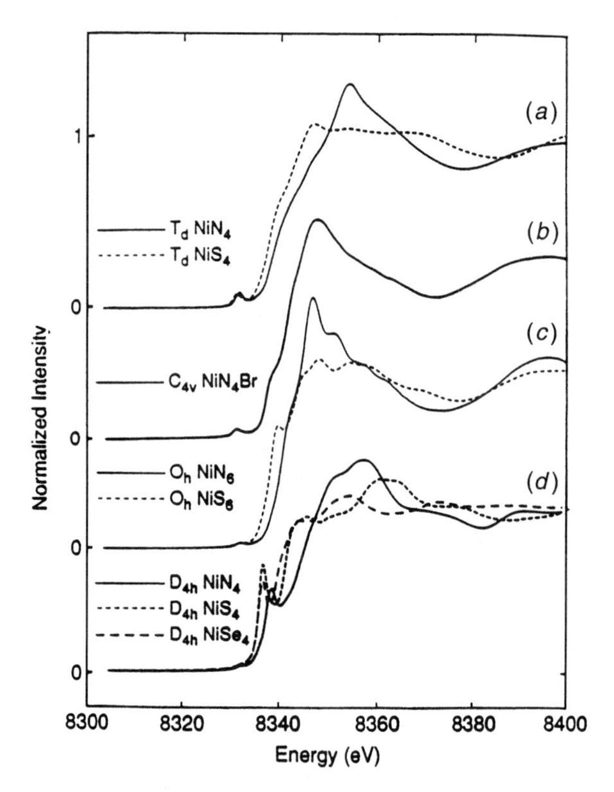

Figure 13

Comparison of Ni K X-ray absorption edge spectra for a series of Ni(II) compounds of varying site symmetry. (*a*) Approximately tetrahedral compounds Ni[tropocoronand–$(CH_2)_6$, $(CH_2)_6$] (solid line)[13] and $[Ni(SPh)_4]^{2-}$ (dotted line).[14] (*b*) Approximately square pyramidal compound $[Ni(tetramethylcyclam)Br]^{+}$.[15,16] (*c*) Approximately octahedral compounds $[Ni(1,4,7\text{-triazacyclononane})_2]^{2+}$ (solid line)[17,18] and $[Ni(1,4,7\text{-trithiacyclononane})_2]^{2+}$ (dotted line).[19] (*d*) Approximately square planar compounds [Ni(phthalocyanine)] (solid line),[20] $[Ni(maleonitriledithiolate)_2]^{2-}$ (dotted line)[21,22] and $\{Ni[Se_2C_2(CF_3)_2]_2\}$ (dashed line).[23] [This figure was adapted from Scott[24]; for an expanded analysis of Ni(II) edges, see Colpas et al.[9]]

the basis of comparison with Cu(II) K-edge spectra, this peak has been assigned to a $1s \rightarrow 4p_z$ transition. (The assignment is actually most likely to be "$1s \rightarrow 4p_z$ plus shakedown",[25] involving a simultaneous promotion of a 1s electron to the $4p_z$ orbital and a ligand–metal charge transfer (LMCT) to a Ni(II) 3d orbital:

$$1s^2 \cdots 3d^8/L^n \rightarrow 1s^1 \cdots 3d^9 4p_z{}^1/L^{n-1})$$

It appears as a resolved peak in the K-edge spectra of square planar Ni(II) complexes owing to the lower energy of the $4p_z$ orbital in the absence of ligands along the molecular *z* axis (perpendicular to the plane of the molecule). Since no discrete peaks are observed at these energies for tetrahedral [Fig. 13(*a*)] or octahedral [Fig. 13(*c*)] Ni(II) sites with the same ligand sets, the presence of a resolved peak in this energy region can be used as a signature for a square planar site.

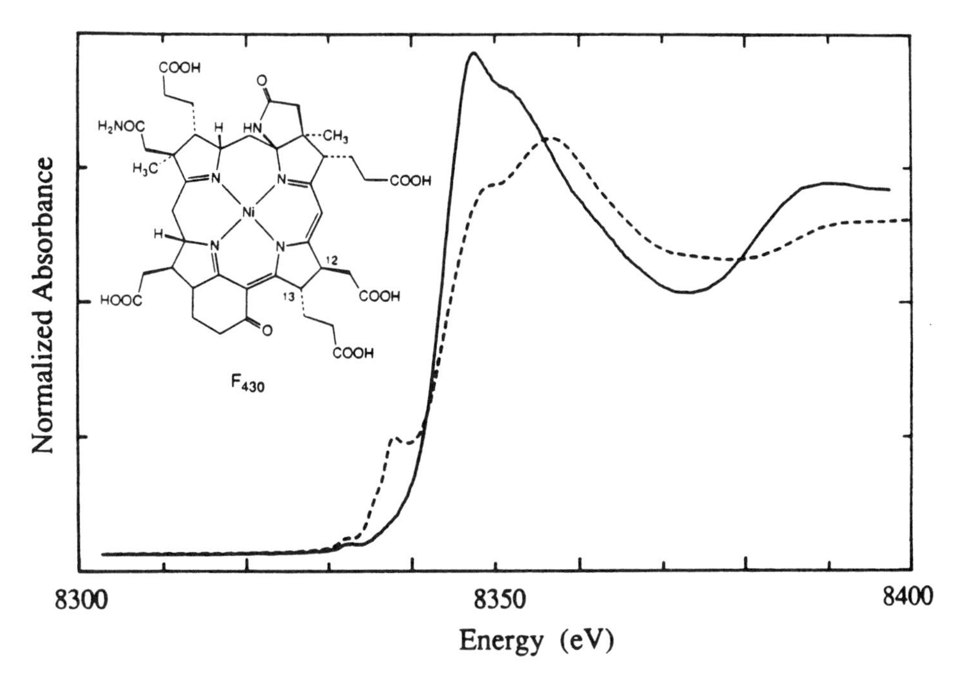

Figure 14
The Ni K X-ray absorption edge spectra of the two isomers of cofactor F_{430} of the methyl reductase enzyme of methanogenic bacteria. The native isomer (solid line) is known to be six coordinate with two axial H_2O ligands and the 12,13-diepimer (dotted line) has a four-coordinate square planar structure. The structure of the unligated native F_{430} cofactor is inset. [This figure was adapted from Shiemke et al.[26]]

For Ni(II) sites with square pyramidal (C_{4v}) symmetry, the K-edge spectrum [Fig. 13(*b*)] exhibits features intermediate between those from square planar and octahedral sites. The 1s → $4p_z$ transition has moved to higher energy (relative to the main edge), becoming a shoulder, as expected from interaction with one axial ligand. The 1s → 3d transition is more intense than that observed for octahedral Ni(II) sites, reflecting the absence of a center of symmetry in the square pyramidal Ni(II) site.

These characteristic edge features can be used to obtain structural information that is complementary to the type of information available from EXAFS. For example, the Ni(II) K edge spectra of two isomers of the Ni containing tetrapyrrole cofactor F_{430} from the methylreductase enzyme of methanogenic bacteria are shown in Figure 14. (This enzyme catalyzes the final step in the production of CH_4 and the F_{430} Ni is thought to be involved.[27]) These data indicated that the native F_{430} isomer was six coordinate with two axial H_2O ligands (giving rise to an octahedral-type edge spectrum) while the 12,13-diepimer was square planar with no axial ligands (giving rise to a square planar-type edge spectrum).[28]

Edge comparisons like the ones discussed above do not always yield a unique answer. Some of the qualitative edge distinctions are not always obvious (cf. solid lines in Figs. 13(*b*) and (*c*). As discussed before for EXAFS analysis, caution is suggested in drawing firm conclusions about site geometry from edge spectra if no other information is available. For example, in F_{430}, it is fairly safe to assume that a NiN_4 equatorial ligand environment is present (from the tetrapyrrole macrocycle) and the question being addressed by edge comparisons is more focused: Does the Ni bind axial ligands?

B. Does a Particular Treatment Generate a Redox Change?

As already mentioned, the position of the X-ray absorption edge is sensitive to the redox state of the metal. The edge energy generally increases with an increase in oxidation of the metal site. However, the presence of pre-edge features arising from bound-state transitions makes the correlation of absolute edge energy with redox state difficult. It is somewhat easier to correlate *changes* in edge position with changes in redox state, although this is also prone to error, especially when a change in molecular symmetry accompanies a redox change. When examining a particular type of metal in selected oxidation states, one must make sure that the spectral changes observed are reliable indicators of redox change and do not just reflect a structural rearrangement.

For example, characteristic differences between the Cu K X-ray absorption edge spectra of Cu(I) and Cu(II) sites have allowed the development of a difference technique for quantitative determination of Cu(I) content in samples with mixed oxidation state composition.[11] Representative Cu K-edge spectra for Cu(I) and Cu(II) sites are shown in Figure 15(*a*). The feature at 8983–8984 eV in the Cu(I) edge spectrum is assigned as a 1s $\rightarrow$ 4p transition and is characteristic of Cu(I). The Cu(I)-minus-Cu(II) difference spectrum gives rise to a peak at this energy, the intensity of which can be correlated with the amount of Cu(I) present in a sample containing a mixture of Cu(II) and Cu(I).[10] [This works well for two- or three-coordinate Cu(I) sites, but tetrahedral Cu(I) does not exhibit the 8984-eV feature.[11]]

This technique was applied to a multi-copper enzyme known as laccase. This polyphenol oxidase contains four copper ions per functional unit, which are spectroscopically distinguishable and are referred to as type 1 (mononuclear), type 2 (mononuclear), and type 3 (dinuclear),[30] the last giving rise to a 330-nm band in the oxidized enzyme. Preparation of a T2D laccase derivative resulted in the absence of a 330-nm UV–vis band. Addition of H_2O_2 (but not other oxidants) regenerated the 330-nm band, a result first attributed to formation of a type-3 Cu–peroxide (O_2^{2-}) complex.[31] However, Cu X-ray absorption edge spectra [Fig. 15(*b*)] unambiguously demonstrated that the type-3 site in T2D laccase remained in the Cu(I) state and could only be reoxidized to Cu(II) by H_2O_2.[11,29]

C. What Types of Atoms Are in the First Coordination Sphere? (Edge Analysis)

Aside from being caused by a redox change, a shift in the X-ray absorption edge energy may also be the result of a change in the ligand composition of the site.

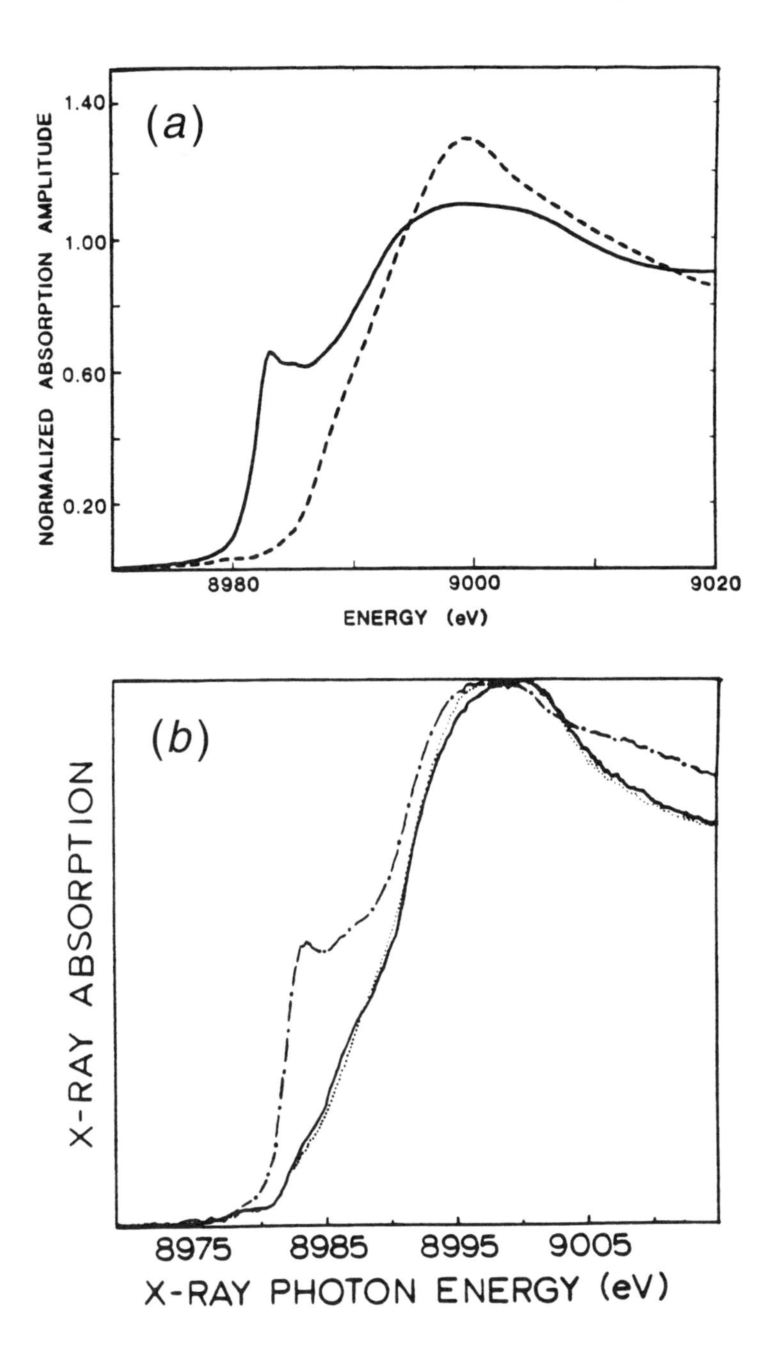

Figure 15
(*a*) Typical Cu(I) (solid line) and Cu(II) (dashed line) K X-ray absorption edge spectra showing the 1s → 4p transition at about 8983 eV in the Cu(I) edge. The Cu(I) complex has CuN_2O ligation and the Cu(II) complex has CuN_4O ligation. [This figure was adapted from Kau et al.[11]] (*b*) The Cu K X-ray absorption edge spectra for native (solid line), type-2 depleted (T2D) (dash–dot line), and H_2O_2 treated T2D (dotted line) forms of the copper enzyme laccase. The native enzyme contains all Cu(II) sites and H_2O_2 treatment of T2D laccase simply reoxidizes a reduced type-3 site (see text). Note that the spectra in (*b*) are not normalized as described in the text. [This figure was adapted from Lu Bien et al.[29]]

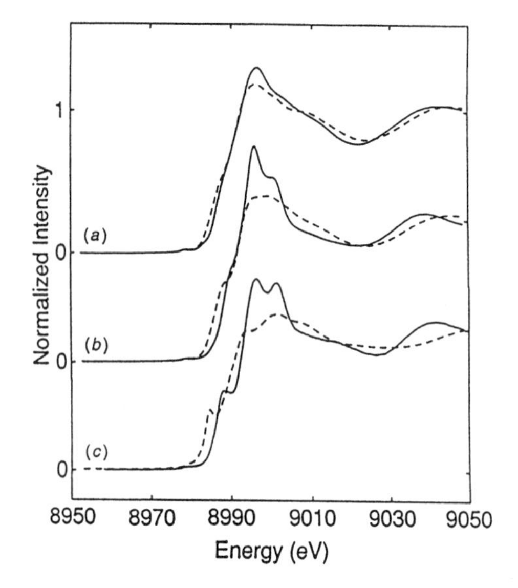

Figure 16
Comparison of Cu(II) K X-ray absorption edge spectra for a series of Cu(II) compounds with (dashed lines) and without (solid lines) sulfur ligation. (*a*) Trigonal bipyramidal complexes with CuN_4O (solid line) or CuN_2S_2O (dashed line) ligation. (*b*) Trigonal bipyramidal complexes with CuN_5 (solid line) or CuN_3S_2 (dashed line) ligation. (*c*) Square planar complexes with CuN_4 (solid line) or CuN_2S_2 (dashed line) ligation. In each case, the lower energy of the edge for the S-containing complexes indicates an increase in charge density at the Cu(II) due to more covalent bonding.

For the higher oxidation states of transition metals, it is often observed that an increase in the proportion of "soft" ligands (donor atoms of P, S, Se, ...) results in a shift of the edge position to lower energy. One possible explanation for this effect is increased covalency of metal–ligand bonds for soft ligands. This increased covalency would presumably involve a significant amount of ligand–metal charge transfer, resulting in an increase in the effective electron density on the metal, decreasing the observed K-edge energy. This sensitivity of edge position to hard/soft ligand composition must be considered along with redox changes as the explanation for an observed edge energy shift (or an abnormally low edge energy).

The alert reader will have already noted that Ni(II) sites obey the rule of thumb that soft ligands result in lower edge energy [Fig. 13(*a*), (*c*), and (*d*)]. It is also generally true that the overall (normalized) edge height is lower for sites with soft ligands (at least those containing M–S bonds) and this is observed routinely in Ni(II) sites, although the explanation for this trend is unknown. Figure 16 illustrates that these trends apply equally well to Cu(II) sites. In each comparison, the $Cu(N,O)_xS_2$ compound displays an edge to lower energy with reduced height compared to that for $Cu(N,O)_{x+2}$ ($x = 2$ or 3). These observed trends can often be used in a "fingerprint" mode to suggest the relative hard/soft character of ligand donor atom composition for sites of unknown structure.

D. What Types of Atoms Are in the First Coordination Sphere? (EXAFS Analysis)

Although the shape and position of the X-ray absorption edge can yield some clues as to the types of ligands bound to the metal site, quantitative information about the structure of the first coordination sphere depends on analysis of the EXAFS data. Usually some information is available on what possible ligand atoms are bound to the metal and this reduces the number of hypothetical compositions that need to be considered. For example, in a metalloenzyme, only biologically relevant ligand atoms (N, O, S, Se, and Cl?) need be considered. For a transition metal complex of unknown structure, the possible ligands are known from other analytical information. For example, mixing a Cu(II) salt with the potential chelating ligand mercaptopropionylglycine (MPG) generates a complex with the proposed structure[32]

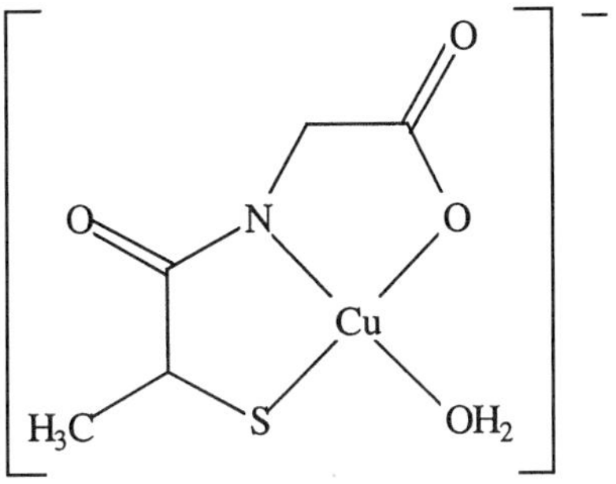

Cu EXAFS can be used to determine whether the thiolate S is in fact bound to the Cu.

The Cu EXAFS data for Na[Cu(MPG)(H_2O)] are shown in Figure 17(*a*). Fourier transformation of these data reveals two separate peaks [Fig. 17(*b*)] originating from two main shells of ligand atoms. Recalling that FT peaks usually appear at distances about 0.4 Å shorter than the true metal–ligand distances, these peaks represent shells with distances of about 1.9 and 2.3 Å. A survey of structures of Cu(II) complexes reveals that typical Cu–(N,O) distances range from about 1.9–2.1 Å. while typical Cu–S distances range from about 2.25–2.35 Å. Thus, the appearance of the FT suggests that the thiolate S is coordinated to the Cu(II). Confirmation of this can be obtained by curve-fitting of the filtered Cu EXAFS data [Fig. 17(*c*)]. In this example, Cu(II)–N and Cu(II)–S scattering functions were empirically determined by examining the EXAFS of a series of structurally characterized (model) Cu(II) compounds. Then, initial guesses for coordination number (N_s), metal–ligand distance (R_{as}), and Debye–Waller ${\sigma_{as}}^2$ parameters were made for each shell, and a subset of these parameters were optimized to match the calculated and observed EXAFS data as well as possible. For each shell, ${\sigma_{as}}^2$ may be fixed at the value found in model compounds and N_s, R_{as} optimized; or N_s may be fixed at selected integer values and ${\sigma_{as}}^2$ and R_{as} optimized. The latter method was used for Na[Cu(MPG)(H_2O)] and the results

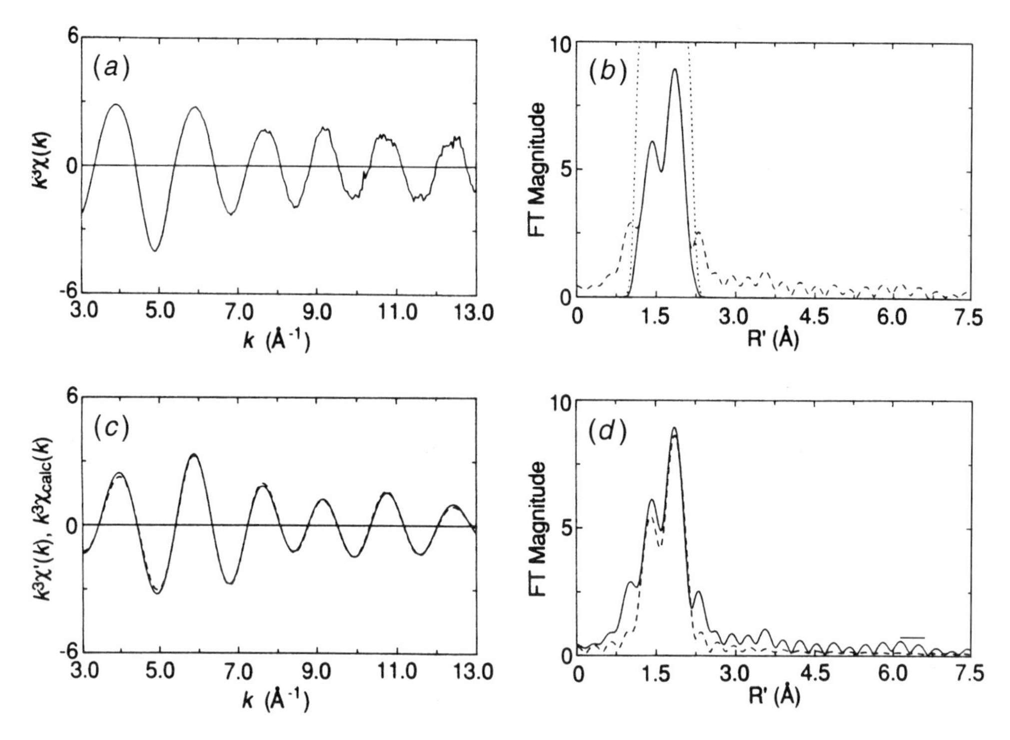

Figure 17

Curve-fitting analysis of the Cu EXAFS data for Na[Cu(MPG)(H_2O)]. The raw EXAFS data in (*a*) give rise to the FT in (*b*). Fourier filtering the first two FT peaks [solid line in (*b*)] gives rise to the filtered data represented by the solid line in (*c*). The best-fit simulation (Table I, Fit 4) is shown as the dashed line in (*c*) and the FT of this simulation is compared to the FT of the raw EXAFS data in (*d*).

shown in Table I. A good fit is one for which the difference between the calculated and observed EXAFS data is minimal (as measured by the f' statistic) and for which the calculated $\sigma_{as}{}^2$ is close to the $\sigma_{as}{}^2$ observed for the model compound ($\Delta\sigma_{as}{}^2 = \sigma_{as}{}^2(\text{sample}) - \sigma_{as}{}^2(\text{model}) \approx 0$).

For Na[Cu(MPG)(H_2O)], Table I shows that the presence of both a Cu–(N,O) and a Cu–S shell is necessary to get an adequate fit to the EXAFS data. For one Cu–S and one Cu–(N,O) shell (Fit 3), the $\sigma_{as}{}^2$ value for the Cu–(N,O) shell is quite large, indicating significant static disorder (spread) in the Cu–(N,O) distances. Alternatively, if two Cu–(N,O) and one Cu–S shells are used (Fits 4 and 5), the $\sigma_{as}{}^2$ values for the two Cu–(N,O) shells more closely approximate the $\sigma_{as}{}^2$ value for the model compound. This improvement in $\sigma_{as}{}^2$ agreement and the reduction in f' lend support for the proposed structure. [Fits 4 and 5 of Table I show that EXAFS cannot distinguish between a site with two long and one short Cu–(N,O) bonds and a site with one long and two short Cu–(N,O) bonds.]

Table I
Curve-Fitting Results for the First Coordination Sphere of $Na[Cu(MPG)(H_2O)]$[a]

	Cu–(N,O)			Cu–S			
Fit	N_s	R_{as}(Å)	$\Delta\sigma_{as}^2$(Å^2)	N_s	R_{as}(Å)	$\Delta\sigma_{as}^2$(Å^2)	f'[b]
1	(4)[c]	2.05	+0.0096				0.154
2				(4)	2.26	+0.0119	0.077
3	(3)	1.99	+0.0097	(1)	2.28	−0.0002	0.026
4	(1) (2)	2.08 1.93	−0.0020 +0.0028	(1)	2.30	−0.0004	0.014
5	(2) (1)	2.04 1.89	+0.0012 −0.0005	(1)	2.29	−0.0003	0.014

[a] The parameter N_s is the number of scatterers per copper; R_{as} is the copper-scatterer distance; $\Delta\sigma_{as}^2$ is a relative mean-square deviation in R_{as}, $\Delta\sigma_{as}^2 = \sigma_{as}^2(\text{sample}) - \sigma_{as}^2(\text{model})$, where the model is $[Cu(imidazole)_4]^{2+}$ at 4 K for Cu–(N,O) and $[Cu(maleonitriledithiolate)_2]^{2-}$ at 4 K for Cu–S. All fits were over the range $k = 3.0$–$13.0\ \text{Å}^{-1}$.

[b] The parameter f' is a goodness-of-fit statistic normalized to the overall magnitude of the $k^3\chi(k)$ data:

$$f' = \frac{\{\Sigma[k^3(\chi_{obsd}(i) - \chi_{calc}(i))]^2/N\}^{1/2}}{(k^3\chi)_{max} - (k^3\chi)_{min}}$$

[c] Numbers in parentheses were not varied during optimization.

E. Does a Particular Treatment Result in a Structural Change?

Since both the X-ray absorption edge and the EXAFS are sensitive to the local structural environment of the absorbing metal atom(s), XAS is a good technique for determining whether the structure of a metal site is affected by a given treatment. The position and shape of the edge can give information about electronic structure (e.g., redox state) and molecular geometry, while the EXAFS can be analyzed to yield the detailed nature of the structural change.

As an example, consider the reductive lithiation (lithium intercalation) of the amorphous material, MoS_3.[33] (The lithiated MS_3 compounds have shown some promise as electrode materials for lithium batteries.) Comparison of the Mo K X-ray absorption edge spectra for the parent MoS_3 and the lithiated material of limiting composition, $Li_{4.0}MoS_3$, showed a shift in the edge to lower energy as expected for reduction of Mo upon Li incorporation [Fig. 18(*a*)][34]. The Mo EXAFS Fourier transforms indicate that a very significant structural change of the Mo coordination environment accompanies this reduction [Fig. 18(*b*)]. The initial structure of MoS_3, involving 5–6 Mo–S bonds at 2.42 Å and about 1 Mo–Mo bond at 2.75 Å, changes into one containing more Mo–Mo bonds upon

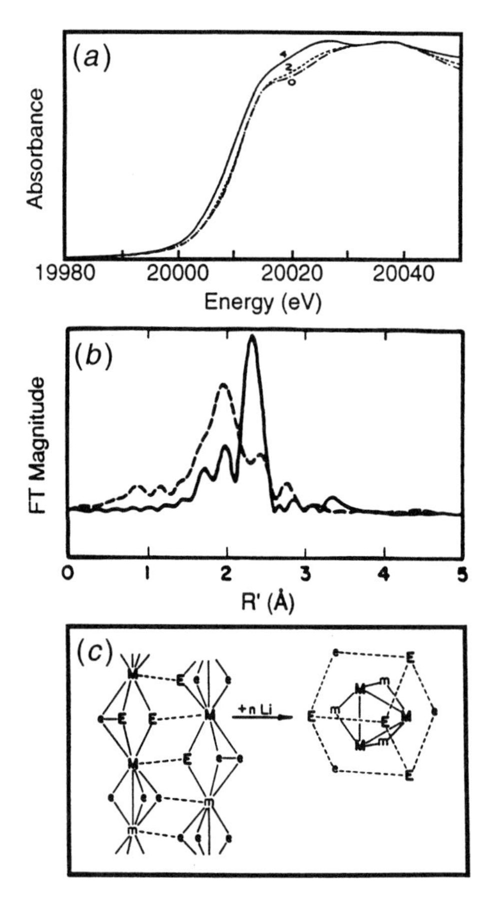

Figure 18

(a) Comparison of the Mo K-edge spectra for untreated MoS_3 (dash–dot line labeled 0) and for $Li_{4.0}MoS_3$ (solid line labeled 4). Note the shift of the Mo edge for the lithiated sample, suggesting a reduction in the average Mo oxidation state. (b) Comparison of the Mo EXAFS Fourier transforms for untreated MoS_3 (dashed line) and for $Li_{4.0}MoS_3$ (solid line). Note the substantial increase in the high R' (2.3–2.4 Å) FT peak in the lithiated material. This indicates an increase in the number of the approximate 2.7-Å Mo–Mo (bonded) interactions upon lithiation, as suggested by the proposed structural change from an extended-chain to a high-nuclearity cluster structure in (c). [These figures were adapted from Scott et al.[34]]

lithiation to $Li_{4.0}$ MoS_3 ($\sim$3 Mo–S bonds at 2.50 Å and about 4 Mo–Mo bonds at 2.66 Å). This increase in Mo neighbors indicates a condensation of Mo atoms into higher nuclearity clusters upon lithiation, as indicated in the proposed conversion of a chain structure to an octahedral Mo_6 cluster [Fig. 18(c)[34]].

In many cases, the treatment being examined is the addition of a potential ligand and XAS is then used to determine whether this compound actually binds to the metal site of interest. In biochemistry, the site of binding of known inhibitors (or substrate analogues) for a metalloenzyme can be tested. This works best if the inhibitor contains a heavy atom (e.g., S, Br, and As) that can easily be detected by EXAFS in the presence of a number of other light-atom (e.g., N and

O) ligands (e.g., see Scott et al.,[35] Clark et al.[36], Scott et al.[37], Holz et al.[38], and Cramer and Hille[39]). Alternatively, inhibitors that generate other unique EXAFS signatures can be used; the most common is cyanide. Figure 19 compares edge, EXAFS, and FT data for two samples of a Cu substituted form of liver alcohol dehydrogenase (LADH), one untreated (solid lines) and the other treated with cyanide (dashed lines). This form of LADH has Cu substituted for Zn in the catalytic sites of each monomer of the dimeric enzyme and is therefore referred to as [Cu_2Zn_2]LADH. The coordination of the Cu(II) in the untreated enzyme is by two cysteine thiolates, one histidine imidazole, and one water. Binding of cyanide to a metal ion usually results in a linear M–C–N arrangement with a fairly short M–C bond distance. The EXAFS scattering amplitude from the remote N is enhanced by a multiple-scattering mechanism (see Fig. 6) through the intervening C, resulting in a large M···N FT peak at about 2.8 Å. This FT peak may be used as a signature for detecting the binding of cyanide to a metal site. Both edge and EXAFS spectra are different for the cyanide-treated [Cu_2Zn_2]LADH compared to the untreated form, suggesting a significant change in the Cu site structure [Fig. 19(*a* and *b*)]. More definitively, a large new FT peak appears at about

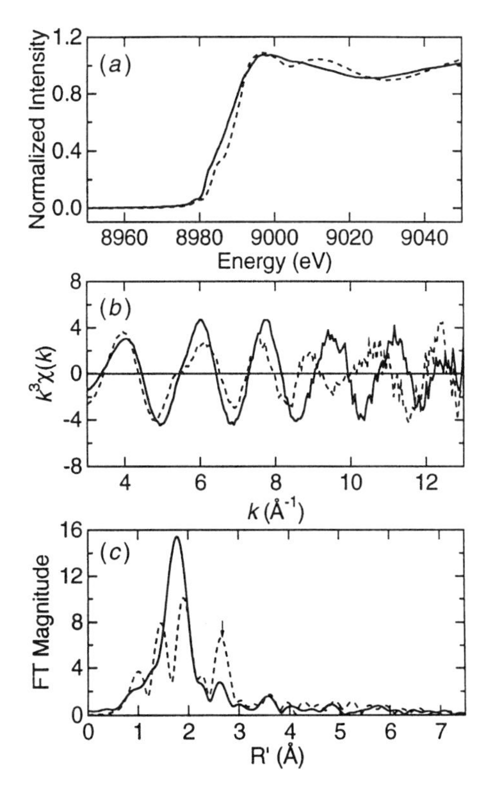

Figure 19

(a) The Cu XAS edge, (b) the EXAFS, and (c) FT data for the Cu substituted form of liver alcohol dehydrogenase ([Cu_2Zn_2]LADH). Solid lines are data from the untreated form and dashed lines are data from the cyanide-treated form. The FT data were generated by k^3-weighted transform over the $k = 3.0$–13.0-Å^{-1} range. [This figure is from unpublished work of Dooley and Scott.]

2.8 Å [arrow in Fig. 19(*c*)], suggesting that cyanide binds to the Cu. Cyanide binding is also suggested by the splitting of the first-shell FT peak into two peaks, implying a larger separation in first-shell distances in the cyanide-bound enzyme [Cu–C at ~1.9 compared to Cu–(N,O) at ~2.0 and Cu–S at ~2.2 Å]. Detailed curve-fitting confirms this conclusion.

F. Is the Metal Being Studied Part of a Metal Cluster?

Since the scattering power of an atom increases with increasing atomic number, additional metal atoms (M′) in the vicinity (~2.5–4.0 Å) of the absorbing metal atom (M) should give rise to characteristic, easily discernible EXAFS contributions. Observation of this M–M′ scattering in the EXAFS could then confirm the presence of a metal cluster $(M, M')_n$ $(n \geq 2)$ in a structurally uncharacterized sample. The M–M′ scattering stands out when M and M′ are directly bonded, when the M–M′ distance is relatively short (<3 Å), or when M′ is a very heavy metal. examples like this include the Mo–Mo clusters in Li_xMoS_3 already discussed, the Fe–Fe clusters in iron–sulfur proteins,[40] and organometallic Mo–Ir clusters.[41]

In all of these examples, a separate peak in the FT can be assigned to the M–M′ scattering [e.g., see Fig. 18(*b*)]. However, even in mononuclear sites, FT peaks often occur in a region corresponding to M–X distances of about 3–4 Å [e.g., see Fig. 19(*c*)]. These usually result from outer-shell scattering atoms that are part of an extended ligand framework (e.g., C and N of imidazole- or pyridine-like rings, C of a porphyrin skeleton). These FT peaks can obscure the presence (or absence) of a M–M′ FT peak, making the detection of a metal cluster difficult.[42] This interference is a frequent problem in metalloenzymes thought to contain dinuclear sites, since histidine imidazole and porphyrin ligation are so common. For example, it has been particularly difficult to measure the Cu···Fe distance in the dinuclear O_2 interaction site of resting state cytochrome *c* oxidase given the presence of histidine and porphyrin ligation. Studies by two different groups have assigned the Cu···Fe distance as 3.0 or 3.8 Å, depending on which FT peak in the Cu EXAFS contains the Cu···Fe contribution.[43] Figure 20 shows the FTs of the Cu and Fe EXAFS of resting state cytochrome *c* oxidase; the arrows indicate the peaks that have been interpreted as containing Cu···Fe (Fe···Cu) scattering; both peaks also contain M···C scattering from histidine imidazoles (M = Cu) or from the porphyrin ring (M = Fe). It is noteworthy that, even with crystal structures now available for two cytochrome oxidases,[45] the Cu_B–Fe_{a3} distance remains a matter of controversy. Similar difficulties were encountered in the successful analysis of the dinuclear Fe–Fe site of hemerythrin.[46–48]

The main problem is that the standard single-scattering EXAFS analysis cannot adequately simulate scattering from outer-shell atoms, since it contains a significant multiple-scattering component. Also, there are often not enough structurally characterized model dinuclear compounds from which to extract empirical M–M′ scattering functions. Now with some of the newer EXAFS analytical techniques that properly treat multiple scattering and rely on more exact ab initio computation of scattering functions[4,5] some of these ambiguities are beginning to be overcome.

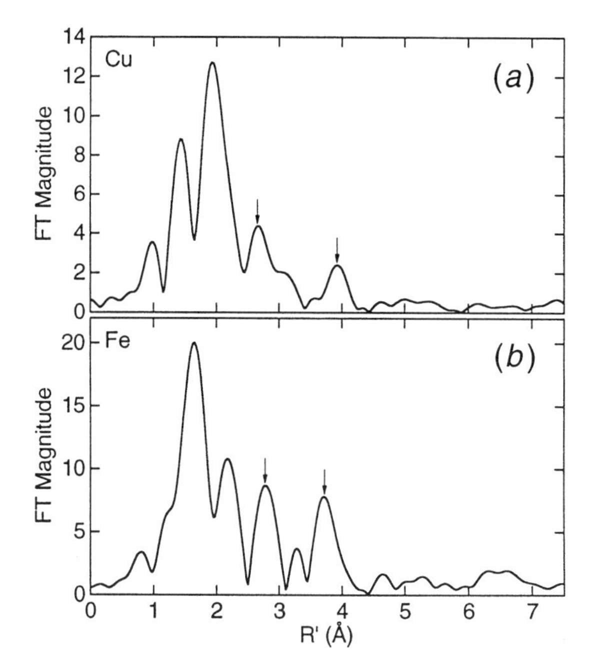

Figure 20
Fourier transforms of Cu EXAFS (*a*) and Fe EXAFS (*b*) for resting state cytochrome *c* oxidase. The arrows in both FTs indicate the peaks that have been assigned by different researchers to Cu···Fe (Fe···Cu) scattering. [Spectrum (*a*) was adapted from Li et al.[44] Spectrum (*b*) was adapted from Scott.[43]]

References

1. Koningsberger, D. C.; Prins, R., Eds.; *X-Ray Absorption. Principles, Applications, Techniques of EXAFS, SEXAFS, and XANES*, Wiley: New York, 1988.
2. Teo, B. K. *EXAFS: Basic Principles and Data Analysis*, Springer-Verlag; Berlin, 1986.
3. Powers, L. "X-ray Absorption Spectroscopy. Application to Biological Molecules" *Biochim. Biophys. Acta* **1982**, *683*, 1–38.
4. Strange, R. W.; Blackburn, N. J.; Knowles, P. F.; Hasnain, S. S. "X-ray Absorption Spectroscopy of Metal–Histidine Coordination in Metalloproteins. Exact Simulation of the EXAFS of Tetrakis(imidazole)copper(II) Nitrate and Other Copper–Imidazole Complexes by the Use of a Multiple-Scattering Treatment" *J. Am. Chem. Soc.* **1987**, *109*, 7157–7162.
5. Rehr, J. J.; Leon, J. M. d.; Zabinsky, S. I.; Albers, R. C. "Theoretical X-ray Absorption Fine Structure Standards" *J. Am. Chem. Soc.* **1991**, *113*, 5135–5140.
6. Epstein, H. M.; Schwerzel, R. E.; Mallozzi, P. J.; Campbell, B. E. "Flash-EXAFS for Structural Analysis of Transient Species: Rapidly Metling Aluminum" *J. Am. Chem. Soc.* **1983**, *105*, 146–148.
7. MacGillavry, C. H.; Rieck, G. D., Eds.; *International Tables for X-Ray Crystallography*, Kynoch Press: Birmingham, UK, 1968.

8. Kutzler, F. W.; Scott, R. A.; Berg, J. M.; Hodgson, K. O.; Doniach, S.; Cramer, S. P.; Chang, C. H. "Single-Crystal Polarized X-ray Absorption Spectroscopy. Observation and Theory for $(MoO_2S_2)^{2-}$" *J. Am. Chem. Soc.* **1981**, *103*, 6083–6088.
9. Colpas, G. J.; Maroney, M. J.; Bagyinka, C.; Kumar, M.; Willis, W. S.; Suib, S. L.; Baidya, N.; Mascharak, P. K. "X-ray Spectroscopic Studies of Nickel Complexes, with Application to the Structure of Nickel Sites in Hydrogenases" *Inorg. Chem.* **1991**, *30*, 920–928.
10. Roe, A. L.; Schneider, D. J.; Mayer, R. J.; Pyrz, J. W.; Widom, J.; Que, L., Jr., "X-ray Absorption Spectroscopy of Iron-Tyrosinate Proteins" *J. Am. Chem. Soc.* **1984**, *106*, 1676–1681.
11. Kau, L.-S.; Spira-Solomon, D. J.; Penner-Hahn, J. E.; Hodgson, K. O.; Solomon, E. I. "X-ray Absorption Edge Determination of the Oxidation State and Coordination Number of Copper: Application to the Type 3 Site in *Rhus* vernicifera Laccase and Its Reaction with Oxygen" *J. Am. Chem. Soc.* **1987**, *109*, 6433–6442.
12. Hahn, J. E.; Scott, R. A.; Hodgson, K. O.; Doniach, S.; Desjardins, S. R.; Solomon, E. I. "Observation of an Electric Quadrupole Transition in the X-ray Absorption Spectrum of a Cu(II) Complex" *Chem. Phys. Lett.* **1982**, *88*, 595–598.
13. Davis, W. M.; Roberts, M. M.; Zask, A.; Nakanishi, K.; Nozoe, T.; Lippard, S. J. "Stereochemical and Electron Spin State Tuning of the Metal Center in Nickel(II) Tropocoronands" *J. Am. Chem. Soc.* **1985**, *107*, 3864–3870.
14. Swenson, D.; Baenziger, N. C.; Coucouvanis, D. "Tetrahedral Mercaptide Complexes. Crystal and Molecular Structures of $[(C_6H_5)_4P]_2M(SC_6H_5)_4$ Complexes (M = Cd(II), Zn(II), Ni(II), Co(II), and Mn(II))" *J. Am. Chem. Soc.* **1978**, *100*, 1932–1934.
15. Ram, M. S.; Bakac, A., Espenson, J. H. "Free-Radical Pathways to Alkyl Complexes of a Nickel Tetraaza Macrocycle" *Inorg. Chem.* **1986**, *25*, 3267–3272.
16. Barefield, E. K.; Wagner, F. "Metal Complexes of 1,4,8,11-Tetramethyl-1,4,8,11-tetraazacyclotetradecane, *N*-Tetramethylcyclam" *Inorg. Chem.* **1973**, *12*, 2435–2439.
17. McAuley, A.; Norman, P. R.; Olubuyide, O. "Preparation, Characterization, and Outer-Sphere Electron-Transfer Reactions of Nickel Complexes of 1,4,7-Triazacyclononane" *Inorg. Chem.* **1984**, *23*, 1938–1943.
18. Zompa, L. J.; Margulis, T. N. "Trigonal Distortion in a Nickel(II) Complex. Structure of $[Ni([9]aneN_3)_2](NO_3)Cl \cdot H_2O$" *Inorg. Chim. Acta* **1978**, *28*, L157–L159.
19. Setzer, W. N.; Ogle, C. A.; Wilson, G. S.; Glass, R. S. "1,4,7-Trithiacyclononane, a Novel Tridentate Thioether Ligand, and the Structures of its Nickel(II), Cobalt(II), and Copper(II) Complexes" *Inorg, Chem.* **1983**, *22*, 266–271.
20. Schramm, C. J.; Scaringe, R. P.; Stojakovic, D. R.; Hoffman, B. M.; Ibers, J. A.; Marks, T. J. "Chemical, Spectral, Structural, and Charge Transport Properties of the "Molecular Metals" Produced by Iodination of Nickel Phthalocyanine" *J. Am. Chem. Soc.* **1980**, *102*, 6702–6713.
21. Davison, A.; Holm, R. H. "Metal Complexes Derived from *cis*-1,2-Dicyano-1,2-Ethylenedithiolate and Bis(trifluoromethyl)-1,2-Dithiete" *Inorg. Synth.* **1967**, *10*, 8–26.
22. Hove, M. J.; Hoffman, B. M.; Ibers, J. A. "Triplet Exciton EPR and Crystal Structure of $[TMPD^+]_2[Ni(mnt)_2]^{-2}$" *J. Chem. Phys.* **1972**, *56*, 3490–3502.
23. Bonamico, M.; Dessy, G. "Structural Studies of Metal Diselenocarbamates. Crystal and Molecular Structures of Nickel(II), Copper(II), and Zinc(II) Diethyldiselenocarbamates" *J. Chem. Soc. A* **1971**, 264–269.
24. Scott, R. A. "Comparative X-ray Absorption Spectropscopic Structural Characterization of Nickel Metalloenzyme Active Sites" *Physica B* **1989**, *158*, 84–86.
25. Kosugi, N.; Yokoyama, T.; Asakura, K.; Kuroda, H. "Polarized Cu K-Edge XANES of Square Planar $CuCl_4{}^{2-}$ Ion. Experimental and Theoretical Evidence for Shake-Down Phenomena" *Chem. Phys.* **1984**, *91*, 249–256.
26. Shiemke, A. K.; Hamilton, C. L.; Scott, R. A. "Structural Heterogeneity and Purification of Protein-free F_{430} from the Cytoplasm of *Methanobacterium thermoautotrophicum*" *J. Biol. Chem.* **1988**, *263*, 5611–5616.

27. Albracht, S. P.; Ankel-Fuchs, D.; Böcher, R.; Ellermann, J.; Moll, J.; van der Zwaan, J. W.; Thauer, R. K. "Five New EPR Signals Assigned to Nickel in Methyl-coenzyme M Reductase from *Methanobacterium thermoautotrophicum*, strain Marburg" *Biochim. Biophys. Acta* **1988**, *955*, 86–102.
28. Shiemke, A. K.; Shelnutt, J. A.; Scott, R. A. "Coordination Chemistry of F_{430}. Axial Ligation Equilibrium Between Square-Planar and Bis-Aquo Species in Aqueous Solution" *J. Biol. Chem.* **1989**, *264*, 11236–11245.
29. LuBien, C. D.; Winkler, M. E.; Thamann, T. J.; Scott, R. A.; Co, M. S.; Hodgson, K. O.; Solomon, E. I. "Chemical and Spectroscopic Properties of the Binuclear Copper Active Site in *Rhus* Laccase; Direct Confirmation of a Reduced Binuclear Type 3 Copper Site in Type 2 Depleted Laccase and Intramolecular Coupling of the Type 3 to the Type 1 and Type 2 Copper Sites" *J. Am. Chem. Soc.* **1981**, *103*, 7014–7016.
30. Fee, J. A. "Copper Proteins. Systems Containing the "Blue" Copper Center" *Structure and Bonding (Berlin)* **1975**, *23*, 1–60.
31. Farver, O.; Frank, P.; Pecht, I. "Peroxide Binding to the Type 3 Site in *Rhus vernicifera* Laccase Depleted of Type 2 Copper" *Biochem. Biophys. Res. Commun.* **1982**, *108*, 273–278.
32. Sugiura, Y.; Hirayama, Y.; Tanaka, H.; Ishizu, K. "Copper(II) Complex of Sulfur-Containing Peptides. Characterization and Similarity of Electron Spin Resonance Spectrum to the Chromophore in Blue Copper Proteins" *J. Am. Chem. Soc.* **1975**, *97*, 5577–5581.
33. Liang, K. S.; deNeufville, J. P.; Jacobson, A. J.; Chianelli, R. R.; Betts, F. "Structure of Amorphous Transition Metal Sulfides" *J. Non-Cryst. Solids* **1980**, *35–36*, 1249–1254.
34. Scott, R. A.; Jacobson, A. J.; Chianelli, R. R.; Pan, W.-H.; Stiefel, E. I.; Hodgson, K. O.; Cramer, S. P. "Reactions of MoS_3, WS_3, WSe_3, and $NbSe_3$ with Lithium. Metal Cluster Rearrangement Revealed by EXAFS" *Inorg. Chem.* **1986**, *25*, 1461–1466.
35. Scott, R. A.; Coté, C. E.; Dooley, D. M. "Copper X-ray Absorption Spectroscopic Studies of the Bovine Plasma Amine Oxidase–Sulfide Complex" *Inorg. Chem.* **1988**, *27*, 3859–3861.
36. Clark, P. A.; Wilcox, D. E.; Scott, R. A. "X-ray Absorption Spectroscopic Evidence for Binding of the Competitive Inhibitor 2-Mercaptoethanol to the Nickel Sites of Jackbean Urease. A New Ni–Ni Interaction in the Inhibited Enzyme" *Inorg. Chem.* **1990**, *29*, 579–581.
37. Scott, R. A.; Wang, S.; Eidsness, M. K.; Kriauciunas, A.; Frolik, C. A.; Chen. V. J. "X-ray Absorption Spectroscopic Studies of the High-Spin Iron(II) Active Site of Isopenicillin N Synthase: Evidence for Fe–S Interaction in the Enzyme–Substrate Complex" *Biochemistry* **1992**, *31*, 4596–4601.
38. Holz, R. C.; Salowe, S. P., Smith, C. K.; Cuca, G. C.; Que, L. Jr., "EXAFS Evidence for a "Cysteine Switch" in the Activation of Prostromelysin" *J. Am. Chem. Soc.* **1992**, *114*, 9611–9614.
39. Cramer, S. P.; Hille, C. R. "Arsenite-Inhibited Xanthine Oxidase—Determination of the Mo–S–As Geometry by EXAFS" *J. Am. Chem. Soc.* **1985**, *107*, 8164–8169.
40. Teo, B.-K.; Shulman, R. G.; Brown, G. S.; Meixner, A. E. "EXAFS Studies of Proteins and Model Compounds Containing Dimeric and Tetrameric Iron–Sulfur Clusters" *J. Am. Chem. Soc.* **1979**, *101*, 5624–5631.
41. Shapley, J.; Uchiyama, W. S.; Scott, R. A. "Bimetallic Catalysts from Alumina-Supported Molybdenum–Irdium Clusters" *J. Phys. Chem.* **1990**, *94*, 1190–1196.
42. Scott, R. A.; Eidsness, M. K. "The Use of X-Ray Absorption Spectroscopy for Detection of Metal–Metal Interactions. Applications to Copper-Containing Enzymes" *Comments Inorg. Chem.* **1988**, *7*, 235–267.
43. Scott, R. A. "X-ray Absorption Spectroscopic Investigations of Cytochrome *c* Oxidase Structure and Function" *Annu. Rev. Biophys. Biophys. Chem.* **1989**, *18*, 137–158.

44. Li, P. M.; Gelles, J.; Chan, S. I.; Sullivan, R. J.; Scott, R. A. "Extended X-Ray Absorption Fine Structure of Copper in Cu_A-Depleted, *p*-(Hydroxymercuri)benzoate-Modified, and Native Cytochrome *c* Oxidase" *Biochemistry* **1987**, *26*, 2091–2095.
45. Iwata, S.; Ostermeier, C.; Ludwig, B.; Michel, H. "Strcuture at 2.8 Å resolution of cytochrome *c* oxidase from *Paracoccus denitrificans*" *Nature (London)* **1995**, 376, 660–669; Tsukihara, T.; Aoyama, H.; Yamashita, E.; Tomizaki, T.; Yamaguchi, H.; Shinzawa-Itoh, K.; Nakashima, R.; Yaono, R.; Yoshikawa, S. "Structures of metal sites of oxidized bovine heart cytochrome *c* oxidase at 2.8 Å" *Science* **1995**, 269, 1069–1074; Tsukihara, T.; Aoyama, H.; Yamashita, E.; Tomizaki, T.; Yamaguchi, H.; Shinzawa-Itoh, K.; Nakashima, R.; Yaono, R.; Yoshikawa, S. "The Whole Structure of the 13-Subunit Oxidized Cytochrome *c* Oxidase at 2.8 Å" *Science* **1996**, 272, 1136–1144.
46. Maroney, M. J.; Scarrow, R. C.; L Que, J.; Roe, A. L.; Lukat, G. S.; D. M. Kurtz, J. "X-ray Absorption Spectroscopic Studies of the Sulfide Complexes of Hemerythrin" *Inorg. Chem.* **1989**, *28*, 1342–1348.
47. Zhang, K.; Stern, E. A.; Ellis, F.; Sanders-Loehr, J.; Shiemke, A. K. "The Active Site of Hemerythrin As Determined by X-ray Absorption Fine Structure" *Biochemistry* **1988**, *27*, 7470–7479.
48. Scarrow, R. C.; Maroney, M. J.; Palmer, S. M.; L. Que, J.; Roe, A. L.; Salowe, S. P.; Stubbe, J. "EXAFS Studies of Binuclear Iron Proteins: Hemerythrin and Ribonucleotide Reductase" *J. Am. Chem. Soc.* **1987**, *109*, 7857–7864.

General References

Teo, B. K. *EXAFS: Basic Principles and Data Analysis*, Springer-Verlag: Berlin, 1986.

A good basic introduction to the background and theory of XAS, but without the modern *ab initio* theoretical treatment of photoelectron scattering.

EXAFS Spectroscopy. Techniques and Applications, Teo, B. K.; Joy, D. C., Ed.; Plenum: New York, 1981.

An early attempt at providing history, background, theory, and applications of EXAFS to mostly materials research. Some discussion of synchrotron sources of the era is also provided.

X-Ray Absorption. Principles, Applications, Techniques of EXAFS, SEXAFS, and XANES, Koningsberger, D. C.; Prins, R., Ed.; J Wiley: New York, 1988.

A more recent attempt to provide a comprehensive review of the background and theory of both EXAFS and XANES, as well as applications ranging from biology to surface science

Lee, P. A.; Citrin, P. H.; Eisenberger, P.; Kincaid, B. M. "Extended x-ray absorption fine structure—its strengths and limitations as a structural tool" *Rev. Mod. Phys.* **1981**, *53*, 769–806.

An early, now classic, review of the EXAFS technique and its application to structural determinations, mostly in solid-state materials.

Crozier, E. D. "A review of the current status of XAFS spectroscopy" *Nucl. Instrum Methods Phys. Res.* **1997**, *B133*, 134–144.

A very recent discussion of the state of the art, including remaining unsolved problems. Most of the examples are from materials science applications.

Shulman, R. G.; Eisenberger, P.; Kincaid, B. M. "X-ray Absorption Spectroscopy of Biological Molecules" *Annu. Rev. Biophys. Bioeng.* **1978**, *7*, 559–578.
Cramer, S. P.; Hodgson, K. O. "X-ray Absorption Spectroscopy: a New Structural Method and Its Applicatios to Bioinorganic Chemistry" *Prog. Inorg. Chem.* **1979**, *25*, 1–39.
Powers, L. "X-ray Absorption Spectroscopy. Application to Biological Molecules" *Biochim. Biophys. Acta* **1982**, *683*, 1–38.
Scott, R. A. "X-Ray Absorption Spectroscopy" In *Structural and Resonance Techniques in Biological Research*; D. L. Rousseau, Ed.; Academic: New York, 1984; pp 295–362.
Scott, R. A. "Measurement of Metal–Ligand Distances by EXAFS" *Methods Enzymol.* **1985**, *117*, 414–459.
Charnock, J. M. "Biological Applications of EXAFS Spectroscopy" *Rad. Phys. Chem.* **1995**, *45*, 385–391.

A chronological series of useful reviews of the biological applications of XAS. Most also have a treatment of the basic theory and methodology of XAS.

Kau, L.-S.; Spira-Solomon, D. J.; Penner-Hahn, J. E.; Hodgson, K. O.; Solomon, E. I. "X-ray Absorption Edge Determination of the Oxidation State and Coordination Number of Copper: Application to the Type 3 Site in *Rhus* vernicifera Laccase and Its Reaction with Oxygen" *J. Am. Chem. Soc.* **1987**, *109*, 6433–6442.
Colpas, G. J.; Maroney, M. J.; Bagyinka, C.; Kumar, M.; Willis, W. S.; Suib, S. L.; Baidya, N.; Mascharak, P. K. "X-ray Spectroscopic Studies of Nickel Complexes, with Application to the Structure of Nickel Sites in Hydrogenases" *Inorg. Chem.* **1991**, *30*, 920–928.
Roe, A. L.; Schneider, D. J.; Mayer, R. J.; Pyrz, J. W.; Widom, J.; L. Que, J. "X-ray Absorption Spectroscopy of Iron-Tyrosinate Proteins" *J. Am. Chem. Soc.* **1984**, *106*, 1676–1681.

Three excellent examples of the use of the edge region of the X-ray absorption spectrum to provide information on the molecular geometry of metal sites.

Mustre de Leon, J.; Rehr, J. J.; Zabinsky, S. I.; Albers, R. C. *Phys Rev B* **1990**, *44*, 4146–4156.
Rehr, J. J.; Zabinsky, S. I.; Albers, R. C. "High-order Multiple Scattering Calculations of X-ray-Absorption Fine Structure" *Phys. Rev. Lett.* **1992**, *69*, 3397–3400.
Fonda, L. "Multiple-Scattering Theory of X-ray Absorption—A Review" *J. Phys. Cond. Matter* **1992**, *4*, 8269–8302.

Recent, in-depth discussions of the importance of multiple scattering and its application to structural applications of EXAFS.

10
Case Studies

CONTENTS

I. The Cu_A Site of Cytochrome *c* Oxidase

JOANN SANDERS-LOEHR
Department of Biochemistry and Molecular Biology
Oregon Graduate Institute of Science and Technology
Portland, OR 97291-1000

Cytochrome *c* oxidase (CCO) is an enzyme that carries out the final step of respiration in mitochondria and aerobic bacteria, utilizing four equivalents from cytochrome *c* to reduce dioxygen to water.[1] This enzyme contains four redox centers, each of which is capable of storing or transmitting one electron (Fig. 1).

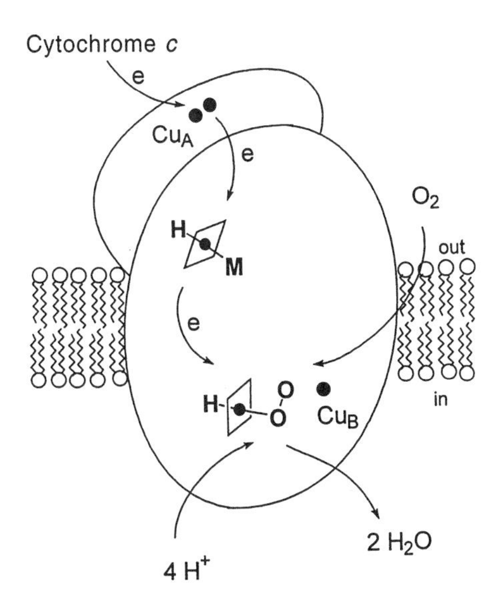

Figure 1
Proposed electron-transfer pathway in cytochrome *c* oxidase, based on the X-ray crystal structure of the enzyme from *Paracoccus denitrificans*.[2]

The Cu_A site accepts one electron from cytochrome *c* and passes it via intramolecular electron transfer to a low-spin heme Fe_a and finally to a dinuclear heme Fe_{a3}/Cu_B site, where the reduction of O_2 to H_2O takes place. Because the spectroscopic properties of CCO are dominated by contributions from the two heme groups, it has been difficult to characterize the Cu_A site. Consequently, the nature of the Cu_A site has been the subject of intense dispute for over 60 years.[3] Elemental analyses revealed only enough copper to create a mononuclear species, but the unusual spectroscopic properties of Cu_A could not be explained by a mononuclear site.

The first major advance in understanding the nature of the Cu_A site came from the prediction that it contained a magnetically interacting pair of copper ions, as a means of explaining its unusual electron paramagnetic resonance (EPR) spectrum.[4] The extremely narrow $A_{\|}$ hyperfine splitting of less than 40 G (Fig. 2) is reminiscent of the small splittings found for mononuclear type 1 copper sites where there is substantial delocalization of electron density onto the cysteine thiolate ligand.[6] However, instead of the four-line hyperfine splitting pattern of mononuclear Cu (see Chapter 3), a seven-line pattern is observed (Fig. 2). The seven-line splitting is best explained as arising from a single unpaired electron interacting with two $I = \frac{3}{2}$ copper nuclei to generate a 1 : 2 : 3 : 4 : 3 : 2 : 1 intensity

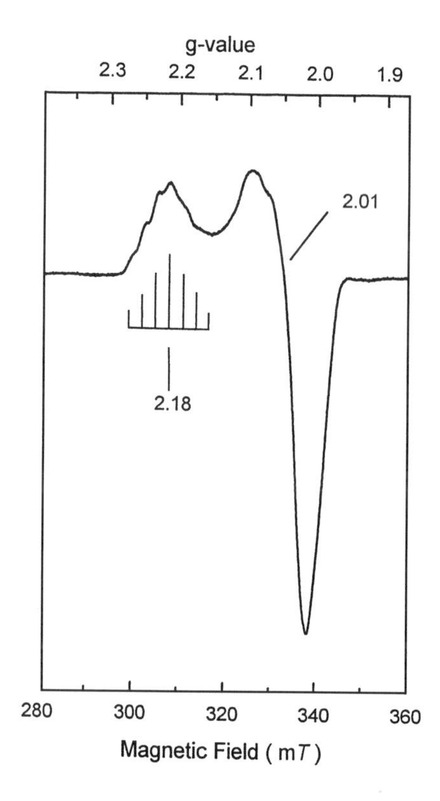

Figure 2
The X-band EPR spectrum of Cu_A domain from *P. denitrificans*. [Reprinted with permission from Farrar et al.[5]]

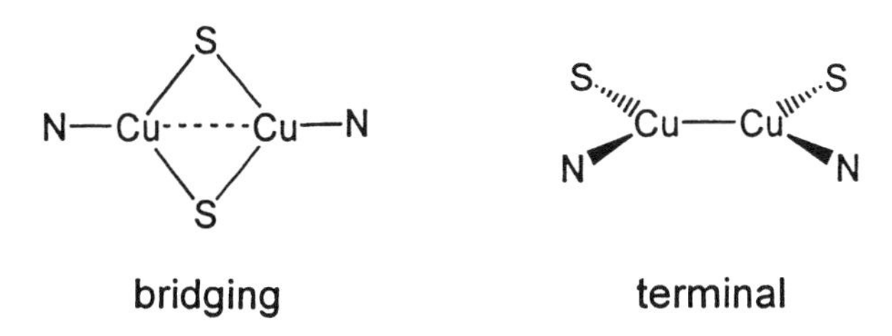

Figure 3
Models for Cu_A site containing either bridging or terminal cysteine thiolates (S) and terminal histidine imidazoles (N).

ratio. Such a seven-line hyperfine splitting was initially recognized in the enzyme N_2O reductase and interpreted as a Cu^I–Cu^{II} pair in which the single unpaired electron was completely delocalized over the two copper ions to generate a Cu(1.5)–Cu(1.5) state.[4] Although the splitting is not as clear in the EPR spectrum of native CCO, the detection of a seven-line pattern in the Cu_A fragment (Fig. 2) provides strong evidence for a similar mixed-valence Cu(1.5)–Cu(1.5) state in all Cu_A sites.

Another major advance was provided by molecular engineering when the gene for the Cu_A-containing domain was truncated from the rest of the membrane protein, allowing the expression of a soluble protein with only the Cu_A site.[7] Electrospray mass spectrometry revealed that Cu-reconstituted protein differed from the apoprotein by exactly two Cu atoms, and the EPR spectrum revealed a clear seven-line hyperfine splitting. Potential ligands to the dinuclear site were proposed as two cysteines and two histidines on the basis of their being totally conserved in all Cu_A domains, their sequence similarity to mononuclear copper sites in homologous proteins such as plastocyanin, and from the loss of Cu_A character when any one of these residues was mutated.[5,7]

Two dicopper models were proposed[5] in which each copper was given an identical ligand set in order to account for the complete delocalization of the unpaired electron (Fig. 3). In addition, a Cu–Cu bond (in the terminal model) or a strong Cu–Cu interaction (in the bridging model) was implicated by extended X-ray absorption fine structure (EXAFS) measurements.[8,9] The EXAFS studies of the Cu_A fragment were performed and compared with a model complex containing a delocalized Cu(1.5)–Cu(1.5) moiety in an octaazacryptand macrocycle with each Cu coordinated to four amines and a second Cu atom. The Fourier transform (FT) for the model compound is well fit by four N scatterers at 2.07 Å and a second Cu at 2.41 Å [Fig. 4(*a*)], in excellent agreement with the X-ray crystal structure. The FT for the Cu_A domain [Fig. 4(*b*)], is well fit by 2.0 N scatterers at 1.95 Å, 1.3 S at 2.30 Å, and 0.7 Cu at 2.5 Å (with low S and Cu occupancies arising from the 30% adventious Cu). This distribution was originally assigned[8] to the terminal model of Figure 3. However, more recent EXAFS data[9] on a Cu_A domain from *Thermus thermophilus* (no adventitious Cu) can be well fit by 1 N, 2 S, and 1 Cu scatterer per Cu, which agrees better with the bridging model of Figure 3. In all cases, the short Cu · · · Cu distance of 2.4–2.5 Å is highly unusual and, thus, suggestive of a direct interaction between the two metal ions.

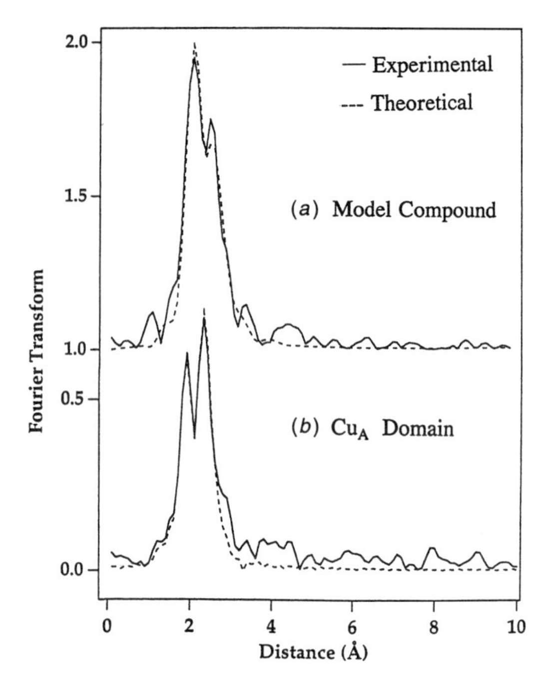

Figure 4
Fourier transformed Cu K-edge EXAFS for (*a*) mixed-valence model compound and (*b*) Cu_A domain from *Bacillus subtilis*. The EXAFS spectra were simulated (dashed line) with a Cu–Cu interaction to account for the resolved shell at 2.4–2.5 Å. [Reprinted with permission from Blackburn, N. J.; de Vries, S.; Barr, M. E.; Houser, R. P.; Tolman, W. B.; Sanders, D.; Fee, J. *J. Am. Chem. Soc.* **1997**, *119*, 6135–6143. Copyright © 1997 American Chemical Society.]

The X-ray crystal structure of CCO[2] revealed that the proposed bridging thiolate structure was indeed correct for the Cu_A site (Fig. 5). The metal–scatterer distances are in good agreement with the EXAFS results, including a short Cu–Cu at about 2.55 Å. However, one striking finding is that each of the Cu

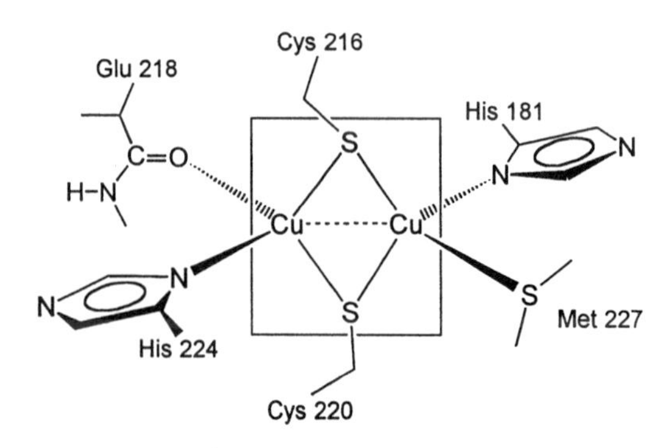

Figure 5
The Cu_A site in *P. denitrificans* CCO from X-ray crystal structure at 2.8-Å resolution. [Redrawn from data in Iwata et al.[2]]

atoms is actually closer to tetrahedral than trigonal planar geometry with the imidazoles offset from the Cu_2S_2 plane and a weakly coordinated amide-backbone carbonyl (Glu 218) or methionine thioether (Met 227) as the fourth ligand. This same ligand set was also observed in the crystal structure of an engineered Cu_A containing protein (purple CyoA) produced by adding two cysteine and two histidine residues to the homologous domain of a quinol oxidase from *Escherichia coli*.[10] The presence of different ligands on the two Cu atoms raises the question of how the complete electron delocalization of the mixed-valence state is achieved.

Further information about the geometry of these dinuclear copper sites has been obtained by resonance Raman (RR) spectroscopy. Copper cysteinates exhibit intense absorption in the visible region due to (Cys)S $\rightarrow$ Cu charge transfer. Excitation within these absorption bands leads to a rich RR spectrum arising from vibrational modes of the Cu–Cys moiety, and these can be identified by frequency shifts to lower energy upon substitution of ^{65}Cu for ^{63}Cu or ^{34}S for ^{32}S in the protein.[11] In mononuclear Cu–Cys proteins, a single Cu–S stretching mode at 300–420 cm^{-1} exhibits a large S-isotope shift of -4 cm^{-1} , as expected for a single Cys ligand. In contrast, in dinuclear Cu_A proteins[12,13] two Cu–S stretching modes are observed at 260 and 340 cm^{-1} with ^{34}S shifts of -4 and -5 cm^{-1}, respectively [Fig. 6(*a*)]. Both the frequencies and isotope shifts are well fit by normal coordinate analysis,[14] assuming a $Cu_2S_2Im_2$ core, where Im = imidazole, as in the bridging thiolate model in Figure 3. The 340-cm^{-1} mode is assigned to the totally symmetric stretch of the Cu_2S_2 rhombus, whereas the 260-cm^{-1} mode is assigned to a combination of Cu–S(Cys) and Cu–Im stretching motions (Fig. 7). The observed RR data cannot be fit using the terminal thiolate model in Figure 3, which predicts two Cu–S stretching modes at 340–350 cm^{-1}, and thus provide definitive evidence against such a structure.

Another result of the normal coordinate analysis is the finding that the predicted vibrational coupling between the Cu–Im and Cu–S stretching modes has a marked angular dependence. The observed S-, Cu-, and N-isotope shifts in the Cu_A site are best fit by a structure in which the Cu–N bond of each of the terminal His ligands is tilted by about 40° above or below the Cu_2S_2 plane.[14] This analysis implies a distorted tetrahedral coordination geometry for each Cu, thereby providing strong spectroscopic support for the structure determined by X-ray crystallography (Fig. 5). The question of a Cu–Cu bonding interaction has also been addressed by RR spectroscopy. Although the Cu sensitive A_g mode at 138 cm^{-1} is predominantly a vibration of the two Cu atoms (Fig. 7), its motion in a normal coordinate analysis can be equally well defined as a Cu–Cu stretch or a Cu–S–Cu bend. Evidence in favor of a Cu–Cu interaction comes from the observation that the 138-cm^{-1} mode is particularly enhanced by excitation within the 780-nm absorption band of Cu_A and that the Cu–Cu bonded octaazacryptand model has a similar Cu–Cu electronic transition near 800 nm.[15]

The RR spectra of Cu_A containing proteins from a variety of different sources are remarkably similar to one another, as can be seen for *P. denitificans, T. thermophilus*, and *B. subtilis* in Figure 6. All of the spectra are dominated by the two Cu–S modes at 260 and 340 cm^{-1}, and the total variability in these frequencies is only ± 4 cm^{-1} in all of the species studied.[12] In contrast, the principal ν(Cu–S) mode of the mononuclear blue copper proteins (cupredoxins) varies by

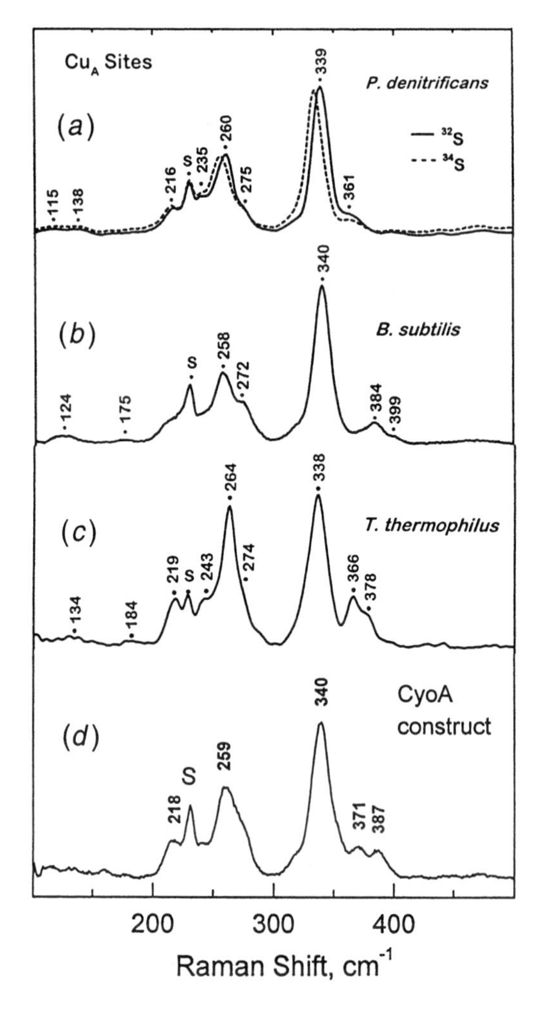

Figure 6
The RR spectra of Cu_A sites in fragments of cytochrome *c* oxidase from (*a*) *P. denitrificans* grown on [^{32}S]- or [^{34}S]-sulfate, (*b*) *B. subtilis*, (*c*) *T. thermophilus*, and (*d*) of reengineered quinol oxidase (CyoA) from *E. coli.* Spectra were obtained with 488-nm excitation at 15 K as described in Andrew et al.[12] (*a–c*) and Andrew et al.[13] (*d*). Frozen solvent = S.

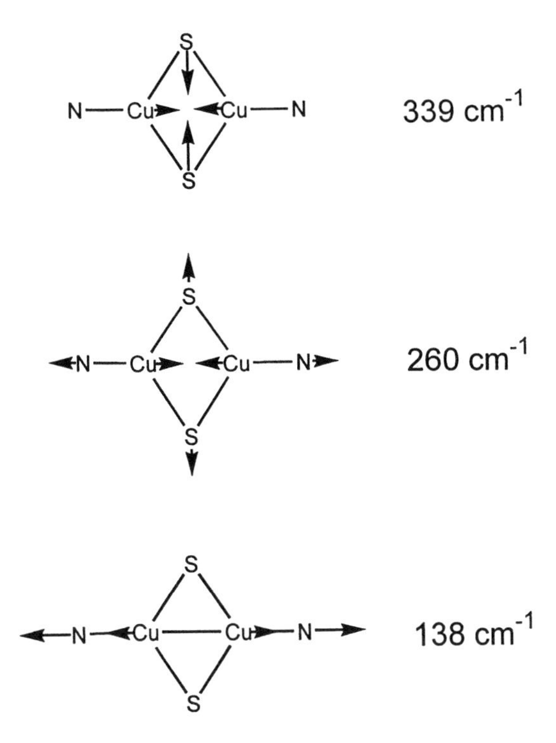

Figure 7
Calculated atomic displacements and frequencies for the symmetric A_g stretching modes of the Cu_A site based on a normal coordinate analysis. [Reprinted with permission from Andrew, C. R.; Fraczkiewicz, R.; Czernuszewicz, R. S.; Lappalainen, P.; Saraste, M.; and Sanders-Loehr, J. *J. Am. Chem. Soc.* **1996**, *118*, 10436–10445, Copyright © 1996 American Chemical Society.]

$\pm 38\ cm^{-1}$, indicating that there is a continuum of geometries from trigonal (His_2Cys ligation) to tetrahedral depending on the strength of the weaker axial ligand (generally methionine).[11,16] The greater structural stability of the dinuclear $Cu_2S_2Im_2$ core is further demonstrated by its ease of self-assembly. Thus, the purple cyoA construct (prepared by inserting two His and two Cys into homologous sites in a non-Cu_A containing quinol oxidase) has the RR signature characteristic of a $Cu_2S_2Im_2$ site with thiolate bridged Cu atoms and tetrahedral coordination geometry [Fig. 6(*d*)].

It would appear that the structure and electronic properties of the Cu_A site, as reflected in vibrational frequencies, are much less affected by the weakly coordinated methionine and carbonyl ligands (Fig. 5) than is the case in the mononuclear cupredoxins. A similar conclusion has been reached from an analysis of the paramagnetically shifted NMR signals of a Cu_A construct where less than 1% of the unpaired spin density is associated with the methionine and carbonyl ligands, and 50–60% is distributed over the Cys and His ligands.[17] The symmetry introduced by the bridging thiolates and terminal imidazoles is apparently great

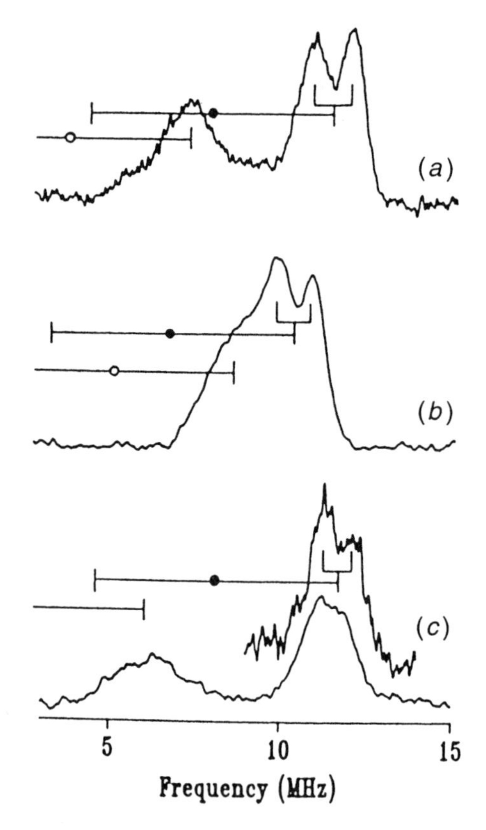

Figure 8
The ^{14}N ENDOR spectra at 35-GHz of (*a*) bovine cytochrome *aa*$_3$ (*b*) *T. thermophilus* cytochrome *ba*$_3$, and (*c*) *T. thermophilus* cytochrome *caa*$_3$. The assignment to A(N1)/2 (•) and A(N2)/2 (○) and twice the Larmor frequency (⊢⊣) are shown for the two nitrogen ligands. [Reprinted with permission from Gurbiel, R. J.; Fann, Y.-C.; Surerus, K. K.; Werst, M. M.; Musser, S. M.; Doan, P. E.; Chan, S. I.; Fee, J. A. Hoffman, B. M. *J. Am. Chem. Soc.* **1993**, *115*, 10888–10894. Copyright © 1993 American Chemical Society.]

enough to ensure a valence-delocalized cluster where the unpaired electron is shared between the two copper atoms. This coupling is preserved despite the fact that the two His ligands in Cu_A (from bovine mitochondria and *T. thermophilis*) are in different enough protein environments to render them distinguishable by $^{14,15}N$ electron–nuclear double resonance (ENDOR) spectroscopy (Fig. 8).[18] The greater stability of the dinuclear cluster toward perturbations from the protein environment could have been the driving force for the evolution of dinuclear Cu_A sites in preference to mononuclear blue Cu sites where the structure of the metal cluster is more variable.[11]

The question arises as to how the unusually stable structure of the Cu_2S_2 core in Cu_A relates to its biological function as a mediator of electron transfer. A key factor in achieving a rapid rate of electron transfer is to minimize the structural reorganization of a metal site as it cycles between different oxidation states.[19]

Preventing reorganization is particularly important for copper, which favors a trigonal array of ligands in the Cu(I) state and a square planar array in the Cu(II) state. The mononuclear Cu sites in cupredoxins already provide a solution to this problem by imposing a distorted tetrahedral coordination geometry that is favorable to both Cu(I) and Cu(II). Such a distorted tetrahedral geometry continues to be a structural feature for the two Cu atoms in Cu_A (Fig. 5). An additional factor for Cu_A is the ability to distribute the bond-length changes from a one-electron redox reaction over two copper centers, thereby enabling an even greater lowering of the reorganization energy. Similar benefits of electron delocalization are observed in other electron-transfer proteins such as the ferredoxins, where a one-electron change in charge is distributed over dinuclear or tetranuclear iron–sulfur clusters.[3]

II. Isopenicillin N Synthase

LAWRENCE QUE, JR.

Department of Chemistry and Center for Metals in Biocatalysis
University of Minnesota, Minneapolis, MN 55455

Isopenicillin N synthase (IPNS) is a nonheme Fe(II) dependent enzyme found in β-lactam antibiotic-producing microorganisms, such as *Cephalosporium*, *Penicillium*, and *Streptomyces*. These organisms catalyzes the formation of isopenicillin N from δ(L-α-aminoadipoyl)-L-cysteinyl-D-valine (ACV). This reaction involves a four-electron oxidation that requires concomitant reduction of dioxygen to water (Scheme 1);[20] thus, the enzyme is technically an oxidase, with dioxygen reduction providing the driving force for the two ring forming steps of the reaction.

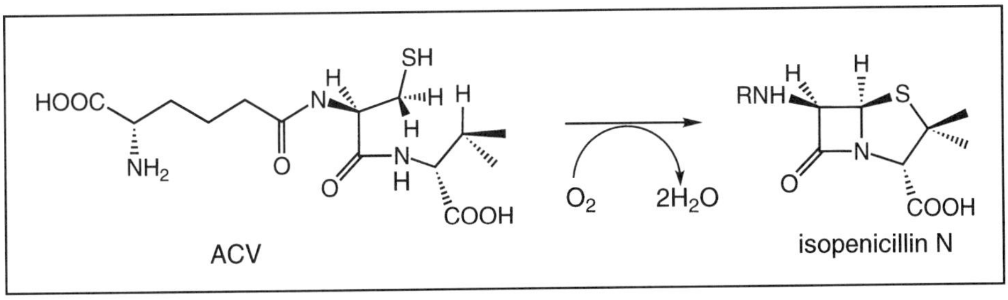

HOOC NH2 O H N SH H H H N H O COOH
ACV
O2 2H2O
RNH H H S N O COOH
isopenicillin N

Scheme 1

Addition of 1 equivalent of Fe(II) to apo-IPNS immediately reconstitutes Fe^{II}IPNS and elicits full enzymatic activity. It exhibits a Mössbauer doublet with $\delta = 1.30\ mm\ s^{-1}$ (Fig. 9; see also Chapter 7), which is associated with a high-spin iron(II) center in an O/N ligand environment.[21] The Fe^{II}IPNS complex has an

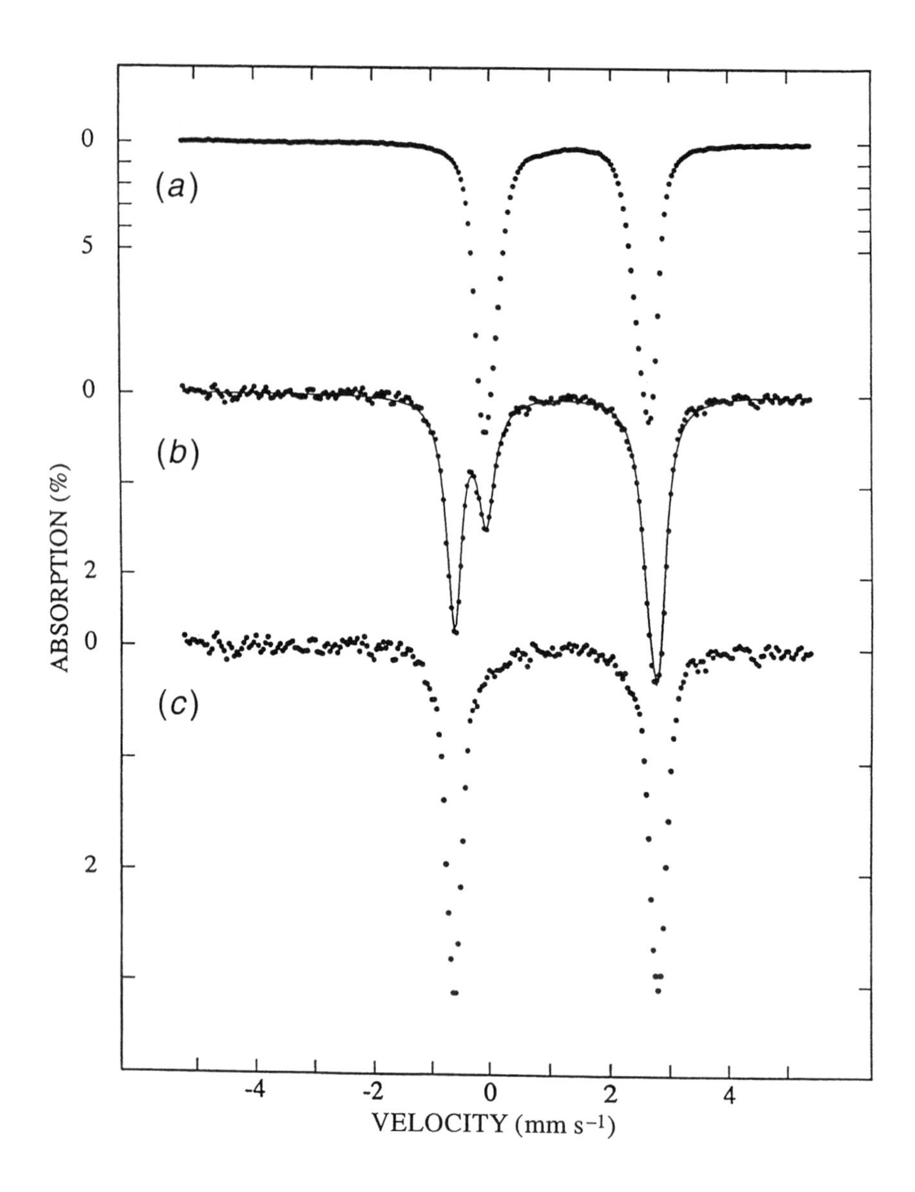

Figure 9
Mössbauer spectra of ^{57}Fe enriched Fe^{II}IPNS as isolated (*a*) and in the presence of 35-mM ACV (substrate) (*b*). Spectrum (*c*) represents the spectrum of the enzyme–substrate complex obtained by removing the contribution of the as isolated enzyme from spectrum (*b*). [Reprinted with permission from the American Society for Biochemistry and Molecular Biology, Chen et al.[21]]

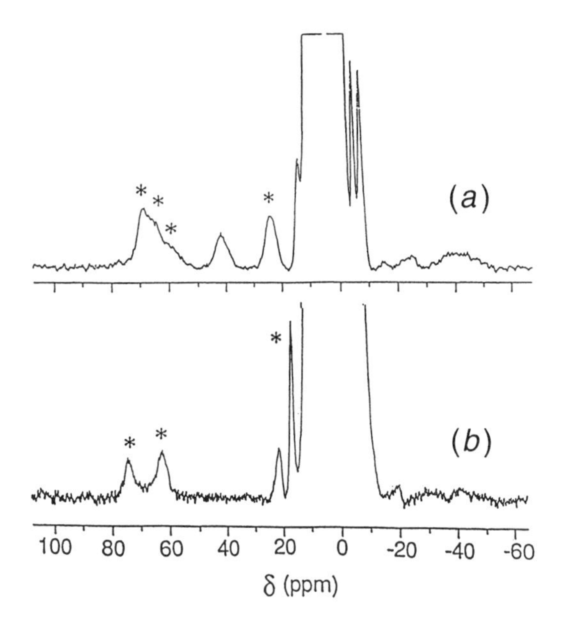

Figure 10
The ^{1}H NMR spectra of Fe^{II}IPNS as isolated (*a*) and its ternary complex with ACV and NO (*b*). Signals marked with asterisks are solvent exchangeable. [Adapted from Ming, L.-J.; Que, L., Jr.; Kriauciunas, A.; Frolik, C. A.; Chen, V. J., Biochemistry, **1991**, *30*, 11653. Copyright © 1991 American Chemical Society.]

X-ray absorption pre-edge absorption peak at about 7112 eV with an intensity of 7.6 units; this feature has been assigned to the 1s → 3d transition (see Chapter 9) and its intensity correlates with the amount of 3d–4p orbital mixing, that is enhanced by a departure from a centrosymmetric coordination environment.[22] Based on a comparison of model complexes with known structure,[23] the pre-edge absorption for Fe^{II}IPNS suggests a distorted six-coordinate environment.[23,24]

The ^{1}H NMR studies of Fe^{II}IPNS show a number of paramagnetically shifted signals (Fig. 10).[25] The high-spin Fe(II) center has four unpaired electrons with a relatively fast spin–lattice relaxation rate; thus, reasonably sharp paramagnetically shifted features can be observed (see Chapter 8). The unsymmetrically shaped signal above 60 ppm [Fig. 10(*a*)] disappears when the solvent is changed to D_2O, which indicates that the protons can exchange with solvent. Since the NH protons of Fe(II) coordinated imidazoles in model complexes are solvent exchangeable and have shifts around 60 ppm, the approximate 60 ppm signals of Fe^{II}IPNS are proposed to arise from His imidazole N–Hs. The asymmetry of the signals implicates the presence of more than one His ligand. The poor spectral resolution prevents the assignment of other features in the spectrum.

Further support for His ligation is found in the Cu(II) substituted derivative.[26] The Cu^{II}IPNS complex exhibits a Type II EPR spectrum with $g_{\parallel} = 2.346$, $A_{\parallel} = 465$ MHz, $g_{\perp} = 2.06$, and $A_{\perp} = 35$ MHz [Fig. 11(*a*)]; these parameters im-

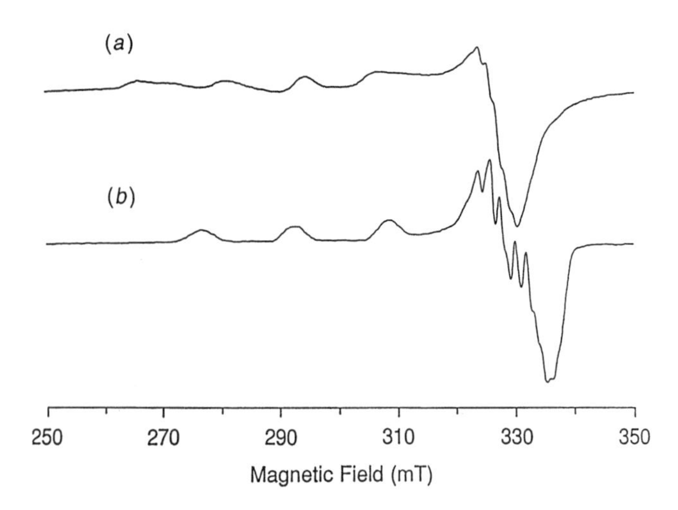

Figure 11
The X-band EPR spectra of Cu^{II}IPNS (*a*) and its ACV complex (*b*). [Reprinted with permission from Ming, L.-J.; Que, L., Jr.; Kriauciunas, A.; Frolik, C. A.; Chen, V. J. *Inorg. Chem.* **1990**, *29*, 1111–1112. Copyright © 1990 American Chemical Society.]

ply a tetragonal environment,[27] consistent with the XAS pre-edge result. Electron spin echo envelope modulation (ESEEM) studies of Cu^{II}IPNS reveal three sharp lines at 0.46, 1.18, and 1.63 MHz, a double quantum feature at about 4.2 MHz, and weaker signals at 2.08, 2.81, and 3.27 MHz [Fig. 12(*a*)].[8] The first four lines are characteristic of the interaction of the unpaired electron of Cu(II) and the remote nitrogen of a coordinated imidazole under the condition of "exact cancellation" (see Chapter 4). The weaker signals arise when more than one imidazole is coordinated to the Cu(II). Simulations favor the presence of two equatorially bound imidazoles.

When the ESEEM experiment is repeated in D_2O, a new peak appears at 2.06 MHz (near the ^2H frequency at the magnetic filed of the measurement), showing that deuterons interact with the Cu(II) unpaired electron.[27] It is not straightforward to ascertain whether these deuterons arise from a coordinated water molecule or from solvent molecules in the vicinity of the active site without assessing the strength of the deuteron coupling. However, the addition of ACV causes the ^2H feature to diminish in intensity and a detailed comparison of the deuteron modulations of the two complexes suggests the displacement of a bound water (or hydroxide) upon ACV binding.

Support for the bound water is also found in EPR studies of Fe^{II}IPNS–NO. The binding of NO converts the high-spin Fe(II) center (typically EPR silent) to an EPR active $S = \frac{3}{2}$ species [Fig. 13(*a*)].[21] The six d electrons of the high-spin Fe(II) interact with the lone unpaired electron of NO to engender an $\{FeNO\}^7$ moiety that has three unpaired electrons. Because of zero-field splitting (zfs) (Scheme 2), the four $S = \frac{3}{2}$ spin states split into two Kramers doublets. Under near axial symmetry, the $\pm\frac{3}{2}$ doublet is EPR silent, while the $\pm\frac{1}{2}$ doublet gives rise to signals near $g = 4$ and 2 (see Chapter 3).

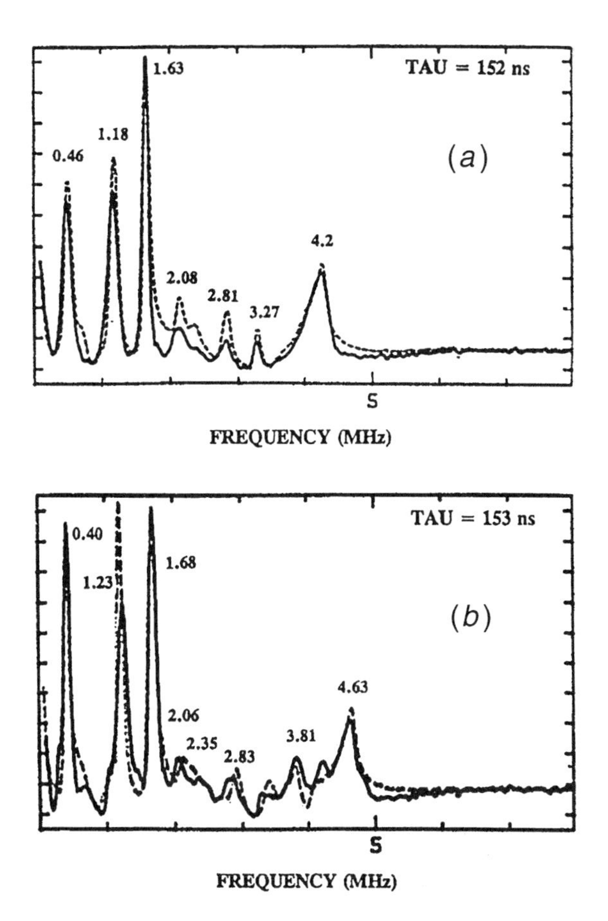

Figure 12
The ESEEM spectra of Cu^{II}IPNS in the absence (pH 6.0) (*a*) and the presence (pH 7.1) of ACV (*b*). [Reprinted with permission from Jiang, F.; Peisach, J.; Ming, L.-J.; Que, L., Jr.; Chen, V. J. *Biochemistry*, **1991**, *30*, 11437–11445. Copyright © 1991 American Chemical Society.]

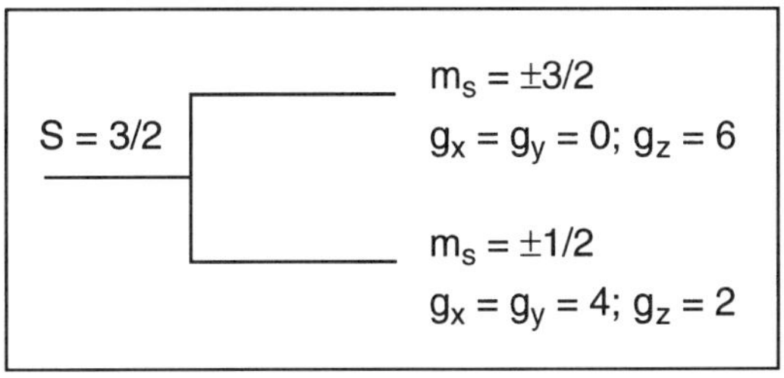

S = 3/2

$m_S = \pm 3/2$
$g_x = g_y = 0;\ g_z = 6$

$m_S = \pm 1/2$
$g_x = g_y = 4;\ g_z = 2$

Scheme 2

The $S = \frac{3}{2}$ EPR signals of Fe^{II}IPNS–NO are quite sharp, indicating an active site of low heterogeneity. The sharp signals have allowed the detection of nuclear hyperfine interactions from bound solvent [Fig. 13(*a*)].[28] When Fe^{II}IPNS–NO is dissolved in $H_2{}^{17}O$ buffer, the signals are noticeably broadened, presumably via interaction with the ^{17}O ($I = \frac{5}{2}$) nucleus. These observations provide further evi-

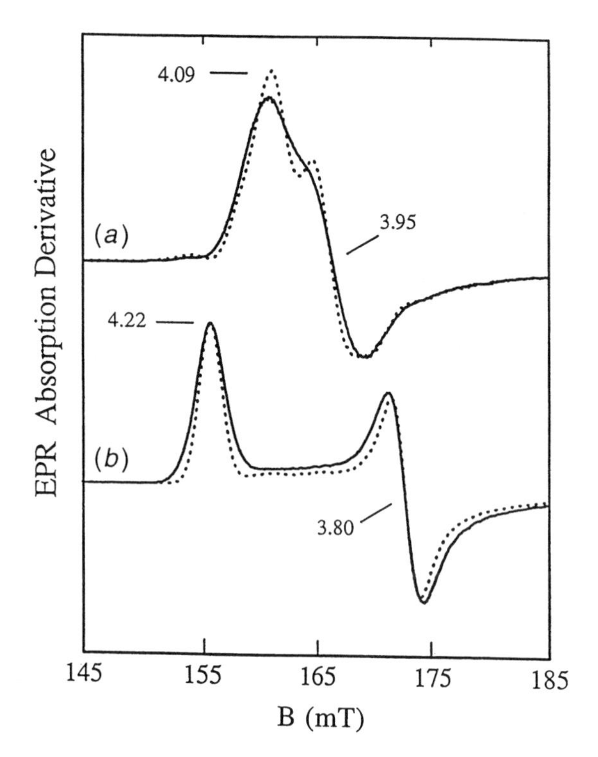

Figure 13
Effect of $H_2{}^{17}O$ on the EPR spectra of Fe^{II}IPNS–NO (*a*) and Fe^{II}IPNS–ACV–NO (*b*) (solid line, $H_2{}^{17}O$; dotted line, $H_2{}^{16}O$). [Reprinted with permission from the American Society for Biochemistry and Molecular Biology, Chen et al.[21]]

dence that the inner-coordination sphere of the active site iron is accessible to solvent. The spectroscopic data discussed thus far provide evidence for two exogenous ligand sites, one for solvent and another for NO (and presumably O_2).

The remaining ligand in the metal coordination sphere that can be identified is a carboxylate, a conclusion derived mainly from NMR[25] studies. As mentioned earlier, the ^1H NMR spectrum of Fe^{II}IPNS is quite broad; however, the substitution of Fe(II) with Co(II) affords a sharper NMR spectrum, particularly in the case of Co^{II}IPNS-ACV (Fig. 14), because of its more favorable electronic relaxation properties (see Chapter 8).[25] The sharpness of the features allows nuclear Overhauser enhancement (NOE) experiments to be performed. These are crucial experiments for establishing proton–proton proximity, especially in spectral regions where the isotropic shift values do not allow for unequivocal assignment of signals to a particular residue. (Typically, the smaller the paramagnetic shift, the more ambiguous the assignment; so additional evidence is helpful.)

Saturation of the signal at 39 ppm engenders a response at 24 and 15 ppm (Fig. 14). By varying the irradiation time and monitoring the responses of the 15 and 24 ppm signals, it can be deduced that the 15 ppm proton is more distant from the 39 ppm proton than the 24 ppm proton. Irradiation experiments at 24 and 15 ppm confirm these conclusions. This trio of signals behaves like an ABX system as seen on an amino acid side chain, that is, $X{-}C_\beta H_2{-}C_\alpha H<$ (Scheme 3).

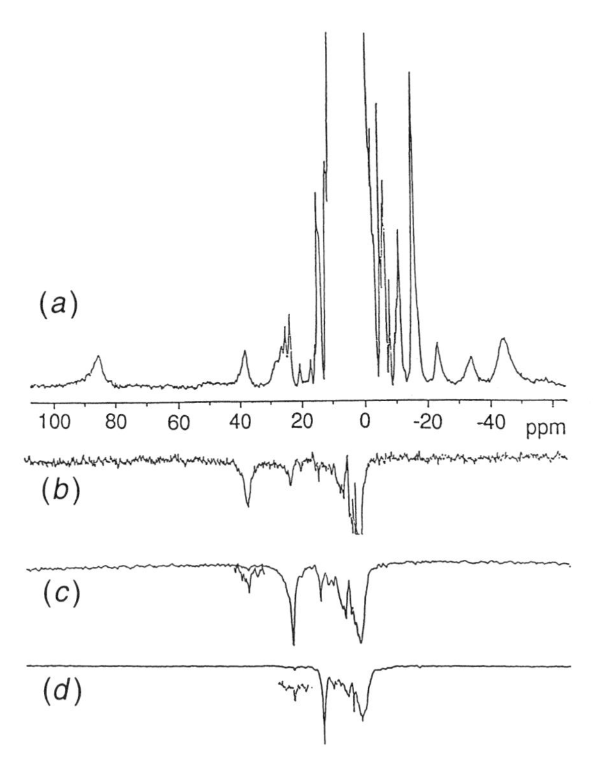

Figure 14
(*a*) The ^{1}H NMR spectra of Co^{II}IPNS–ACV in D_2O buffer. (*b–d*) the NOE difference spectra obtained by subtraction of the reference spectrum with those obtained in which the resonances at 38.6 (*b*), 24.3 (*c*), and 14.6 ppm (*d*) had been irradiated. [Reprinted with permission from Ming, L.-J.; Que, L., Jr.; Kriauciunas, A.; Frolik, C. A.; Chen, V. J., *Biochemistry*, **1991**, *30*, 11653. Copyright © 1991 American Chemical Society.]

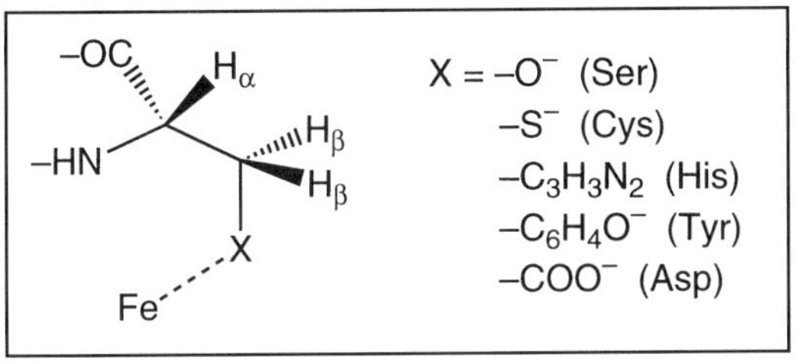

–OC, H_α, –HN, H_β, H_β, X, Fe

X = $-O^-$ (Ser)
$-S^-$ (Cys)
$-C_3H_3N_2$ (His)
$-C_6H_4O^-$ (Tyr)
$-COO^-$ (Asp)

Scheme 3

Without consideration of the chemical shift values, X may be O, S, phenolate, imidazole, or carboxylate. However, from model compound studies, the $C_\beta H_2$ protons of Ser and Cys are expected to have much larger shifts, while those of His will have smaller shifts. The Tyr residue can be eliminated, because its ring protons are expected to exhibit significant isotropic shifts, and such significantly shifted resonances are not observed. A Glu residue ($^{-}OOC{-}C_\gamma H_2{-}C_\beta H_2{-}C_\alpha H$<) is not considered likely, because the NOE experiments would have provided evidence for an ABMNX system, as has been found for the bridging Glu in the

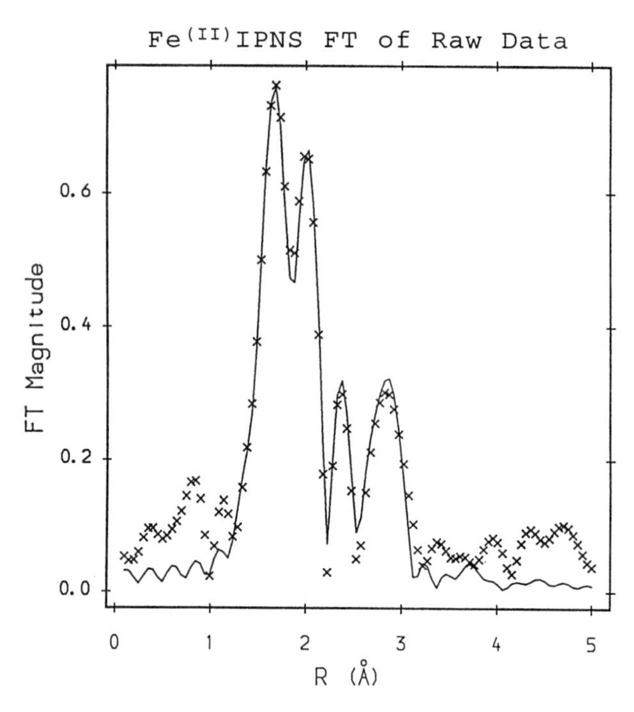

Figure 15
Fourier transformed unfiltered Fe K-edge EXAFS of Fe^{II}IPNS. [Reprinted with permission from Randall, C. R.; Zang, Y.; True, A. E.; Que, L., Jr.; Chamock, J. M.; Garner, C. D.; Fujishima, Y.; Schofield, C. J.; Baldwin, J. E. *Biochemistry* **1993**, *32*, 6664–6673. Copyright © 1993 American Chemical Society.]

diiron site of myohemerythrin.[29] An aspartate is thus the most likely residue that gives rise to the ABX systems with the 39 and 24 ppm signals assigned to the $C_{\beta}H_2$ protons and the 15 ppm resonance to the $C_{\alpha}H$ proton. In support, the Asp residue coordinated to the Co in E_2Co_2-superoxide dismutase has its $C_{\beta}H_2$ protons at 37 and 43 ppm.[30]

Extended X-ray absorption fine structure studies of Fe^{II}IPNS support the spectroscopic conclusions drawn so far.[23] The Fourier transformed (FT) data show a split first coordination shell consisting of O/N scatterers at 2.01 and 2.15 Å (Fig. 15). The 2.15-Å distance is typical of Fe^{II}–$N_{imidazole}$ distances and is thus assigned to the proposed His ligands. The 2.01-Å distance is associated with the carboxylate and a solvent hydroxide. Furthermore, the EXAFS data cannot be simulated well without the inclusion of a scatterer at 2.6–2.7 Å. A scatterer at such a long distance must be fixed quite rigidly to contribute to the fit. Typically, the second shell carbon atoms on the coordinated imidazoles can be found at about 3 Å, while the carboxylate carbon of a chelated carboxylate can be found at about 2.5 Å.[31,32] The 2.7-Å distance required in the fit of Fe^{II}IPNS is in between these distances and has been ascribed to an unsymmetrically chelated carboxylate. A proposed structure for the active site of Fe^{II}IPNS is shown in Scheme 4.

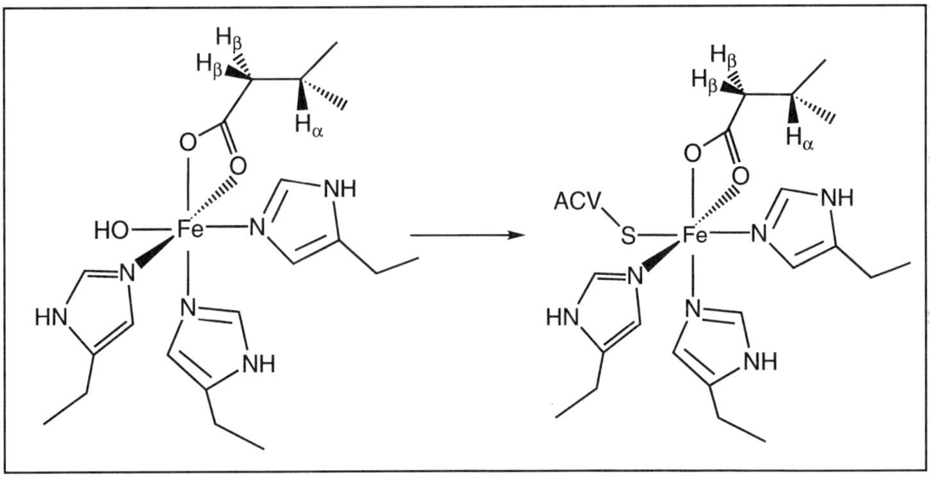

Scheme 4

The addition of the substrate ACV engenders a number of spectroscopic changes that indicate the coordination of its thiolate to the active site Fe(II) center. The Mössbauer isomer shift of the ES complex ($\delta = 1.10$ mm s^{-1}) decreases relative to the substrate-free enzyme (Fig. 9), suggesting increased covalency of the ligand environment.[21,28] The EXAFS analysis of the ES complex requires the inclusion of an Fe–S component at 2.34–2.35 Å, a distance consistent with Fe^{II}-thiolate bonds.[23,24] Furthermore, the Fe^{II}IPNS–ACV–NO complex exhibits a visible spectrum with absorption features at 508 and 720 nm (Fig. 16).[2] These

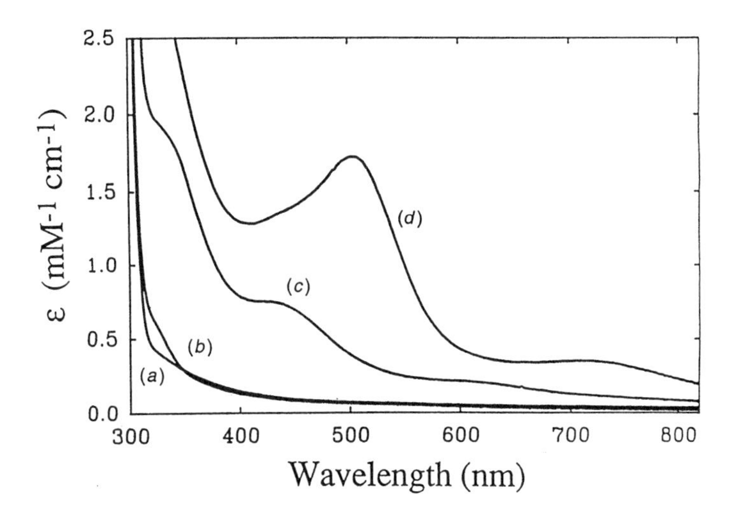

Figure 16
The UV–vis spectra of Fe^{II}IPNS as isolated (*a*), Fe^{II}IPNS–ACV (*b*), Fe^{II}IPNS–NO (*c*), and Fe^{II}IPNS–ACV–NO (*d*). [Reprinted with permission from the American Society for Biochemistry and Molecular Biology, Chen et al.[21]]

features are not observed for the Fe^{II}IPNS–NO and Fe^{II}IPNS–ASV–NO (Ser in place of Cys) complexes, both of which have maxima at 430 and 630 nm.[28] The ACV thiolate thus appears essential to elicit these features. The possibility that a protein Cys residue may be responsible for the spectral change has been eliminated by the observation that the characteristic visible bands of the ternary complex persist in mutant proteins wherein the two Cys residues are replaced by serines.[28]

For Cu^{II}IPNS, ACV binding results in a change in the Cu^{II} signal ([Fig. 11(*b*)], $g_{\parallel} = 2.234$, $A_{\parallel} = 498$ MHz, $g_{\perp} = 2.04$, $A_{\perp} = 54$ MHz) and the growth of a charge-transfer transition at 385 nm,[26,27] as previously found for the binding of sulfur anions to Type II Cu(II) centers.[33–35] The ESEEM data for Cu^{II}IPNS-ACV retains the three sharp NQI lines and the double quantum line found in CU^{II}IPNS, but these lines are shifted in response to ACV binding [Fig. 12(*b*)]. A second double quantum line is observed at 3.81 MHz, indicating the presence of two inequivalent imidazoles. Thus ACV binding causes the two equivalent equatorial His residues to be distinguished. A reasonable explanation for this effect is the coordination of the ACV thiolate trans to one of the His residues, thereby lengthening that Cu–N_{His} bond and weakening the interaction of the unpaired electron with the distal ^{14}N of the trans His residue. The ACV binding further causes a decrease in the signal associated with deuteron modulation in Cu^{II}IPNS. It is likely that the ACV thiolate replaces a coordinated solvent molecule (H_2O or HO^-) trans to one His upon enzyme–substrate complex formation. The effect of ACV on the Fe^{II}IPNS active site is shown in Scheme 4.

The active enzyme does not react with O_2 in the absence of substrate. Once the ACV binds to the active site, the Fe(II) center is now presumably poised to interact with O_2. The introduction of the thiolate ligand undoubtedly lowers the redox potential of the iron and acts as a switch to initiate catalysis. Such a switch has the advantage of utilizing the oxidative power of O_2 only when the proper substrate is present. To date, no oxygenated intermediate such as Fe^{II}IPNS–ACV–O_2 has been observed. The NO complexes discussed above represent the closest analogues for such an intermediate and demonstrate that there is a site available for O_2 binding.[21,28]

Extended X-ray absorption fine structure studies of Fe^{II}IPNS–ACV–NO reveal a short (1.71 Å) Fe–N bond, expected for an NO ligand, and an Fe–S bond at 2.32 Å, providing evidence for at least two exogenous ligands.[23] A third exogenous ligand, a solvent molecule, is indicated by the broadening effect of $H_2{}^{17}O$ on the EPR spectrum of Fe^{II}IPNS–ACV–NO.[28] The presence of three exogenous ligands indicates that there can only be three endogenous ligands in the ternary complex. NMR studies of Fe^{II}IPNS–ACV–NO show that there are only two His residues in this complex [Fig. 10(*b*)],[25] with the Asp residue likely to be the third endogenous ligand.

Subsequent to these spectroscopic studies, crystal structures of Mn^{II}IPNS, the Fe^{II}IPNS–ACV complex, and the ternary Fe^{II}IPNS–ACV–NO complex were reported (Figure 17).[36,37] These structures substantiate many of the conclusions based on the spectroscopic data. Three (1 Asp and 2 His) of the four endogenous ligands were correctly identified. These three ligands occupy one face of the oc-

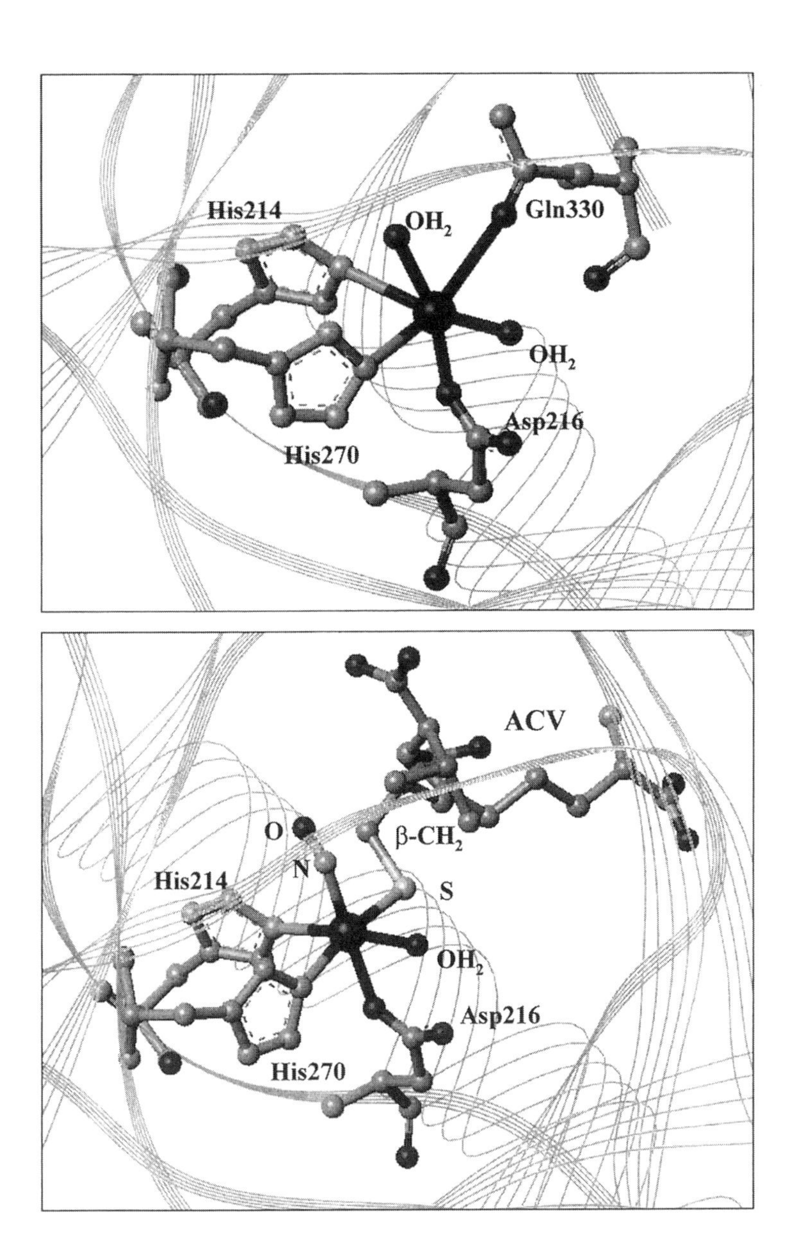

Figure 17
Crystal Structures of Mn(II)IPNS and the Fe(II)IPNS · ACV · NO complex derived from coordinates reported in refs 36 and 37.

tahedron and serve to hold the metal center in place during the catalytic cycle. Indeed this 2-His–1-carboxylate facial triad is emerging as a common motif in the structures of mononuclear nonheme iron(II) enzymes.[38] The fourth endogenous ligand, originally thought to be a histidine, turns out to be a glutamine. Since this residue has only recently been associated with iron active sites,[39,40] its involvement in metal coordination was not considered and spectroscopic signatures for such a ligand have not been established. The spectroscopic studies also correctly anticipated the changes in the coordination environment upon formation of the enzyme–substrate complex and the ternary ES–NO complex.

With many of the structural features of the IPNS active site now clarified, focus now shifts to the nature of the iron species responsible for the oxidative cyclization of the substrate ACV. A plausible mechanism is shown in Scheme 5, which is proposed on the basis of the products derived from the oxidation of a variety of substrate analogues by Fe^{II}IPNS.[20] Though no intermediates have thus far been observed in the catalytic cycle, the two species of interest are the iron(II)–peroxo intermediate that may be generated in the formation of the *β*-lactam ring and the iron(IV)–oxo species that is proposed to be responsible for initiating the formation of the thiazolidine ring. Obtaining insights into how this enzyme works will be useful not only for understanding penicillin and cephalosporin biosynthesis but also for understanding the mechanisms of oxygen activation by nonheme iron(II) enzymes in general.

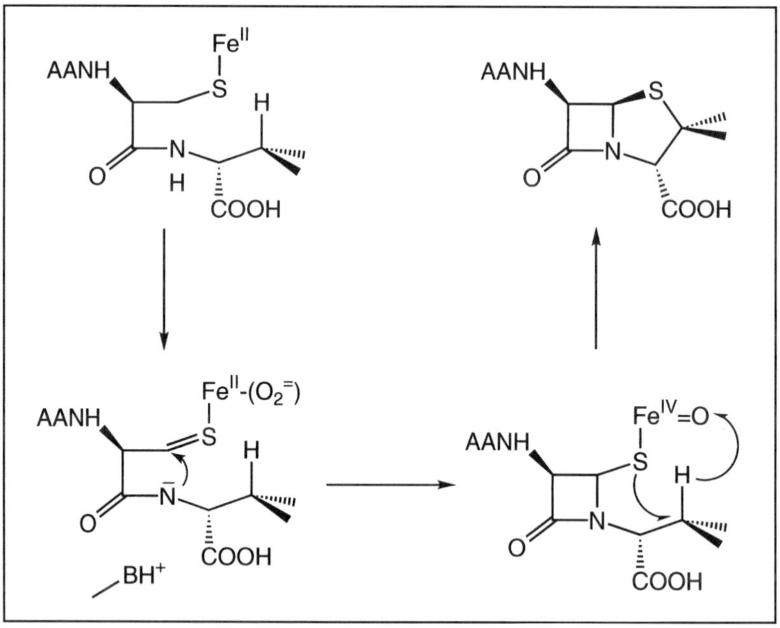

Scheme 5

References to The Cu_A Site of Cytochrome *c* Oxidase Case Study

1. Babcock, G. T.; Wikström, M., "Oxygen Activation and the Conservation of Energy in Cell Respiration." *Nature* (*London*) **1992**, *356*, 301–309.
2. Iwata, S.; Ostermeier, C.; Ludwig, B.; Michel, H., "Structure at 2.8 Å Resolution of Cytochrome *c* Oxidase from *Paracoccus denitrificans*." *Nature* (*London*) **1995**, *376*, 660–669.
3. Beinert, H., "Copper A of Cytochrome *c* Oxidase, a Novel, Long-Embattled, Biological Electron-Transfer Site." Eur. J. Biochem. **1997**, *245*, 521–532.
4. Kroneck, P. M. H., Antholine; W. E.; Kastrau, D. H. W.; Buse, G.; Steffens, G. C. M.; Zumft, W. G., "Multifrequency EPR Evidence for a Bimetallic Centre at the Cu_A site in Cytochrome *c* Oxidase." *Eur. J. Biochem.* **1992**, *209*, 875–881.
5. Farrar, J. A.; Lappalainen, P.; Zumft, W. G.; Saraste, M.; Thomson, A. J., "Spectroscopic and Mutagenesis Studies on the Cu_A Centre from the Cytochrome-*c* Oxidase complex of *Paracoccus denitrificans*." *Eur. J. Biochem.* **1995**, *232*, 294–303.
6. Farrar, J. A.; Neese, F.; Lappalainen, P.; Kroneck, P. M. H.; Saraste, M.; Zumft, W. G.; Thomson, A. J., "The Electronic Structure of Cu_A: A Novel Mixed-Valence Dinuclear Copper Electron-Transfer Center." *J. Am. Chem. Soc.* **1996**, *118*, 11501–11514.
7. Lappalainen, P.; Aasa, R.; Malmström, B. G.; Saraste, M., "Soluble Cu_A-binding Domain from the *Paracoccus Cytochrome c* Oxidase." *J. Biol. Chem.* **1993**, *268*, 26416–26421.
8. Blackburn, N. J.; Barr, M. E.; Woodruff, W. H.; van der Oost, J.; de Vries, S., "Metal–Metal Bonding in Biology: EXAFS Evidence for a 2.5 Å Copper–Copper Bond in the Cu_A Center of Cytochrome *c* Oxidase." *Biochemistry* **1994**, *33*, 10401–10407.
9. Blackburn, N. J.; de Vries, S.; Barr, M. E.; Houser, R. P.; Tolman, W. B.; Sanders, D.; Fee, J. A. "X-ray Absorption Studies on the Mixed-Valence and Fully Reduced Forms of the Soluble Cu_A Domain of Cytochrome *c* Oxidase." *J. Am. Chem. Soc.* **1997**, *119*, 6135–6143.
10. Wilmanns, M.; Lappalainen, P.; Kelly, M.; Sauer-Eriksson, E.; Saraste, M., "Crystal Structure of the Membrane-Exposed Domain from a Respiratory Quinol Oxidase Complex with an Engineered Dinuclear Copper Center." *Proc. Natl. Acad. Sci. USA* **1995**, *92*, 11955–11959.
11. Andrew, C. R.; Sanders-Loehr, J., "Copper–Sulfur Proteins: Using Raman Spectroscopy to Predict Coordination Geometry." *Acc Chem. Res.* **1996**, *29*, 365–372.
12. Andrew, C. R.; Lappalainen, P.; Saraste, M.; Hay, M. T.; Lu, Y.; Dennison, C.; Canters, G. W.; Fee, J. A.; Slutter, C. E.; Nakamura, N.; Sanders-Loehr, J., "Engineered Cupredoxins and Bacterial Cytochrome *c* Oxidases Have Similar Cu_A Sites: Evidence from Resonance Raman Spectroscopy." *J. Am. Chem. Soc.* **1995**, *117*, 10759–10760.
13. Andrew, C. R.; Nakamura, N.; Sanders-Loehr, J.; Saraste, M., unpublished results.
14. Andrew, C. R.; Fraczkiewicz, R.; Czernuszewicz, R. S.; Lappalainen, P.; Saraste, M.; and Sanders-Loehr, J., "Identification and Description of Copper–Thiolate Vibrations in the Dinuclear Cu_A Site of Cytochrome *c* Oxidase." *J. Am. Chem. Soc.* **1996**, *118*, 10436–10445.
15. Wallace-Williams, S. E.; James, C. A.; de Vries, S.; Saraste, M.; Lappalainen, P.; van der Oost, J.; Fabian, M.; Palmer, G.; Woodruff, W. H., "Far-Red Resonance Raman Study of Copper A in Subunit II of Cytochrome *c* Oxidase." *J. Am. Chem. Soc.* **1996**, *118*, 3986–3987.
16. Andrew, C. R.; Yeom, H.; Valentine, J. S.; Karlsson, B. G.; Bonander, N.; van Pouderoyen, G.; Canters, G. W.; Loehr, T. M.; Sanders-Loehr, J., "Raman Spectroscopy as an Indicator of Cu–S Bond Length in Type 1 and type 2 Copper Cysteinate Proteins." *J. Am. Chem. Soc.* **1994**, *116*, 11489–11498.
17. Dennison, C.; Berg, A.; Canters, G. W., "An ^1H NMR Study of the Paramagnetic Active Site of the Cu_A Variant of Amicyanin." *Biochemistry* **1997**, *36*, 3262–3269.
18. Gurbiel, R. J.; Fann, Y.-C.; Surerus, K. K.; Werst, M. M.; Musser, S. M.; Doan, P. E.; Chan, S. I.; Fee, J. A.; Hoffman, B. M., "Detection of Two Histidyl Ligands to Cu_A of Cytochrome Oxidase by 35-CHz ENDOR." *J. Am. Chem. Soc.* **1993**, *115*, 10888–10894.

19. Ramirez, B. E., Malmström, B. G., Winkler, J. R., and Gray, H. B., "The Currents of Life: The Terminal Electron-Transfer Complex of Respiration." *Proc. Natl Acad. Sci. USA*, **1995**, *92*, 11949–11951.

References to the Isopenicillin N Synthase Case Study

20. Baldwin, J. E.; Bradley, M., "Isopenicillin N Synthase: Mechanistic Studies" *Chem. Rev.* **1990**, *90*, 1079–1088.
21. Chen, V. J.; Orville, A. M.; Harpel, M. R.; Frolik, C. A.; Surerus, K. K.; Münck, E.; Lipscomb, J. D., "Spectroscopic Studies of Isopenicillin N Synthase. A Mononuclear Nonheme Fe^{2+} Oxidase with Metal Coordination Sites for Small Molecules and Substrate" *J. Biol. Chem.* **1989**, *264*, 21677–21681.
22. Roe, A. L.; Schneider, D. J.; Mayer, R.; Pyrz, J. W.; Widom, J.; Que, L., Jr., "X-ray Absorption Spectroscopy of Iron-Tyrosinate Proteins" *J. Am. Chem. Soc.* **1984**, *106*, 1676–1681.
23. Randall, C. R.; Zang, Y.; True, A. E.; Que, L., Jr.; Charnock, J. M.; Garner, C. D.; Fujishima, Y.; Schofield, C. J.; Baldwin, J. E., "X-ray Absorption Studies of the Ferrous Active Site of Isopenicillin N Synthase and Related Model Complexes" *Biochemistry* **1993**, *32*, 6664–6673.
24. Scott, R. A.; Wang, S.; Eidsness, M. K.; Kriauciunas, A.; Frolik, C. A.; Chen, V. J., "X-ray Absorption Spectroscopic Studies of the High-Spin Iron(II) Active Site of Isopenicillin N Synthase; Evidence for an Fe–S Interaction in the Enzyme–Substrate Complex" *Biochemistry* **1992**, *31*, 4596–4601.
25. Ming, L.-J.; Que, L., Jr.; Kriauciunas, A.; Frolik, C. A.; Chen, V. J., "NMR Studies of the Active Site of Isopenicillin N Synthase, a Non-Heme Iron(II) Enzyme" *Biochemistry* **1991**, *30*, 11653–11659.
26. Ming, L.-J.; Que, L., Jr.; Kriauciunas, A.; Frolik, C. A.; Chen, V. J., "Coordination Chemistry of the Metal Binding Site of Isopenicillin N Synthase" *Inorg. Chem.* **1990**, *29*, 1111–1112.
27. Jiang, F.; Peisach, J.; Ming, L.-J.; Que, L., Jr.; Chen, V. J., "Electron Spin Echo Envelope Modulation Studies of the Cu(II)-Substituted Derivative of Isopenicillin N Synthase: A Structural and Spectroscopic Model" *Biochemistry* **1991**, *30*, 11437–11445.
28. Orville, A. M.; Chen, V. J.; Kriauciunas, A.; Harpel, M. R.; Fox, B. G.; Münck, E.; Lipscomb, J. D., "Thiolate Ligation of the Active Site Fe^{2+} of Isopenicillin N Synthase Derives from Substrate Rather Than Endogenous Cysteine: Spectroscopic Studies of Site-Specific Cys → Ser Mutated Enzymes" *Biochemistry* **1992**, *31*, 4602–4612.
29. Wang, Z.; Martins, L.; Ellis, W. R., Jr.; Que, L., Jr., "Proton NMR Studies of Myohemerythrin from *Themiste zostericola*" *J. Biol. Inorg. Chem.* **1997**, *2*, 56–64.
30. Banci, L.; Bertini, I.; Luchinat, C.; Viezzoli, M. S., "A Comment on the ^1H NMR Spectra of Cobalt(II)-Substituted Superoxide Dismutases with Histidines Deuteriated in the ε1-Position" *Inorg. Chem.* **1990**, *29*, 1438–1440.
31. Kitajima, N.; Fukui, H.; Moro-oka, Y.; Mizutani, Y.; Kitagawa, T., "Synthetic Model for Dioxygen Binding Sites of Non-Heme Iron Proteins. X-ray Structure of Fe(OBz)(MeCN)·(HB(3,5-$iPr_2pz)_3$) and Resonance Raman Evidence for Reversible Formation of a Peroxo Adduct" *J. Am. Chem. Soc.* **1990**, *112*, 6402–6403.
32. Zang, Y.; Elgren, T. E.; Dong, Y.; Que, L., Jr., "A High-Potential Ferrous Complex and Its Conversion to an Alkylperoxoiron(III) Intermediate. A Lipoxygenase Model" *J. Am. Chem. Soc.* **1993**, *115*, 811–813.
33. Dooley, D. M.; Coté, C. E., "Inactivation of Beef Plasma Amine Oxidase by Sulfide" *J. Biol. Chem.* **1984**, *259*, 2923–2926.
34. Hughey, J. L., IV; Fawcett, T. G.; Rudich, S. M.; Lalancette, R. A.; Potenza, J. A.; Schugar, H. G., "Preparation and Characterization of [rac-5,7,7,1,2,14,14-Hexamethyl-1,4,8,11-tetraazocyclotetradecane]copper(II)*o*-Mercaptobenzoate Hydrate [Cu(tet b)·

(o-$SC_6H_4CO_2$)] · H_2O, a Complex with a CuN_4S(Mercaptide) Chromophore" *J. Am. Chem. Soc.* **1979**, *101*, 2617–2623.

35. Morpurgo, L.; Desideri, A.; Rigo, A.; Viglino, P.; Rotilio, G., "Reaction of *N,N*-Diethyldithiocarbamate and Other Bidendate Ligands with Zn, Co and Cu Bovine Carbonic Anhydrases. Inhibition of the Enzyme Activity and Evidence for Stable Ternary Enzyme–Metal–Ligand Complexes." *Biochim. Biophys. Acta* **1983**, *746*, 168–175.
36. Roach, P. L.; Clifton, I. J.; Fülöp, V.; Harlos, K.; Barton, G. J.; Hajdu, J.; Andersson, I.; Schofield, C. J.; Baldwin, J. E., "Crystal Structure of Isopenicillin *N* Synthase is the First From a New Structural Family of Enzymes" *Nature* (*London*) **1995**, *375*, 700–704.
37. Roach, P. L.; Clifton, I. J.; Hensgens, C. M. H.; Shibata, N.; Schofield, C. J.; Hajdu, J.; Baldwin, J. E., "Structure of Isopenicillin *N* Synthase Complexed with Substrate and the Mechanism of Penicillin Formation" *Nature* (*London*) **1997**, *387*, 827–830.
38. Hegg, E. L.; Que, L., Jr., "The 2-His-1-Carboxylate Facial Triad: An Emerging Structural Motif in Mononuclear Non-Heme Iron(II) Enzymes" *Eur. J. Biochem.*, **1997**, *250*, 625–629.
39. Minor, W.; Steczko, J.; Bolin, J. T.; Otwinowski, Z.; Axelrod, B., "Crystallographic Determination of the Active Site Iron and Its Ligands in Soybean Lipoxygenase L-1" *Biochemistry* **1993**, *32*, 6320–6323.
40. Sträter, N.; Klabunde, T.; Tucker, P.; Witzel, H.; Krebs, B., "Crystal Structure of a Purple Acid Phosphatase Containing a Dinuclear Fe(III)–Zn(II) Active Site" *Science* **1995**, *268*, 1489–1492.

APPENDIX

Problems for Selected Chapters

Problems for Chapter 1: Electronic Absorption Spectroscopy

1. **(a)** Show that a $^4A_2 \rightarrow {}^4T_1$ transition in a tetrahedral Co(II) complex is electronically allowed but that the $^4A_2 \rightarrow {}^4T_2$ transition is forbidden. **(b)** If the normal modes of vibration span the a_1, e and t_2 representations of the T_d point group, show that the $^4A_2 \rightarrow {}^4T_2$ transition is vibronically allowed.

2. Assume that the $Co(NH_3)_5Cl^{2+}$ ion can be classified as C_{4v}. Show that the $a_1(\sigma) \rightarrow a_1(z^2)$ Cl $\rightarrow$ Co CT transition is polarized parallel to the z (C_4) axis. Then prove that the $e(\pi) \rightarrow a_1(z^2)$ transition is perpendicularly polarized.

3. If a solution containing compound X has an absorbance of 0.50 at the wavelength of irradiation, what fraction of the incident light is absorbed? (0.684) What fraction of the incident light is absorbed by compound X if it contributes an absorbance of 0.50 to a total solution absorbance of 2.00? (0.248)

4. If the $t_1(L) \rightarrow e^*(M)$ transition occurs at 530 nm in the CT spectrum of MnO_4^- and at 350 nm in the case of CrO_4^{2-}, show that the difference in the optical electronegativities of the metal centers is about 0.3.

5. The lowest energy $p\pi \rightarrow t_{2g}$ CT transition in $TaCl_6^-$ occurs around 34,800 cm^{-1}. Estimate the optical electronegativity of the d^0 Ta(V) ion in octahedral geometry.

6. **(a)** The visible spectrum of $[(H_3N)_5Ru(py)]^{2+}$, where py denotes pyridine, exhibits a broad MLCT transition that maximizes at 407 nm in acetonitrile. Explain why the corresponding transition occurs at a longer wavelength (555 nm) for the Os(II) analogue. **(b)** If the symmetry of the osmium complex is approximated as C_{2v} with the pyridine ligand located in the yz plane, what are the symmetries of the filled 6d orbitals of the osmium? **(c)** A second, weaker MLCT transition is observed with a maximum at 428 nm in the spectrum of the osmium complex. How many MLCT transitions to the LUMO of the pyridine ligand are orbitally allowed if the LUMO forms a basis for the b_1 representation of C_{2v}?

7. **(a)** The near UV/visible spectra of $[Ru(NH_3)_6]^{2+}$ **(A)**, $[(H_3N)_5Ru(4\text{-}NH_2\text{-}py)]^{2+}$ **(B)** and $[(H_3N)_5Ru(4\text{-}NH_2\text{-}py)]^{3+}$ **(C)** are presented below in the form of semilog plots. The

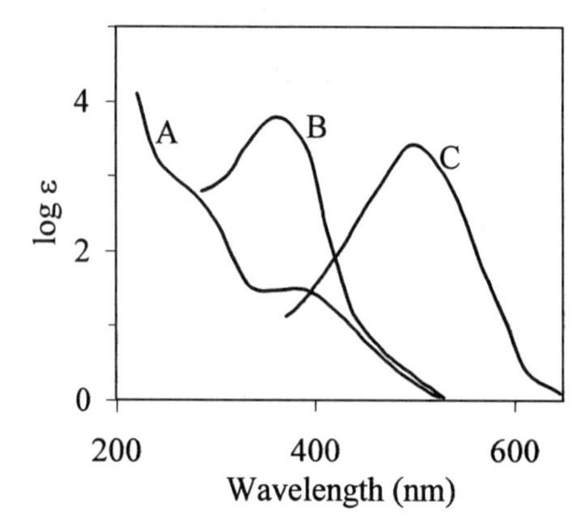

hexammine complex exhibits two weak transitions that are poorly resolved. Propose assignments for the bands. **(b)** The intense band in the absorption spectrum of the Ru(II) complex with the aminopyridine ligand is blue shifted relative to that of the analogous compound described in Problem 6. Rationalize the influence that the amino group has on the transition. **(c)** The intense transition in the Ru(III) complex has a completely different type of orbital parentage. Propose an assignment. (*Hint:* No such transition occurs in the absense of the amino group.)

8. The terms associated with the ground configuration of O_2 are determined by the π^{*2}_{2p} configuration where the π^*_{2p} orbitals form a basis for the π_g representation of $D_{\infty h}$. **(a)** Show that there are six independent ways to assign two electrons to the π^*_{2p} shell. **(b)** Use the correlation table below and the fact that

$$\pi_g \times \pi_g = \Delta_g + \Sigma_g{}^+ + \Sigma_g{}^-$$

to derive the terms associated with the ground configuration of O_2.

$D_{\infty h}$	C_{2v}
$\Sigma_g{}^+$ or $\Sigma_u{}^+$	A_1
$\Sigma_g{}^-$ or $\Sigma_u{}^-$	A_2
π_g or π_u	$B_1 + B_2$
Δ_g or Δ_u	$A_1 + A_2$

9. The $[Cr(oxalate)_3]^{3-}$ system exhibits low-lying, spin-allowed d–d bands that maximize at 420 and at 570 nm. **(a)** Estimate $10Dq$ on the basis of the $^4A_{2g} \rightarrow {}^4T_{2g}$ transition. **(b)** What values of Dq and B give the best fit of the spectrum according to the Tanabe–Sugano diagram?

10. The 0–0 transitions associated with the $^4A_2 \rightarrow {}^2E$ absorption and $^4A_2 \leftarrow {}^2E$ emission of an octahedral Cr(III) complex are usually nearly superposed on each other. The figure below shows the $^4A_2 \leftarrow {}^2E$ emission spectra of $[Cr(CN)_{6-n}(OH_2)_n]^{n-3}$ complexes for $n = 0$, 3, and 6. [Reprinted with permission from A. Ghaith; Forster, L. S.; Rund,

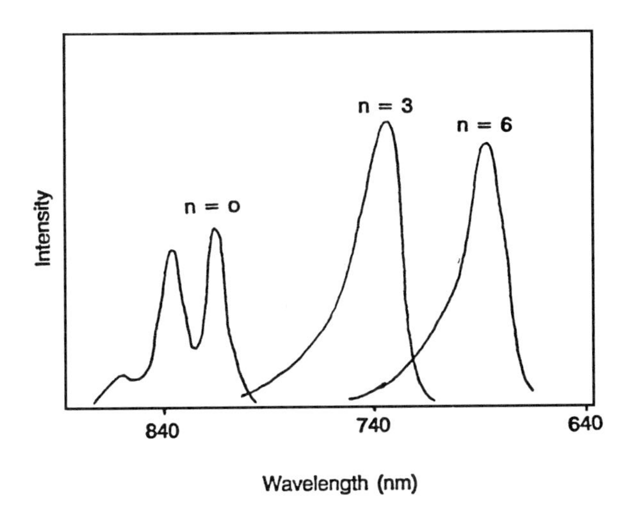

J. V. *Inorg. Chim. Acta* **1986**, *116*, 11–13.] **(a)** To a good approximation, the energy of the 0–0 transition is given as 21 B. Describe the configuration involved and explain why the energy is independent of Δ. **(b)** Estimate the value of B for the $[Cr(CN)_6]^{3-}$ complex. (This value is sometimes called B_{55}, and it is specific for electron–electron repulsions between electrons in the t_{2g} subshell.) Compare your answer with the gas-phase value for Cr^{3+} which is 1030 cm^{-1}.

11. The first band of appreciable intensity in the d–d spectrum of $[Co(en)_3]^{3+}$ appears at 21,400 cm^{-1}. The figure below shows that [*trans*-$Co(en)_2Cl_2]^+$ has absorption in the same region, but it is split into two components. **(a)** Approximate the symmetry of

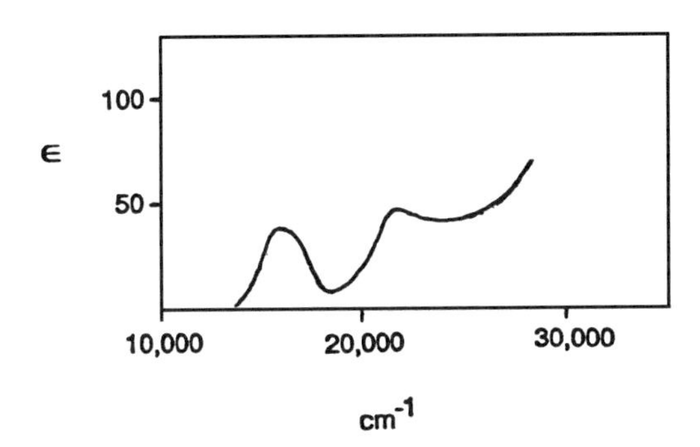

$[Co(en)_3]^{3+}$ as O_h and the symmetry of the mixed-ligand complex as D_{4h}. Use a correlation table and explain why the $^1A_{1g} \rightarrow {}^1T_{1g}$ band is split in the D_{4h} system. **(b)** Single-crystal studies of the mixed-ligand complex show that the lower energy component is not strongly polarized while the higher energy component is polarized in the *xy* plane (⊥ to Cl–Co–Cl). Assign D_{4h} symmetry labels to the two transitions.

12. In view of the relative intensity, the 555-nm transition of $[(H_3N)_5Os(py)]^{2+}$, described in Problem 6, can be assigned as the $b_1(xz) \rightarrow b_1(\pi^*)$ transition. Why?

13. When the CT term determines the absorption intensity, the intensity depends in a predictable way on the dipole length of the transition Phifer and McMillin.[18]

Rationalize the MLCT intensities for the following series of bis(terpyridine)Ru(II) systems.

R_1	R_2	ε_{max} (M^{-1} cm^{-1})
H	C_6H_5	38,400
H	H	16,100
C_6H_5	H	6,850

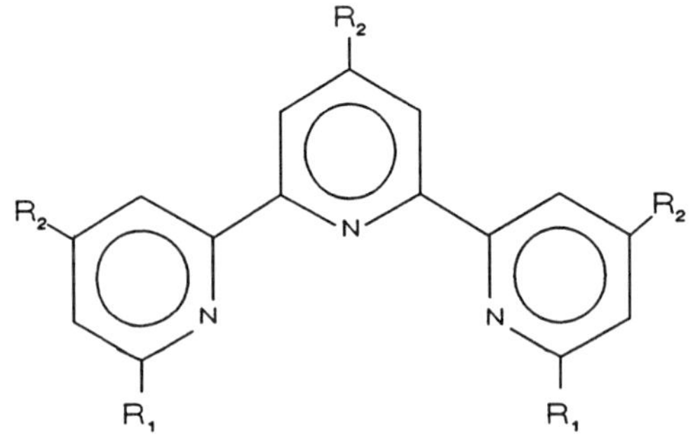

14. In the D_{4h} ion $[CuCl_4]^{2-}$, the energies of the d orbitals of the copper center are believed to vary as follows:

$$z^2 < xz, yz < xy \ll x^2 - y^2$$

(a) The ground state is therefore a $^2B_{1g}$ state. Identify the terms associated with the three d–d excited states of the ion. **(b)** Only two of the d–d transitions are vibronically allowed in z polarization, that is, when the electric vector is normal to the molecular plane. Determine which two are z polarized if the normal modes of vibration are $a_{1g} + a_{2u} + b_{1g} + b_{2g} + b_{1u} + 2e_u$. **(c)** Show that all three d–d transitions are vibronically allowed with x, y polarization.

15. Oxygenated hemocyanin exhibits an intense peroxide-to-copper CT band at 345 nm and a weaker one at 575 nm. **(a)** Use the MO scheme in Figure 12 to determine the polarization of each of the CT transitions. **(b)** Explain why only one of the two transitions derives intensity from the CT term.

Problems for Chapter 4: ESEEM and ENDOR Spectroscopy

1. From the first-order energy equation (3), derive Eqs. (6a and b) for the ENDOR frequencies of an $I = \frac{1}{2}$ nucleus coupled with an electron spin $S = \frac{1}{2}$.

2. **(a)** Sketch the energy level diagram for a system with one unpaired electron ($S = \frac{1}{2}$, $m_S = \pm\frac{1}{2}$) and two equivalent protons ($I_1 = I_2 = \frac{1}{2}$, $m_{I1} = m_{I2} = \pm\frac{1}{2}$) each with the same coupling constant A. Label each level with the appropriate quantum numbers and energies, as is done in Figure 1. The first-order energies for this system are given by extending Eq. (3) to include a second nucleus, that is, $E(m_S m_I) = g\beta B m_S - g_n\beta_n B m_{I1} -$

$g_n\beta_n Bm_{I2} + hAm_S m_{I1} + hAm_S m_{I2}$. **(b)** From the energy level diagram obtained in (a) show that the EPR spectrum (selection rule: $\Delta m_S = \pm 1$, $\Delta m_I = 0$) will consist of three lines at the positions ν_e, and $\nu_e \pm A$ ($\nu_e = gg_n\beta_n B/h$), separated by A, and having relative intensities 1:2:1. Show that the ENDOR spectrum (selection rule: $\Delta m_S = 0$, $\Delta m_I = \pm 1$) will consist of only two lines, separated by A, at the positions $\nu_n \pm A/2$. Verify that this is an example of the $2nI + 1$ rule for the number of lines in the EPR spectrum where $n = 2$ and the $4I$ rule for the number of lines in the ENDOR spectrum.

3. Show that the Hamiltonian operator for the quadrupole interaction given by Eq. (8) is equivalent to Eq. (9b). (*Hint:* Equation (8) in expanded form is given by $\mathscr{H}_Q = P_x\hat{I}_x{}^2 + P_y\hat{I}_y{}^2 + P_z\hat{I}_z{}^2$. Make substitutions for the quantities in Eq. (9b) to reduce it to Eq. (8). Note that $\eta = (P_x - P_y)/P_z$, $\hat{I}^2 = \hat{I}_x{}^2 + \hat{I}_y{}^2 + \hat{I}_z{}^2 = I(I+1)$ (the eigenvalue of the $\hat{I}^2$ operator) and $P_x + P_y + P_z = 0$).

4. Show that the Eqs. (12a–c) for the ENDOR frequencies of the upper electron spin state ($m_S = +\frac{1}{2}$) reduce to the pure quadrupole frequencies given by Eqs. (21a–c) at the exact cancellation condition $\nu_n = a/2$.

5. The frequency spectrum from the two-pulse ESEEM of Cu^{2+}–transferrin–CO_3 shows peaks from ^{14}N at 0.74, 0.86, 1.60, and 4.05 MHz at a field of 3185 G (Fig. 10). **(a)** From the three quadrupole lines, calculate K, the quadrupole coupling constant e^2qQ, the asymmetry parameter η, and the principal values of the quadrupole tensor P_x, P_y, and P_z. **(b)** Calculate the ^{14}N isotropic nuclear hyperfine coupling constant a from the "double quantum" peak ν_{13} at 4.05 MHz of the lower $m_S = -\frac{1}{2}$ manifold of nuclear states. **(c)** Does the value obtained for a confirm that the spectrum is obtained at a field corresponding to exact cancellation, that is, $\nu_n \approx a/2$?

6. The echo amplitude in Figure 6(*a*) as a function of time is given in the following table:

Time (τ) (μs)	Amplitude[a]
1.0	1.55
1.5	0.97
2.0	0.57
2.5	0.29
3.0	0.17

[a] Relative units after "smoothing out" the modulation.

(a) By assuming that the decay of the echo is exponential, determine the value of the phase memory time T_m of the electron spin from the appropriate first-order plot. **(b)** From the value of the phase memory time, estimate the minimum line width $\Delta\nu_{1/2}$ expected for the frequency spectrum [Fig. 6(*b*)]. Is the result of your calculation born out by the experimental line widths?.

7. The ^{15}N spectrum of [^{15}N]imidazole of the vanadyl imidazole complex, $[VO(Im)_4]^{2+}$, along the z axis (VO bond axis) shows couplings from two different nitrogens with coupling constants $A_z = 9.05$ and 9.89 MHz, respectively (Mulks, C. F.; Kirste, B.; van Willigen, H. "ENDOR Study of VO^{2+}-Imidazole Complexes in Frozen Solution", *J. Am. Chem. Soc.* **1982**, *104*, 5906–5911). **(a)** What do you predict for the corresponding coupling constants for [^{14}N]imidazole? **(b)** The [^{14}N]imidazole complex spectrum consists of two overlapping sets of four lines from each of the two ^{14}N nuclei. Their positions are 1.26, 2.98, 3.66, and 5.38 MHz and 2.50, 3.19, 4.21, and 4.90 MHz at a field of 2780 G in the EPR spectrum. Determine A_z and P_z from these data.

8. To identify the nucleus responsible for a particular line in the frequency spectrum of an ESEEM experiment, one can make measurements at more than one EPR frequency and follow the frequency shift of the line as a function of magnetic field. If a line appears at 1.43 MHz at 3150 G, at 1.82 MHz at 4421 G and at 2.11 MHz at 5359 G in the ESEEM frequency spectrum, what is the nucleus? Note that the line may not, and often does not, appear at the Zeeman or Larmor frequency of the nucleus because of hyperfine and/or quadrupolar couplings.

9. The X-band ^{15}N ENDOR spectrum of phthalate dioxygenase (PDO) at 3588 G consists of two doublets from two ^{15}N nuclei with lines occurring at 1.00 and 4.10 MHz, and 1.80 and 4.90 MHz, respectively (Fig. 18). **(a)** Determine the coupling constants. **(b)** Calculate where the four lines should appear in the Q-band ENDOR spectrum at a field of 13,137 G. How well does your answer agree with the positions of the two high-frequency lines at 8.2 and 9.2 MHz in Figure 19(*a*)?

10. Given that the value of $P_x = 0.85$ MHz and $P_y = 0.45$ MHz for the site 1 ^{14}N spectrum of phthalate dioxygenase, what must P_z be?

11. VO^{2+} complexes typically display EPR spectra of axial symmetry (e.g., Fig. 13) where the direction of $g_{\|}$ corresponds to the short VO bond axis (≈ 1.6 Å). In water/methanol solvent a proton is hydrogen bonded to the vanadyl VO oxygen atom and has couplings of $A_{\|} = 4.34$ MHz and $A_{\perp} = (-)1.40$ MHz as determined by ENDOR (See Mustafi and Makinen[17]). The g-factors of the complex are approximately $g_{\|} = 1.93$ and $g_{\perp} = 1.98$. **(a)** From Eq. (5), calculate the value of the proton coupling $A(\theta)$ for the magnetic field vector making an angle 45° to the VO bond axis. Assume that the proton lies on the principal axis. **(b)** Compute the value of the isotropic g_0 value, the isotropic nuclear hyperfine coupling a and the dipolar coupling F in MHz. **(c)** From the value of F compute the V–H distance assuming a simple magnetic dipole model.

12. For protons at distances of 4.0–5.0 Å from the metal, show that the modulation depth parameter k given by Eq. (17b) reduces to $k = (B/\omega_n)^2$ at magnetic field strengths of an X-band experiment (~3300 G), and therefore that $k \propto (\sin\theta\cos\theta)^2/r^6$. At what angle θ of the magnetic field vector with the metal–proton direction is modulation maximum? [*Hint:* Note that at 3300G, ω_n, the Larmor frequency of the proton, is much larger than either the A or B terms given by Eqs. (17e and f).]

13. For weakly coupled protons, the ENDOR frequencies ω_a and ω_b of the upper $m_S = +\frac{1}{2}$ and lower $m_S = -\frac{1}{2}$ electron spin manifold are essentially equal to ω_n the Larmor frequency of the proton. [See Eqs. (17c and d).] Thus resonances from weakly coupled protons can be τ suppressed with a single value of all. If a proton line appears at 15.2 MHz in the frequency spectrum from a three-pulse ESEEM experiment, what choices of τ can be employed to suppress this line?

14. An ESEEM study of a number of complexes of the enzyme pyruvate kinase has been carried out at various microwave frequencies (Tipton, P. A., McCracken, J. Cornelius, J. B.; Peisach, J. "Electron Spin Echo Envelope Modulation Studies of Pyruvate Kinase Active-Site Complexes" *Biochemistry* **1989**, *28*, 5720–5728). Given below are the frequencies of the ESEEM peaks obtained at the designated magnetic field strengths. Identify the nucleus by its Larmor frequency and determine the coupling constant where possible.

Complex[a]	B(G)	ν_{obs} (MHz)
(a) VO · PK · [1-^{13}C]pyruvate	4064	3.4, 5.6
(b) VO · PK · [2-^{13}C]pyruvate	4050	1.7, 7.1
(c) Mn · PK · MgATP · oxalate	3404	3.9, 8.0
(d) VO · PK · Cs · pyruvate	3148	1.8
(e) VO · PK · Na · pyruvate	3110	3.5
(f) VO · PK · [2-^{13}C]pyruvate-d_3	3416	2.2

Adensosine triphosphate = ATP.

15. Often ESEEM measurements on metalloproteins are made in both D_2O and H_2O solutions to detect exchangeable protons from water in the first and second coordination spheres or from protons on the protein. The ratio of the ESEEM obtained in D_2O to that in H_2O effectively cancels all modulations from nonexchangeable protons and other nuclei [see Eq. (18)]. The resultant ESEEM quotient contains modulations from the exchanged protons *and* deuterons, $V_{quotient} = V_{mod}(D)/V_{mod}(H)$, but because deuterium modulations are deeper, they dominate the ESEEM (see, e.g., McCracken et al.[9]). The numbers of deuteriums and their distances from the metal are usually estimated from simulation of the observed modulation depth. The three-pulse ESEEM quotient spectrum (D_2O/H_2O) of the low-spin $S = \frac{1}{2}$ cyanide adduct of $Fe^3{}_2$-transferrin has been observed at fields of 2677, 2906, and 3226 G, corresponding to the field positions of g_x, g_y, and g_z in the X-band EPR spectrum. The ESEEM frequency spectra show peaks at 1.78, 1.93, and 2.09 MHz for these three field settings, respectively. **(a)** Verify that these peaks are due to deuterium by computing the nuclear g_n value. **(b)** The peak line width $\Delta\nu_{1/2}$ is approximately 250 KHz. Estimate the minimum distance of a single deuteron from the metal assuming that the line width is due solely to unresolved electron–deuteron hyperfine coupling, which is point dipolar in nature. Is the deuteron located in the first coordination sphere?

16. At a field of 3226 G corresponding to g_z, the frequency spectrum from the three-pulse ESEEM of the low-spin [^{15}N]cyanide complex of $Fe^3{}_2$-transferrin shows strong sharp peaks at 1.11 and 1.76 MHz and much weaker and broader peaks at 2.70 and 3.53 MHz. (Snetsinger, P. A.; Chasteen, N. D.: Cornelius, J. B., and Singel, D. J. "Probing the Iron Center of the Low-Spin Cyanide Adduct of Transferrin by ESEEM Spectroscopy" *J. Chem Phys.* **1992**, *96*, 7917–7922.) Calculate the ^{15}N coupling constant. What can you conclude about the number of CN^- coordinated to the iron?

17. Starting with the Hamiltonian given by Eq. (9a), derive energies of the pure quadrupole states and show that the pure quadrupole transition frequencies are given by Eq. (21). (*Hint:* Set up the 3 × 3 secular determinant using the basis functions $|m_I\rangle = |-1\rangle$, $|0\rangle$ and $|+1\rangle$ and solve for the energies. See, for example, Section B8 of Wertz and Bolton, 1972, for formulas for the various matrix elements.)

18. Derive the energies given in Figure 3 starting with the Hamiltonian $\mathscr{H} = g\beta B_z\hat{S}_z - g_n\beta_n B_z\hat{I}_z + a\hat{S}_z\hat{I}_z + e^2Qq/4I(2I-1)[3\hat{\mathbf{I}}_z{}^2 - I(I+1) + \eta(\hat{\mathbf{I}}_x{}^2 - \hat{\mathbf{I}}_y{}^2)]$. Assume an isotropic g factor and an isotropic hyperfine coupling constant a with the magnetic field along the z axis. (*Hint:* Set up the two 3 × 3 secular determinants for the $m_S = \frac{1}{2}$ and $m_S = -\frac{1}{2}$ electron spin states. Use the basis functions $|M_I\rangle = |-1\rangle, |0\rangle, |+1\rangle$ for each determinant and solve for the energies (see Problem 17).

PROBLEM ANSWERS

5. **(a)** $K = 0.41$ MHz, $e^2qQ = 1.64$ MHz, $\eta = 0.90$, $P_x = -0.041$ MHz, $P_y = -0.779$ MHz, $P_z = 0.820$ MHz. (Signs of P_x, P_y, and P_z are relative only.) **(b)** $a = 1.86$ MHz; **(c)** Yes, $\nu_n = 0.98$ MHz at 3185 G, which is about $\frac{a}{2} = 0.9$ MHz.

6. **(a)** From the slope of a plot of ln(amplitude) versus time, a value of $T_m = 0.89$ μs is obtained [see Eq. (16b)]. **(b)** Yes, $\Delta\nu_{1/2} \approx 0.36$ MHz. The width of the ^1H line at 16.3 MHz is about 0.5 MHz. The widths of the other lines appear to be broader but their widths are difficult to measure because of overlap. Note that if the modulation itself decays away faster than the amplitude of the echo, then the lines will be broader than 0.36 MHz. That is, the widths of the individuals lines in the frequency spectrum are determined by the persistence of the corresponding modulations in the time domain.

7. **(a)** $A_z = 6.45$ and 7.05 MHz for the two ^{14}N nuclei; **(b)** $A_z = 6.64$ MHz and $P_z = 0.57$ MHz (or 0.80 MHz depending on which pair of lines is used in the calculation) and $A_z = 7.40$ and $P_z = 0.23$ MHz for the two ^{14}N nuclei, respectively. Note that the experimental values of A_z for ^{14}N from **(b)** do not perfectly agree with the calculated ones from **(a)**.

8. To first-order, a plot of ν_{obs} versus B has a slope $g_n\beta_n/h$ from which $g_n = 0.404$ is obtained. Therefore the nucleus is ^{14}N.

9. **(a)** Case III, Eq. (24c) applies giving $A = 5.10$ and 6.7 MHz; **(b)** Case I, Eq. (24a) applies giving the line positions 3.12 and 8.22 MHz, and 2.32 and 9.02 MHz. Agreement is satisfactory for the purposes of making spectral assignments.

10. $P_z = -1.30$ MHz.

11. **(a)** $A(45°) = 3.19$ MHz; **(b)** $g_0 = 1.96$, $a = 0.51$ MHz, $F = 1.91$ MHz; **(c)** $r = 3.43$ Å.

12. 45°.

13. 65.8 ns, 131.6 ns, 197.4 ns, 263.2 ns, and so on.

14. [Answer: **(a)** ^{13}C ($g_n = 1.45$ measured versus 1.405 literature), $A = 2.2$ MHz; **(b)** ^{13}C ($g_n = 1.42$ measured versus 1.405 literature), $A = 5.4$ MHz; **(c)** ^{31}P ($g_n = 2.29$ measured versus 2.263 literature), $A = 4.1$ MHz; **(d)** ^{133}Cs ($g_n = 0.750$ measured versus 0.7378 literature); **(e)** ^{23}Na ($g_n = 1.47$ MHz measured versus 1.4784 literature); **(f)** ^2H ($g_n = 0.845$ measured versus 0.8574 literature).

15. **(a)** The deuterium Larmor frequencies at these fields are 1.75, 1.90, and 2.11 MHz, which are in good agreement with the observed frequencies of 1.78, 1.93, and 2.09 MHz; the measured g_n values are 0.872, 0.871, and 0.850 compared with the literature value of 0.8574. **(b)** Setting the line width equal to F given by Eq. (17h), a value of $r = 3.6$ Å is estimated. This distance is too large for the D_2O to be in the first-coordination sphere.

16. The peaks at 1.11 and 1.76 are centered about the ^{15}N Larmor frequency of 1.43 MHz so the coupling constant is 0.65 MHz. The combination peak at 2.70 MHz ($\approx 1.11 + 1.76$) and the harmonic peak at 3.53 MHz ($\approx 2 \times 1.76$) in the three-pulse frequency spectrum indicate that there are more than one similarly coupled CN^- groups coordinated to the iron.

17. $E = K(1 + \eta), K(1 - \eta), -2K$.

Problems for Chapter 5: CD and MCD Spectroscopy

1. An optically active sample of $[Ni(en)_3]Cl_2$ (en = ethylenediamine) has d–d bands centered at 345, 545, and 890 nm in the UV-vis absorption spectrum. **(a)** Assign the transitions and determine which would be expected to exhibit CD under D_3 symmetry. **(b)** Would temperature-dependent MCD be expected for these d–d bands? Indicate the likely origin of the low-temperature MCD intensity for these transitions (i.e., MCD A,

B, or C terms). **(c)** Outline how you could use variable temperature MCD measurements to assess ground-state zero-field splitting.

2. Using Eqs. (40a), (40c), and (44) evaluate the ratio of the Faraday parameters, A_1/D_0 or C_0/D_0 in terms of the ground- or excited-state g values for a ${}^1A_{1g} \rightarrow {}^1T_{1u}$ and ${}^1T_{1u} \rightarrow {}^1A_{1g}$ transitions under idealized octahedral symmetry. (The worked solution is give in Stephens.[8])

3. The MCD intensity in arbitrary units for three well-resolved bands (labeled A, B, and C) of a complex multicluster metalloenzyme are tabulated below as function of the applied magnetic field. The data were collected at 1.6 K with applied magnetic fields between 0 and 7 T. Assess the spin state of the ground state from which each transition arises making the following three assumptions: (a) the MCD intensity arises solely from C terms; (b) only the lowest doublet of the ground-state manifold is significantly populated at 1.6 K; (c) if $S > \frac{1}{2}$, the system is axial with $D < 0$ such that the ground-state doublet has $g_{\parallel} = 4S$ and $g_{\perp} = 0$.

Magnetic Field (T)	A(au)	B(au)	C(au)
0.00	0.0	0.0	0.0
0.14	13.5	9.6	2.0
0.29	24.7	18.7	4.0
0.43	32.7	26.6	5.9
0.71	41.6	38.7	9.7
0.95	44.9	45.4	12.7
1.43	47.7	52.8	17.9
1.91	48.7	56.2	22.1
2.86	49.4	58.8	27.8
3.81	49.7	59.8	30.7
4.76	49.8	60.2	32.1
5.72	49.8	60.5	32.8
6.67	49.9	60.6	33.1

Index

D

F

G

H

I

J

K

L

N

O

S

T

U

V